£84.99

DC/AC FUNDAMENTALS

A SYSTEMS APPROACH

THOMAS L. FLOYD
DAVID M. BUCHLA

PEARSON

Boston Columbus Indianapolis New York San Francisco Upper Saddle River
Amsterdam Cape Town Dubai London Madrid Milan Munich Paris Montreal Toronto
Delhi Mexico City São Paulo Sydney Hong Kong Seoul Singapore Taipei Tokyo

Editorial Director: Vernon R. Anthony
Senior Acquisitions Editor: Lindsey Prudhomme
Development Editor: Dan Trudden
Editorial Assistant: Yvette Schlarman
Director of Marketing: David Gesell
Marketing Manager: Harper Coles
Senior Marketing Coordinator: Alicia Wozniak
Marketing Assistant: Les Roberts
Senior Managing Editor: JoEllen Gohr
Senior Project Manager: Rex Davidson
Senior Operations Supervisor: Pat Tonneman
Creative Director: Andrea Nix
Art Director: Diane Y. Ernsberger
Text and Cover Designer: Candace Rowley
Cover Image: takito/Shutterstock.com
Media Project Manager: Karen Bretz
Full-Service Project Management: Penny Walker/Aptara®, Inc.
Composition: Aptara®, Inc
Printer/Binder: Courier Kendallville
Cover Printer: Lehigh/Phoenix Color Hagerstown
Text Font: Times Roman

Credits and acknowledgments for materials borrowed from other sources and reproduced, with permission, in this textbook appear on the appropriate page within text.

Library of Congress Cataloging-in-Publication Data

Floyd, Thomas L.
  DC/AC fundamentals : a systems approach / Thomas L. Floyd,
David M. Buchla.
    p. cm.
  ISBN-13: 978-0-13-293393-3
  ISBN-10: 0-13-293393-4
  1. Electric circuits.   I. Buchla, David M.   II. Title.
TK454.F5565 2013
621.319'2—dc23

                                    2012020326

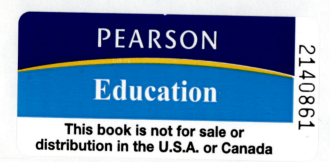

10 9 8 7 6 5 4 3 2 1

ISBN 10:   0-13-293393-4
ISBN 13: 978-0-13-293393-3

CONTENTS

PART II AC CIRCUITS

PREFACE

This first edition of *DC/AC Fundamentals: A Systems Approach* takes a broader view of DC/AC circuits than most standard texts. This textbook provides relevance to basic theory by stressing applications of dc/ac circuits in actual systems. As electronics has evolved, the need for troubleshooting to the component level has lessened, and the need to understand the relationship of system blocks, interfaces, and input/output signals has increased. We have addressed these changes by including basic system examples and system notes throughout every chapter. Examples and notes are specifically selected to complement and illustrate the basic dc/ac concepts for the chapter in which they appear. Most chapters include a troubleshooting section that emphasizes testing and measurements necessary for troubleshooting a system.

The mathematical level in this text is basic algebra and understanding the sine, cosine, and tangent functions for a right triangle. Complex arithmetic is not required; basic phasors are shown graphically. This approach shows the reason and importance of phase differences in circuits without bogging down in the mathematics of complex examples.

Features

- Fundamental dc/ac laws are presented with specific system examples.
- All chapters feature system examples and system notes that are coordinated with chapter objectives.
- The oscilloscope and DMM are described in detail, and tips are given on using these instruments.
- Multisim is used in selected examples, figures, and problems to provide practice in simulating circuits and systems and in troubleshooting.
- Worked examples illustrate basic concepts. Related problems in each worked example provide additional practice. Many examples include a Multisim exercise.
- System notes present interesting facts and information about system-related issues.
- Many chapters have a troubleshooting section that relates to topics covered in the chapter and emphasizes troubleshooting techniques and the use of instrumentation.
- Hands-on-tips (HOT) provide useful and practical information.
- Each chapter begins with a list of sections (outline) chapter, objectives, introduction, key terms list, and website reference.
- Each section within a chapter begins with an introduction and objectives for that section.
- Each section concludes with Checkup exercises that emphasize the main concepts presented in the section.
- Each chapter ends with a summary, key term glossary, formula list, true/false quiz, self-test, troubleshooting quiz, and sectionalized problem set.
- Answers to section checkups, related problems for examples, true/false quiz, self-test, and troubleshooting quiz are at the end of each chapter.
- An end-of-book glossary contains all bold and color terms in the text.
- Answers to odd-numbered problems are at the end of the book.
- Website (www.pearsonhighered.com) includes Multisim files for selected examples and Troubleshooting problems.

Student Resources

- *Experiments in DC/AC Fundamentals: A Systems Approach* (ISBN 0132989867) by David Buchla. Lab exercises are coordinated with the text and solutions are provided in the Instructor's Resource Manual.
- *Multisim Experiments for the DC/AC, Digital, and Devices Courses* (ISBN 0132113880) by Gary Snyder and David Buchla. Students take data, analyze results, and write a conclusion to simulate an actual laboratory experience.
- **Multisim Files Available on the Website**   Circuit files coordinated with this text in Version 11 and 12 of Multisim are located on the website at www.pearsonhighered.com/floyd. Circuit files with prefix E are example circuits and files with prefix P are Multisim troubleshooting circuits.

In order to use the Multisim circuit files, you must have Multisim software installed on your computer. Multisim software is available at **www.ni.com/Multisim**. Although the Multisim circuit files are intended to complement classroom, textbook, and laboratory study, these files are not essential to successfully using this text.

Instructor Resources

Instructor resources are available from Pearson's **Instructor's Resource Center**.

- PowerPoint® slides (ISBN 0132987708) support the topics in each chapter.
- Instructor's Resource Manual (ISBN 0132989875) contains the solutions to the text problems and the solutions to the lab manual.
- TestGen (ISBN 0132989883) This electronic bank of test questions can be used to develop customized quizzes, tests, and/or exams.

To access supplementary materials online, instructors need to request an instructor access code. Go to **www.pearsonhighered.com/irc**, where you can register for an instructor access code. Within 48 hours after registering, you will receive a confirming e-mail, including an instructor access code. Once you have received your code, go to the site and log on for full instructions on downloading the materials you wish to use.

Illustrations of Textbook Features

Chapter Opener A typical chapter opener is shown in Figure P–1.

Worked Example, Related Problem, Multisim Exercise A typical worked-out example with a Related Problem and Multisim exercise is shown in Figure P–2.

Section Opener Each section in a chapter begins with a brief introduction that includes a general overview and section objectives, as shown in Figure P–2.

Section Checkup A typical Section Checkup is shown in Figure P–3. (Answers to the Section Checkups are at the end of the chapter.)

System Example Many System Examples are used to illustrate how the basic concepts are applied in a system application. A typical system example is shown in Figure P–3.

System Notes Many System Notes reinforce the material and give added insight to how it relates to systems. A typical system note is shown in Figure P–4.

Troubleshooting Section A portion of a typical troubleshooting section is shown in Figure P–5.

Hands On Tip A typical Hands On Tip is shown in Figure P–5.

CHAPTER 4

SERIES CIRCUITS

OUTLINE

4–1 Resistors in Series
4–2 Total Series Resistance
4–3 Current in a Series Circuit
4–4 Application of Ohm's Law
4–5 Voltage Sources in Series
4–6 Kirchhoff's Voltage Law
4–7 Voltage Dividers
4–8 Power in Series Circuits
4–9 Voltage Measurements
4–10 Troubleshooting

OBJECTIVES

- Identify a series resistive circuit
- Determine total series resistance
- Determine the current throughout a series circuit
- Apply Ohm's law in series circuits
- Determine the total effect of voltage sources connected in series
- Apply Kirchhoff's voltage law
- Use a series circuit as a voltage divider
- Determine power in a series circuit
- State how to measure voltage with respect to ground
- Troubleshoot series circuits

KEY TERMS

Series
Series-aiding
Series-opposing
Kirchhoff's voltage law

Voltage divider
Reference ground
Open
Short

INTRODUCTION

Resistive circuits can be of two basic forms: series or parallel. In this chapter series circuits are discussed. Parallel circuits are covered in Chapter 5, and combinations of series and parallel circuits are examined in Chapter 6. In this chapter you will see how Ohm's law is used in series circuits; and you will study another important law, Kirchhoff's voltage law. Also, important system applications of series circuits are presented.

VISIT THE WEBSITE
Study aids for this chapter are available at
http://pearsonhighered.com/floyd

FIGURE P–1   Chapter opener.

12–3 ANALYSIS OF SERIES *RL* CIRCUITS

Ohm's law and Kirchhoff's voltage law are used in the analysis of series *RL* circuits to determine voltage, current, and impedance. Also, in this section *RL* lead and lag circuits are examined.

After completing this section, you should be able to

- Analyze a series *RL* circuit
  - Apply Ohm's law and Kirchhoff's voltage law to series *RL* circuits
  - Determine the phase relationships of the voltages and current
  - Show how impedance and phase angle vary with frequency
  - Discuss and analyze the *RL* lag circuit
  - Discuss and analyze the *RL* lead circuit

Ohm's Law

The application of Ohm's law to series *RL* circuits involves the use of the quantities of Z, V, and I. The three equivalent forms of Ohm's law were stated in Chapter 10 for *RC* circuits. They apply also to *RL* circuits and are restated here:

$$V = IZ \qquad I = \frac{V}{Z} \qquad Z = \frac{V}{I}$$

The following example illustrates the use of Ohm's law.

EXAMPLE 12–2

The current in Figure 12–6 is 200 μA. Determine the source voltage.

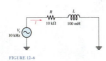

FIGURE 12–6

SOLUTION

From Equation 11–10, the inductive reactance is

$$X_L = 2\pi f L = 2\pi(10\ \text{kHz})(100\ \text{mH}) = 6.28\ \text{k}\Omega$$

The impedance is

$$Z = \sqrt{R^2 + X_L^2} = \sqrt{(10\ \text{k}\Omega)^2 + (6.28\ \text{k}\Omega)^2} = 11.8\ \text{k}\Omega$$

Applying Ohm's law yields

$$V_s = IZ = (200\ \mu\text{A})(11.8\ \text{k}\Omega) = \textbf{2.36 V}$$

RELATED PROBLEM

If the source voltage in Figure 12–6 is 5 V, what would be the current?

MULTISIM

Open Multisim file E12-02; files are found at www.pearsonhighered.com/floyd. Measure the current at 10 kHz, 5 kHz, and 20 kHz. Explain the results of your measurement.

FIGURE P–2   Worked example with Related Problem and Multisim exercise. Example of section opener

FIGURE P–3   Section Checkup and System Example.

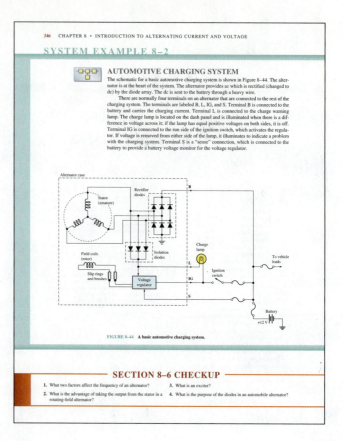

FIGURE P–4   System Note.

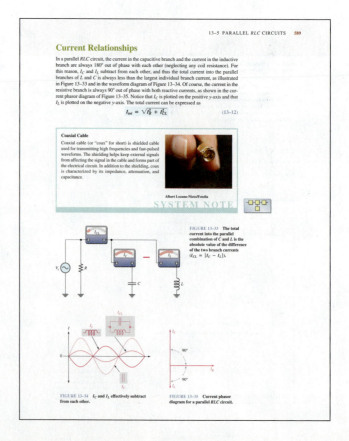

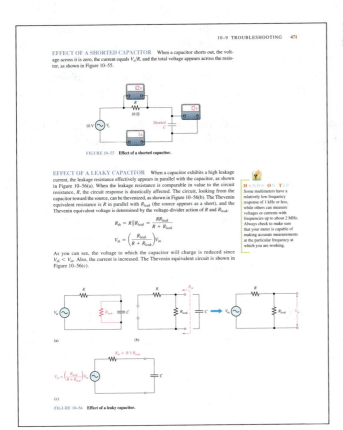

FIGURE P–5   **Troubleshooting section and Hands On Tip.**

Other Features

End of Chapter The following features are at the end of each chapter.

- Summary
- Key term glossary
- Formulas
- True/False quiz
- Self-test
- Troubleshooting quiz
- Sectionalized and categorized problem set
- Answers to section checkups, related problems for examples, true/false quiz, and self-test, and Troubleshooting quiz

End of Book The following features are at the end of book.

- Appendices: Table of Standard Resistor Values, Capacitor Color Coding and Marking, Norton's Theorem and Millman's Theorem, NI Multisim for Circuit Simulation
- Answers to odd-numbered problems
- Comprehensive glossary
- Index

To the Student

Any career training requires effort, and the electrical/electronics field is no exception. The best way for you to learn new material is by reading, thinking, and doing. This text is designed to help you along the way and illustrate how basic laws are used in the real world.

Read each section of the text carefully and think about what you have read. Sometimes you may need to read the section more than once. Work through each example problem step-by-step before you try the related problem that goes with the example. Read the System Examples and System Notes to see how the material relates to practical situations. After each section, answer the checkup questions. Answers to the related problems and the section checkup questions are at the end of the chapter.

Review the chapter summary, the key term definitions, and the formula list. Take the true/false quiz, the multiple-choice self-test, and the troubleshooting quiz. Check your answers against those at the end of the chapter. Finally, work the problems and verify your answers to the odd-numbered problems with those provided at the end of the book.

The importance of obtaining a thorough understanding of the basic DC/AC laws and concepts contained in this text cannot be overemphasized. These will prove to be invaluable when you are dealing with complex circuits and systems. Most employers prefer to hire people who have both a thorough grounding in the basics and the ability and eagerness to grasp new concepts and techniques. If you have a good training in the basics, an employer will train you in the specifics of the job to which you are assigned.

Careers in Electronics

The fields of electricity and electronics are very diverse, and career opportunities are available in many areas. There are many types of job classifications for which a person with training in electricity and electronics technology may qualify. Many contemporary industries such as biotechnology and medical fields require technical skills. The recent emphasis on renewable energy systems has increased the need for electrical and electronic skills to assist in developing, manufacturing, installing, and maintaining these systems. A newer field is that of the "mech-tech", a synergetic combination of mechanical and electronic technician. Mech-techs work on robotic assemblies and other systems in which both mechanical and electronic skills are required. Other workers that require technical skills are those people working in quality control, customer service or training, technical writers, and technical sales.

Milestones in Electronics

Let's briefly look at some of the important developments that led to the electronics technology we have today. The names of many of the early pioneers in electricity and electromagnetics still live on in terms of familiar units and quantities. Names such as Ohm, Ampere, Volta, Farad, Henry, Coulomb, Oersted, and Hertz are some of the better known examples. More widely known names such as Franklin and Edison are also significant in the history of electricity and electronics because of their tremendous contributions.

The Beginning of Electronics   Early experiments with electronics involved electric currents in vacuum tubes. Heinrich Geissler (1814–1879) removed most of the air from a glass tube and found that the tube glowed when there was current through it. Later, Sir William Crookes (1832–1919) found the current in vacuum tubes seemed to consist of particles. Thomas Edison (1847–1931) experimented with carbon filament bulbs with plates and discovered that there was a current from the hot filament to a positively charged plate. He patented the idea but never used it. Other early experimenters measured the properties of the particles that flowed in vacuum tubes. Sir Joseph Thompson (1856–1940) measured properties of these particles, later called *electrons*.

Although wireless telegraphic communication dates back to 1844, electronics is basically a 20th century concept that began with the invention of the vacuum tube amplifier. An early vacuum tube that allowed current in only one direction was constructed by John A. Fleming in 1904. Called the Fleming valve, it was the forerunner of vacuum tube diodes. In 1907, Lee deForest added a grid to the vacuum tube. The new device, called the audiotron, could amplify a weak signal. By adding the control element, deForest ushered in the electronics revolution. It was with an improved version of his device that made transcontinental telephone service and radios possible. In 1912, a radio amateur in San Jose, California, was regularly broadcasting music!

In 1921, the secretary of commerce, Herbert Hoover, issued the first license to a broadcast radio station; within two years over 600 licenses were issued. By the end of the 1920s radios were in many homes. A new type of radio, the superheterodyne radio, invented by Edwin Armstrong, solved problems with high-frequency communication. In 1923, Vladimir Zworykin, an American researcher, invented the first television picture tube, and in 1927 Philo T. Farnsworth applied for a patent for a complete television system.

The 1930s saw many developments in radio, including metal tubes, automatic gain control, "midget sets," directional antennas, and more. Also started in this decade was the development of the first electronic computers. Modern computers trace their origins to the work of John Atanasoff at Iowa State University. Beginning in 1937, he envisioned a binary machine that could do complex mathematical work. By 1939, he and graduate student Clifford Berry had constructed a binary machine called ABC, (for Atanasoff-Berry Computer) that used vacuum tubes for logic and condensers (capacitors) for memory. In 1939, the magnetron, a microwave oscillator, was invented in Britain by Henry Boot and John Randall. In the same year, the klystron microwave tube was invented in America by Russell and Sigurd Varian.

During World War II, electronics developed rapidly. Radar and very high-frequency communication were made possible by the magnetron and klystron. Cathode ray tubes were improved for use in radar. Computer work continued during the war. By 1946, John von Neumann had developed the first stored program computer, the Eniac, at the University of Pennsylvania. The decade ended with one of the most important inventions ever, the transistor.

Solid-State Electronics   The crystal detectors used in early radios were the forerunners of modern solid-state devices. However, the era of solid-state electronics began with the invention of the transistor in 1947 at Bell Labs. The inventors were Walter Brattain, John Bardeen, and William Shockley. PC (printed circuit) boards were introduced in 1947, the year the transistor was invented. Commercial manufacturing of transistors began in Allentown, Pennsylvania, in 1951.

The most important invention of the 1950s was the integrated circuit. On September 12, 1958, Jack Kilby, at Texas Instruments, made the first integrated circuit (IC). This invention literally created the modern computer age and brought about sweeping changes in medicine, communication, manufacturing, and the entertainment industry. Many billions of "chips"—as integrated circuits came to be called—have since been manufactured.

The 1960s saw the space race begin and spurred work on miniaturization and computers. The space race was the driving force behind the rapid changes in electronics that followed. The first successful "op-amp" was designed by Bob Widlar at Fairchild Semiconductor in 1965. Called the μA709, it was very successful but suffered from "latch-up" and other problems. Later, the most popular op-amp ever, the 741, was taking shape at Fairchild. This op-amp became the industry standard and influenced design of op-amps for years to come.

By 1971, a new company that had been formed by a group from Fairchild introduced the first microprocessor. The company was Intel and the product was the 4004 chip, which had the same processing power as the Eniac computer. Later in the same year, Intel announced the first 8-bit processor, the 8008. In 1975, the first personal computer was introduced by Altair, and Popular Science magazine featured it on the cover of the January, 1975, issue. The 1970s also saw the introduction of the pocket calculator and new developments in optical integrated circuits.

By the 1980s, half of all U.S. homes were using cable hookups instead of television antennas. The reliability, speed, and miniaturization of electronics continued throughout the 1980s, including automated testing and calibrating of PC boards. The computer became a part of instrumentation and the virtual instrument was created. Computers became a standard tool on the workbench.

The 1990s saw a widespread application of the Internet. In 1993, there were 130 websites, and now there are millions. Companies scrambled to establish a home page and many of the early developments of radio broadcasting had parallels with the Internet. In 1995, the FCC allocated spectrum space for a new service called Digital Audio Radio Service. Digital television standards were adopted in 1996 by the FCC for the nation's next generation of broadcast television.

One of the major technology stories of the 21st century has been the continuous and explosive growth of the Internet. Wireless broadband access has helped fuel the growth tremendously. The processing speed of computers is increasing at a steady rate and data storage media capacity is increasing at an amazing pace. Carbon nanotubes are seen to be the next step forward for computer chips, eventually replacing transistor technology.

During the first decade of the 21st century, communication networks made major shifts to fiber optics to obtain the highest data throughput possible. Underwater fiber optic cables continue to be an important part of worldwide communication for television, phones, and the Internet. Computer speed and power increased and new Bluetooth technology enabled high-speed, short-range radio communication between various devices, such as hands-free cell phones, computers, GPS receivers, and other wireless networked devices.

In the second decade of the 21st century, the advances in electronics technology continue with amazing new devices and applications of technology, especially in robotic assembly and automation. In recent years, the search for alternative sources of energy has led to intensive research and development in the areas of batteries, solar cells, fuel cells, wind energy, as well as advances in automotive technology and efficiency. Advances in medicine have led to the development of prosthetics that may one day allow artificial limbs to be controlled by their wearers and add sensations of touch and feeling by providing a biocompatible material through which nerve fibers can grow. Consumers continue to demand the latest gadgets and manufacturers are constantly striving to meet that demand with the latest technical toy. Advances will continue into the foreseeable future in this exciting field.

Acknowledgments

The concept of this series of systems-oriented textbooks is credited to suggestions and discussions with senior instructional staff at ITT Schools and Vern Anthony at Pearson Education. The staff and others at Pearson Education, by their hard work and dedication, have helped make the textbook a reality. Lois Porter, who did the manuscript editing, has done a great job of suggesting improvements and has added her usual amazing attention to detail. Rex Davidson skillfully guided the work through its many detailed phases of production to create the end product that you are now looking at. Lindsey Prudhomme, acquisitions editor, and Dan Trudden, development editor, have provided effective overall guidance for this project. We also thank Mark Walters at National Instruments Corporation for his assistance in preparing the Multisim appendix and to Gary Snyder for help in preparing the Multisim files.

TOM FLOYD
DAVID BUCHLA

CHAPTER 1

SYSTEMS, QUANTITIES, AND UNITS

OUTLINE

OBJECTIVES

- Describe the electronics industry in general terms
- Describe attributes of a system
- Describe attributes of circuits in general
- Use scientific notation to represent quantities
- Work with electrical units and metric prefixes
- Convert from one unit with a metric prefix to another
- Express measured data with the proper number of significant digits
- Recognize electrical hazards and practice proper safety procedures

KEY TERMS

Circuit	System
Vertical organization	Boundary
Horizontal organization	Input
Integrated circuit	Output
Mechatronics	Block diagram
Transfer curve	Exponent
Gain	Engineering notation
Passive component	SI
Active component	Metric prefix
Oscillator	Error
Alternator	Accuracy
Digital	Precision
Analog	Significant digit
Transducer	Round off
Scientific notation	Electrical shock
Power of ten	

INTRODUCTION

Technical electronics has changed from less emphasis on troubleshooting discrete circuits at the component level into more work whereby technicians install and test systems. Technicians need to become more oriented to system integration than in the past. In order to understand how systems operate, anyone entering a technical field should be grounded in fundamental laws. This chapter begins with an overview of the electronics industry and attributes of systems and circuits. It then covers the basic units, electrical quantities, and engineering and scientific notation used in the field of electronics.

VISIT THE WEBSITE
Study aids for this chapter are available at
http://pearsonhighered.com/floyd

The electronics industry is made up of companies that are competing for business in manufacturing and marketing electronics products. Because of this fierce competition, the level of innovation and new products has been spectacular and continues to this day. One change in the industry has been the growth of small specialized companies with a narrow market area.

After completing this section, you should be able to

- **Describe the electronics industry in general terms**
  - **Describe briefly the processes engineers use to design and test a new circuit**
  - **Explain the difference between vertical and horizontal company organization**
  - **Cite an example of skills that service technicians need to apply to systems work**
  - **Discuss advantages to having a certification in a particular field**

The electronics industry has witnessed revolutionary technological advances, especially in semiconductors. These advances have changed the industry and the way circuits are developed, constructed, and repaired. A **circuit**[*] is an interconnection of electrical components designed to produce a desired result. Circuits have become more compact and with more reliability than in the past. Today, engineers use circuit design software at a workstation to begin the design process for a new circuit. The software simulates the performance of the circuit and looks for potential problems (such as timing issues, and thermal or noise problems). Computer simulations can enable a circuit to be optimized with minimal cost. When the engineer is satisfied that the simulation meets the requirements for the circuit, the computer automatically lays out the circuit on a printed circuit (PC) board and forms a complete parts list. From this, a prototype is constructed in 48 hours or less either in-house or by a company that specializes in prototyping PC boards. An engineer tests the prototype circuit, and the design is either modified or approved for production by a review team.

In addition to computer-aided design and simulation, in recent years companies have tended to shift their focus away from **vertical organization** with a definite hierarchy to a focus that is more specialized. A vertical organization may go from raw material to finished product in one facility. A company with a **horizontal organization** is one that has a structure where decisions are decentralized. The focus can be more on a specialty. Companies tend to outsource any work not directly related to their specialty. One aspect of horizontal organization is that many smaller companies are involved in producing intermediate components that go into the final product. A specialized production company then assembles the final product from the components. These smaller specialized companies are often centered near each other in regional areas to save on shipping and to lower inventory costs. For certain goods, many specialized suppliers and assembly companies are located in countries where manufacturing costs are low. The result is that many products do not have a single country of origin.

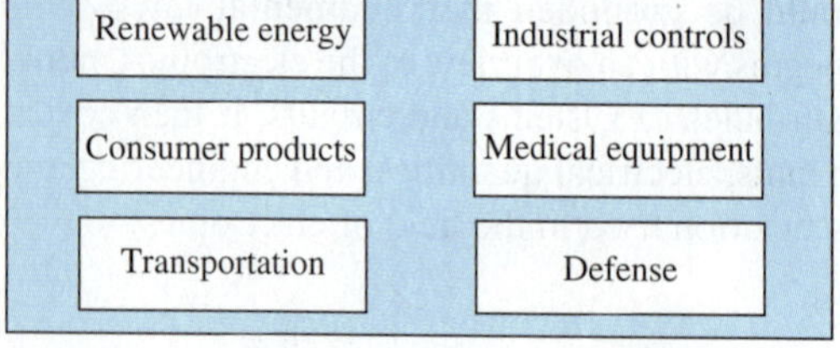

Major Divisions

Two major divisions of the electronics industry are the service sector and the manufacturing sector. The service sector supports all of the manufacturing sectors as well as customers. The manufacturing sector is illustrated in Figure 1–1. Fundamental to all manufacturing is component and semiconductor manufacturers and printed circuit board manufacturers. System manufacturers make systems from fundamental components and/or printed circuit boards. System manufacturers (shown in the blue area) assemble these into completed systems. These manufacturers include companies making communications equipment (including broadcast), computers, renewable energy systems, and more.

Component and board manufacturers

FIGURE 1–1 **The electronics manufacturing industry.**

---

[*]All bold terms are in the end-of-book glossary. The bold terms in color are key terms and are also defined at the end of the chapter.

All of the electronic products that manufacturers produce today involve highly automated processes. These processes can assemble a range of items from integrated circuits to printed circuit boards. An **integrated circuit** is a complex circuit consisting of resistors, transistors, and other components fabricated as a single unit that performs the function of multiple discrete components. The result of automation for consumers has been cheaper products and constant improvements in technology (speed and reliability as well as new features).

As mentioned, the increased reliability of electronic products has caused a shift in the service sector from troubleshooting and repair to a system approach. In the service sector, most technicians need to be able to identify the faulty board, make a quick verification that it is bad, and replace it. Skills that technicians need tend to be broader than the traditional skills required in the past. In the manufacturing sector, technicians may have to work on PLCs (programmable logic controllers), computers, robotics, and mechanical assemblies. Skilled technicians are required to be able to work on a system as a whole for installation, maintenance, and troubleshooting regardless of where a fault may lie. These jobs often involve the new applied engineering classification of **mechatronics**, which is a synergistic combination of mechanics and electronics.

Perhaps surprisingly, the service sector provides more jobs than the manufacturing sector, although there is overlap (as in servicing a machine in a manufacturing plant). Many service sector jobs require a broad range of skills. For example, consider the skills required for the installation and maintenance of solar electrical systems. The technician must have knowledge of electrical and electronic systems and perform electrical and mechanical installations of modules, junction boxes, overcurrent devices, grounding equipment, and pumps. He or she must be able to test and troubleshoot the systems and work safely with both the electrical and nonelectrical hazards associated with these systems.

Manufacturers require people to work on the machines and systems that are used in their industries and reprogram them if necessary. For example, in the food-processing industry, companies commonly utilize many specialized machines for sorting, weighing, and packaging that are controlled by electronic systems. Figure 1–2 shows a check-weight scale that is common in the food-processing industry. The scale is only one element in the overall packaging system. A subassembly may be connected to the scale to acquire data and send it to a computer to determine if the product is within a certain specified range. If the package size were changed, a technician would set the new limits of an acceptable range in the controller.

FIGURE 1–2   An automated check-weight scale in the food-processing industry. (Courtesy of Cornerstone Automation Systems, LLC.)

The service sector also requires technicians with electronics skills for a diverse range of jobs that includes the installation and repair of highly technical communication equipment (such as found in the broadcast industry), system integrators, and solar and wind energy systems to name a few. Other related positions in the service field are for technical trainers and writers, customer service representatives, and sales.

Certification

In virtually every field, there are professional organizations that issue certifications to show competency in that field. Certification allows individuals to showcase their skills in a given technical area and at the same time offers a means for employers and/or the public for judging that competency. In the broadcast field, a certification agency is iNARTE (interNational Association for Radio, Telecommunications, and Electromagnetics). This organization also administers FCC Commercial Operator License Examinations. In the energy field, the North American Board of Certified Energy Practitioners (NABCEP) is a certifying agency for several technical jobs. It can issue installer certification for PV systems, solar thermal systems, or small wind systems. Certification in any field does not replace any license requirements required by a state or governing agency but is useful for potential job applicants to show skill in that field.

SECTION 1–1 CHECKUP*

1. What are the advantages to having a computer simulation of a new circuit before it is constructed?

2. What are the two major divisions of the electronics industry?

3. Name six product categories in the electronics systems manufacturing industry.

4. What is the purpose of certification?

*Answers are at the end of the chapter.

1–2 INTRODUCTION TO ELECTRONIC SYSTEMS

An electronic system is an assembly of components and circuits designed to accomplish a specific function. Examples of electronic systems range from a simple garage door opener to a complex system such as a radar system.

After completing this section, you should be able to

- **Describe attributes of a system**
  - **Define the word *system* as it applies to electrical and electronic systems**
  - **Explain the purpose of a block diagram and read a basic block diagram**
  - **Give an example of a transfer curve as applied to a block within an electronic system.**

System Concepts

A **system** is a group of interrelated parts that perform a specific function. A **boundary** is the dividing line between what is part of the system and its environment, which is everything else, as illustrated in Figure 1–3.

A system communicates to the outside world through its inputs and outputs. An **input** is the voltage, current, or power that is applied to an electrical circuit to achieve a desired result. An **output** is the result obtained from the system after processing one or more inputs. Systems can have multiple inputs and multiple outputs. No system is completely isolated from its environment; however, for simplifying the analysis of a system, it is often treated as if it were isolated. Even a system that seems isolated, such as the International Space Station (ISS), receives solar energy, radiates heat into space, and is affected by the gravitational field of the earth.

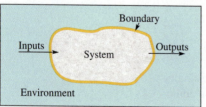

FIGURE 1–3   A system and its environment.

ELECTRICAL SYSTEMS   Electrical systems deal with electric power. As an example of an electrical system, consider the wiring for a house. Normally, the outside walls and roof are defined as the boundary; and the interior space, including the basement and attic, is defined as the system. The connection to the utility grid is done at a panel and is considered to be an input to the system. The outputs represent specific points within the house (or outside) where loads are connected. The boundary can be changed to suit the analysis. For the house-wiring system, a portion of this (such as just the kitchen circuits) could be considered a system or a subsystem.

ELECTRONIC SYSTEMS   Generally, electronic systems deal with signals rather than power. In the context of an electronic system, a *signal* is a changing electrical or electromagnetic quantity that carries information. If a particular input to a system never changes, it is completely predictable. In this case, the input is not considered a signal because no information is present. Most electronic systems process the information contained in a signal. The distinction between electronic and electrical circuits is blurred by the fact that electrical systems frequently use electronic components for regulation. Electronic systems will normally have an electrical subsystem for supplying necessary power.

Block Diagrams

Electronic systems will generally involve a logical sequence of processes. To simplify complicated systems and indicate the order of processing, systems are frequently drawn in block diagram form. A **block diagram** is a model of a system that represents its structure in a graphical format using labeled blocks to represent functions and lines to represent the signal flow. For example, the block diagram of a digital thermometer system is shown in Figure 1–4. Assume that the temperature sensor is one that changes its output voltage based on the temperature of its environment. The signal from the sensor is increased by the amplifier and sent to the ADC. (ADC stands for analog-to-digital converter, which changes the continuous voltage from the amplifier into a digital number.) In addition to basic processing, digital data can be stored and recovered accurately and is much less subject to noise than analog data. The processor converts the digital data to a temperature reading that is displayed. The display for this system is a lighted number. The block diagram shows the key parts of the system and the signal flow without showing the details of the circuits.

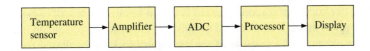

FIGURE 1–4   **A block diagram of a digital thermometer system.**

SYSTEM EXAMPLE 1–1

BLOCK DIAGRAMS AND FLOWCHARTS

Data acquisition systems are used for a variety of applications. Usually, these systems need to convert an analog quantity to digital data for processing. The analog-to-digital converter (ADC) is a key part of this type of system. System analysts can explain the relationship of the ADC, called a successive approximation ADC, in a system by using both a block diagram and a flowchart. A block diagram would be used to illustrate signal flow for the overall system; whereas a flowchart would be used to show the logical process that the control logic block must perform.

The difference between a block diagram and a flowchart is illustrated in Figure 1–5. Part (a) shows a block diagram of the hardware parts, showing the signal flow; it does not show details of the process that the control logic must perform. The control logic block is

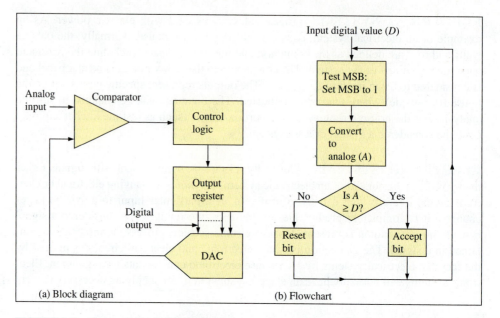

FIGURE 1–5 Comparison of a block diagram and a flowchart.

one part of the overall system. Figure 1–5(b) shows a flowchart diagram that breaks out the logic that the control logic block performs.

Block diagrams show the flow of a signal. The circuit uses a control logic circuit to determine an action to take based on the result of comparing the input (analog) and output (digital) signals. Flowcharts show the logic required to process the signal. The total system is best explained by showing both types of diagrams to clarify the signal flow and the logic.

Transfer Curve

A **transfer curve** is a plot that shows the ratio of the output to the input. In electronic systems, the transfer curve (or response curve) is useful because it describes how the system behaves for a given input. It can be used to illustrate the behavior of a given block or group of blocks. For example, the purpose of the amplifier block in the digital thermometer in Figure 1–4 is to make the small signal from the temperature sensor larger for the ADC. Ideally, the transfer curve for a linear amplifier is a straight line as shown in Figure 1–6. The input is a small voltage that is plotted along the x-axis. The output is a proportionally larger voltage that is plotted on the y-axis. For any small input voltage, the output voltage is larger by a factor known as the **gain**. In the example shown, the gain is 10 because for any given input voltage, the output is 10× larger.

The amplifier transfer curve is a ratio of two voltages; hence it is dimensionless and represents the voltage gain of the amplifier (V_{out}/V_{in}). Ideally, the transfer curve for a linear amplifier is a straight line, indicating constant gain for any input. In practice, amplifiers often approach this ideal for a certain range of inputs, but no amplifier is "perfect."

In some cases, the input and output may have different units; in this case, the transfer curve will not be dimensionless. For example, the transfer curve for a sensor might show a resistance change (a change in opposition to current) as a function of temperature. This type of transfer is shown in Figure 1–7 for a typical thermistor (a type of temperature sensor). Notice that the sensor has a nonlinear response, a point that is easy to spot on the transfer curve. If it is used as a sensor in the digital thermometer system, the data would need some form of processing to convert its output to a temperature reading that can then be displayed.

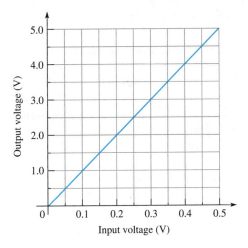

FIGURE 1–6 **Ideal transfer curve for an amplifier that has a gain of 10.**

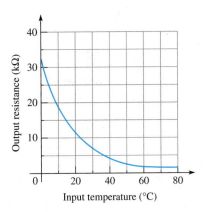

FIGURE 1–7 **Transfer curve for a typical thermistor.**

SECTION 1–2 CHECKUP

1. What is the input and output of a digital thermometer system?

2. Consider the International Space Station as a system. What constitutes the environment of this system?

3. What is the purpose of a block diagram?

4. What name is given to a plot of the ratio of the output to the input for a given block?

1–3 TYPES OF CIRCUITS

Electrical systems deal with power, whereas electronic systems deal with signals. Electrical systems can be either dc (direct current) or ac (alternating current). Electronic systems generally require some form of power, so it is important to have an understanding of both electrical and electronic systems. Fortunately, most of the fundamental laws for both types of systems are the same. This section focuses on the types of circuits used in electrical and electronic systems.

After completing this section, you should be able to

- **Describe circuits in general terms**
  - **Explain the difference between active and passive components**
  - **Name applications for ac and dc circuits in electrical power distribution**
  - **Compare digital and analog electronic circuits**
  - **Define transducer**

Components

The word *circuit* is based on the Latin word *circuitus*, meaning to go around. Electrical circuits must have a complete path (like a circle) that starts at a source, includes a load, and returns to the source. When a circuit has a complete path, it is said to be closed; if the path is broken, it is said to be open. In addition to one or more sources, circuits use components, which are devices that alter the electrical properties of the circuit. Components that do not require electrical power to function are called **passive components**; those that require a source of power are called **active components**. Passive components cannot increase the power in a signal; active components can. Most systems contain both types of components.

Three fundamental passive components are resistors, capacitors, and inductors. Other passive components are diodes, transformers, batteries, and motors. A battery is considered to be a passive component because it cannot increase the power in a signal (although it does supply electrical power).

Active components are generally classified as those that can increase signal power. They include transistors, operational amplifiers, and microprocessors. Sometimes the signal is not an input to the system but is produced internally. An electronic **oscillator** circuit is an example where the output is internally generated, but the device generating the output is considered an active device. Oscillators are circuits that generate a continuous output that can be one of several different shapes. They are commonly used in telecommunications, test instruments, and computers.

Electrical Circuits

Electrical circuits generate, deliver, and control power, which can be in the form of ac (alternating current) or dc (direct current).

AC CIRCUITS

Alternating current (ac) changes polarity a certain number of times each second (the frequency). The power grids in the world are generally ac (with some exceptions). The most common way to produce ac for power delivery is to use a device called an **alternator** (an ac generator). Alternators are discussed in detail in Section 8–6.

AC can be transformed from low-to-high voltage and back easily and with lower expense than dc (however, some very high voltage systems use dc.) High voltages are much more efficient to transmit over long distances, so nearly all power distribution systems use ac. Historically, power companies developed and supplied power to their customers with no thought to interconnections. As a result, a hodgepodge of systems developed with different frequencies and voltages. Eventually, different parts of the world selected one of two frequencies as a standard. For most of the world, the standard frequency is 50 Hz (hertz is the unit of frequency); in the U.S. and North America, it is 60 Hz. In aircraft and certain military applications where weight is critical, 400 Hz is used because it allows smaller and lighter weight components to be used.

DC CIRCUITS

Direct current (dc) is the form of electrical current that does not change polarity (unlike ac). Although most utility power is ac, there is high voltage dc (HVDC) technology used for some long distance transmission and underwater transmission. Although the conversion of dc to the very high voltages required for efficient transmission is more costly than ac, the tower and transmission lines can be designed for a lower cost per mile. For very long runs, dc can be cost effective. It is also useful for interfacing independent ac networks that are not synchronized. This occurs in Japan, where two completely different frequencies are in use (both 50 Hz and 60 Hz). DC is also generated for some low-voltage military and aircraft systems for supplying electrical power; typically 28 V is used.

Electronic Circuits

As you know, electronic circuits work with signals. Pure dc does not have signal information but is very important in electronic applications because it is required by active electronic circuits. For this reason, the study of dc circuits forms a foundation for anyone who wants to have an understanding of electrical or electronic systems. DC sources include batteries, fuel cells, solar cells, and generators. These sources are discussed in Section 2–3. A common method of obtaining dc is to convert ac to dc with a power supply. Power supplies are discussed in Section 3–7.

A signal carries information because of an intentional change in a parameter such as voltage or frequency. There are two major divisions for signal types. Signals that have discrete levels are digital signals; continuous signals are analog signals. In some cases, the signal may be internally generated in the system. For example, Figure 1–8 shows a bicycle warning light. This is an example of a simple digital system in which a flashing signal is internally generated.

Analog signals vary continuously over an allowed range. As an example of an analog signal, consider an AM radio signal. The radio station transmits a high-frequency signal (called the *carrier*) whose amplitude is varied (or modulated) by a lower frequency carrying the information of interest. Figure 1–9 illustrates a simulated AM signal that is an example of an analog signal. In this case, the information is contained in the envelope.

Frequently, you will see electronic circuits that have analog and digital portions to the same circuit (such as the digital thermometer described in the previous section). The conversion from an analog signal to a digital signal is done with a converter (ADC), which often receives its input from a transducer. A **transducer** is a device that transforms energy from one form to another; for electronic systems one of the forms is an electrical signal. The temperature sensor in the digital thermometer system is an example of an input transducer. The transducer converts the temperature to a voltage, which is then converted to a digital signal for processing.

In some cases, such as audio, a digital signal may need to be converted to an analog signal. This is accomplished with a digital-to-analog converter (DAC). For example, a CD player takes a digital signal from the CD, processes it, and converts it to an analog output. The analog signal is amplified and sent to a speaker. The speaker is an output transducer, converting electrical energy to sound.

FIGURE 1–8   **Bicycle warning light.**   (natenn/iStockphoto.com)

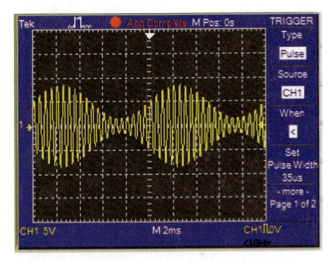

FIGURE 1–9   **A simulated AM radio wave.**

SECTION 1–3 CHECKUP

1. What is the difference between a passive and an active component?

2. Why is a battery not considered to be an electronic device?

3. What are the two frequencies that are used in most of the world's power grids?

4. Why is dc important to active circuits?

5. What is the difference between a digital and an analog circuit?

6. What is a transducer?

1–4 SCIENTIFIC AND ENGINEERING NOTATION

Working with electrical and electronics systems, you will encounter both very small and very large quantities. For example, electrical current can range from hundreds of amperes in power applications to a few thousandths or millionths of an ampere in many electronic circuits. This range of values is typical of many other electrical quantities also. Engineering notation is a specialized form of scientific notation. It is used widely in technical fields to express large and small quantities. In electronics, engineering notation is used to express values of voltage, current, power, resistance, and other quantities.

After completing this section, you should be able to

- **Use scientific notation to represent quantities**
  - **Express any number using a power of ten**
  - **Perform calculations with powers of ten**

Scientific notation provides a convenient method for expressing large and small numbers and for performing calculations involving such numbers. In scientific notation, a quantity is expressed as a product of a number between 1 and 10 (one digit to the left of the decimal point) and a power of ten. For example, the quantity 150,000 is expressed in scientific notation as 1.5×10^5, and the quantity 0.00022 is expressed as 2.2×10^{-4}.

Powers of Ten

Table 1–1 lists some powers of ten, both positive and negative, and the corresponding decimal numbers. The **power of ten** is expressed as an *exponent* of the *base* 10 in each case.

Base   Exponent

$$10^x$$

An **exponent** is a number to which a base number is raised. The exponent indicates the number of places that the decimal point is moved to the right or left to produce the decimal number. For a positive power of ten, move the decimal point to the right to get the equivalent decimal number. As an example, for an exponent of 4,

$$10^4 = 1 \times 10^4 = 1.0000. = 10,000.$$

TABLE 1–1 • Some positive and negative powers of ten.	
$10^6 = 1,000,000$	$10^{-6} = 0.000001$
$10^5 = 100,000$	$10^{-5} = 0.00001$
$10^4 = 10,000$	$10^{-4} = 0.0001$
$10^3 = 1,000$	$10^{-3} = 0.001$
$10^2 = 100$	$10^{-2} = 0.01$
$10^1 = 10$	$10^{-1} = 0.1$
$10^0 = 1$	

For a negative power of ten, move the decimal point to the left to get the equivalent decimal number. As an example, for an exponent of −4,

$$10^{-4} = 1 \times 10^{-4} = .0001. = 0.0001$$

The negative exponent does not indicate that a number is negative; it simply moves the decimal point to the left.

EXAMPLE 1–1

Express each number in scientific notation:

(a) 240   (b) 5100   (c) 85,000   (d) 3,350,000

SOLUTION

In each case, move the decimal point an appropriate number of places to the left to determine the positive power of ten.

(a) $240 = 2.4 \times 10^2$ (b) $5100 = 5.1 \times 10^3$

(c) $85,000 = 8.5 \times 10^4$ (d) $3,350,000 = 3.35 \times 10^6$

RELATED PROBLEM*

Express 750,000,000 in scientific notation.

*Answers are at the end of the chapter.

EXAMPLE 1–2

Express each number in scientific notation:

(a) 0.24   (b) 0.005   (c) 0.00063   (d) 0.000015

SOLUTION

In each case, move the decimal point an appropriate number of places to the right to determine the negative power of ten.

(a) $0.24 = 2.4 \times 10^{-1}$ (b) $0.005 = 5 \times 10^{-3}$

(c) $0.00063 = 6.3 \times 10^{-4}$ (d) $0.000015 = 1.5 \times 10^{-5}$

RELATED PROBLEM

Express 0.00000093 in scientific notation.

EXAMPLE 1–3

Express each of the following numbers as a normal decimal number:

(a) 1×10^5 (b) 2.9×10^3 (c) 3.2×10^{-2} (d) 2.5×10^{-6}

SOLUTION

Move the decimal point to the right or left a number of places indicated by the positive or the negative power of ten respectively.

(a) $1 \times 10^5 = 100,000$ (b) $2.9 \times 10^3 = 2900$

(c) $3.2 \times 10^{-2} = 0.032$ (d) $2.5 \times 10^{-6} = 0.0000025$

RELATED PROBLEM

Express 8.2×10^8 as a normal decimal number.

Calculations With Powers of Ten

The advantage of scientific notation is in addition, subtraction, multiplication, and division of very small or very large numbers.

ADDITION   The steps for adding numbers in powers of ten are as follows:

1. Express the numbers to be added in the same power of ten.
2. Add the numbers without their powers of ten to get the sum.
3. Bring down the common power of ten, which becomes the power of ten of the sum.

EXAMPLE 1–4

Add 2×10^6 and 5×10^7 and express the result in scientific notation.

SOLUTION

1. Express both numbers in the same power of ten: $(2 \times 10^6) + (50 \times 10^6)$.
2. Add $2 + 50 = 52$.
3. Bring down the common power of ten (10^6); the sum is $52 \times 10^6 =$ $\mathbf{5.2 \times 10^7}$.

RELATED PROBLEM

Add 4.1×10^3 and 7.9×10^2.

SUBTRACTION   The steps for subtracting numbers in powers of ten are as follows:

1. Express the numbers to be subtracted in the same power of ten.
2. Subtract the numbers without their powers of ten to get the difference.
3. Bring down the common power of ten, which becomes the power of ten of the difference.

EXAMPLE 1–5

Subtract 2.5×10^{-12} from 7.5×10^{-11} and express the result in scientific notation.

SOLUTION

1. Express each number in the same power of ten: $(7.5 \times 10^{-11}) -$ (0.25×10^{-11}).
2. Subtract $7.5 - 0.25 = 7.25$.
3. Bring down the common power of ten (10^{-11}); the difference is $\mathbf{7.25 \times 10^{-11}}$.

RELATED PROBLEM

Subtract 3.5×10^{-6} from 2.2×10^{-5}.

MULTIPLICATION   The steps for multiplying numbers in powers of ten are as follows:

1. Multiply the numbers directly without their powers of ten.
2. Add the powers of ten algebraically (the exponents do not have to be the same).

EXAMPLE 1–6

Multiply 5×10^{12} by 3×10^{-6} and express the result in scientific notation.

SOLUTION

Multiply the numbers, and algebraically add the powers.

$$(5 \times 10^{12})(3 \times 10^{-6}) = 15 \times 10^{12+(-6)} = 15 \times 10^6 = \mathbf{1.5 \times 10^7}$$

RELATED PROBLEM

Multiply 1.2×10^3 by 4×10^2.

DIVISION   The steps for dividing numbers in powers of ten are as follows:

1. Divide the numbers directly without their powers of ten.
2. Subtract the power of ten in the denominator from the power of ten in the numerator (the exponents do not have to be the same).

EXAMPLE 1–7

Divide 5.0×10^8 by 2.5×10^3 and express the result in scientific notation.

SOLUTION

Write the division problem with a numerator and denominator.

$$\frac{5.0 \times 10^8}{2.5 \times 10^3}$$

Divide the numbers and subtract the powers of ten (3 from 8).

$$\frac{5.0 \times 10^8}{2.5 \times 10^3} = 2 \times 10^{8-3} = \mathbf{2 \times 10^5}$$

RELATED PROBLEM

Divide 8×10^{-6} by 2×10^{-10}.

SCIENTIFIC NOTATION ON A CALCULATOR   Entering a number in scientific notation is accomplished on most calculators using the EE key as follows: Enter the number with one digit to the left of the decimal point, press EE, and enter the power of ten. This method requires that the power of ten be determined before entering the number. Some calculators can be placed in a mode that will automatically convert any decimal number entered into scientific notation.

EXAMPLE 1–8

Enter 23,560 in scientific notation using the EE key.

SOLUTION

Move the decimal point four places to the left so that it comes after the digit 2. This results in the number expressed in scientific notation as

$$2.3560 \times 10^4$$

Enter this number on your calculator as follows:

| 2 | • | 3 | 5 | 6 | EE | 4 | 2.3560E4 |

Note that it is not necessary to enter the zero.

RELATED PROBLEM

Enter the number 573,946 using the EE key.

Engineering Notation

Engineering notation is similar to scientific notation. However, in **engineering notation** a number can have from one to three digits to the left of the decimal point and the power-of-ten exponent must be a multiple of three. For example, the number 33,000 expressed in engineering notation is 33×10^3. In scientific notation, it is expressed as 3.3×10^4. As another example, the number 0.045 is expressed in engineering notation as 45×10^{-3}. In scientific notation, it is expressed as 4.5×10^{-2}. Engineering notation is useful in electrical and electronic calculations that use metric prefixes (discussed in Section 1–5).

EXAMPLE 1–9

Express the following numbers in engineering notation:

(a) 82,000   **(b)** 243,000   **(c)** 1,956,000

SOLUTION

In engineering notation,

(a) 82,000 is expressed as **82×10^3**.
(b) 243,000 is expressed as **243×10^3**.
(c) 1,956,000 is expressed as **1.956×10^6**.

RELATED PROBLEM

Express 36,000,000,000 in engineering notation.

EXAMPLE 1–10

Convert each of the following numbers to engineering notation:

(a) 0.0022   **(b)** 0.000000047   **(c)** 0.00033

SOLUTION

In engineering notation,

(a) 0.0022 is expressed as **2.2×10^{-3}**.
(b) 0.000000047 is expressed as **47×10^{-9}**.
(c) 0.00033 is expressed as **330×10^{-6}**.

RELATED PROBLEM

Express 0.0000000000056 in engineering notation.

ENGINEERING NOTATION ON A CALCULATOR   Use the EE key to enter the number with one, two, or three digits to the left of the decimal point, press EE, and enter the power of ten that is a multiple of three. This method requires that the appropriate power of ten be determined before entering the number.

EXAMPLE 1–11

Enter 51,200,000 in engineering notation using the EE key.

SOLUTION

Move the decimal point six places to the left so that it comes after the digit 1. This results in the number expressed in engineering notation as

$$51.2 \times 10^6$$

Enter this number on your calculator as follows:

         51.2E6

RELATED PROBLEM

Enter the number 273,900 in engineering notation using the EE key.

SECTION 1–4 CHECKUP

1. Scientific notation uses powers of ten. (True or False)

2. Express 100 as a power of ten.

3. Express the following numbers in scientific notation:

  (a) 4350

  (b) 12,010

  (c) 29,000,000

4. Express the following numbers in scientific notation:

  (a) 0.760

  (b) 0.00025

  (c) 0.000000597

5. Do the following operations:

  (a) $(1 \times 10^5) + (2 \times 10^5)$

  (b) $(3 \times 10^6)(2 \times 10^4)$

  (c) $(8 \times 10^3) \div (4 \times 10^2)$

  (d) $(2.5 \times 10^{-6}) - (1.3 \times 10^{-7})$

6. Enter the numbers expressed in scientific notation in Problem 3 into your calculator.

7. Express the following numbers in engineering notation:

  (a) 0.0056

  (b) 0.0000000283

  (c) 950,000

  (d) 375,000,000,000

8. Enter the numbers in Problem 7 into your calculator using engineering notation.

In electronics, you must deal with measurable quantities. For example, you must be able to express how many volts are measured at a certain test point in a circuit, how much current there is through a conductor, or how much power a certain amplifier delivers. In this section, you are introduced to the units and symbols for most of the electrical quantities that are used throughout the book. Metric prefixes are used in conjunction with engineering notation as a "shorthand" for the certain powers of ten that commonly are used.

After completing this section, you should be able to

- Work with electrical units and metric prefixes
  - Name the units for twelve electrical quantities
  - Specify the symbols for the electrical units
  - List the metric prefixes
  - Change a power of ten in engineering notation to a metric prefix
  - Use metric prefixes to express electrical quantities

Electrical Units

Letter symbols are used in electronics to represent both quantities and their units. One symbol is used to represent the name of the quantity, and another is used to represent the unit of measurement of that quantity. Table 1–2 lists the most important electrical quantities, along with their SI units and symbols. For example, italic P stands for *power* and nonitalic (roman) W stands for *watt*, which is the unit of power. In general, italic letters represent quantities and nonitalic letters represent units. Notice that energy is abbreviated with an italic W that represents *work*; and both *energy* and *work* have the same unit (the joule). The term **SI** is the French abbreviation for *International System* (*Système International* in French).

TABLE 1–2 • Electrical quantities and their corresponding units with SI symbols.			
QUANTITY	**SYMBOL**	**SI UNIT**	**SYMBOL**
capacitance	C	farad	F
charge	Q	coulomb	C
conductance	G	siemens	S
current	I	ampere	A
energy or work	W	joule	J
frequency	f	hertz	Hz
impedance	Z	ohm	Ω
inductance	L	henry	H
power	P	watt	W
reactance	X	ohm	Ω
resistance	R	ohm	Ω
voltage	V	volt	V

In addition to the common electrical units shown in Table 1–2, the SI system has many other units that are defined in terms of certain fundamental units. In 1954, by international agreement, *meter, kilogram, second, ampere, degree kelvin,* and *candela* were adopted as the basic SI units (*degree kelvin* was later changed to just *kelvin*). These units form the basis of the mks (for meter-kilogram-second) units that are used for derived quantities and have become the preferred units for nearly all scientific and engineering work. An older metric system, called the cgs system, was based on the centimeter, gram, and second as fundamental units. There are still a number of units in common use based on the cgs system; for example, the gauss is a magnetic flux unit in the cgs system and is still in common usage. In keeping with preferred practice, this text uses mks units, except when otherwise noted.

Metric Prefixes

In engineering notation **metric prefixes** represent each of the most commonly used powers of ten. These metric prefixes are listed in Table 1–3 with their symbols and corresponding powers of ten.

TABLE 1–3 • Metric prefixes with their symbols and corresponding powers of ten and values.			
METRIC PREFIX	**SYMBOL**	**POWER OF TEN**	**VALUE**
femto	f	10^{-15}	one-quadrillionth
pico	p	10^{-12}	one-trillionth
nano	n	10^{-9}	one-billionth
micro	μ	10^{-6}	one-millionth
milli	m	10^{-3}	one-thousandth
kilo	k	10^{3}	one thousand
mega	M	10^{6}	one million
giga	G	10^{9}	one billion
tera	T	10^{12}	one trillion

Metric prefixes are used only with numbers that have a unit of measure, such as volts, amperes, and ohms, and precede the unit symbol. For example, 0.025 amperes can be expressed in engineering notation as 25×10^{-3} A. This quantity expressed using a metric prefix is 25 mA, which is read 25 milliamps. The metric prefix *milli* has replaced 10^{-3}. As another example, the Alpha Ventus offshore wind power system has 12 wind generators, each capable of delivering 5,000,000 W, which is expressed as 5.0 MW. The system generates 60,000,000 W, which is 60 MW. The metric prefix *mega* has replaced 10^{6} in expressing the power in both cases.

EXAMPLE 1–12

Express each quantity using a metric prefix:

(a) 50,000 V   (b) 25,000,000 Ω   (c) 0.000036 A

SOLUTION

(a) 50,000 V $= 50 \times 10^{3}$ V $= \textbf{50 kV}$

(b) 25,000,000 Ω $= 25 \times 10^{6}$ Ω $= \textbf{25 M}\boldsymbol{\Omega}$

(c) 0.000036 A $= 36 \times 10^{-6}$ A $= \textbf{36 } \boldsymbol{\mu}\textbf{A}$

RELATED PROBLEM

Express each quantity using metric prefixes:

(a) 56,000,000 Ω   (b) 0.000470 A

SECTION 1–5 CHECKUP

1. List the metric prefix for each of the following powers of ten: 10^6, 10^3, 10^{-3}, 10^{-6}, 10^{-9}, and 10^{-12}.

2. Use a metric prefix to express 0.000001 A.

3. Use a metric prefix to express 250,000 W.

1–6 METRIC UNIT CONVERSIONS

It is sometimes necessary or convenient to convert a quantity from one unit with a metric prefix to another, such as from milliamperes (mA) to microamperes (μA). Moving the decimal point in the number an appropriate number of places to the left or to the right, depending on the particular conversion, results in a metric unit conversion.

After completing this section, you should be able to

- **Convert from one unit with a metric prefix to another**
  - **Convert between milli, micro, nano, and pico**
  - **Convert between kilo and mega**

The following basic rules apply to metric unit conversions:

1. When converting from a larger unit to a smaller unit, move the decimal point to the right.

2. When converting from a smaller unit to a larger unit, move the decimal point to the left.

3. Determine the number of places to move the decimal point by finding the difference in the powers of ten of the units being converted.

For example, when converting from milliamperes (mA) to microamperes (μA), move the decimal point three places to the right because there is a three-place difference between the two units (mA is 10^{-3} A and μA is 10^{-6} A). The following examples illustrate a few conversions.

EXAMPLE 1–13

Convert 0.15 milliampere (0.15 mA) to microamperes (μA).

SOLUTION

Move the decimal point three places to the right.

$$0.15\,\text{mA} = 0.15 \times 10^{-3}\,\text{A} = 150 \times 10^{-6}\,\text{A} = \mathbf{150\,\mu A}$$

RELATED PROBLEM

Convert 1 mA to microamperes.

EXAMPLE 1–14

Convert 4500 microvolts (4500 μV) to millivolts (mV).

SOLUTION

Move the decimal point three places to the left.

$$4500\,\mu\text{V} = 4500 \times 10^{-6}\,\text{V} = 4.5 \times 10^{-3}\,\text{V} = \textbf{4.5 mV}$$

RELATED PROBLEM

Convert 1000 μV to millivolts.

EXAMPLE 1–15

Convert 5000 nanoamperes (5000 nA) to microamperes (μA).

SOLUTION

Move the decimal point three places to the left.

$$5000\,\text{nA} = 5000 \times 10^{-9}\,\text{A} = 5 \times 10^{-6}\,\text{A} = \textbf{5}\,\boldsymbol{\mu}\textbf{A}$$

RELATED PROBLEM

Convert 893 nA to microamperes.

EXAMPLE 1–16

Convert 47,000 picofarads (47,000 pF) to microfarads (μF).

SOLUTION

Move the decimal point six places to the left.

$$47,000\,\text{pF} = 47,000 \times 10^{-12}\,\text{F} = 0.047 \times 10^{-6}\,\text{F} = \textbf{0.047}\,\boldsymbol{\mu}\textbf{F}$$

RELATED PROBLEM

Convert 10,000 pF to microfarads.

EXAMPLE 1–17

Convert 0.00022 microfarad (0.00022 μF) to picofarads (pF).

SOLUTION

Move the decimal point six places to the right.

$$0.00022\,\mu\text{F} = 0.00022 \times 10^{-6}\,\text{F} = 220 \times 10^{-12}\,\text{F} = \textbf{220 pF}$$

RELATED PROBLEM

Convert 0.0022 μF to picofarads.

EXAMPLE 1–18

Convert 1800 kilohms (1800 kΩ) to megohms (MΩ).

SOLUTION

Move the decimal point three places to the left.

$$1800 \text{ k}\Omega = 1800 \times 10^3 \, \Omega = 1.8 \times 10^6 \, \Omega = \textbf{1.8 M}\boldsymbol{\Omega}$$

RELATED PROBLEM

Convert 2.2 kΩ to megohms.

When adding (or subtracting) quantities with different metric prefixes, first convert one of the quantities to the same prefix as the other quantity.

EXAMPLE 1–19

Add 15 mA and 8000 μA and express the result in milliamperes.

SOLUTION

Convert 8000 μA to 8 mA and add.

$$15 \text{ mA} + 8000 \, \mu\text{A} = 15 \text{ mA} + 8 \text{ mA} = \textbf{23 mA}$$

RELATED PROBLEM

Add 2873 mA and 10,000 μA.

SECTION 1–6 CHECKUP

1. Convert 0.01 MV to kilovolts (kV).

2. Convert 250,000 pA to milliamperes (mA).

3. Add 0.05 MW and 75 kW and express the result in kW.

4. Add 50 mV and 25,000 μV and express the result in mV.

1–7 MEASURED NUMBERS

Whenever a quantity is measured, there is uncertainty in the result due to limitations of the instruments used. When a measured quantity contains approximate numbers, the digits known to be correct are called significant digits. When reporting measured quantities, the number of digits that should be retained are the significant digits and no more than one uncertain digit.

After completing this section, you should be able to

- **Express measured data with the proper number of significant digits**
  - **Define** *accuracy, error,* **and** *precision*
  - **Round numbers properly**

Error, Accuracy, and Precision

Data taken in experiments are not perfect because the accuracy of the data depends on the accuracy of the test equipment and the conditions under which the measurement was made. In order to properly report measured data, the error associated with the measurement should be taken into account. Experimental error should not be thought of as a mistake. All measurements that do not involve counting are approximations of the true value. The difference between the true or best-accepted value of some quantity and the measured value is the **error**. A measurement is said to be accurate if the error is small. **Accuracy** is an indication of the range of error in a measurement. For example, if you measure thickness of a 10.00 mm gauge block with a micrometer and find that it is 10.8 mm, the reading is not accurate because a gauge block is considered to be a working standard. If you measure 10.02 mm, the reading is accurate because it is in reasonable agreement with the standard.

Calibrating Equipment

Precision and accuracy are important to understand if you are required to calibrate any piece of electronic equipment. *Calibration* is the process of comparing a given instrument to a standard. The standard is generally required to be more *precise* by a factor of 4 over the instrument being calibrated, but this alone does not assure accuracy. To assure accuracy, the calibration should be certifiable, and it needs to be traceable to national standards (in the U.S., this is National Institute of Standards and Technology). The standard requires periodic recertifications to assure its accuracy.

SYSTEM NOTE

Another term associated with the quality of a measurement is *precision*. **Precision** is a measure of the repeatability (or consistency) of a measurement of some quantity. It is possible to have a precise measurement in which a series of readings are not scattered, but each measurement is inaccurate because of an instrument error. For example, a meter may be out of calibration and produce inaccurate but consistent (precise) results. However, it is not possible to have an accurate instrument unless it is also precise.

Significant Digits

The digits in a measured number that are known to be correct are called **significant digits**. Most measuring instruments show the proper number of significant digits, but some instruments can show digits that are not significant, leaving it to the user to determine what should be reported. This may occur because of an effect called *loading* (discussed in Section 6–4). A meter can change the actual reading in a circuit by its very presence. It is important to recognize when a reading may be inaccurate; you should not report digits that are known to be inaccurate.

Another problem with significant digits occurs when you perform mathematical operations with numbers. The number of significant digits should never exceed the number in the original measurement. For example, if 1.0 V is divided by 3.0 Ω, a calculator will show 0.33333333. Since the original numbers each contain 2 significant digits, the answer should be reported as 0.33 A, the same number of significant digits.

The rules for determining if a reported digit is significant are

1. Nonzero digits are always considered to be significant.
2. Zeros to the left of the first nonzero digit are never significant.
3. Zeros between nonzero digits are always significant.
4. Zeros to the right of the decimal point for a decimal number are significant.
5. Zeros to the left of the decimal point with a whole number may or may not be significant depending on the measurement. For example, the number 12,100 Ω can have 3, 4, or 5 significant digits. To clarify the significant digits, scientific notation (or a metric prefix) should be used. For example, 12.10 kΩ has 4 significant digits.

When a measured value is reported, one uncertain digit may be retained but other uncertain digits should be discarded. To find the number of significant digits in a number, ignore the decimal point, and count the number of digits from left to right starting with the first nonzero digit and ending with the last digit to the right. All of the digits counted are significant except zeros to the right end of the number, which may or may not be significant.

In the absence of other information, the significance of the right-hand zeros is uncertain. Generally, zeros that are placeholders, and not part of a measurement, are considered to be not significant. To avoid confusion, numbers should be shown using scientific or engineering notation if it is necessary to show the significant zeros.

EXAMPLE 1–20

Express the measured number 4300 with 2, 3, and 4 significant digits.

SOLUTION

Zeros to the right of the decimal point in a decimal number are significant. Therefore, to show two significant digits, write

$$4.3 \times 10^3$$

To show three significant digits, write

$$4.30 \times 10^3$$

To show four significant digits, write

$$4.300 \times 10^3$$

RELATED PROBLEM

How would you show the number 10,000 showing three significant digits?

EXAMPLE 1–21

Underline the significant digits in each of the following measurements:

(a) 40.0   **(b)** 0.3040   **(c)** 1.20×10^5 **(d)** 120,000   **(e)** 0.00502

SOLUTION

(a) <u>40.0</u> has three significant digits; see rule 4.

(b) 0.<u>3040</u> has four significant digits; see rules 2 and 3.

(c) <u>1.20</u> $\times 10^5$ has three significant digits; see rule 4.

(d) <u>12</u>0,000 has at least two significant digits. Although the number has the same value as in (c), zeros in this example are uncertain; see rule 5. This is *not* a recommended method for reporting a measured quantity; use scientific notation or a metric prefix in this case. See Example 1–20.

(e) 0.00<u>502</u> has three significant digits; see rules 2 and 3.

RELATED PROBLEM

What is the difference between a measured quantity of 10 and 10.0?

Rounding Off Numbers

Since they always contain approximate numbers, measurements should be shown only with those digits that are significant plus no more than one uncertain digit. The number of digits shown is indicative of the precision of the measurement. For this reason, you should **round off** a number by dropping one or more digits to the right of the last significant digit. Use only the most significant dropped digit to decide how to round off. The rules for rounding off are

1. If the most significant digit dropped is greater than 5, increase the last retained digit by 1.

2. If the digit dropped is less than 5, do not change the last retained digit.

3. If the digit dropped is 5, increase the last retained digit *if* it makes it an even number, otherwise do not. This is called the "round-to-even" rule.

EXAMPLE 1–22

Round each of the following numbers to three significant digits:

(a) 10.071   (b) 29.961   (c) 6.3948   (d) 123.52   (e) 122.52

SOLUTION

(a) 10.071 rounds to **10.1**.   (b) 29.961 rounds to **30.0**.

(c) 6.3948 rounds to **6.39**.   (d) 123.52 rounds to **124**.

(e) 122.52 rounds to **122**.

RELATED PROBLEM

Round 3.2850 to three significant digits using the round-to-even rule.

In most electrical and electronics systems and circuits, components have tolerances greater than 1% (5% and 10% are common). Most measuring instruments have accuracy specifications better than this, but it is unusual for measurements to be made with higher accuracy than 1 part in 1000. For this reason, three significant digits are appropriate for numbers that represent measured quantities in all but the most exacting work. If you are working with a problem with several intermediate results, keep all digits in your calculator, but round the answers to three when reporting a result.

SECTION 1–7 CHECKUP

1. What is the rule for showing zeros to the right of the decimal point?

2. What is the round-to-even rule?

3. On schematics, you will frequently see a 1000 Ω resistor listed as 1.0 kΩ. What does this imply about the value of the resistor?

4. If a power supply is required to be set to 10.00 V, what does this imply about the accuracy needed for the measuring instrument?

5. How can scientific or engineering notation be used to show the correct number of significant digits in a measurement?

1–8 ELECTRICAL SAFETY

Safety is a major concern when working with electrical systems. The possibility of an electric shock or a burn is always present, so caution should always be used. You provide a current path when voltage is applied across two points on your body, and current produces electrical shock. Electrical components often operate at high temperatures, so you can sustain skin burns when you come in contact with them. Also, the presence of electricity creates a potential fire hazard.

After completing this section, you should be able to

- **Recognize electrical hazards and practice proper safety procedures**
  - **Describe the cause of electrical shock**
  - **List the groups of current paths through the body**
  - **Discuss the effects of current on the human body**
  - **List the safety precautions that you should observe when you work with electricity**

Electrical Shock

Current through your body, not the voltage, is the cause of **electrical shock**. Of course, it takes voltage across a resistance to produce current. When a point on your body comes in contact with a voltage and another point comes in contact with a different voltage or with ground, such as a metal chassis, there will be current through your body from one point to the other. The path of the current depends on the points across which the voltage occurs. The severity of the resulting electrical shock depends on the amount of voltage and the path that the current takes through your body.

The current path through the body determines which tissues and organs will be affected. The current paths can be placed into three groups which are referred to as *touch potential, step potential,* and *touch/step potential.* These are illustrated in Figure 1–10.

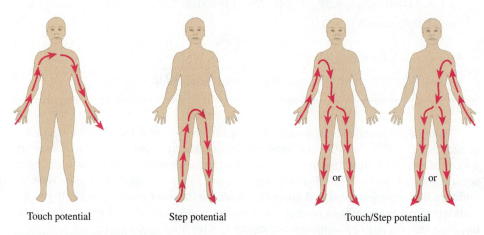

Touch potential   Step potential   Touch/Step potential

FIGURE 1–10   **Shock hazard in terms of three basic current path groups.**

EFFECTS OF CURRENT ON THE HUMAN BODY   The amount of current is dependent on voltage and resistance. The human body has resistance that depends on many factors, which include body mass, skin moisture, and points of contact of the body with a voltage potential. Table 1–4 shows the effects for various values of current in milliamperes.

TABLE 1–4 • Physical effects of electrical current. Values vary depending on body mass.	
CURRENT (mA)	**PHYSICAL EFFECT**
0.4	Slight sensation
1.1	Perception threshold
1.8	Shock, no pain, no loss of muscular control
9	Painful shock, no loss of muscular control
16	Painful shock, let-go threshold
23	Severe painful shock, muscular contractions, breathing difficulty
75	Ventricular fibrillation, threshold
235	Ventricular fibrillation, usually fatal for duration of 5 seconds or more
4,000	Heart paralysis (no ventricular fibrillation)
5,000	Tissue burn

BODY RESISTANCE   Resistance of the human body is typically between $10\,k\Omega$ and $50\,k\Omega$ and depends on the two points between which it is measured. The moisture of the skin also affects the resistance between two points. The resistance determines the amount of voltage required to produce each of the effects listed in Table 1–4. For example, if you have a resistance of $10\,k\Omega$ between two given points on your body, 90 V across those two points will produce enough current (9 mA) to cause painful shock.

Utility Voltages

We tend to take utility voltages for granted, but they can be and have been lethal. It is best to be careful around any source of voltage (even low voltages can present a serious burn hazard). As a general rule, you should avoid working on any energized circuit, and check that the power is off with a known good meter. Most work in educational labs uses low voltages, but you should still avoid touching any energized circuit. If you are working on a circuit that is connected to utility voltages, the service should be disconnected, a notice should be placed on the equipment or place where the service is disconnected, and a padlock should be used to prevent someone from accidentally turning on the power. This procedure is called *lockout/tagout* and is widely used in industry. There are specific OSHA and industry standards for lockout/tagout.

Most laboratory equipment is connected to the utility line ("ac") and in North America, this is 120 V rms (rms is discussed in Section 8–2). A faulty piece of equipment can cause the "hot" lead to inadvertently become exposed. You should inspect cords for exposed wires and check equipment for missing covers or other potential safety problems. The single-phase utility lines in homes and electrical laboratories use three insulated wires that are referred to as the "hot" (black or red wire), neutral (white wire), and safety ground (green wire). The hot and neutral wires will have current, but the green safety line should never have current in normal operation. The safety wire is connected to the metal exterior of encased equipment and is also connected to conduit and the metal boxes for housing receptacles. Figure 1–11 shows the location of these conductors on a standard receptacle. Notice on the receptacle that the neutral lead is larger than the hot lead.

The safety ground should be connected to the neutral at the service panel. The metal chassis of an instrument or appliance is also connected to ground. In the event that the hot wire is accidentally in contact with ground, the resulting high current should trip the circuit breaker or open a fuse to remove the hazard. However, a broken or missing ground lead may not have high current until it is contacted by a person. This danger is one obvious reason for ensuring that line cords have not been altered by removing the ground pin.

Many circuits are further protected with a special device called a ground-fault circuit interrupter (GFCI, which is sometimes called just GFI). If a fault occurs in a GFCI circuit, a sensor detects that the current in the hot line and the neutral line are not equal as they should be and trips the circuit breaker. The GFCI breaker is very fast acting and can trip faster than the breaker on the main panel. GFCI breakers are required in areas where a shock hazard exists such as wherever there is water or moisture. Pools, bathrooms, kitchens, basements, and garages should all have GFCI outlets. Figure 1–12 shows a ground-fault receptacle with reset and test buttons. When the test button is pressed, the circuit should immediately open. The reset button restores power.

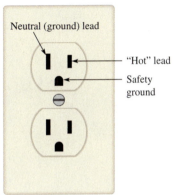

FIGURE 1–11   Standard receptacle and connections.

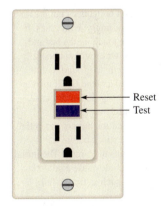

FIGURE 1–12   GFCI receptacle.

Safety Precautions

There are many practical things that you should do when you work with electrical and electronic equipment. Some important precautions are listed here.

- Avoid contact with any voltage source. Turn power off before you work on circuits when you need to touch circuit parts.
- Do not work alone. A telephone should be available for emergencies.

- Do not work when tired or taking medications that make you drowsy.
- Remove rings, watches, and other metallic jewelry when you work on circuits.
- Do not work on equipment until you know proper procedures and are aware of potential hazards.
- Make sure power cords are in good condition and grounding pins are not missing or bent.
- Keep your tools properly maintained. Make sure the insulation on metal tool handles is in good condition.
- Handle tools properly and maintain a neat work area.
- Wear safety glasses when appropriate, particularly when soldering, clipping wires, or working with power tools.
- Always shut off power and discharge capacitors before you touch any part of a circuit with your hands.
- Know the location of the emergency power-off switch and emergency exits.
- Never try to override or tamper with safety devices such as an interlock switch or ground pin on a three-prong plug.
- Always wear shoes and keep them dry. Do not stand on metal or wet floors when working on electrical circuits.
- Never handle instruments when your hands are wet.
- Never assume that a circuit is off. Double-check it with a reliable meter before handling.
- Set the limiter on electronic power supplies to prevent currents larger than necessary to supply the circuit under test.
- Some devices such as capacitors can store a lethal charge for long periods after power is removed. They must be properly discharged before you work with them.
- When making circuit connections, always make the connection to the point with the highest voltage as your last step.
- Avoid contact with the terminals of power supplies.
- Always use wires with insulation and connectors or clips with insulating shrouds.
- Keep cables and wires as short as possible. Connect polarized components properly.
- Report any unsafe condition.
- Be aware of and follow all workplace and laboratory rules. Do not have drinks or food near equipment.
- If another person cannot let go of an energized conductor, switch the power off immediately. If that is not possible, use any available nonconductive material to try to separate the body from the contact.
- Use a lockout/tagout procedure to avoid someone turning power on while you are working on a circuit.

SAFETY NOTE

A GFCI outlet does not prevent shock or injury in all cases. If you are touching the hot and neutral wires without being grounded, no ground fault is detected and the GFCI breaker will not trip. In another case, the GFCI may prevent electrocution but not the initial electric shock before it interrupts the circuit. The initial shock could cause a secondary injury, such as from a fall.

SECTION 1–8 CHECKUP

1. What causes physical pain and/or damage to the body when electrical contact is made?

2. It's OK to wear a ring when working on an electrical circuit. (T or F)

3. Standing on a wet floor presents no safety hazard when working with electricity.   (T or F)

4. A circuit can be rewired without removing the power if you are careful.   (T or F)

5. Electrical shock can be extremely painful or even fatal. (T or F)

6. What does *GFCI* stand for?

SUMMARY

- Steps for design of a new product include using circuit design software, computer lay out, prototype construction, and testing.
- Many companies have tended to move away from vertical organization (raw material to finished product in one facility) to horizontal organization (focus on a specialty).
- The electronics industry can be divided into the manufacturing sector and the service sector.
- Many technical service sector jobs also require a broad range of skills including skills related to both electrical/electronic and mechanical installations.
- A system is a group of interrelated parts that perform a specific function. It is characterized by a boundary that separates the system from its environment. The system communicates to the outside world through inputs and outputs.
- A block diagram is a model of a system that represents the system's structure in a graphical format using labeled blocks to represent functions and lines to represent the signal flow.
- A transfer curve is the ratio of the output to the input for a system or circuit.
- Electrical and electronic circuits must have a complete path, one or more sources, and a load. Electrical and electronic circuits contain either passive components or active components.
- Passive components cannot increase the power in a signal; active components can.
- Signals that have discrete levels are called digital signals; continuous signals are analog signals.
- A transducer is a device that transforms energy from one form to another.
- Scientific notation is a method for expressing very large and very small numbers as a number between one and ten (one digit to left of decimal point) times a power of ten.
- Engineering notation is a form of scientific notation in which quantities are expressed with one, two, or three digits to the left of the decimal point times a power of ten that is a multiple of three.
- Metric prefixes are symbols used to represent powers of ten that are multiples of three.
- The uncertainty of a measured quantity depends on the accuracy and precision of the measurement.
- The number of significant digits in the result of a mathematical operation should never exceed the significant digits in the original numbers.
- Standard connections to electrical plugs include a hot wire, a neutral, and a safety ground.
- GFCI breakers sense the current in the hot wire and in the neutral wire and trip the breaker if they are different, indicating a ground fault.

KEY TERMS

Key terms and other bold terms in the chapter are defined in the end-of-book glossary.

Accuracy   An indication of the range of error in a measurement.

Active component   A component that requires electrical power to function correctly; active components can deliver more signal power than they receive from the input signal.

Alternator   An ac generator; alternators convert mechanical energy to electrical energy.

Analog   A continuous signal.

Block diagram   A model of a system that represents its structure in a graphical format using labeled blocks to represent functions and lines to represent the signal flow.

Boundary   The dividing line between what is part of the system and its environment.

Circuit   An interconnection of electrical components designed to produce a desired result.

Digital   A signal that has discrete levels.

Electrical shock   The physical sensation resulting from current through the body.

Engineering notation   A system for representing any number as a one-, two-, or three-digit number times a power of ten with an exponent that is a multiple of 3.

Error   The difference between the true or best-accepted value of some quantity and the measured value.

Exponent   The number to which a base number is raised.

Gain   The ratio of the output to the input for an amplifier.

Horizontal organization   A business structure that is decentralized, to allow managers to focus on a specialty and streamline decision making.

Input   The voltage, current, or power applied to an electrical circuit to achieve a desired result.

Integrated circuit   An active complex circuit consisting of resistors, transistors, and other components fabricated as a single unit that performs the function of multiple discrete components.

Mechatronics   A synergistic combination of mechanics and electronics that includes instrumentation and control systems.

Metric prefix   A symbol that is used to replace the power of ten in numbers expressed in engineering notation.

Oscillator   A circuit where the output signal is internally generated; the output is a continuous signal that can be one of several different shapes.

Output   The result obtained from the system after processing the inputs.

Passive component   A component that does not require power; passive components cannot increase signal power.

Power of ten   A numerical representation consisting of a base of 10 and an exponent; the number 10 raised to a power.

Precision   A measure of the repeatability (or consistency) of a series of measurements.

Round off   The process of dropping one or more digits to the right of the last significant digit in a number.

Scientific notation   A system for representing any number as a number between 1 and 10 times an appropriate power of ten.

SI   Standardized international system of units used for all engineering and scientific work; abbreviation for French *Le Systeme International d'Unites*.

Significant digit   A digit known to be correct in a number.

System   A group of interrelated parts that perform a specific function.

Transducer   A device that transforms energy from one form to another.

Transfer curve   A plot showing the ratio of the output to the input.

Vertical organization   A centralized business structure that has an organized top-down approach. The tendency is less toward a specialization.

TRUE/FALSE QUIZ

Answers are at the end of the chapter.

1. A vertical organization is characterized by decentralized decision making.

2. Technicians working in service and repair usually need to identify a faulty component and replace it.

3. Mechatronics is an applied engineering classification that is a synergistic combination of mechanics and electronics.

4. It is not possible to completely isolate a system from its environment.

5. A block diagram is the same thing as a flowchart.

6. The units for a transfer curve must be the same for input and output.

7. A passive device cannot have power gain.

8. An example of a transducer is a loudspeaker.

9. The number 3300 is written as 3.3×10^3 in both scientific and engineering notation.

10. A negative number that is expressed in scientific notation will always have a negative exponent.

11. When you multiply two numbers written in scientific notation, the exponents need to be the same.

12. When you divide two numbers written in scientific notation, the exponent of the denominator is subtracted from the exponent of the numerator.

13. The metric prefix *micro* has an equivalent power of ten equal to 10^6.

14. To express 56×10^6 with a metric prefix, the result is 56 M.

15. 0.047 μF is equal to 47 nF.

16. The number of significant digits in the number 0.0102 is three.

17. When you apply the *round-to-even* rule to round off 26.25 to three digits, the result is 26.3.

18. The white neutral lead for ac power should have the same current as the hot lead.

SELF-TEST

Answers are at the end of the chapter.

1. A PLC is a
(a) printed logic circuit
(b) power limiting circuit
(c) programmable logic controller
(d) peak limit controller

2. The boundary of a system separates
(a) inputs from outputs
(b) the system from its environment
(c) only the inputs from the environment
(d) only the outputs from the environment

3. For a system, a block diagram is a graphical tool to illustrate
(a) functions and signal flow
(b) details of the system
(c) power supplies
(d) logical processes

4. The transfer curve always
(a) is a straight line
(b) is the ratio of output to input
(c) is dimensionless
(d) all of the above

5. An active component is a device that
(a) supplies its own power
(b) supplies power to other components
(c) does not have an output
(d) can increase signal power

6. An example of an ac source is a
(a) battery   (b) alternator   (c) solar cell   (d) fuel cell

7. The quantity 4.7×10^3 is the same as
(a) 470   (b) 4700   (c) 47,000   (d) 0.0047

8. The quantity 56×10^{-3} is the same as
(a) 0.056   (b) 0.560   (c) 560   (d) 56,000

9. The number 3,300,000 can be expressed in engineering notation as
(a) 3300×10^3 (b) 3.3×10^{-6} (c) 3.3×10^6 (d) either (a) or (c)

10. Ten milliamperes can be expressed as
(a) 10 MA   (b) 10 μA   (c) 10 kA   (d) 10 mA

11. Five thousand volts can be expressed as
(a) 5000 V   (b) 5 MV   (c) 5 kV   (d) either (a) or (c)

12. Twenty million ohms can be expressed as
(a) 20 mΩ (b) 20 MW   (c) 20 MΩ (d) 20 $\mu\Omega$

13. 15,000 W is the same as
(a) 15 mW   (b) 15 kW   (c) 15 MW   (d) 15 μW

14. Which of the following is not an electrical quantity?
(a) current   (b) voltage   (c) time   (d) power

15. The unit of current is
(a) volt   (b) watt   (c) ampere   (d) joule

16. The unit of voltage is
    (a) ohm   **(b)** watt   **(c)** volt   **(d)** farad

17. The unit of resistance is
    (a) ampere   **(b)** henry   **(c)** hertz   **(d)** ohm

18. Hertz is the unit of
    (a) power   **(b)** inductance   **(c)** frequency   **(d)** time

19. The number of significant digits in the number 0.1050 is
    (a) two   **(b)** three   **(c)** four   **(d)** five

PROBLEMS

Answers to odd-numbered problems are at the end of the book.

BASIC PROBLEMS AND QUESTIONS

SECTION 1–1   The Electronics Industry

1. If an engineer has an idea for improving a circuit board, what steps should be taken before the change is implemented?

2. Cite examples of products made by system manufacturers.

3. Cite the fundamental reasons that technician work has shifted from troubleshooting and repair at the component level to a system approach.

SECTION 1–2   Introduction to Electronic Systems

4. What are key differences between an electrical system and an electronic system?

5. What are two advantages to converting an analog signal into a digital signal?

6. What is the primary difference between a block diagram and a flowchart?

SECTION 1–3   Types of Circuits

7. (a) What is an oscillator?   (b) How does it differ from most circuits?

8. (a) What do the initials HVDC stand for?   (b) Where is it used?

9. What is a carrier?

10. The transfer function for a device is a plot of the output to the input. What does the transfer function look like for an ADC?

SECTION 1–4   Scientific and Engineering Notation

11. Express each of the following numbers in scientific notation:
    (a) 3000   **(b)** 75,000   **(c)** 2,000,000

12. Express each fractional number in scientific notation:
    (a) 1/500   **(b)** 1/2000   **(c)** 1/5,000,000

13. Express each of the following numbers in scientific notation:
    (a) 8400   **(b)** 99,000   **(c)** 0.2×10^6

14. Express each of the following numbers in scientific notation:
    (a) 0.0002   **(b)** 0.6   **(c)** 7.8×10^{-2}

15. Express each of the following as a regular decimal number:
    (a) 2.5×10^{-6} **(b)** 5.0×10^2 **(c)** 3.9×10^{-1}

16. Express each number in regular decimal form:
    (a) 4.5×10^{-6} **(b)** 8×10^{-9} **(c)** 4.0×10^{-12}

17. Add the following numbers:
    (a) $(9.2 \times 10^6) + (3.4 \times 10^7)$ **(b)** $(5 \times 10^3) + (8.5 \times 10^{-1})$
    (c) $(5.6 \times 10^{-8}) + (4.6 \times 10^{-9})$

18. Perform the following subtractions:
    (a) $(3.2 \times 10^{12}) - (1.1 \times 10^{12})$ **(b)** $(2.6 \times 10^8) - (1.3 \times 10^7)$
    (c) $(1.5 \times 10^{-12}) - (8 \times 10^{-13})$

19. Perform the following multiplications:
   (a) $(5 \times 10^3)(4 \times 10^5)$ (b) $(1.2 \times 10^{12})(3 \times 10^2)$
   (c) $(2.2 \times 10^{-9})(7 \times 10^{-6})$

20. Divide the following:
   (a) $(1.0 \times 10^3) \div (2.5 \times 10^2)$ (b) $(2.5 \times 10^{-6}) \div (5.0 \times 10^{-8})$
   (c) $(4.2 \times 10^8) \div (2 \times 10^{-5})$

21. Express each number in engineering notation:
   (a) 89,000   (b) 450,000   (c) 12,040,000,000,000

22. Express each number in engineering notation:
   (a) 2.35×10^5 (b) 7.32×10^7 (c) 1.333×10^9

23. Express each number in engineering notation:
   (a) 0.000345   (b) 0.025   (c) 0.00000000129

24. Express each number in engineering notation:
   (a) 9.81×10^{-3} (b) 4.82×10^{-4} (c) 4.38×10^{-7}

25. Add the following numbers and express each result in engineering notation:
   (a) $2.5 \times 10^{-3} + 4.6 \times 10^{-3}$ (b) $68 \times 10^6 + 33 \times 10^6$
   (c) $1.25 \times 10^6 + 250 \times 10^3$

26. Multiply the following numbers and express each result in engineering notation:
   (a) $(32 \times 10^{-3})(56 \times 10^3)$ (b) $(1.2 \times 10^{-6})(1.2 \times 10^{-6})$
   (c) $100(55 \times 10^{-3})$

27. Divide the following numbers and express each result in engineering notation:
   (a) $50 \div (2.2 \times 10^3)$ (b) $(5 \times 10^3) \div (25 \times 10^{-6})$
   (c) $(560 \times 10^3) \div (660 \times 10^3)$

SECTION 1–5   Units and Metric Prefixes

28. Express each number in Problem 11 in ohms using a metric prefix.

29. Express each number in Problem 13 in amperes using a metric prefix.

30. Express each of the following as a quantity having a metric prefix:
   (a) 31×10^{-3} A   (b) 5.5×10^3 V   (c) 20×10^{-12} F

31. Express the following using metric prefixes:
   (a) 3×10^{-6} F   (b) $3.3 \times 10^6 \ \Omega$ (c) 350×10^{-9} A

32. Express each quantity with a power of ten:
   (a) $5 \ \mu$A   (b) 43 mV   (c) 275 kΩ (d) 10 MW

SECTION 1–6   Metric Unit Conversions

33. Perform the indicated conversions:
   (a) 5 mA to microamperes   (b) 3200 μW to milliwatts
   (c) 5000 kV to megavolts   (d) 10 MW to kilowatts

34. Determine the following:
   (a) The number of microamperes in 1 milliampere
   (b) The number of millivolts in 0.05 kilovolt
   (c) The number of megohms in 0.02 kilohm
   (d) The number of kilowatts in 155 milliwatts

35. Add the following quantities:
   (a) 50 mA + 680 μA   (b) 120 kΩ + 2.2 MΩ (c) 0.02 μF + 3300 pF

36. Do the following operations:
   (a) 10 kΩ ÷ (2.2 kΩ + 10 kΩ)   (b) 250 mV ÷ 50 μV   (c) 1 MW ÷ 2 kW

SECTION 1–7   Measured Numbers

37. How many significant digits are in each of the following numbers:
   (a) 1.00×10^3 (b) 0.0057   (c) 1502.0
   (d) 0.000036   (e) 0.105   (f) 2.6×10^2

38. Round each of the following numbers to three significant digits. Use the "round-to-even" rule.
   (a) 50,505   (b) 220.45   (c) 4646   (d) 10.99   (e) 1.005

ANSWERS TO SECTION CHECKUPS

SECTION 1–1   The Electronics Industry

1. Simulation software can enable a circuit to be optimized with minimal cost. It can look for potential problems such as timing, noise, or thermal problems.

2. Manufacturing and service

3. Communications, computers, renewable energy, industrial controls, consumer products, medical equipment, transportation, defense.

4. Certification allows individuals to showcase their skills and gives employers a means of judging a person's competency.

SECTION 1–2   Introduction to Electronic Systems

1. The digital thermometer extracts heat from its surroundings, so the input is heat; the output is the displayed temperature.

2. The environment of the ISS includes the gravitational field of the earth and the other orbiting material around earth, and the radiated heat and light from the sun.

3. A block diagram shows the relationship of functional blocks and the signal flow paths.

4. A transfer curve is a plot that shows the ratio of the output to the input.

SECTION 1–3   Types of Circuits

1. A passive component does not require electrical power to function and cannot increase the power in a signal. An active component does require electrical power to function correctly and can increase the power in a signal.

2. A battery can only supply electrical power; it cannot increase the power in a signal.

3. 50 Hz and 60 Hz

4. Normally it is dc that is converted to signal power by an active circuit.

5. A digital circuit is one that processes discrete signals; an analog circuit is one that processes continuous signals.

6. A transducer is a device that transforms energy from one form to another.

SECTION 1–4   Scientific and Engineering Notation

1. True

2. 10^2

3. (a) 4.35×10^3 (b) 1.201×10^4 (c) 2.9×10^7

4. (a) 7.6×10^{-1} (b) 2.5×10^{-4} (c) 5.97×10^{-7}

5. (a) 3×10^5 (b) 6×10^{10} (c) 2×10^1 (d) 2.37×10^{-6}

6. Enter the digits, press EE, and enter the power of ten.

7. (a) 5.6×10^{-3} (b) 28.3×10^{-9} (c) 950×10^3 (d) 375×10^9

8. Enter the digits, press EE, and enter the power of ten.

SECTION 1–5   Units and Metric Prefixes

1. Mega (M), kilo (k), milli (m), micro (μ), nano (n), and pico (p)

2. 1 μA (one microampere)

3. 250 kW (250 kilowatts)

SECTION 1–6   Metric Unit Conversions

1. 0.01 MV = 10 kV

2. 250,000 pA = 0.00025 mA

3. 125 kW

4. 75 mV

SECTION 1–7   Measured Numbers

1. Zeros should be retained only if they are significant because if they are shown, they are considered significant.

2. If the digit dropped is 5, increase the last retained digit *if* it makes it even, otherwise do not.

3. A zero to the right of the decimal point implies that the resistor is accurate to the nearest 100 Ω (0.1 kΩ).

4. The instrument must be accurate to four significant digits.

5. Scientific and engineering notation can show any number of digits to the right of a decimal. Numbers to the right of the decimal are always considered significant.

SECTION 1–8   Electrical Safety

1. Current

2. F

3. F

4. F

5. T

6. Ground-fault circuit interrupter

ANSWERS TO RELATED PROBLEMS FOR EXAMPLES

1–1 7.5×10^8

1–2 9.3×10^{-7}

1–3 820,000,000

1–4 4.89×10^3

1–5 1.85×10^{-5}

1–6 4.8×10^5

1–7 4×10^4

1–8 Enter 5.73946; press EE, enter 5.

1–9 36×10^9

1–10 5.6×10^{-12}

1–11 Enter 273.9, press EE, enter 3.

1–12 **(a)** 56 MΩ   **(b)** 470 μA

1–13 1000 μA

1–14 1 mV

1–15 0.893 μA

1–16 0.01 μF

1–17 2200 pF

1–18 0.0022 MΩ

1–19 2883 mA

1–20 10.0×10^3

1–21 The number 10 has two significant digits; the number 10.0 has three.

1–22 3.28

ANSWERS TO TRUE/FALSE QUIZ

1. F   **2.** F   **3.** T   **4.** T   **5.** F   **6.** F   **7.** T   **8.** T   **9.** T
10. F   **11.** F   **12.** T   **13.** F   **14.** T   **15.** T   **16.** T   **17.** F   **18.** T

ANSWERS TO SELF-TEST

1. (c)   **2.** (b)   **3.** (a)   **4.** (b)   **5.** (d)   **6.** (b)   **7.** (b)
8. (a)   **9.** (c)   **10.** (d)   **11.** (d)   **12.** (c)   **13.** (b)   **14.** (c)
15. (c)   **16.** (c)   **17.** (d)   **18.** (c)   **19.** (c)

CHAPTER 2

VOLTAGE, CURRENT, AND RESISTANCE

OUTLINE

OBJECTIVES

- Describe the basic structure of an atom
- Explain the concept of electrical charge
- Define *voltage* and discuss its characteristics
- Define *current* and discuss its characteristics
- Define *resistance* and discuss its characteristics
- Describe a basic electric circuit
- Make basic circuit measurements

KEY TERMS

Atom	Voltage
Electron	Volt (V)
Free electron	Voltage source
Conductor	Fuel cell
Semiconductor	Current
Insulator	Ampere (A)
Charge	Current source
Coulomb's law	Resistance
Coulomb (C)	Ohm (Ω)
Conductance	Switch
Siemens (S)	Fuse
Resistor	Circuit breaker
Potentiometer	AWG
Rheostat	Ground
Load	Voltmeter
Schematic	Ammeter
Closed circuit	Ohmmeter
Open circuit	DMM

INTRODUCTION

Three basic electrical quantities presented in this chapter are voltage, current, and resistance. No matter what type of electrical or electronic system you may work with, these quantities will always be of primary importance. This is true for dc and ac circuits, but our focus will be dc circuits in Part 1 of this book. Because of its importance in electrical systems, an ac circuit may be used occasionally to illustrate a particular concept; however, in these special cases the analysis and calculations are the same as for an equivalent dc circuit.

To help you understand voltage, current, and resistance, the basic structure of the atom is discussed and the concept of charge is introduced. The basic electric circuit is studied, along with techniques for measuring voltage, current, and resistance.

VISIT THE WEBSITE
Study aids for this chapter are available at
http://pearsonhighered.com/floyd

2–1 ATOMS

All matter is made of atoms; and all atoms consist of electrons, protons, and neutrons. The configuration of certain electrons in an atom is the key factor in determining how well a conductive or semiconductive material conducts electric current.

After completing this section, you should be able to

- **Describe the basic structure of an atom**
  - **Define *nucleus, proton, neutron,* and *electron***
  - **Define *atomic number***
  - **Define *shell***
  - **Explain what a valence electron is**
  - **Describe ionization**
  - **Explain what a free electron is**
  - **Define *conductor, semiconductor,* and *insulator***

An **atom** is the smallest particle of an **element** that retains the characteristics of that element. Each of the known 118 elements has atoms that are different from the atoms of all other elements. This gives each element a unique atomic structure. According to the classic Bohr model, an atom is visualized as having a planetary type of structure that consists of a central nucleus surrounded by orbiting electrons, as illustrated in Figure 2–1. The **nucleus** consists of positively charged particles called **protons** and uncharged particles called **neutrons**. The basic particles of negative charge are called **electrons**. Electrons orbit the nucleus.

Each type of atom has a certain number of protons that distinguishes it from the atoms of all other elements. For example, the simplest atom is that of hydrogen, which has one proton and one electron, as pictured in Figure 2–2(a). As another example, the helium atom, shown in Figure 2–2(b), has two protons and two neutrons in the nucleus and two electrons orbiting the nucleus.

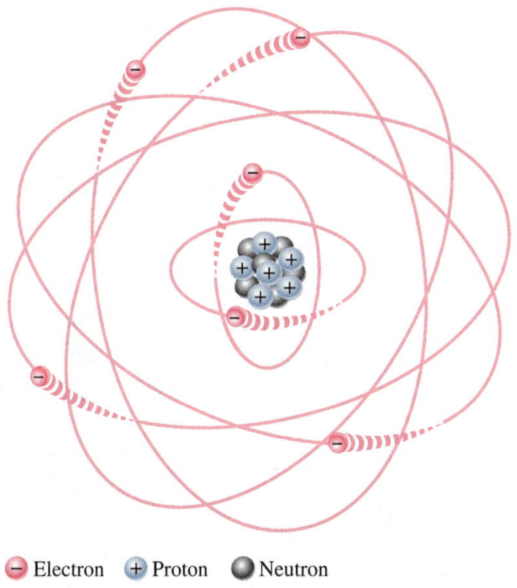

Atomic Number

All elements are arranged in the periodic table of the elements in order according to their **atomic number**. The atomic number equals the number of protons in the nucleus. For example, hydrogen has an atomic number of 1 and helium has an atomic number of 2. In their normal (or neutral) state, all atoms of a given element have the same number of electrons as protons; the positive charges cancel the negative charges, and the atom has a net charge of zero, making it electrically balanced.

⊖ Electron   ⊕ Proton   ⬤ Neutron

FIGURE 2–1   **The Bohr model of an atom showing electrons in circular orbits around the nucleus. The "tails" on the electrons indicate they are moving.**

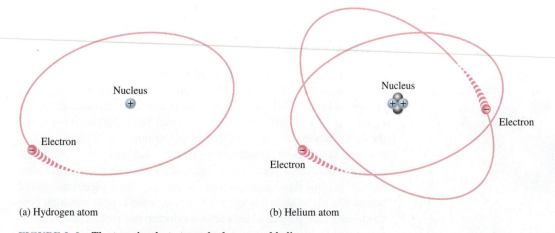

(a) Hydrogen atom   (b) Helium atom

FIGURE 2–2   **The two simplest atoms, hydrogen and helium.**

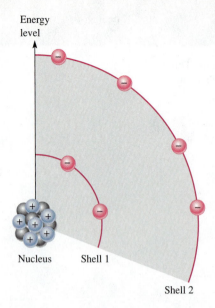

FIGURE 2–3   **Energy levels increase as the distance from the nucleus increases.**

Electron Shells and Orbits

Electrons orbit the nucleus of an atom at certain distances from the nucleus. Electrons near the nucleus have less energy than those in more distant **orbits**. It is known that only discrete (separate and distinct) values of electron energies exist within atomic structures. Therefore, electrons must orbit only at discrete distances from the nucleus.

ENERGY LEVELS   Each discrete distance (orbit) from the nucleus corresponds to a certain energy level. In an atom, the orbits are grouped into energy bands known as **shells**. A given atom has a fixed number of shells. Each shell has a fixed maximum number of electrons at permissible energy levels (orbits). The shells are designated 1, 2, 3, and so on, with 1 being closest to the nucleus. This energy band concept is illustrated in Figure 2–3, which shows two energy levels. Additional shells may exist in other types of atoms, depending on the element.

The number of electrons in each shell follows a predictable pattern according to the formula, $2N^2$, where N is the number of the shell. The first shell of any atom ($N = 1$) can have up to two electrons, the second shell ($N = 2$) up to eight electrons, the third shell up to 18 electrons, and the fourth shell up to 32 electrons. In many elements, electrons start filling the fourth shell after eight electrons are in the third shell.

Valence Electrons

Electrons that are in orbits farther from the nucleus have higher energy and are less tightly bound to the atom than those closer to the nucleus. This is because the force of attraction between the positively charged nucleus and the negatively charged electron decreases with increasing distance from the nucleus. Electrons with the highest energy levels exist in the outermost shell of an atom and are relatively loosely bound to the atom. This outermost shell is known as the **valence** shell, and electrons in this shell are called **valence electrons**. These valence electrons contribute to chemical reactions and bonding within the structure of a material, and they determine a material's electrical properties.

Free Electrons and Ions

If an electron absorbs a **photon** of sufficient energy, the electron escapes from the atom and becomes a **free electron**. Any time an atom or group of atoms is left with a net charge, the atom or group of atoms is called an **ion**. When an electron escapes from the neutral hydrogen atom (designated H), the atom is left with a net positive charge and becomes a *positive ion* (designated H^+). In other cases, an atom or group of atoms can acquire one or more electrons, in which case it is called a *negative ion*.

The Copper Atom

Copper is the most commonly used metal in electrical systems. The copper atom has 29 electrons that orbit the nucleus in four shells, as shown in Figure 2–4. Notice that the fourth or outermost shell, the valence shell, has only one valence electron. The inner shells are called the core. When the valence electron in the outer shell of the copper atom gains sufficient thermal energy, it can break away from the parent atom and become a free electron. In a piece of copper at room temperature, a "sea" of these free electrons is present. These electrons are not bound to a given atom but are free to move in the copper material. Free electrons make copper an excellent conductor and make electrical current possible.

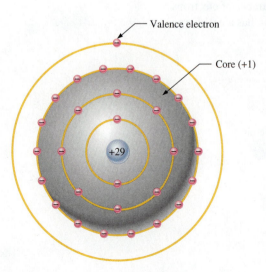

FIGURE 2–4   **The copper atom.**

Categories of Materials

Three categories of materials are used in electronics: conductors, semiconductors, and insulators.

CONDUCTORS   **Conductors** are materials that readily allow current. They have a large number of free electrons and are characterized by one to three valence electrons in their structure. Most metals are good conductors. Silver is the best conductor, and copper is next. Copper is the most widely used conductive material because it is less expensive than silver. Copper wire is commonly used as a conductor in electric circuits.

SEMICONDUCTORS   **Semiconductors** are classed below the conductors in their ability to carry current because they have fewer free electrons than do conductors. Semiconductors have four valence electrons in their atomic structures. However, because of their unique characteristics, certain semiconductor materials are the basis for electronic devices such as the diode, transistor, and integrated circuit. Silicon and germanium are common semiconductive materials.

INSULATORS   **Insulators** are nonmetallic materials that are poor conductors of electric current; they are used to prevent current where it is not wanted. Insulators have no free electrons in their structure. The valence electrons are bound to the nucleus and not considered "free." Although nonmetal elements are generally considered to be insulators, most practical insulators used in electrical and electronic systems are compounds such as glass, porcelain, Teflon, and polyethylene, to name a few.

SECTION 2–1 CHECKUP*

1. What is the basic particle of negative charge?

2. Define *atom.*

3. What does an atom consist of?

4. Define *atomic number.*

5. Do all elements have the same types of atoms?

6. What is a free electron?

7. What is a shell in the atomic structure?

8. Name two conductive materials.

*Answers are at the end of the chapter.

2–2 ELECTRICAL CHARGE

As you know, an electron is the smallest particle that exhibits negative electrical charge. When an excess of electrons exists in a material, there is a net negative electrical charge. When a deficiency of electrons exists, there is a net positive electrical charge.

After completing this section, you should be able to

- **Explain the concept of electrical charge**
  - **Name the unit of charge**
  - **Name the types of charge**
  - **Describe the forces between charges**
  - **Determine the amount of charge on a given number of electrons**

The charge of an electron and that of a proton are equal in magnitude and opposite in sign. Electrical **charge** is an electrical property of matter that exists because of an excess or deficiency of electrons. Charge is symbolized by Q. Static electricity is the presence of a net positive or negative charge in a material. Everyone has experienced the effects of static electricity from time to time, for example, when attempting to touch a metal surface or another person or when the clothes in a dryer cling together.

Materials with charges of opposite polarity are attracted to each other; materials with charges of the same polarity are repelled, as indicated symbolically in Figure 2–5. A force acts between charges, as evidenced by the attraction or repulsion. This force, called an *electric field,* consists of invisible lines of force as represented in Figure 2–6.

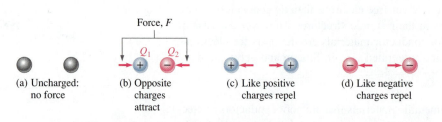

(a) Uncharged: no force

(b) Opposite charges attract

(c) Like positive charges repel

(d) Like negative charges repel

FIGURE 2–5 **Attraction and repulsion of electrical charges.**

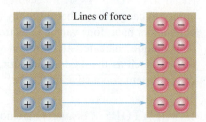

FIGURE 2–6 **Electric field between two oppositely charged surfaces.**

Coulomb's law states

A force (F) exists between two point-source charges (Q_1, Q_2) that is directly proportional to the product of the two charges and inversely proportional to the square of the distance (d) between the charges.

Coulomb: The Unit of Charge

Electrical charge is measured in coulombs, symbolized by C.

One coulomb is the total charge possessed by 6.25×10^{18} electrons.

A single electron has a charge of 1.6×10^{-19} C. The total charge Q, expressed in coulombs, for a given number of electrons is found by the following formula:

$$Q = \frac{\textbf{number of electrons}}{\textbf{6.25} \times \textbf{10}^{18} \textbf{ electrons/C}} \qquad (2\text{–}1)$$

Positive and Negative Charge

Consider a neutral atom—that is, one that has the same number of electrons and protons and thus has no net charge. As you know, when a valence electron is pulled away from the atom by the application of energy, the atom is left with a net positive charge (more protons than electrons) and becomes a positive ion. A **positive ion** is defined as an atom or group of atoms with a net positive charge. If an atom acquires an extra electron in its outer shell, it has a net negative charge and becomes a negative ion. A **negative ion** is defined as an atom or group of atoms with a net negative charge.

The amount of energy required to free a valence electron is related to the number of electrons in the outer shell. An atom can have up to eight valence electrons. The more complete the outer shell, the more stable the atom and thus the more energy is required to remove an electron. Figure 2–7 illustrates the creation of a positive ion and a negative ion when a hydrogen atom gives up its single valence electron to a chlorine atom, forming gaseous hydrogen chloride (HCl). When the gaseous HCl is dissolved in water, hydrochloric acid is formed.

Hydrogen atom
(1 proton, 1 electron)

Chlorine atom
(17 protons, 17 electrons)

(a) The neutral hydrogen atom has a single valence
electron.

(b) The atoms combine by sharing the
valence electron to form gaseous
hydrogen chloride (HCl).

Positive hydrogen ion
(1 proton, no electrons)

Negative chloride ion
(17 protons, 18 electrons)

(c) When dissolved in water, hydrogen chloride gas separates into positive hydrogen ions
and negative chloride ions. The chlorine atom retains the electron given up by the
hydrogen atom forming both positive and negative ions in the same solution.

FIGURE 2–7   **Example of the formation of positive and negative ions.**

EXAMPLE 2–1

How many coulombs of charge do 93.8×10^{16} electrons represent?

SOLUTION

$$Q = \frac{\text{number of electrons}}{6.25 \times 10^{18} \text{ electrons/C}} = \frac{93.8 \times 10^{16} \text{ electrons}}{6.25 \times 10^{18} \text{ electrons/C}} = 15 \times 10^{-2}\,\text{C} = \mathbf{0.15\ C}$$

RELATED PROBLEM*

How many electrons does it take to have 3 C of charge?

*Answers are at the end of the chapter.

SECTION 2–2 CHECKUP

1. What is the symbol used for charge?

2. What is the unit of charge, and what is the unit symbol?

3. What are the two types of charge?

4. How much charge, in coulombs, is there in 10×10^{12} electrons?

2–3 VOLTAGE

As you have learned, a force of attraction exists between a positive and a negative charge. A certain amount of energy must be exerted in the form of work to overcome the force and move the charges a given distance apart. All opposite charges possess a certain potential energy because of the separation between them. The difference in potential energy of the charges is the potential difference or *voltage*. Voltage is the driving force in electric circuits and is what establishes current.

After completing this section, you should be able to

- **Define *voltage* and discuss its characteristics**
  - **State the formula for voltage**
  - **Name and define the unit of voltage**
  - **Describe the basic sources of voltage**

Voltage is defined as energy per unit of charge and is expressed as

$$V = \frac{W}{Q} \tag{2–2}$$

where V is voltage in volts (V), W is energy in joules (J), and Q is charge in coulombs (C). As a simple analogy, you can think of voltage as corresponding to the pressure difference created by a pump that causes water to flow through a pipe in a closed water system.

Volt: The Unit of Voltage

The unit of voltage is the **volt**, symbolized by V.

> **One volt is the potential difference (voltage) between two points when one joule of energy is used to move one coulomb of charge from one point to the other.**

EXAMPLE 2–2

If 50 J of energy are required to move 10 C of charge, what is the voltage?

SOLUTION

$$V = \frac{W}{Q} = \frac{50\ \text{J}}{10\ \text{C}} = \textbf{5 V}$$

RELATED PROBLEM

How much energy is required to move 50 C from one point in a circuit to another when the voltage between the two points is 12 V?

The DC Voltage Source

A **voltage source** provides electrical energy or electromotive force (emf), more commonly known as voltage. Voltage is produced by means of chemical energy, light energy, and magnetic energy combined with mechanical motion.

THE IDEAL DC VOLTAGE SOURCE   An ideal dc voltage source can provide a constant voltage for any current required by a circuit. The ideal voltage source does not exist but can be closely approximated in practice. We will assume ideal unless otherwise specified.

Voltage sources can be either dc or ac. A common symbol for a dc voltage source is shown in Figure 2–8.

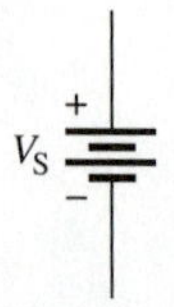

FIGURE 2–8   **Symbol for dc voltage sources.**

A graph showing current versus voltage for an ideal dc voltage source is shown in Figure 2–9. As you can see, the voltage is constant for any current from the source. For a practical (nonideal) voltage source connected in a circuit, the voltage decreases slightly as the current increases as shown by the dashed line. Current is always drawn from a voltage source when a load such as a resistance is connected to it.

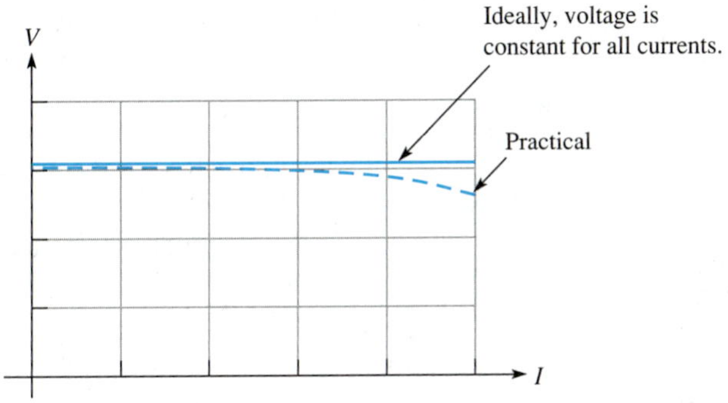

Types of DC Voltage Sources

BATTERIES   A **battery** is a type of voltage source that is composed of one or more cells that convert chemical energy directly into electrical energy. As you know, work (or energy) per charge is the basic unit for voltage, and a battery adds energy to each unit of charge. It is something of a misnomer to talk about "charging a battery" because a battery does not store charge but rather stores chemical potential energy. All batteries use a specific type of chemical reaction called an *oxidation-reduction* reaction. In this type of reaction, electrons are transferred from one reactant to the other. If the chemicals used in the reaction are separated, it is possible to cause the electrons to travel in the external circuit, creating current. As long as there is an external path for the electrons, the reaction can proceed, and stored chemical energy is converted to electrical current. If the path is broken, the reaction stops and the battery is said to be in equilibrium. In a battery, the terminal that supplies electrons has a surplus of electrons and is the negative electrode or **anode**. The electrode that acquires electrons has a positive potential and is the **cathode**.

Figure 2–10 shows a nonrechargeable single-cell copper-zinc battery that we will use for illustration of battery operation. The copper-zinc cell is simple to construct and illustrates concepts common to all nonrechargeable batteries. A zinc electrode and a copper electrode are immersed in solutions of zinc sulfate ($ZnSO_4$) and copper sulfate ($CuSO_4$), which are separated by a salt bridge that prevents the Cu^{2+} ions from reacting directly with the Zn metal. The zinc metal electrode supplies Zn^{2+} ions to the solution and electrons to the external circuit, so this electrode is constantly eaten away as the reaction proceeds. The salt bridge allows ions to pass through it to maintain charge balance in the cell. There are no free electrons in the solutions, so an external path for electrons is provided through an ammeter (in our case) or other load. On the cathode side, the electrons that were given up by the zinc combine with copper ions from the solution to form copper metal, which deposits on the copper electrode. The chemical reactions (shown in the diagram) occur at

FIGURE 2–9   Voltage source graph.

Ideally, voltage is constant for all currents.

Practical

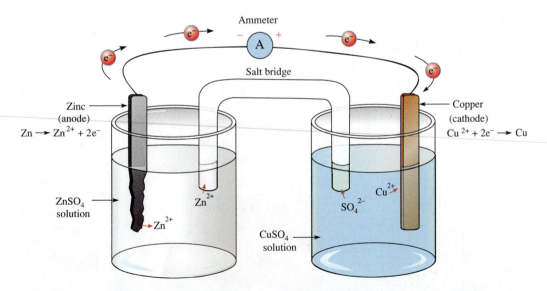

FIGURE 2–10   A copper-zinc battery. The reaction can only occur if an external path is provided for the electrons. As the reaction proceeds, the Zn anode is eaten away and Cu^{2+} ions combine with electrons to form copper metal on the cathode.

the electrode. Different types of batteries have different reactions, but all involve transfer of electrons in the external circuit and migration of ions internally as the battery discharges.

A single cell will have a certain fixed voltage. In the copper-zinc cell, the voltage is 1.1 V. In a lead-acid cell, the kind used in car batteries, a potential difference of about 2.1 V is between the anode and cathode. A typical car battery has six such cells connected in series. The voltage of any cell depends on the cell chemistry. Nickel-cadmium cells are about 1.2 V and lithium cells can be as high as almost 4 V, depending on the second reactant. Cell chemistry also determines the shelf life and discharge characteristic for a battery. For example, a lithium-MnO_2 battery typically has five times the shelf life as a comparable carbon-zinc battery.

Although the voltage of a battery cell is fixed by its chemistry, the capacity is variable and depends on the quantity of materials in the cell. Essentially, the *capacity* of a cell is the number of electrons that can be obtained from it and is measured by the amount of current (defined in Section 2–4) that can be supplied over time.

Battery Backup

Batteries are used in many systems to provide redundancy, such as in the case of alarm systems where the main power may be off at a critical time, or to back up utility power in case of failure. In solar-energy systems, batteries are used to supply power when the sun is not available. Deep-cycle batteries are constructed with heavier plates than normal batteries and are designed to be discharged further than regular automotive batteries. However, deep-cycle batteries tend to be expensive, inefficient, and require regular maintenance.

Photo courtesy of NREL.

SYSTEM NOTE

Batteries normally consist of multiple cells that are electrically connected together internally. The way that the cells are connected and the type of cells determine the voltage and current capacity of the battery. If the positive electrode of one cell is connected to the negative electrode of the next and so on, as illustrated in Figure 2–11(a), the battery voltage is the sum of the individual cell voltages. This is called a *series connection*. To increase

(a) Series-connected cells increase voltage.

(b) Parallel-connected cells increase current capacity.

FIGURE 2–11   Cells connected to form batteries.

battery current capacity, the positive electrodes of several cells are connected together and all the negative electrodes are connected together, as illustrated in Figure 2–11(b). This is called a *parallel connection*.

There are many sizes of batteries; large batteries, with more material, can supply more current. In addition to the many sizes and shapes, batteries are classified according to their chemical makeup and if they are rechargeable or not. Primary batteries are not rechargeable and are discarded when they run down because their chemical reactions are irreversible; secondary batteries are reusable because their chemical reactions are reversible. The following are some important types of batteries:

- **Alkaline-MnO₂.** This is a primary battery that is commonly used in palm-type computers, photographic equipment, toys, radios, and recorders. It has a longer shelf life and a higher power density than a carbon-zinc battery.

- **Carbon-zinc.** This is a primary multiuse battery for flashlights and small appliances. It is available in a variety of sizes such as AAA, AA, C, and D.

- **Lead-acid.** This is a secondary (rechargable) battery that is commonly used in automotive, marine, and other similar applications. Deep-cycle lead-acid batteries are commonly used as system backup types.

- **Lithium-ion.** This is a secondary battery that is commonly used in all types of portable electronics. This type of battery is increasingly being used in defense, aerospace, and automotive applications.

- **Lithium-MnO₂.** This is a primary battery that is commonly used in photographic and electronic equipment, smoke alarms, personal organizers, memory backup, and communications equipment.

- **Nickel-metal hydride.** This is a secondary (rechargeable) battery that is commonly used in portable computers, cell phones, camcorders, and other portable consumer electronics.

- **Silver oxide.** This is a primary battery that is commonly used in watches, photographic equipment, hearing aids, and electronics requiring high-capacity batteries.

- **Zinc air.** This is a primary battery that is commonly used in hearing aids, medical monitoring instruments, pagers, and similar applications.

FUEL CELLS A fuel cell is a device that converts electrochemical energy into dc voltage directly. Fuel cells combine a fuel (usually hydrogen) with an oxidizing agent (usually oxygen). In the hydrogen fuel cell, hydrogen and oxygen react to form water, which is the only by-product. The process is clean, quiet, and more efficient than burning. Fuel cells and batteries are similar in that they both use an oxidation-reduction chemical reaction in which electrons are forced to travel in the external circuit. However, a battery is a closed system with all its chemicals stored inside, whereas in a fuel cell, the hydrogen and oxygen constantly flow into the cell where they combine and produce electricity.

Hydrogen fuel cells are usually classified by their operating temperature and the type of electrolyte they use. Some types work well for use in stationary power generation plants. Others may be useful for small portable applications or for powering cars. For example, the type that holds the most promise for automotive applications is the polymer exchange membrane fuel cell (PEMFC), which is a type of hydrogen fuel cell. A simplified diagram is shown in Figure 2–12 to illustrate the basic operation.

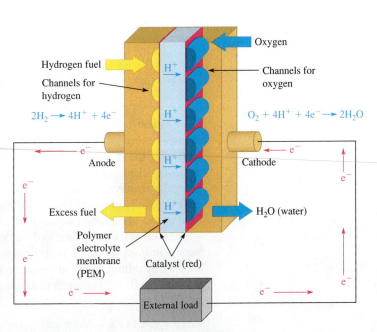

FIGURE 2–12 Simplified diagram of a fuel cell.

The channels disperse pressurized hydrogen gas and oxygen gas equally over the surface of the catalyst, which facilitates the reaction of the hydrogen and oxygen. When an H_2 molecule comes in contact with the platinum catalyst on the anode side of the fuel cell, it splits into two H^+ ions and two electrons (e^-). The hydrogen ions are passed through the polymer electrolyte membrane (PEM) onto the cathode. The electrons pass through the anode and into the external circuit to create current.

When an O_2 molecule comes in contact with the catalyst on the cathode side, it breaks apart, forming two oxygen ions. The negative charge of these ions attracts two H^+ ions through the electrolyte membrane and together they combine with electrons from the external circuit to form a water molecule (H_2O), which is passed from the cell as a by-product. In a single fuel cell, this reaction produces only about 0.7 V. To get higher voltages, multiple fuel cells are connected in series.

Current research on fuel cells is ongoing and is focused on developing reliable, smaller, and cost-effective components for vehicles and other applications. The conversion to fuel cells also requires research on how best to obtain and provide hydrogen fuel where it is needed. Potential sources for hydrogen include using solar, geothermal, or wind energy to break apart water. Hydrogen can also be obtained by breaking down coal or natural gas molecules, which are rich in hydrogen. Various demonstration projects for using hydrogen fuel cells have been completed. For example, the mail-processing center at Anchorage, AK, uses five 200 kW fuel cells to power the building.

SOLAR CELLS   The operation of solar cells is based on the **photovoltaic effect**, which is the process whereby light energy is converted directly into electrical energy. A basic solar cell consists of two layers of different types of semiconductive materials joined together to form a junction. When one layer is exposed to light, many electrons acquire enough energy to break away from their parent atoms and cross the junction. This process forms negative ions on one side of the junction and positive ions on the other, and thus a potential difference (voltage) is developed. Figure 2–13 shows the construction of a basic solar cell.

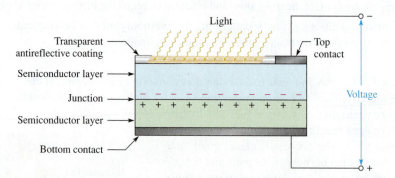

FIGURE 2–13   **Construction of a basic solar cell.**

Although solar cells can be used in room light for powering a calculator, research is focusing more on converting sunlight to electricity. There is considerable research in increasing the efficiency of solar cells and photovoltaic (PV) modules today because they are a very clean source of energy using sunlight. A complete system for continuous power generally requires a battery backup to provide energy when the sun is not shining. Solar cells are well suited for remote locations where energy sources are unavailable and are used in providing power to satellites.

Scientists are working to develop flexible solar cells that can be printed by a banknote printing process. The technology uses organic cells that can be mass-produced cheaply by printing them in the same manner as money is printed. "Printable electronics" is at the forefront of polymer technology research.

GENERATOR   Electrical **generators** convert mechanical energy into electrical energy using a principle called *electromagnetic induction* (see Chapter 7). A coil is rotated

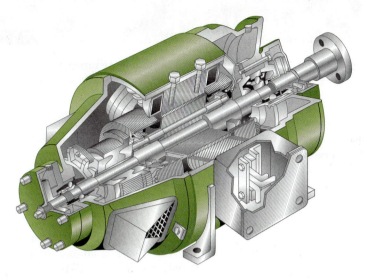

FIGURE 2–14   Cutaway view of a dc voltage generator.

through a magnetic field, and a voltage is produced across the coil. A typical generator is pictured in Figure 2–14.

THE POWER SUPPLY   **Power supplies** convert the ac voltage from the wall outlet to a constant (dc) voltage that is available across two terminals. A basic commercial power supply is shown in Figure 2–15. Power supplies are covered in more detail in Section 3–7.

FIGURE 2–15   **A basic power supply.**   (Courtesy of B+K Precision)

THERMOCOUPLES   The **thermocouple** is a thermoelectric type of voltage source that is commonly used to sense temperature. A thermocouple is formed by the junction of two dissimilar metals, and its operation is based on the **Seebeck effect** that describes the voltage generated at the junction of the metals as a function of temperature.

    Standard types of thermocouples are characterized by the specific metals used. These standard thermocouples produce predictable output voltages for a range of temperatures. The most common is type K, made of chromel and alumel. Other types are also designated by letters as E, J, N, B, R, and S. Most thermocouples are available in wire or probe form.

PIEZOELECTRIC SENSORS   These sensors act as very low-power voltage sources and are based on the **piezoelectric effect** where a voltage is generated when a piezoelectric material is mechanically deformed by an external force. Quartz and ceramic are two types of piezoelectric material. Piezoelectric sensors are used in various systems, such as pressure sensors, force sensors, accelerometers, microphones, and ultrasonic devices.

SECTION 2–3 CHECKUP

1. Define *voltage*.

2. What is the unit of voltage?

3. What is the voltage when 24 J of energy are required to move 10 C of charge?

4. List seven sources of voltage.

5. What type of chemical reaction occurs in all batteries and fuel cells?

2–4 CURRENT

Voltage provides energy to electrons that allows them to move through a circuit. This movement of electrons is the current, which results in work being done in an electric circuit.

After completing this section, you should be able to

- **Define** *current* **and discuss its characteristics**
  - **Explain the movement of electrons**
  - **State the formula for current**
  - **Name and define the unit of current**

As you have learned, free electrons are available in all conductive and semiconductive materials. These outer-shell electrons drift randomly in all directions, from atom to atom, within the structure of the material, as indicated in Figure 2–16. These electrons are loosely bound to the positive metal ions in the material; but because of thermal energy, they are free to move about the crystalline structure of the metal.

FIGURE 2–16 **Random motion of free electrons in a material.**

If a voltage is placed across a conductive or semiconductive material, one end becomes positive and the other negative, as indicated in Figure 2–17. The repulsive force produced by the negative voltage at the left end causes the free electrons (negative charges) to move toward the right. The attractive force produced by the positive voltage at the right end pulls the free electrons to the right. The result is a net movement of the free electrons from the negative end of the material to the positive end, as shown in Figure 2–17.

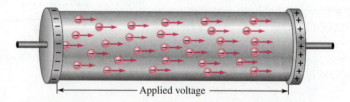

Applied voltage

FIGURE 2–17 **Electrons flow from negative to positive when a voltage is applied across a conductive or semiconductive material.**

The movement of the free electrons from the negative end of a material to the positive end is the electrical current, symbolized by *I*.

Electrical current is the rate of flow of charge.

Current in a conductive material is measured by the number of electrons (amount of charge) that flow past a point in a unit of time.

$$I = \frac{Q}{t}$$

(2–3)

where *I* is current in amperes (A), *Q* is the charge of the electrons in coulombs (C), and *t* is time in seconds (s). As a simple analogy, you can think of current as corresponding to water flowing through a pipe in a water system when pressure (corresponding to voltage) is applied by a pump (corresponding to a voltage source). *Voltage causes current.*

Ampere: The Unit of Current

Current is measured in a unit called the *ampere* or *amp* for short, symbolized by A.

> **One ampere (1 A) is the amount of current that exists when a number of electrons having a total charge of one coulomb (1 C) move through a given cross-sectional area in one second (1 s).**

See Figure 2–18. Remember, one coulomb is the charge carried by 6.25×10^{18} electrons.

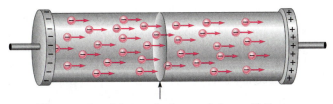

When a number of electrons having a total charge of 1 C pass through a cross-sectional area in 1 s, there is 1 A of current.

FIGURE 2–18   **Illustration of 1 A of current (1 C/s) in a material.**

E X A M P L E   2 – 3

Ten coulombs of charge flow past a given point in a wire in 2 s. What is the current in amperes?

S O L U T I O N

$$I = \frac{Q}{t} = \frac{10\ \text{C}}{2\ \text{s}} = \textbf{5 A}$$

R E L A T E D   P R O B L E M

If there are 8 A of direct current through the filament of a light bulb, how many coulombs have moved through the filament in 1.5 s?

The Current Source

THE IDEAL CURRENT SOURCE   As you know, an ideal voltage source can provide a constant voltage for any load. An ideal current source can provide a constant current in any load. Just as in the case of a voltage source, the ideal current source does not exist but can be approximated in practice. We will assume ideal unless otherwise specified.

The symbol for a current source is shown in Figure 2–19(a). The graph for an ideal current source is a horizontal line as illustrated in Figure 2–19(b). This graph is called the *IV* characteristic curve. Notice that the current is constant for any voltage across the current source. In a nonideal current source, the current drops off as shown by the dashed line in Figure 2–19(b).

ACTUAL CURRENT SOURCES   Power supplies are normally thought of as voltage sources because they are the most common type of source in the laboratory. However, current sources are another type of power supply. Current sources may be "stand-alone"

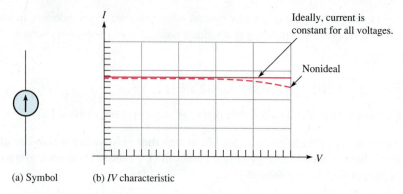

(a) Symbol   (b) *IV* characteristic

FIGURE 2–19   **The current source.**

instruments or may be combined with other instruments, such as a voltage source, DMM, or function generator. Examples of combination instruments are the source-measurement units shown in Figure 2–20. These units can be set up as voltage or current sources and include a built-in DMM, as well as other instruments. They are used primarily for testing transistors and other semiconductors.

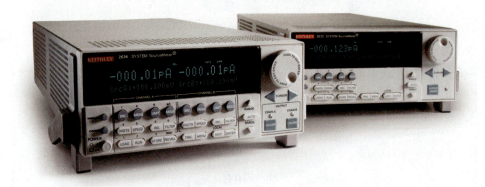

FIGURE 2–20   **Source-measurement units that have current and voltage sources.**   (Courtesy of Keithley Instruments)

In most transistor circuits, the transistor acts as a current source because part of the *IV* characteristic curve is a horizontal line as shown by the transistor characteristic in Figure 2–21. The flat part of the graph indicates where the transistor current is constant over a range of voltages. The constant-current region is used to form a constant-current source.

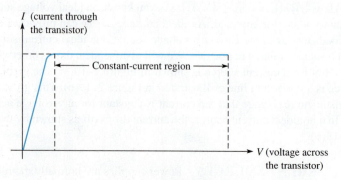

FIGURE 2–21   **Characteristic curve of a transistor showing the constant-current region.**

SYSTEM EXAMPLE 2–1

A CURRENT SOURCE

Current sources are useful in a variety of automotive lighting requirements (taillights, turn signals, brakes, daytime running lights, interior lights) and in requirements like studio lighting where constant illumination is required. High brightness light-emitting diodes (LEDs) are popular in these applications because they are energy efficient, reliable, small, and can turn on rapidly. Many integrated circuits are specifically designed to provide the high currents needed for constant illumination. An example is the BP5843A constant current LED driver for illumination. The BP5843A can accept either a dc source (from 113 V to 170 V) or an ac source (from 80 V to 120 V) and can supply a range of currents from 250 mA to 350 mA at a constant level as set by the user.

A basic constant-current illumination lighting system is shown in Figure 2–22. The input ac is filtered, transformed, and rectified (converted to pulsing dc). The components labeled C are capacitors, which are for filtering and noise removal. (Capacitors are discussed in Chapter 9.) The resistor, labeled R, is used to set the amount of current through the high brightness LEDs. (Resistors are covered in Section 2–5.)

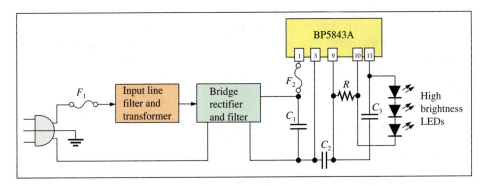

FIGURE 2–22   A basic constant-current illumination lighting system.

SECTION 2–4 CHECKUP

1. Define *current* and state its unit.

2. How many electrons make up one coulomb of charge?

3. What is the current in amperes when 20 C flow past a point in a wire in 4 s?

2–5 RESISTANCE

When there is current through a material, the free electrons move through the material and occasionally collide with atoms. These collisions cause the electrons to lose some of their energy, and thus their movement is restricted. The more collisions, the more the flow of electrons is restricted. This restriction varies and is determined by the type of material. The property of a material that restricts the flow of electrons is called *resistance*, designated with an *R*.

After completing this section, you should be able to

- Define *resistance* and discuss its characteristics
  - Name and define the unit of resistance
  - Describe the basic types of resistors
  - Determine resistance value by color code or labeling

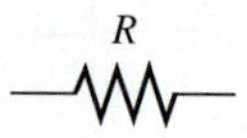

FIGURE 2–23 Resistance/resistor symbol.

Resistance is the opposition to current.

The schematic symbol for resistance is shown in Figure 2–23.

When there is current through any material that has resistance, heat is produced by the collisions of free electrons and atoms. Therefore, wire, which typically has a very small resistance, can become warm or even hot when there is sufficient current through it.

As a simple analogy, you can think of a resistor as corresponding to a partially open valve in a closed water system that restricts the amount of water flowing through a pipe. If the valve is opened more (corresponding to less resistance), the water flow (corresponding to current) increases. If the valve is closed a little (corresponding to more resistance), the water flow (corresponding to current) decreases.

Ohm: The Unit of Resistance

Resistance, *R*, is expressed in the unit of ohms, which is symbolized by the Greek letter omega (Ω).

One ohm (1 Ω) of resistance exists when there is one ampere (1 A) of current in a material with one volt (1 V) applied across the material.

CONDUCTANCE The reciprocal of resistance is **conductance**, symbolized by *G*. It is a measure of the ease with which current is established. The formula is

$$G = \frac{1}{R} \tag{2–4}$$

The unit of conductance is the **siemens**, symbolized by S. For example, the conductance of a 22 kΩ resistor is

$$G = \frac{1}{22\ \text{k}\Omega} = 45.5\ \mu\text{S}$$

Occasionally, the obsolete unit of *mho* is still used for conductance.

Resistors

Components that are specifically designed to have a certain amount of resistance are called **resistors**. The principal applications of resistors are to limit current, divide voltage, and, in certain cases, generate heat. Although different types of resistors come in many shapes and sizes, they can all be placed in one of two main categories: fixed or variable.

FIXED RESISTORS Fixed resistors are available with a large selection of resistance values that are set during manufacturing and cannot be changed easily. Fixed resistors are constructed using various methods and materials. Figure 2–24 shows several common types.

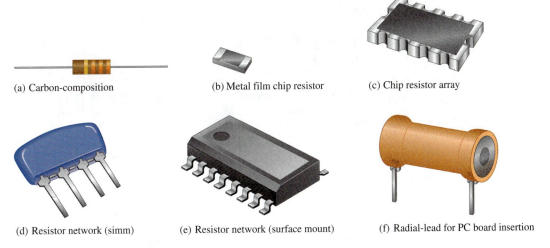

(a) Carbon-composition

(b) Metal film chip resistor

(c) Chip resistor array

(d) Resistor network (simm)

(e) Resistor network (surface mount)

(f) Radial-lead for PC board insertion

FIGURE 2–24   **Typical fixed resistors.**

One common fixed resistor is the carbon-composition type, which is made with a mixture of finely ground carbon, insulating filler, and a resin binder. The ratio of carbon to insulating filler sets the resistance value. The mixture is formed into rods which are cut into short lengths, and lead connections are made. The entire resistor is then encapsulated in an insulated coating for protection. Figure 2–25(a) shows the construction of a typical carbon-composition resistor.

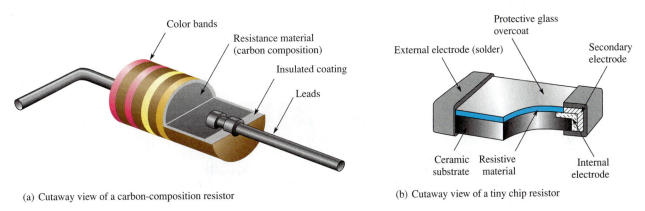

(a) Cutaway view of a carbon-composition resistor

(b) Cutaway view of a tiny chip resistor

FIGURE 2–25   **Two types of fixed resistors (not to scale).**

The chip resistor is another type of fixed resistor and is in the category of SMT (surface-mount technology) components. It has the advantage of a very small size for compact assemblies. Small-value chip resistors ($<1 \ \Omega$) can be made with very close tolerances ($\pm 0.5\%$) and have application as current-sensing resistors. Figure 2–25(b) shows the construction of a chip resistor.

Other types of fixed resistors include carbon film, metal film, metal-oxide film, and wirewound. In film resistors, a resistive material is deposited evenly onto a high-grade ceramic rod. The resistive film may be carbon (carbon film) or nickel chromium (metal film). In these types of resistors, the desired resistance value is obtained by removing part of the resistive material in a helical pattern along the rod using a spiraling technique, as shown in Figure 2–26(a). Very close **tolerance** can be achieved with this method. Film resistors are also available in the form of resistor networks, as shown in Figure 2–26(b).

Wirewound resistors are constructed with resistive wire wound around an insulating rod and then sealed. Normally, wirewound resistors are used because of their relatively high power ratings. Since they are constructed with a coil of wire, wirewound resistors have significant inductance and are not used at higher frequencies. Some typical wirewound resistors are shown in Figure 2–27.

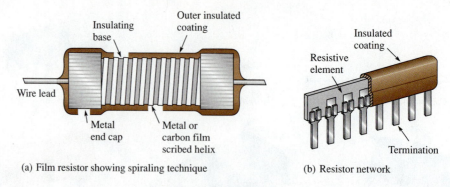

(a) Film resistor showing spiraling technique   (b) Resistor network

FIGURE 2–26   Construction views of typical film resistors.

FIGURE 2–27   Typical wirewound power resistors.

RESISTOR COLOR CODES   Some types of fixed resistors with value tolerances of 5% or 10% are color coded with four bands to indicate the resistance value and the tolerance. This color-code band system is shown in Figure 2–28, and the **color code** is listed in Table 2–1. The bands are always located closer to one end.

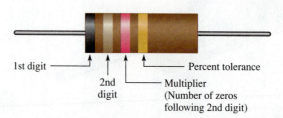

FIGURE 2–28   Color-code bands on a 4-band resistor.

The 4-band color code is read as follows:

1. Start with the band closest to one end of the resistor. The first band is the first digit of the resistance value. If it is not clear which is the banded end, start from the end that does not begin with a gold or silver band.

2. The second band is the second digit of the resistance value.

3. The third band is the number of zeros following the second digit, or the *multiplier*.

4. The fourth band indicates the percent tolerance and is usually gold or silver. If there is no fourth band, the tolerance is ±20%.

For example, a 5% tolerance means that the *actual* resistance value is within ±5% of the color-coded value. Thus, a 100 Ω resistor with a tolerance of ±5% can have an acceptable range of values from a minimum of 95 Ω to a maximum of 105 Ω.

As indicated in the table, for resistance values less than 10 Ω, the third band is either gold or silver. Gold in the third band represents a multiplier of 0.1, and silver represents 0.01. For example, a color code of red, violet, gold, and silver represents 2.7 Ω with a tolerance of ±10%. A table of standard resistance values is in Appendix A.

TABLE 2–1 • 4-band resistor color code.

	Digit	Color	
Resistance value, first three bands: First band—1st digit Second band—2nd digit *Third band—multiplier (number of zeros following the 2nd digit)	0 1 2 3 4 5 6 7 8 9		Black Brown Red Orange Yellow Green Blue Violet Gray White
Fourth band—tolerance	±5% ±10%		Gold Silver

* For resistance values less than 10 Ω, the third band is either gold or silver. Gold is for a multiplier of 0.1 and silver is for a multiplier of 0.01.

EXAMPLE 2–4

Find the resistance values in ohms and the percent tolerance for each of the color-coded resistors shown in Figure 2–29.

(a)　　　　　　　　　　(b)　　　　　　　　　　(c)

FIGURE 2–29

SOLUTION

(a) First band is red = 2, second band is violet = 7, third band is orange = 3 zeros, fourth band is silver = ±10% tolerance.

$$R = 27{,}000 \ \Omega \ \pm \ 10\%$$

(b) First band is brown = 1, second band is black = 0, third band is brown = 1 zero, fourth band is silver = ±10% tolerance.

$$R = 100 \ \Omega \ \pm \ 10\%$$

(c) First band is green = 5, second band is blue = 6, third band is green = 5 zeros, fourth band is gold = ±5% tolerance.

$$R = 5{,}600{,}000 \ \Omega \ \pm \ 5\%$$

RELATED PROBLEM

A certain resistor has a yellow first band, a violet second band, a red third band, and a gold fourth band. Determine the value in ohms and its percent tolerance.

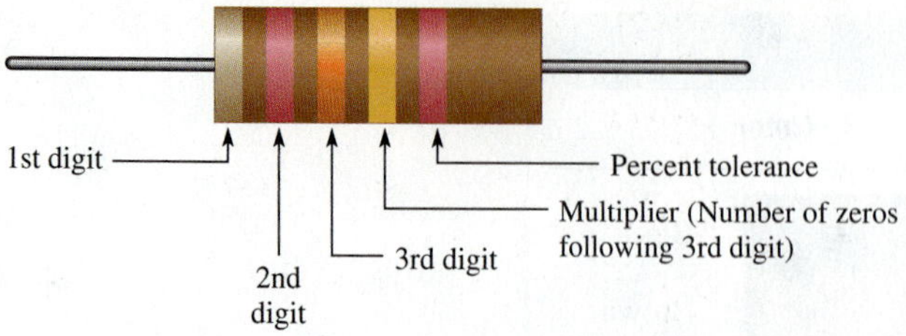

1st digit

2nd digit

3rd digit

Multiplier (Number of zeros following 3rd digit)

Percent tolerance

FIGURE 2–30 **Color-code bands on a 5-band resistor.**

FIVE-BAND COLOR CODE Certain precision resistors with tolerances of 2%, 1%, or less are generally color coded with five bands, as shown in Figure 2–30. Begin at the band closest to one end. The first band is the first digit of the resistance value, the second band is the second digit, the third band is the third digit, the fourth band is the multiplier (number of zeros after the third digit), and the fifth band indicates the tolerance. Table 2–2 shows the 5-band color code.

TABLE 2–2 • 5-band resistor color code.

	DIGIT	**COLOR**
Resistance value, first four bands:	0	Black
	1	Brown
	2	Red
First band—1st digit	3	Orange
Second band—2nd digit	4	Yellow
Third band—3rd digit	5	Green
Fourth band—multiplier	6	Blue
(number of zeros following 3rd digit)	7	Violet
	8	Gray
	9	White
Fourth band—multiplier	0.1	Gold
	0.01	Silver
	±2%	Red
	±1%	Brown
Fifth band—tolerance	±0.5%	Green
	±0.25%	Blue
	±0.1%	Violet

EXAMPLE 2–5

Find the resistance value in ohms and the percent tolerance for each of the color-coded resistors shown in Figure 2–31.

(a) (b) (c)

FIGURE 2–31

SOLUTION

(a) First band is red = 2, second band is violet = 7, third band is black = 0, fourth band is gold = ×0.1, fifth band is red = ±2% tolerance.

$$R = 270 \times 0.1 = \textbf{27 } \mathbf{\Omega} \textbf{ ± 2\%}$$

(b) First band is yellow = 4, second band is black = 0, third band is red = 2, fourth band is black = 0, fifth band is brown = ±1% tolerance.

$$R = 402\ \Omega \pm 1\%$$

(c) First band is orange = 3, second band is orange = 3, third band is red = 2, fourth band is orange = 3, fifth band is green = ±0.5% tolerance.

$$R = 332{,}000\ \Omega \pm 0.5\%$$

RELATED PROBLEM

A certain resistor has a yellow first band, a violet second band, a green third band, a gold fourth band, and a red fifth band. Determine its value in ohms and its percent tolerance.

RESISTOR LABEL CODES   Not all types of resistors are color coded. Many, including surface-mount resistors, use typographical marking to indicate the resistance value and tolerance. These label codes consist of either all numbers (numeric) or a combination of numbers and letters (alphanumeric). In some cases when the body of the resistor is large enough, the entire resistance value and tolerance are stamped on it in standard form. For example, a 33,000 Ω resistor may be labeled as 33 kΩ.

Numeric labeling uses three digits to indicate the resistance value, as shown in Figure 2–32 using a specific example. The first two digits give the first two digits of the resistance value, and the third digit gives the multiplier or number of zeros that follow the first two digits. This code is limited to values of 10 Ω or greater.

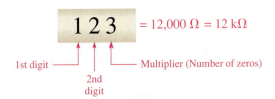

FIGURE 2–32   **Example of three-digit labeling for a resistor.**

Another common type of marking is a three- or four-character label that uses both digits and letters. An alphanumeric label typically consists of only three digits or two or three digits and one of the letters R, K, or M. The letter is used to indicate the multiplier, and the position of the letter indicates the decimal point placement. The letter R indicates a multiplier of 1 (no zeros after the digits), the K indicates a multiplier of 1000 (three zeros after the digits), and the M indicates a multiplier of 1,000,000 (six zeros after the digits). In this format, values from 100 to 999 consist of three digits and no letter to represent the three digits in the resistance value. Figure 2–33 shows three examples of this type of resistor label.

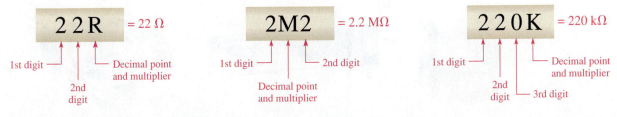

FIGURE 2–33   **Examples of the alphanumeric resistor label.**

EXAMPLE 2–6

Interpret the following alphanumeric resistor labels:

(a) 470   (b) 5R6   (c) 68K   (d) 10M   (e) 3M3

SOLUTION

(a) 470 = **470 Ω**   (b) 5R6 = **5.6 Ω**   (c) 68K = **68 kΩ**

(d) 10M = **10 MΩ**   (e) 3M3 = **3.3 MΩ**

RELATED PROBLEM

What is the resistance indicated by 1K2?

One system of labels for resistance tolerance values uses the letters F, G, and J:

$$F = \pm1\% \qquad G = \pm2\% \qquad J = \pm5\%$$

For example, 620F indicates a 620 Ω resistor with a tolerance of ±1%, 4R6G is a 4.6 Ω ±2% resistor, and 56KJ is a 56 kΩ ±5% resistor.

VARIABLE RESISTORS   Variable resistors are designed so that their resistance values can be changed easily. Two basic uses for variable resistors are to divide voltage and to control current. The variable resistor used to divide voltage is called a **potentiometer**. The variable resistor used to control current is called a **rheostat**. Schematic symbols for these types are shown in Figure 2–34. The potentiometer is a three-terminal device, as indicated in part (a). Terminals 1 and 2 have a fixed resistance between them, which is the total resistance. Terminal 3 is connected to a moving contact (**wiper**). You can vary the resistance between 3 and 1 or between 3 and 2 by moving the contact.

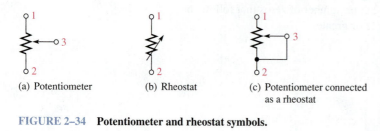

(a) Potentiometer   (b) Rheostat   (c) Potentiometer connected
                                               as a rheostat

FIGURE 2–34   Potentiometer and rheostat symbols.

Figure 2–34(b) shows the rheostat as a two-terminal variable resistor. Part (c) shows how you can use a potentiometer as a rheostat by connecting terminal 3 to either terminal 1 or terminal 2. Parts (b) and (c) are equivalent symbols. Some typical potentiometers are pictured in Figure 2–35. The construction view is shown under the normal view.

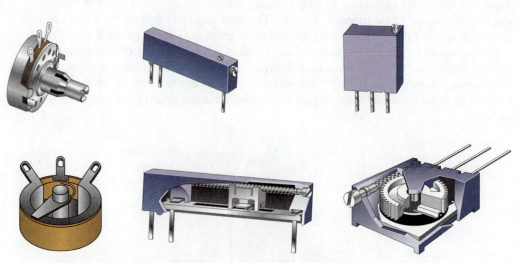

FIGURE 2–35   Typical potentiometers and construction views.

Potentiometers and rheostats can be classified as linear or tapered, as shown in Figure 2–36, where a potentiometer with a total resistance of 100 Ω is used as an example. As shown in part (a), in a linear potentiometer, the resistance between either terminal and the moving contact varies linearly with the position of the moving contact. For example, one-half of a turn results in one-half the total resistance. Three-quarters of a turn results in three-quarters of the total resistance between the moving contact and one terminal, or one-quarter of the total resistance between the other terminal and the moving contact.

In a **tapered** potentiometer, the resistance varies nonlinearly with the position of the moving contact, so that one-half of a turn does not necessarily result in one-half the total resistance. This concept is illustrated in Figure 2–36(b), where the nonlinear values are arbitrary.

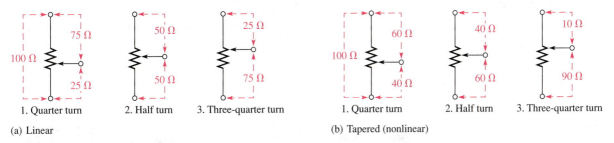

| 1. Quarter turn | 2. Half turn | 3. Three-quarter turn | | 1. Quarter turn | 2. Half turn | 3. Three-quarter turn |

(a) Linear   (b) Tapered (nonlinear)

FIGURE 2–36   **Examples of (a) linear and (b) tapered potentiometers.**

The potentiometer is used as a voltage-control device. When a fixed voltage is applied across the end terminals, a variable voltage is obtained at the wiper contact with respect to either end terminal. The rheostat is used as a current-control device; the current can be changed by changing the wiper position.

VARIABLE RESISTANCE SENSORS   Many sensors operate on the concept of a variable resistance, in which a physical quantity alters the electrical resistance. Depending on the sensor and the measurement requirements, the change in resistance may be determined directly or indirectly using the resistance change to alter a voltage or current.

Examples of resistance sensors include **thermistors** that change resistance as a function of temperature, **photoconductive cells** that change resistance as a function of light, and **strain gages** that change resistance when a force is applied to them. Thermistors are commonly used in thermostats. Photocells have many applications; for example, they are used to turn on street lights at dusk and turn them off at dawn. Strain gages are widely used in scales and applications where mechanical motion needs to be sensed. The measuring instruments for strain gages need to be very sensitive because the change in resistance is very small. Figure 2–37 shows symbols for these types of resistance sensors.

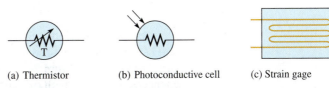

(a) Thermistor   (b) Photoconductive cell   (c) Strain gage

FIGURE 2–37   **Symbols for resistance devices with sensitivities to temperature, light, and force.**

Truck Scales

Many large scales, such as those that are used for weighing trucks, use strain gages as the basic sensor to detect the force. The strain gage is part of a complete electromechanical assembly consisting of four gages. Typically, the strain gages are composed of a back-and forth pattern of thin wires that are bonded to a steel beam or plate. The steel is bent by the weight, stretching or compressing the wire in the strain gage and changing its resistance. The change is detected and converted to a number that represents the weight.

Courtesy of Cardinal Scale
Manufacturing Co.
www.CardinalScale.com

SYSTEM NOTE

SECTION 2–5 CHECKUP

1. Define *resistance* and name its unit.

2. What are the two main categories of resistors? Briefly explain the difference between them.

3. In the 4-band resistor color code, what does each band represent?

4. Determine the resistance and percent tolerance for each of the following color codes:

   (a) yellow, violet, red, gold

   (b) blue, red, orange, silver

   (c) brown, gray, black, gold

   (d) red, red, blue, red, green

5. What resistance value is indicated by each alphanumeric label:

   (a) 33R

   (b) 5K6

   (c) 900

   (d) 6M8

6. What is the basic difference between a rheostat and a potentiometer?

7. Name three resistance sensors and the physical quantity that affects their resistance.

2–6 THE ELECTRIC CIRCUIT

A basic electric circuit is an arrangement of physical components that use voltage, current, and resistance to perform some useful function.

After completing this section, you should be able to

- **Describe a basic electric circuit**
  - **Relate a schematic to a physical circuit**
  - **Define *open circuit* and *closed circuit***
  - **Describe various types of protective devices**
  - **Describe various types of switches**
  - **Explain how wire sizes are related to gauge numbers**
  - **Define *ground* or *common***

Basically, an electric circuit consists of a voltage source, a load, and a path for current between the source and the load. The **load** is a device on which work is done by the current through it. Figure 2–38 shows an example of a simple electric circuit: a battery connected to a lamp with two conductors (wires). The battery is the voltage source, the lamp is the load on the battery because it draws current from the battery, and the two wires provide the current path from the negative terminal of the battery to the lamp and back to the positive terminal of the battery, as indicated by the red arrows. There is current through the filament of the lamp (which has a resistance), causing it to become hot enough to emit visible light. Current through the battery is produced by chemical action.

In many practical cases, one terminal of the battery is connected to a ground (common) point. For example, in automobiles, the negative battery terminal is generally connected to the metal chassis of the car. The chassis is the ground for the automobile electrical system and provides a current path for the circuit. (The concept of *ground* is covered later in this chapter.)

As you know, block diagrams are a basic way to show the signal path and functional blocks in a system. When details of a circuit are required for troubleshooting or analysis, in large systems, for example,

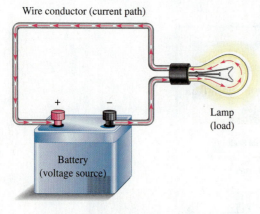

Wire conductor (current path)

Lamp (load)

Battery (voltage source)

FIGURE 2–38 A simple electric circuit.

it is useful to have a schematic of the circuit. A **schematic** is a diagram that uses standard symbols and designators to represent components and lines to represent conductors; it shows the circuit in a logical organized manner so that circuit operation can be determined. The schematic may be quite different from the physical layout and is usually drawn to show signal flow from left to right. In general, the symbols used in schematics have been standardized by various standards organizations. For the U.S., the standard is IEEE STD 315-1975, available from the American National Standards Institute (ANSI). In this textbook, symbols are introduced as they are needed. Figure 2–39 is a basic schematic for the circuit in Figure 2–38.

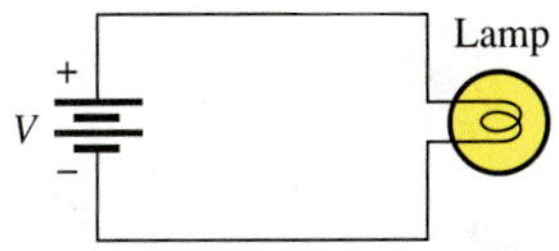

FIGURE 2–39 **A schematic for the circuit in Figure 2–38.**

SAFETY NOTE

To avoid electrical shock, never touch a circuit while it is connected to a voltage source. If you need to handle a circuit, remove a component, or change a component, first make sure the voltage source is disconnected.

Circuit Current Control and Protection

The circuit in Figure 2–40(a) illustrates a **closed circuit**—that is, a circuit in which the current has a complete path. An **open circuit** is a circuit in which the current path is broken so that there is no current, as shown in part (b). An open circuit is considered to have infinite resistance (*infinite* means immeasurably large).

MECHANICAL SWITCHES **Switches** are commonly used for controlling the opening or closing of circuits. For example, a switch is used to turn a lamp on or off, as illustrated in Figure 2–40. Each circuit pictorial is shown with its associated schematic. The type of switch indicated is a *single-pole–single-throw* (SPST) toggle switch. The term *pole* refers to the movable arm in a switch, and the term *throw* indicates the number of contacts that are affected (either opened or closed) by a single switch action (a single movement of a pole).

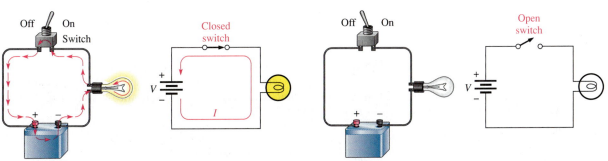

(a) There is current in a *closed* circuit because there is a complete current path (switch is ON or in the *closed* position). Current is always indicated by a red arrow in this text.

(b) There is no current in an *open* circuit because the path is broken (switch is OFF or in the *open* position).

FIGURE 2–40 **Illustration of closed and open circuits using an SPST switch for control.**

Figure 2–41 shows a somewhat more complicated circuit using a *single-pole–double-throw* (SPDT) type of switch to control the current to two different lamps. When one lamp is on, the other is off, and vice versa, as illustrated by the two schematics that represent each of the switch positions.

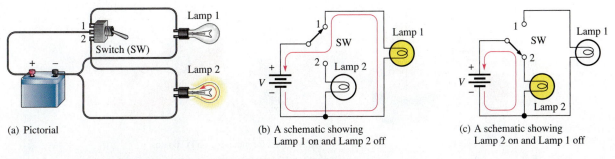

(a) Pictorial

(b) A schematic showing Lamp 1 on and Lamp 2 off

(c) A schematic showing Lamp 2 on and Lamp 1 off

FIGURE 2–41 **An example of an SPDT switch controlling two lamps.**

In addition to the SPST and the SPDT switches (symbols are shown in Figure 2–42(a) and (b)), the following other types of switches are also important:

- **Double-pole–single-throw (DPST).** The DPST switch permits simultaneous opening or closing of two sets of contacts. The symbol is shown in Figure 2–42(c). The dashed line indicates that the contact arms are mechanically linked so that both move with a single switch action.

- **Double-pole–double-throw (DPDT).** The DPDT switch provides connection from one set of contacts to either of two other sets. The schematic symbol is shown in Figure 2–42(d).

- **Push button (PB).** In the normally open push-button switch (NOPB), shown in Figure 2–42(e), connection is made between two contacts when the button is depressed, and connection is broken when the button is released. In the normally closed push-button switch (NCPB), shown in Figure 2–42(f), connection between the two contacts is broken when the button is depressed.

- **Rotary.** In a rotary switch, a knob is turned to make a connection between one contact and any one of several others. A symbol for a simple six-position rotary switch is shown in Figure 2–42(g).

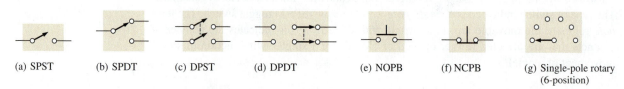

(a) SPST   (b) SPDT   (c) DPST   (d) DPDT   (e) NOPB   (f) NCPB   (g) Single-pole rotary (6-position)

FIGURE 2–42   **Switch symbols.**

Figure 2–43 shows several varieties of switches, and Figure 2–44 shows a construction view of a typical toggle switch. In addition to mechanical switches, a transistor can be used as the equivalent of a single-pole–single-throw switch in certain applications.

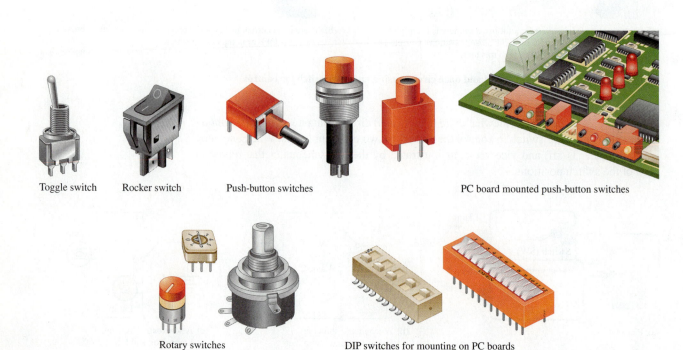

Toggle switch   Rocker switch   Push-button switches   PC board mounted push-button switches

Rotary switches   DIP switches for mounting on PC boards

FIGURE 2–43   **Typical mechanical switches.**

PROTECTIVE DEVICES   **Fuses** and **circuit breakers** are placed in the current path and are used to deliberately create an open circuit when the current exceeds a specified number of amperes due to a malfunction or other abnormal condition in a circuit. For example, a fuse or circuit breaker with a 20 A rating will open a circuit when the current exceeds 20 A.

FIGURE 2–44   **Construction view of a typical toggle switch.**

The basic difference between a fuse and a circuit breaker is that when a fuse is "blown," it must be replaced; but when a circuit breaker opens, it can be reset and reused repeatedly. Both of these devices protect against damage to a circuit due to excess current or prevent a hazardous condition created by the overheating of wires and other components when the current is too great. Because fuses cut off excess current more quickly than circuit breakers, fuses are used whenever delicate electronic equipment needs to be protected. Several typical fuses and circuit breakers, along with their schematic symbols, are shown in Figure 2–45.

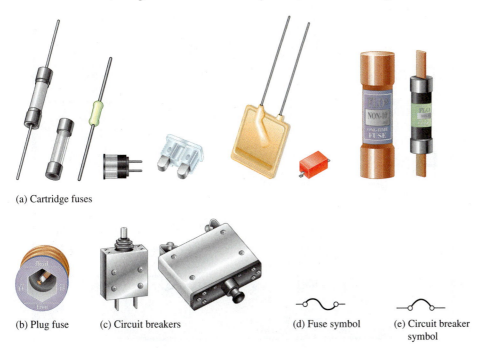

(a) Cartridge fuses

(b) Plug fuse   (c) Circuit breakers   (d) Fuse symbol   (e) Circuit breaker symbol

FIGURE 2–45   Typical fuses and circuit breakers and their symbols.

Two basic categories of fuses in terms of their physical configuration are cartridge type and plug type (screw in). Cartridge type fuses have various shaped housings with leads or other types of contacts as shown in Figure 2–45(a). A typical plug type fuse is shown in part (b). Fuse operation is based on the melting temperature of a wire or other metal element. As current increases, the fuse element heats up and when the rated current is exceeded, the element reaches its melting temperature and opens, thus removing power from the circuit.

Two common types of fuses are the fast-acting and the time-delay (slow blow). Fast-acting fuses are type F and time-delay fuses are type T. In normal operation, fuses are often subjected to intermittent current surges that may exceed the rated current, such as when power to a circuit is turned on. Over time, this reduces the fuse's ability to withstand short surges or even current at the rated value. A slow blow fuse can tolerate greater and longer duration surges of current than the typical fast-acting fuse. A fuse symbol is shown in Figure 2–45(d).

Typical circuit breakers are shown in Figure 2–45(c) and the symbol is shown in part (e). Generally, a circuit breaker detects excess current either by the heating effect of the current or by the magnetic field it creates. In a circuit breaker based on the heating effect, a bimetallic spring opens the contacts when the rated current is exceeded. Once opened, the contact is held open by mechanical means until manually reset. In a circuit breaker based on a magnetic field, the contacts are opened by a sufficient magnetic force created by excess current and must be mechanically reset.

SAFETY NOTE

Always use fully insulated fuse pullers to remove and replace fuses in an electrical box. Even if the disconnect switch is in the off position, line voltage is still present in the box. Never use metal tools to remove and replace fuses.

SYSTEM EXAMPLE 2–2

A QUIZ BOARD

Figure 2–46 shows a quiz board for a science fair project; it is an example of a small system that uses components introduced in this chapter. The operation of the circuit is for the user to rotate the switch into one of four positions to select a battery type (shown with a light). After selecting the battery type, the user presses a pushbutton corresponding to the correct cell voltage for that battery. The "correct" light will illuminate only if the selected pushbutton is correct. The quiz board is constructed with the sequence of correct answers as B, D, A, C for batteries 1, 2, 3, and 4. A rheostat is added to control the brightness of all lights and the circuit should be fused.

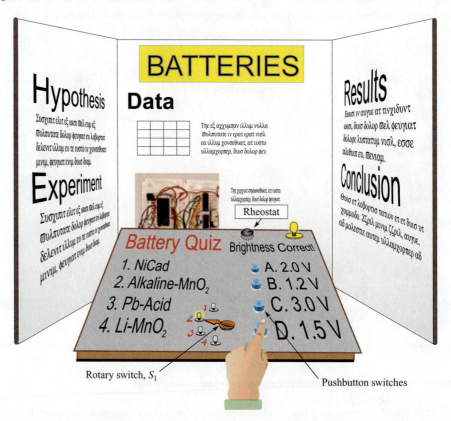

FIGURE 2–46 **A quiz board.**

The schematic for the circuit is shown in Figure 2–47. Notice that each position of the rotary switch goes to a light and that only one pushbutton switch corresponds to the correct answer.

FIGURE 2–47 **Quiz board schematic.**

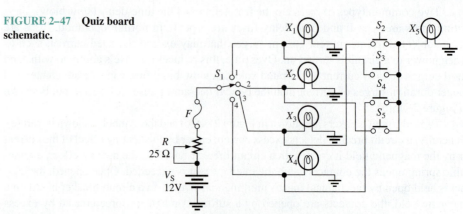

Wires

Wires are the most common form of conductive material used in electrical applications. They vary in diameter size and are arranged according to standard gauge numbers, called **AWG** (American Wire Gauge) sizes. As the gauge number increases, the wire diameter decreases. The size of a wire is also specified in terms of its cross-sectional area, as illustrated in Figure 2–48. The unit of cross-sectional area is the **circular mil,** abbreviated CM. One circular mil is the area of a wire with a diameter of 0.001 inch (0.001 in., or 1 mil). You can find the cross-sectional area in circular mils by expressing the diameter in thousandths of an inch (mils) and squaring it, as follows:

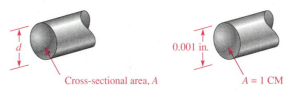
Cross-sectional area, A 0.001 in. $A = 1$ CM

FIGURE 2–48   Cross-sectional area of a wire.

$$A = d^2 \qquad\qquad (2\text{–}5)$$

where A is the cross-sectional area in circular mils and d is the diameter in mils. Table 2–3 lists the AWG sizes with their corresponding cross-sectional area and resistance in ohms per 1000 ft at 20°C.

TABLE 2–3 • AWG (American Wire Gauge) sizes and resistances for solid round copper.					
AWG #	AREA (CM)	RESISTANCE (Ω/1000 FT AT 20°C)	AWG #	AREA (CM)	RESISTANCE (Ω/1000 FT AT 20°C)
0000	211,600	0.0490	19	1,288.1	8.051
000	167,810	0.0618	20	1,021.5	10.15
00	133,080	0.0780	21	810.10	12.80
0	105,530	0.0983	22	642.40	16.14
1	83,694	0.1240	23	509.45	20.36
2	66,373	0.1563	24	404.01	25.67
3	52,634	0.1970	25	320.40	32.37
4	41,742	0.2485	26	254.10	40.81
5	33,102	0.3133	27	201.50	51.47
6	26,250	0.3951	28	159.79	64.90
7	20,816	0.4982	29	126.72	81.83
8	16,509	0.6282	30	100.50	103.2
9	13,094	0.7921	31	79.70	130.1
10	10,381	0.9989	32	63.21	164.1
11	8,234.0	1.260	33	50.13	206.9
12	6,529.0	1.588	34	39.75	260.9
13	5,178.4	2.003	35	31.52	329.0
14	4,106.8	2.525	36	25.00	414.8
15	3,256.7	3.184	37	19.83	523.1
16	2,582.9	4.016	38	15.72	659.6
17	2,048.2	5.064	39	12.47	831.8
18	1,624.3	6.385	40	9.89	1049.0

EXAMPLE 2–7

What is the cross-sectional area of a wire with a diameter of 0.005 inch?

SOLUTION

$$d = 0.005 \text{ in.} = 5 \text{ mils}$$
$$A = d^2 = (5 \text{ mils})^2 = \mathbf{25\ CM}$$

RELATED PROBLEM

What is the cross-sectional area of a 0.0201 in. diameter wire? What is the AWG # for this wire from Table 2–3?

WIRE RESISTANCE Although copper wire conducts current extremely well, it still has some resistance, as do all conductors. The resistance of a wire depends on three physical characteristics: (a) type of material, (b) length of wire, and (c) cross-sectional area. In addition, temperature can also affect the resistance.

Each type of conductive material has a characteristic called its *resistivity,* which is represented by the Greek letter rho (ρ). For each material, ρ is a constant value at a given temperature. The formula for the resistance of a wire of length l and cross-sectional area A is

$$R = \frac{\rho l}{A} \tag{2–6}$$

Connectors

Connectors route wiring to and from system assemblies. The size and type of connector needed depends on the number and type of signals, power requirements, environment, and physical requirements. Connectors should be mechanically strong, reliable, and provide a low-resistance contact. A loose or corroded connection will have higher than normal resistance and can lead to internal arcing and breakdown.

In most systems, connectors are critical for reliability. NASA has experienced delays in missions due to connector problems and has developed certain reliability tests for connectors for space missions. For this reason, life support systems, space, and military systems all have unique requirements and specifications for system connectors.

SYSTEM NOTE

This formula shows that resistance increases with an increase in resistivity and length and decreases with an increase in cross-sectional area. For resistance to be calculated in ohms, the length must be in feet (ft), the cross-sectional area in circular mils (CM), and the resistivity in CM-Ω/ft.

EXAMPLE 2–8

Find the resistance of a 100 ft length of copper wire with a cross-sectional area of 810.1 CM. The resistivity of copper is 10.37 CM-Ω/ft at 20°C.

SOLUTION

$$R = \frac{\rho l}{A} = \frac{(10.37 \text{ CM-}\Omega/\text{ft})\,(100 \text{ ft})}{810.1 \text{ CM}} = \mathbf{1.280\ \Omega}$$

Use Table 2–3 to determine the resistance of 100 ft of copper wire with a cross-sectional area of 810.1 CM. Compare with the calculated value.

As mentioned, Table 2–3 lists the resistance of the various standard wire sizes in ohms per 1000 ft at 20°C. For example, a 1000 ft length of 14-gauge copper wire has a resistance of 2.525 Ω. A 1000 ft length of 22-gauge wire has a resistance of 16.14 Ω. For a given length, the smaller-gauge wire has more resistance. Thus, for a given voltage, larger-gauge wires can carry more current than smaller ones.

SPLICING WIRES   In some systems it is necessary to splice wires, especially when the run is long. Specific rules for splicing of electrical cables are given in the National Electric Code (NEC). Spicing is generally permissible provided that the wire is clean and a good electrical and mechanical connection is made. Most of the time, the splice is within an electrical box to protect it. For small-diameter wires, the crimp-type connectors are useful and quick. A crimping tool is used to make the connections.

Soldering the wires together in a splice is also permitted except on conductors used for grounding, provided that there is a good mechanical connection prior to soldering. After splicing, the splice should be encased in an insulation that is as good as the wire insulation.

One problem to avoid is splicing wires made from different conductors such as aluminum and copper. In this case, a chemical reaction called electrolysis can occur at the junction, leading to a high resistance connection that fails. Another problem with splicing different conductors is that they can expand differently if they are warmed, causing the splice to loosen and fail.

Ground

Ground is the reference point in an electric circuit. The term *ground* originated from the fact that one conductor of a circuit was typically connected with an 8-foot long metal rod driven into the earth itself. Today, this type of connection is referred to as an *earth ground*. In household wiring, earth ground is indicated with a green or bare copper wire. Earth ground is normally connected to the metal chassis of an appliance or a metal electrical box for safety. Unfortunately, there have been exceptions to this rule, which can present a safety hazard if a metal chassis is not at earth ground. It is a good idea to confirm that a metal chassis is actually at earth ground potential before doing any work on an instrument or appliance.

Another type of ground is called a *reference ground*. Voltages are always specified with respect to another point. If that point is not stated explicitly, the reference ground is understood. Reference ground defines 0 V for the circuit. The reference ground can be at a completely different potential than the earth ground. Reference ground is also called **common** and labeled COM or COMM because it represents a common conductor. When you are wiring a protoboard in the laboratory, you will normally reserve one of the bus strips (a long line along the length of the board) for this common conductor.

Three ground symbols are shown in Figure 2–49. Unfortunately, there is not a separate symbol to distinguish between earth ground and reference ground. The symbol in (a) represents either an earth ground or a reference ground, (b) shows a chassis ground, and (c) is an alternate reference symbol typically used when there is more than one common connection (such as analog and digital ground in the same circuit). In this book, the symbol in part (a) will be used throughout.

Figure 2–50 illustrates a simple circuit with ground connections. The current is from the negative terminal of the 12 V source, through the common ground connection, through the lamp, and through the wire back to the positive terminal of the source. Ground provides a path for the current from the source because all of the ground points are electrically the same point and provide a zero resistance (ideally) current path. The voltage at the top of the circuit is +12 V with respect to ground. You can think of all the ground points in a circuit as being connected together by a conductor.

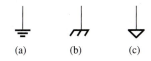

(a)   (b)   (c)

FIGURE 2–49   Symbols for ground.

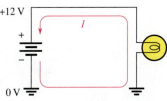

FIGURE 2–50   A simple circuit with ground connections.

SECTION 2–6 CHECKUP

1. What are the basic elements of an electric circuit?

2. Define *open circuit.*

3. Define *closed circuit.*

4. What is the resistance of an open switch? Ideally, what is the resistance of a closed switch?

5. What is the purpose of a fuse?

6. What is the difference between a fuse and a circuit breaker?

7. Which wire is larger in diameter, AWG #3 or AWG #22?

8. What is ground in an electric circuit?

2–7 BASIC CIRCUIT MEASUREMENTS

In working on electrical or electronic circuits, you will frequently need to measure voltage, current, and resistance and to use meters safely and correctly.

After completing this section, you should be able to

- **Make basic circuit measurements**
  - **Properly measure voltage in a circuit**
  - **Properly measure current in a circuit**
  - **Properly measure resistance**
  - **Set up and read basic meters**

Voltage, current, and resistance measurements are commonly required in electrical and electronics work. The instrument used to measure voltage is a **voltmeter**, the instrument used to measure current is an **ammeter**, and the instrument used to measure resistance is an **ohmmeter**. Commonly, all three instruments are combined into a single instrument known as a **multimeter,** in which you can choose the specific quantity to measure by selecting the appropriate function with a switch.

Typical portable multimeters are shown in Figure 2–51. Part (a) shows a digital multimeter (DMM), which provides a digital readout of the measured quantity; and part (b) shows an analog meter with a needle pointer. Many digital multimeters also include a bar graph display.

FIGURE 2–51 Typical portable multimeters.   (a) Courtesy of Fluke Corporation. Reproduced with permission. (b) Courtesy of B+K Precision.

(a) Digital multimeter   (b) Analog multimeter

Meter Symbols

Throughout this book certain symbols will be used in circuits to represent meters, as shown in Figure 2–52. You may see any of four types of symbols for voltmeters, ammeters, and ohmmeters, depending on which symbol most effectively conveys the information required. The digital meter symbol is used when specific values are to be indicated in a

(a) Digital   (b) Bar graph   (c) Analog   (d) Generic

FIGURE 2–52 **Examples of meter symbols used in this book. Each of the symbols can be used to represent either an ammeter (A), a voltmeter (V), or an ohmmeter (Ω).**

circuit. The bar graph meter symbol and sometimes the analog meter symbol are used to illustrate the operation of a circuit when *relative* measurements or changes in quantities, rather than specific values, need to be depicted. A changing quantity may be indicated by an arrow in the display showing an increase or decrease. The generic symbol is used to indicate placement of meters in a circuit when no values or value changes need to be shown.

Measuring Current

Figure 2–53 illustrates how to measure current with an ammeter. Part (a) shows a simple circuit in which the current through a resistor is to be measured. Connect an ammeter in the current path by first opening the circuit, as shown in part (b). Then insert the meter as shown in part (c). Such a connection is a *series* connection. The polarity of the meter must be such that the current is in at the negative terminal and out at the positive terminal. If you are using a DMM to measure current, the leads must be moved to a special current jack.

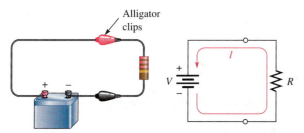

(a) Circuit in which the current is to be measured

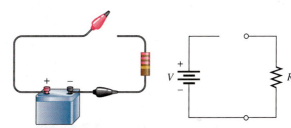

(b) Open the circuit either between the resistor and the positive terminal or between the resistor and the negative terminal of source.

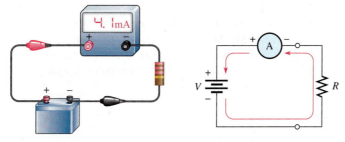

(c) Install the ammeter in the current path with polarity as shown (negative to negative, positive to positive).

FIGURE 2–53 **Example of an ammeter connection to measure current in a simple circuit.**

Measuring Voltage

To measure voltage, connect a voltmeter across the component for which the voltage is to be found. Such a connection is a *parallel* connection. The negative terminal of the meter must be connected to the negative side of the circuit, and the positive terminal of the meter must be connected to the positive side of the circuit. Figure 2–54 shows a voltmeter connected to measure the voltage across the resistor.

SAFETY NOTE

Never wear rings or any type of metallic jewelry while you work on a circuit. These items may accidentally come in contact with the circuit, causing shock and/or damage to the circuit. With high-energy sources, such as a car battery, a short across jewelry (a watch or ring) can become hot instantly, causing burns to the wearer.

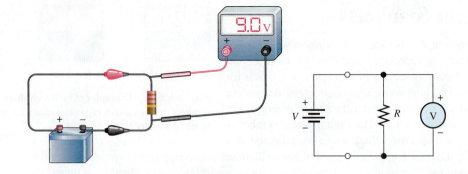

Measuring Resistance

To measure resistance, connect an ohmmeter across the resistor. *The resistor must first be removed or disconnected from the circuit.* This procedure is shown in Figure 2–55.

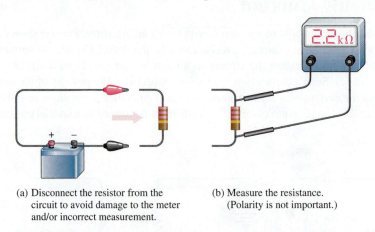

(a) Disconnect the resistor from the circuit to avoid damage to the meter and/or incorrect measurement.

(b) Measure the resistance. (Polarity is not important.)

FIGURE 2–55   **Example of using an ohmmeter to measure resistance.**

Digital Multimeters (DMMs)

A **DMM** is a multifunction electronic instrument that can measure voltage, current, or resistance. DMMs are the most widely used type of electronic measuring instrument. Generally, DMMs provide more functions, better accuracy, greater ease of reading, and greater reliability than do analog meters, which are covered next. Analog meters have at least one advantage over DMMs, however. They can track short-term variations and trends in a measured quantity that many DMMs are too slow to respond to. Typical DMMs are shown in Figure 2–56. Many DMMs are autoranging types in which the proper range is automatically selected by internal circuitry.

FIGURE 2–56   **Typical digital multimeters (DMMs).**
(Courtesy of B+K Precision)

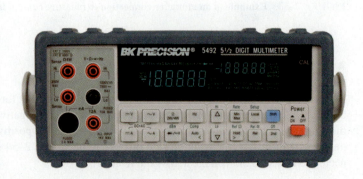

DMM FUNCTIONS   The basic functions found on most DMMs are

- Ohms
- DC voltage and current
- AC voltage and current

Some DMMs provide additional functions such as analog bar graph displays, transistor or diode tests, power measurement, and decibel measurement for audio amplifier tests.

DMM DISPLAYS   DMMs are available with either LCD (liquid-crystal display) or LED (light-emitting diode) readouts. The LCD is the most commonly used readout in battery-powered instruments because it requires only very small amounts of current. A typical battery-powered DMM with an LCD readout operates on a 9 V battery that will last from a few hundred hours to 2000 hours and more. The disadvantages of LCD readouts are that (a) they are difficult or impossible to see in low-light conditions and (b) they are relatively slow to respond to measurement changes. LEDs, on the other hand, can be seen in the dark and respond quickly to changes in measured values. LED displays require much more current than LCD displays; and, therefore, battery life is shortened when LEDs are used in portable equipment.

Both LCD and LED DMM displays are in a seven-segment format. Each digit in the display consists of seven separate segments, as shown in Figure 2–57(a). Each of the ten decimal digits is formed by the activation of appropriate segments, as illustrated in Figure 2–57(b). In addition to the seven segments, there is also a decimal point.

(a)   (b)

FIGURE 2–57   Seven-segment display.

RESOLUTION   The **resolution** of a DMM is the smallest increment of a quantity that the DMM can measure. The smaller the increment, the better the resolution. One factor that determines the resolution of a meter is the number of digits in the display.

Because many DMMs have 3½ digits in their display, we will use this case for illustration. A 3½-digit multimeter has three digit positions that can indicate from 0 through 9, and one digit position that can indicate only a value of 1. This latter digit, called the *half-digit,* is always the most significant digit in the display. For example, suppose that a DMM is reading .999 V, as shown in Figure 2–58(a). If the voltage increases by 0.001 V to 1 V, the display correctly shows 1.000 V, as shown in part (b). The "1" is the half-digit. Thus, with 3½ digits, a variation of 0.001 V, which is the resolution, can be observed.

(a) Resolution: 0.001 V   (b) Resolution: 0.001 V   (c) Resolution: 0.001 V   (d) Resolution: 0.01 V

FIGURE 2–58   A 3½-digit DMM illustrates how the resolution changes with the number of digits in use.

Now, suppose that the voltage increases to 1.999 V. This value is indicated on the meter as shown in Figure 2–58(c). If the voltage increases by 0.001 V to 2 V, the half-digit cannot display the "2," so the display shows 2.00. The half-digit is blanked and only three digits are active, as indicated in part (d). With only three digits active, the resolution is 0.01 V rather than 0.001 V as it is with 3½ active digits. The resolution remains 0.01 V up to 19.99 V. The resolution goes to 0.1 V for readings of 20.0 V to 199.9 V. At 200 V, the resolution goes to 1 V, and so on.

The resolution capability of a DMM is also determined by the internal circuitry and the rate at which the measured quantity is sampled. DMMs with displays of 4½ through 8½ digits are also available.

ACCURACY   As defined in Chapter 1, accuracy is an indication, usually expressed in percentage, of the range of error which is the difference in the measured and true or accepted value of a quantity. The accuracy of a DMM is established strictly by its internal circuitry and calibration. For typical meters, accuracies range from 0.01% to 0.5%, with some precision laboratory-grade meters going to 0.002%.

Reading Analog Multimeters

Although the DMM is predominately used, you may have to use on analog meter on occasion.

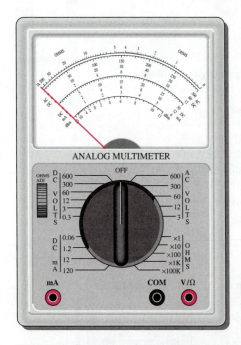

FIGURE 2–59   A typical analog multimeter.

FUNCTIONS   The face of a typical analog (needle-type) multimeter is represented in Figure 2–59. This particular instrument can be used to measure both direct current (dc) and alternating current (ac) quantities as well as resistance values. It has four selectable functions: dc volts (DC VOLTS), dc milliamperes (DC mA), ac volts (AC VOLTS), and OHMS. Many analog multimeters are similar to this one, although range selections and scales may vary.

RANGES   Within each function there are several ranges, as indicated by the brackets around the selector switch. For example, the DC VOLTS function has 0.3 V, 3 V, 12 V, 60 V, 300 V, and 600 V ranges. Thus, dc voltages from 0.3 V full-scale to 600 V full-scale can be measured. On the DC mA function, direct currents from 0.06 mA full-scale to 120 mA full-scale can be measured. On the ohm scale, the range settings are ×1, ×10, ×100, ×1000, and ×100,000.

THE OHM SCALE   Ohms are read on the top scale of the meter. This scale is nonlinear; that is, the values represented by each division (large or small) vary as you go across the scale. In Figure 2–59, notice how the scale becomes more compressed as you go from right to left.

To read the actual value in ohms, multiply the number on the scale as indicated by the pointer by the factor selected by the switch. For example, when the switch is set at ×100 and the pointer is at 20, the reading is 20 × 100 = 2000 Ω.

As another example, assume that the switch is at ×10 and the pointer is at the seventh small division between the 1 and 2 marks, indicating 17 Ω (1.7 × 10). Now, if the meter remains connected to the same resistance and the switch setting is changed to ×1, the pointer will move to the second small division between the 15 and 20 marks. This, of course, is also a 17 Ω reading, illustrating that a given resistance value can often be read at more than one switch setting. However, the meter should be *zeroed* each time the range is changed by touching the leads together and adjusting the needle.

THE AC-DC AND DC mA SCALES   The second, third, and fourth scales from the top (labeled "AC" and "DC") are used in conjunction with the DC VOLTS and AC VOLTS functions. The upper ac-dc scale ends at the 300 mark and is used with range settings such as 0.3, 3, and 300. For example, when the switch is at 3 on the DC VOLTS function, the

300 scale has a full-scale value of 3 V; at the range setting of 300, the full-scale value is 300 V. The middle ac-dc scale ends at 60. This scale is used in conjunction with range settings such as 0.06, 60, and 600. For example, when the switch is at 60 on the DC VOLTS function, the full-scale value is 60 V. The lower ac-dc scale ends at 12 and is used in conjunction with switch settings such as 1.2, 12, and 120. The three DC mA scales are used in a similar way to measure current.

HANDS ON TIP

When you are using a multimeter, such as the analog multimeter illustrated in Figure 2–59, where you manually select the voltage and current ranges, it is good practice to always set the multimeter on the maximum range before you measure an unknown voltage or current. You can then reduce the range until you get an acceptable reading.

EXAMPLE 2–9

In Figure 2–60, determine the quantity (voltage, current, or resistance) that is being measured and its value for each of the following specified switch settings on the meter in Figure 2–59.

(a) DC volts: 60   **(b)** DC mA: 12   **(c)** OHMS: ×1K

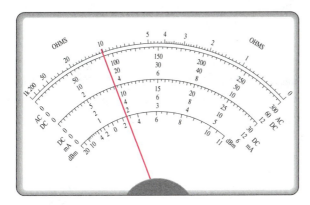

FIGURE 2–60

SOLUTION

(a) The reading taken from the middle AC-DC scale is **18 V**.

(b) The reading taken from the lower AC-DC scale is **3.8 mA**.

(c) The reading taken from the ohms scale (top) is **10 kΩ**.

RELATED PROBLEM

For part (c) in this example, the switch is moved to the ×100 ohms setting. Assuming that the same resistance is being measured, what will the needle do?

SECTION 2–7 CHECKUP

1. Name the multimeter functions for measuring

   (a) current

   (b) voltage

   (c) resistance

2. Show how to place two ammeters in the circuit of Figure 2–41 to measure the current through either lamp (be sure to observe the polarities). How can the same measurements be accomplished with only one ammeter?

3. Show how to place a voltmeter to measure the voltage across lamp 2 in Figure 2–41.

4. List two common types of DMM displays, and discuss the advantages and disadvantages of each.

5. Define *resolution* in a DMM.

6. The analog multimeter in Figure 2–59 is set on the 3 V range to measure dc voltage. Assume the pointer is at 150 on the upper ac-dc scale. What voltage is being measured?

7. How do you set up the multimeter in Figure 2–59 to measure 275 V dc, and on what scale do you read the voltage?

8. If you expect to measure a resistance in excess of 20 kΩ with the multimeter in Figure 2–59, where do you set the switch?

SUMMARY

- An atom is the smallest particle of an element that retains the characteristics of that element.
- The electron is the basic particle of negative electrical charge.
- The proton is the basic particle of positive charge.
- An ion is an atom that has gained or lost an electron and is no longer neutral.
- When electrons in the outer orbit of an atom (valence electrons) break away, they become free electrons.
- Free electrons make current possible.
- Like charges repel each other, and opposite charges attract each other.
- Voltage must be applied to a circuit before there can be current.
- Fuel cells and batteries convert chemical energy to electrical energy using oxidation-reduction reaction.
- Resistance limits the current.
- Basically, an electric circuit consists of a source, a load, and a current path.
- An open circuit is one in which the current path is broken.
- A closed circuit is one which has a complete current path.
- An ammeter is connected in line (series) with the current path to measure current.
- A voltmeter is connected across (parallel) the current path to measure voltage.
- An ohmmeter is connected across a resistor to measure resistance. The resistor must be disconnected from the circuit.
- Figure 2–61 shows the electrical symbols introduced in this chapter.

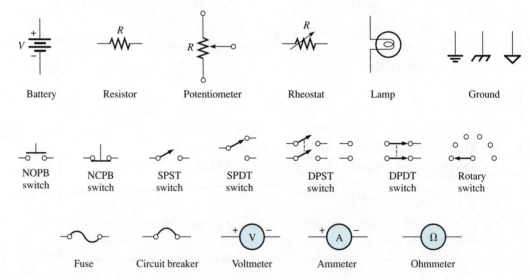

FIGURE 2–61

- One coulomb is the charge on 6.25×10^{18} electrons.
- One volt is the potential difference (voltage) between two points when one joule of energy is used to move one coulomb of charge from one point to the other.
- One ampere is the amount of current that exists when one coulomb of charge moves through a given cross-sectional area of a material in one second.
- One ohm is the resistance when there is one ampere of current in a material with one volt applied across the material.

KEY TERMS

Key terms and other bold terms in the chapter are defined in the end-of-book glossary.

Ammeter   An electrical instrument used to measure current.

Ampere (A)   The unit of electrical current.

Atom   The smallest particle of an element possessing the unique characteristics of that element.

AWG   American Wire Gauge; a standardization based on wire diameter.

Charge   An electrical property of matter that exists because of an excess or a deficiency of electrons. Charge can be either positive or negative.

Circuit breaker   A resettable protective device used for interrupting excessive current in an electric circuit.

Closed circuit   A circuit with a complete current path.

Conductance   The ability of a circuit to allow current. The unit is the siemens (S).

Conductor   A material in which electric current is easily established. An example is copper.

Coulomb (C)   The unit of electrical charge; the total charge possessed by 6.25×10^{18} electrons.

Coulomb's law   A law that states a force exists between two charged bodies that is directly proportional to the product of the two charges and inversely proportional to the square of the distance between them.

Current   The rate of flow of charge (free electrons).

Current source   A device that produces a constant current for a varying load.

DMM   Digital multimeter; an electronic instrument that combines meters for measurement of voltage, current, and resistance.

Electron   A basic particle of electrical charge in matter. The electron possesses negative charge.

Free electron   A valence electron that has broken away from its parent atom and is free to move from atom to atom within the atomic structure of a material.

Fuel cell   A device that converts electrochemical energy from an external source into dc voltage. The hydrogen fuel cell is the most common type.

Fuse   A protective device that burns open when there is excessive current in a circuit.

Ground   The common or reference point in a circuit.

Insulator   A material that does not allow current under normal conditions.

Load   An element (resistor or other component) connected across the output terminals of a circuit that draws current from the source and upon which work is done.

Ohm (Ω)   The unit of resistance.

Ohmmeter   An instrument for measuring resistance.

Open circuit   A circuit in which there is not a complete current path.

Potentiometer   A three-terminal variable resistor.

Resistance   Opposition to current. The unit is the ohm (Ω).

Resistor   An electrical component designed specifically to have a certain amount of resistance.

Rheostat   A two-terminal variable resistor.

Schematic   A symbolized diagram of an electrical or electronic circuit.

Semiconductor   A material that has a conductance value between that of a conductor and an insulator. Silicon and germanium are examples.

Siemens (S)   The unit of conductance.

Switch   An electrical or electronic device for opening and closing a current path.

Volt (V)   The unit of voltage or electromotive force.

Voltage   The amount of energy per charge available to move electrons from one point to another in an electric circuit.

Voltage source   A device that produces a constant voltage for a varying load.

Voltmeter   An instrument used to measure voltage.

KEY FORMULAS

(2–1) $Q = \dfrac{\text{number of electrons}}{6.25 \times 10^{18} \text{ electrons/C}}$ Charge

(2–2) $V = \dfrac{W}{Q}$ Voltage in volts equals energy in joules divided by charge in coulombs.

(2–3) $I = \dfrac{Q}{t}$ Current in amperes equals charge in coulombs divided by time in seconds.

(2–4) $G = \dfrac{1}{R}$ Conductance in siemens is the reciprocal of resistance in ohms.

(2–5) $A = d^2$ Cross-sectional area in circular mils equals the diameter in mils squared.

(2–6) $R = \dfrac{\rho l}{A}$ Resistance is resistivity in CM Ω/ft times length in feet divided by cross-sectional area in circular mils.

TRUE/FALSE QUIZ

Answers are at the end of the chapter.

1. The number of neutrons in the nucleus is the atomic number of that element.

2. The unit of charge is the ampere.

3. Energy in a battery is stored in the form of chemical energy.

4. A volt can be defined in terms of energy per charge.

5. In a five-band precision resistor, the fourth band is the tolerance band.

6. A rheostat performs the same function as a potentiometer.

7. A strain gage changes resistance in response to an applied force.

8. Soldering is never allowed on a splice.

9. All circuits must have a complete path for current.

10. A circular mil is a unit of area.

11. The three basic measurements that can be done by a DMM are voltage, current, and power.

12. To measure current with a meter, the meter should be placed in series.

SELF-TEST

Answers are at the end of the chapter.

1. A neutral atom with an atomic number of three has how many electrons?
   (a) 1   (b) 3   (c) none   (d) depends on the type of atom

2. Electron orbits are called
   (a) shells   (b) nuclei   (c) waves   (d) valences

3. Materials in which current cannot be established are called
   (a) filters   (b) conductors   (c) insulators   (d) semiconductors

4. When placed close together, a positively charged material and a negatively charged material will
   (a) repel   (b) become neutral   (c) attract   (d) exchange charges

5. The charge on a single electron is
   (a) 6.25×10^{-18} C   (b) 1.6×10^{-19} C   (c) 1.6×10^{-19} J   (d) 3.14×10^{-6} C

6. *Potential difference* is another term for
   (a) energy   (b) voltage   (c) distance of an electron from the nucleus   (d) charge

7. The unit of energy is the
   (a) watt   (b) coulomb   (c) joule   (d) volt

8. Which one of the following is not a type of energy source?
   (a) battery   (b) solar cell   (c) generator   (d) potentiometer

9. Which one of the following is the byproduct of a hydrogen fuel cell?
   (a) oxygen   (b) carbon dioxide   (c) hydrochloric acid   (d) water

10. Which one of the following is generally not a possible condition in an electric circuit?

    (a) voltage and no current   (b) current and no voltage

    (c) voltage and current   (d) no voltage and no current

11. Electrical current is defined as

    (a) free electrons   (b) the rate of flow of free electrons

    (c) the energy required to move electrons   (d) the charge on free electrons

12. There is no current in a circuit when

    (a) a series switch is closed   (b) a series switch is open   (c) there is no source voltage

    (d) both (a) and (c)   (e) both (b) and (c)

13. The primary purpose of a resistor is to

    (a) increase current   (b) limit current   (c) produce heat   (d) resist current change

14. Potentiometers and rheostats are types of

    (a) voltage sources   (b) variable resistors   (c) fixed resistors   (d) circuit breakers

15. The current in a given circuit is not to exceed 22 A. Which value of fuse is best?

    (a) 10 A   (b) 25 A   (c) 20 A   (d) a fuse is not necessary

PROBLEMS

Answers to odd-numbered problems are at the end of the book.

BASIC PROBLEMS

SECTION 2–2   Electrical Charge

1. How many coulombs of charge do 50×10^{31} electrons possess?

2. How many electrons does it take to make 80 μC of charge?

3. What is the charge in coulombs of the nucleus of a copper atom?

4. What is the charge in coulombs of the nucleus of a chlorine atom?

SECTION 2–3   Voltage

5. Determine the voltage in each of the following cases:
       (a) 10 J/C   (b) 5 J/2 C   (c) 100 J/25 C

6. Five hundred joules of energy are used to move 100 C of charge through a resistor. What is the voltage across the resistor?

7. What is the voltage of a battery that uses 800 J of energy to move 40 C of charge through a resistor?

8. How much energy does a 12 V battery in your car use to move 2.5 C through the electrical circuit?

9. Assume that a solar battery charger delivers 2.5 J of energy when 0.2 C of charge has been moved. What is the voltage?

SECTION 2–4   Current

10. If the solar cell in Problem 9 has moved the charge in 10 s, what is the current?

11. Determine the current in each of the following cases:
        (a) 75 C in 1 s   (b) 10 C in 0.5 s   (c) 5 C in 2 s

12. Six-tenths coulomb passes a point in 3 s. What is the current in amperes?

13. How long does it take 10 C to flow past a point if the current is 5 A?

14. How many coulombs pass a point in 0.1 s when the current is 1.5 A?

SECTION 2–5   Resistance

15. Figure 2–62(a) shows color-coded resistors. Determine the resistance value and the tolerance of each.

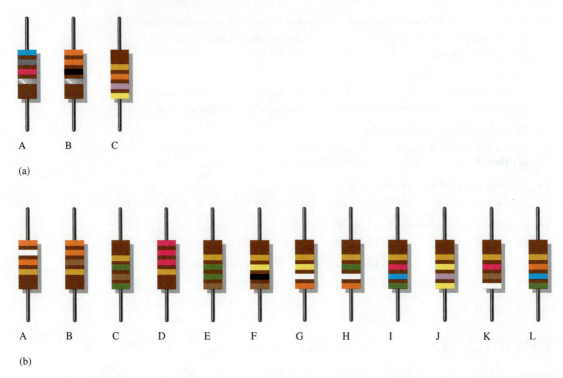

FIGURE 2–62

16. Find the minimum and the maximum resistance within the tolerance limits for each resistor in Figure 2–62(a).

17. (a) If you need a 270 Ω resistor with 5% tolerance, what color bands should you look for?
    (b) From the selection of resistors in Figure 2–62(b), choose the following values: 330 Ω, 2.2 kΩ, 39 kΩ, 56 kΩ, and 100 kΩ.

18. Determine the resistance value and tolerance for each resistor in Figure 2–63.

(a)   (b)   (c)

FIGURE 2–63

19. Determine the resistance and tolerance of each of the following 4-band resistors:
    (a) brown, black, black, gold   (b) green, brown, green, silver   (c) blue, gray, black, gold

20. Determine the color bands for each of the following 4-band resistors. Assume each has a 5% tolerance.
    (a) 0.47 Ω   (b) 270 kΩ   (c) 5.1 MΩ

21. Determine the resistance and tolerance of each of the following 5-band resistors:
    (a) red, gray, violet, red, brown   (b) blue, black, yellow, gold, brown
    (c) white, orange, brown, brown, brown

22. Determine the color bands for each of the following 5-band resistors. Assume each has a 1% tolerance.
    (a) 14.7 kΩ   (b) 39.2 Ω   (c) 9.76 kΩ

23. Determine the resistance values represented by the following labels:
    (a) 220   (b) 472   (c) 823   (d) 3K3   (e) 560   (f) 10M

24. The adjustable contact of a linear potentiometer is set at the mechanical center of its adjustment. If the total resistance is 1000 Ω, what is the resistance between each end terminal and the adjustable contact?

SECTION 2–6 The Electric Circuit

25. Trace the current path in the lamp circuit of Figure 2–41(a) with the switch making contact between the middle and lower pins.

26. With the switch in either position, redraw the lamp circuit in Figure 2–41(b) with a fuse connected to protect the circuit against excessive current.

SECTION 2–7   Basic Circuit Measurements

27. Show the placement of an ammeter and a voltmeter to measure the current and the source voltage in Figure 2–64.

28. Show how you would measure the resistance of R_2 in Figure 2–64.

29. In Figure 2–65 what does each voltmeter indicate when the switch (SW) is in position 1? In position 2?

30. In Figure 2–65, show how to connect an ammeter to measure the current from the voltage source regardless of the switch (SW) position.

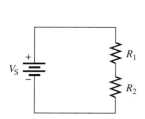

FIGURE 2–64

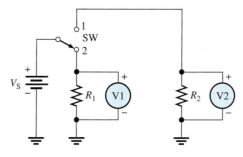

FIGURE 2–65

31. What is the voltage reading of the meter in Figure 2–66?

32. How much resistance is the meter in Figure 2–67 measuring?

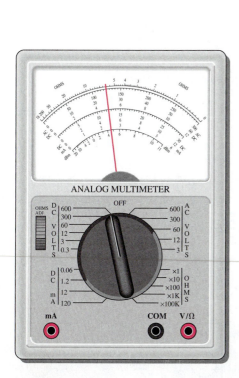

FIGURE 2–66

FIGURE 2–67

33. Determine the resistance indicated by each of the following ohmmeter readings and range settings:
   (a) pointer at 2, range setting at R × 100   **(b)** pointer at 15, range setting at R × 10M
   (c) pointer at 45, range setting at R × 100

34. A multimeter has the following ranges: 1 mA, 10 mA, 100 mA; 100 mV, 1 V, 10 V; R × 1, R × 10, R × 100. Indicate schematically how you would connect the multimeter in Figure 2–68 to measure the following quantities:

(a) I_{R1} (b) V_{R1} (c) R_1

In each case indicate the *function* to which you would set the meter and the *range* that you would use.

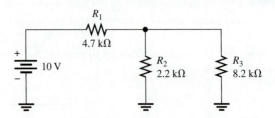

FIGURE 2–68

ADVANCED PROBLEMS

35. A resistor with a current of 2 A through it in an amplifier circuit converts 1000 J of electrical energy to heat energy in 15 s. What is the voltage across the resistor?

36. If 574×10^{15} electrons flow through a wire in 250 ms, what is the current in amperes?

37. A 120 V source is to be connected to a 1500 Ω resistive load by two lengths of wire as shown in Figure 2–69. The voltage source is to be located 50 ft from the load. Using Table 2–3, determine the gauge number of the *smallest* wire that can be used if the total resistance of the two lengths of wire is not to exceed 6 Ω.

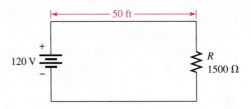

FIGURE 2–69

38. Determine the resistance and tolerance of each resistor labeled as follows:

(a) 4R7J   (b) 560KF   (c) 1M5G

39. There is only one circuit in Figure 2–70 in which it is possible to have all lamps on at the same time. Determine which circuit it is.

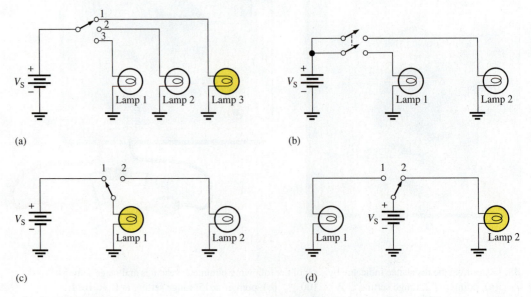

FIGURE 2–70

40. Through which resistor in Figure 2–71 is there always current, regardless of the position of the switches?

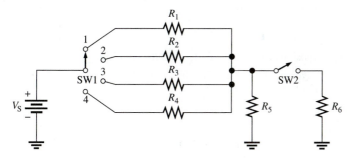

FIGURE 2–71

41. In Figure 2–71, show the proper placement of ammeters to measure the current through each resistor and the current out of the battery.

42. Show the proper placement of voltmeters to measure the voltage across each resistor in Figure 2–71.

43. Devise a switch arrangement whereby two voltage sources V_{S1} and V_{S2} can be connected simultaneously to either of two resistors (R_1 and R_2) as follows:

$$V_{S1} \text{ connected to } R_1 \text{ and } V_{S2} \text{ connected to } R_2$$
$$V_{S1} \text{ connected to } R_2 \text{ and } V_{S2} \text{ connected to } R_1$$

44. Show how to wire the quiz board in Figure 2–46 if the correct sequence of answers is D-B-C-A.

ANSWERS TO SECTION CHECKUPS

SECTION 2–1   Atoms

1. The electron is the basic particle of negative charge.

2. An atom is the smallest particle of an element that retains the unique characteristics of the element.

3. An atom is a positively charged nucleus surrounded by orbiting electrons.

4. Atomic number is the number of protons in a nucleus.

5. No, each element has a different type of atom.

6. A free electron is an outer-shell electron that has drifted away from the parent atom.

7. Shells are energy bands in which electrons orbit the nucleus of an atom.

8. Copper and silver

SECTION 2–2   Electrical Charge

1. Q is the symbol for charge.

2. Coulomb is the unit of charge, and C is the symbol for coulomb.

3. The two types of charge are positive and negative.

4. $Q = \dfrac{10 \times 10^{12} \text{ electrons}}{6.25 \times 10^{18} \text{ electrons/C}} = 1.6 \times 10^{-6}\,\text{C} = 1.6\,\mu\text{C}$

SECTION 2–3   Voltage

1. Voltage is energy per unit charge.

2. The unit of voltage is the volt.

3. $V = W/Q = 24\,\text{J}/10\,\text{C} = 2.4\,\text{V}$

4. Battery, fuel cell, power supply, solar cell, generator, thermocouple, and piezoelectric sensors are voltage sources.

5. Oxidation-reduction reactions

SECTION 2–4   Current

1. Current is the rate of flow of charge; **the** unit of current is the ampere (A).

2. There are 6.25×10^{18} electrons in one coulomb.

3. $I = Q/t = 20 \text{ C}/4 \text{ s} = 5 \text{ A}$

SECTION 2–5   Resistance

1. Resistance is opposition to current and its unit is the ohm (Ω).

2. Two resistor categories are fixed and variable. The value of a fixed resistor cannot be changed, but that of a variable resistor can.

3. *First band:* first digit of resistance value.

   Second band: second digit of resistance value.

   Third band: number of zeros following the 2nd digit.

   Fourth band: percent tolerance.

4. **(a)** Yellow, violet, red, gold = 4700 Ω $\pm$ 5%
   (b) Blue, red, orange, silver = 62,000 Ω $\pm$ 10%
   (c) Brown, gray, black, gold = 18 Ω $\pm$ 5%
   (d) Red, red, blue, red, green = 22.6 kΩ $\pm$ 0.5%

5. **(a)** 33R = 33 Ω **(b)** 5K6 = 5.6 kΩ **(c)** 900 = 900 Ω **(d)** 6M8 = 6.8 MΩ

6. A rheostat has two terminals; a potentiometer has three terminals.

7. Thermistor—temperature; photoconductor cell—light; strain gage—force.

SECTION 2–6   The Electric Circuit

1. A basic electric circuit consists of source, load, and current path between source and load.

2. An open circuit is one that has no path for current.

3. A closed circuit is one that has a complete path for current.

4. $R = \infty$ (infinite); $R = 0 \ \Omega$

5. A fuse protects a circuit against excessive current.

6. A fuse must be replaced once blown. A circuit breaker can be reset once tripped.

7. AWG #3 is larger than AWG #22.

8. Ground is the reference point with zero voltage with respect to other points.

SECTION 2–7   Basic Circuit Measurements

1. **(a)** Ammeter measures current.
   (b) Voltmeter measures voltage.
   (c) Ohmmeter measures resistance.

2. See Figure 2–72.

3. See Figure 2–73.

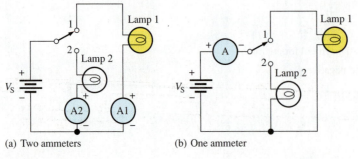

(a) Two ammeters   (b) One ammeter

FIGURE 2–72

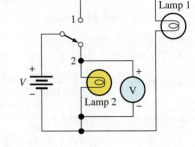

FIGURE 2–73

4. Two types of DMM displays are LED and LCD. The LCD requires little current, but it is difficult to see in low light and is slow to respond. The LED can be seen in the dark, and it responds quickly; however, it requires much more current than does the LCD.

5. Resolution is the smallest increment of a quantity that a meter can measure.

6. The voltage being measured is 1.5 V.

7. DC VOLTS, 600 setting; 275 V is read on the 60 scale near midpoint.

8. OHMS $\times$ 1000

ANSWERS TO RELATED PROBLEMS FOR EXAMPLES

2–1 1.88×10^{19} electrons

2–2 600 J

2–3 12 C

2–4 4700 Ω $\pm$ 5%

2–5 47.5 Ω $\pm$ 2%

2–6 1.2 kΩ

2–7 404.01 CM; #24

2–8 1.280 Ω; same as calculated result

2–9 The needle will indicate 100 on the top scale.

ANSWERS TO TRUE/FALSE QUIZ

1. F **2.** F **3.** T **4.** T **5.** F **6.** F **7.** T **8.** F **9.** T
10. T **11.** F **12.** T

ANSWERS TO SELF-TEST

1. (b) **2.** (a) **3.** (c) **4.** (c) **5.** (b) **6.** (b) **7.** (c) **8.** (d)
9. (d) **10.** (b) **11.** (b) **12.** (e) **13.** (b) **14.** (b) **15.** (c)

CHAPTER 3

OHM'S LAW, ENERGY, AND POWER

OBJECTIVES

- Explain Ohm's law
- Use Ohm's law to determine voltage, current, or resistance
- Define *energy* and *power*
- Calculate power in a circuit
- Properly select resistors based on power consideration
- Explain energy conversion and voltage drop
- Discuss characteristics of power supplies and batteries
- Describe a basic approach to troubleshooting

KEY TERMS

Ohm's law	Power
Linear	Joule (J)
Energy	Watt (W)
Kilowatt-hour (kWh)	Efficiency
Watt's law	Ah rating
Power rating	Troubleshooting
Voltage drop	Half-splitting
Power supply	

INTRODUCTION

Georg Simon Ohm (1787–1854) experimentally found that voltage, current, and resistance are all related in a specific way. This basic relationship, known as *Ohm's law*, is one of the most fundamental and important laws in the fields of electricity and electronics. In this chapter, Ohm's law is examined, and its use in practical circuit applications is discussed and demonstrated by numerous examples.

In addition to Ohm's law, the concepts and definitions of energy and power in electric circuits and systems are introduced, and the Watt's law power formulas are given. A general approach to troubleshooting using the analysis, planning, and measurement (APM) method is also introduced.

VISIT THE WEBSITE
Study aids for this chapter are available at
http://pearsonhighered.com/floyd

3–1 OHM'S LAW

Ohm's law describes mathematically how voltage, current, and resistance in a circuit are related. Ohm's law can be written in three equivalent forms; the formula you use depends on the quantity you need to determine.

After completing this section, you should be able to

- **Explain Ohm's law**
  - **Describe how voltage (*V*), current (*I*), and resistance (*R*) are related**
  - **Express *I* as a function of *V* and *R***
  - **Express *V* as a function of *I* and *R***
  - **Express *R* as a function of *V* and *I***

Ohm determined experimentally that if the voltage across a resistor is increased, the current through the resistor will increase; and, likewise, if the voltage is decreased, the current will decrease. For example, if the voltage is doubled, the current will double. If the voltage is halved, the current will also be halved. This relationship is illustrated in Figure 3–1, with relative meter indications of voltage and current.

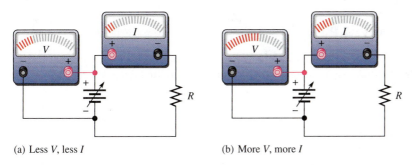

(a) Less *V*, less *I*   (b) More *V*, more *I*

FIGURE 3–1   Effect on the current of changing the voltage with the resistance at a constant value.

Ohm also determined that if the voltage is held constant, less resistance results in more current, and more resistance results in less current. For example, if the resistance is halved, the current doubles. If the resistance is doubled, the current is halved. This concept is illustrated by the meter indications in Figure 3–2, where the resistance is increased and the voltage is held constant.

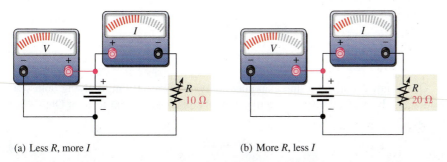

(a) Less *R*, more *I*   (b) More *R*, less *I*

FIGURE 3–2   Effect on the current of changing the resistance with the voltage at a constant value.

Ohm's law states that current is directly proportional to voltage and inversely proportional to resistance.

$$I = \frac{V}{R}$$

(3–1)

where I is current in amperes (A), V is voltage in volts (V), and R is resistance in ohms (Ω). This formula describes the relationship illustrated by the action in the circuits of Figures 3–1 and 3–2.

For a constant resistance, if the voltage applied to a circuit is increased, the current will increase; and if the voltage is decreased, the current will decrease.

$$I = \frac{V}{R} \qquad\qquad I = \frac{V}{R} \qquad\qquad R \text{ constant}$$

Increase V, I increases   Decrease V, I decreases

For a constant voltage, if the resistance in a circuit is increased, the current will decrease; and if the resistance is decreased, the current will increase.

$$I = \frac{V}{R} \qquad\qquad I = \frac{V}{R} \qquad\qquad V \text{ constant}$$

Increase R, I decreases   Decrease R, I increases

EXAMPLE 3–1

Using the Ohm's law formula in Equation 3–1, verify that the current through a 10 Ω resistor increases when the voltage is increased from 5 V to 20 V.

SOLUTION

For $V = 5$ V,

$$I = \frac{V}{R} = \frac{5 \text{ V}}{10 \text{ }\Omega} = \textbf{0.5 A}$$

For $V = 20$ V,

$$I = \frac{V}{R} = \frac{20 \text{ V}}{10 \text{ }\Omega} = \textbf{2 A}$$

RELATED PROBLEM*

Show that the current decreases when the resistance is increased from 5 Ω to 20 Ω and the voltage is a constant 10 V.

*Answers are at the end of the chapter.

Ohm's law can also be stated another equivalent way. By multiplying both sides of Equation 3–1 by R and transposing terms, you obtain an equivalent form of Ohm's law, as follows:

$$V = IR \qquad\qquad\qquad (3–2)$$

With this formula, you can calculate voltage in volts if you know the current in amperes and resistance in ohms.

EXAMPLE 3–2

Use the Ohm's law formula in Equation 3–2 to calculate the voltage across a 1.0 kΩ resistor when the current is 5.0 mA.

SOLUTION

$$V = IR = (5.0\,\text{mA})(1.0\,\text{k}\Omega) = \mathbf{5.0\,V}$$

RELATED PROBLEM

Find the voltage across a 1.0 kΩ resistor when the current is 1 mA.

There is a third equivalent way to state Ohm's law. By dividing both sides of Equation 3–2 by *I* and transposing terms, you obtain the following formula:

$$R = \frac{V}{I} \qquad\qquad (3\text{–}3)$$

This expression of the Ohm's law formula is used to determine resistance in ohms if you know the values of voltage in volts and current in amperes.

Remember, the three expressions—Equations 3–1, 3–2, and 3–3—are all equivalent. They are simply three ways of stating Ohm's law.

EXAMPLE 3–3

Use the Ohm's law formula in Equation 3–3 to calculate the resistance of a rear window defroster grid in a certain vehicle. When it is connected to 12.6 V, it draws 15.0 A from the battery. What is the resistance of the defroster grid?

SOLUTION

$$R = \frac{V}{I} = \frac{12.6\,\text{V}}{15.0\,\text{A}} = \mathbf{840\,m\Omega}$$

RELATED PROBLEM

If one of the grid wires opens, the current drops to 13.0 A. What is the new resistance?

The Linear Relationship of Current and Voltage

In resistive circuits, current and voltage are linearly proportional. **Linear** means that if one is increased or decreased by a certain percentage, the other will increase or decrease by the same percentage, assuming that the resistance is constant in value. For example, if the voltage across a resistor is tripled, the current will triple. If the voltage is reduced by half, the current will decrease by half.

EXAMPLE 3–4

Show that if the voltage in the circuit of Figure 3–3 is increased to three times its present value, the current will triple in value.

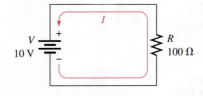

FIGURE 3–3

SOLUTION

With 10 V, the current is

$$I = \frac{V}{R} = \frac{10 \text{ V}}{100 \ \Omega} = 0.1 \text{ A}$$

If the voltage is increased to 30 V, the current will be

$$I = \frac{V}{R} = \frac{30 \text{ V}}{100 \ \Omega} = 0.3 \text{ A}$$

The current went from 0.1 A to 0.3 A when the voltage was tripled to 30 V.

RELATED PROBLEM

If the voltage in Figure 3–3 is quadrupled, will the current also quadruple?

Let's take a constant value of resistance, for example, 10 Ω, and calculate the current for several values of voltage ranging from 10 V to 100 V in the circuit in Figure 3–4(a). The current values obtained are shown in Figure 3–4(b). The graph of the I values versus the V values is shown in Figure 3–4(c). Note that it is a straight line graph. This graph shows that a change in voltage results in a linearly proportional change in current. No matter what value R is, assuming that R is constant, the graph of I versus V will always be a straight line.

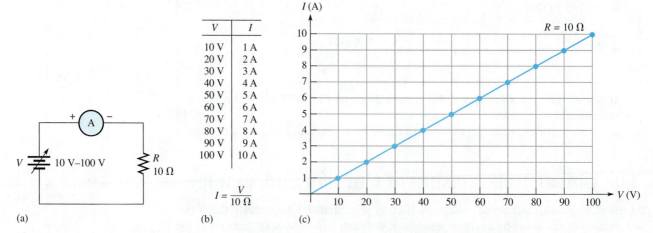

(a)   (b)   (c)

FIGURE 3–4   **Graph of current versus voltage for the circuit in part (a).**

A Graphic Aid for Ohm's Law

You may find the graphic aid in Figure 3–5 helpful for applying Ohm's law. It is a way to remember the formulas.

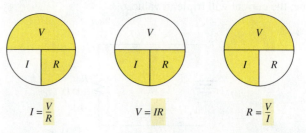

FIGURE 3–5   A graphic aid for the Ohm's law formulas.

SECTION 3–1 CHECKUP*

1. Briefly state Ohm's law in words.

2. Write the Ohm's law formula for calculating current.

3. Write the Ohm's law formula for calculating voltage.

4. Write the Ohm's law formula for calculating resistance.

5. If the voltage across a resistor is tripled, does the current increase or decrease? By how much?

6. There is a fixed voltage across a variable resistor, and you measure a current of 10 mA. If you double the resistance, how much current will you measure?

7. What happens to the current in a linear circuit where both the voltage and the resistance are doubled?

*Answers are at the end of the chapter.

3–2 APPLICATION OF OHM'S LAW

This section provides examples of the application of Ohm's law for calculating voltage, current, and resistance in electric circuits. You will also see how to use quantities expressed with metric prefixes in circuit calculations.

After completing this section, you should be able to

- **Use Ohm's law to determine voltage, current, or resistance**
  - **Use Ohm's law to find current when you know voltage and resistance**
  - **Use Ohm's law to find voltage when you know current and resistance**
  - **Use Ohm's law to find resistance when you know voltage and current**
  - **Use quantities with metric prefixes**

Current Calculations

In these examples you will learn to determine current values when you know the values of voltage and resistance. In these problems, the formula $I = V/R$ is used. In order to get current in amperes, you must express the value of V in volts and the value of R in ohms.

EXAMPLE 3–5

An indicator light requires a 330 Ω resistor to limit current. The voltage across the current limiting resistor is 3 V. What is the current in the resistor?

SOLUTION

$$I = \frac{V}{R} = \frac{3.0 \text{ V}}{330 \text{ Ω}} = \textbf{9.09 mA}$$

RELATED PROBLEM

How will the current change if a 270 Ω resistor is used instead and 3.0 V still is across the resistor?

In electronics, resistance values of thousands or millions of ohms are common. The metric prefixes *kilo* (k) and *mega* (M) are used to indicate large values. Thus, thousands of ohms are expressed in kilohms (kΩ), and millions of ohms are expressed in megohms (MΩ). The following examples illustrate how to use kilohms and megohms when you use Ohm's law to calculate current.

EXAMPLE 3–6

Calculate the current in milliamperes for the circuit of Figure 3–6.

FIGURE 3–6

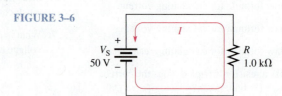

SOLUTION

Remember that $1.0 \, \text{k}\Omega$ is the same as $1.0 \times 10^3 \, \Omega$. Use the formula $I = V/R$ and substitute 50 V for V and $1.0 \times 10^3 \, \Omega$ for R.

$$I = \frac{V_S}{R} = \frac{50 \, \text{V}}{1.0 \, \text{k}\Omega} = \frac{50 \, \text{V}}{1.0 \times 10^3 \, \Omega} = 50 \times 10^{-3} \, \text{A} = \mathbf{50 \, mA}$$

RELATED PROBLEM

If the resistance in Figure 3–6 is increased to $10 \, \text{k}\Omega$, what is the current?

In Example 3–6, the current is expressed as 50 mA. Thus, *when volts (V) are divided by kilohms (kΩ), the current is in milliamperes (mA).*

When volts (V) are divided by megohms (MΩ), the current is in microamperes (μA), as Example 3–7 illustrates.

EXAMPLE 3–7

Determine the amount of current in microamperes for the circuit of Figure 3–7.

FIGURE 3–7

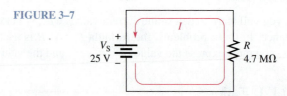

SOLUTION

Recall that $4.7 \, \text{M}\Omega$ equals $4.7 \times 10^6 \, \Omega$. Use the formula $I = V/R$ and substitute 25 V for V and $4.7 \times 10^6 \, \Omega$ for R.

$$I = \frac{V_S}{R} = \frac{25 \, \text{V}}{4.7 \, \text{M}\Omega} = \frac{25 \, \text{V}}{4.7 \times 10^6 \, \Omega} = 5.32 \times 10^{-6} \, \text{A} = \mathbf{5.32 \, \mu A}$$

RELATED PROBLEM

If the resistance in Figure 3–7 is decreased to $1.0 \, \text{M}\Omega$, what is the current?

Small voltages, usually less than 50 V, are common in electronic circuits. Occasionally, however, large voltages are encountered. For example, the high-voltage supply in radio transmitters, plasma guns, ion motors, and x-ray machines are commonly well above 1000 V. Transmission voltages generated by the power companies may be as high as 345,000 V (345 kV).

EXAMPLE 3–8

How much current in microamperes is there through a 100 MΩ resistor when 50 kV are applied across it?

SOLUTION

Divide 50 kV by 100 MΩ to get the current. Substitute 50×10^3 V for 50 kV and 100×10^6 Ω for 100 MΩ in the formula for current. V_R is the voltage across the resistor.

$$I = \frac{V_R}{R} = \frac{50 \text{ kV}}{100 \text{ MΩ}} = \frac{50 \times 10^3 \text{ V}}{100 \times 10^6 \text{ Ω}} = 0.5 \times 10^{-3} \text{ A}$$

$$= 500 \times 10^{-6} = \mathbf{500 \; \mu A}$$

RELATED PROBLEM

How much current is there through 10 MΩ when 2 kV are applied?

Voltage Calculations

In these examples you will learn how to determine voltage values when you know the current and resistance using the formula $V = IR$. To obtain voltage in volts, you must express the value of I in amperes and the value of R in ohms.

EXAMPLE 3–9

In the circuit of Figure 3–8, how much voltage is needed to produce 5 A of current?

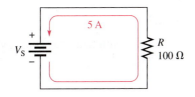

FIGURE 3–8

SOLUTION

Substitute 5 A for I and 100 Ω for R into the formula $V = IR$.

$$V_S = IR = (5 \text{ A})(100 \text{ Ω}) = \mathbf{500 \; V}$$

Thus, 500 V are required to produce 5 A of current through a 100 Ω resistor.

RELATED PROBLEM

How much voltage is required to produce 8 A in the circuit of Figure 3–8?

EXAMPLE 3–10

How much voltage will be measured across the resistor in Figure 3–9?

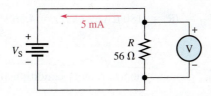

FIGURE 3–9

SOLUTION

Note that 5 mA equals 5×10^{-3} A. Substitute the values for I and R into the formula $V = IR$.

$$V_R = IR = (5 \text{ mA})(56 \ \Omega) = (5 \times 10^{-3} \text{ A})(56 \ \Omega) = \textbf{280 mV}$$

When milliamperes are multiplied by ohms, the result is millivolts.

RELATED PROBLEM

Change the resistor in Figure 3–9 to 22 Ω and determine the voltage required to produce 10 mA.

EXAMPLE 3–11

The circuit in Figure 3–10 has a current of 10 mA. What is the source voltage?

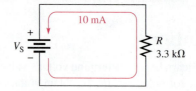

FIGURE 3–10

SOLUTION

Note that 10 mA equals 10×10^{-3} A and that 3.3 kΩ equals 3.3×10^3 Ω. Substitute these values into the formula $V = IR$.

$$V_S = IR = (10 \text{ mA})(3.3 \text{ k}\Omega) = (10 \times 10^{-3} \text{ A})(3.3 \times 10^3 \ \Omega) = \textbf{33 V}$$

When milliamperes and kilohms are multiplied, the result is volts.

RELATED PROBLEM

What is the voltage in Figure 3–10 if the current is 5 mA?

EXAMPLE 3–12

A small solar cell is connected to a 27 kΩ resistor. In bright sunlight, the solar cell looks like a current source that can supply 180 μA to the resistor, as shown in Figure 3–11. What is the voltage across the resistor?

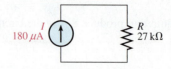

FIGURE 3–11

SOLUTION

$$V_R = IR = (180 \ \mu\text{A})(27 \text{ k}\Omega) = \textbf{4.86 V}$$

RELATED PROBLEM

How does the voltage change when in cloudy conditions the current drops to 40 μA?

MULTISIM

Open Multisim file E03-12. Connect a voltmeter across the resistor and confirm the voltage calculated in this example.

Resistance Calculations

In these examples you will learn how to determine resistance values when you know the voltage and current using the formula $R = V/I$. To find resistance in ohms, you must express the value of V in volts and the value of I in amperes.

EXAMPLE 3–13

An automotive lamp draws 2 A from the 13.2 V battery. What is the resistance of the bulb?

SOLUTION

$$R = \frac{V}{I} = \frac{13.2 \text{ V}}{2.0 \text{ A}} = \mathbf{6.6 \ \Omega}$$

RELATED PROBLEM

When operated at 6.6 V, the same bulb has a current of 1.1 A. What is the resistance of the bulb in this case?

EXAMPLE 3–14

A cadmium sulfide cell (CdS cell) is a photo-sensitive resistor that changes resistance when light strikes it. It is used in applications such as turning on lights at dusk. You could monitor the resistance indirectly from an ammeter when the cell is in an active circuit with the setup shown in Figure 3–12. What resistance is implied by the readings shown?

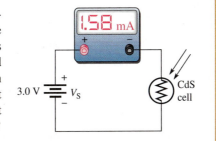

FIGURE 3–12

SOLUTION

$$R = \frac{V}{I} = \frac{3.0 \text{ V}}{1.58 \text{ mA}} = \mathbf{1.90 \ k\Omega}$$

RELATED PROBLEM

In the dark, the current drops to 76 μA. What resistance does this imply?

SYSTEM EXAMPLE 3–1

CURRENT-SENSING RESISTORS

Current-sensing resistors are precision low-value resistors that are widely used in various systems, including power supplies, battery-charging circuits, motor controllers, automotive systems, telecommunications, and computers. As stated by Ohm's law, there is a voltage drop across the resistor when current is in it. The current-sensing resistor converts this current into a small voltage that can be easily measured and monitored. The majority of applications use precision surface-mount resistors such as the one shown in Figure 3–13.

Current-sensing resistors have low-resistance values in order to minimize power loss and self-heating effects. Low values also ensure there is a minimum effect on the circuit due to insertion loss. Resistance values of 20–25 mΩ are common.

One application for current-sensing resistors is for a battery charger, which is a type of power supply. Figure 3–14 shows a simplified drawing of a type of battery charger that

is designed to regulate current. The current-sensing resistor is shown in red and is the focus of this example. The current from the supply is converted to a small voltage that is monitored by the controller, which is a small integrated circuit. The controller can adjust the amount of current based on an optimum charging profile. The controller enables the charger to be highly efficient and reliable.

FIGURE 3–13 Surface mount current-sensing resistor.

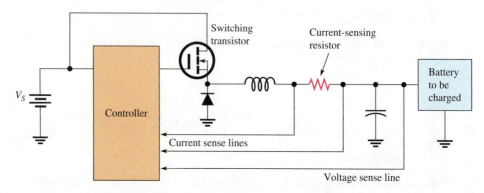

FIGURE 3–14 A power supply system for charging batteries that uses a current-sensing resistor to monitor the charging current.

SECTION 3–2 CHECKUP

1. $V = 10$ V and $R = 4.7$ Ω. Find I.

2. If a 4.7 MΩ resistor has 20 kV across it, how much current is there?

3. How much current will 10 kV across a 2 kΩ resistance produce?

4. $I = 1$ A and $R = 10$ Ω. Find V.

5. What voltage do you need to produce 3 mA of current in a 3 kΩ resistance?

6. A battery produces 2 A of current through a 6 Ω resistive load. What is the battery voltage?

7. $V = 10$ V and $I = 2$ A. Find R.

8. In a stereo amplifier circuit there is a resistor across which you measure 25 V, and your ammeter indicates 50 mA of current through the resistor. What is the resistor's value in kilohms? In ohms?

3–3 ENERGY AND POWER

When there is current through a resistance, electrical energy is converted to heat or other form of energy, such as light. A common example of this is an incandescent light bulb that becomes too hot to touch. The current through the filament that produces light also produces unwanted heat because the filament has resistance. Electrical components must be able to dissipate a certain amount of energy in a given period of time.

After completing this section, you should be able to

- **Define** *energy* **and** *power*
  - **Express power in terms of energy and time**
  - **State the unit of power**
  - **State the common units of energy**
  - **Perform energy and power calculations**

Energy is the ability to do work, and power is the rate at which energy is used.

In other words, power, P, is a certain amount of energy, W, used in a certain length of time (t), expressed as follows:

$$P = \frac{W}{t} \qquad (3\text{–}4)$$

where P is power in watts (W), W is energy in joules (J), and t is time in seconds (s). Note that an italic W is used to represent energy in the form of work and a nonitalic W is used for watts, the unit of power. The **joule** is the SI unit for energy.

Energy in joules divided by time in seconds gives power in watts. For example, if 50 J of energy are used in 2 s, the power is 50 J/2 s = 25 W. By definition,

One watt is the amount of power when one joule of energy is used in one second.

Thus, the number of joules used in 1 s is always equal to the number of watts. For example, if 75 J are used in 1 s, the power is

$$P = \frac{W}{t} = \frac{75 \text{ J}}{1 \text{ s}} = 75 \text{ W}$$

Amounts of power much less than one watt are common in certain areas of electronics. As with small current and voltage values, metric prefixes are used to designate small amounts of power. Thus, milliwatts (mW) and microwatts (μW) are commonly found in some applications.

In the electrical utilities field, kilowatts (kW) and megawatts (MW) are common units. Radio and television stations also use large amounts of power to transmit signals. Electric motors are commonly rated in horsepower (hp) where 1 hp = 746 W.

Since power is the rate at which energy is used, power utilized over a period of time represents energy consumption. If you multiply power in watts and time in seconds, you have energy in joules, symbolized by W.

$$W = Pt$$

EXAMPLE 3–15

An amount of energy equal to 100 J is used in 5 s. What is the power in watts?

SOLUTION

$$P = \frac{\text{energy}}{\text{time}} = \frac{W}{t} = \frac{100 \text{ J}}{5 \text{ s}} = \textbf{20 W}$$

RELATED PROBLEM

If 100 W of power occurs for 30 s, how much energy in joules is used?

The Kilowatt-hour (kWh) Unit of Energy

The joule has been defined as the unit of energy. However, there is another way to express energy. Since power is expressed in watts and time can be expressed in hours, a unit of energy called the kilowatt-hour (kWh) can be used.

When you pay your electric bill, you are charged on the basis of the amount of energy you use. Because power companies deal in huge amounts of energy, the most practical unit is the kilowatt-hour. You use a **kilowatt-hour** of energy when you use the equivalent of 1000 W of power for 1 h. For example, a 100 W for television that is on 10 h uses 1 kWh of energy.

$$W = Pt = (100 \text{ W})(10 \text{ h}) = 1000 \text{ Wh} = 1 \text{ kWh}$$

EXAMPLE 3-16

Determine the number of kilowatt-hours (kWh) for each of the following energy consumptions:

(a) 1400 W for 1 hr   **(b)** 2500 W for 2 h   **(c)** 100,000 W for 5 h

SOLUTION

(a) 1400 W = 1.4 kW
$$W = Pt = (1.4\,\text{kW})(1\,\text{h}) = \textbf{1.4 kWh}$$

(b) 2500 W = 2.5 kW
$$\text{Energy} = (2.5\,\text{kW})(2\,\text{h}) = \textbf{5 kWh}$$

(c) 100,000 W = 100 kW
$$\text{Energy} = (100\,\text{kW})(5\,\text{h}) = \textbf{500 kWh}$$

RELATED PROBLEM

How many kilowatt-hours of energy are used by a 250 W light bulb burning for 8 h?

Table 3–1 lists the typical power rating in watts for several household appliances. You can determine the maximum kWh for various appliances by using the power rating in Table 3–1 converted to kilowatts times the number of hours it is used.

TABLE 3–1

APPLIANCE	POWER RATING (WATTS)
Air conditioner	860
Blow dryer	1000
Clock	2
Clothes dryer	4000
Dishwasher	1200
Heater	1322
Microwave oven	800
Range	12,200
Refrigerator	500
Television	250
Washing machine	400
Water heater	2500

EXAMPLE 3-17

During a typical 24-hour period, you use the following appliances for the specified lengths of time:

air conditioner: 15 hours

blow dryer: 10 minutes

clock: 24 hours

clothes dryer: 1 hour

dishwasher: 45 minutes

microwave oven: 15 minutes

refrigerator: 12 hours

television: 2 hours

water heater: 8 hours

Determine the total kilowatt-hours and the electric bill for the time period. The rate is 11 cents per kilowatt-hour.

SOLUTION

Determine the kWh for each appliance used by converting the watts in Table 3–1 to kilowatts and multiplying by the time in hours:

air conditioner: $0.860 \, kW \times 15 \, h = 12.9 \, kWh$

blow dryer: $1.0 \, kW \times 0.167 \, h = 0.167 \, kWh$

clock: $0.002 \, kW \times 24 \, h = 0.048 \, kWh$

clothes dryer: $4.0 \, kW \times 1 \, h = 4.0 \, kWh$

dishwasher: $1.2 \, kW \times 0.75 \, h = 0.9 \, kWh$

microwave: $0.8 \, kW \times 0.25 \, h = 0.2 \, kWh$

refrigerator: $0.5 \, kW \times 12 \, h = 6 \, kWh$

television: $0.25 \, kW \times 2 \, h = 0.5 \, kWh$

water heater: $2.5 \, kW \times 8 \, h = 20 \, kWh$

Now, add up all the kilowatt-hours to get the total energy for the 24-hour period.

Total energy = $(12.9 + 0.167 + 0.048 + 4.0 + 0.9 + 0.2 + 6.0 + 0.5 + 20) \, kWh =$ **44.7 kWh**

At 11 cents/kilowatt-hour, the cost of energy to run the appliances for the 24-hour period is

Energy cost = $44.7 \, kWh \times 0.11 \, \$/kWh =$ **\$4.92**

RELATED PROBLEM

In addition to the appliances, suppose you used one 200 W humidifier for 2 hours and one 75 W heating pad for 3 hours. Calculate your cost for the 24-hour period for everything.

SECTION 3–3 CHECKUP

1. Define *power*.

2. Write the formula for power in terms of energy and time.

3. Define *watt*.

4. Express each of the following values of power in the most appropriate units:

   (a) 68,000 W   (b) 0.005 W   (c) 0.000025 W

5. If you use 100 W of power for 10 h, how much energy (in kilowatt-hours) have you used?

6. Convert 2000 W to kilowatts.

7. How much does it cost to run a heater (1322 W) for 24 hours if energy cost is 11¢ per kilowatt-hour?

3–4 POWER IN AN ELECTRIC CIRCUIT

The generation of heat, which occurs when electrical energy is converted to heat energy, in an electric circuit is often an unwanted by-product of current through the resistance in the circuit. In some cases, however, the generation of heat is the primary purpose of a circuit as, for example, in an electric resistive heater. In any case, you must frequently deal with power in electrical and electronic circuits.

After completing this section, you should be able to

- **Calculate power in a circuit**
  - **Determine power when you know** I **and** R
  - **Determine power when you know** V **and** I
  - **Determine power when you know** V **and** R

FIGURE 3–15 **Power dissipation in an electric circuit is seen as heat given off by the resistance. The power dissipation is equal to the power supplied by the voltage source.**

Heat produced by current through resistance is a result of energy conversion.

When there is current through a resistance, the collisions of the electrons as they move through the resistance give off heat, resulting in a conversion of electrical energy to thermal energy as indicated in Figure 3–15. The amount of power dissipated in an electrical circuit is dependent on the amount of resistance and on the amount of current, expressed as follows:

$$P = I^2R \tag{3–5}$$

where P is power in watts (W), V is voltage in volts (V), and I is current in amperes (A). You can get an equivalent expression for power in terms of voltage and current by substituting $I \times I$ for I^2 and V for IR.

$$P = I^2R = (I \times I)R = I(IR) = (IR)I$$

$$P = VI \tag{3–6}$$

You can obtain another equivalent expression by substituting V/R for I (Ohm's law).

$$P = VI = V\left(\frac{V}{R}\right)$$

$$P = \frac{V^2}{R} \tag{3–7}$$

The three power expressions in Equations 3–5, 3–6, and 3–7 are known as **Watt's law**. To calculate the power in a resistance, you can use any one of the three equivalent Watt's law power formulas, depending on what information you have. For example, assume that you know the values of current and voltage; in this case, calculate the power with the formula $P = VI$. If you know I and R, use the formula $P = I^2R$. If you know V and R, use the formula $P = V^2/R$.

An aid for using both Ohm's law and Watt's law is found in the chapter summary, Figure 3–27.

EXAMPLE 3–18

Calculate the power in each of the three circuits of Figure 3–16.

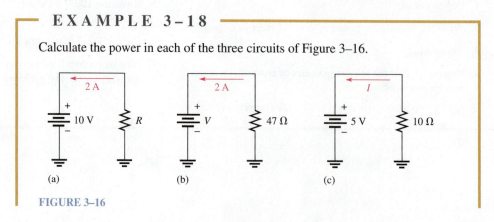

(a) (b) (c)

FIGURE 3–16

SOLUTION

In circuit (a), you know V and I. The power is determined as follows:

$$P = VI = (10\text{ V})(2\text{ A}) = \mathbf{20\ W}$$

In circuit (b), you know I and R. Therefore,

$$P = I^2R = (2\text{ A})^2(47\ \Omega) = \mathbf{188\ W}$$

In circuit (c), you know V and R. Therefore,

$$P = \frac{V^2}{R} = \frac{(5\text{ V})^2}{10\ \Omega} = \mathbf{2.5\ W}$$

RELATED PROBLEM

Determine the power in each circuit of Figure 3–16 for the following changes: In circuit (a), I is doubled and V remains the same; in circuit (b), R is doubled and I remains the same; in circuit (c), V is halved and R remains the same.

EXAMPLE 3–19

A solar yard light such as the one shown in Figure 3–17 has a solar collector that can provide 1.0 W of power for charging the 3.0 V batteries. What is the maximum charging current the solar collector could supply to fully discharged 3.0 V batteries?

FIGURE 3–17

SOLUTION

$$I = \frac{P}{V} = \frac{1.0\text{ W}}{3.0\text{ V}} = \mathbf{0.33\ A}$$

RELATED PROBLEM

What is the power dissipated by the light at night if the current is 30 mA?

SECTION 3–4 CHECKUP

1. Assume a car window defroster is connected to 13.0 V and has a current of 12 A. What power is dissipated in the defroster?

2. If there is a current of 5 A through a 47 Ω resistor, what is the power dissipated?

3. Many oscilloscopes have a 50 Ω input position that places a 2 W, 50 Ω resistor between the input and ground. What is the maximum voltage that could be applied to the input before exceeding the power rating of this resistor?

4. Assume a car seat heater has an internal resistance of 3.0 Ω. If the battery voltage is 13.4 V, what power is dissipated by the heater when it is on?

5. How much power does a 2.2 kΩ resistor with 8 V across it produce?

6. What is the resistance of a 55 W bulb that draws 0.5 A?

As you know, a resistor gives off heat when there is current through it. There is a limit to the amount of heat that a resistor can give off, which is specified by its power rating.

After completing this section, you should be able to

- Properly select resistors based on power consideration
- Define *power rating*
- Explain how physical characteristics of resistors determine their power rating
- Check for resistor failure with an ohmmeter

The **power rating** is the maximum amount of power that a resistor can dissipate without being damaged by excessive heat buildup. The power rating is not related to the ohmic value (resistance) but rather is determined mainly by the physical composition, size, and shape of the resistor. All else being equal, the larger the surface area of a resistor, the more power it can dissipate. *The surface area of a cylindrically shaped resistor is equal to the length (l) times the circumference (c),* as indicated in Figure 3–18. The area of the ends is not included.

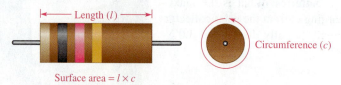

Length (l)

Surface area = $l \times c$

Circumference (c)

FIGURE 3–18   The power rating of a resistor is directly related to its surface area.

Metal-film resistors are available in standard power ratings from $\frac{1}{8}$ W to 1 W, as shown in Figure 3–19. Available power ratings for other types of resistors vary. For example, wirewound resistors have ratings up to 225 W or greater. Figure 3–20 shows some of these resistors.

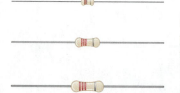

FIGURE 3–19   Relative sizes of metal-film resistors with standard power ratings of $\frac{1}{8}$ W, $\frac{1}{4}$ W, $\frac{1}{2}$ W, and 1 W.

Braking Resistors

Power resistors that are rated for several thousand watts are used as braking resistors for motors that need to be stopped completely. When a motor is turning, it has kinetic energy. For example, there is a large amount of energy in a moving elevator that must be dissipated when the elevator is stopped at a floor. When the electrical input is removed from an elevator motor, it would try to continue spinning due to inertia. To stop the motor quickly and bring it to a smooth stop, a special high-power braking resistor is switched across the motor. The braking resistor dissipates the energy from the moving system as heat. The braking resistor is normally mounted to a heat sink to help remove the heat.

Photo courtesy of Glover Resistors, Inc.

SYSTEM NOTE

(a) Axial-lead wirewound   (b) Adjustable wirewound   (c) Radial-lead for PC board insertion   (d) Thick-film power

FIGURE 3–20   Typical resistors with high power ratings.

When a resistor is used in a circuit, its power rating should be greater than the maximum power that it will have to handle to create a safety margin. Generally, the next higher standard value is used. For example, if a metal-film resistor is to dissipate 0.75 W in a circuit application, its rating should be the next higher standard value which is 1 W.

SYSTEM EXAMPLE 3–2

A LOAD TEST BOX

Most manufacturers run some quality control on a sample of their products to assure each tested product meets specifications before shipping it to a customer. In many cases, a specialized test fixture is designed to perform the quality test. Assume a manufacturer needs to test four different supplies at their maximum rated output, which are as follows: 5.0 V @ 1.0 A, 8.0 V at 500 mA, 10 V @ 100 mA, and 15 V at 100 mA. A set of selectable resistors is needed that will allow a technician to check that the power supply can operate properly under load. For this example, a resistor test box is designed that will meet the requirements, using resistors selected from stock.

Step 1 of the design process is to determine the resistance values that will provide the required loads and the power dissipated by the resistor when the power supply is connected to it. From Ohm's law, the required resistances for each supply are

5.0 V supply: $R = V/I = 5.0 \text{ V}/1.0 \text{ A} = 5.0 \ \Omega$

8.0 V supply: $R = V/I = 8.0 \text{ V}/0.50 \text{ A} = 16 \ \Omega$

10 V supply: $R = V/I = 10 \text{ V}/0.10 \text{ A} = 100 \ \Omega$

15 V supply: $R = V/I = 15 \text{ V}/0.10 \text{ A} = 150 \ \Omega$

Step 2 is to determine the amount of power expected in each resistor. From Watt's law, the expected power dissipation for each power supply is

5.0 V supply: $P = VI = (5.0 \text{ V})(1.0 \text{ A}) = 5.0 \text{ W}$

8.0 V supply: $P = VI = (8.0 \text{ V})(0.50 \text{ A}) = 4.0 \text{ W}$

10 V supply: $P = VI = (10 \text{ V})(0.10 \text{ A}) = 1.0 \text{ W}$

15 V supply: $P = VI = (15 \text{ V})(0.10 \text{ A}) = 1.5 \text{ W}$

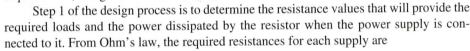

To be safe, specify resistors that have at least a 20% margin above their rated power in case a power supply has more voltage than expected. The power rating of the resistors depends on the type of resistor and the power tolerance required. Assume the choices for power ratings are 1.0 W, 2.0 W, 5.0 W, or 10 W. The selected resistors are then

5.0 V supply: $R = 5.0 \ \Omega$; $P = 5.0 \text{ W}$, choose $P_{(rating)} = 10 \text{ W}$

8.0 V supply: $R = 16 \ \Omega$; $P = 4.0 \text{ W}$, choose $P_{(rating)} = 5.0 \text{ W}$

10 V supply: $R = 100 \ \Omega$; $P = 1.0 \text{ W}$, choose $P_{(rating)} = 2.0 \text{ W}$

15 V supply: $R = 150 \ \Omega$; $P = 1.5 \text{ W}$, choose $P_{(rating)} = 2.0 \text{ W}$

The final step is to specify the fuse and list the parts. The fuse needs to be large enough to accept a more than the maximum expected current (1A). A reasonable fuse is 1.5 A. The final design is shown in Figure 3–21. To complete the box, a parts list is prepared including the pc board, components, and hardware items (box, screws, knob, etc.).

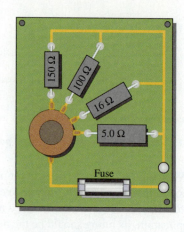

FIGURE 3–21   A load test box showing top view and the pc board.

When the power dissipated in a resistor is greater than its rating, the resistor will become excessively hot. As a result, the resistor may burn open or its resistance value may be greatly altered.

A resistor that has been damaged because of overheating can often be detected by the charred or altered appearance of its surface. If there is no visual evidence, a resistor that is suspected of being damaged can be checked with an ohmmeter for an open or increased resistance value. Recall that a resistor should be disconnected from the circuit to measure resistance. Sometimes an overheated resistor is due to another failure in the circuit. After replacing an overheated resistor, the underlying cause should be investigated before restoring power to the circuit.

EXAMPLE 3–20

Determine whether the resistor in each circuit of Figure 3–22 has possibly been damaged by overheating.

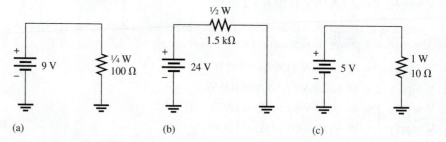

(a)   (b)   (c)

FIGURE 3–22

SOLUTION

For the circuit in Figure 3–22(a),

$$P = \frac{V^2}{R} = \frac{(9\ \text{V})^2}{100\ \Omega} = 0.81\ \text{W}$$

The rating of the resistor is ¼ W (0.25 W), which is insufficient to handle the power. The resistor has been overheated and may be burned out, making it an open.

For the circuit in Figure 3–22(b),

$$P = \frac{V^2}{R} = \frac{(24\ \text{V})^2}{1.5\ \text{k}\Omega} = 0.384\ \text{W}$$

The rating of the resistor is ½ W (0.5 W), which is sufficient to handle the power.

For the circuit in Figure 3–22(c),

$$P = \frac{V^2}{R} = \frac{(5\text{ V})^2}{10\ \Omega} = 2.5\text{ W}$$

The rating of the resistor is 1 W, which is insufficient to handle the power. The resistor has been overheated and may be burned out, making it an open.

RELATED PROBLEM

A ¼ W, 1.0 kΩ resistor is connected across a 12 V battery. Will it overheat?

SECTION 3–5 CHECKUP

1. Name two important parameters associated with a resistor.

2. How does the physical size of a resistor determine the amount of power that it can handle?

3. List the standard power ratings of metal-film resistors.

4. A resistor must handle 0.3 W. What standard size metal-film resistor should be used to dissipate the energy properly?

5. What is the maximum voltage that can be applied to a ¼ W, 100 Ω resistor if the power rating is not to be exceeded?

3–6 ENERGY CONVERSION AND VOLTAGE DROP IN A RESISTANCE

As you have learned, when there is current through a resistance, electrical energy is converted to heat energy. This heat is caused by collisions of the free electrons within the atomic structure of the resistive material. When a collision occurs, heat is given off; and the electron gives up some of its acquired energy as it moves through the material.

After completing this section, you should be able to

- **Explain energy conversion and voltage drop**
  - **Discuss the cause of energy conversion in a circuit**
  - **Define *voltage drop***
  - **Explain the relationship between energy conversion and voltage drop**

Figure 3–23 illustrates charge in the form of electrons flowing from the negative terminal of a battery, through a circuit, and back to the positive terminal. As they emerge from the negative terminal, the electrons are at their highest energy level. The electrons flow through each of the resistors that are connected together to form a current path (this type of connection is called series, as you will learn in Chapter 4). As the electrons flow through each resistor, some of their energy is given up in the form of heat. Therefore, the electrons have more energy when they enter a resistor than when they exit the resistor, as illustrated in the figure by the decrease in the intensity of the red color. When they have traveled through the circuit back to the positive terminal of the battery, the electrons are at their lowest energy level.

Recall that voltage equals energy per charge ($V = W/Q$) and that charge is a property of electrons. Based on the voltage of the battery, a certain amount of energy is imparted to all of the electrons that flow out of the negative terminal. The same number of electrons flow at each point throughout the circuit, but their energy decreases as they move through the resistance of the circuit.

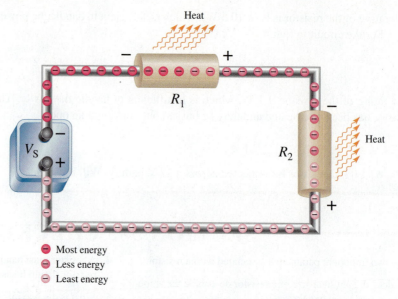

Most energy

Less energy

Least energy

FIGURE 3–23   A loss of energy by electrons (charge) as they flow through a resistance creates a voltage drop because voltage equals energy divided by charge.

In Figure 3–23, the voltage at the left end of R_1 is equal to W_{enter}/Q, and the voltage at the right end of R_1 is equal to W_{exit}/Q. The same number of electrons that enter R_1 also exit R_1, so Q is constant. However, the energy W_{exit} is less than W_{enter}, so the voltage at the right end of R_1 is less than the voltage at the left end. This decrease in voltage across the resistor due to a loss of energy is called a **voltage drop**. The voltage at the right end of R_1 is less negative (or more positive) than the voltage at the left end. The voltage drop is indicated by − and + signs (the + implies a more positive voltage).

The electrons have lost some energy in R_1 and now they enter R_2 with a reduced energy level. As they flow through R_2, they lose more energy, resulting in another voltage drop across R_2.

Integrated Circuit Thermal Sensors

Heat is the bane of many electronic systems. Noncontact integrated circuit (IC) temperature sensors that measure infrared energy are available for applications where heat can be a problem and space is limited. Temperature sensors are available that include the sensor, signal conditioning, analog-to-digital converter (ADC) and a reference all in a single IC package. This simplifies data collection and recording of temperature data over a continuous time period.

Photo/image provided courtesy of Fluke Corporation.

SYSTEM NOTE

SECTION 3–6 CHECKUP

1. What is the basic reason for energy conversion in a resistor?

2. What is a voltage drop?

3. What is the polarity of a voltage drop in relation to current direction?

3–7 POWER SUPPLIES AND BATTERIES

Recall that power supplies and batteries were briefly introduced as types of voltage sources in Section 2–3. A **power supply** is generally defined as a device that converts ac (alternating current) from the utility lines to a dc (direct current) voltage that virtually all electronic circuits and some transducers require. Batteries are also capable of supplying dc; in fact, many systems, such as laptop computers, can run from a power supply or internal battery. In this section, both types of voltage sources are described.

After completing this section, you should be able to

- Discuss characteristics of power supplies and batteries
    - Describe controls on typical laboratory power supplies
    - Determine the efficiency of a power supply given the input and output power
    - Define ampere-hour rating of batteries

Utilities universally have adopted ac for transmitting electricity from the generating station to the user because it can be readily transformed to high voltages for transmission and low voltages for the end user. High voltages are much more efficient and cost-effective to transmit over long distances. In the United States, the standard voltage supplied to outlets is approximately 120 V or 240 V at 60 Hz, but in Europe and other countries, the outlet voltage is 240 V at 50 Hz.

Virtually all electronic systems require stable dc for the integrated circuits and other devices to work properly. Power supplies fulfill this function by converting ac to stable dc and are usually built into the product. Many electronic systems have a recessed and protected switch that allows the internal power supply to be set for either the 120 V standard or for the 240 V standard. That switch must be set correctly, or serious damage can occur to the equipment.

In the laboratory, circuits are developed and tested. The purpose of a laboratory power supply is to provide the required stable dc to the circuit under test. The test circuit can be anything from a simple resistive network to a complex amplifier or logic circuit. To meet the requirement for a constant voltage, with almost no noise or ripple, laboratory power supplies are **regulated power supplies**, meaning the output is constantly sensed and automatically adjusted if it tries to change because of a change in the line voltage or the load.

Many circuits require multiple voltages, as well as the ability to set the voltage to a precise value or change it a small amount for testing. For this reason, laboratory power supplies usually have two or three outputs that are independent of each other and can be controlled separately. Output metering is normally part of a good laboratory power supply, in order to set and monitor the output voltage or current. Control may include fine and coarse controls or digital inputs to set precise voltages.

Figure 3–24 shows a triple output bench supply such as the type used in many electronic laboratories. The model shown has two 0–30 V independent supplies and a 4 V–6.5 V high current supply (commonly referred to a logic supply). Voltages can be precisely set using coarse and fine controls. The 0–30 V supplies have floating outputs, meaning they are not referenced to ground. This allows the user to set them up as a positive or negative supply or even connect them to another external supply. Another feature of this supply is that it can be set up as a current source, with a maximum voltage set for constant-current applications.

As in many power supplies, there are three output banana jacks for each of the 0–30 V supplies. The output is taken between the red (more positive) and black terminals. The green jack is referenced to the chassis, which is earth ground. Ground can be connected to either the red or black

HANDS ON TIP
A power supply provides both output voltage and current. You should make sure the voltage range is sufficient for your applications. Also, you must have sufficient current capacity to assure proper circuit operation. The current capacity is the maximum current that a power supply can provide to a load at a given voltage.

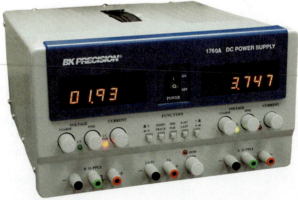

FIGURE 3–24 A triple output power supply. (Photo courtesy of B&K Precision Corp.)

jacks, which are normally "floating" (not referenced to ground). In addition, current and voltage can be monitored using the built-in digital meters.

The power delivered by a power supply is the product of the absolute voltage and current. For example, if a power supply is providing −15.0 V at 3.0 A, the supplied power is 45 W. For a triple output supply, the total power supplied by all three supplies is the sum of the power from each one individually.

EXAMPLE 3–21

What is the total power delivered by a triple output power supply if the output voltage and current are as follows?

Source 1: 18 V at 2.0 A

Source 2: −18 V at 1.5 A

Source 3: 5.0 V at 1.0 A

SOLUTION

Power delivered from each supply is the product of voltage and current (ignoring the sign).

Source 1: $P_1 = V_1 I_1 = (18\ \text{V})(2.0\ \text{A}) = 36\ \text{W}$

Source 2: $P_2 = V_2 I_2 = (18\ \text{V})(1.5\ \text{A}) = 27\ \text{W}$

Source 3: $P_3 = V_3 I_3 = (5.0\ \text{V})(1.0\ \text{A}) = 5.0\ \text{W}$

The total power is

$$P_T = P_1 + P_2 + P_3 = 36\ \text{W} + 27\ \text{W} + 5.0\ \text{W} = \textbf{68 W}$$

RELATED PROBLEM

How will the total power delivered change if the current from Source 1 increases to 2.5 A?

Power Supply Efficiency

An important characteristic of power supplies is efficiency. **Efficiency** is the ratio of the output power, P_{OUT}, to the input power, P_{IN}.

$$\text{Efficiency} = \frac{P_{\text{OUT}}}{P_{\text{IN}}} \qquad (3\text{–}8)$$

Efficiency is often expressed as a percentage. For example, if the input power is 100 W and the output power is 50 W, the efficiency is (50 W/100 W) × 100% = 50%.

All electronic power supplies are energy converters and require that power be put into them in order to get power out. For example, an electronic dc power supply might use the ac power from a wall outlet as its input. Its output is usually regulated dc voltage. The output power is *always* less than the input power because some of the total power is used internally to operate the power supply circuitry. This internal power dissipation is normally called the *power loss*. The output power is the input power minus the power loss.

$$P_{\text{OUT}} = P_{\text{IN}} - P_{\text{LOSS}} \qquad (3\text{–}9)$$

High efficiency means that very little power is dissipated in the power supply and there is a higher proportion of output power for a given input power.

EXAMPLE 3–22

A certain electronic power supply requires 25 W of input power. It can produce an output power of 20 W. What is its efficiency, and what is the power loss?

SOLUTION

$$\text{Efficiency} = \frac{P_{OUT}}{P_{IN}} = \left(\frac{20\text{ W}}{25\text{ W}}\right) = \textbf{0.8}$$

Expressed as a percentage,

$$\text{Efficiency} = \frac{20\text{ W}}{25\text{ W}} = 80\%$$

$$P_{LOSS} = P_{IN} - P_{OUT} = 25\text{ W} - 20\text{ W} = \textbf{5 W}$$

RELATED PROBLEM

A power supply has an efficiency of 92%. If P_{IN} is 50 W, what is P_{OUT}?

Ampere-Hour Ratings of Batteries

Batteries convert stored chemical energy to electrical energy. They are widely used to power small systems, such as laptop computers and cell phones, to supply the stable dc required. The batteries used in these small systems are normally rechargeable, meaning that the chemical reaction can be reversed from an external source. The capacity for any battery is measured in ampere-hours (Ah). For a rechargeable battery, the Ah rating is the capacity before it needs to be recharged. The **Ah rating** determines the length of time a battery can deliver a certain amount of current at the rated voltage.

A rating of one ampere-hour means that a battery can deliver an average of one ampere of current to a load for one hour at the rated voltage output. This same battery can deliver an average of two amperes for one-half hour. The more current the battery is required to deliver, the shorter the life of the battery. In practice, a battery usually is rated for a specified current level and output voltage. For example, a 12 V automobile battery may be rated for 70 Ah at 3.5 A. This means that it can supply an average of 3.5 A for 20 h at the rated voltage.

EXAMPLE 3–23

For how many hours can a battery deliver 2 A if it is rated at 70 Ah?

SOLUTION

The ampere-hour rating is the current times the number of hours, x.

$$70\text{ Ah} \times (2\text{ A})(x\text{ h})$$

Solving for the number of hours, x, yields

$$x = \frac{70\text{ Ah}}{2\text{ A}} = \textbf{35 h}$$

RELATED PROBLEM

A certain battery delivers 10 A for 6 h. What is its minimum Ah rating?

SECTION 3–7 CHECKUP

1. When a loading device draws an increased amount of current from a power supply, does this change represent a greater or a smaller load on the supply?

2. A power supply produces an output voltage of 10 V. If the supply provides 0.5 A to a load, what is the power output?

3. If a battery has an ampere-hour rating of 100 Ah, how long can it provide 5 A to a load?

4. If the battery in Question 3 is a 12 V device, what is its power output for the specified value of current?

5. A power supply used in the lab operates with an input power of 1 W. It can provide an output power of 750 mW. What is its efficiency?

3–8 INTRODUCTION TO TROUBLESHOOTING

Technicians must be able to diagnose and repair malfunctioning circuits and systems. In this section, you learn a general approach to troubleshooting using a simple example. Troubleshooting coverage is an important part of this textbook, so you will find a troubleshooting section in many of the chapters as well as troubleshooting problems for skill building.

After completing this section, you should be able to

- **Describe a basic approach to troubleshooting**
  - **List three steps in troubleshooting**
  - **Explain what is meant by half-splitting**
  - **Discuss and compare the three basic measurements of voltage, current, and resistance**

Troubleshooting is the application of logical thinking combined with a thorough knowledge of circuit or system operation to correct a malfunction. The basic approach to troubleshooting consists of three steps: *analysis, planning,* and *measuring.* We will refer to this 3-step approach as APM.

Analysis

HANDS ON TIP
When you troubleshoot a newly manufactured circuit on a pc board, certain failures are possible that are not likely on a circuit that has been working. For example, a wrong part may have accidentally been used in the new circuit or a pin may have been bent when it was installed. Cases occur where a series of newly manufactured boards have identical faults because of a wrong part.

The first step in troubleshooting a circuit is to analyze clues or symptoms of the failure. The analysis can begin by determining the answer to certain questions:

1. Has the circuit ever worked?

2. If the circuit once worked, under what conditions did it fail?

3. What are the symptoms of the failure?

4. What are the possible causes of failure?

Planning

The second step in the troubleshooting process, after analyzing the clues, is formulating a logical plan of attack. Much time can be saved by proper planning. A working knowledge of the circuit is a prerequisite to a plan for troubleshooting. If you are not certain how the circuit is supposed to operate, take time to review circuit diagrams (schematics), operating instructions, and other pertinent information. A schematic with proper voltages marked at various test points is particularly useful. Although logical thinking is perhaps the most important tool in troubleshooting, it rarely can solve the problem by itself.

Measuring

The third step is to narrow the possible failures by making carefully thought out measurements. These measurements usually confirm the direction you are taking in solving the problem, or they may point to a new direction that you should take. Occasionally, you may find a totally unexpected result.

An APM Example

The thought process that is part of the APM approach can be illustrated with a simple example. Suppose you have a string of eight decorative 15 V bulbs connected in series to a 120 V source V_S, as shown in Figure 3–25. Assume that this circuit worked properly at one time but stopped working after it was moved to a new location. When plugged in at the new location, the bulbs fail to turn on. How do you go about finding the trouble?

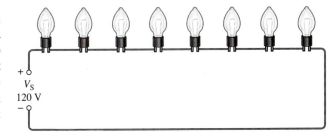

FIGURE 3–25   **A string of bulbs connected to a voltage source.**

THE ANALYSIS THOUGHT PROCESS   You may think like this as you proceed to analyze the situation:

- Since the circuit worked before it was moved, the problem could be that there is no voltage at the new location.
- Perhaps the wiring was loose and pulled apart when moved.
- It is possible that a bulb is burned out or loose in its socket.

With this reasoning, you have considered possible causes and failures that may have occurred. The thought process continues:

- The fact that the circuit once worked eliminates the possibility that the original circuit was improperly wired.
- If the fault is due to an open path, it is unlikely that there is more than one break which could be either a bad connection or a burned out bulb.

You have now analyzed the problem and are ready to plan the process of finding the fault in the circuit.

THE PLANNING THOUGHT PROCESS   The first part of your plan is to measure for voltage at the new location. If the voltage is present, then the problem is in the string of lights. If voltage is not present, check the circuit breaker at the distribution box in the house. Before resetting breakers, you should think about why a breaker may be tripped. Let's assume that you find the voltage is present. This means that the problem is in the string of lights.

   The second part of your plan is to measure either the resistance in the string of lights or to measure voltages across the bulbs. The decision whether to measure resistance or voltage is a toss-up and can be made based on the ease of making the test. Seldom is a troubleshooting plan developed so completely that all possible contingencies are included. You will frequently need to modify the plan as you go along.

THE MEASUREMENT PROCESS   You proceed with the first part of your plan by using a multimeter to check the voltage at the new location. Assume the measurement shows a voltage of 120 V. Now you have eliminated the possibility of no voltage. You know that, since you have voltage across the string and there is no current because no bulb is on, there must be an open in the current path. Either a bulb is burned out, a connection at the lamp socket is broken, or the wire is broken.

   Next, you decide to locate the break by measuring resistance with your multimeter. Applying logical thinking, you decide to measure the resistance of each half of the string instead of measuring the resistance of each bulb. By measuring the resistance of half the bulbs at once, you can usually reduce the effort required to find the open. This technique is a type of troubleshooting procedure called **half-splitting**.

Once you have identified the half in which the open occurs, as indicated by an infinite resistance, you use half-splitting again on the faulty half and continue until you narrow the fault down to a faulty bulb or connection. This process is shown in Figure 3–26, assuming for purposes of illustration that the seventh bulb is burned out.

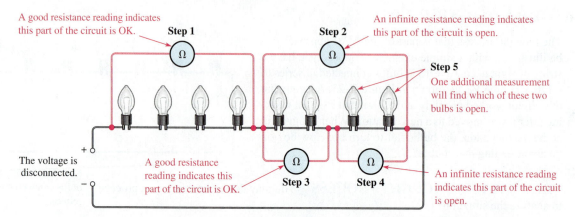

FIGURE 3–26   Illustration of the half-splitting method of troubleshooting. The numbered steps indicate the sequence in which the multimeter is moved from one position to another.

As you can see in the figure, the half-splitting approach in this particular case takes a maximum of five measurements to identify the open bulb. If you had decided to measure each bulb individually and had started at the left, you would have needed seven measurements. So, sometimes half-splitting saves steps; sometimes it doesn't. The number of steps required depends on where you make your measurements and in what sequence.

Unfortunately, most troubleshooting is more difficult than this example. However, analysis and planning are essential for effective troubleshooting in any situation. As measurements are made, the plan is often modified; the experienced troubleshooter narrows the search by fitting the symptoms and measurements into a probable cause. In some cases, low-cost equipment is simply discarded or recycled when troubleshooting and repair costs are comparable to replacement costs.

Comparison of *V*, *R*, and *I* Measurements

As you know from Section 2–7, you can measure voltage, current, or resistance in a circuit. To measure voltage, place the voltmeter in parallel across the component; that is, place one lead on each side of the component. This makes voltage measurements the easiest of the three types of measurements.

To measure resistance, connect the ohmmeter across a component; however, the voltage must be first disconnected, and sometimes the component itself must be removed from the circuit. Therefore, resistance measurements are generally more difficult than voltage measurements.

To measure current, place the ammeter in series with the component; that is, the ammeter must be in line with the current path. To do this you must disconnect a component lead or a wire before you connect the ammeter. This usually makes a current measurement the most difficult to perform.

SECTION 3–8 CHECKUP

1. Name the three steps in the APM approach to troubleshooting.

2. Explain the basic idea of the half-splitting technique.

3. Why are voltages easier to measure than currents in a circuit?

SUMMARY

- Voltage and current are linearly proportional.
- Ohm's law gives the relationship of voltage, current, and resistance.
- Current is inversely proportional to resistance.
- A kilohm (kΩ) is one thousand ohms.
- A megohm (MΩ) is one million ohms.
- A microampere (μA) is one-millionth of an ampere.
- A milliampere (mA) is one-thousandth of an ampere.
- Use $I = V/R$ to calculate current.
- Use $V = IR$ to calculate voltage.
- Use $R = V/I$ to calculate resistance.
- One watt equals one joule per second.
- Watt is the unit of power, and joule is the unit of energy.
- The power rating of a resistor determines the maximum power that it can handle safely.
- Resistors with a larger physical size can dissipate more power in the form of heat than smaller ones.
- A resistor should have a power rating as high or higher than the maximum power that it is expected to handle in the circuit.
- Power rating is not related to resistance value.
- A resistor usually opens when it overheats and fails.
- Energy is equal to power multiplied by time.
- The kilowatt-hour is a unit of energy.
- An example of one kilowatt-hour is one thousand watts used for one hour.
- A power supply is an energy source used to operate electrical and electronic devices.
- A battery is one type of power supply that converts chemical energy into electrical energy.
- A power supply converts commercial energy (ac from the power company) to regulated dc at various voltage levels.
- The output power of a power supply is the output voltage times the load current.
- A load is a device that draws current from the power supply.
- The capacity of a battery is measured in ampere-hours (Ah).
- One ampere-hour equals one ampere used for one hour, or any other combination of amperes and hours that has a product of one.
- A power supply with a high efficiency has a smaller percentage power loss than one with a lower efficiency.
- The formula wheel in Figure 3–27 gives the Ohm's law and Watt's law relationships.
- APM (analysis, planning, and measurement) provides a logical approach to troubleshooting.
- The half-splitting method of troubleshooting generally results in fewer measurements.

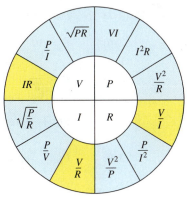

◼ Ohm's law   ◻ Watt's law

FIGURE 3–27

KEY TERMS

Key terms and other bold terms in the chapter are defined in the end-of-book glossary.

Ah rating   A capacity rating for batteries determined by multiplying the current (A) times the length of time (h) a battery can deliver that current to a load.

Efficiency   The ratio of the output power to the input power of a circuit, usually expressed as a percent.

Energy   The ability to do work. The unit is the joule (J).

Half-splitting A troubleshooting procedure where one starts in the middle of a circuit or system and, depending on the first measurement, works toward the output or toward the input to find the fault.

Joule (J) The SI unit of energy.

Kilowatt-hour (kWh) A large unit of energy used mainly by utility companies.

Linear Characterized by a straight-line relationship.

Ohm's law A law stating that current is directly proportional to voltage and inversely proportional to resistance.

Power The rate of energy usage. The unit is the watt (W).

Power rating The maximum amount of power that a resistor can dissipate without being damaged by excessive heat buildup.

Power supply A device that converts ac from the utility lines to a dc voltage.

Troubleshooting A systematic process of isolating, identifying, and correcting a fault in a circuit or system.

Voltage drop The decrease in voltage across a resistor due to a loss of energy.

Watt (W) The unit of power. One watt is the power when 1 J of energy is used in 1 s.

Watt's law A law that states the relationships of power to current, voltage, and resistance.

KEY FORMULAS

(3–1) $I = \dfrac{V}{R}$ Form of Ohm's law for calculating current

(3–2) $V = IR$ Form of Ohm's law for calculating voltage

(3–3) $R = \dfrac{V}{I}$ Form of Ohm's law for calculating resistance

(3–4) $P = \dfrac{W}{t}$ Power equals energy divided by time.

(3–5) $P = I^2R$ Power equals current squared times resistance.

(3–6) $P = VI$ Power equals voltage times current.

(3–7) $P = \dfrac{V^2}{R}$ Power equals voltage squared divided by resistance.

(3–8) $\text{Efficiency} = \dfrac{P_{\text{OUT}}}{P_{\text{IN}}}$ Power supply efficiency

(3–9) $P_{\text{OUT}} = P_{\text{IN}} - P_{\text{LOSS}}$ Output power

TRUE/FALSE QUIZ

Answers are at the end of the chapter.

1. If the total resistance of a circuit increases, current decreases.

2. Ohm's law for finding resistance is $R = I/V$.

3. When milliamps and kilohms are multiplied together, the result is volts.

4. If a $10\,\text{k}\Omega$ resistor is connected to a 10 V source, the current in the resistor will be 1 A.

5. The kilowatt-hour is a unit of power.

6. One watt is equal to one joule per second.

7. The power rating of a resistor should always be less than the required power dissipation in the circuit.

8. Within limits, a regulated power supply can automatically keep the output voltage constant even if the load changes.

9. A power supply that has a negative output voltage absorbs power from the load.

10. When analyzing a circuit problem, you should consider the conditions under which it failed.

SELF-TEST

Answers are at the end of the chapter.

1. Ohm's law states that
   (a) current equals voltage times resistance
   (b) voltage equals current times resistance
   (c) resistance equals current divided by voltage
   (d) voltage equals current squared times resistance

2. When the voltage across a resistor is doubled, the current will
   (a) triple   (b) halve   (c) double   (d) not change

3. When 10 V are applied across a 20 Ω resistor, the current is
   (a) 10 A   (b) 0.5 A   (c) 200 A   (d) 2 A

4. When there are 10 mA of current through a 1.0 kΩ resistor, the voltage across the resistor is
   (a) 100 V   (b) 0.1 V   (c) 10 kV   (d) 10 V

5. If 20 V are applied across a resistor and there are 6.06 mA of current, the resistance is
   (a) 3.3 kΩ   (b) 33 kΩ   (c) 330 Ω   (d) 3.03 kΩ

6. A current of 250 μA through a 4.7 kΩ resistor produces a voltage drop of
   (a) 53.2 V   (b) 1.175 mV   (c) 18.8 V   (d) 1.175 V

7. A resistance of 2.2 MΩ is connected across a 1 kV source. The resulting current is approximately
   (a) 2.2 mA   (b) 455 μA   (c) 45.5 μA   (d) 0.455 A

8. Power can be defined as
   (a) energy   (b) heat
   (c) the rate at which energy is used   (d) the time required to use energy

9. For 10 V and 50 mA, the power is
   (a) 500 mW   (b) 0.5 W
   (c) 500,000 μW   (d) answers (a), (b), and (c)

10. When the current through a 10 kΩ resistor is 10 mA, the power is
    (a) 1 W   (b) 10 W   (c) 100 mW   (d) 1 mW

11. A 2.2 kΩ resistor dissipates 0.5 W. The current is
    (a) 15.1 mA   (b) 227 μA   (c) 1.1 mA   (d) 4.4 mA

12. A 330 Ω resistor dissipates 2 W. The voltage is
    (a) 2.57 V   (b) 660 V   (c) 6.6 V   (d) 25.7 V

13. The power rating of a resistor that is to handle up to 1.1 W should be
    (a) 0.25 W   (b) 1 W   (c) 2 W   (d) 5 W

14. A 22 Ω half-watt resistor and a 220 Ω half-watt resistor are connected across a 10 V source. Which one(s) will overheat?
    (a) 22 Ω   (b) 220 Ω   (c) both   (d) neither

15. When the needle of an analog ohmmeter indicates infinity, the resistor being measured is
    (a) overheated   (b) shorted   (c) open   (d) reversed

TROUBLESHOOTING: SYMPTOM AND CAUSE

The purpose of these exercises is to help develop thought processes essential to troubleshooting. Answers are at the end of the chapter.

Determine the cause for each set of symptoms. Refer to Figure 3–28.

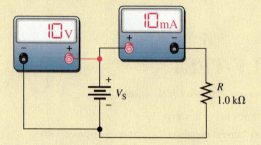

FIGURE 3–28 **The meters indicate the correct readings for this circuit.**

1. *Symptom:* The ammeter reading is zero, and the voltmeter reading is 10 V.
   Cause:
   (a) R is shorted.
   (b) R is open.
   (c) The voltage source is faulty.

2. *Symptom:* The ammeter reading is zero, and the voltmeter reading is 0 V.
   Cause:
   (a) R is open.
   (b) R is shorted.
   (c) The voltage source is turned off or faulty.

3. *Symptom:* The ammeter reading is 10 mA, and the voltmeter reading is 0 V.
   Cause:
   (a) The voltmeter is defective.
   (b) The ammeter is faulty.
   (c) The voltage source is turned off or faulty.

4. *Symptom:* The ammeter reading is 1 mA, and the voltmeter reading is 10 V.
   Cause:
   (a) The voltmeter is defective.
   (b) The resistor value is higher than it should be.
   (c) The resistor value is lower than it should be.

5. *Symptom:* The ammeter reading is 100 mA, and the voltmeter reading is 10 V.
   Cause:
   (a) The voltmeter is defective.
   (b) The resistor value is higher than it should be.
   (c) The resistor value is lower than it should be.

PROBLEMS

Answers to odd-numbered problems are at the end of the book.

BASIC PROBLEMS

SECTION 3–1   Ohm's Law

1. The current in a circuit is 1 A. Determine what the current will be when
   (a) the voltage is tripled
   (b) the voltage is reduced by 80%
   (c) the voltage is increased by 50%

2. The current in a circuit is 100 mA. Determine what the current will be when
   (a) the resistance is increased by 100%
   (b) the resistance is reduced by 30%
   (c) the resistance is quadrupled

3. The current in a circuit is 10 mA. What will the current be if the voltage is tripled and the resistance is doubled?

SECTION 3–2   Application of Ohm's Law

4. Determine the current in each case.
   (a) $V = 5\,V, R = 1.0\,\Omega$ (b) $V = 15\,V, R = 10\,\Omega$
   (c) $V = 50\,V, R = 100\,\Omega$ (d) $V = 30\,V, R = 15\,k\Omega$
   (e) $V = 250\,V, R = 4.7\,M\Omega$

5. Determine the current in each case.
   (a) $V = 9\,V, R = 2.7\,k\Omega$ (b) $V = 5.5\,V, R = 10\,k\Omega$
   (c) $V = 40\,V, R = 68\,k\Omega$ (d) $V = 1\,kV, R = 2\,k\Omega$
   (e) $V = 66\,kV, R = 10\,M\Omega$

6. A 10 Ω resistor is connected across a 12 V battery. How much current is there through the resistor?

7. A resistor is connected across the terminals of a dc voltage source in each part of Figure 3–29. Determine the current in each resistor.

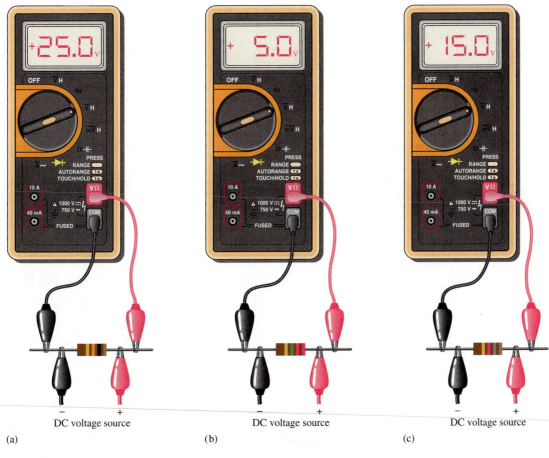

(a)   (b)   (c)

FIGURE 3–29

8. A 5-band resistor is connected across a 12 V source. Determine the current if the resistor color code is orange, violet, yellow, gold, brown.

9. If the voltage in Problem 8 is doubled, will a 0.5 A fuse blow? Explain your answer.

10. Calculate the voltage for each value of I and R.
    (a) $I = 2\,A, R = 18\,\Omega$ (b) $I = 5\,A, R = 47\,\Omega$
    (c) $I = 2.5\,A, R = 620\,\Omega$ (d) $I = 0.6\,A, R = 47\,\Omega$
    (e) $I = 0.1\,A, R = 470\,\Omega$

11. Calculate the voltage for each value of *I* and *R*.
    (a) $I = 1$ mA, $R = 10$ Ω   (b) $I = 50$ mA, $R = 33$ Ω
    (c) $I = 3$ A, $R = 4.7$ kΩ   (d) $I = 1.6$ mA, $R = 2.2$ kΩ
    (e) $I = 250$ μA, $R = 1.0$ kΩ   (f) $I = 500$ mA, $R = 1.5$ MΩ
    (g) $I = 850$ μA, $R = 10$ MΩ   (h) $I = 75$ μA, $R = 47$ Ω

12. A 20 mΩ current-sensing resistor has 3 A. What is the voltage across the current-sensing resistor?

13. Assign a voltage value to each source in the circuits of Figure 3–30 to obtain the indicated amounts of current.

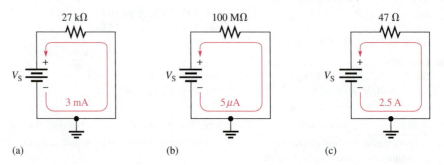

(a)   (b)   (c)

FIGURE 3–30

14. Calculate the resistance for each value of *V* and *I*.
    (a) $V = 10$ V, $I = 2$ A   (b) $V = 90$ V, $I = 45$ A
    (c) $V = 50$ V, $I = 5$ A   (d) $V = 5.5$ V, $I = 10$ A
    (e) $V = 150$ V, $I = 0.5$ A

15. Calculate *R* for each set of *V* and *I* values.
    (a) $V = 10$ kV, $I = 5$ A   (b) $V = 7$ V, $I = 2$ mA
    (c) $V = 500$ V, $I = 250$ mA   (d) $V = 50$ V, $I = 500$ μA
    (e) $V = 1$ kV, $I = 1$ mA

16. Six volts are applied across a resistor. A current of 2 mA is measured. What is the value of the resistor?

17. Choose the correct value of resistance to get the current values indicated in each circuit of Figure 3–31.

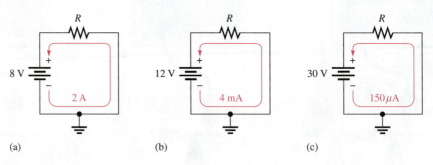

(a)   (b)   (c)

FIGURE 3–31

18. A flashlight is operated from 3.2 V and has a resistance of 3.9 Ω when the filament is hot. What current is supplied by the batteries?

SECTION 3–3   Energy and Power

19. The flashlight in Problem 18 uses 26 J in 10 s. What is the power in watts?

20. What is the power when energy is used at the rate of 350 J/s?

21. What is the power in watts when 7500 J of energy are used in 5 h?

22. Convert the following to kilowatts:
    (a) 1000 W   (b) 3750 W   (c) 160 W   (d) 50,000 W

23. Convert the following to megawatts:
    (a) 1,000,000 W   (b) 3×10^6 W   (c) 15×10^7 W   (d) 8700 kW

24. Convert the following to milliwatts:
    (a) 1 W   (b) 0.4 W
    (c) 0.002 W   (d) 0.0125 W

25. Convert the following to microwatts:
    (a) 2 W   (b) 0.0005 W
    (c) 0.25 mW   (d) 0.00667 mW

26. Convert the following to watts:
    (a) 1.5 kW   (b) 0.5 MW
    (c) 350 mW   (d) 9000 μW

27. Prove that the unit for power (the watt) is equivalent to 1 V $\times$ 1 A.

28. Show that there are 3.6×10^6 joules in a kilowatt-hour.

SECTION 3–4   Power in an Electric Circuit

29. If a resistor has 5.5 V across it and 3 mA through it, what is the power dissipation?

30. An electric heater works on 115 V and draws 3 A of current. How much power does it use?

31. What is the power when there are 500 mA of current through a 4.7 kΩ resistor?

32. A 25 mΩ current-sensing resistor has 5 A of current. What power is dissipated?

33. If there are 60 V across a 620 Ω resistor, what is the power dissipation?

34. A 56 Ω resistor is connected across the terminals of a 1.5 V battery. What is the power dissipation in the resistor?

35. If a resistor is to carry 2 A of current and handle 100 W of power, how many ohms must it be? Assume that the voltage can be adjusted to any required value.

36. Convert 5×10^6 watts used for 1 minute to kWh.

37. Convert 6700 watts used for 1 second to kWh.

38. How many kilowatt-hours do 50 W used for 12 h equal?

39. Assume that an alkaline D-cell battery can maintain an average voltage of 1.25 V for 90 h in a 10 Ω load before becoming unusable. What average power is delivered to the load during the life of the battery?

40. What is the total energy in joules that is delivered during the 90 hours for the battery in Problem 39?

SECTION 3–5   The Power Rating of Resistors

41. A 6.8 kΩ resistor has burned out in a circuit. You must replace it with another resistor with the same resistance value. If the resistor carries 10 mA, what should its power rating be? Assume that you have available resistors in all the standard power ratings.

42. A certain type of power resistor comes in the following ratings: 3 W, 5 W, 8 W, 12 W, 20 W. Your particular application requires a resistor that can handle approximately 8 W. Which rating would you use for a minimum safety margin of 20% above the rated value? Why?

SECTION 3–6   Energy Conversion and Voltage Drop in a Resistance

43. For each circuit in Figure 3–32, assign the proper polarity for the voltage across the resistor.

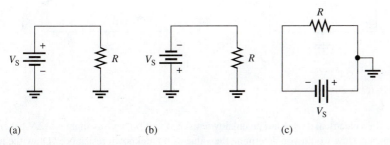

(a)   (b)   (c)

FIGURE 3–32

SECTION 3–7   Power Supplies and Batteries

44. A 50 Ω load dissipates 1 W of power. What is the output voltage of the power supply?

45. A battery can provide an average of 1.5 A of current for 24 h. What is its ampere-hour rating?

46. How much average current can be drawn from an 80 Ah battery for 10 h?

47. If a battery is rated at 650 mAh, how much average current will it provide for 48 h?

48. If the input power is 500 mW and the output power is 400 mW, what is the power loss? What is the efficiency of this power supply?

49. To operate at 85% efficiency, how much output power must a source produce if the input power is 5 W?

SECTION 3–8   Introduction to Troubleshooting

50. In the light circuit of Figure 3–33, identify the faulty bulb based on the series of ohmmeter readings shown.

51. Assume you have a 32-light string and one of the bulbs is burned out. Using the half-splitting approach and starting in the left half of the circuit, how many resistance measurements will it take to find the faulty bulb if it is seventeenth from the left.

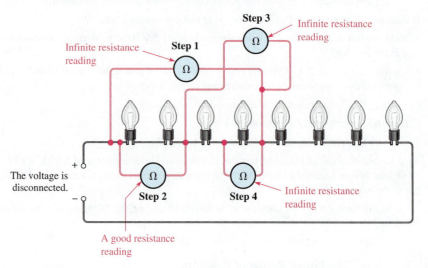

FIGURE 3–33

ADVANCED PROBLEMS

52. A certain power supply provides a continuous 2 W to a load. It is operating at 60% efficiency. In a 24 h period, how many kilowatt-hours does the power supply use?

53. The filament of a light bulb in the circuit of Figure 3–34(a) has a certain amount of resistance, represented by an equivalent resistance in Figure 3–34(b). If the bulb operates with 120 V and 0.8 A of current, what is the resistance of its filament?

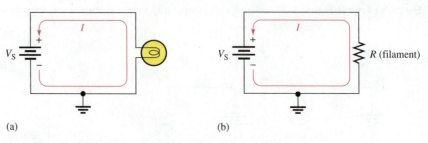

(a)   (b)

FIGURE 3–34

54. A certain electrical device has an unknown resistance. You have available a 12 V battery and an ammeter. How would you determine the value of the unknown resistance? Draw the necessary circuit connections.

55. A variable voltage source is connected to the circuit of Figure 3–35. Start at 0 V and increase the voltage in 10 V steps up to 100 V. Determine the current at each voltage value, and plot a graph of V versus I. Is the graph a straight line? What does the graph indicate?

56. In a certain circuit, $V_S = 1$ V and $I = 5$ mA. Determine the current for each of the following voltages in a circuit with the same resistance:
   (a) $V_S = 1.5$ V (b) $V_S = 2$ V (c) $V_S = 3$ V
   (d) $V_S = 4$ V (e) $V_S = 10$ V

57. Figure 3–36 is a graph of current versus voltage for three resistance values. Determine R_1, R_2, and R_3.

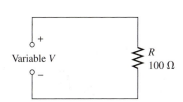

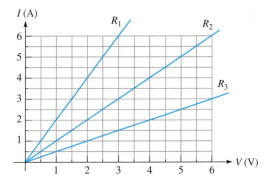

Variable V

R
$100\ \Omega$

FIGURE 3–35

FIGURE 3–36

58. You are measuring the current in a circuit that is operated on a 10 V battery. The ammeter reads 50 mA. Later, you notice that the current has dropped to 30 mA. Eliminating the possibility of a resistance change, you must conclude that the voltage has changed. How much has the voltage of the battery changed, and what is its new value?

59. If you wish to increase the amount of current in a resistor from 100 mA to 150 mA by changing the 20 V source, by how many volts should you change the source? To what new value should you set it?

60. A 6 V source is connected to a 100 Ω resistor by two 12 ft lengths of 18-gauge copper wire. Refer to Table 2–3 to determine the following:
   (a) current (b) resistor voltage (c) voltage across each length of wire

61. If a 300 W bulb is allowed to burn continuously for 30 days, how many kilowatt-hours of energy does it use?

62. At the end of a 31-day period, your utility bill shows that you have used 1500 kWh. What is the average daily power?

63. A certain type of power resistor comes in the following ratings: 3 W, 5 W, 8 W, 12 W, 20 W. Your particular application requires a resistor that can handle approximately 10 W. Which rating would you use? Why?

64. A 12 V source is connected across a 10 Ω resistor for 2 min.
   (a) What is the power dissipation?
   (b) How much energy is used?
   (c) If the resistor remains connected for an additional minute, does the power dissipation increase or decrease?

65. The rheostat in Figure 3–37 is used to control the current to a heating element. When the rheostat is adjusted to a value of 8 Ω or less, the heating element can burn out. What is the rated value of the fuse needed to protect the circuit if the voltage across the heating element at the point of maximum current is 100 V?

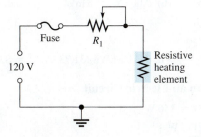

FIGURE 3–37

66. A 30 mΩ current-sensing resistor is rated for ½ W. What is the maximum current it can have before the rating is exceeded?

67. What happens to the power dissipated in a resistor if the current is doubled?

68. Develop a complete materials list for the load test box in Figure 3–21.

69. Draw the schematic for the load test box in Figure 3–21.

MULTISIM TROUBLESHOOTING PROBLEMS

70. Open file P03-70; files are found at www.pearsonhighered.com/floyd. Determine whether or not the circuit is operating properly. If not, find the fault.

71. Open file P03-71. Determine whether or not the circuit is operating properly. If not, find the fault.

72. Open file P03-72. Determine whether or not the circuit is operating properly. If not, find the fault.

73. Open file P03-73. Determine whether or not the circuit is operating properly. If not, find the fault.

74. Open file P03-74. Determine whether or not the circuit is operating properly. If not, find the fault.

ANSWERS TO SECTION CHECKUPS

SECTION 3–1   Ohm's Law

1. Ohm's law states that current varies directly with voltage and inversely with resistance.

2. $I = V/R$

3. $V = IR$

4. $R = V/I$

5. The current increases; three times when V is tripled

6. Doubling R cuts I in half to 5 mA.

7. No change in I if V and R are doubled.

SECTION 3–2   Application of Ohm's Law

1. $I = 10\,\text{V}/4.7\,\Omega = 2.13\,\text{A}$

2. $I = 20\,\text{kV}/4.7\,\text{M}\Omega = 4.26\,\text{mA}$

3. $I = 10\,\text{kV}/2\,\text{k}\Omega = 5\,\text{A}$

4. $V = (1\,\text{A})(10\,\Omega) = 10\,\text{V}$

5. $V = (3\,\text{mA})(3\,\text{k}\Omega) = 9\,\text{V}$

6. $V = (2\,\text{A})(6\,\Omega) = 12\,\text{V}$

7. $R = 10\,\text{V}/2\,\text{A} = 5\,\Omega$

8. $R = 25\,\text{V}/50\,\text{mA} = 0.5\,\text{k}\Omega = 500\,\Omega$

SECTION 3–3   Energy and Power

1. Power is the rate at which energy is used.

2. $P = W/t$

3. Watt is the unit of power. One watt is the power when 1 J of energy is used in 1 s.

4. **(a)** 68,000 W = 68 kW   **(b)** 0.005 W = 5 mW   **(c)** 0.000025 = 25 μW

5. (100 W)(10 h) = 1 kWh

6. 2000 W = 2 kW

7. (1.322 kW)(24 h) = 31.73 kWh; (0.11 $/kWh)(31.73 kWh) = $3.49

SECTION 3–4   Power in an Electric Circuit

1. $P = IV = (12\,\text{A})(13\,\text{V}) = 156\,\text{W}$

2. $P = (5\,\text{A})^2(47\,\Omega) = 1175\,\text{W}$

3. $V = \sqrt{PR} = \sqrt{(2\,\text{W})(50\,\Omega)} = 10\,\text{V}$

4. $P = \dfrac{V^2}{R} = \dfrac{(13.4\ \text{V})^2}{3.0\ \Omega} = 60\ \text{W}$

5. $P = (8\ \text{V})^2/2.2\ \text{k}\Omega = 29.1\ \text{mW}$

6. $R = 55\ \text{W}/(0.5\ \text{A})^2 = 220\ \Omega$

SECTION 3–5   The Power Rating of Resistors

1. Two resistor parameters are resistance and power rating.

2. A larger resistor physical size dissipates more energy.

3. Standard ratings of metal-film resistors are 0.125 W, 0.25 W, 0.5 W, and 1 W.

4. At least a 0.5 W rating to handle 0.3 W

5. 5.0 V

SECTION 3–6   Energy Conversion and Voltage Drop in a Resistance

1. Energy conversion in a resistor is caused by collisions of free electrons with the atoms in the material.

2. Voltage drop is a decrease in voltage across a resistor due to a loss of energy.

3. Voltage drop is negative to positive in the direction of current.

SECTION 3–7   Power Supplies and Batteries

1. An increased amount of current represents a greater load.

2. $P_{\text{OUT}} = (10\ \text{V})(0.5\ \text{A}) = 5\ \text{W}$

3. $100\ \text{Ah}/5\ \text{A} = 20\ \text{h}$

4. $P_{\text{OUT}} = (12\ \text{V})(5\ \text{A}) = 60\ \text{W}$

5. Efficiency $= (750\ \text{mW}/1000\ \text{mW})100\% = 75\%$

SECTION 3–8   Introduction to Troubleshooting

1. Analysis, Planning, and Measuring

2. Half-splitting identifies the fault by successively isolating half of the remaining circuit.

3. Voltage is measured across a component. Current is measured in series with the component.

ANSWERS TO RELATED PROBLEMS FOR EXAMPLES

3–1 $I_1 = 10\ \text{V}/5\ \Omega = 2\ \text{A}; I_2 = 10\ \text{V}/20\ \Omega = 0.5\ \text{A}$

3–2 1 V

3–3 970 mΩ

3–4 Yes

3–5 11.1 mA

3–6 5 mA

3–7 25 μA

3–8 200 μA

3–9 800 V

3–10 220 mV

3–11 16.5 V

3–12 Voltage drops to 1.08 V.

3–13 6.0 Ω

3–14 39.5 kΩ

3–15 3000 J

3–16 2 kWh

3–17 $4.92 + $0.07 = $4.99

3–18 **(a)** 40 W   **(b)** 376 W   **(c)** 625 mW

3–19 26 W

3–20 No

3–21 77 W

3–22 46 W

3–23 60 Ah

ANSWERS TO TRUE/FALSE QUIZ

1. T   **2.** F   **3.** T   **4.** F   **5.** F   **6.** T   **7.** F   **8.** T

9. F   **10.** T

ANSWERS TO SELF-TEST

1. (b)   **2.** (c)   **3.** (b)   **4.** (d)   **5.** (a)   **6.** (d)   **7.** (b)   **8.** (c)

9. (d)   **10.** (a)   **11.** (a)   **12.** (d)   **13.** (c)   **14.** (a)   **15.** (c)

ANSWERS TO TROUBLESHOOTING: SYMPTOM AND CAUSE

1. (b)

2. (c)

3. (a)

4. (b)

5. (c)

CHAPTER 4

SERIES CIRCUITS

OUTLINE

OBJECTIVES

- Identify a series resistive circuit
- Determine total series resistance
- Determine the current throughout a series circuit
- Apply Ohm's law in series circuits
- Determine the total effect of voltage sources connected in series
- Apply Kirchhoff's voltage law
- Use a series circuit as a voltage divider
- Determine power in a series circuit
- State how to measure voltage with respect to ground
- Troubleshoot series circuits

KEY TERMS

- Series
- Series-aiding
- Series-opposing
- Kirchhoff's voltage law
- Voltage divider
- Reference ground
- Open
- Short

INTRODUCTION

Resistive circuits can be of two basic forms: series or parallel. In this chapter series circuits are discussed. Parallel circuits are covered in Chapter 5, and combinations of series and parallel circuits are examined in Chapter 6. In this chapter you will see how Ohm's law is used in series circuits; and you will study another important law, Kirchhoff's voltage law. Also, important system applications of series circuits are presented.

VISIT THE WEBSITE
Study aids for this chapter are available at
http://pearsonhighered.com/floyd

When connected in series, resistors form a "string" in which there is only one path for current.

After completing this section, you should be able to

- Identify a series resistive circuit
  - Translate a physical arrangement of resistors into a schematic diagram

Figure 4–1(a) shows two resistors connected in series between point *A* and point *B*. Part (b) shows three resistors in series, and part (c) shows four in series. Of course, there can be any number of resistors in series.

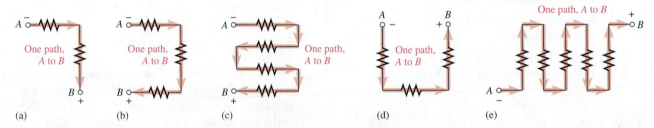

FIGURE 4–1   Resistors in series.

When a voltage source is connected from point *A* to point *B*, the only way for current to get from one point to the other in any of the connections of Figure 4–1 is to go through each of the resistors. A series circuit is identified as follows:

> A **series** circuit provides only one path for current between two points so that the current is the same through each series resistor.

In an actual circuit diagram, a series circuit may not always be as easy to identify as those in Figure 4–1. For example, Figure 4–2 shows series resistors drawn in other ways with a voltage applied. Remember, if there is only one current path between two points, the resistors between those two points are in series, no matter how they appear in a diagram.

FIGURE 4–2   Some examples of series connections of resistors. Notice that the current must be the same at all points because the current has only one path.

EXAMPLE 4–1

Five resistors are positioned on a protoboard as shown in Figure 4–3. Wire them together in series so that, starting from the negative (−) terminal, R_1 is first, R_2 is second, R_3 is third, and so on. Draw a schematic showing this connection.

FIGURE 4–3

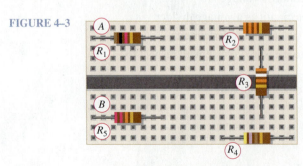

SOLUTION

The wires are connected as shown in Figure 4–4(a), which is the assembly diagram. The schematic is shown in Figure 4–4(b). Note that the schematic does not necessarily show the actual physical arrangement of the resistors as does the assembly diagram. The schematic shows how components are connected electrically; the assembly diagram shows how components are arranged and interconnected physically.

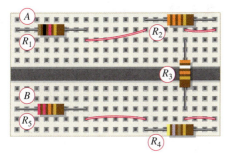

(a) Assembly diagram

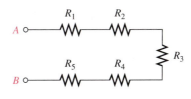

(b) Schematic

FIGURE 4–4

RELATED PROBLEM*

(a) Show how you would rewire the circuit board in Figure 4–4(a) so that all the odd-numbered resistors come first starting at point A followed by the even-numbered ones.

(b) Determine the value of each resistor.

Answers are at the end of the chapter.

EXAMPLE 4–2

Describe how the resistors on the printed circuit (PC) board in Figure 4–5 are related electrically. Determine the resistance value of each resistor.

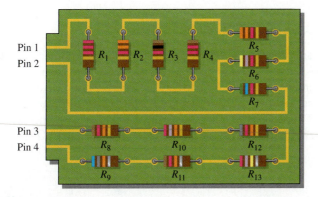

FIGURE 4–5

SOLUTION

Resistors R_1 through R_7 are in series with each other. This series combination is connected between pins 1 and 2 on the PC board.

Resistors R_8 through R_{13} are in series with each other. This series combination is connected between pins 3 and 4 on the PC board.

The values of the resistors are $R_1 = 2.2\,\text{k}\Omega$, $R_2 = 3.3\,\text{k}\Omega$, $R_3 = 1.0\,\text{k}\Omega$, $R_4 = 1.2\,\text{k}\Omega$, $R_5 = 3.3\,\text{k}\Omega$, $R_6 = 4.7\,\text{k}\Omega$, $R_7 = 5.6\,\text{k}\Omega$, $R_8 = 12\,\text{k}\Omega$, $R_9 = 68\,\text{k}\Omega$, $R_{10} = 27\,\text{k}\Omega$, $R_{11} = 12\,\text{k}\Omega$, $R_{12} = 82\,\text{k}\Omega$, and $R_{13} = 270\,\text{k}\Omega$.

RELATED PROBLEM

How is the circuit in Figure 4–5 changed when pin 2 and pin 3 are connected?

SECTION 4–1 CHECKUP*

1. How are the resistors connected in a series circuit?

2. How can you identify a series circuit?

3. Complete the schematics for the circuits in each part of Figure 4–6 by connecting the resistors in series in numerical order from A to B.

4. Now connect each group of series resistors in Figure 4–6 in series.

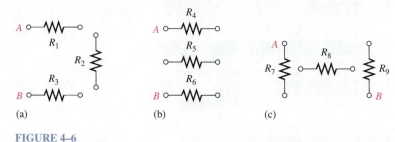

(a)   (b)   (c)

FIGURE 4–6

*Answers are at the end of the chapter.

4–2   TOTAL SERIES RESISTANCE

The total resistance of a series circuit is equal to the sum of the resistances of each individual resistor.

After completing this section, you should be able to

- **Determine total series resistance**
  - **Explain why resistance values add when resistors are connected in series**
  - **Apply the series resistance formula**

Series Resistor Values Add

When resistors are connected in series, the resistor values add because each resistor offers opposition to the current in direct proportion to its resistance. A greater number of resistors connected in series creates more opposition to current. More opposition to current implies a higher value of resistance. Thus, for every resistor that is added in series, the total resistance increases.

Figure 4–7 illustrates how series resistances add together to increase the total resistance. Part (a) has a single 10 Ω resistor. Part (b) shows another 10 Ω resistor connected in

FIGURE 4–7 Total resistance increases with each additional series resistor. The ground symbol used here was introduced in Section 2–6.

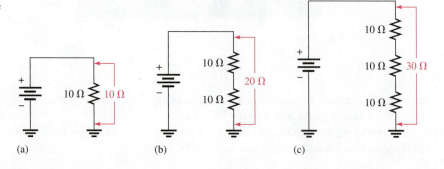

(a)   (b)   (c)

series with the first one, making a total resistance of 20 Ω. If a third 10 Ω resistor is connected in series with the first two, as shown in part (c), the total resistance becomes 30 Ω.

Series Resistance Formula

For any number of individual resistors connected in series, the total resistance is the sum of each of the individual values.

$$R_T = R_1 + R_2 + R_3 + \cdots + R_n \tag{4–1}$$

where R_T is the total resistance and R_n is the last resistor in the series string (n can be any positive integer equal to the number of resistors in series). For example, if there are four resistors in series ($n = 4$), the total resistance formula is

$$R_T = R_1 + R_2 + R_3 + R_4$$

If there are six resistors in series ($n = 6$), the total resistance formula is

$$R_T = R_1 + R_2 + R_3 + R_4 + R_5 + R_6$$

To illustrate the calculation of total series resistance, let's determine R_T of the circuit of Figure 4–8, where V_S is the source voltage. This circuit has five resistors in series. To find the total resistance, simply add the values.

$$R_T = 56\ \Omega + 100\ \Omega + 27\ \Omega + 10\ \Omega + 47\ \Omega = 240\ \Omega$$

Note in Figure 4–8 that the order in which the resistances are added does not matter. Also, you can physically change the positions of the resistors in the circuit without affecting the total resistance or the current.

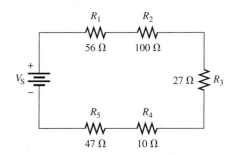

FIGURE 4–8 Example of five resistors in series. V_S stands for source voltage.

EXAMPLE 4–3

Connect the resistors on the protoboard in Figure 4–9 in series, and determine the total resistance, R_T, from the color codes.

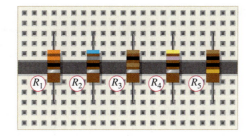

FIGURE 4–9

SOLUTION

The resistors are connected as shown in Figure 4–10. Find the total resistance by adding all the values.

$$R_T = R_1 + R_2 + R_3 + R_4 + R_5 = 33\ \Omega + 68\ \Omega + 100\ \Omega + 47\ \Omega + 10\ \Omega = \mathbf{258\ \Omega}$$

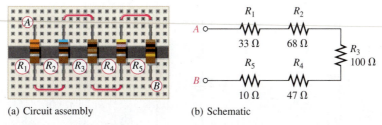

(a) Circuit assembly   (b) Schematic

FIGURE 4–10

RELATED PROBLEM

Determine the total resistance in Figure 4–10(a) if the positions of R_2 and R_4 are interchanged.

EXAMPLE 4–4

Calculate the total resistance, R_T, for each circuit in Figure 4–11.

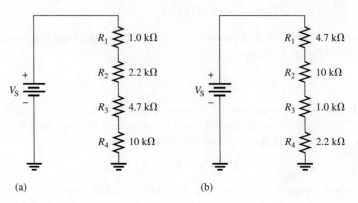

(a)   (b)

FIGURE 4–11

SOLUTION

For circuit (a),

$$R_T = 1.0\,k\Omega + 2.2\,k\Omega + 4.7\,k\Omega + 10\,k\Omega = \mathbf{17.9\,k\Omega}$$

For circuit (b),

$$R_T = 4.7\,k\Omega + 10\,k\Omega + 1.0\,k\Omega + 2.2\,k\Omega = \mathbf{17.9\,k\Omega}$$

Notice that the total resistance does not depend on the position of the resistors. Both circuits are identical in terms of total resistance.

RELATED PROBLEM

What is the total resistance for the following series resistors: $1.0\,k\Omega$, $2.2\,k\Omega$, $3.3\,k\Omega$, and $5.6\,k\Omega$?

EXAMPLE 4–5

Determine the value of R_4 in the circuit of Figure 4–12.

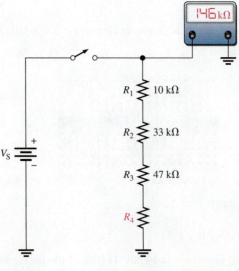

FIGURE 4–12

SOLUTION

From the ohmmeter reading, $R_T = 146\,\text{k}\Omega$.

$$R_T = R_1 + R_2 + R_3 + R_4$$

Solving for R_4 yields

$$R_4 = R_T - (R_1 + R_2 + R_3) = 146\,\text{k}\Omega - (10\,\text{k}\Omega + 33\,\text{k}\Omega + 47\,\text{k}\Omega) = \mathbf{56\,k\Omega}$$

RELATED PROBLEM

Determine the value of R_4 in Figure 4–12 if the ohmmeter reading is $112\,\text{k}\Omega$.

Equal-Value Series Resistors

When a circuit has more than one resistor of the same value in series, there is a shortcut method to obtain the total resistance: Simply multiply the resistance value of the resistors having the same value by the number of equal-value resistors that are in series. This method is essentially the same as adding the values. For example, five $100\,\Omega$ resistors in series have an R_T of $5(100\,\Omega) = 500\,\Omega$. In general, the formula is expressed as

$$R_T = nR \tag{4–2}$$

where n is the number of equal-value resistors and R is the resistance value.

EXAMPLE 4–6

Find the R_T of eight $22\,\Omega$ resistors in series.

SOLUTION

Find R_T by adding the values.

$$R_T = 22\,\Omega + 22\,\Omega + 22\,\Omega + 22\,\Omega + 22\,\Omega + 22\,\Omega + 22\,\Omega + 22\,\Omega$$
$$= \mathbf{176\,\Omega}$$

However, it is much easier to multiply.

$$R_T = 8(22\,\Omega) = \mathbf{176\,\Omega}$$

RELATED PROBLEM

Find R_T for three $1.0\,\text{k}\Omega$ and two $680\,\Omega$ resistors in series.

SYSTEM EXAMPLE 4–1

A MOTOR STARTER

Shunt dc motors (discussed in Section 7–7) are a type of motor that has a torque-speed characteristic that can be varied over a wide range while still retaining high efficiency. For variable loads, such as an electric hoist motor, a dc shunt motor has an advantage of allowing control of both speed and torque. Speed is controlled by voltage; torque is controlled by current.

The shunt motor has two interacting electromagnets: one is called the field coil and the other is called the rotor, which is a rotating electromagnet. If full voltage is applied without a current-limiting device to the motor, the initial current can be very high, which runs the risk of burning out the motor or overloading the line and blowing a fuse or circuit breaker. To avoid this, a shunt dc motor should be started by limiting current with resistors (or other means) until the motor begins to speed up. After that, the resistance can be dropped out to adjust speed and torque to the load requirement.

A motor starter circuit, such as the one shown in Figure 4–13, provides the means to limit current. The resistors are effectively removed as required so that there is less resistance as the rotary switch is moved clockwise. For this example, a manual rotary switch is shown, but automated methods are also used. To start the motor, the rotary switch is moved from the off position to position 1. This applies dc to the field coil and the series solenoid (an electromagnet, which keeps the switch in place). If power drops out for any reason, the solenoid releases its pull and the switch is pulled by the spring back to the off position. This prevents the motor from restarting with no resistance in place.

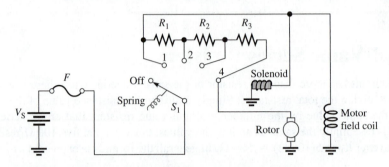

FIGURE 4–13   A manual motor starter for a dc shunt motor.

SECTION 4–2 CHECKUP

1. Calculate R_T between points A and B for each circuit in Figure 4–14.

2. The following resistors are in series: one 100 Ω, two 47 Ω, four 12 Ω, and one 330 Ω. What is the total resistance?

3. Suppose that you have one resistor for each of the following values: 1.0 kΩ, 2.7 kΩ, 3.3 kΩ, and 1.8 kΩ. To get a total resistance of 10 kΩ, you need one more resistor. What should its value be?

4. What is the R_T for twelve 47 Ω resistors in series?

FIGURE 4–14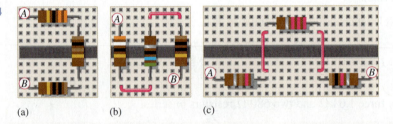

(a)   (b)   (c)

4–3 CURRENT IN A SERIES CIRCUIT

The current is the same through all points in a series circuit. The current through each resistor in a series circuit is the same as the current through all the other resistors that are in series with it.

After completing this section, you should be able to

• **Determine the current throughout a series circuit**
  • **Show that the current is the same at all points in a series circuit**

Figure 4–15 shows three resistors connected in series to a voltage source. *At any point in this circuit, the current into that point must equal the current out of that point,* as illustrated by the current directional arrows. Notice also that the current out of each of the resistors must equal the current in because there is no place where part of the current can branch off and go somewhere else. Therefore, the current in each section of the circuit is the same as the current in all other sections. It has only one path going from the negative (−) side of the source to the positive (+) side.

In Figure 4–16, the battery supplies 1.82 mA of current to the series resistors. There are 1.82 mA from the battery's negative terminal. As shown, the same current is measured at several points in the series circuit.

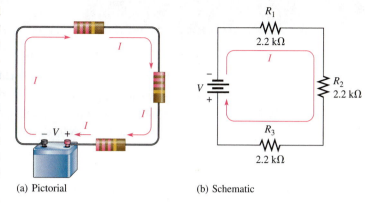

(a) Pictorial   (b) Schematic

FIGURE 4–15   Current entering any point in a series circuit is the same as the current leaving that point.

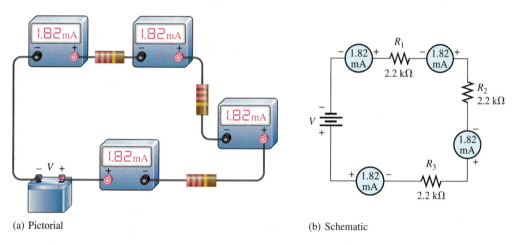

(a) Pictorial   (b) Schematic

FIGURE 4–16   Current is the same at all points in a series circuit.

SECTION 4–3 CHECKUP

1. What statement can you make about the amount of current at any point in a series circuit?

2. In a series circuit with a 100 Ω and a 47 Ω resistor, there are 20 mA through the 100 Ω resistor. How much current is there through the 47 Ω resistor?

3. A milliammeter is connected between points *A* and *B* in Figure 4–17. It measures 50 mA. If you move the meter and connect it between points *C* and *D*, how much current will it indicate? Between *E* and *F*?

4. In Figure 4–18, how much current does ammeter 1 indicate? Ammeter 2?

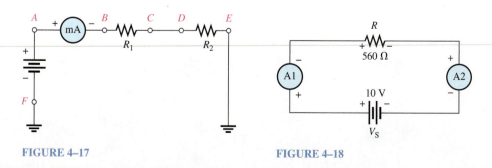

FIGURE 4–17   **FIGURE 4–18**

4–4 APPLICATION OF OHM'S LAW

The basic concepts of series circuits and Ohm's law can be applied to series circuit analysis.

After completing this section, you should be able to

- **Apply Ohm's law in series circuits**
  - **Find the current in a series circuit**
  - **Find the voltage across each resistor in series**

The following are key points to remember when you analyze series circuits:

1. Current through any of the series resistors is the same as the total current.
2. If you know the total applied voltage and the total resistance, you can determine the total current by Ohm's law.

$$I_T = \frac{V_T}{R_T}$$

3. If you know the voltage drop across one of the series resistors, R_x, you can determine the total current by Ohm's law.

$$I_T = \frac{V_x}{R_x}$$

4. If you know the total current, you can find the voltage drop across any of the series resistors by Ohm's law.

$$V_x = I_T R_x$$

5. The polarity of a voltage drop across a resistor is positive at the end of the resistor that is closest to the positive terminal of the voltage source.
6. The current through a resistor is defined to be in a direction from the negative end of the resistor to the positive end.
7. An open in a series circuit prevents current; and, therefore, there is zero voltage drop across each series resistor. The total voltage appears across the points between which there is an open.

Now let's look at several examples that use Ohm's law for series circuit analysis.

EXAMPLE 4–7

Find the current in the circuit of Figure 4–19.

FIGURE 4–19

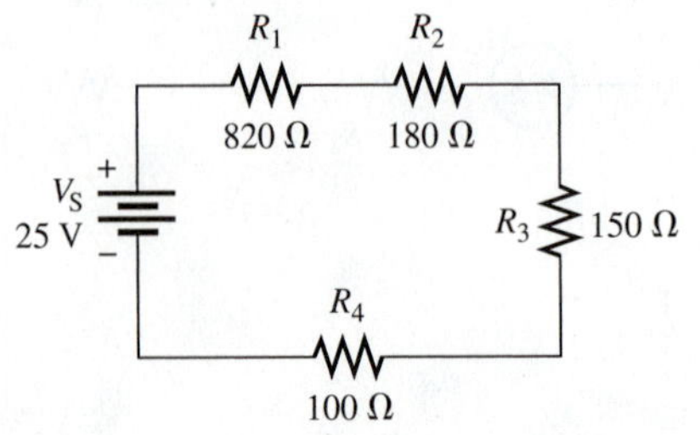

SOLUTION

The current is determined by the source voltage and the total resistance. First, calculate the total resistance.

$$R_T = R_1 + R_2 + R_3 + R_4 = 820 \ \Omega + 180 \ \Omega + 150 \ \Omega + 100 \ \Omega = 1.25 \ k\Omega$$

Next, use Ohm's law to calculate the current.

$$I = \frac{V_S}{R_T} = \frac{25\text{ V}}{1.25\text{ k}\Omega} = \mathbf{20\text{ mA}}$$

Remember, the same current exists at all points in the circuit. Thus, each resistor has 20 mA through it.

RELATED PROBLEM

What is the current in the circuit of Figure 4–19 if R_4 is changed to 200 Ω?

EXAMPLE 4–8

The current in the circuit of Figure 4–20 is 1 mA. For this amount of current, what must the source voltage V_S be?

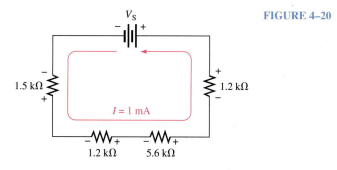

FIGURE 4–20

SOLUTION

In order to calculate V_S, first determine R_T.

$$R_T = 1.2\text{ k}\Omega + 5.6\text{ k}\Omega + 1.2\text{ k}\Omega + 1.5\text{ k}\Omega = 9.5\text{ k}\Omega$$

Now use Ohm's law to determine V_S.

$$V_S = IR_T = (1\text{ mA})(9.5\text{ k}\Omega) = \mathbf{9.5\text{ V}}$$

RELATED PROBLEM

If the 5.6 kΩ resistor is changed to 3.9 kΩ, what value of V_S is necessary to keep the current at 1 mA?

EXAMPLE 4–9

Calculate the voltage drop across each resistor in Figure 4–21, and find the value of V_S. To what value can V_S be raised if the current is to be limited to 5 mA?

FIGURE 4–21

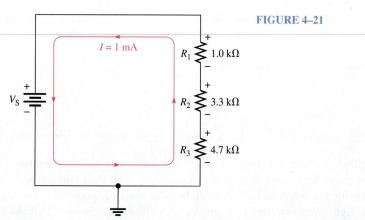

SOLUTION

By Ohm's law, the voltage drop across each resistor is equal to its resistance multiplied by the current through it. Use the Ohm's law formula $V = IR$ to determine the voltage drop across each of the resistors. Keep in mind that there is the same current through each series resistor. The voltage drop across R_1 (designated V_1) is

$$V_1 = IR_1 = (1 \text{ mA})(1.0 \text{ k}\Omega) = \textbf{1 V}$$

The voltage drop across R_2 is

$$V_2 = IR_2 = (1 \text{ mA})(3.3 \text{ k}\Omega) = \textbf{3.3 V}$$

The voltage drop across R_3 is

$$V_3 = IR_3 = (1 \text{ mA})(4.7 \text{ k}\Omega) = \textbf{4.7 V}$$

To calculate the value of V_S, first determine the total resistance.

$$R_T = 1.0 \text{ k}\Omega + 3.3 \text{ k}\Omega + 4.7 \text{ k}\Omega = 9 \text{ k}\Omega$$

The source voltage V_S is equal to the current times the total resistance.

$$V_S = IR_T = (1 \text{ mA})(9 \text{ k}\Omega) = \textbf{9 V}$$

Notice that if you add the voltage drops of the resistors, they total 9 V, which is the same as the source voltage.

V_S can be increased to a maximum value where $I = 5$ mA. Therefore the maximum value of V_S is

$$V_{S(max)} = IR_T = (5 \text{ mA})(9 \text{ k}\Omega) = \textbf{45 V}$$

RELATED PROBLEM

Repeat the calculations for V_1, V_2, V_3, V_S, and $V_{S(max)}$ if $R_3 = 2.2 \text{ k}\Omega$ and I is maintained at 1 mA.

MULTISIM

Open Multisim file E04-09. Using the multimeter, verify that the voltages across the resistors agree with the values calculated in this example.

Indicator Lights

Indicator lights are a common way to ascertain a specific condition of a machine or relay, such as a motor starter. Incandescent indicator lights are available for common voltages used in industrial plants (120 V, 240 V, 480 V, and 600 V). In small systems, low voltages are common, and LEDs are a good choice for indicating a condition (such as a power supply is *on*). A typical indicating LED will drop about 2 V and has an *on* current of 20 mA. To prevent current from exceeding the specification, a current-limiting resistor is used in series (in-line) with the LED. Example 4–10 shows how to calculate the size of current limiting resistors when a variable brightness is needed to account for different ambient conditions.

(yukosourov/Fotolia.com)

SYSTEM NOTE

EXAMPLE 4–10

LEDs are commonly used in systems to indicate a certain condition. The LED indicators have specific current requirements. Too little current, and the LED is too dim; too much and it burns out. In some systems, a control is needed to adjust the light intensity based on ambient conditions. Figure 4–22 shows this basic application, where the red LED is used as a system indicator light. In this case, a rheostat and series fixed resistor limit current to the LED. The rheostat is included to dim the LED, depending on ambient conditions. We will focus on just these two current-limiting resistors.

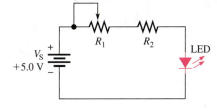

FIGURE 4–22

A red LED will always have a voltage across it of about +1.7 V when it is on and working within its normal operating range. The remaining voltage from the power supply will be across the two series resistors. Together, the rheostat and the fixed resistor will have a total of 3.3 V across them.

Assume you want current in the LED to range from a minimum of 2.5 mA (dim) to a maximum of 10 mA (bright). What values of R_1 and R_2 would you choose to accomplish this?

SOLUTION

Start with the brightest condition when the resistance of the rheostat is adjusted for 0 Ω. In this case, there will be no voltage across R_1 and 3.3 V will be across R_2. Because it is a series circuit, the same current is in R_2 as the LED. Therefore,

$$R_2 = \frac{V}{I} = \frac{3.3\text{ V}}{10\text{ mA}} = \textbf{330 }\boldsymbol{\Omega}$$

Now determine the total resistance required to limit the current to 2.5 mA. The total resistance is $R_T = R_1 + R_2$, and the voltage drop across R_T is 3.3 V. From Ohm's law,

$$R_T = \frac{V}{I} = \frac{3.3\text{ V}}{2.5\text{ mA}} = 1.32\text{ k}\Omega$$

To find R_1, subtract the value of R_2 from the total resistance.

$$R_1 = R_T - R_2 = 1.32\text{ k}\Omega - 330\ \Omega = 990\ \Omega$$

Choose a **1.0 kΩ** rheostat as the nearest standard value.

RELATED PROBLEM

What is the value of R_2 if the highest current is 12 mA?

SECTION 4–4 CHECKUP

1. A 6 V battery is connected across three 100 Ω resistors in series. What is the current through each resistor?

2. How much voltage is required to produce 5 mA through the circuit of Figure 4–23?

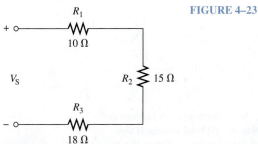

FIGURE 4–23

3. How much voltage is dropped across each resistor in Figure 4–23 when the current is 5 mA?

4. Four equal-value resistors are connected in series with a 5 V source. The measured current is 4.63 mA. What is the value of each resistor?

5. What size should a series current-limiting resistor be to limit the current to a red LED to 10 mA from a 3.0 V source? Assume the LED drops 1.7 V.

4–5 VOLTAGE SOURCES IN SERIES

Recall that a voltage source is an energy source that provides a constant voltage to a load. Batteries and power supplies are practical examples of dc voltage sources. When two or more voltage sources are in series, the total voltage is equal to the algebraic sum of the individual source voltages.

After completing this section, you should be able to

- Determine the total effect of voltage sources connected in series
  - Determine the total voltage of series sources with the same polarities
  - Determine the total voltage of series sources with opposite polarities

When batteries are placed in a flashlight, they are connected in a **series-aiding** arrangement to produce a larger voltage, as illustrated in Figure 4–24. In this example, three 1.5 V batteries are placed in series to produce a total voltage ($V_{S(tot)}$).

$$V_{S(tot)} = V_{S1} + V_{S2} + V_{S3} = 1.5\ V + 1.5\ V + 1.5\ V = 4.5\ V$$

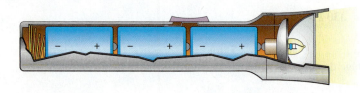

(a) Flashlight with series batteries

(b) Schematic of flashlight circuit

FIGURE 4–24   Example of series-aiding voltage sources.

Series voltage sources (batteries in this case) are added when their polarities are in the same direction, or series-aiding, and are subtracted when their polarities are in opposite directions, or **series-opposing**. For example, if one of the batteries in the flashlight is turned around, as indicated in the schematic of Figure 4–25, its voltage subtracts because it has a negative value and reduces the total voltage.

$$V_{S(tot)} = V_{S1} - V_{S2} + V_{S3} = 1.5\ V - 1.5\ V + 1.5\ V = 1.5\ V$$

There is no valid reason for reversing a battery, and it can lead to high current and reduced life if it happens accidentally. However, an opposing source does occur naturally in the case of motors. An opposing voltage to the source voltage is generated within the motor, which reduces the current as you will see in Section 7–7.

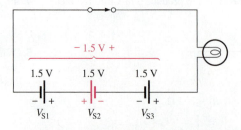

FIGURE 4–25   When batteries are connected in opposite directions, the total voltage is the algebraic sum of the voltages. As noted, this is not a valid configuration for batteries.

EXAMPLE 4–11

What is the total source voltage ($V_{S(tot)}$) in Figure 4–26?

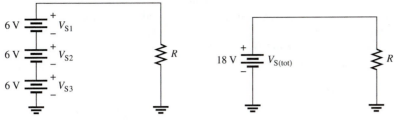

FIGURE 4–26   FIGURE 4–27

SOLUTION

The polarity of each source is the same (the sources are connected in the same direction in the circuit). Therefore, add the three voltages to get the total.

$$V_{S(tot)} = V_{S1} + V_{S2} + V_{S3} = 6\,V + 6\,V + 6\,V = \mathbf{18\,V}$$

The three individual sources can be replaced by a single equivalent source of 18 V with its polarity as shown in Figure 4–27.

RELATED PROBLEM

If the battery V_{S3} in Figure 4–26 is accidentally installed backwards, what is the total voltage?

MULTISIM

Open Multisim file E04-11.
Verify the total source voltage.
Repeat for the related problem.

SYSTEM EXAMPLE 4–2

SOLAR PANELS IN SERIES

Solar panels (modules) are composed of many solar cells in an array, wired to produce a certain voltage in full sunlight. Within the module, each solar cell produces only about 0.5 V in full sun. Therefore, an 18 V panel requires at least 36 solar cells wired in series, and a 24 V panel requires at least 48 solar cells in series. For battery charging, this is sufficient voltage for most applications. In general, module voltages range from 16.5 V to 72.3 V as specified under standard test conditions (STC), which allow comparisons between modules. Actual operating conditions are different.

Figure 4–28 is a block diagram of a basic solar electric system that is independent of the utility grid (in this case). It has a battery bank for storing energy when there is no solar

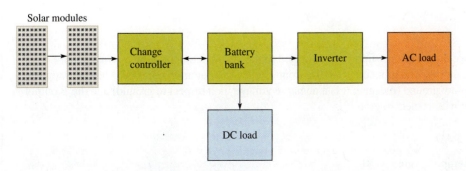

FIGURE 4–28   **Typical small solar electric system.**

input. The solar modules convert sunlight to dc electrical power, which is sent to a charge controller. The charge controller regulates charging current to prevent overcharging the batteries or draining the batteries back through the solar modules when they are needed for the load. The inverter converts the dc output to ac for powering small appliances. The inverter is a basic one in this system rather than a more expensive grid-tied inverter. Grid-tied inverters are special inverters that are required for any renewable energy system that is connected to a utility grid.

For most systems, a number of modules are connected together to provide a high voltage dc for converting to ac. The voltage rating of the module and the inverter voltage ratings determine how many modules can be put in series. All components in the system (including modules, wiring, and inverters) must be rated for the maximum system voltage. In the United States, the maximum permitted voltage is 600 V for most installations.

For this example, assume a single panel has a maximum voltage of 34.6 V and is rated for 250 W. The number of these panels that can be installed in series depends on the maximum permitted voltage. For a permitted voltage of 600 V, 17 panels in series is the maximum that can be connected without exceeding 600 V. In this case, the maximum voltage will be 588 V. The advantage of the series connection for these panels is that the higher voltage is more efficient and can use smaller wire because the current is the same going into and from each panel. The maximum rated power from the array is 17 panels $\times$ 250 W/panel = 4250 W.

EXAMPLE 4–12

Many circuits use positive and negative supply voltages. A dual-power supply will normally have two independent outputs, such as those shown in Figure 4–29. Show how to connect the two 12 V outputs from the power supply so that there is both a positive and a negative output and the two sources are series-aiding.

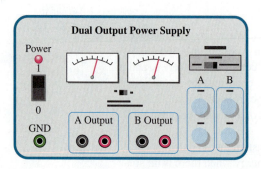

FIGURE 4–29

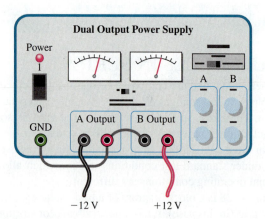

FIGURE 4–30

SOLUTION

See Figure 4–30. The positive output of one supply is connected to the negative output of the second supply. This makes the connection series-aiding. The ground terminal is connected to the common point between the supplies, forcing the A output to be below ground (meaning it is a negative voltage with respect to ground) and the B output to be above ground (positive with respect to ground).

RELATED PROBLEM

Draw the schematic of the setup in Figure 4–30.

SECTION 4–5 CHECKUP

1. How many 12 V batteries must be connected in series to produce 60 V? Draw a schematic that shows the battery connections.

2. Four 1.5 V flashlight batteries are connected in series-aiding. What is the total voltage across the bulb?

3. The resistive circuit in Figure 4–31 is used to bias a transistor amplifier. Show how to connect the resistors to two 15 V power supplies in order to get 30 V across the series resistors.

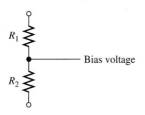

FIGURE 4–31

4. Determine the total source voltage for the circuit in Figure 4–32.

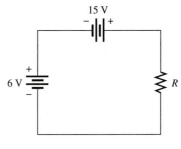

FIGURE 4–32

5. Assume that in a four-cell flashlight one of the four 1.5 V batteries was accidentally installed in the wrong direction. In this case, what voltage would be across the bulb when the light was turned on?

4–6 KIRCHHOFF'S VOLTAGE LAW

Kirchhoff's voltage law is a fundamental circuit law that states that the algebraic sum of all the voltages around a single closed path is zero or, in other words, the sum of the voltage drops equals the total source voltage.

After completing this section, you should be able to

- **Apply Kirchhoff's voltage law**
  - **State Kirchhoff's voltage law**
  - **Determine the source voltage by adding the voltage drops**
  - **Determine an unknown voltage drop**

In an electric circuit, the voltages across the resistors (voltage drops) *always* have polarities opposite to the source voltage polarity. For example, in Figure 4–33, follow a counterclockwise loop around the circuit. Note that the source polarity is plus-to-minus and each voltage drop is minus-to-plus.

In Figure 4–33, the current is out of the negative side of the source and through the resistors as the arrows indicate. The current is into the negative side of each resistor and out the positive side. As you learned in Chapter 3, when electrons flow through a resistor, they lose energy and are therefore at a lower energy level when they emerge. The lower energy side is less negative (more positive) than the higher energy side. The drop in energy level across a resistor creates a potential difference, or voltage drop, with a minus-to-plus polarity in the direction of the current.

The voltage from point A to point B in the circuit of Figure 4–33 is the source voltage, V_S. Also, the voltage from A to B is the sum of the series resistor voltage drops. Therefore, the source voltage is equal to the sum of the three voltage drops, as stated by Kirchhoff's voltage law.

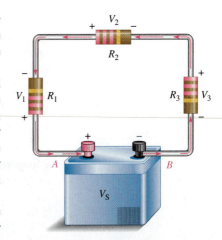

FIGURE 4–33 Illustration of voltage polarities in a closed-loop circuit.

The sum of all the voltage drops around a single closed path in a circuit is equal to the total source voltage in that closed path.

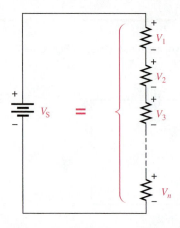

FIGURE 4–34   Sum of *n* voltage drops equals the source voltage.

Kirchhoff's voltage law applied to a series circuit is illustrated in Figure 4–34. For this case, Kirchhoff's voltage law can be expressed by the equation

$$V_S = V_1 + V_2 + V_3 + \cdots + V_n \tag{4–3}$$

where the subscript *n* represents the number of voltage drops.

If all the voltage drops around a closed path are added and then this total is subtracted from the source voltage, the result is zero. This result occurs because the sum of the voltage drops in a series circuit always equals the source voltage.

Equation 4–3 can be applied to any circuit that has a voltage source and resistors or other loads. The key idea is that a single closed path is followed from an arbitrary starting point and back to this point. In a series circuit, the closed path will always include a voltage source and one or more resistors or other loads. In this case, the source represents a voltage *rise,* and each load represents a voltage *drop.* Another way of stating Kirchhoff's voltage law for a series circuit is that the sum of all voltage rises is equal to the sum of all voltage drops.

You can verify Kirchhoff's voltage law in a series circuit by connecting the circuit and measuring each resistor voltage and the source voltage, as illustrated in Figure 4–35. When the resistor voltages are added together, their sum will equal the source voltage. Any number of resistors can be used.

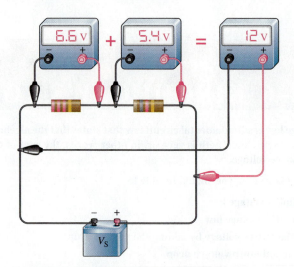

FIGURE 4–35   Illustration of a verification of Kirchhoff's voltage law.

Kirchhoff's voltage law was developed here for a series circuit, but the concept can be applied to any circuit. In complex circuits, you can still write Kirchhoff's voltage law, but there are cases where there is no source voltage in a given closed loop. Even so, Kirchhoff's voltage law still applies. This leads to a more general statement of Kirchhoff's voltage law:

> **The algebraic sum of the voltages around any closed path in a circuit is equal to zero.**

If a voltage source is present, it is just treated as one of the terms in the summation. It is important to assign the correct algebraic sign to the terms, noting if the voltage is rising or dropping as one traverses the loop. In equation form,

$$V_1 + V_2 + V_3 + \cdots + V_n = 0 \tag{4–4}$$

Any of the variables in Equation 4–4 can represent a voltage rise or a voltage drop. This more-general form of Kirchhoff's voltage law can be expressed more concisely as

$$\sum_{i=1}^{n} V_i = 0$$

This equation is a shorthand way of writing Equation 4–4. The capital sigma (Σ) means to add the voltages from the first ($i = 1$) to the last ($i = n$).

This form of Kirchhoff's voltage law can be applied to circuits other than the series circuits, provided you follow a single closed path. In applying Equation 4–4, you need to assign an algebraic sign to each voltage in the path. In circuits other than series circuits, a voltage across a resistor can appear as either a rise or a drop, depending on the path chosen. As you traverse the path, you will need to write voltage rises and voltage drops in a consistent manner. For series circuits such as given in Figure 4–34, the main idea is that the source voltage (rise) is equal to the sum of the voltages (drops) across the loads (resistors).

EXAMPLE 4–13

Determine the source voltage, V_S, in Figure 4–36 where the two voltage drops are given.

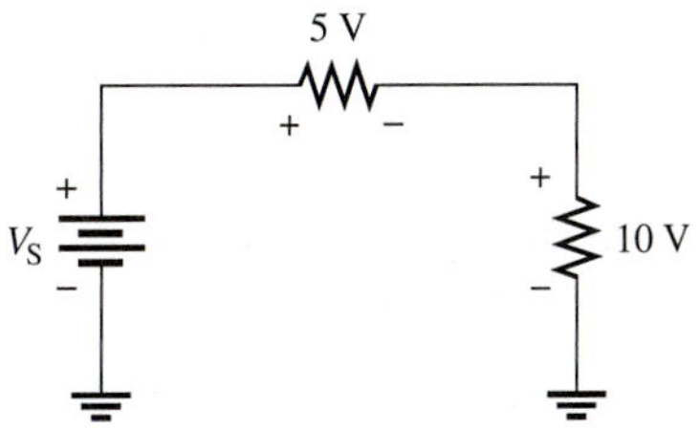

SOLUTION

By Kirchhoff's voltage law (Eq. 4–3), the source voltage (applied voltage) must equal the sum of the voltage drops. Adding the voltage drops gives the value of the source voltage.

$$V_S = 5\text{ V} + 10\text{ V} = \textbf{15 V}$$

FIGURE 4–36

RELATED PROBLEM

If V_S is increased to 30 V in Figure 4–36, what are the two voltage drops?

MULTISIM

Open Multisim file E04-13. Verify that the voltage drops equal to V_S. Repeat for the related problem.

EXAMPLE 4–14

Determine the unknown voltage drop, V_3, in Figure 4–37.

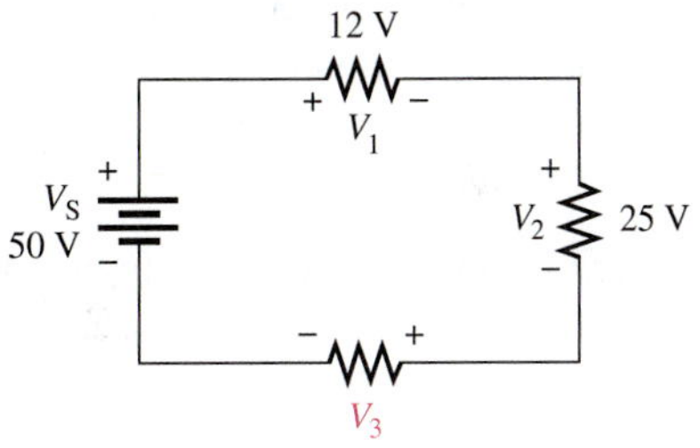

SOLUTION

By Kirchhoff's voltage law (Equation 4–4), the algebraic sum of the voltages around the circuit is zero.

$$V_1 + V_2 + V_3 + V_4 = 0$$

FIGURE 4–37

Substitute $-V_S$ for V_4. Then,

$$V_1 + V_2 + V_3 - V_S = 0$$

Solve for V_3.

$$V_3 = V_S - V_1 - V_2 = 50\text{ V} - 12\text{ V} - 25\text{ V} = \textbf{13 V}$$

The voltage drop, V_3, across R_3 is 13 V, and its polarity is as shown in Figure 4–37.

RELATED PROBLEM

Determine V_3 in Figure 4–37 if the source is changed to 25 V.

MULTISIM

Open Multisim file E04-14. Verify that V_3 agrees with the calculated value. Repeat for the related problem.

EXAMPLE 4–15

Find the value of R_4 in Figure 4–38.

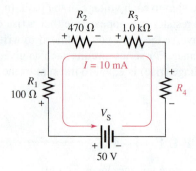

FIGURE 4–38

SOLUTION

In this problem you will use both Ohm's law and Kirchhoff's voltage law. First, use Ohm's law to find the voltage drop across each of the known resistors.

$$V_1 = IR_1 = (10\,\text{mA})(100\,\Omega) = 1.0\,\text{V}$$
$$V_2 = IR_2 = (10\,\text{mA})(470\,\Omega) = 4.7\,\text{V}$$
$$V_3 = IR_3 = (10\,\text{mA})(1.0\,\text{k}\Omega) = 10\,\text{V}$$

Next, use Kirchhoff's voltage law to find V_4, the voltage drop across the unknown resistor.

$$V_S - V_1 - V_2 - V_3 - V_4 = 0\,\text{V}$$
$$50\,\text{V} - 1.0\,\text{V} - 4.7\,\text{V} - 10\,\text{V} - V_4 = 0\,\text{V}$$
$$34.3\,\text{V} - V_4 = 0\,\text{V}$$
$$V_4 = 34.3\,\text{V}$$

Now that you know V_4, you can use Ohm's law to calculate R_4.

$$R_4 = \frac{V_4}{I} = \frac{34.3\,\text{V}}{10\,\text{mA}} = \mathbf{3.43\,k\Omega}$$

R_4 is most likely a $3.3\,\text{k}\Omega$ standard-value resistor because $3.43\,\text{k}\Omega$ is within a standard tolerance range ($\pm 5\%$) of $3.3\,\text{k}\Omega$.

RELATED PROBLEM

Determine the value of R_4 in Figure 4–38 if $V_S = 20\,\text{V}$ and $I = 10\,\text{mA}$.

MULTISIM

Open Multisim file E04-15. Verify that the calculated value of R_4 produces the current in Figure 4–38. Repeat for the related problem.

SECTION 4–6 CHECKUP

1. State Kirchhoff's voltage law in two ways.

2. A 50 V source is connected to a series resistive circuit. What is the sum of the voltage drops in this circuit?

3. Two equal-value resistors are connected in series across a 10 V battery. What is the voltage drop across each resistor?

4. In a series circuit with a 25 V source, there are three resistors. One voltage drop is 5 V, and the other is 10 V. What is the value of the third voltage drop?

5. The individual voltage drops in a series circuit are 1 V, 3 V, 5 V, 7 V, and 8 V. What is the total voltage applied across the series circuit?

A series circuit acts as a voltage divider. The voltage divider is an important application of series circuits.

After completing this section, you should be able to

- **Use a series circuit as a voltage divider**
  - **Apply the voltage-divider formula**
  - **Use the potentiometer as an adjustable voltage divider**
  - **Describe some voltage-divider applications**

A circuit consisting of a series string of resistors connected to a voltage source acts as a **voltage divider**. Figure 4–39(a) shows a circuit with two resistors in series, although there can be any number. As you already know, there are two voltage drops: one across R_1 and one across R_2. These voltage drops are labeled V_1 and V_2, respectively, as indicated in the schematic. Since each resistor has the same current through it, the voltage drops are proportional to the resistance values. For example, if the value of R_2 is twice that of R_1, then the value of V_2 is twice that of V_1.

The total voltage drop around a single closed path divides among the series resistors in amounts directly proportional to the resistance values. The smallest resistance has the least voltage, and the largest resistance has the most voltage ($V = IR$). For example, in Figure 4–39(b), if V_S is 10 V, R_1 is 100 Ω, and R_2 is 200 Ω, then V_1 is one-third the total voltage, or 3.33 V, because R_1 is one-third the total resistance. Likewise, V_2 is two-thirds V_S, or 6.67 V, because R_2 is two-thirds the total resistance.

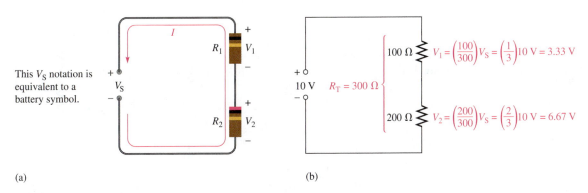

This V_S notation is equivalent to a battery symbol.

(a)

(b)

FIGURE 4–39 Example of a two-resistor voltage divider.

Voltage-Divider Formula

With a few calculations, you can develop a formula for determining how the voltages divide among series resistors. Assume a circuit with n resistors in series as shown in Figure 4–40, where n can be any number.

Let V_x represent the voltage drop across any one of the resistors and R_x represent the number of a particular resistor or combination of resistors. By Ohm's law, you can express the voltage drop across R_x as follows:

$$V_x = IR_x$$

The current through the circuit is equal to the source voltage divided by the total resistance ($I = V_S/R_T$). In the circuit of Figure 4–40, the total resistance is

$$R_T = R_1 + R_2 + R_3 + \cdots + R_n$$

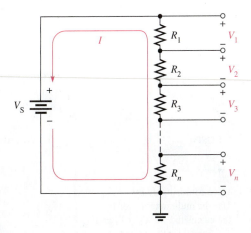

FIGURE 4–40 Generalized voltage divider with n resistors.

By substitution of V_S/R_T for I in the expression for V_x,

$$V_x = \left(\frac{V_S}{R_T}\right)R_x$$

Rearranging the terms you get

$$V_x = \left(\frac{R_x}{R_T}\right)V_S \tag{4–5}$$

Equation 4–5 is the general voltage-divider formula, which can be stated as follows:

The voltage drop across any resistor or combination of resistors in a series circuit is equal to the ratio of that resistance value to the total resistance, multiplied by the source voltage.

Flash Converters

In many systems, an analog signal is converted to digital for processing. This process uses an analog-to-digital converter (ADC). For applications where speed is critical, a flash converter is the fastest type of ADC.

The standard flash converter uses a large voltage-divider string composed of hundreds of resistors to set up a series of reference voltages. For example, an 8-bit flash ADC uses 255 resistors in a divider string. The resistors must be very accurate, so they are carefully controlled within the integrated circuit. The input analog signal is compared to all of the different voltages simultaneously, and the level closest to the signal level is digitized.

SYSTEM NOTE

EXAMPLE 4–16

Determine V_1 (the voltage across R_1) and V_2 (the voltage across R_2) in the circuit in Figure 4–41.

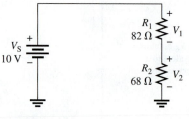

FIGURE 4–41

SOLUTION

To determine V_1, use the voltage-divider formula, $V_x = (R_x/R_T)V_S$, where $x = 1$. The total resistance is

$$R_T = R_1 + R_2 = 82\ \Omega + 68\ \Omega = 150\ \Omega$$

R_1 is 82 Ω and V_S is 10 V. Substitute these values into the voltage-divider formula.

$$V_1 = \left(\frac{R_1}{R_T}\right)V_S = \left(\frac{82\ \Omega}{150\ \Omega}\right)10\ \text{V} = \textbf{5.47 V}$$

There are two ways to find the value of V_2: Kirchhoff's voltage law or the voltage-divider formula. If you use Kirchhoff's voltage law ($V_S = V_1 + V_2$), substitute the values for V_S and V_1 and solve for V_2.

$$V_2 = V_S - V_1 = 10\ \text{V} - 5.47\ \text{V} = \textbf{4.53 V}$$

To determine V_2, use the voltage-divider formula where $x = 2$.

$$V_2 = \left(\frac{R_2}{R_T}\right)V_S = \left(\frac{68\ \Omega}{150\ \Omega}\right)10\ \text{V} = \textbf{4.53 V}$$

RELATED PROBLEM

Find the voltage drops across R_1 and R_2 if R_2 is changed to 180 Ω in Figure 4–41.

MULTISIM

Open Multisim file E04-16. Use the multimeter to verify the calculated values of V_1 and V_2. Repeat for the related problem.

EXAMPLE 4–17

Calculate the voltage drop across each resistor in the circuit of Figure 4–42.

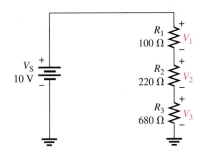

SOLUTION

Look at the circuit and consider the following: The total resistance is 1000 Ω. Ten percent of the total voltage is across R_1 because it is 10% of the total resistance (100 Ω is 10% of 1000 Ω). Likewise, 22% of the total voltage is dropped across R_2 because it is 22% of the total resistance (220 Ω is 22% of 1000 Ω).

FIGURE 4–42

Finally, R_3 drops 68% of the total voltage (680 Ω is 68% of 1000 Ω).

Because of the convenient values in this problem, it is easy to mentally calculate the voltage drops ($V_1 = 0.10 \times 10$ V $= 1$ V, $V_2 = 0.22 \times 10$ V $= 2.2$ V, and $V_3 = 0.68 \times 10$ V $= 6.8$ V). Such is usually not the case, but sometimes a little thinking will produce a result more efficiently.

Although you have already reasoned through this problem, the calculations are

$$V_1 = \left(\frac{R_1}{R_T}\right)V_S = \left(\frac{100\ \Omega}{1000\ \Omega}\right)10\ \text{V} = \mathbf{1\ V}$$

$$V_2 = \left(\frac{R_2}{R_T}\right)V_S = \left(\frac{220\ \Omega}{1000\ \Omega}\right)10\ \text{V} = \mathbf{2.2\ V}$$

$$V_3 = \left(\frac{R_3}{R_T}\right)V_S = \left(\frac{680\ \Omega}{1000\ \Omega}\right)10\ \text{V} = \mathbf{6.8\ V}$$

Notice that the sum of the voltage drops is equal to the source voltage, in accordance with Kirchhoff's voltage law. This check is a good way to verify your results.

RELATED PROBLEM

If R_1 and R_2 are both changed to 680 Ω in Figure 4–42, what are all the voltage drops?

MULTISIM

Open Multisim file E04-17. Verify the values of V_1, V_2, and V_3. Repeat for the related problem.

EXAMPLE 4–18

Determine the voltages between the following points in the circuit of Figure 4–43:

(a) *A* to *B*   **(b)** *A* to *C*   **(c)** *B* to *C*   **(d)** *B* to *D*   **(e)** *C* to *D*

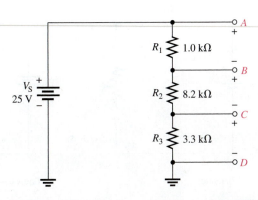

FIGURE 4–43

SOLUTION

First determine R_T.

$$R_T = 1.0\,k\Omega + 8.2\,k\Omega + 3.3\,k\Omega = 12.5\,k\Omega$$

Then apply the voltage-divider formula to obtain each required voltage.

(a) The voltage A to B is also the voltage drop across R_1.

$$V_{AB} = \left(\frac{R_1}{R_T}\right)V_S = \left(\frac{1.0\,k\Omega}{12.5\,k\Omega}\right)25\,V = \textbf{2 V}$$

(b) The voltage from A to C is the combined voltage drop across both R_1 and R_2. In this case, R_x in the general formula given in Equation 4–5 is $R_1 + R_2$.

$$V_{AC} = \left(\frac{R_1 + R_2}{R_T}\right)V_S = \left(\frac{9.2\,k\Omega}{12.5\,k\Omega}\right)25\,V = \textbf{18.4 V}$$

(c) The voltage from B to C is the voltage drop across R_2.

$$V_{BC} = \left(\frac{R_2}{R_T}\right)V_S = \left(\frac{8.2\,k\Omega}{12.5\,k\Omega}\right)25\,V = \textbf{16.4 V}$$

(d) The voltage from B to D is the combined voltage drop across both R_2 and R_3. In this case, R_x in the general formula is $R_2 + R_3$.

$$V_{BD} = \left(\frac{R_2 + R_3}{R_T}\right)V_S = \left(\frac{11.5\,k\Omega}{12.5\,k\Omega}\right)25\,V = \textbf{23 V}$$

(e) Finally, the voltage from C to D is the voltage drop across R_3.

$$V_{CD} = \left(\frac{R_3}{R_T}\right)V_S = \left(\frac{3.3\,k\Omega}{12.5\,k\Omega}\right)25\,V = \textbf{6.6 V}$$

If you connect this voltage divider, you can verify each of the calculated voltages by connecting a voltmeter between the appropriate points in each case.

RELATED PROBLEM

Determine each of the previously calculated voltages if V_S is doubled.

MULTISIM

Open Multisim file E04-18. Verify the values of V_{AB}, V_{AC}, V_{BC}, V_{BD}, and V_{CD}. Repeat for the related problem.

The Potentiometer as an Adjustable Voltage Divider

Recall from Chapter 2 that a potentiometer is a variable resistor with three terminals. A linear potentiometer connected to a voltage source is shown in Figure 4–44. Notice that the two end terminals are labeled 1 and 2. The adjustable terminal or wiper is labeled 3. The

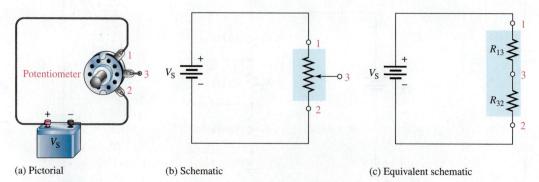

(a) Pictorial   (b) Schematic   (c) Equivalent schematic

FIGURE 4–44   **The potentiometer used as a voltage divider.**

potentiometer acts as a voltage divider, which can be illustrated by separating the total resistance into two parts as shown in Figure 4–44(c). The resistance between terminal 1 and terminal 3 (R_{13}) is one part, and the resistance between terminal 3 and terminal 2 (R_{32}) is the other part. The potentiometer acts like a two-resistor voltage divider that can be manually adjusted.

Figure 4–45 shows what happens when the wiper contact (3) is moved. In part (a), the wiper is exactly centered, making the two resistances equal. If you measure the voltage across terminals 3 to 2 as indicated, you have one-half of the total source voltage. When the wiper is moved up from the center position, as in part (b), the resistance between terminals 3 and 2 increases, and the voltage across it increases proportionally. When the wiper is moved down from the center position, as in part (c), the resistance between terminals 3 and 2 decreases, and the voltage decreases proportionally.

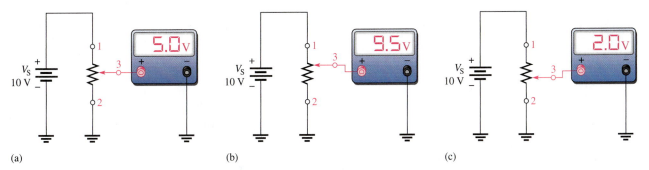

(a)   (b)   (c)

FIGURE 4–45   **Adjusting the voltage divider.**

Applications

There are many applications of voltage dividers in systems. A radio receiver can be thought of as a small system. The volume control is a common application of a potentiometer used as a voltage divider. Since the loudness of the sound is dependent on the amount of voltage associated with the audio signal, you can increase or decrease the volume by adjusting the potentiometer, that is, by turning the knob of the volume control. The block diagram in Figure 4–46 shows how a potentiometer can be used for volume control.

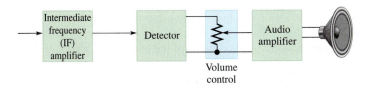

FIGURE 4–46   **A variable voltage divider used for volume control in a radio receiver.**

Figure 4–47 illustrates how a potentiometer voltage divider can be used as a level sensor in a storage tank. As shown in part (a), the float moves up as the tank is filled and moves down as the tank empties. The float is mechanically linked to the wiper arm of a potentiometer, as shown in part (b). The output voltage varies proportionally with the position of the wiper arm. As the liquid in the tank decreases, the sensor output voltage also decreases. The output voltage goes to the indicator circuitry, which controls a digital readout to show the amount of liquid in the tank. The schematic of this system is shown in part (c).

Another application for voltage dividers is in setting the gain of operational amplifiers and setting up reference voltages in power supplies. In certain precision applications, voltage dividers are available in IC form. In addition, voltage dividers are used for setting the dc operating voltage (bias) in transistor amplifiers. Figure 4–48 shows a voltage divider used for this purpose.

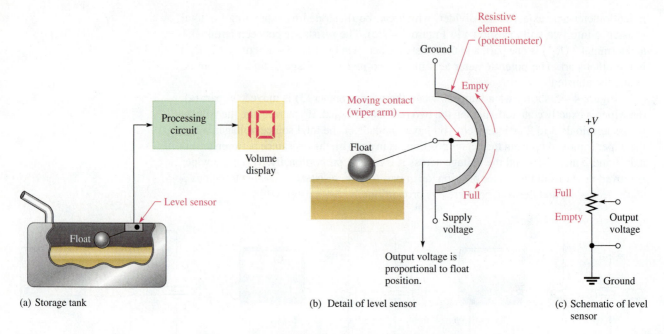

(a) Storage tank

(b) Detail of level sensor

(c) Schematic of level sensor

FIGURE 4–47   A potentiometer voltage divider used as a level sensor.

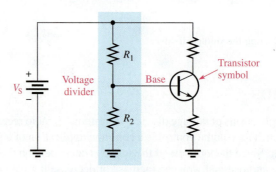

FIGURE 4–48   The voltage divider used as a bias circuit for a transistor amplifier.

A voltage divider is useful for converting a resistance sensor output to a voltage. Resistance sensors were described in Section 2–5 and include thermistors, photoconductive cells, and strain gages. To convert a change in resistance to an output voltage, the resistance sensor can be used in place of one of the resistors of a voltage divider.

EXAMPLE 4–19

Assume you have a CdS cell (a type of photoconductive cell) configured as shown in Figure 4–49 that uses three AA batteries for a voltage source (4.5 V). At dusk, the resistance of the cell rises from a low resistance to above 90 kΩ. The cell is used to trigger a logic circuit that will turn on lights if V_{OUT} is greater than approximately 1.5 V. What value of R will produce an output voltage of 1.5 V when the cell resistance is 90 kΩ?

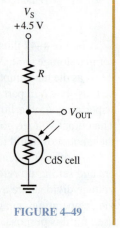

FIGURE 4–49

SOLUTION

Notice that the threshold voltage (1.5 V) is one-third of the supply voltage. You know that 90 kΩ represents one-third of the total resistance at this point. Therefore, the total resistance is

$$R_T = 3(90 \text{ k}\Omega) = 270 \text{ k}\Omega$$

The resistance needed to produce an output voltage of 1.5 V is

$$R = R_T - 90\,\text{k}\Omega = 270\,\text{k}\Omega - 90\,\text{k}\Omega = \mathbf{180\,k\Omega}$$

RELATED PROBLEM

Starting with Equation 4–5, prove that the required resistance is $180\,\text{k}\Omega$ to produce an output of 1.5 V when the cell resistance is $90\,\text{k}\Omega$.

SECTION 4–7 CHECKUP

1. What is a voltage divider?

2. How many resistors can there be in a series voltage-divider circuit?

3. Write the general voltage-divider formula.

4. If two series resistors of equal value are connected across a 20 V source, how much voltage is there across each resistor?

5. A $56\,\text{k}\Omega$ resistor and an $82\,\text{k}\Omega$ resistor are connected as a voltage divider. The source voltage is 10 V. Draw the circuit, and determine the voltage across each of the resistors.

6. The circuit of Figure 4–50 is an adjustable voltage divider. If the potentiometer is linear, where would you set the wiper in order to get 5 V from B to A and 5 V from C to B?

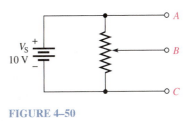

FIGURE 4–50

4–8 POWER IN SERIES CIRCUITS

The power dissipated by each individual resistor in a series circuit contributes to the total power in the circuit. The individual powers are additive.

After completing this section, you should be able to

- • Determine power in a series circuit
  - • Apply any of the power formulas

The total amount of power in a series resistive circuit is equal to the sum of the powers in each resistor in series.

$$P_T = P_1 + P_2 + P_3 + \cdots + P_n \qquad (4\text{–}6)$$

where P_T is the total power and P_n is the power in the last resistor in series (n can be any positive integer equal to the number of resistors in series).

The power formulas that you learned in Chapter 3 are, applicable to series circuits. Since each resistor in series has the same current through it, the following formulas are used to calculate the total power:

$$P_T = V_S I$$
$$P_T = I^2 R_T$$
$$P_T = \frac{V_S^2}{R_T}$$

where I is the current through the circuit, V_S is the source voltage across the series circuit, and R_T is the total resistance.

EXAMPLE 4–20

Determine the total amount of power in the series circuit in Figure 4–51.

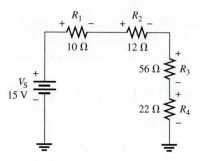

FIGURE 4–51

SOLUTION

The source voltage is 15 V. The total resistance is

$$R_T = 10\ \Omega + 12\ \Omega + 56\ \Omega + 22\ \Omega = 100\ \Omega$$

The easiest formula to use is $P_T = V_S^2/R_T$ since you know both V_S and R_T.

$$P_T = \frac{V_S^2}{R_T} = \frac{(15\ \text{V})^2}{100\ \Omega} = \frac{225\ \text{V}^2}{100\ \Omega} = 2.25\ \text{W}$$

If you determine the power of each resistor separately and add all these powers, you obtain the same result. Another calculation will illustrate. First, find the current.

$$I = \frac{V_S}{R_T} = \frac{15\ \text{V}}{100\ \Omega} = 150\ \text{mA}$$

Next, calculate the power for each resistor using $P = I^2R$.

$$P_1 = (150\ \text{mA})^2(10\ \Omega) = 225\ \text{mW}$$
$$P_2 = (150\ \text{mA})^2(12\ \Omega) = 270\ \text{mW}$$
$$P_3 = (150\ \text{mA})^2(56\ \Omega) = 1.26\ \text{W}$$
$$P_4 = (150\ \text{mA})^2(22\ \Omega) = 495\ \text{mW}$$

Now, add these powers to get the total power.

$$P_T = 225\ \text{mW} + 270\ \text{mW} + 1.260\ \text{W} + 495\ \text{mW} = \textbf{2.25 W}$$

This result shows that the sum of the individual powers is equal to the total power as determined by the formula $P_T = V_S^2/R_T$.

RELATED PROBLEM

What is the total power in the circuit of Figure 4–51 if V_S is increased to 30 V?

SECTION 4–8 CHECKUP

1. If you know the power in each resistor in a series circuit, how can you find the total power?

2. The resistors in a series circuit have the following powers: 1 W, 2 W, 5 W, and 8 W. What is the total power in the circuit?

3. A circuit has a 100 Ω, a 330 Ω, and a 680 Ω resistor in series. There is a current of 1 mA through the circuit. What is the total power?

The concept of *reference ground* was introduced in Chapter 2 and designated as the 0 V reference point for a circuit. Voltage is always measured with respect to another point in the circuit. *Ground* is discussed in more detail in this section.

After completing this section, you should be able to

- **State how to measure voltage with respect to ground**
  - **Define the term *reference ground***
  - **Explain the use of single and double subscripts for indicating voltages**

The term *ground* has its origin in telephone systems in which one of the conductors was the earth itself. The term was also used in early radio receiving antennas (called *aerials*) where one part of the system was connected to a metal pipe driven into the earth. Today, *ground* can mean different things and is not necessarily at the same potential as the earth. In electronic systems, **reference ground** (or **common**) refers to a conductor that is the comparison point for voltage measurements in a circuit. Frequently, it is the conductor that carries the power supply return current. Most electronic circuit boards have a larger conducting surface area for ground. For many multilayer boards the surface area is a separate internal layer, which is referred to as a **ground plane**.

In electrical wiring, the reference ground is usually the same as the earth potential because neutral and earth ground are connected at the entrance point to a building. In this case, reference ground and earth ground are at the same potential. (Section 517 of the National Electric Code (NEC) specifies certain exceptions to this grounding method for hospital operating rooms and health care facilities.)

The concept of *reference ground* is also used in automotive electrical systems. In most automotive systems, the chassis of the automobile is the ground reference (even though the tires insulate it from the earth, which can be a different potential). In nearly all modern automobiles, the negative post of the battery is connected to the chassis with a solid low-resistance connection. This makes the chassis of the vehicle serve as the return path for all of the electrical circuits in the vehicle, as illustrated in the simplified drawing in Figure 4–52. In some vintage cars, the positive terminal was connected to the chassis in an arrangement called *positive ground*. In both cases, the chassis represents the reference ground point.

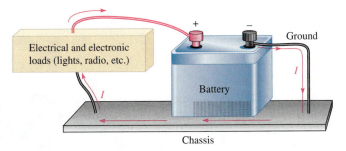

FIGURE 4–52   **The chassis serves as a return path for the electrical circuits in a vehicle.**

Measuring Voltages with Respect to Ground

When voltages are measured with respect to ground, they are indicated with a single-letter subscript. For example, V_A is voltage at point A with respect to ground. Each circuit in Figure 4–53 consists of three 1.0 kΩ series resistors and four lettered reference points. Reference ground represents a potential of 0 V with respect to all other points in each circuit as illustrated. In part (a), the reference point is D, and all voltages are positive with respect to D. In part (b), the reference point is A in the circuit, and all other points have a negative voltage.

Many circuits use both positive and negative voltages and as mentioned, the return path for these supplies is designated as reference ground. Figure 4–53(c) shows the same

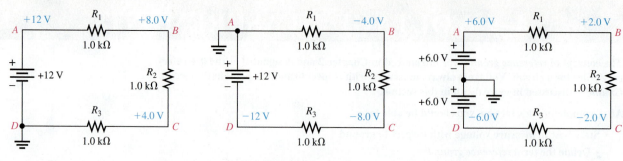

(a) Positive voltages with respect to ground

(b) Negative voltages with respect to ground

(c) Positive and negative voltages with respect to ground

FIGURE 4–53   **The ground point does not affect the current in the circuit or the voltage drops across the resistors.**

circuit but with two 6 V series-aiding sources replacing the 12 V source. In this case, the reference point is arbitrarily designated between the two voltage sources. There is exactly the same current in all three circuits, but now voltages are referenced to the new ground point. As you can see from these examples, the reference ground point is arbitrary and does not change the current.

Not all voltages are measured with respect to ground. If you wish to specify the voltage drop across an ungrounded resistor, you can name the resistor in the subscript or use two-lettered subscripts. When two different subscripts are used, the voltage represents the difference between the points. For example, V_{BC} means $V_B - V_C$. In Figure 4–53, V_{BC} is the same in all three circuits (+4.0 V) as you can confirm by performing the subtraction in each case. Another way of designating V_{BC} is to write simply V_{R2}.

There is one more convention commonly used to express voltages using subscripts. Power supply voltages are usually given with a double-letter subscript. The reference point is ground or common. For example, a voltage written as V_{CC} is a positive power supply voltage with respect to ground. Other common power supply voltages are V_{DD} (positive), V_{EE} (negative), and V_{SS} (negative).

To measure voltages with a digital meter, the meter leads can be connected across any two points and the meter will indicate the voltage, either positive or negative. The meter reference jack is labeled "COM" (normally black). This is common only to the meter, not to the circuit. Figure 4–54 shows a DMM used to measure the voltage across the ungrounded

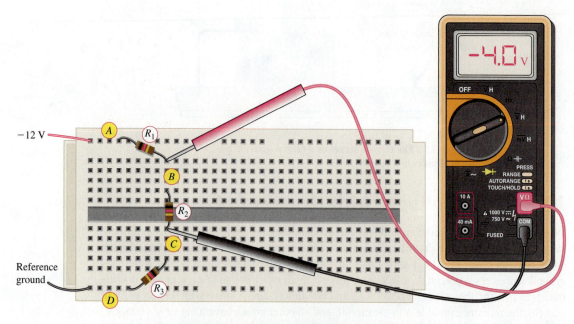

FIGURE 4–54   **A DMM has a "floating" common, so the leads can be connected to any points in the circuit and it will indicate the correct voltage between the two leads.**

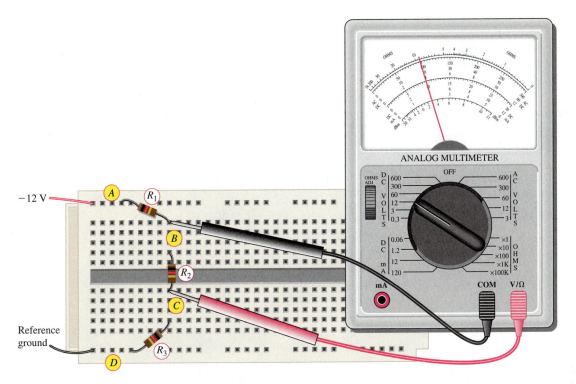

FIGURE 4–55   An analog meter needs to be connected so that the positive lead goes to the more positive point in the circuit.

(floating) resistor R_2. The circuit is the same one given in Figure 4–53(b), which has a negative power supply. It is shown as it might be constructed in a lab. Notice that the meter indicates a negative voltage, which means that the meter's COM lead is the more positive lead. If you wanted to measure voltages with respect to the circuit's reference ground, you would connect the COM on the meter to the circuit's reference ground. The voltage with respect to the reference ground will be indicated.

If you are using an analog meter to make circuit measurements, you must connect the meter so that its common lead is connected to the most negative point in the circuit, or else the meter movement will try to move backward. Figure 4–55 shows an analog meter connected to the same circuit as before; notice that the meter leads are reversed. To measure voltage across R_2, the leads must be connected so that the meter will deflect in the positive direction. The positive lead on the meter is connected to the more positive point in the circuit, which in this case is the circuit's ground. The user appends a minus sign to the reading when recording the reading.

MULTISIM

Open Multisim file E04-21. For each circuit, verify the values of the voltages at each point with respect to ground. Repeat for the related problem.

EXAMPLE 4–21

Determine the voltages at each of the indicated points in each circuit of Figure 4–56 with respect to ground. Since each of the four resistors has the same value, 25 V are dropped across each one.

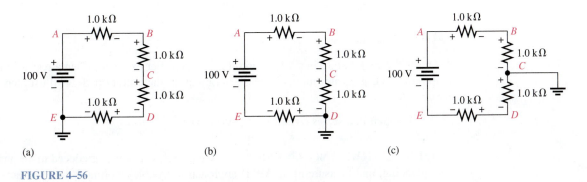

(a)   (b)   (c)

FIGURE 4–56

SOLUTION

For the circuit in Figure 4–56(a), the voltage polarities are as shown. Point E is ground. Single-letter subscripts denote voltage at a point with respect to ground. The voltages with respect to ground are as follows:

$$V_E = 0\ \text{V}, \quad V_D = +25\ \text{V}, \quad V_C = +50\ \text{V}, \quad V_B = +75\ \text{V}, \quad V_A = +100\ \text{V}$$

For the circuit in Figure 4–56(b), the voltage polarities are as shown. Point D is ground. The voltages with respect to ground are as follows:

$$V_E = -25\ \text{V}, \quad V_D = 0\ \text{V}, \quad V_C = +25\ \text{V}, \quad V_B = +50\ \text{V}, \quad V_A = +75\ \text{V}$$

For the circuit in Figure 4–56(c), the voltage polarities are as shown. Point C is ground. The voltages with respect to ground are as follows:

$$V_E = -50\ \text{V}, \quad V_D = -25\ \text{V}, \quad V_C = 0\ \text{V}, \quad V_B = +25\ \text{V}, \quad V_A = +50\ \text{V}$$

RELATED PROBLEM

If the ground is moved to point A in Figure 4–56(a), what are the voltages at each of the other points with respect to ground?

SECTION 4–9 CHECKUP

1. What is the reference point in a circuit called?

2. If V_{AB} in a circuit is +5.0 V, what is V_{BA}?

3. Voltages in a circuit are generally referenced to ground. (True or False)

4. The housing or chassis can be used as reference ground. (True or False)

4–10 TROUBLESHOOTING

Open components or contacts and shorts between conductors are common problems in all circuits. An open produces an infinite resistance. A short produces nearly zero resistance.

After completing this section, you should be able to

- **Troubleshoot series circuits**
  - **Check for an open circuit**
  - **Check for a short circuit**
  - **Identify primary causes of opens and shorts**

Open Circuit

The most common failure in a series circuit is an **open**. For example, when a resistor or a lamp burns out, it creates an open, causing a break in the current path, as illustrated in Figure 4–57.

An open in a series circuit prevents current.

TROUBLESHOOTING AN OPEN   In Chapter 3, you were introduced to the analysis, planning, and measurement (APM) approach to troubleshooting. You also learned

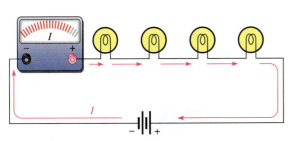

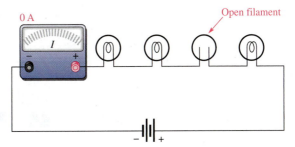

(a) A complete series circuit has current.   (b) An open series circuit has no current.

FIGURE 4–57   Current ceases when an open occurs.

about the half-splitting method and saw an example using an ohmmeter. Now, the same principles will be applied using voltage measurements instead of resistance measurements. As you know, voltage measurements are generally the easiest to make because you do not have to disconnect anything.

As a beginning step, prior to analysis, it is a good idea to make a visual check of the faulty circuit. Occasionally, you can find a charred resistor, a broken lamp filament, a loose wire, or a loose connection this way. However, it is possible (and probably more common) for a resistor or other component to open without showing visible signs of damage. When a visual check reveals nothing, then proceed with the APM approach.

When an open occurs in a series circuit, all of the source voltage appears across the open. The reason for this is that the open condition prevents current through the series circuit. With no current, there can be no voltage drop across any of the other resistors (or other component). Since $IR = (0 \text{ A})R = 0 \text{ V}$, the voltage on each end of a good resistor is the same. Therefore, the voltage applied across a series string also appears across the open component because there are other voltage drops in the circuit, as illustrated in Figure 4–58. The source voltage will appear across the open resistor in accordance with Kirchhoff's voltage law as follows:

$$V_S = V_1 + V_2 + V_3 + V_4 + V_5 + V_6$$
$$V_4 = V_S - V_1 - V_2 - V_3 - V_5 - V_6$$
$$= 10 \text{ V} - 0 \text{ V} - 0 \text{ V} - 0 \text{ V} - 0 \text{ V} - 0 \text{ V}$$
$$V_4 = V_S = 10 \text{ V}$$

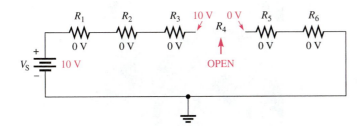

FIGURE 4–58   The source voltage appears across the open series resistor.

EXAMPLE OF HALF-SPLITTING USING VOLTAGE MEASUREMENTS

Suppose a circuit has four resistors in series. You have determined by *analyzing* the symptoms (there is voltage but no current) that one of the resistors is open, and you are *planning* to find the open resistor using a voltmeter for *measuring* by the half-splitting method. A sequence of measurements for this particular case is illustrated in Figure 4–59.

Step 1: Measure across R_1 and R_2 (the left half of the circuit). A 0 V reading indicates that neither of these resistors is open.

Step 2: Move the meter to measure across R_3 and R_4; the reading is 10 V. This indicates there is an open in the right half of the circuit, so either R_3 or R_4 is the faulty resistor (assume no bad connections).

HANDS ON TIP

When measuring a resistance, make sure that you do not touch the meter leads or the resistor leads. If you hold both ends of a high-value resistor in your fingers along with the meter probes, the measurement will be inaccurate because your body resistance can affect the measured value. When body resistance is placed in parallel with a high-value resistor, the measured value will be less than the actual value of the resistor.

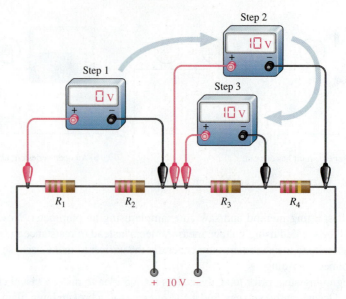

FIGURE 4–59   **Troubleshooting a series circuit for an open using half-splitting.**

Step 3:   Move the meter to measure across R_3. A measurement of 10 V across R_3 identifies it as the open resistor. If you had measured across R_4, it would have indicated 0 V. This would have also identified R_3 as the faulty component because it would have been the only one left that could have 10 V across it.

Short Circuit

Sometimes an unwanted short circuit occurs when two conductors touch or a foreign object such as solder or a wire clipping accidentally connects two sections of a circuit together. This situation is more common in circuits with a high component density. Three potential causes of short circuits are illustrated on the PC board in Figure 4–60.

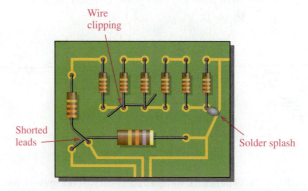

FIGURE 4–60   **Examples of shorts on a PC board.**

When there is a **short**, a portion of the series resistance is bypassed (all of the current goes through the short), thus reducing the total resistance as illustrated in Figure 4–61. Notice that the current increases as a result of the short.

A short in a series circuit causes more current.

TROUBLESHOOTING A SHORT   A short is generally very difficult to troubleshoot. As in any troubleshooting situation, it is a good idea to make a visual check of the faulty circuit. In the case of a short in the circuit, a wire clipping, solder splash, or touching leads is often found to be the culprit. In terms of component failure, shorts are less common than opens in many types of components. Furthermore, a short in one part of a circuit can

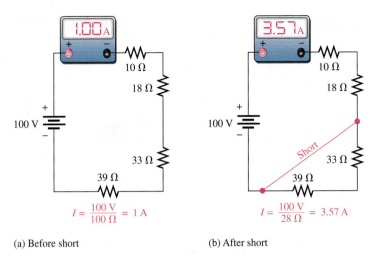

(a) Before short

$$I = \frac{100 \text{ V}}{100 \text{ } \Omega} = 1 \text{ A}$$

(b) After short

$$I = \frac{100 \text{ V}}{28 \text{ } \Omega} = 3.57 \text{ A}$$

FIGURE 4–61   **The effect of a short in a series circuit.**

cause overheating in another part due to the higher current caused by the short. As a result two failures, an open and a short, may occur together.

When a short occurs in a series circuit, there is essentially no voltage across the shorted part. A short has zero or near zero resistance, although shorts with significant resistance values can occur from time to time. These are called *resistive shorts.* For purposes of illustration, zero resistance is assumed for all shorts.

In order to troubleshoot a short, measure the voltage across each resistor until you get a reading of 0 V. This is the straightforward approach and does not use half-splitting. In order to apply the half-splitting method, you must know the correct voltage at each point in the circuit and compare it to measured voltages. Example 4–22 illustrates using half-splitting to find a short.

EXAMPLE 4–22

Assume you have determined that there is a short in a circuit with four series resistors because the current is higher than it should be. You know that the voltage at each point in the circuit should be as shown in Figure 4–62 if the circuit is working properly. The voltages are shown relative to the negative terminal of the source. Find the location of the short.

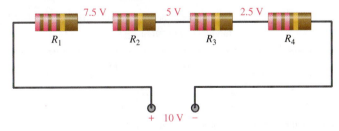

FIGURE 4–62   **Series circuit (without a short) with correct voltages marked.**

SOLUTION

Use the half-splitting method to troubleshoot the short.

Step 1:   Measure across R_1 and R_2. The meter shows a reading of 6.67 V, which is higher than the normal voltage (it should be 5 V). Look elsewhere for a voltage that is lower than normal because a short will make the voltage less across that part of the circuit.

Step 2:   Move the meter and measure across R_3 and R_4; the reading of 3.33 V is incorrect and lower than normal (it should be 5 V). This shows that the short is in the right half of the circuit and that either R_3 or R_4 is shorted.

Step 3:   Again move the meter and measure across R_3. A reading of 3.3 V across R_3 tells you that R_4 is shorted because it must have 0 V across it. Figure 4–63 illustrates this troubleshooting technique.

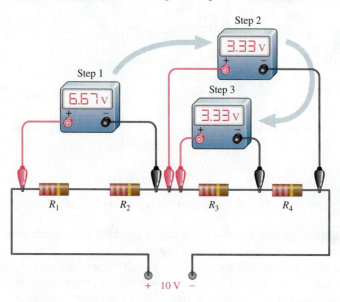

FIGURE 4–63   Troubleshooting a series circuit for a short using half-splitting.

RELATED PROBLEM

Assume that R_1 is shorted in Figure 4–63. What would the Step 1 measurement be?

SECTION 4–10 CHECKUP

1. Define *open.*

2. Define *short.*

3. What happens when a series circuit opens?

4. Name two general ways in which an open circuit can occur in practice. How can a short circuit occur?

5. When a resistor fails, it will normally open.   (True or False)

6. The total voltage across a string of series resistors is 24 V. If one of the resistors is open, how much voltage is there across it? How much is there across each of the good resistors?

7. Explain why the voltage measured in Step 1 of Figure 4–63 is higher than normal.

SUMMARY

- The total series resistance is the sum of all resistors in the series circuit.
- The total resistance between any two points in a series circuit is equal to the sum of all resistors connected in series between those two points.
- If all of the resistors in a series circuit are of equal value, the total resistance is the number of resistors multiplied by the resistance value of one resistor.
- The current is the same at all points in a series circuit.
- Voltage sources in series add algebraically.
- Kirchhoff's voltage law: The sum of the voltage drops in a series circuit equals the total source voltage.
- Kirchhoff's voltage law: the algebraic sum of all the voltages around a closed single path is zero.
- The voltage drops in a circuit are always opposite in polarity to the total source voltage.
- Current is defined to be out of the negative side of a source and into the positive side.
- Current is defined to be into the negative side of each resistor and out of the positive side.

- A voltage divider is a series arrangement of resistors connected to a voltage source.
- A voltage divider is so named because the voltage drop across any resistor in the series circuit is divided down from the total voltage by an amount proportional to that resistance value in relation to the total resistance.
- A potentiometer can be used as an adjustable voltage divider.
- The total power in a resistive circuit is the sum of all the individual powers of the resistors making up the series circuit.
- Voltages given with a single-lettered subscript are referenced to ground. When two different letters are used in the subscript, the voltage is the difference in the two points.
- Ground (common) is zero volts with respect to all points referenced to it in the circuit.
- *Negative ground* is the term used when the negative side of the source is grounded.
- *Positive ground* is the term used when the positive side of the source is grounded.
- The voltage across an open component always equals the source voltage.
- The voltage across a shorted component is always 0 V.

KEY TERMS

Key terms and other bold terms in the chapter are defined in the end-of-book glossary.

Kirchhoff's voltage law   A law stating that (1) the sum of the voltage drops around a single closed path equals the source voltage in that loop or (2) the algebraic sum of all the voltages around a single closed path in a circuit is zero.

Open   A circuit condition in which the current path is interrupted.

Reference ground   The metal chassis that houses the assembly or a large conductive area on a printed circuit board used as the common or reference point; also called common (com).

Series   In an electric circuit, a relationship of components in which the components are connected such that they provide a single current path between two points.

Series-aiding   An arrangement of two or more series voltage sources with polarities in the same direction.

Series-opposing   An arrangement of two series voltage sources with polarities in the opposite direction; not a normal configuration.

Short   A circuit condition in which there is a zero or abnormally low resistance path between two points: usually an inadvertent condition.

Voltage divider   A circuit consisting of series resistors across which one or more output voltages are taken.

KEY FORMULAS

(4–1)	$R_T = R_1 + R_2 + R_3 + \cdots + R_n$	Total resistance of n resistors in series
(4–2)	$R_T = nR$	Total resistance of n equal-value resistors in series
(4–3)	$V_S = V_1 + V_2 + V_3 + \cdots + V_n$	Kirchhoff's voltage law in a series circuit
(4–4)	$V_1 + V_2 + V_3 + \cdots + V_n = 0$	Kirchhoff's voltage law
(4–5)	$V_x = \left(\dfrac{R_x}{R_T}\right)V_S$	Voltage-divider formula
(4–6)	$P_T = P_1 + P_2 + P_3 + \cdots + P_n$	Total power

TRUE/FALSE QUIZ

Answers are at the end of the chapter.

1. A series circuit can have more than one path for current.

2. The total resistance of a series circuit can be less than the largest resistor in that circuit.

3. If two series resistors are different sizes, the larger resistor will have the larger current.

4. If two series resistors are different sizes, the larger resistor will have the larger voltage.

5. If three equal resistors are used in a voltage divider, the voltage across each one will be one-third of the source voltage.

6. There is no valid electrical reason for installing flashlight batteries so that they are series-opposing.

7. Kirchhoff's voltage law is valid only if a loop contains a voltage source.

8. The voltage-divider equation can be written as $V_x = (R_x/R_T)V_S$.

9. The power dissipated by the resistors in a series circuit is the same as the power supplied by the source.

10. If point A in a circuit has a voltage of $+10$ V and point B has a voltage of -2 V, then V_{AB} is $+8$ V.

SELF-TEST

Answers are at the end of the chapter.

1. Five equal-value resistors are connected in series and there is a current of 2 mA into the first resistor. The amount of current out of the second resistor is
   (a) 2 mA   **(b)** 1 mA   **(c)** 4 mA   **(d)** 0.4 mA

2. To measure the current out of the third resistor in a circuit consisting of four series resistors, an ammeter can be placed
   (a) between the third and fourth resistors   **(b)** between the second and third resistors
   (c) at the positive terminal of the source   **(d)** at any point in the circuit

3. When a third resistor is connected in series with two series resistors, the total resistance
   (a) remains the same   **(b)** increases   **(c)** decreases   **(d)** increases by one-third

4. When one of four series resistors is removed from a circuit and the circuit reconnected, the current
   (a) decreases by the amount of current through the removed resistor
   (b) decreases by one-fourth   **(c)** quadruples   **(d)** increases

5. A series circuit consists of three resistors with values of 100 Ω, 220 Ω, and 330 Ω. The total resistance is
   (a) less than 100 Ω   **(b)** the average of the values   **(c)** 550 Ω   **(d)** 650 Ω

6. A 9 V battery is connected across a series combination of 68 Ω, 33 Ω, 100 Ω, and 47 Ω resistors. The amount of current is
   (a) 36.3 mA   **(b)** 27.6 A   **(c)** 22.3 mA   **(d)** 363 mA

7. While putting four 1.5 V batteries in a flashlight, you accidentally put one of them in backward. The light will be
   (a) brighter than normal   **(b)** dimmer than normal   **(c)** off   **(d)** the same

8. If you measure all the voltage drops and the source voltage in a series circuit and add them together, taking into consideration the polarities, you will get a result equal to
   (a) the source voltage   **(b)** the total of the voltage drops
   (c) zero   **(d)** the total of the source voltage and the voltage drops

9. There are six resistors in a given series circuit and each resistor has 5 V dropped across it. The source voltage is
   (a) 5 V   **(b)** 30 V
   (c) dependent on the resistor values   **(d)** dependent on the current

10. A series circuit consists of a 4.7 kΩ, a 5.6 kΩ, and a 10 kΩ resistor. The resistor that has the most voltage across it is
    (a) the 4.7 kΩ   **(b)** the 5.6 kΩ
    (c) the 10 kΩ   **(d)** impossible to determine from the given information

11. Which of the following series combinations dissipates the most power when connected across a 100 V source?
    (a) one 100 Ω resistor   **(b)** two 100 Ω resistors
    (c) three 100 Ω resistors   **(d)** four 100 Ω resistors

12. The total power in a certain circuit is 1 W. Each of the five equal-value series resistors making up the circuit dissipates
    (a) 1 W   (b) 5 W   (c) 0.5 W   (d) 0.2 W

13. When you connect an ammeter in a series resistive circuit and turn on the source voltage, the meter reads zero. You should check for
    (a) a broken wire   (b) a shorted resistor   (c) an open resistor   (d) both (a) and (c)

14. While checking out a series resistive circuit, you find that the current is higher than it should be. You should look for
    (a) an open circuit   (b) a short   (c) a low resistor value   (d) both (b) and (c)

TROUBLESHOOTING: SYMPTOM AND CAUSE

The purpose of these exercises is to help develop thought processes essential to troubleshooting. Answers are at the end of the chapter.

Determine the cause for each set of symptoms. Refer to Figure 4–64.

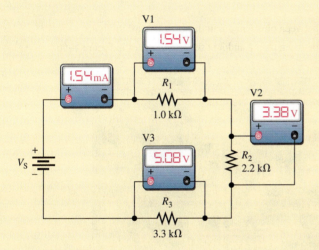

FIGURE 4–64   The meters indicate the correct readings for this circuit.

1. *Symptom:* The ammeter reading is zero, the voltmeter 1 and voltmeter 3 readings are zero, and the voltmeter 2 reading is 10 V.
   Cause:
   (a) R_1 is open.   (b) R_2 is open.   (c) R_3 is open.

2. *Symptom:* The ammeter reading is zero, and all the voltmeter readings are zero.
   Cause:
   (a) A resistor is open.   (b) The voltage source is turned off or faulty.
   (c) One of the resistor values is too high.

3. *Symptom:* The ammeter reading is 2.33 mA, and the voltmeter 2 reading is zero.
   Cause:
   (a) R_1 is shorted.   (b) The voltage source is set too high.   (c) R_2 is shorted.

4. *Symptom:* The ammeter reading is zero, voltmeter 1 reads 0 V, voltmeter 2 reads 5 V, and voltmeter 3 reads 5 V.
   Cause:
   (a) R_1 is shorted.   (b) R_1 and R_2 are open.   (c) R_2 and R_3 are open.

5. *Symptom:* The ammeter reading is 0.645 mA, the voltmeter 1 reading is too high, and the other two voltmeter readings are too low.
   Cause:
   (a) R_1 has an incorrect value of 10 kΩ.   (b) R_2 has an incorrect value of 10 kΩ.
   (c) R_3 has an incorrect value of 10 kΩ.

PROBLEMS

Answers to odd-numbered problems are at the end of the book.

BASIC PROBLEMS

SECTION 4–1   Resistors in Series

1. Connect each set of resistors in Figure 4–65 in series between points *A* and *B*.

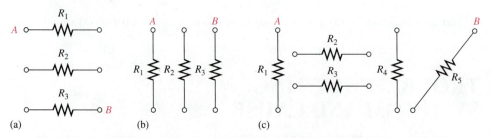

FIGURE 4–65

2. Determine which resistors in Figure 4–66 are in series. Show how to interconnect the pins to put all the resistors in series.

3. Determine the resistance between pins 1 and 8 in the circuit board in Figure 4–66.

4. Determine the resistance between pins 2 and 3 in the circuit board in Figure 4–66.

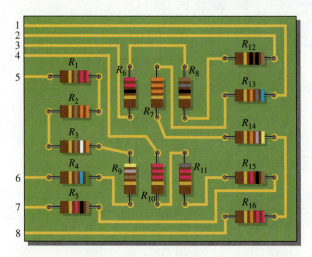

FIGURE 4–66

SECTION 4–2   Total Series Resistance

5. An 82 Ω resistor and a 56 Ω resistor are connected in series. What is the total resistance?

6. Find the total resistance of each group of series resistors shown in Figure 4–67.

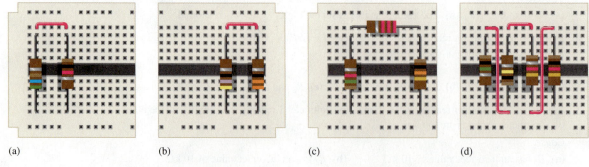

FIGURE 4–67

7. Determine R_T for each circuit in Figure 4–68. Show how to measure R_T with an ohmmeter.

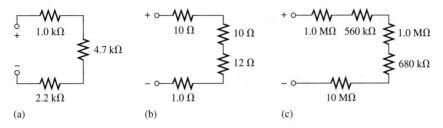

(a)   (b)   (c)

FIGURE 4–68

8. What is the total resistance of twelve 5.6 kΩ resistors in series?

9. Six 47 Ω resistors, eight 100 Ω resistors, and two 22 Ω resistors are in series. What is the total resistance?

10. The total resistance in Figure 4–69 is 20 kΩ. What is the value of R_5?

11. Determine the resistance between each of the following sets of pins on the PC board in Figure 4–66.
    (a) pin 1 and pin 8   **(b)** pin 2 and pin 3
    (c) pin 4 and pin 7   **(d)** pin 5 and pin 6

12. If all the resistors in Figure 4–66 are connected in series, what is the total resistance?

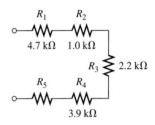

FIGURE 4–69

SECTION 4–3   Current in a Series Circuit

13. What is the current through each of four resistors in a series circuit if the source voltage is 12 V and the total resistance is 120 Ω?

14. The current from the source in Figure 4–70 is 5 mA. How much current does each milliammeter in the circuit indicate?

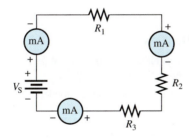

FIGURE 4–70

SECTION 4–4   Application of Ohm's Law

15. What is the current in each circuit of Figure 4–71? Show how to connect an ammeter in each case.

16. Determine the voltage across each resistor in Figure 4–71.

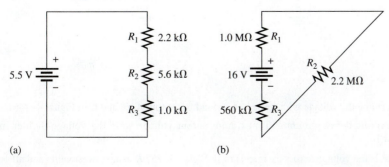

(a)   (b)

FIGURE 4–71

17. Three 470 Ω resistors are in series with a 48 V source.
    (a) How much current is there?
    (b) What is the voltage across each resistor?
    (c) What is the minimum power rating of the resistors?

18. Four equal-value resistors are in series with a 5 V source, and a current of 1 mA is measured. What is the value of each resistor?

SECTION 4–5   Voltage Sources in Series

19. Show how to connect four 6 V batteries to achieve a voltage of 24 V.

20. What happens if one of the batteries in Problem 19 is accidentally connected in reverse?

SECTION 4–6   Kirchhoff's Voltage Law

21. The following voltage drops are measured across each of three resistors in series: 5.5 V, 8.2 V, and 12.3 V. What is the value of the source voltage to which these resistors are connected?

22. Five resistors are in series with a 20 V source. The voltage drops across four of the resistors are 1.5 V, 5.5 V, 3 V, and 6 V. How much voltage is across the fifth resistor?

23. Determine the unspecified voltage drop(s) in each circuit of Figure 4–72. Show how to connect a voltmeter to measure each unknown voltage drop.

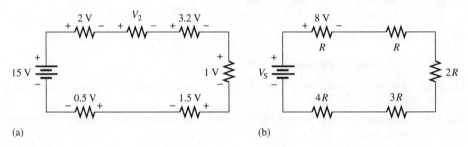

(a)   (b)

FIGURE 4–72

SECTION 4–7   Voltage Dividers

24. The total resistance of a series circuit is 500 Ω. What percentage of the total voltage appears across a 22 Ω resistor in the series circuit?

25. Find the voltage between A and B in each voltage divider of Figure 4–73.

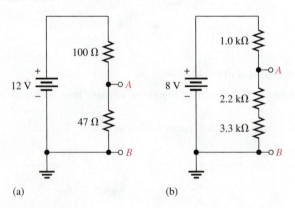

(a)   (b)

FIGURE 4–73

26. Determine the voltage with respect to ground for output A, B, and C in Figure 4–74(a).

27. Determine the minimum and maximum output voltage from the voltage divider in Figure 4–74(b).

28. What is the voltage across each resistor in Figure 4–75? R is the lowest value and all others are multiples of that value as indicated.

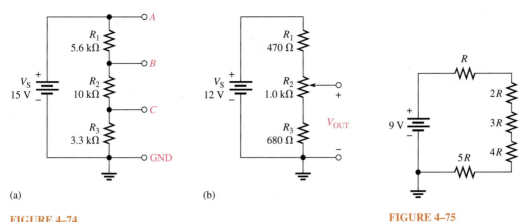

FIGURE 4–74

FIGURE 4–75

29. What is the voltage across each resistor on the protoboard in Figure 4–76(b)?

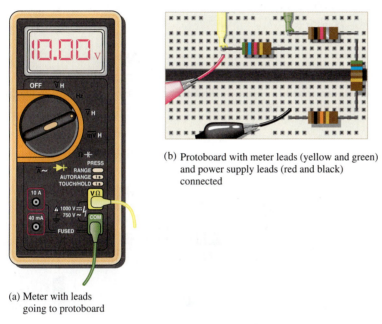

(a) Meter with leads
   going to protoboard

(b) Protoboard with meter leads (yellow and green)
   and power supply leads (red and black)
   connected

FIGURE 4–76

SECTION 4–8   Power in Series Circuits

30. Five series resistors each dissipate 50 mW of power. What is the total power?

31. Find the total power in Figure 4–76.

SECTION 4–9   Voltage Measurements

32. Determine the voltage at each point with respect to ground in Figure 4–77.

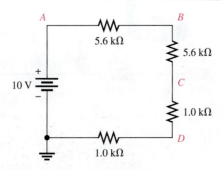

FIGURE 4–77

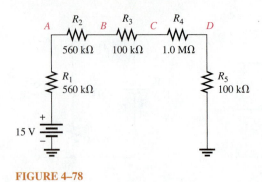

FIGURE 4–78

33. In Figure 4–78, how would you determine the voltage across R_2 by measurement, without connecting a meter directly across the resistor?

34. Determine the voltage at each point with respect to ground in Figure 4–78.

35. In Figure 4–78, what is V_{AC}?

36. In Figure 4–78, what is V_{CA}?

SECTION 4–10   Troubleshooting

37. By observing the meters in Figure 4–79, determine the types of failures in the circuits and which components have failed.

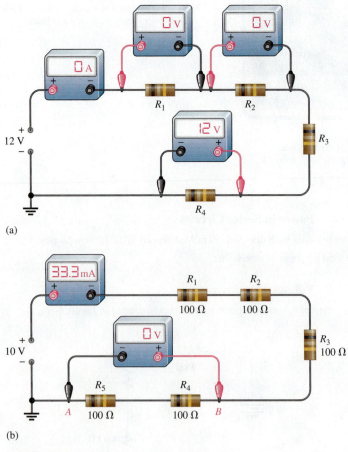

(a)

(b)

FIGURE 4–79

38. Is the multimeter reading in Figure 4–80 correct? If not, what is wrong?

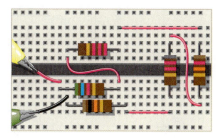

(b) Protoboard with meter leads connected

(a) Meter with leads
going to protoboard

FIGURE 4–80

ADVANCED PROBLEMS

39. Determine the unknown resistance (R_3) in the circuit of Figure 4–81.

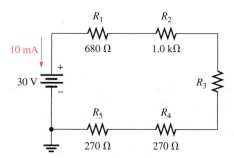

FIGURE 4–81

40. You have the following resistor values available to you in the lab in unlimited quantities: 10 Ω, 100 Ω, 470 Ω, 560 Ω, 680 Ω, 1.0 kΩ, 2.2 kΩ, and 5.6 kΩ. All of the other standard values are out of stock. A project that you are working on requires an 18 kΩ resistance. What combination of available values can you use to obtain the needed value?

41. Determine the voltage at each point in Figure 4–82 with respect to ground.

FIGURE 4–82

42. Find all the unknown quantities (shown in red) in Figure 4–83.

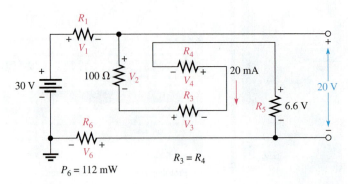

FIGURE 4–83

43. There are 250 mA in a series circuit with a total resistance of 1.5 kΩ. The current must be reduced by 25%. Determine how much resistance must be added in order to accomplish this reduction in current.

44. Four ½ W resistors are in series: 47 Ω, 68 Ω, 100 Ω, and 120 Ω. To what maximum value can the current be raised before the power rating of one of the resistors is exceeded? Which resistor will burn out first if the current is increased above the maximum?

45. A certain series circuit is made up of a ⅛ W resistor, a ¼ W resistor, and a ½ W resistor. The total resistance is 2400 Ω. If each of the resistors is operating at its maximum power level, determine the following:
    (a) I (b) V_S (c) the value of each resistor

46. Using 1.5 V batteries, a switch, and three lamps, devise a circuit to apply 4.5 V across one lamp, two lamps in series, or three lamps in series with a single control switch. Draw the schematic.

47. Develop a variable voltage divider to provide output voltages ranging from a minimum of 10 V to a maximum of 100 V using a 120 V source. The maximum voltage must be at the maximum resistance setting of the potentiometer. The minimum voltage must be at the minimum resistance (zero ohms) setting. The current is to be 10 mA.

48. Using the standard resistor values given in Appendix A, design a voltage divider to provide the following approximate voltages with respect to the negative terminal of a 30 V source: 8.18 V, 14.7 V, and 24.6 V. The current drain on the source must be limited to no more than 1 mA. The number of resistors, their resistance values, and their power ratings must be specified. Draw a schematic showing the circuit with all resistor values indicated.

49. On the double-sided PC board in Figure 4–84, identify each group of series resistors and determine its total resistance. Note that many of the interconnections feed through the board from the top side to the bottom side.

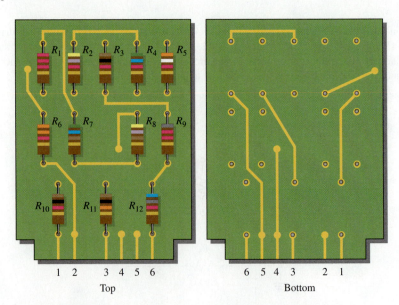

Top   Bottom

FIGURE 4–84

50. What is the total resistance from *A* to *B* for each switch position in Figure 4–85?

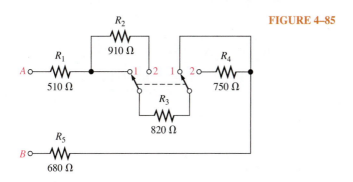

FIGURE 4–85

51. Determine the current measured by the meter in Figure 4–86 for each switch position.

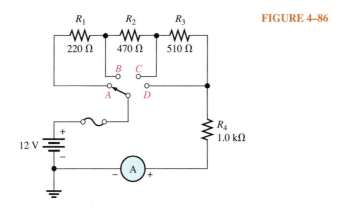

FIGURE 4–86

52. Determine the current measured by the meter in Figure 4–87 for each position of the ganged switch.

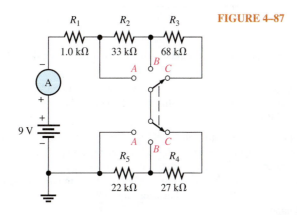

FIGURE 4–87

53. Determine the voltage across each resistor in Figure 4–88 for each switch position if the current through R_5 is 6 mA when the switch is in the *D* position.

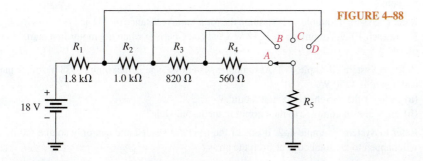

FIGURE 4–88

54. Table 4–1 shows the results of resistance measurements on the PC circuit board in Figure 4–84. Are these results correct? If not, identify the possible problems.

TABLE 4–1	
BETWEEN PINS	**RESISTANCE**
1 and 2	∞
1 and 3	∞
1 and 4	4.23 kΩ
1 and 5	∞
1 and 6	∞
2 and 3	23.6 kΩ
2 and 4	∞
2 and 5	∞
2 and 6	∞
3 and 4	∞
3 and 5	∞
3 and 6	∞
4 and 5	∞
4 and 6	∞
5 and 6	19.9 kΩ

55. You measure 15 kΩ between pins 5 and 6 on the PC board in Figure 4–84. Does this indicate a problem? If so, identify it.

56. In checking out the PC board in Figure 4–84, you measure 17.83 kΩ between pins 1 and 2. Also, you measure 13.6 kΩ between pins 2 and 4. Does this indicate a problem on the PC board? If so, identify the fault.

57. The three groups of series resistors on the PC board in Figure 4–84 are connected in series with each other to form a single series circuit by connecting pin 2 to pin 4 and pin 3 to pin 5. A voltage source is connected across pins 1 and 6 and an ammeter is placed in series. As you increase the source voltage, you observe the corresponding increase in current. Suddenly, the current drops to zero and you smell smoke. All resistors are $1/2$ W.
   (a) What has happened?
   (b) Specifically, what must you do to fix the problem?
   (c) At what voltage did the failure occur?

58. Refer to System Example 4–1. If the field coil should open while the motor is running, what happens?

59. Refer to System Example 4–1. Assume the rotor has a resistance of 2 Ω and resistors R_1, R_2, and R_3 are each 10 Ω. Assuming V_S is 12 V, what is the current when the motor first starts (in position 1)?

60. Refer to System Example 4–2. Assume the system voltage is maximum (588 V for 17 panels) and power is 4250 W.
   (a) What is the current in the first module?
   (b) How does it compare to the current in the last module?

61. Refer to System Example 4–2. If one of the panels is shaded and can only source 5.0 A, what will happen to the total current from the array?

MULTISIM TROUBLESHOOTING PROBLEMS

62. Open file P04-62: Files are found at www.pearsonhighered.com/floyd. Determine if there is a fault and, if so, identify the fault.

63. Open file P04-63. Determine if there is a fault and, if so, identify the fault.

64. Open file P04-64. Determine if there is a fault and, if so, identify the fault.

65. Open file P04-65. Determine if there is a fault and, if so, identify the fault.

66. Open file P04-66. Determine if there is a fault and, if so, identify the fault.

67. Open file P04-67. Determine if there is a fault and, if so, identify the fault.

ANSWERS TO SECTION CHECKUPS

SECTION 4–1   Resistors in Series

1. Series resistors are connected end-to-end in a "string."

2. There is a single current path in a series circuit.

3. See Figure 4–89.

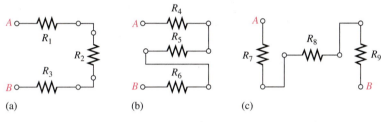

(a)   (b)   (c)

FIGURE 4–89

4. See Figure 4–90.

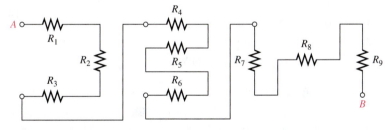

FIGURE 4–90

SECTION 4–2   Total Series Resistance

1. **(a)** $R_T = 33\ \Omega + 100\ \Omega + 10\ \Omega = 143\ \Omega$
   (b) $R_T = 39\ \Omega + 56\ \Omega + 10\ \Omega = 105\ \Omega$
   (c) $R_T = 820\ \Omega + 2200\ \Omega + 1000\ \Omega = 4020\ \Omega$

2. $R_T = 100\ \Omega + 2(47\ \Omega) + 4(12\ \Omega) + 330\ \Omega = 572\ \Omega$

3. $10\ k\Omega - 8.8\ k\Omega = 1.2\ k\Omega$

4. $R_T = 12(47\ \Omega) = 564\ \Omega$

SECTION 4–3   Current in a Series Circuit

1. The current is the same at all points in a series circuit.

2. There are 20 mA through the 47 Ω resistor.

3. 50 mA between C and D, 50 mA between E and F

4. Ammeter 1 indicates 17.9 mA. Ammeter 2 indicates 17.9 mA.

SECTION 4–4   Ohm's Law in Series Circuits

1. $I = 6\,V/300\,\Omega = 0.020\,A = 20\,mA$
2. $V = (5\,mA)(43\,\Omega) = 125\,mV$
3. $V_1 = (5\,mA)(10\,\Omega) = 50\,mV$,
   $V_2 = (5\,mA)(15\,\Omega) = 75\,mV$,
   $V_3 = (5\,mA)(18\,\Omega) = 90\,mV$
4. $R = 1.25\,V/4.63\,mA = 270\,\Omega$
5. $R = 130\,\Omega$

SECTION 4–5   Voltage Sources in Series

1. $60\,V/12\,V = 5$; see Figure 4–91.
2. $V_T = (4)(1.5\,V) = 6.0\,V$
3. See Figure 4–92.

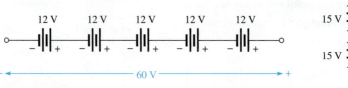

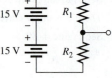

FIGURE 4–91   **FIGURE 4–92**

4. $V_{S(tot)} = 6\,V + 15\,V = 21\,V$
5. $3.0\,V$.

SECTION 4–6   Kirchhoff's Voltage Law

1. Kirchhoff's voltage law states:
   (a) The algebraic sum of the voltages around a closed path is zero.
   (b) The sum of the voltage drops equals the total source voltage.
2. $V_{R(tot)} = V_S = 50\,V$
3. $V_{R1} = V_{R2} = 10\,V/2 = 5\,V$
4. $V_{R3} = 25\,V - 5\,V - 10\,V = 10\,V$
5. $V_S = 1\,V + 3\,V + 5\,V + 7\,V + 8\,V = 24\,V$

SECTION 4–7   Voltage Dividers

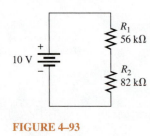

FIGURE 4–93

1. A voltage divider is a series circuit with two or more resistors in which the voltage taken across any resistor or combination of resistors is proportional to the value of that resistance.
2. Two or more resistors form a voltage divider.
3. $V_x = (R_x/R_T)V_S$ is the general voltage-divider formula.
4. $V_R = 20\,V/2 = 10\,V$
5. See Figure 4–93. $V_{R1} = (56\,k\Omega/138\,k\Omega)10\,V = 4.06\,V$,
   $V_{R2} = (82\,k\Omega/138\,k\Omega)10\,V = 5.94\,V$.
6. Set the potentiometer at the midpoint.

SECTION 4–8   Power in Series Circuits

1. Add the powers in each resistor to get total power.
2. $P_T = 1\,W + 2\,W = 5\,W + 8\,W = 16\,W$
3. $P_T = (1\,mA)^2(100\,\Omega + 330\,\Omega + 680\,\Omega) = 1.11\,mW$

SECTION 4–9   Voltage Measurements

1. Ground
2. $-5.0\,V$

3. True

4. True

SECTION 4–10   Troubleshooting

1. An open is a break in the current path.

2. A short is a zero resistance path that bypasses a portion of a circuit.

3. Current ceases when a series circuit opens.

4. An open can be created by a component failure or a faulty contact. A short can be created by wire clippings, solder splashes, etc.

5. True

6. 24 V across the open R; 0 V across the other resistors.

7. Because R_4 is shorted, more voltage is dropped across the other resistors than normal. The total voltage is divided across three equal-value resistors.

ANSWERS TO RELATED PROBLEMS FOR EXAMPLES

4–1   (a) Left end of R_1 to terminal A, right end of R_1 to top end of R_3, bottom end of R_3 to right end of R_5, left end of R_5 to left end of R_2, right end of R_2 to right end of R_4, left end of R_4 to terminal B

     (b) $R_1 = 1.0\ \text{k}\Omega, R_2 = 33\ \text{k}\Omega, R_3 = 39\ \text{k}\Omega, R_4 = 470\ \Omega, R_5 = 22\ \text{k}\Omega$

4–2   The two series circuits are connected in series, so all the resistors on the board are in series.

4–3 $258\ \Omega$ (No change)

4–4 $12.1\ \text{k}\Omega$

4–5 $22\ \text{k}\Omega$

4–6 $4.36\ \text{k}\Omega$

4–7 $18.5\ \text{mA}$

4–8 $7.8\ \text{V}$

4–9 $V_1 = 1\ \text{V}, V_2 = 3.3\ \text{V}, V_3 = 2.2\ \text{V}, V_S = 6.5\ \text{V}, V_{S(max)} = 32.5\ \text{V}$

4–10 $275\ \Omega$

4–11 $6\ \text{V}$

4–12   See Figure 4–94.

4–13 $10\ \text{V}, 20\ \text{V}$

4–14 $6.5\ \text{V}$

4–15 $430\ \Omega$

4–16 $V_1 = 3.13\ \text{V}; V_2 = 6.87\ \text{V}$

4–17 $V_1 = V_2 = V_3 = 3.33\ \text{V}$

4–18 $V_{AB} = 4\ \text{V}; V_{AC} = 36.8\ \text{V}; V_{BC} = 32.8\ \text{V}; V_{BD} = 46\ \text{V}; V_{CD} = 13.2\ \text{V}$

4–19 $V_x = \left(\dfrac{R_x}{R_T}\right)V_S$

$$\frac{V_x}{V_S} = \frac{R_x}{R_T} = \frac{R_x}{R + R_x}$$

$$\frac{1.5\ \text{V}}{4.5\ \text{V}} = \frac{90\ \text{k}\Omega}{R + 90\ \text{k}\Omega}$$

$$1.5\ \text{V}(R + 90\ \text{k}\Omega) = (4.5\ \text{V})(90\ \text{k}\Omega)$$

$$1.5R = 270\ \Omega$$

$$R = 180\ \Omega$$

4–20 $9\ \text{W}$

4–21 $V_A = 0\ \text{V}; V_B = -25\ \text{V}; V_C = -50\ \text{V}; V_D = -75\ \text{V}; V_E = -100\ \text{V}$

4–22 $3.33\ \text{V}$

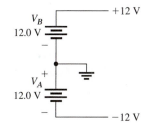

FIGURE 4–94

ANSWERS TO TRUE/FALSE QUIZ

1. F   2. F   3. F   4. T   5. T
6. T   7. F   8. T   9. T   10. F

ANSWERS TO SELF-TEST

1. (a)   2. (d)   3. (b)   4. (d)   5. (d)   6. (a)   7. (b)
8. (c)   9. (b)   10. (c)   11. (a)   12. (d)   13. (d)   14. (d)

ANSWERS TO TROUBLESHOOTING: SYMPTOM AND CAUSE

1. (b)   2. (b)   3. (c)   4. (c)   5. (a)

CHAPTER 5

PARALLEL CIRCUITS

OUTLINE

OBJECTIVES

- Identify a parallel resistive circuit
- Determine total parallel resistance
- Determine the voltage across each parallel branch
- Apply Ohm's law in a parallel circuit
- Apply Kirchhoff's current law
- Use a parallel circuit as a current divider
- Determine power in a parallel circuit
- Troubleshoot parallel circuits

KEY TERMS

Parallel

Branch

Kirchhoff's current law

Node

Current divider

Radiometric image

INTRODUCTION

In this chapter, you will see how Ohm's law is used in parallel circuits; and you will learn Kirchhoff's current law. Also, several system applications of parallel circuits, including automotive lighting, residential wiring, and control circuits are presented. You will learn how to determine total parallel resistance and how to troubleshoot for open resistors.

When resistors are connected in parallel and a voltage is applied across the parallel circuit, each resistor provides a separate path for current. The total resistance of a parallel circuit is reduced as more resistors are connected in parallel. The voltage across each of the parallel resistors is equal to the voltage applied across the entire parallel circuit.

VISIT THE WEBSITE
Study aids for this chapter are available at
http://pearsonhighered.com/floyd

5–1 RESISTORS IN PARALLEL

When two or more resistors are individually connected between the same two points, they are in **parallel** with each other. A parallel circuit provides more than one path for current.

After completing this section, you should be able to

- Identify a parallel resistive circuit
- Translate a physical arrangement of parallel resistors into a schematic

___

Each parallel path in a circuit is called a **branch**. Two resistors connected in parallel are shown in Figure 5–1(a). As shown in part (b), the current out of the source (I_T) divides when it gets to point *B*. I_1 goes through R_1 and I_2 goes through R_2. The two currents come back together at point *A*. If additional resistors are connected in parallel with the first two, more current paths are provided, as shown in Figure 5–1(c). All points along the top shown in blue are electrically the same as point *A*, and all the points along the bottom shown in green are electrically the same as point *B*.

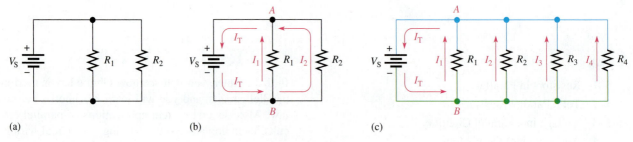

(a) (b) (c)

FIGURE 5–1 Resistors in parallel.

In Figure 5–1, it is obvious that the resistors are connected in parallel. However, in actual circuit diagrams, the parallel relationship often is not as clear. It is important that you learn to recognize parallel circuits regardless of how they may be drawn.

A rule for identifying parallel circuits is as follows:

> **If there is more than one current path (branch) between two points, and if the voltage between those two points also appears across each of the branches, then there is a parallel circuit between those two points.**

Figure 5–2 shows parallel resistors drawn in different ways between two points labeled *A* and *B*. Notice that in each case, the current has two paths going from *A* to *B*, and the voltage across each branch is the same. Although these figures show only two parallel paths, there can be any number of resistors in parallel.

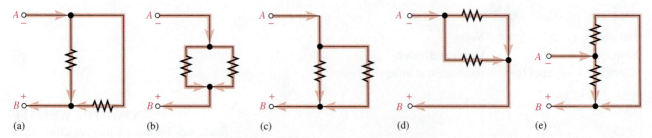

(a) (b) (c) (d) (e)

FIGURE 5–2 Examples of circuits with two parallel paths.

EXAMPLE 5–1

Five resistors are positioned on a protoboard as shown in Figure 5–3. Show the wiring required to connect all the resistors in parallel between A and B. Draw a schematic and label each of the resistors with its value.

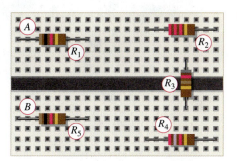

FIGURE 5–3

SOLUTION

Wires are connected as shown in the assembly wiring diagram of Figure 5–4(a). The schematic is shown in Figure 5–4(b) with resistance values from the color bands. Again, note that the schematic does not necessarily have to show the actual physical arrangement of the resistors. The schematic shows how components are connected electrically.

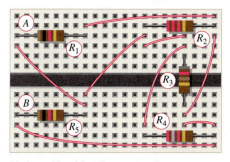

(a) Assembly wiring diagram

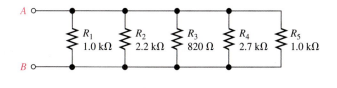

(b) Schematic

FIGURE 5–4

RELATED PROBLEM*

Would the circuit have to be rewired if R_2 is removed?

*Answers are at the end of the chapter.

EXAMPLE 5–2

Determine the parallel groupings in Figure 5–5 and the value of each resistor.

SOLUTION

Resistors R_1 through R_4 and R_{11} and R_{12} are all in parallel. This parallel combination is connected to pins 1 and 4. Each resistor in this group is 56 kΩ.

Resistors R_5 through R_{10} are all in parallel. This combination is connected to pins 2 and 3. Each resistor in this group is 100 kΩ.

RELATED PROBLEM

How would you connect all the resistors on the PC board in parallel?

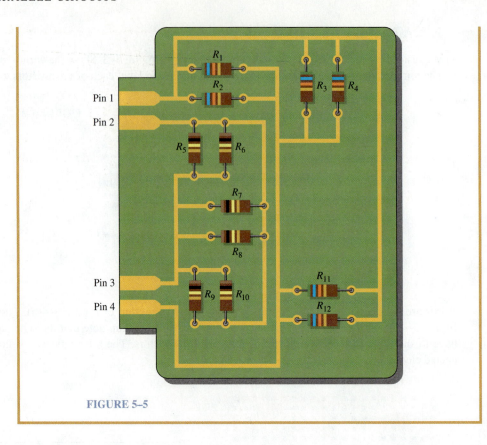

FIGURE 5–5

SECTION 5–1 CHECKUP*

1. How are the resistors connected in a parallel circuit?

2. How do you identify a parallel circuit?

3. Complete the schematics for the circuits in each part of Figure 5–6 by connecting the resistors in parallel between points *A* and *B*.

4. Connect each group of parallel resistors in Figure 5–6 in parallel with each other.

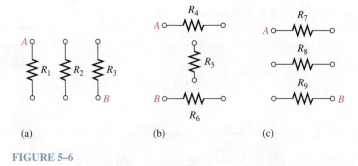

(a)   (b)   (c)

FIGURE 5–6

*Answers are at the end of the chapter.

When resistors are connected in parallel, the total resistance of the circuit decreases. The total resistance of a parallel circuit is always less than the value of the smallest resistor. For example, if a 10 Ω resistor and a 100 Ω resistor are connected in parallel, the total resistance is less than 10 Ω.

After completing this section, you should be able to

- **Determine total parallel resistance**
  - **Explain why resistance decreases as resistors are connected in parallel**
  - **Apply the parallel-resistance formula**
  - **Describe two applications of a parallel circuit**

When resistors are connected in parallel, the current has more than one path. The number of current paths is equal to the number of parallel branches.

In Figure 5–7(a), there is only one current path because it is a series circuit. There is a certain amount of current, I_1, through R_1. If resistor R_2 is connected in parallel with R_1, as shown in Figure 5–7(b), there is an additional amount of current, I_2, through R_2. The total current coming from the source has increased with the addition of the parallel branch. Assuming that the source voltage is constant, an increase in the total current from the source means that the total resistance has decreased, in accordance with Ohm's law. Additional resistors connected in parallel will further reduce the resistance and increase the total current.

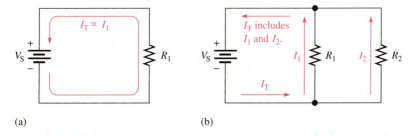

(a)   (b)

FIGURE 5–7   Connecting resistors in parallel reduces total resistance and increases total current.

Formula for Total Parallel Resistance, R_T

The circuit in Figure 5–8 shows a general case of n resistors in parallel (n can be any integer greater than 1). As resistors are added, there are more paths for current; hence, there is increased conductance. Recall that conductance (G) was defined in Section 2–5 as the reciprocal of resistance, $(1/R)$. It has the unit of siemens (S).

With parallel resistors, it is simple to think in terms of the conductive paths; each resistor adds to the total conductance as given by the following equation:

$$G_T = G_1 + G_2 + G_3 + \cdots + G_n$$

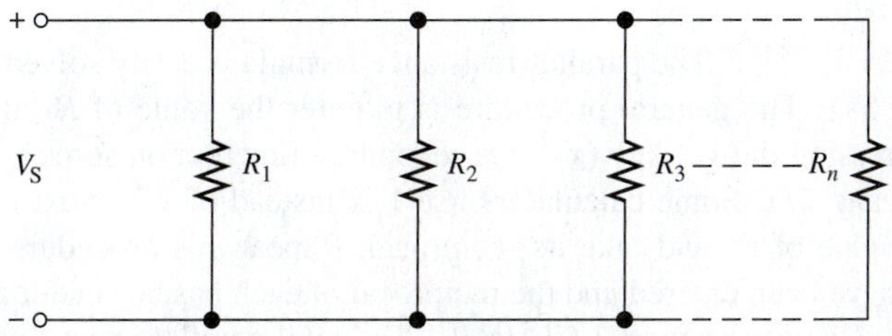

FIGURE 5–8   Circuit with n resistors in parallel.

Substituting $1/R$ for G,

$$\frac{1}{R_T} = \frac{1}{R_1} + \frac{1}{R_2} + \frac{1}{R_3} + \cdots + \frac{1}{R_n}$$

Solve for R_T by taking the reciprocal of (that is, by inverting) both sides of the equation.

$$R_T = \frac{1}{\dfrac{1}{R_1} + \dfrac{1}{R_2} + \dfrac{1}{R_3} + \cdots + \dfrac{1}{R_n}} \qquad\qquad \textbf{(5–1)}$$

Equation 5–1 shows that to find the total parallel resistance, add all the $1/R$ (or conductance, G) terms and then take the reciprocal of the sum.

$$R_T = \frac{1}{G_T}$$

EXAMPLE 5–3

Calculate the total parallel resistance between points A and B of the circuit in Figure 5–9.

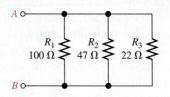

SOLUTION

Use Equation 5–1 to calculate the total resistance when you know the individual resistances.

FIGURE 5–9

First, find the conductance, which is the reciprocal of the resistance, of each of the three resistors.

$$G_1 = \frac{1}{R_1} = \frac{1}{100\ \Omega} = 10\ \text{mS}$$

$$G_2 = \frac{1}{R_2} = \frac{1}{47\ \Omega} = 21.3\ \text{mS}$$

$$G_3 = \frac{1}{R_3} = \frac{1}{22\ \Omega} = 45.5\ \text{mS}$$

Next, calculate R_T by adding G_1, G_2, and G_3 and taking the reciprocal of the sum.

$$R_T = \frac{1}{\dfrac{1}{R_1} + \dfrac{1}{R_2} + \dfrac{1}{R_3}} = \frac{1}{G_1 + G_2 + G_3}$$

$$= \frac{1}{10\ \text{mS} + 21.3\ \text{mS} + 45.5\ \text{mS}} = \frac{1}{76.8\ \text{mS}} = \textbf{13.0 }\boldsymbol{\Omega}$$

For a rough accuracy check, notice that the value of R_T (13.0 Ω) is smaller than the smallest value in parallel, which is R_3 (22 Ω), as it should be.

RELATED PROBLEM

If a 33 Ω resistor is connected in parallel in Figure 5–9, what is R_T?

CALCULATOR TIP   The parallel-resistance formula is easily solved on a calculator using Equation 5–1. The general procedure is to enter the value of R_1 and then take its reciprocal by pressing the x^{-1} key (x^{-1} is a secondary function on some calculators). The notation x^{-1} means $1/x$. Some calculators use $1/x$ instead of x^{-1}. Next press the $+$ key; then enter the value of R_2 and take its reciprocal. Repeat this procedure until all of the resistor values have been entered and the reciprocal of each has been added. The final step is to press the x^{-1} key to convert $1/R_T$ to R_T. The total parallel resistance is now on the display. The display format may vary, depending on the particular calculator.

EXAMPLE 5–4

Show the steps required for a typical calculator solution of Example 5–3.

SOLUTION

1. Enter 100. Display shows 100.
2. Press the x^{-1} key (or the 2nd then x^{-1} keys).
3. Press the + key.
4. Enter 47.
5. Press the x^{-1} key.
6. Press the + key.
7. Enter 22.
8. Press the x^{-1} key.
9. Press the ENTER key. Display shows $76.7311411992\text{E}^{-3}$.
10. Press the x^{-1} and ENTER keys. Display shows $13.0325182758\text{E}^{0}$.

The number displayed in Step 10 is the total resistance, which is rounded to 13.0 Ω.

RELATED PROBLEM

Show the additional calculator steps for R_T when a 33 Ω resistor is placed in parallel in the circuit of Example 5–3.

THE CASE OF TWO RESISTORS IN PARALLEL   Equation 5–1 is a general formula for finding the total resistance of any number of resistors in parallel. It is often useful to consider only two resistors in parallel because this situation occurs commonly in practice.

Derived from Equation 5–1, the formula for two resistors in parallel is

$$R_T = \frac{R_1 R_2}{R_1 + R_2} \tag{5–2}$$

Equation 5–2 states

The total resistance for two resistors in parallel is equal to the product of the two resistors divided by the sum of the two resistors.

This equation is sometimes referred to as the "product over the sum" formula.

EXAMPLE 5–5

Calculate the total resistance connected to the voltage source of the circuit in Figure 5–10.

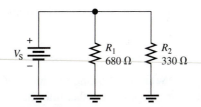

FIGURE 5–10

SOLUTION

Use Equation 5–2.

$$R_T = \frac{R_1 R_2}{R_1 + R_2} = \frac{(680\ \Omega)(330\ \Omega)}{680\ \Omega + 330\ \Omega} = \frac{224{,}000\ \Omega^2}{1{,}010\ \Omega} = \mathbf{222\ \Omega}$$

RELATED PROBLEM

Find R_T if a 220 Ω resistor replaces R_1 in Figure 5–10.

THE CASE OF EQUAL-VALUE RESISTORS IN PARALLEL   Another special case of parallel circuits is the parallel connection of several resistors having the same value. The following is a shortcut method of calculating R_T when this case occurs:

$$R_T = \frac{R}{n}$$ (5–3)

Equation 5–3 says that when any number of resistors (n), all having the same resistance (R), are connected in parallel, R_T is equal to the resistance divided by the number of resistors in parallel.

EXAMPLE 5–6

Find the total resistance between points A and B in Figure 5–11.

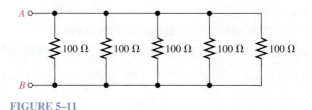

FIGURE 5–11

SOLUTION

There are five 100 Ω resistors in parallel. Use Equation 5–3.

$$R_T = \frac{R}{n} = \frac{100\ \Omega}{5} = \mathbf{20\ \Omega}$$

RELATED PROBLEM

Find R_T for three 100 kΩ resistors in parallel.

NOTATION FOR PARALLEL RESISTORS   Sometimes, for convenience, parallel resistors are designated by two parallel vertical marks. For example, R_1 in parallel with R_2 can be written as $R_1 \| R_2$. Also, when several resistors are in parallel with each other, this notation can be used. For example,

$$R_1 \| R_2 \| R_3 \| R_4 \| R_5$$

indicates that R_1 through R_5 are all in parallel.

This notation is also used with resistance values. For example,

$$10\,\text{k}\Omega \| 5\,\text{k}\Omega$$

means that a 10 kΩ resistor is in parallel with a 5 kΩ resistor.

Applications of Parallel Circuits

AUTOMOTIVE   One advantage of a parallel circuit over a series circuit is that when one branch opens, the other branches are not affected. For example, Figure 5–12 shows a simplified diagram of an automobile lighting system. When one headlight on your car goes out, it does not cause the other lights to go out because they are all in parallel.

Notice that the brake lights are switched on independently of the headlights and taillights. They come on only when the driver closes the brake light switch by depressing the brake pedal. When the lights switch is closed, both headlights and both taillights are on. When the headlights are on, the parking lights are off and vice versa, as indicated by the dashed line connecting the switches. If any one of the lights burns out (opens), there is still current in each of the other lights. The back-up lights are switched on when the reverse gear is engaged.

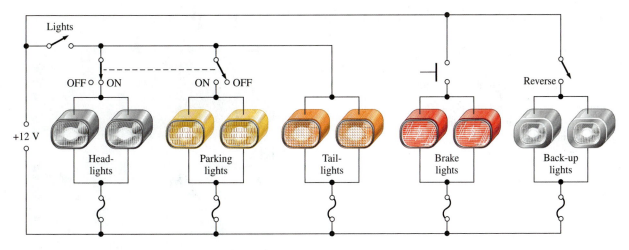

FIGURE 5–12   Simplified diagram of the exterior light system of an automobile.

Another automotive application of parallel resistances is the rear window defroster. As you know, power is dissipated in resistance in the form of heat. The defroster consists of a group of parallel resistance wires that warm the glass when power is applied. A typical defroster can dissipate over 100 W across the window. Although resistance heating is not as efficient as other forms of heat, in applications like this it is simple and cost effective.

RESIDENTIAL   Another common use of parallel circuits is in residential electrical systems. All the lights and appliances in a home are wired in parallel. Figure 5–13 shows a typical room wiring arrangement with two switch-controlled lights and three wall outlets in parallel.

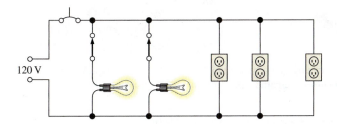

FIGURE 5–13   Example of a parallel circuit in residential wiring.

CONTROL CIRCUITS   Many control systems use parallel circuits or equivalent circuits for control and monitoring an industrial process such as a production line. Most complex control applications are implemented on a dedicated computer called a **programmable logic controller (PLC)**. A PLC shows an equivalent parallel circuit, which is displayed on a computer screen. Internally, the circuit may exist only as a computer code, written in a computer programming language. However, the displayed circuit is one that could in principle be constructed with hardware. These circuits can be drawn to resemble a ladder—with rungs representing the load (and a source) and the rails representing the two conductors that are connected to the voltage source. (For example, notice that Figure 5–13 resembles a ladder on its side, with the lights and wall outlets as loads.) Parallel control circuits use **ladder diagrams** but add additional control elements (switches, relays, timers, and more). The result of adding the control elements creates a logical diagram, which is referred to as **ladder logic**. Ladder logic is easy to understand, so it is popular for showing control logic in industrial environments, such as factories or food processors. Entire books are written about ladder logic, but a ladder diagram (parallel circuit) is at the core of all ladder logic. The most important aspect is that ladder logic shows the basic functionality of the circuit in an easy-to-read format.

In addition to ladder diagrams used in industrial control problems, many auto repair manuals use these diagrams for showing circuits for troubleshooting purposes. Ladder diagrams are logical, so a technician can organize a test procedure by reading the diagram.

SECTION 5–2 CHECKUP

1. Does the total resistance increase or decrease as more resistors are connected in parallel?

2. The total parallel resistance is always less than what value?

3. Determine R_T (between pin 1 and pin 4) for the circuit in Figure 5–14. Note that pins 1 and 2 are connected and pins 3 and 4 are connected.

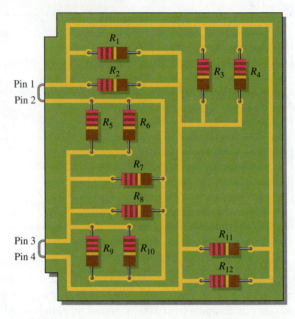

FIGURE 5–14

5–3 VOLTAGE IN A PARALLEL CIRCUIT

The voltage across any given branch of a parallel circuit is equal to the voltage across each of the other branches in parallel. As you know, each current path in a parallel circuit is called a *branch*.

After completing this section, you should be able to

- **Determine the voltage across each branch in a parallel circuit**
    - **Explain why the voltage is the same across all parallel resistors**

To illustrate voltage in a parallel circuit, let's examine Figure 5–15(a). Points *A*, *B*, *C*, and *D* along the left side of the parallel circuit are electrically the same point because the voltage is the same along this line. You can think of all of these points as being connected

FIGURE 5–15 **Voltage across parallel branches is the same.**

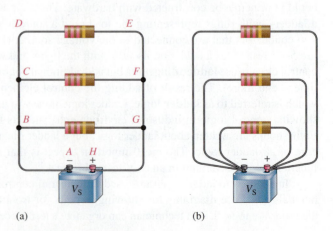

(a)   (b)

by a single wire to the negative terminal of the battery. The points *E, F, G,* and *H* along the right side of the circuit are all at a voltage equal to that of the positive terminal of the source. Thus, voltage across each parallel resistor is the same, and each is equal to the source voltage. Notice that the parallel circuit in Figure 5–15 resembles a ladder as described earlier.

Figure 5–15(b) is the same circuit as in part (a), drawn in a slightly different way. Here the left side of each resistor is connected to a single point, which is the negative battery terminal. The right side of each resistor is connected to a single point, which is the positive battery terminal. The resistors are still all in parallel across the source.

In Figure 5–16, a 12 V battery is connected across three parallel resistors. When the voltage is measured across the battery and then across each of the resistors, the readings are the same. As you can see, the same voltage appears across each branch in a parallel circuit.

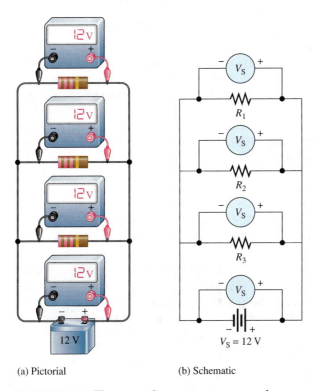

(a) Pictorial   (b) Schematic

FIGURE 5–16   **The same voltage appears across each resistor in parallel.**

E X A M P L E   5 – 7

Determine the voltage across each resistor in Figure 5–17.

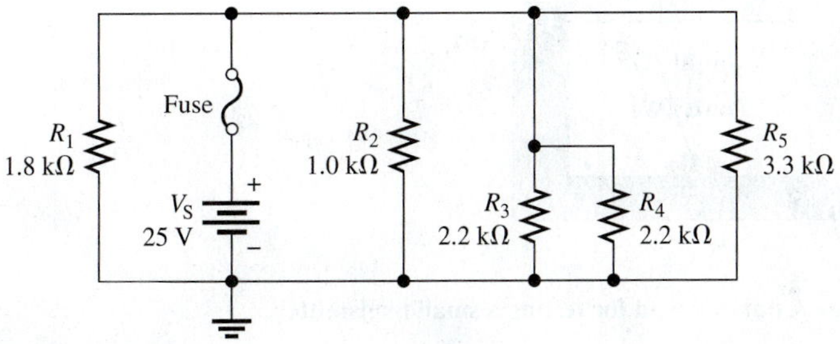

FIGURE 5–17

184   CHAPTER 5 • PARALLEL CIRCUITS

Open Multisim file E05-07; files are found at www. pearsonhighered.com/floyd. Verify that each resistor voltage equals the source voltage. Repeat for the related problem.

SOLUTION

The five resistors are in parallel; so the voltage across each one is equal to the source voltage, V_S. There is no voltage across the fuse.

$$V_1 = V_2 = V_3 = V_4 = V_5 = V_S = 25 \text{ V}$$

RELATED PROBLEM

If R_4 is removed from the circuit, what is the voltage across R_3?

SYSTEM EXAMPLE 5–1

DUMMY LOADS AND LOAD BANKS

A **dummy load** is a precision high-power resistor and may include multiple resistors in one enclosure. A dummy load is used to test a system such as a radio transmitter without the actual load (the antenna) in place. A *load bank* performs the same function electronically, but it absorbs a certain amount of current, which is controlled to represent a specific resistance. Both a dummy load and a load bank absorb energy from the system. For radio transmitters, the energy from the transmitter that is normally routed to the antenna system as electromagnetic energy is instead routed to the dummy load, where it is dissipated in the form of heat. Without an antenna or a dummy load, the transmitter could be damaged. The dummy load enables the transmitter to be tested under simulated conditions without causing interference to other systems as adjustments are made. In addition to testing radio transmitters, dummy loads can allow testing of other systems. For example, load cells can be tested with a dummy load for calibration purposes.

Figure 5–18(a) shows the schematic for a dummy load for testing a small radio transmitter and is composed of 1 W resistors (R_{11} and R_{12} can be 0.5 W each). Resistors R_{11} and R_{12} form a voltage divider, which is in parallel with the other resistors in the array. The output from the divider is rectified and measured with a voltmeter that can be calibrated for power. To match a 50 Ω antenna system, the total resistance is 50 Ω as you can determine from the values given. The entire assembly is contained in a shielded enclosure to avoid radiating any electromagnetic energy. The enclosure contains a meter to read the power in the load, as shown in Figure 5–18(b).

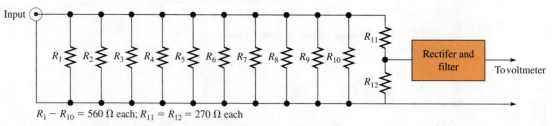

$R_1 - R_{10} = 560 \text{ Ω each}; R_{11} = R_{12} = 270 \text{ Ω each}$

(a) Schematic

(b) Front panel of dummy load

FIGURE 5–18   A dummy load for testing a small transmitter.

SECTION 5–3 CHECKUP

1. A 100 Ω and a 220 Ω resistor are connected in parallel with a 5 V source. What is the voltage across each of the resistors?

2. A voltmeter connected across R_1 in Figure 5–19 measures 118 V. If you move the meter and connect it across R_2, how much voltage will it indicate? What is the source voltage?

3. In Figure 5–20, how much voltage does voltmeter 1 indicate? Voltmeter 2?

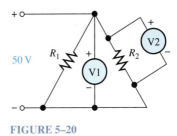

FIGURE 5–20

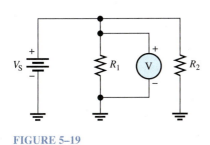

FIGURE 5–19

4. How are voltages across each branch of a parallel circuit related?

5–4 APPLICATION OF OHM'S LAW

Ohm's law can be applied to parallel circuit analysis.

After completing this section, you should be able to

- **Apply Ohm's law in a parallel circuit**
  - **Find the total current in a parallel circuit**
  - **Find the branch currents, voltage, and resistance in a parallel circuit**

The following examples illustrate how to apply Ohm's law in parallel circuits.

EXAMPLE 5–8

Find the total current from the battery in Figure 5–21.

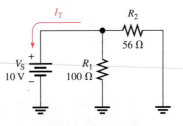

FIGURE 5–21

SOLUTION

The battery "sees" a total parallel resistance that determines the amount of current. First, calculate R_T.

$$R_T = \frac{R_1 R_2}{R_1 + R_2} = \frac{(100\ \Omega)(56\ \Omega)}{100\ \Omega + 56\ \Omega} = \frac{5600\ \Omega^2}{156\ \Omega} = 35.9\ \Omega$$

The battery voltage is 10 V. Use Ohm's law to find I_T.

$$I_T = \frac{V_S}{R_T} = \frac{10\ \text{V}}{35.9\ \Omega} = \textbf{279 mA}$$

RELATED PROBLEM

Find the currents through R_1 and R_2 in Figure 5–21. Show that the sum of the currents through R_1 and R_2 equals the total current.

EXAMPLE 5–9

Determine the current through each resistor in the parallel circuit of Figure 5–22.

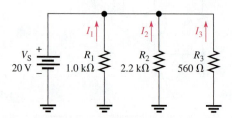

FIGURE 5–22

SOLUTION

The voltage across each resistor (branch) is equal to the source voltage. That is, the voltage across R_1 is 20 V, the voltage across R_2 is 20 V, and the voltage across R_3 is 20 V. Determine the current through each resistor as follows:

$$I_1 = \frac{V_S}{R_1} = \frac{20\text{ V}}{1.0\text{ k}\Omega} = \textbf{20.0 mA}$$

$$I_2 = \frac{V_S}{R_2} = \frac{20\text{ V}}{2.2\text{ k}\Omega} = \textbf{9.09 mA}$$

$$I_3 = \frac{V_S}{R_3} = \frac{20\text{ V}}{560\ \Omega} = \textbf{35.7 mA}$$

RELATED PROBLEM

If an additional resistor of 910 Ω is connected in parallel to the circuit in Figure 5–22, determine all the branch circuits.

EXAMPLE 5–10

Find the voltage across the parallel circuit shown in Figure 5–23.

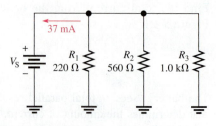

FIGURE 5–23

SOLUTION

The total current into the parallel circuit is 37 mA. If you know the total resistance, then you can apply Ohm's law to get the voltage. The total resistance is

$$R_T = \cfrac{1}{\dfrac{1}{R_1} + \dfrac{1}{R_2} + \dfrac{1}{R_3}} = \cfrac{1}{\dfrac{1}{220\ \Omega} + \dfrac{1}{560\ \Omega} + \dfrac{1}{1.0\text{ k}\Omega}}$$

$$= \frac{1}{4.55\text{ mS} + 1.79\text{ mS} + 1\text{ mS}} = \frac{1}{7.34\text{ mS}} = 136\ \Omega$$

Therefore, the source voltage and the voltage across each branch is

$$V_S = I_T R_T = (37 \text{ mA})(136 \text{ } \Omega) = \textbf{5.05 V}$$

RELATED PROBLEM

Find the total current if R_3 opens in Figure 5–23. Assume V_S remains the same.

EXAMPLE 5–11

Sometimes a direct measurement of resistance is not practical. For example, tungsten filament bulbs get hot after turning them on, causing the resistance to increase. An ohmmeter can only measure the cold resistance. Assume you want to know the equivalent hot resistance of the two headlight bulbs and the two taillight bulbs in a car. The two headlights normally operate on 12.6 V and draw 2.8 A each when they are on.

(a) What is the total equivalent hot resistance when the two headlights are on?

(b) Assume the total current for headlights and taillights is 8.0 A when all four bulbs are on. What is the equivalent resistance of each taillight?

SOLUTION

(a) Use Ohm's law to calculate the equivalent resistance of one headlight.

$$R_{\text{HEAD}} = \frac{V}{I} = \frac{12.6 \text{ V}}{2.8 \text{ A}} = 4.5 \text{ } \Omega$$

Because the two bulbs are in parallel and have equal resistances,

$$R_{\text{T(HEAD)}} = \frac{R_{\text{HEAD}}}{n} = \frac{4.5 \text{ } \Omega}{2} = \textbf{2.25 } \boldsymbol{\Omega}$$

(b) Use Ohm's law to find the total resistance with two taillights and two headlights on.

$$R_{\text{T(HEAD+TAIL)}} = \frac{12.6 \text{ V}}{8.0 \text{ A}} = 1.58 \text{ } \Omega$$

Apply the formula for parallel resistors to find the resistance of the two taillights together.

$$\frac{1}{R_{\text{T(HEAD+TAIL)}}} = \frac{1}{R_{\text{T(HEAD)}}} + \frac{1}{R_{\text{T(TAIL)}}}$$

$$\frac{1}{R_{\text{T(TAIL)}}} = \frac{1}{R_{\text{T(HEAD+TAIL)}}} - \frac{1}{R_{\text{T(HEAD)}}} = \frac{1}{1.58 \text{ } \Omega} - \frac{1}{2.25 \text{ } \Omega}$$

$$R_{\text{T(TAIL)}} = 5.25 \text{ } \Omega$$

The two taillights are in parallel, so each light has a resistance of

$$R_{\text{TAIL}} = n R_{\text{T(TAIL)}} = 2(5.25 \text{ } \Omega) = \textbf{10.5 } \boldsymbol{\Omega}$$

RELATED PROBLEM

What is the total equivalent resistance of two headlights that each draw 3.15 A?

SECTION 5–4 CHECKUP

1. A 12 V battery is connected across three 680 Ω resistors that are in parallel. What is the total current from the battery?

2. How much voltage is required to produce 20 mA through the circuit of Figure 5–24?

3. How much current is there through each resistor of Figure 5–24?

4. There are four equal-value resistors in parallel with a 12 V source and 6 mA of current from the source. What is the value of each resistor?

5. A 1.0 kΩ and a 2.2 kΩ resistor are connected in parallel. There is a total of 100 mA through the parallel combination. How much voltage is dropped across the resistors?

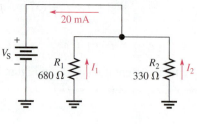

FIGURE 5–24

5–5 KIRCHHOFF'S CURRENT LAW

Kirchhoff's voltage law deals with voltages in a single closed path. Kirchhoff's current law applies to currents in multiple paths.

After completing this section, you should be able to

- **Apply Kirchhoff's current law**
  - **State Kirchhoff's current law**
  - **Define** *node*
  - **Determine the total current by adding the branch currents**
  - **Determine an unknown branch current**

Kirchhoff's current law, often abbreviated KCL, can be stated as follows:

> **The sum of the currents into a node (total current in) is equal to the sum of the currents out of that node (total current out).**

A **node** is any point or junction in a circuit where two or more components are connected. In a parallel circuit, a node is a point where the parallel branches come together. For example, in the circuit of Figure 5–25, there are two nodes as indicated, node *A* and node *B*. Let's

FIGURE 5–25 **Kirchhoff's current law: The current into a node equals the current out of that node.**

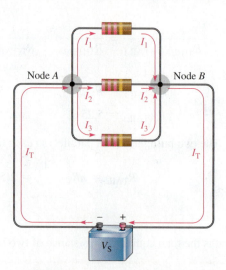

start at the negative terminal of the source and follow the current. The total current I_T from the source is *into* node A. At this point, the current splits up among the three branches as indicated. Each of the three branch currents (I_1, I_2 and I_3) is *out of* node A. Kirchhoff's current law says that the total current into node A is equal to the total current out of node A; that is,

$$I_T = I_1 + I_2 + I_3$$

Now, following the currents in Figure 5–25 through the three branches, you see that they come back together at node B. Currents I_1, I_2, and I_3 are into node B, and I_T is out of node B. Kirchhoff's current law formula at node B is therefore the same as at node A.

$$I_1 + I_2 + I_3 = I_T$$

Kirchhoff's current law can be applied to anything that acts as a load for an electrical system. For example, the power connections to a circuit board will have the same current supplied to the board as leaving the board. Likewise, a building has the same current on the incoming "hot" line as leaving on the neutral. The supply and return currents are always the same, but a fault in a system can cause return current to follow a different path other than the neutral line (a condition known as a "ground fault").

 SYSTEM NOTE

By Kirchhoff's current law, the sum of the currents into a node must equal the sum of the currents out of that node. Figure 5–26 illustrates the general case of Kirchhoff's current law and can be written as a mathematical relationship.

$$I_{IN(1)} + I_{IN(2)} + I_{IN(3)} + \cdots + I_{IN(n)}$$
$$= I_{OUT(1)} + I_{OUT(2)} + I_{OUT(3)} + \cdots + I_{OUT(m)} \qquad (5\text{--}4)$$

Moving the terms on the right side to the left side and changing the sign results in the following equivalent equation:

$$I_{IN(1)} + I_{IN(2)} + I_{IN(3)} + \cdots + I_{IN(n)} - I_{OUT(1)} - I_{OUT(2)} - I_{OUT(3)} - \cdots - I_{OUT(m)} = 0$$

This equation shows that all current into and out of the junction sums to zero. It is equivalent to "balancing the books" by accounting for all current into and out of a node.

An equivalent way of writing Kirchhoff's current law can be expressed using the mathematical summation shorthand as was done for Kirchhoff's voltage law in Section 4–6. To express Kirchhoff's current law with this method, we will assign all currents with a sequential subscript (1, 2, 3, and so on) no matter if the current is into or out of the node.

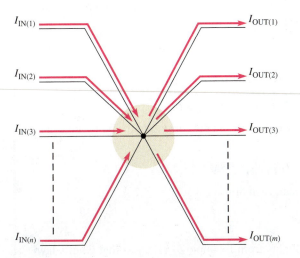

FIGURE 5–26 Generalized circuit node illustrates Kirchhoff's current law.

$$I_{IN(1)} + I_{IN(2)} + I_{IN(3)} + \cdots + I_{IN(n)} = I_{OUT(1)} + I_{OUT(2)} + I_{OUT(3)} + \cdots I_{OUT(m)}$$

Currents into the node are assigned a positive sign and currents leaving the node are assigned a negative sign. Using this notation, Kirchhoff's current law can be stated this way:

$$\sum_{i=1}^{n} I_i = 0$$

This equation uses the mathematical symbol for summation, $\sum$, which means to add the terms from $i = 1$ to $i = n$, taking into account their signs, and set them equal to zero. It can be expressed as a statement as

The algebraic sum of all of the currents entering and leaving a node is equal to zero.

You can verify Kirchhoff's current law by connecting a circuit and measuring each branch current and the total current from the source, as illustrated in Figure 5–27. *When the branch currents are added together, their sum will equal the total current.* This rule applies for any number of branches.

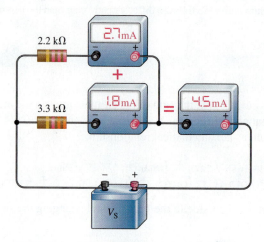

FIGURE 5–27   An illustration of Kirchhoff's current law.

EXAMPLE 5–12

In Example 5–11, you found the equivalent resistance of the headlights and taillights of a car. Using Kirchhoff's current law, find the current in each of the two taillights, given that the total current is 8.0 A and together the headlights draw 5.6 A. Assume the lights are the only load on the battery.

SOLUTION

The current from the battery equals the current to the lights.

$$I_{BAT} = I_{T(HEAD)} + I_{T(TAIL)}$$

The total current from the battery is +8.0 A and the currents to the bulbs are negative. Rearranging,

$$I_{T(TAIL)} = I_{BAT} - I_{T(HEAD)} = 8.0\,A - 5.6\,A = 2.4\,A$$

Because the taillights are identical, each draws 2.4 A/2 = **1.2 A**.

RELATED PROBLEM

Using the resistance of the taillights from Example 5–11, show that Ohm's law gives the same result.

EXAMPLE 5–13

You know the branch currents in the circuit of Figure 5–28. Determine the total current entering node A and the total current leaving node B.

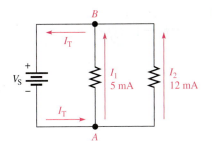

FIGURE 5–28

SOLUTION

The total current out of node A is the sum of the two branch currents. Therefore, the total current into node A is

$$I_T = I_1 + I_2 = 5\,\text{mA} + 12\,\text{mA} = \textbf{17 mA}$$

The total current entering node B is the sum of the two branch currents. Thus, the total current out of node B is

$$I_T = I_1 + I_2 = 5\,\text{mA} + 12\,\text{mA} = \textbf{17 mA}$$

RELATED PROBLEM

If a third resistor is connected in parallel to the circuit of Figure 5–28 and its current is 3 mA, what is the total current into node A and out of node B?

MULTISIM

Open Multisim file E05-13. Verify that the total current is the sum of the branch currents. Repeat for the related problem.

EXAMPLE 5–14

Determine the current through R_2 in Figure 5–29.

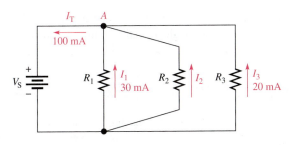

FIGURE 5–29

SOLUTION

The total current out of node A is $I_T = I_1 + I_2 + I_3$. From Figure 5–29, you know the total current and the branch currents through R_1 and R_3. Solve for I_2.

$$I_2 = I_T - I_1 - I_3 = 100\,\text{mA} - 30\,\text{mA} - 20\,\text{mA} = \textbf{50 mA}$$

RELATED PROBLEM

Determine I_T and I_2 if a fourth branch is added to the circuit in Figure 5–29 and it has 12 mA through it.

MULTISIM

Open Multisim file E05-14. Verify the calculated value for I_2. Repeat for the related problem.

EXAMPLE 5–15

Use Kirchhoff's current law to find the current measured by ammeters A3 and A5 in Figure 5–30.

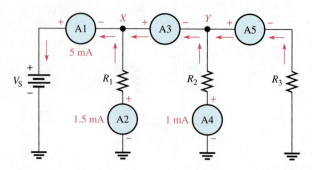

FIGURE 5–30

SOLUTION

The total current out of node X is 5 mA. Two currents are into node X: 1.5 mA through resistor R_1 and the current through A3. Kirchhoff's current law at node X is

$$5\text{ mA} = 1.5\text{ mA} + I_{A3}$$

Solving for I_{A3} yields

$$I_{A3} = 5\text{ mA} - 1.5\text{ mA} = \textbf{3.5 mA}$$

The total current out of node Y is $I_{A3} = 3.5$ mA. Two currents are into node Y: 1 mA through resistor R_2 and the current through A5 and R_3. Kirchhoff's current law applied to node Y gives

$$3.5\text{ mA} = 1\text{ mA} + I_{A5}$$

Solving for I_{A5} yields

$$I_{A5} = 3.5\text{ mA} - 1\text{ mA} = \textbf{2.5 mA}$$

RELATED PROBLEM

How much current will an ammeter measure when it is placed in the circuit in Figure 5–30 right below R_3? Below the negative battery terminal?

SECTION 5–5 CHECKUP

1. State Kirchhoff's current law in two ways.

2. A total current of 2.5 A is into a node and the current out of the node divides into three parallel branches. What is the sum of all three branch currents?

3. In Figure 5–31, 100 mA and 300 mA are into the node. What is the amount of current out of the node?

4. Each of the two taillights on a trailer draws 1 A and the two brake lights each draw another 1 A. What is the current in the ground lead to the trailer when all lights are on?

5. A underground pump has 10 A of current on the hot line. What should be the current on the neutral line?

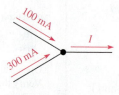

FIGURE 5–31

5–6 CURRENT DIVIDERS

A parallel circuit acts as a **current divider** because the current entering the junction of parallel branches "divides" up into several individual branch currents.

After completing this section, you should be able to

- **Use a parallel circuit as a current divider**
  - **Apply the current-divider formula**
  - **Determine an unknown branch current**

In a parallel circuit, the total current into the junction of the parallel branches (node) divides among the branches. Thus, a parallel circuit acts as a current divider. This current-divider principle is illustrated in Figure 5–32 for a two-branch parallel circuit in which part of the total current I_T goes through R_1 and part through R_2.

Since the same voltage is across each of the resistors in parallel, the branch currents are inversely proportional to the values of the resistors. For example, if the value of R_2 is twice that of R_1, then the value of I_2 is one-half that of I_1. In other words,

> **The total current divides among parallel resistors into currents with values inversely proportional to the resistance values.**

The branches with higher resistance have less current, and the branches with lower resistance have more current, in accordance with Ohm's law. If all the branches have the same resistance, the branch currents are all equal.

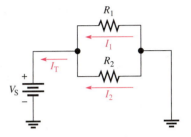

FIGURE 5–32   Total current divides between the two branches.

Power Resistors in the TO Case

In high-current systems (up to about 40 A), a shunt resistor normally has a very small resistance. Small-value resistors with a large power rating are available in various case configurations that are designed to be mounted to a heat sink. The resistor shown is in a TO style case. Resistors as small as 5 mΩ with 1% tolerance are available in the TO case.

The TO case configuration allows heat to be dissipated by the heat sink and away from the sensitive electronics assembly. Power specifications depends on the thermal design and range from 1 W to 100 W. An advantage to resistors used in TO cases is that they can operate in fast-switching circuits and in high-frequency applications because there is little effect on the signal.

SYSTEM NOTE

Figure 5–33 shows specific values to demonstrate how the currents divide according to the branch resistances. Notice that in this case the resistance of the upper branch is one-tenth the resistance of the lower branch, but the upper branch current is ten times the lower branch current.

Current-Divider Formula

You can develop a formula for determining how currents divide among any number of parallel resistors, as shown in Figure 5–34, where n is the total number of resistors.

The current through any one of the parallel resistors is I_x, where x represents the number of a particular resistor (1, 2, 3, and so on). By Ohm's law, you can express the current through any one of the resistors in Figure 5–34 as follows:

$$I_x = \frac{V_S}{R_x}$$

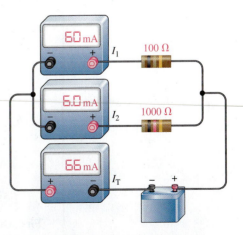

FIGURE 5–33   The branch with the lowest resistance has the most current, and the branch with the highest resistance has the least current.

I_T
+
V_S
−
I_T
I_1
R_1
I_2
R_2
I_3
R_3
R_n
I_n

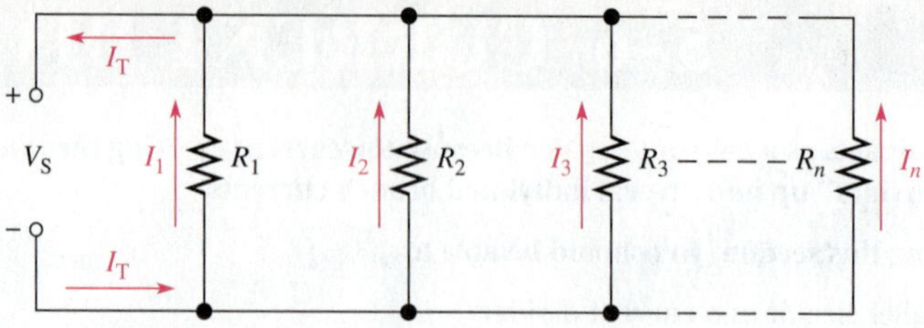

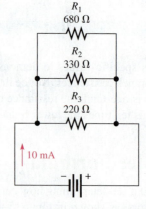

FIGURE 5–34 **A parallel circuit with *n* branches.**

The source voltage, V_S, appears across each of the parallel resistors, and R_x represents any one of the parallel resistors. The total source voltage, V_S, is equal to the total current times the total parallel resistance.

$$V_S = I_T R_T$$

Substituting $I_T R_T$ for V_S in the expression for I_x results in

$$I_x = \frac{I_T R_T}{R_x}$$

Rearranging terms yields

$$I_x = \left(\frac{R_T}{R_x}\right) I_T \quad (5\text{–}5)$$

where $x = 1, 2, 3$, etc. Equation 5–5 is the general current-divider formula and applies to a parallel circuit with any number of branches.

The current (I_x) through any branch equals the total parallel resistance (R_T) divided by the resistance (R_x) of that branch, and then multiplied by the total current (I_T) into the junction of parallel branches.

EXAMPLE 5–16

Determine the current through each resistor in the circuit of Figure 5–35.

FIGURE 5–35

R_1
680 Ω
R_2
330 Ω
R_3
220 Ω
10 mA
−
+

SOLUTION

First, calculate the total parallel resistance.

$$R_T = \frac{1}{\dfrac{1}{R_1} + \dfrac{1}{R_2} + \dfrac{1}{R_3}} = \frac{1}{\dfrac{1}{680\ \Omega} + \dfrac{1}{330\ \Omega} + \dfrac{1}{220\ \Omega}} = 111\ \Omega$$

The total current is 10 mA. Use Equation 5–5 to calculate each branch current.

$$I_1 = \left(\frac{R_T}{R_1}\right)I_T = \left(\frac{111\ \Omega}{680\ \Omega}\right)10\ \text{mA} = \textbf{1.63 mA}$$

$$I_2 = \left(\frac{R_T}{R_2}\right)I_T = \left(\frac{111\ \Omega}{330\ \Omega}\right)10\ \text{mA} = \textbf{3.36 mA}$$

$$I_3 = \left(\frac{R_T}{R_3}\right)I_T = \left(\frac{111\ \Omega}{220\ \Omega}\right)10\ \text{mA} = \textbf{5.05 mA}$$

RELATED PROBLEM

Determine the current through R_1 and R_2 in Figure 5–35 if R_3 is removed. Assume the source voltage remains the same.

CURRENT-DIVIDER FORMULAS FOR TWO BRANCHES   You already know how to use Ohm's law ($I = V/R$) to determine the current in any branch in a parallel circuit when you know the voltage and resistance. When you do not know the voltage but you know total current, you can find both of the branch currents (I_1 and I_2) by using the following formulas:

$$I_1 = \left(\frac{R_2}{R_1 + R_2}\right)I_T \qquad (5\text{–}6)$$

$$I_2 = \left(\frac{R_1}{R_1 + R_2}\right)I_T \qquad (5\text{–}7)$$

These formulas show that the current in either branch is equal to the opposite branch resistance divided by the sum of the two resistances and then multiplied by the total current.

EXAMPLE 5–17

Find I_1 and I_2 in Figure 5–36.

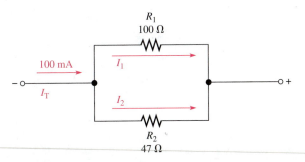

FIGURE 5–36

SOLUTION

Use Equation 5–6 to determine I_1.

$$I_1 = \left(\frac{R_2}{R_1 + R_2}\right)I_T = \left(\frac{47\ \Omega}{147\ \Omega}\right)100\ \text{mA} = \textbf{32.0 mA}$$

Use Equation 5–7 to determine I_2.

$$I_2 = \left(\frac{R_1}{R_1 + R_2}\right)I_T = \left(\frac{100\ \Omega}{147\ \Omega}\right)100\ \text{mA} = \textbf{68.0 mA}$$

> **RELATED PROBLEM**
>
> If $R_1 = 56 \ \Omega$ and $R_2 = 82 \ \Omega$ in Figure 5–36 and I_T stays the same, what will each branch current be?

SYSTEM EXAMPLE 5-2

AMMETER SHUNTS

Many systems incorporate basic analog ammeters; they are built into a control panel to give a visual indication of current. Most ammeters are set up to read more than one scale. Often, the meter is sensitive and will be destroyed with large currents. In order to avoid this and to read higher currents, parallel resistors called *shunts* are switched in, allowing most of the current to go around the meter.

The mechanism in an analog ammeter that causes the pointer to move in proportion to the current is called the *meter movement*. A meter movement has a certain resistance and a maximum current. This maximum current, called the *full-scale deflection current,* causes the pointer to go all the way to the right side of the scale. The meter in this example has 50 Ω internal resistance and a full-scale deflection current of 1 mA. A basic meter with this particular movement can measure currents of 1 mA or less, but if a current greater than 1 mA is in the meter movement, the pointer will "peg" (or stop) at full scale, which can damage the sensitive movement.

Parallel resistors are used to extend the range of an ammeter. The resistor values can be determined by applying the current-divider rule or Ohm's law. Figure 5–37 illustrates a 50 Ω, 1 mA meter with three ranges: 1 mA, 10 mA, and 100 mA. In the 1 mA setting, there is no shunt. In the 10 mA setting, the shunt resistance is 1/9 of the meter resistance or 5.56 Ω.

The maximum current in the meter and maximum shunt current are shown in Table 5–1 for each setting of the range switch. Notice that the maximum meter current never exceeds its rated 1.0 mA. The meter in Figure 5–37 is shown measuring the maximum current. In this case, 99 mA goes through the shunt and 1.0 mA goes through the meter movement. If the meter reads less than full scale, the currents in the meter and the shunt are proportionately less.

TABLE 5–1 • Meter current.		
RANGE SWITCH SETTING	**MAXIMUM METER CURRENT**	**MAXIMUM SHUNT CURRENT**
1 mA	1.0 mA	0
10 mA	1.0 mA	9.0 mA
100 mA	1.0 mA	99.0 mA

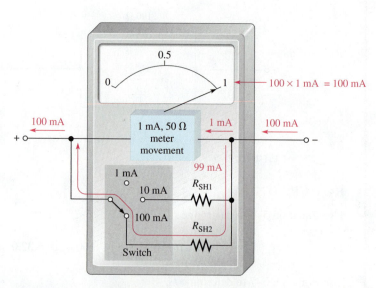

FIGURE 5–37   A milliammeter with three ranges.

SECTION 5–6 CHECKUP

1. A circuit has the following resistors in parallel with a voltage source: 220 Ω, 100 Ω, 68 Ω, 56 Ω, and 22 Ω. Which resistor has the most current through it? The least current?

2. Determine the current through R_3 in Figure 5–38.

3. Find the currents through each resistor in the circuit of Figure 5–39.

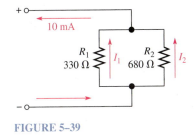

FIGURE 5–39

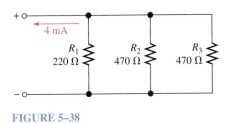

FIGURE 5–38

5–7 POWER IN PARALLEL CIRCUITS

Total power in a parallel circuit is found by adding up the powers of all the individual resistors, the same as for series circuits.

After completing this section, you should be able to

- Determine power in a parallel circuit

Equation 5–8 states the formula for total power for any number of resistors in parallel.

$$P_T = P_1 + P_2 + P_3 + \cdots + P_n \tag{5–8}$$

where P_T is the total power and P_n is the power in the last resistor in parallel. As you can see, the powers are additive, just as in a series circuit.

The power formulas from Chapter 3 are directly applicable to parallel circuits. The following formulas are used to calculate the total power P_T:

$$P_T = V_S I_T$$
$$P_T = I_T^2 R_T$$
$$P_T = \frac{V_S^2}{R_T}$$

where V_S is the voltage across the parallel circuit, I_T is the total current into the parallel circuit, and R_T is the total resistance of the parallel circuit. Examples 5–18 and 5–19 show how total power can be calculated in a parallel circuit.

EXAMPLE 5–18

Determine the total amount of power in the parallel circuit in Figure 5–40.

SOLUTION

The total current is 200 mA. The total resistance is

$$R_T = \cfrac{1}{\cfrac{1}{68\ \Omega} + \cfrac{1}{33\ \Omega} + \cfrac{1}{22\ \Omega}} = 11.1\ \Omega$$

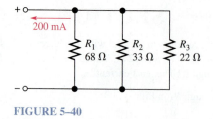

FIGURE 5–40

The easiest formula to use is $P_T = I_T^2 R_T$ since you know both I_T and R_T. Thus,

$$P_T = I_T^2 R_T = (200 \text{ mA})^2 (11.1 \text{ } \Omega) = \textbf{444 mW}$$

Let's demonstrate that if you determine the power in each resistor and if you add all of these values together, you get the same result. First, find the voltage across each branch of the circuit.

$$V_S = I_T R_T = (200 \text{ mA})(11.1 \text{ } \Omega) = 2.22 \text{ V}$$

Remember that the voltage across all branches is the same.

Next, use $P = V_S^2/R$ to calculate the power for each resistor.

$$P_1 = \frac{(2.22 \text{ V})^2}{68 \text{ } \Omega} = 72.5 \text{ mW}$$

$$P_2 = \frac{(2.22 \text{ V})^2}{33 \text{ } \Omega} = 149 \text{ mW}$$

$$P_3 = \frac{(2.22 \text{ V})^2}{22 \text{ } \Omega} = 224 \text{ mW}$$

Now, add these powers to get the total power.

$$P_T = 72.5 \text{ mW} + 149 \text{ mW} + 224 \text{ mW} = \textbf{446 mW}$$

This calculation shows that the sum of the individual powers is equal (approximately) to the total power as determined by one of the power formulas. Rounding to three significant figures accounts for the difference.

RELATED PROBLEM

Find the total power in Figure 5–40 if the total voltage is doubled.

EXAMPLE 5–19

The amplifier in one channel of a stereo system as shown in Figure 5–41 drives two parallel speakers. If the maximum voltage* to the speakers is 15 V, how much power must the amplifier be able to deliver to the speakers?

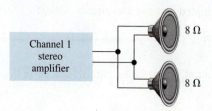

FIGURE 5–41

---

*Voltage is ac in this case; but power is determined the same for ac voltage as for dc voltage, as you will learn later.

SOLUTION

The speakers are connected in parallel to the amplifier output, so the voltage across each is the same. The maximum power to each speaker is

$$P_{max} = \frac{V_{max}^2}{R} = \frac{(15\ V)^2}{8\ \Omega} = 28.1\ W$$

The total power that the amplifier must be capable of delivering to the speaker system is twice the power in an individual speaker because the total power is the sum of the individual powers.

$$P_{T(max)} = P_{max} + P_{max} = 2P_{max} = 2(28.1\ W) = \textbf{56.2 W}$$

RELATED PROBLEM

If the amplifier can produce a maximum of 18 V, what is the maximum total power to the speakers?

SECTION 5–7 CHECKUP

1. If you know the power in each resistor in a parallel circuit, how can you find the total power?

2. The resistors in a parallel circuit dissipate the following powers: 1 W, 2 W, 5 W, and 8 W. What is the total power in the circuit?

3. A circuit has a 1.0 kΩ, a 2.7 kΩ, and a 3.9 kΩ resistor in parallel. There is a total current of 1 mA into the parallel circuit. What is the total power?

4. Normally, circuits are protected with a fuse that is at least 120% of the maximum current expected. What size of fuse should be installed for an automotive rear window defroster that is rated for 100 W? Assume the battery voltage is 12.6 V.

5–8 TROUBLESHOOTING

Recall that an open circuit is one in which the current path is interrupted and there is no current. In this section, you will see how an open in a parallel branch affects the parallel circuit.

After completing this section, you should be able to

- **Troubleshoot parallel circuits**
  - **Check for an open in a circuit**

Open Branches

If a switch is connected in a branch of a parallel circuit, as shown in Figure 5–42, an open or a closed path can be made by the switch. When the switch is closed, as in Figure 5–42(a), R_1 and R_2 are in parallel. The total resistance is 50 Ω (two 100 Ω resistors in parallel). Current is through both resistors. If the switch is opened, as in Figure 5–42(b), R_1 is effectively removed

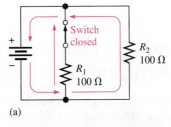

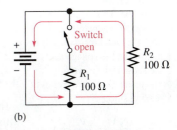

(a)

(b)

FIGURE 5–42 **When the switch opens, total current decreases and current through R_2 remains unchanged.**

from the circuit, and the total resistance is 100 Ω. There is still the same voltage across R_2 and the same current through it although the total current from the source is reduced by the amount of the R_1 current.

In general,

> **When an open occurs in a parallel branch, the total resistance increases, the total current decreases, and the same current continues through each of the remaining parallel paths.**

Consider the lamp circuit in Figure 5–43. There are four bulbs in parallel with a 12 V source. In part (a), there is current through each bulb. Now suppose that one of the bulbs burns out, creating an open path as shown in Figure 5–43(b). This light will go out because there is no current through the open path. Notice, however, that current continues through all the other parallel bulbs, and they continue to glow. The open branch does not change the voltage across the remaining parallel branches; it remains at 12 V, and the current through each branch remains the same.

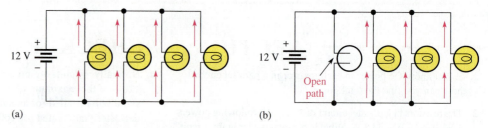

(a)   (b)

FIGURE 5–43   When a lamp filament opens, total current decreases by the amount of current in the lamp that opened. The other branch currents remain unchanged.

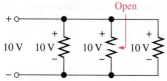

FIGURE 5–44   All parallel branches (open or not) have the same voltage.

You can see that a parallel circuit has an advantage over a series connection in lighting systems because if one or more of the parallel bulbs burn out, the others will stay on. In a series circuit, when one bulb goes out, all of the others go out also because the current path is completely interrupted.

When a resistor in a parallel circuit opens, the open resistor cannot be located by measurement of the voltage across the branches because the same voltage exists across all the branches. Thus, there is no way to tell which resistor is open by simply measuring voltage (the half-splitting method is not applicable). The good resistors will always have the same voltage as the open one, as illustrated in Figure 5–44 (note that the middle resistor is open). However, a temperature measurement with a thermal-imaging camera (see System Note on page 202) can reveal which resistors have current (good) and which do not (bad).

If a visual inspection does not reveal the open resistor, it can be located by resistance or current measurements. In practice, measuring resistance or current is more difficult than measuring voltage because you must disconnect a component to measure the resistance and you must insert an ammeter in series to measure the current. Thus, a wire or a printed circuit connection must usually be cut or disconnected, or one end of a component must be lifted off the circuit board, in order to connect a DMM to measure resistance and current. This procedure, of course, is not required when voltage measurements are made because the meter leads are simply connected across a component.

HANDS ON TIP
An alternate way to measure current without having to break a circuit to connect an ammeter is to place, as part of the original circuit, a 1 Ω series resistor in each line through which current is to be measured. This small "sense" resistor will generally not affect the total resistance. By measuring the voltage across the sense resistor, you automatically have the current reading since

$$I = \frac{V}{R} = \frac{V}{1\ \Omega} = V$$

Finding an Open Branch by Current Measurement

In a parallel circuit with a suspected open branch, the total current can be measured to find the open. *When a parallel resistor opens, I_T is always less than its normal value.* Once you know I_T and the voltage across the branches, a few calculations will determine the open resistor when all the resistors are of different values.

Consider the two-branch circuit in Figure 5–45(a). If one of the resistors opens, the total current will equal the current in the good resistor. Ohm's law quickly tells you what the current in each resistor should be.

$$I_1 = \frac{50\ V}{560\ \Omega} = 89.3\ mA$$

$$I_2 = \frac{50\ V}{100\ \Omega} = 500\ mA$$

$$I_T = I_1 + I_2 = 589.3\ mA$$

If R_2 is open, the total current is 89.3 mA, as indicated in Figure 5–45(b). If R_1 is open, the total current is 500 mA, as indicated in Figure 5–45(c).

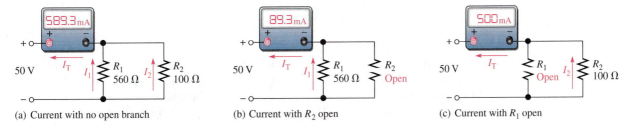

(a) Current with no open branch   (b) Current with R_2 open   (c) Current with R_1 open

FIGURE 5–45   **Finding an open path by current measurement.**

This procedure can be extended to any number of branches having unequal resistances. If the parallel resistances are all equal, the current in each branch must be checked until a branch is found with no current.

EXAMPLE 5–20

In Figure 5–46, there is a total current of 31.09 mA, and the voltage across the parallel branches is 20 V. Is there an open resistor, and if so, which one is it?

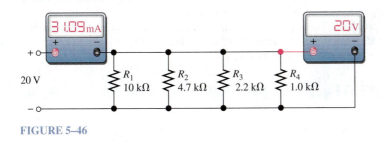

FIGURE 5–46

SOLUTION

Calculate the current in each branch.

$$I_1 = \frac{V}{R_1} = \frac{20\ V}{10\ k\Omega} = 2\ mA$$

$$I_2 = \frac{V}{R_2} = \frac{20\ V}{4.7\ k\Omega} = 4.26\ mA$$

$$I_3 = \frac{V}{R_3} = \frac{20\ V}{2.2\ k\Omega} = 9.09\ mA$$

$$I_4 = \frac{V}{R_4} = \frac{20\ V}{1.0\ k\Omega} = 20\ mA$$

MULTISIM

Open Multisim file E05-20. Measure the total current and the current in each resistor. There are no faults in the circuit.

The total current should be

$$I_T = I_1 + I_2 + I_3 + I_4 = 2\,\text{mA} + 4.26\,\text{mA} + 9.09\,\text{mA} + 20\,\text{mA}$$
$$= 35.35\,\text{mA}$$

The actual measured current is 31.09 mA, as stated, which is 4.26 mA less than normal, indicating that the branch with 4.26 mA is open. Thus, R_2 **must be open**.

RELATED PROBLEM

What is the total current measured in Figure 5–46 if R_4 and not R_2 is open?

Shorted Branches

When a branch in a parallel circuit shorts, the current increases to an excessive value, usually causing the resistor to burn open or a fuse or circuit breaker to blow. This results in a difficult troubleshooting problem because it is hard to isolate the shorted branch.

A pulser and a current tracer are tools often used to find shorts in a circuit. They are not restricted to use in digital circuits but can be effective in any type of circuit. The pulser is a pen-shaped tool that applies pulses to a selected point in a circuit, causing pulses of current through the shorted path. The current tracer is also a pen-shaped tool that senses pulses of current. By following the current with the tracer, the current path can be identified.

Thermal Imaging

Components are always designed to handle a certain amount of power as specified by the manufacturer. Sometimes a fault occurs (such as a loose or oxidized connection) and a component or junction overheats. A tool for detecting overheating in any device is thermal imaging, which is sensitive to the infrared radiation emitted by all objects. Thermal imaging is useful in a number of fields (e.g., mechanical, electrical, medical).

Courtesy of Fluke Corporation

Many manufacturers routinely do thermal imaging of equipment as part of a preventive maintenance program. Thermal imaging has advantages in that it can often spot potential problems before they cause a complete malfunction. A thermal inspection is noncontacting and very sensitive to temperature differences.

SYSTEM NOTE

Thermography

Figure 5–47 shows the main part of the electromagnetic spectrum. Electromagnetic energy is a continuous band of frequencies that includes visible light. We respond to the visible region with our eyes, but this is only a small fraction of the electromagnetic spectrum. The infrared region has longer wavelengths than the visible region. The infrared region is subdivided into the *near-infrared, mid-infrared,* and the *far-infrared.* It is this far-infrared region that the temperature-sensitive nerve endings in our skin sense as thermal energy when we are in sunlight or stand next to a hot stove. At wavelengths that are longer than the far-infrared are microwave frequencies and radio waves. At wavelengths shorter than the visible light region are the ultraviolet and x-ray regions.

The surfaces of all bodies radiate electromagnetic energy. If all bodies are constantly emitting energy, why don't they eventually radiate all of their energy and cool to absolute zero? The answer is because they continuously absorb energy from their surroundings. If

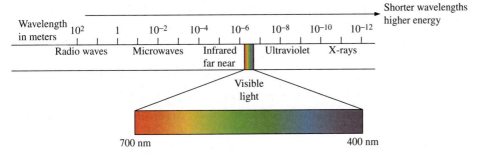

Shorter wavelengths
higher energy

Wavelength in meters

10^2 $\quad$ 1 $\quad$ 10^{-2} $\quad$ 10^{-4} $\quad$ 10^{-6} $\quad$ 10^{-8} $\quad$ 10^{-10} $\quad$ 10^{-12}

Radio waves $\quad$ Microwaves $\quad$ Infrared $\quad$ Ultraviolet $\quad$ X-rays
far near

Visible light

700 nm $\qquad\qquad$ 400 nm

FIGURE 5–47 **The electromagnetic spectrum.**

the amount of energy they absorb is greater than what is radiated, the temperature rises; if it is less, the temperature falls. The wavelength that is emitted depends on the temperature of the body. If a body is hot enough, it radiates energy in the visible spectrum. If it is cooler, the radiated energy is primarily in the infrared region.

Although our skin is very sensitive to heat, humans have physical limitations on detecting specific direction and quantizing the amount of heat. Thermal imaging instruments extend our senses and enable us to measure heat and provide a radiometric image of a hot source. A **radiometric image** is a thermal image that contains temperature-measurement calculations for various points within the image. The image is displayed on a screen where colors correspond to the temperature of the infrared radiation emitted by the target. Figure 5–48 shows a thermal imager that is viewing the infrared energy emitted by three fuses. The color corresponds to the temperature within the image. From this, it is obvious which fuses are carrying the largest current because they are warmer.

Thermal imaging is used in a variety of applications and is widely used in electronics and industrial operations for both troubleshooting and preventive maintenance. It can immediately show a component that is too hot or one that is not carrying current because it is cooler than normal. In the case of the dummy load in System Example 5–1, the set of parallel resistors needs to all be functioning correctly to present the required resistance that simulates an antenna. If one resistor were to open, the total resistance would increase and there would be a reflected signal back toward the transmitter. A thermal imager will show the open resistor as a cooler component than the others with no need to disconnect it in the circuit to test it. The test is fast and noncontacting.

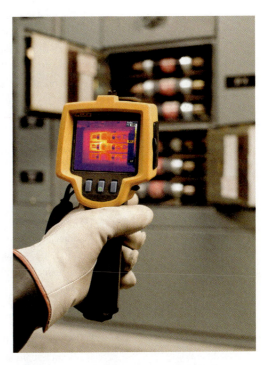

FIGURE 5–48 **Thermal imager showing the temperature profile of the target (a set of three fuses).** (Courtesy of Fluke Corporation)

SECTION 5–8 CHECKUP

1. If a parallel branch opens, what changes can be detected in the circuit's voltage and the currents, assuming that the parallel circuit is across a constant voltage source?

2. What happens to the total resistance if one branch opens?

3. If several light bulbs are connected in parallel and one of the bulbs opens (burns out), will the others continue to glow?

4. There is one ampere of current in each branch of a parallel circuit. If one branch opens, what is the current in each of the remaining branches?

5. Assume a load draws a hot current of 1.00 A, and the neutral return current is 0.90 A. If the source voltage is 120 V, what is the apparent faulty resistance in the ground path?

6. What do the colors represent in a radiometric image?

SUMMARY

- Resistors in parallel are connected between two nodes in a circuit.
- A parallel circuit provides more than one path for current.
- The total parallel resistance is less than the lowest-value parallel resistor.
- The voltages across all branches of a parallel circuit are the same.
- Kirchhoff's current law: The sum of the currents into a node (total current in) equals the sum of the currents out of the node (total current out).

  The algebraic sum of all the currents entering and leaving a node is zero.
- A parallel circuit is a current divider, so called because the total current entering a node divides up into each of the branches connected to the node.
- If all of the branches of a parallel circuit have equal resistance, the currents through all of the branches are equal.
- The total power in a parallel-resistive circuit is the sum of all of the individual powers of the resistors making up the parallel circuit.
- The total power for a parallel circuit can be calculated with the power formulas using values of total current, total resistance, or total voltage.
- If one of the branches of a parallel circuit opens, the total resistance increases, and therefore the total current decreases.
- If a branch of a parallel circuit opens, there is still the same current through the remaining branches.
- A radiometric image of a circuit is useful for preventive maintenance and for troubleshooting.

KEY TERMS

Key terms and other bold terms in the chapter are defined in the end-of-book glossary.

Branch   One current path in a parallel circuit.

Current divider   A parallel circuit in which the currents divide inversely proportional to the parallel branch resistances.

Kirchhoff's current law   A law stating that the total current into a node equals the total current out of the node. Equivalently, it can be stated that the algebraic sum of all the currents entering and leaving a node is zero.

Node   A point or junction in a circuit at which two or more components are connected.

Parallel   The relationship between two circuit components that exists when they are connected between the same pair of nodes.

Radiometric image   A thermal picture in which colors represent temperature.

KEY FORMULAS

(5–1)	$R_T = \dfrac{1}{\dfrac{1}{R_1} + \dfrac{1}{R_2} + \dfrac{1}{R_3} + \cdots + \dfrac{1}{R_n}}$	Total parallel resistance
(5–2)	$R_T = \dfrac{R_1 R_2}{R_1 + R_2}$	Special case for two resistors in parallel
(5–3)	$R_T = \dfrac{R}{n}$	Special case for n equal-value resistors in parallel
(5–4)	$\begin{aligned} I_{\text{IN}(1)} + I_{\text{IN}(2)} + I_{\text{IN}(3)} + \cdots + I_{\text{IN}(n)} \\ = I_{\text{OUT}(1)} + I_{\text{OUT}(2)} + I_{\text{OUT}(3)} \\ + \cdots + I_{\text{OUT}(m)} \end{aligned}$	Kirchhoff's current law
(5–5)	$I_x = \left(\dfrac{R_T}{R_x} \right) I_T$	General current-divider formula
(5–6)	$I_1 = \left(\dfrac{R_2}{R_1 + R_2} \right) I_T$	Two-branch current-divider formula

(5–7) $I_2 = \left(\dfrac{R_1}{R_1 + R_2}\right)I_T$ Two-branch current-divider formula

(5–8) $P_T = P_1 + P_2 + P_3 + \cdots + P_n$ Total power

TRUE/FALSE QUIZ

Answers are at the end of the chapter.

1. To find the total conductance of parallel resistors, you can add the conductance of each of the resistors.

2. The total resistance of parallel resistors is always smaller than the smallest resistor.

3. The product-over-sum rule works for any number of parallel resistors.

4. In a parallel circuit, the voltage is larger on a larger resistor and smaller on a smaller resistor.

5. When a new path is added to a parallel circuit, the total resistance goes up.

6. When a new path is added to a parallel circuit, the total current goes up.

7. The total current entering a node is always equal to the total current leaving the node.

8. In the current-divider formula, $I_x = (R_T/R_x)I_T$, the quantity in parentheses is a fraction that is never greater than 1.

9. When two resistors are in parallel, the smaller resistor will have the smaller power dissipation.

10. The total power dissipated by parallel resistors can be larger than the power supplied by the source.

SELF-TEST

Answers are at the end of the chapter.

1. In a parallel circuit, each resistor has
   - (a) the same current
   - (b) the same voltage
   - (c) the same power
   - (d) all of the above

2. When a 1.2 kΩ resistor and a 100 Ω resistor are connected in parallel, the total resistance is
   - (a) greater than 1.2 kΩ
   - (b) greater than 100 Ω but less than 1.2 kΩ
   - (c) less than 100 Ω but greater than 90 Ω
   - (d) less than 90 Ω

3. A 330 Ω resistor, a 270 Ω resistor, and a 68 Ω resistor are all in parallel. The total resistance is approximately
   - (a) 668 Ω
   - (b) 47 Ω
   - (c) 68 Ω
   - (d) 22 Ω

4. Eight resistors are in parallel. The two lowest-value resistors are both 1.0 kΩ. The total resistance
   - (a) cannot be determined
   - (b) is greater than 1.0 kΩ
   - (c) is less than 1.0 kΩ
   - (d) is less than 500 Ω

5. When an additional resistor is connected across an existing parallel circuit, the total resistance
   - (a) decreases
   - (b) increases
   - (c) remains the same
   - (d) increases by the value of the added resistor

6. If one of the resistors in a parallel circuit is removed, the total resistance
   - (a) decreases by the value of the removed resistor
   - (b) remains the same
   - (c) increases
   - (d) doubles

7. The currents into a node are along two paths. One current is 5 A and the other is 3 A. The total current out of the node is
   - (a) 2 A
   - (b) unknown
   - (c) 8 A
   - (d) the larger of the two

8. The following resistors are in parallel across a voltage source: 390 Ω, 560 Ω, and 820 Ω. The resistor with the least current is
   - (a) 390 Ω
   - (b) 560 Ω
   - (c) 820 Ω
   - (d) impossible to determine without knowing the voltage

9. A sudden decrease in the total current into a parallel circuit may indicate
   (a) a short   (b) an open resistor
   (c) a drop in source voltage   (d) either (b) or (c)

10. In a four-branch parallel circuit, there are 10 mA of current in each branch. If one of the branches opens, the current in each of the other three branches is
    (a) 13.33 mA   (b) 10 mA   (c) 0 A   (d) 30 mA

11. In a certain three-branch parallel circuit, R_1 has 10 mA through it, R_2 has 15 mA through it, and R_3 has 20 mA through it. After measuring a total current of 35 mA, you can say that
    (a) R_1 is open   (b) R_2 is open
    (c) R_3 is open   (d) the circuit is operating properly

12. If there are a total of 100 mA into a parallel circuit consisting of three branches and two of the branch currents are 40 mA and 20 mA, the third branch current is
    (a) 60 mA   (b) 20 mA   (c) 160 mA   (d) 40 mA

13. A complete short develops across one of five parallel resistors on a PC board. The most likely result is
    (a) the smallest-value resistor will burn out
    (b) one or more of the other resistors will burn out
    (c) the fuse in the power supply will blow
    (d) the resistance values will be altered

14. The power dissipation in each of four parallel branches is 1 mW. The total power dissipation is
    (a) 1 mW   (b) 4 mW   (c) 0.25 mW   (d) 16 mW

TROUBLESHOOTING: SYMPTOM AND CAUSE

The purpose of these exercises is to help develop thought processes essential to troubleshooting. Answers are at the end of the chapter.

Determine the cause for each set of symptoms. Refer to Figure 5–49.

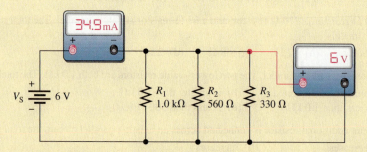

FIGURE 5–49   **The meters indicate the correct readings for this circuit.**

1. *Symptom:* The ammeter and voltmeter readings are zero.
   Cause:
   (a) R_1 is open.   (b) The voltage source is turned off or faulty.   (c) R_3 is open.

2. *Symptom:* The ammeter reading is 16.7 mA, and the voltmeter reading is 6 V.
   Cause:
   (a) R_1 is open.   (b) R_2 is open.   (c) R_3 is open.

3. *Symptom:* The ammeter reading is 28.9 mA, and the voltmeter reading is 6 V.
   Cause:
   (a) R_1 is open.   (b) R_2 is open.   (c) R_3 is open.

4. *Symptom:* The ammeter reading is 24.2 mA, and the voltmeter reading is 6 V.

   Cause:

   (a) R_1 is open.
   (b) R_2 is open.
   (c) R_3 is open.

5. *Symptom:* The ammeter reading is 34.9 mA, and the voltmeter reading is 0 V.

   Cause:

   (a) A resistor is shorted.
   (b) The voltmeter is faulty.
   (c) The voltage source is turned off or faulty.

PROBLEMS

Answers to odd-numbered problems are at the end of the book.

BASIC PROBLEMS

SECTION 5–1   Resistors in Parallel

1. Connect the resistors in Figure 5–50 in parallel across the battery.

2. Determine whether or not all the resistors in Figure 5–51 are connected in parallel on the printed circuit board. Draw the schematic including resistor values.

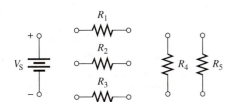

FIGURE 5–50

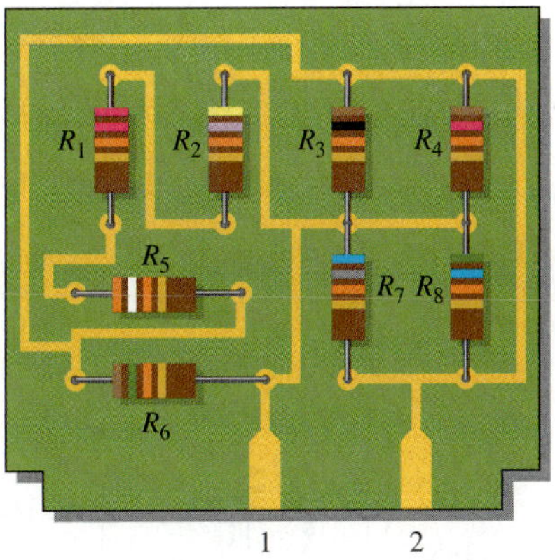

FIGURE 5–51

SECTION 5–2   Total Parallel Resistance

3. Determine the total resistance between pins 1 and 2 in Figure 5–51.

4. The following resistors are connected in parallel: 1.0 MΩ, 2.2 MΩ, 4.7 MΩ, 12 MΩ, and 22 MΩ. Determine the total resistance.

5. Find the total resistance between nodes *A* and *B* for each group of parallel resistors in Figure 5–52.

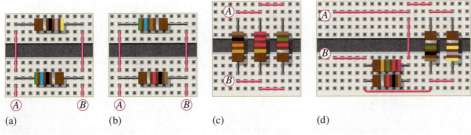

(a)   (b)   (c)   (d)

FIGURE 5–52

6. Calculate R_T for each circuit in Figure 5–53.

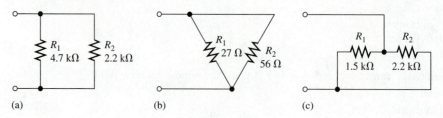

(a) (b) (c)

FIGURE 5–53

7. What is the total resistance of eleven 22 kΩ resistors in parallel?

8. Five 15 Ω, ten 100 Ω, and two 10 Ω resistors are all connected in parallel. What is the total resistance?

SECTION 5–3 Voltage in a Parallel Circuit

9. Determine the voltage across and the current through each parallel resistor if the total voltage is 12 V and the total resistance is 600 Ω. There are four resistors, all of equal value.

10. The source voltage in Figure 5–54 is 100 V. How much voltage does each of the three meters read?

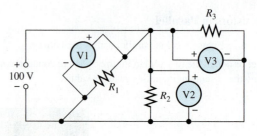

FIGURE 5–54

SECTION 5–4 Application of Ohm's Law

11. What is the total current, I_T, in each circuit of Figure 5–55?

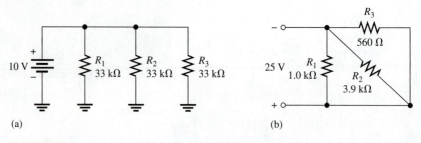

(a) (b)

FIGURE 5–55

12. The resistance of a 60 W bulb is approximately 240 Ω. What is the current from the source when three bulbs are on in a 120 V parallel circuit?

13. What is the current in each resistor for the circuits shown in Figure 5–56?

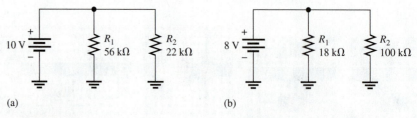

(a) (b)

FIGURE 5–56

14. Four equal-value resistors are connected in parallel. Five volts are applied across the parallel circuit, and 2.5 mA are measured from the source. What is the value of each resistor?

SECTION 5–5   Kirchhoff's Current Law

15. The following currents are measured in the same direction in a three-branch parallel circuit: 250 mA, 300 mA, and 800 mA. What is the value of the current into the node to which these three branches are connected?

16. There is a total of 500 mA of current into five parallel resistors. The currents through four of the resistors are 50 mA, 150 mA, 25 mA, and 100 mA. What is the current through the fifth resistor?

17. How much current is through R_2 and R_3 in Figure 5–57 if R_2 and R_3 have the same resistance? Show how to connect ammeters to measure these currents.

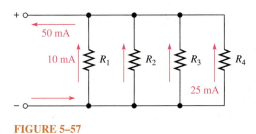

FIGURE 5–57

18. A trailer has four running lights that draw 0.5 A each and two taillights that draw 1.2 A each. What is the current supplied to the trailer when the taillights and running lights are on?

19. Assume the trailer in Problem 18 has two brake lights that draw 1 A each.
   (a) When all lights are on, what current is supplied to the trailer?
   (b) For this condition, what is the ground return current from the trailer?

SECTION 5–6   Current Dividers

20. A 10 kΩ resistor and a 15 kΩ resistor are in parallel with a voltage source. Which resistor has the highest current?

21. How much branch current should each meter in Figure 5–58 indicate?

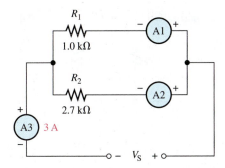

FIGURE 5–58

22. Using the current-divider formula, determine the current in each branch of the circuits of Figure 5–59.

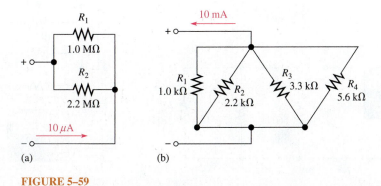

(a)   (b)

FIGURE 5–59

SECTION 5–7   Power in Parallel Circuits

23. Five parallel resistors each handle 40 mW. What is the total power?

24. Determine the total power in each circuit of Figure 5–59.

25. Six light bulbs are connected in parallel across 120 V. Each bulb is rated at 75 W. How much current is through each bulb, and what is the total current?

SECTION 5–8   Troubleshooting

26. If one of the bulbs burns out in Problem 25, how much current will be through each of the remaining bulbs? What will be the total current?

27. In Figure 5–60, the current and voltage measurements are indicated. Has a resistor opened, and, if so, which one?

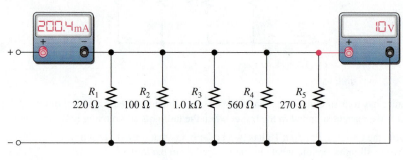

FIGURE 5–60

28. What is wrong with the circuit in Figure 5–61?

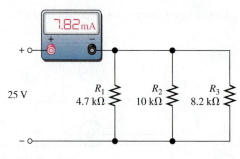

FIGURE 5–61

29. Find the open resistor in Figure 5–62.

30. From the ohmmeter reading in Figure 5–63, can you tell if a resistor is open? If so, can you identify it?

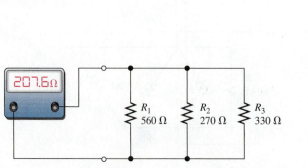

FIGURE 5–62

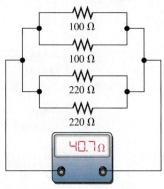

FIGURE 5–63

ADVANCED PROBLEMS

31. In the circuit of Figure 5–64, determine resistances R_2, R_3, and R_4.

32. The total resistance of a parallel circuit is 25 Ω. If the total current is 100 mA, how much current is through a 220 Ω resistor that makes up part of the parallel circuit?

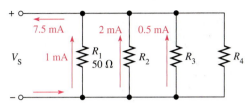

FIGURE 5–64

33. What is the current through each resistor in Figure 5–65? R is the lowest-value resistor, and all others are multiples of that value as indicated.

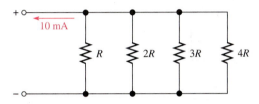

FIGURE 5–65

34. A certain parallel circuit consists of only $\frac{1}{2}$ W resistors each having the same resistance. The total resistance is 1 kΩ, and the total current is 50 mA. If each resistor is operating at one-half its maximum power level, determine the following:
(a) the number of resistors   **(b)** the value of each resistor
(c) the current in each branch   **(d)** the applied voltage

35. Find the values of the unspecified quantities (shown in color) in each circuit of Figure 5–66.

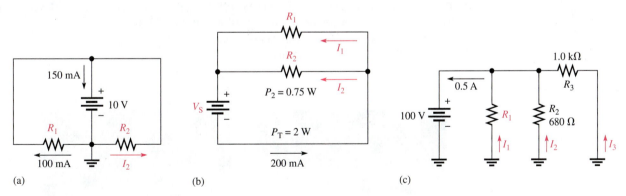

(a)   (b)   (c)

FIGURE 5–66

36. What is the total resistance between terminal A and ground in Figure 5–67 for the following conditions?
(a) SW1 and SW2 open   **(b)** SW1 closed, SW2 open
(c) SW1 open, SW2 closed   **(d)** SW1 and SW2 closed

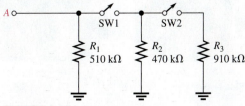

FIGURE 5–67

37. What value of R_2 in Figure 5–68 will cause excessive current?

38. Determine the total current from the source and the current through each resistor for each switch position in Figure 5–69.

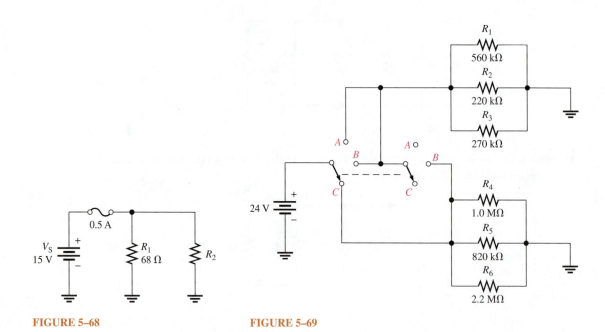

FIGURE 5–68

FIGURE 5–69

39. The electrical circuit in a room is protected with a 15 A circuit breaker. A space heater that draws 8.0 A is plugged into one wall outlet, and two table lamps that draw 0.833 A each are plugged into another outlet. Can a vacuum cleaner that uses 5.0 A be plugged into this same circuit without exceeding the breaker current? Explain your answer.

40. The total resistance of a parallel circuit is 25 Ω. What is the current through a 220 Ω resistor that makes up part of the parallel circuit if the total current is 100 mA?

41. Identify which groups of resistors are in parallel on the double-sided PC board in Figure 5–70. Determine the total resistance of each group.

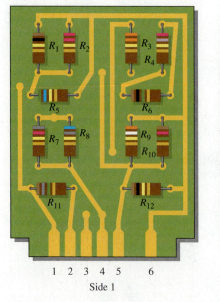
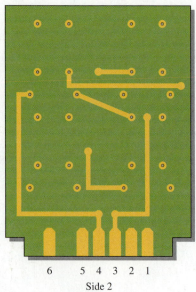

FIGURE 5–70

42. If the total resistance in Figure 5–71 is 200 Ω, what is the value of R_2?

43. Determine the unknown resistances in Figure 5–72.

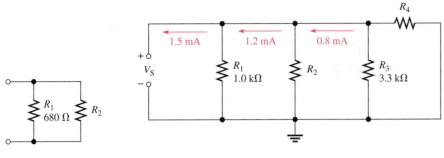

FIGURE 5–71   **FIGURE 5–72**

44. There is a total of 250 mA into a parallel circuit with a total resistance of 1.5 kΩ. The current must be increased by 25%. Determine how much resistance to add in parallel to accomplish this increase in current.

45. Draw the schematic for the setup in Figure 5–73 and determine what is wrong with the circuit if 25 V are applied across the red and black leads.

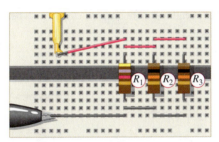

(b) Protoboard with leads connected. Yellow lead is from meter and gray lead is from 25 V power supply ground. The red meter lead goes to +25 V.

(a) Meter with yellow lead going to protoboard and red lead going to the positive terminal of the 25 V power supply.

FIGURE 5–73

46. Refer to System Example 5–1. If you needed to design the load for a 10 W, 75 Ω resistance using standard values of 1 W, 1% resistors, what values would you choose? Draw your completed design.

47. A dummy load has the values shown in Figure 5–74. All resistors are rated for 2 W except R_{14} and R_{15} are rated for ½ W each. (a) What is the resistance of the load? (b) What is its power rating?

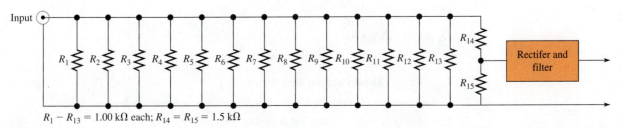

$R_1 - R_{13} = 1.00$ kΩ each; $R_{14} = R_{15} = 1.5$ kΩ

FIGURE 5–74

48. Refer to System Example 5–2. For this example, what is the correct value of R_{SH2}?

49. Assume you have a meter that has a 100 μA full-scale deflection current and an internal resistance of 80 Ω. What shunt resistance would you specify to cause the meter to read 1 mA full scale?

MULTISIM TROUBLESHOOTING PROBLEMS

50. Open file P05-50; files are found at www.pearsonhighered.com/floyd. Using current measurements, determine if there is a fault in the circuit. If so, identify the fault.

51. Open file P05-51. Using current measurements, determine if there is a fault in the circuit. If so, identify the fault.

52. Open file P05-52. Using resistance measurements, determine if there is a fault in the circuit. If so, identify the fault.

53. Open file P05-53. Measure the total resistance of each circuit and compare to the calculated values.

ANSWERS TO SECTION CHECKUPS

SECTION 5–1 Resistors in Parallel

1. Parallel resistors are connected between the same two points.

2. A parallel circuit has more than one current path between two given points.

3. See Figure 5–75.

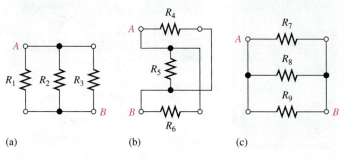

(a)   (b)   (c)

FIGURE 5–75

4. See Figure 5–76.

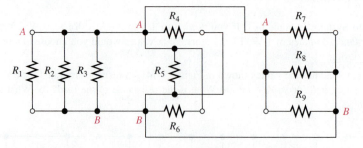

FIGURE 5–76

SECTION 5–2 Total Parallel Resistance

1. As more resistors are connected in parallel, the total resistance decreases.

2. R_T is always less than the smallest resistance value.

3. $R_T = 2.2 \text{ k}\Omega/12 = 183 \ \Omega$

SECTION 5–3   Voltage in a Parallel Circuit

1. 5 V
2. $V_{R2} = 118$ V; $V_S = 118$ V
3. $V_{R1} = 50$ V; $V_{R2} = 50$ V
4. Voltage is the same across all parallel branches.

SECTION 5–4   Application Ohm's Law

1. $I_T = 12$ V/$(680 \ \Omega/3) = 53$ mA
2. $V_S = 20$ mA$(680 \ \Omega \| 330 \ \Omega) = 4.44$ V
3. $I_1 = 4.44$ V/$680 \ \Omega = 6.53$ mA; $I_2 = 4.44$ V/$330 \ \Omega = 13.5$ mA
4. $R_T = 4(12$ V/6 mA$) = 8 \ k\Omega$
5. $V = (1.0 \ k\Omega \| 2.2 \ k\Omega)100$ mA $= 68.8$ V

SECTION 5–5   Kirchhoff's Current Law

1. Kirchhoff's current law: The algebraic sum of all currents at a node is zero; The sum of the currents entering a node equals the sum of all the currents leaving that node.
2. $I_1 + I_2 + I_3 = 2.5$ A
3. 100 mA $+ 300$ mA $= 400$ mA
4. 4 A
5. 10 A

SECTION 5–6   Current Dividers

1. The $22 \ \Omega$ has most current; the $220 \ \Omega$ has least current.
2. $I_3 = (R_T/R_3)4$ mA $= (113.6 \ \Omega/470 \ \Omega)4$ mA $= 967 \ \mu A$
3. $I_2 = (R_T/680 \ \Omega)10$ mA $= 3.27$ mA; $I_1 = (R_T/330 \ \Omega)10$ mA $= 6.73$ mA

SECTION 5–7   Power in Parallel Circuits

1. Add the powers in each resistor to get P_T.
2. $P_T = 1$ W $+ 2$ W $+ 5$ W $+ 8$ W $= 16$ W
3. $P_T = (1$ mA$)^2 R_T = 615 \ \mu W$
4. The normal current is $I = P/V = 100$ W/12.6 V $= 7.9$ A. Choose a 10 A fuse.

SECTION 5–8   Troubleshooting

1. When a parallel branch opens, there is no change in voltage and the total current decreases.
2. When the total resistance is greater than it should be, a branch is open.
3. Yes, all other bulbs continue to glow.
4. The current in each nonopen branch remains 1 A.
5. By Kirchhoff's current law, the ground return current is 0.10 A and represents a parallel return path. From Ohm's law, $R = 120$ V/0.1 A $= 1200 \ \Omega$.
6. Temperature

ANSWERS TO RELATED PROBLEMS FOR EXAMPLES

5–1   No rewiring is necessary.

5–2   Connect pin 1 to pin 2 and pin 3 to pin 4.

5–3 $9.34 \ \Omega$

5–4   Replace Step 9 with "Press the $+$ key."

Display shows $100^{-1} + 47^{-1} + 22^{-1} +$.

Replace Step 10 with "Enter 33."

Display shows $100^{-1} + 47^{-1} + 22^{-1} + 33$.

Step 11: Press the x^{-1} key.

Display shows $100^{-1} + 47^{-1} + 22^{-1} + 33^{-1}$.

Step 12: Press the ENTER key. Display shows $107.034171502e^{-3}$.

Step 13: Press the x^{-1} and ENTER keys. Display shows 9.34281067406E0.

5–5 132 Ω

5–6 33.3 kΩ

5–7 25 V

5–8 $I_1 = 100$ mA; $I_2 = 179$ mA; 100 mA + 179 mA = 279 mA

5–9 $I_1 = 20.0$ mA; $I_2 = 9.09$ mA; $I_3 = 35.7$ mA; $I_4 = 22.0$ mA

5–10 31.9 mA

5–11 2.0 Ω

5–12 $I = V/R_{TAIL} = 12.6$ V/10.5 Ω = 1.2 A

5–13 20 mA

5–14 $I_T = 112$ mA; $I_2 = 50$ mA

5–15 2.5 mA; 5 mA

5–16 $I_1 = 1.63$ mA; $I_2 = 3.35$ mA

5–17 $I_1 = 59.4$ mA; $I_2 = 40.6$ mA

5–18 1.78 W

5–19 81 W

5–20 15.4 mA

ANSWERS TO TRUE/FALSE QUIZ

1. T **2.** T **3.** F **4.** F **5.** F **6.** T **7.** T **8.** T **9.** F **10.** F

ANSWERS TO SELF-TEST

1. (b) **2.** (c) **3.** (b) **4.** (d) **5.** (a) **6.** (c) **7.** (c)
8. (c) **9.** (d) **10.** (b) **11.** (a) **12.** (d) **13.** (c) **14.** (b)

ANSWERS TO TROUBLESHOOTING: SYMPTOM AND CAUSE

1. (b)

2. (c)

3. (a)

4. (b)

5. (b)

CHAPTER 6

SERIES-PARALLEL CIRCUITS

OUTLINE

OBJECTIVES

- Identify series-parallel relationships
- Analyze series-parallel circuits
- Analyze loaded voltage dividers
- Determine the loading effect of a voltmeter on a circuit
- Analyze and apply a Wheatstone bridge
- Apply Thevenin's theorem to simplify a circuit for analysis
- Apply the maximum power transfer theorem
- Apply the superposition theorem to circuit analysis
- Troubleshoot series-parallel circuits

KEY TERMS

- Loading
- Load current
- Bleeder current
- Wheatstone bridge
- Balanced bridge
- Unbalanced bridge
- Thevenin's theorem
- Terminal equivalency
- Maximum power transfer
- Superposition

INTRODUCTION

Various combinations of both series and parallel resistors are often found in electronic systems. In this chapter, examples of such series-parallel arrangements are examined and analyzed. An important circuit called the *Wheatstone bridge* is introduced, which is widely used in measurement systems. You will learn how complex circuits can be simplified using Thevenin's theorem. The maximum power transfer theorem, used in applications where it is important for a given circuit to provide maximum power to a load, is discussed. Also, circuits with more than one voltage source are analyzed in simple steps using the superposition theorem. Troubleshooting series-parallel circuits for shorts and opens is also covered.

VISIT THE WEBSITE
Study aids for this chapter are available at
http://pearsonhighered.com/floyd

6–1 IDENTIFYING SERIES-PARALLEL RELATIONSHIPS

A series-parallel circuit consists of combinations of both series and parallel current paths. It is important to be able to identify how the components in a circuit are arranged in terms of their series and parallel relationships.

After completing this section, you should be able to

- **Identify series-parallel relationships**
  - **Recognize how each resistor in a given circuit is related to the other resistors**
  - **Determine series and parallel relationships on a PC board**

Figure 6–1(a) shows an example of a simple series-parallel combination of resistors. Notice that the resistance from A to B is R_1. The resistance from B to C is R_2 and R_3 in parallel ($R_2 \| R_3$) because these resistors are connected to the same pair of nodes (node B and node C). The total resistance from A to C is R_1 in series with the parallel combination of R_2 and R_3, as indicated in Figure 6–1(b).

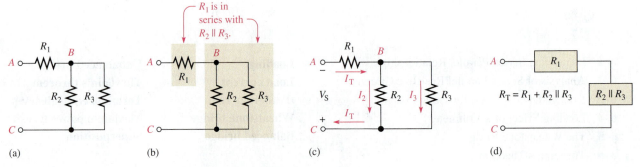

(a) (b) (c) (d)

FIGURE 6–1 A simple series-parallel resistive circuit.

When the resistors in Figure 6–1 are connected to a voltage source as shown in part (c) the total current through R_1 divides at point B into the two parallel paths. These two branch currents then recombine. The total current is into the positive source terminal as shown. The resistor relationships are shown in block form in part (d).

Now, to further illustrate series-parallel relationships, let's increase the complexity of the arrangement in Figure 6–1(a) step-by-step.

1. In Figure 6–2(a), R_4 is connected in series with R_1. The resistance between points A and B is now $R_1 + R_4$. This series combination is in series with the parallel combination of R_2 and R_3, as illustrated in Figure 6–2(b). Part (c) shows the resistor relationships in block form.

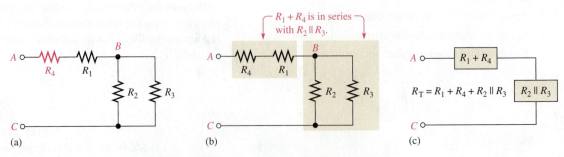

(a) (b) (c)

FIGURE 6–2 R_4 is added to the circuit in series with R_1.

2. In Figure 6–3(a), R_5 is connected in series with R_2. The series combination of R_2 and R_5 is in parallel with R_3. This entire series-parallel combination is in series with the $R_1 + R_4$ combination, as illustrated in Figure 6–3(b). The block diagram is shown in part (c).

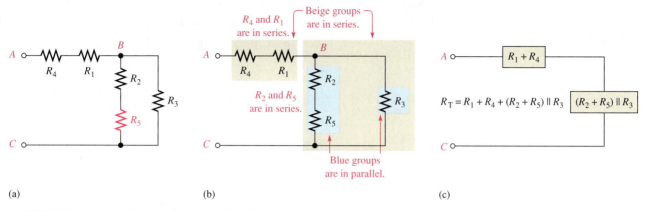

(a)   (b)   (c)

FIGURE 6–3 R_5 **is added to the circuit in series with** R_2**.**

3. In Figure 6–4(a), R_6 is connected in parallel with the series combination of R_1 and R_4. The series-parallel combination of R_1, R_4, and R_6 is in series with the series-parallel combination of R_2, R_3, and R_5, as indicated in Figure 6–4(b). The block diagram is shown in part (c).

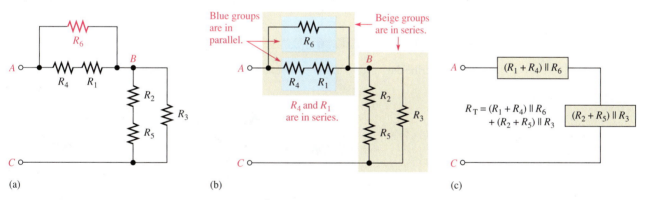

(a)   (b)   (c)

FIGURE 6–4 R_6 **is added to the circuit in parallel with the series combination of** R_1 **and** R_4**.**

EXAMPLE 6–1

Identify the series-parallel relationships in Figure 6–5.

SOLUTION

Starting at the negative terminal of the source, follow the current paths.

1. All the current produced by the source must go through R_1, which is in series   **FIGURE 6–5**
with the rest of the circuit.

2. The total current takes two paths when it gets to node A. Part of it is through R_2, and part of it is through R_3.

3. Resistors R_2 and R_3 are in parallel with each other because they are connected to the same pair of nodes. This parallel combination is in series with R_1.

4. At node B, the currents through R_2 and R_3 come together again into a single path. Thus, the total current is through R_4.

5. Resistor R_4 is in series with both R_1 and the parallel combination of R_2 and R_3.

The currents are shown in Figure 6–6, where I_T is the total current. In summary, R_1 and R_4 are in series with the parallel combination of R_2 and R_3.

$$R_1 + R_4 + R_2 \| R_3$$

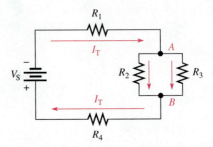

FIGURE 6–6

RELATED PROBLEM*

If another resistor, R_5, is connected from node A to the positive side of the source in Figure 6–6, what is its relationship to the other resistors?

*Answers are at the end of the chapter.

EXAMPLE 6–2

Describe the series-parallel combination between terminals A and D in Figure 6–7.

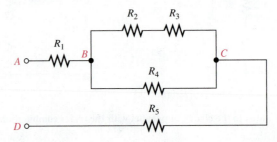

FIGURE 6–7

SOLUTION

Between nodes B and C, there are two parallel paths.

1. The lower path consists of R_4.

2. The upper path consists of a series combination of R_2 and R_3.

This parallel combination is in series with both R_1 and R_5. In summary, R_1 and R_5 are in series with the parallel combination of R_4 and $(R_2 + R_3)$.

$$R_1 + R_5 + R_4 \| (R_2 + R_3)$$

RELATED PROBLEM

If a resistor is connected from node C to node D in Figure 6–7, describe its relationship in the circuit.

EXAMPLE 6–3

Describe the total resistance between each pair of terminals in Figure 6–8.

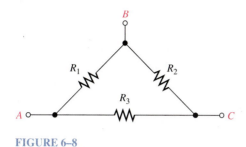

FIGURE 6–8

SOLUTION

1. **From *A* to *B*:** R_1 is in parallel with the series combination of R_2 and R_3.
2. **From *A* to *C*:** R_3 is in parallel with the series combination of R_1 and R_2.
3. **From *B* to *C*:** R_2 is in parallel with the series combination of R_1 and R_3.

RELATED PROBLEM

In Figure 6–8, describe the total resistance between each terminal and an added ground if a new resistor, R_4, is connected from terminal *C* to ground. None of the existing resistors connect directly to the ground.

When connecting a circuit on a protoboard from a schematic, it is easier to check the circuit if the resistors and connections on the protoboard are oriented so that they approximately match the way the schematic is drawn. In some cases, it is difficult to see the series-parallel relationships on a schematic because of the way in which it is drawn. In such a situation, it helps to redraw the diagram so that the relationships become clear. The physical arrangement of components on a PC or protoboard usually bears no resemblance to the actual electrical relationships. By tracing out the circuit and rearranging the components on paper into a recognizable form, you can determine the series-parallel relationships.

EXAMPLE 6–4

Identify the series-parallel relationships in Figure 6–9.

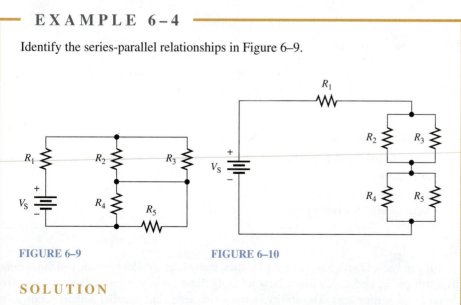

FIGURE 6–9   **FIGURE 6–10**

SOLUTION

The circuit schematic is redrawn in Figure 6–10 to better illustrate the series-parallel relationships. Now you can see that R_2 and R_3 are in parallel with each other and also that R_4 and R_5 are in parallel with each other. Both parallel combinations are in series with each other and with R_1.

$$R_1 + R_2 \| R_3 + R_4 \| R_5$$

RELATED PROBLEM

If a resistor is connected from the bottom end of R_3 to the top end of R_5 in Figure 6–10, what effect does it have on the circuit? Explain.

SECTION 6–1 CHECKUP*

1. A certain series-parallel circuit is described as follows: R_1 and R_2 are in parallel. This parallel combination is in series with another parallel combination of R_3 and R_4. Draw the circuit.

2. In the circuit of Figure 6–11, describe the series-parallel relationships of the resistors.

3. Which resistors are in parallel in Figure 6–12?

4. Identify the parallel resistors in Figure 6–13.

5. Are the parallel combinations in Figure 6–13 in series?

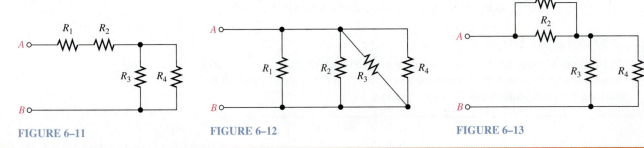

FIGURE 6–11

FIGURE 6–12

FIGURE 6–13

*Answers are at the end of the chapter.

6–2 ANALYSIS OF SERIES-PARALLEL RESISTIVE CIRCUITS

The analysis of series-parallel circuits can be approached in many ways, depending on what information you need and what circuit values you know. The examples in this section do not represent an exhaustive coverage, but they give you an idea of how to approach series-parallel circuit analysis.

After completing this section, you should be able to

- **Analyze series-parallel circuits**
  - **Determine total resistance**
  - **Determine all the currents**
  - **Determine all the voltage drops**

If you know Ohm's law, Kirchhoff's laws, the voltage-divider formula, and the current-divider formula, and if you know how to apply these laws, you can solve most resistive circuit analysis problems. The ability to recognize series and parallel combinations is, of course, essential. There is no standard "cookbook" approach that can be applied to all situations. Logical thought is the most powerful tool you can apply to problem solving.

Total Resistance

In Chapter 4, you learned how to determine total series resistance. In Chapter 5, you learned how to determine total parallel resistance. To find the total resistance (R_T) of a series-parallel combination, first identify the series and parallel relationships, and then simplify the circuit by finding the total resistance. The following two examples illustrate the general approach.

EXAMPLE 6–5

Determine R_T between terminals A and B of the circuit in Figure 6–14.

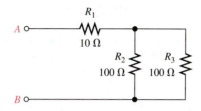

FIGURE 6–14

SOLUTION

Resistors R_2 and R_3 are in parallel, and this parallel combination is in series with R_1. First find the parallel resistance of R_2 and R_3. Since R_2 and R_3 are equal in value, divide the value by 2.

$$R_{2\|3} = \frac{R}{n} = \frac{100 \ \Omega}{2} = 50 \ \Omega$$

Now, since R_1 is in series with $R_{2\|3}$, add their values.

$$R_T = R_1 + R_{2\|3} = 10 \ \Omega + 50 \ \Omega = \mathbf{60 \ \Omega}$$

RELATED PROBLEM

Determine R_T in Figure 6–14 if R_3 is changed to 82 Ω.

EXAMPLE 6–6

Find R_T of the circuit in Figure 6–15.

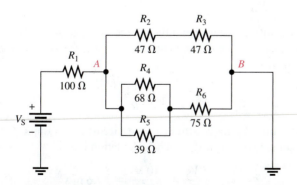

FIGURE 6–15

SOLUTION

1. In the upper branch between nodes A and B, R_2 is in series with R_3. The series combination is designated R_{2+3} and is equal to $R_2 + R_3$.

$$R_{2+3} = R_2 + R_3 = 47 \ \Omega + 47 \ \Omega = 94 \ \Omega$$

2. In the lower branch, R_4 and R_5 are in parallel with each other. This parallel combination is designated $R_{4\|5}$.

$$R_{4\|5} = \frac{R_4 R_5}{R_4 + R_5} = \frac{(68\ \Omega)(39\ \Omega)}{68\ \Omega + 39\ \Omega} = 24.8\ \Omega$$

3. Also in the lower branch, the parallel combination of R_4 and R_5 is in series with R_6. This series-parallel combination is designated $R_{4\|5+6}$.

$$R_{4\|5+6} = R_6 + R_{4\|5} = 75\ \Omega + 24.8\ \Omega = 99.8\ \Omega$$

Figure 6–16 shows the original circuit in a simplified equivalent form.

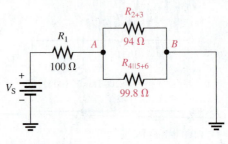

FIGURE 6–16

4. Now you can find the resistance between nodes A and B. It is R_{2+3} in parallel with $R_{4\|5+6}$. Calculate the equivalent resistance as follows:

$$R_{AB} = \frac{1}{\dfrac{1}{R_{2+3}} + \dfrac{1}{R_{4\|5+6}}} = \frac{1}{\dfrac{1}{94\ \Omega} + \dfrac{1}{99.8\ \Omega}} = 48.4\ \Omega$$

5. Finally, the total circuit resistance is R_1 in series with R_{AB}.

$$R_T = R_1 + R_{AB} = 100\ \Omega + 48.4\ \Omega = \mathbf{148\ \Omega}$$

MULTISIM

Open Multisim file E06-06. Verify the total calculated resistance. Remove R_5 from the circuit and measure the total resistance. Then check your measured value by calculating the total resistance.

RELATED PROBLEM

Determine R_T if a 68 Ω resistor is connected to the circuit in Figure 6–16 from node A to node B.

Total Current

Once you know the total resistance and the source voltage, you can apply Ohm's law to find the total current in a circuit. Total current is the source voltage divided by the total resistance.

$$I_T = \frac{V_S}{R_T}$$

For example, let's find the total current in the circuit of Example 6–6 (Figure 6–15). Assume that the source voltage is 10 V. The calculation is

$$I_T = \frac{V_S}{R_T} = \frac{10\ V}{148\ \Omega} = 67.6\ mA$$

Branch Currents

Using the current-divider formula, Kirchhoff's current law, Ohm's law, or combinations of these, you can find the current in any branch of a series-parallel circuit. In some cases, it may take repeated application of the formula to find a given current.

EXAMPLE 6–7

Determine the current through R_4 in Figure 6–17 if $V_S = 5.0$ V.

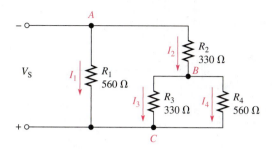

FIGURE 6–17

SOLUTION

First, find the current (I_2) into node B. Once you know this current, you can use the current-divider formula to find I_4, the current through R_4.

Notice that there are two main branches in the circuit. The left-most branch consists of only R_1. The right-most branch has R_2 in series with the parallel combination of R_3 and R_4. The voltage across both of these main branches is the same and equal to 5.0 V. Calculate the equivalent resistance ($R_{2+3\|4}$) of the right-most main branch and then apply Ohm's law; I_2 is the total current through this main branch. Thus,

$$R_{2+3\|4} = R_2 + \frac{R_3 R_4}{R_3 + R_4} = 330 \text{ } \Omega + \frac{(330 \text{ } \Omega)(560 \text{ } \Omega)}{890 \text{ } \Omega} = 538 \text{ } \Omega$$

$$I_2 = \frac{V_S}{R_{2+3\|4}} = \frac{5.0 \text{ V}}{538 \text{ } \Omega} = 9.29 \text{ mA}$$

Use the two-resistor current-divider formula to calculate I_4.

$$I_4 = \left(\frac{R_3}{R_3 + R_4}\right) I_2 = \left(\frac{330 \text{ } \Omega}{890 \text{ } \Omega}\right) 9.29 \text{ mA} = \textbf{3.45 mA}$$

RELATED PROBLEM

Find I_1, I_3, and I_T in Figure 6–17.

MULTISIM

Open Multisim file E06-07. Measure the current in each resistor. Compare the measurements with the calculated values.

Voltage Relationships

The circuit in Figure 6–18 illustrates voltage relationships in a series-parallel circuit. Voltmeters are connected to measure each of the resistor voltages, and the readings are indicated.

Some general observations about Figure 6–18 are as follows:

1. V_{R1} and V_{R2} are equal because R_1 and R_2 are in parallel. (Recall that voltages across parallel branches are the same.) V_{R1} and V_{R2} are the same as the voltage from A to B.

2. V_{R3} is equal to $V_{R4} + V_{R5}$ because R_3 is in parallel with the series combination of R_4 and R_5. (V_{R3} is the same as the voltage from B to C.)

3. V_{R4} is about one-third of the voltage from B to C because R_4 is about one-third of the resistance $R_4 + R_5$ (by the voltage-divider principle).

4. V_{R5} is about two-thirds of the voltage from B to C because R_5 is about two-thirds of $R_4 + R_5$.

5. $V_{R1} + V_{R3} - V_S = 0$ because, by Kirchhoff's voltage law, the algebraic sum of the voltage drops around a single closed path must equal zero.

FIGURE 6–18 **Illustration of voltage relationships.**

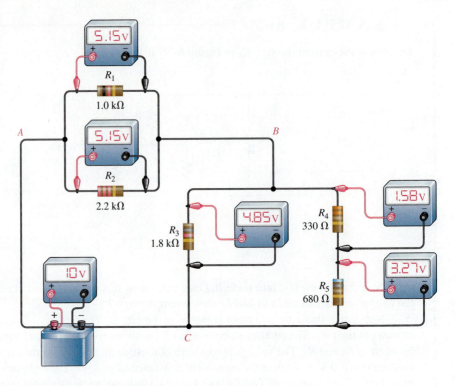

Example 6–8 will verify the meter readings in Figure 6–18.

EXAMPLE 6–8

Verify that the voltmeter readings in Figure 6–18 are correct. The circuit is redrawn as a schematic in Figure 6–19.

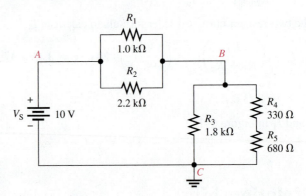

FIGURE 6–19

SOLUTION

The resistance from A to B is the parallel combination of R_1 and R_2.

$$R_{AB} = \frac{R_1 R_2}{R_1 + R_2} = \frac{(1.0 \text{ k}\Omega)(2.2 \text{ k}\Omega)}{3.2 \text{ k}\Omega} = 688 \ \Omega$$

The resistance from B to C is R_3 in parallel with the series combination of R_4 and R_5.

$$R_4 + R_5 = 330 \ \Omega + 680 \ \Omega = 1010 \ \Omega = 1.01 \text{ k}\Omega$$

$$R_{BC} = \frac{R_3(R_4 + R_5)}{R_3 + R_4 + R_5} = \frac{(1.8 \text{ k}\Omega)(1.01 \text{ k}\Omega)}{2.81 \text{ k}\Omega} = 647 \ \Omega$$

The resistance from A to B is in series with the resistance from B to C, so the total circuit resistance is

$$R_{\text{T}} = R_{AB} + R_{BC} = 688 \ \Omega + 647 \ \Omega = 1335 \ \Omega$$

Use the voltage-divider principle to calculate the voltages.

$$V_{AB} = \left(\frac{R_{AB}}{R_T}\right)V_S = \left(\frac{688\ \Omega}{1335\ \Omega}\right)10\ V = 5.15\ V$$

$$V_{BC} = \left(\frac{R_{BC}}{R_T}\right)V_S = \left(\frac{647\ \Omega}{1335\ \Omega}\right)10\ V = 4.85\ V$$

$$V_{R1} = V_{R2} = V_{AB} = \mathbf{5.15\ V}$$
$$V_{R3} = V_{BC} = \mathbf{4.85\ V}$$
$$V_{R4} = \left(\frac{R_4}{R_4 + R_5}\right)V_{BC} = \left(\frac{330\ \Omega}{1010\ \Omega}\right)4.85\ V = \mathbf{1.58\ V}$$
$$V_{R5} = \left(\frac{R_5}{R_4 + R_5}\right)V_{BC} = \left(\frac{680\ \Omega}{1010\ \Omega}\right)4.85\ V = \mathbf{3.27\ V}$$

RELATED PROBLEM

Determine each voltage drop in Figure 6–19 if the source voltage is doubled.

EXAMPLE 6–9

Determine the voltage drop across each resistor in Figure 6–20.

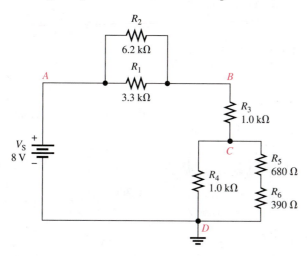

FIGURE 6–20

SOLUTION

Because you know the total voltage, you can solve this problem by forming an equivalent series circuit and applying the voltage-divider formula.

Step 1: Reduce each parallel combination to an equivalent resistance. Since R_1 and R_2 are in parallel between nodes A and B, combine their values.

$$R_{AB} = \frac{R_1 R_2}{R_1 + R_2} + \frac{(3.3\ k\Omega)(6.2\ k\Omega)}{9.5\ k\Omega} = 2.15\ k\Omega$$

Since R_4 is in parallel with the R_5 and R_6 series combination (R_{5+6}) between nodes C and D, combine these values.

$$R_{CD} = \frac{R_4(R_{5+6})}{R_4 + R_{5+6}} + \frac{(1.0\ k\Omega)(1.07\ k\Omega)}{2.07\ k\Omega} = 517\ \Omega$$

Step 2: Draw the equivalent circuit as shown in Figure 6–21. The total circuit resistance is

$$R_T = R_{AB} + R_3 + R_{CD} = 2.15\ k\Omega + 1.0\ k\Omega + 517\ \Omega = 3.67\ k\Omega$$

FIGURE 6–21

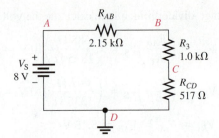

Step 3: Apply the voltage-divider formula to solve for the voltages in the equivalent series circuit.

$$V_{AB} = \left(\frac{R_{AB}}{R_T}\right)V_S = \left(\frac{2.15 \text{ k}\Omega}{3.67 \text{ k}\Omega}\right)8 \text{ V} = 4.69 \text{ V}$$

$$V_{BC} = \left(\frac{R_3}{R_T}\right)V_S = \left(\frac{1.0 \text{ k}\Omega}{3.67 \text{ k}\Omega}\right)8 \text{ V} = 2.18 \text{ V}$$

$$V_{CD} = \left(\frac{R_{CD}}{R_T}\right)V_S = \left(\frac{517 \text{ }\Omega}{3.67 \text{ k}\Omega}\right)8 \text{ V} = 1.13 \text{ V}$$

Refer to Figure 6–20. V_{AB} equals the voltage across both R_1 and R_2.

$$V_{R1} = V_{R2} = V_{AB} = \textbf{4.69 V}$$

V_{BC} is the voltage across R_3.

$$V_{R3} = V_{BC} = \textbf{2.18 V}$$

V_{CD} is the voltage across R_4 and also across the series combination of R_5 and R_6.

$$V_{R4} = V_{CD} = \textbf{1.13 V}$$

Step 4: Apply the voltage-divider formula to the series combination of R_5 and R_6 to get V_{R5} and V_{R6}.

$$V_{R5} = \left(\frac{R_5}{R_5 + R_6}\right)V_{CD} = \left(\frac{680 \text{ }\Omega}{1070 \text{ }\Omega}\right)1.13 \text{ V} = \textbf{718 mV}$$

$$V_{R6} = \left(\frac{R_6}{R_5 + R_6}\right)V_{CD} = \left(\frac{390 \text{ }\Omega}{1070 \text{ }\Omega}\right)1.13 \text{ V} = \textbf{412 mV}$$

RELATED PROBLEM

Determine the current and power in each resistor in Figure 6–20.

MULTISIM

Open Multisim file E06-09. Measure the voltage across each resistor and compare to the calculated values. If R_4 is increased to 2.2 kΩ, specify which voltage drops increase and which ones decrease. Verify this by measurement.

SECTION 6–2 CHECKUP

1. Find the total resistance between A and B in the circuit of Figure 6–22.

2. Find the current through R_3 in Figure 6–22.

3. Find V_{R2} in Figure 6–22.

4. Determine R_T and I_T in Figure 6–23.

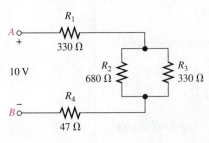

FIGURE 6–22

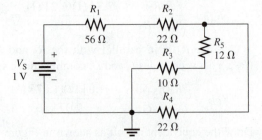

FIGURE 6–23

6–3 VOLTAGE DIVIDERS WITH RESISTIVE LOADS

Voltage dividers were introduced in Chapter 4. In this section, you will learn how resistive loads affect the operation of voltage-divider circuits.

After completing this section, you should be able to

- **Analyze loaded voltage dividers**
  - **Determine the effect of a resistive load on a voltage-divider circuit**
  - **Define** *bleeder current*

The voltage divider in Figure 6–24(a) produces an output voltage (V_{OUT}) of 5 V because the input is 10 V and the two resistors are of equal value. This voltage is the unloaded output voltage. When a load resistor, R_L, is connected from the output to ground as shown in Figure 6–24(b), the output voltage is reduced by an amount that depends on the value of R_L. This effect is called **loading**. The load resistor is in parallel with R_2, reducing the resistance from node A to ground and, as a result, also reducing the voltage across the parallel combination. This is one effect of loading a voltage divider. Another effect of a load is that more current is drawn from the source because the total resistance of the circuit is reduced.

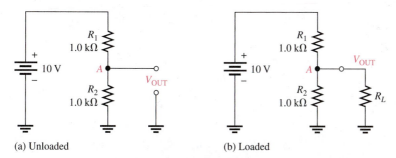

(a) Unloaded   (b) Loaded

FIGURE 6–24   A voltage divider with both unloaded and loaded outputs.

The effect of loading on a voltage divider is important to take into account when choosing resistors for a divider. When R_L is large compared to the divider resistors, the loading effect is small and the output voltage will change only a small amount from its unloaded value. If the loading effect is small, the divider is said to be a *stiff voltage divider*. This term is relative but generally means that the difference between the unloaded and loaded output voltage is small. As a rule of thumb, a **stiff voltage divider** is one in which the load resistor is at least ten times larger than the divider resistors. Stiff voltage dividers are more stable but use more power, so the choice of divider resistors is a trade-off between these two parameters. Figure 6–25 illustrates the effect of a load resistor on the output voltage.

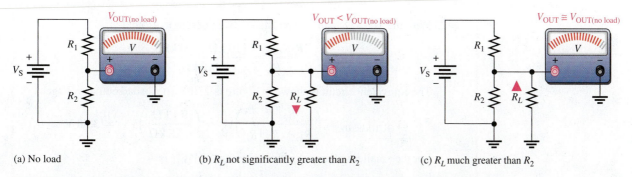

(a) No load   (b) R_L not significantly greater than R_2 (c) R_L much greater than R_2

FIGURE 6–25   The effect of a load resistor. The circuit in part (c) illustrates a stiff voltage divider.

EXAMPLE 6–10

(a) Determine the unloaded output voltage of the voltage divider in Figure 6–26.

(b) Find the loaded output voltages of the voltage divider in Figure 6–26 for the following two values of load resistance: $R_L = 10\,\text{k}\Omega$ and $R_L = 100\,\text{k}\Omega$.

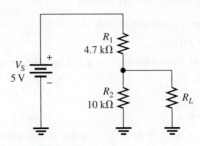

FIGURE 6–26

SOLUTION

(a) The unloaded output voltage is

$$V_{OUT(unloaded)} = \left(\frac{R_2}{R_1 + R_2}\right)V_S = \left(\frac{10\,\text{k}\Omega}{14.7\,\text{k}\Omega}\right)5\,\text{V} = \mathbf{3.40\,V}$$

(b) With the 10 kΩ load resistor connected, R_L is in parallel with R_2, which gives

$$R_2\,\|\,R_L = \frac{R_2 R_L}{R_2 + R_L} = \frac{(10\,\text{k}\Omega)(10\,\text{k}\Omega)}{20\,\text{k}\Omega} = 5.0\,\text{k}\Omega$$

The equivalent circuit is shown in Figure 6–27(a). The loaded output voltage is

$$V_{OUT(loaded)} = \left(\frac{R_2\,\|\,R_L}{R_1 + R_2\,\|\,R_L}\right)V_S = \left(\frac{5.0\,\text{k}\Omega}{9.7\,\text{k}\Omega}\right)5\,\text{V} = \mathbf{2.58\,V}$$

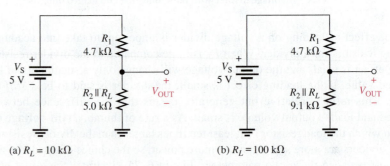

(a) $R_L = 10\,\text{k}\Omega$ (b) $R_L = 100\,\text{k}\Omega$

FIGURE 6–27

With the 100 kΩ load, the resistance from output to ground is

$$R_2\,\|\,R_L = \frac{R_2 R_L}{R_2 + R_L} = \frac{(10\,\text{k}\Omega)(100\,\text{k}\Omega)}{110\,\text{k}\Omega} = 9.1\,\text{k}\Omega$$

The equivalent circuit is shown in Figure 6–27(b). The loaded output voltage is

$$V_{OUT(loaded)} = \left(\frac{R_2\,\|\,R_L}{R_1 + R_2\,\|\,R_L}\right)V_S = \left(\frac{9.1\,\text{k}\Omega}{13.8\,\text{k}\Omega}\right)5\,\text{V} = \mathbf{3.30\,V}$$

For the smaller value of R_L, the reduction in V_{OUT} is

$$3.40\,\text{V} - 2.58\,\text{V} = 0.82\,\text{V} \qquad \text{(a 24\% drop in output voltage)}$$

For the larger value of R_L, the reduction in V_{OUT} is

$$3.40 \text{ V} - 3.30 \text{ V} = 0.10 \text{ V} \qquad \text{(a 3\% drop in output voltage)}$$

This illustrates the loading effect of R_L on the voltage divider.

RELATED PROBLEM

Determine V_{OUT} in Figure 6–26 for a 1.0 MΩ load resistance.

MULTISIM

Open Multisim file E06-10. Measure the voltage at the output terminal with respect to ground. Connect a 10 kΩ load resistor from the output to ground and measure the output voltage. Change the load resistor to 100 kΩ and measure the output voltage. Do these measurements agree closely with your calculated values?

Load Current and Bleeder Current

In a multiple-tap loaded voltage-divider circuit, the total current drawn from the source consists of currents through the load resistors, called **load currents**, and the divider resistors. Figure 6–28 shows a voltage divider with two voltage outputs or taps. Notice that the total current, I_T, is through R_1. The total current is composed of the two branch currents, I_{RL1} and I_2. The current I_2 is composed of two additional branch currents, I_{RL2} and I_3. Current I_3 is called the **bleeder current**, which is the current left after the total load current is subtracted from the total current in the circuit.

$$I_{\text{BLEEDER}} = I_T - I_{RL1} - I_{RL2} \qquad (6\text{--}1)$$

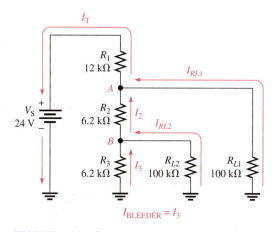

FIGURE 6–28   **Currents in a two-tap loaded voltage divider.**

EXAMPLE 6–11

Determine the load currents I_{RL1} and I_{RL2} and the bleeder current I_3 in the two-tap loaded voltage divider in Figure 6–28.

SOLUTION

The equivalent resistance from node A to ground is the 100 kΩ load resistor R_{L1} in parallel with the combination of R_2 in series with the parallel combination of R_3 and R_{L2}. Determine the resistance values first. R_3 in parallel with R_{L2} is designated R_B. The resulting equivalent circuit is shown in Figure 6–29(a).

$$R_B = \frac{R_3 R_{L2}}{R_3 + R_{L2}} = \frac{(6.2 \text{ k}\Omega)(100 \text{ k}\Omega)}{106.2 \text{ k}\Omega} = 5.84 \text{ k}\Omega$$

R_2 in series with R_B is designated R_{2+B}. The resulting equivalent circuit is shown in Figure 6–29(b).

$$R_{2+B} = R_2 + R_B = 6.2 \text{ k}\Omega + 5.84 \text{ k}\Omega = 12.0 \text{ k}\Omega$$

R_{L1} in parallel with R_{2+B} is designated R_A. The resulting equivalent circuit is shown in Figure 6–29(c).

$$R_A = \frac{R_{L1} R_{2+B}}{R_{L1} + R_{2+B}} = \frac{(100 \text{ k}\Omega)(12.0 \text{ k}\Omega)}{112 \text{ k}\Omega} = 10.7 \text{ k}\Omega$$

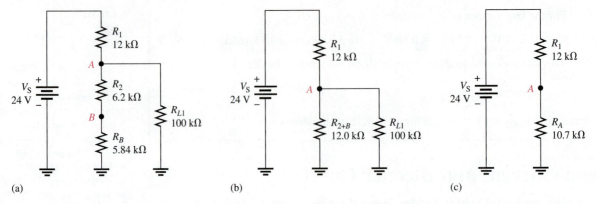

FIGURE 6–29

R_A is the total resistance from node A to ground. The total resistance for the circuit is

$$R_T = R_A + R_1 = 10.7 \, \text{k}\Omega + 12 \, \text{k}\Omega = 22.7 \, \text{k}\Omega$$

Determine the voltage across R_{L1} as follows, using the equivalent circuit in Figure 6–29(c):

$$V_{RL1} = V_A = \left(\frac{R_A}{R_T}\right)V_S = \left(\frac{10.7 \, \text{k}\Omega}{22.7 \, \text{k}\Omega}\right)24 \, \text{V} = 11.3 \, \text{V}$$

The load current through R_{L1} is

$$I_{RL1} = \frac{V_{RL1}}{R_{L1}} = \left(\frac{11.3 \, \text{V}}{100 \, \text{k}\Omega}\right) = \mathbf{113 \, \mu A}$$

Determine the voltage at node B by using the equivalent circuit in Figure 6–29(a) and the voltage at node A.

$$V_B = \left(\frac{R_B}{R_{2+B}}\right)V_A = \left(\frac{5.84 \, \text{k}\Omega}{12.0 \, \text{k}\Omega}\right)11.3 \, \text{V} = 5.50 \, \text{V}$$

The load current through R_{L2} is

$$I_{RL2} = \frac{V_{RL2}}{R_{L2}} = \frac{V_B}{R_{L2}} = \frac{5.50 \, \text{V}}{100 \, \text{k}\Omega} = \mathbf{55 \, \mu A}$$

The bleeder current is

$$I_3 = \frac{V_B}{R_3} = \frac{5.50 \, \text{V}}{6.2 \, \text{k}\Omega} = \mathbf{887 \, \mu A}$$

MULTISIM

Open Multisim file E06-11. Measure the voltage across and the current through each load resistor, R_{L1} and R_{L2}.

RELATED PROBLEM

What will happen to the load current in R_{L2} if R_{L1} is disconnected?

SYSTEM EXAMPLE 6–1

DISCRETE AMPLIFIERS WITH VOLTAGE-DIVIDER BIAS

Amplifiers are fundamental to electronic systems because they are the basis of many other circuits. In linear circuits, bipolar transistors are often selected because of their linear characteristics. For a small-signal bipolar transistor amplifier, a voltage-divider is commonly used to develop the correct dc conditions (called *bias*) in order for the transistor to amplify. Figure 6–30 shows a simplified bipolar transistor voltage amplifier. The transistor requires a certain amount of base current to operate, which is set up by the voltage-divider bias network composed of R_1 and R_2. In an amplifier in which bias is set by a voltage divider, the base current should normally be much smaller (<10%) than the bleeder current. When

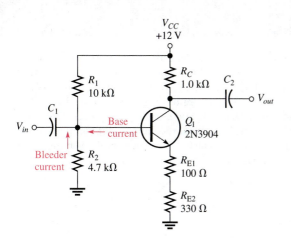

FIGURE 6–30 A voltage amplifier.

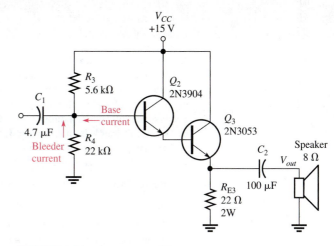

FIGURE 6–31 A power amplifier.

this condition is met, the basic calculation of base voltage is simple; simply apply the unloaded voltage-divider rule you already learned, substituting V_{CC} for the source voltage, V_S, in the original equation:

$$V_{BASE} = \left(\frac{R_2}{R_1 + R_2}\right)V_{CC} = \left(\frac{4.7\text{ k}\Omega}{10\text{ k}\Omega + 4.7\text{ k}\Omega}\right)12\text{ V} = 3.84\text{ V}$$

The actual base voltage in this circuit is 3.7 V, showing a small loading effect.

Figure 6–31 shows another example with voltage-divider bias, but the load is a speaker and the amplifier is a power amplifier. The divider for a power amplifier will usually have different values for the divider string depending on the type of amplifier and the bias current required for proper operation. If the bias current requirement is large, the bias resistors will be smaller. The simplifying assumption of an unloaded voltage divider can be used whenever the base current is small compared to the bleeder current, and the result is a reasonable estimate of the base voltage.

SECTION 6–3 CHECKUP

1. A load resistor is connected to an output on a voltage divider. What effect does the load resistor have on the output voltage?

2. A larger-value load resistor will cause the output voltage of a voltage divider to change less than a small-value one will. (True or False)

3. For the voltage divider in Figure 6–32, determine the unloaded output voltage. Also determine the output voltage with a 10 MΩ load resistor connected from the output to ground.

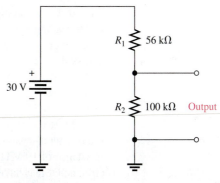

FIGURE 6–32

As you have learned, voltmeters must be connected in parallel with a resistor in order to measure the voltage across the resistor. Because of its internal resistance, a voltmeter, or any other measuring instrument for that matter, puts a load on the circuit and will affect, to a certain extent, the voltage that is being measured. Until now, we have ignored the loading effect because the internal resistance of a voltmeter is very high, and normally it has negligible effect on the circuit that is being measured. However, if the internal resistance of the voltmeter is not sufficiently greater than the circuit resistance across which it is connected, the loading effect will cause the measured voltage to be less than its actual value.

After completing this section, you should be able to

- **Determine the loading effect of a voltmeter on a circuit**
  - **Explain why a voltmeter can load a circuit**
  - **Discuss the internal resistance of a voltmeter**

When a voltmeter is connected to a circuit as shown, for example, in Figure 6–33(a), its internal resistance appears in parallel with R_3, as shown in part (b). The resistance from A to B is altered by the loading effect of the voltmeter's internal resistance, R_M, and is equal to $R_3 \| R_M$, as indicated in part (c).

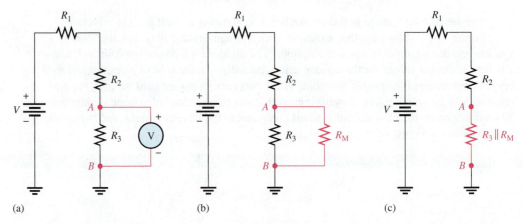

(a)   (b)   (c)

FIGURE 6–33   The loading effect of a voltmeter.

If R_M is much greater than R_3, the resistance from A to B changes very little, and the meter reading is very close to the actual voltage when no meter is present. If R_M is not sufficiently greater than R_3, the resistance from A to B is reduced significantly, and the voltage across R_3 is altered by the loading effect of the meter. A good rule of thumb for troubleshooting work is that *if the meter resistance is at least ten times greater than the resistance across which it is connected, the loading effect can be neglected (measurement error is less than 10%).*

Most voltmeters are part of a multifunction instrument such as the DMM or the analog multimeter discussed in Section 2–7. The voltmeter in a DMM will typically have an internal resistance of 10 MΩ or more, so the loading effect is important only in very high-resistance circuits. DMMs have a constant resistance on all ranges because the input is connected to an internal fixed voltage divider. For analog multimeters, the internal resistance depends on the range selected for making a measurement. To determine the loading effect, you need to know the **sensitivity** of the meter, a value given by the manufacturer on the meter or in the manual. Sensitivity is expressed in ohms/volt and is typically about 20,000 Ω/V. To determine the internal series resistance, multiply the sensitivity by the maximum voltage on the range selected. For example, a 20,000 Ω/V meter will have an internal resistance of 20,000 Ω on the 1 V range and 200,000 Ω on the 10 V range. As you can see, there is a smaller loading effect for higher voltage ranges than for lower ones on the analog multimeter.

EXAMPLE 6-12

How much does the digital voltmeter affect the voltage being measured for each circuit indicated in Figure 6–34? Assume the meter has an input resistance (R_M) of 10 MΩ.

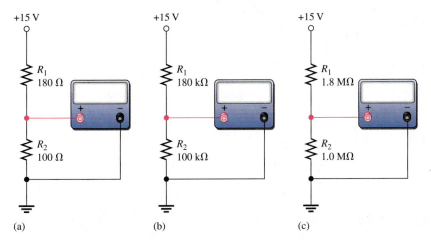

FIGURE 6–34

SOLUTION

To show the small differences more clearly, the results are expressed in more than three significant digits in this example.

(a) Refer to Figure 6–34(a). The unloaded voltage across R_2 in the voltage-divider circuit is

$$V_{R2} = \left(\frac{R_2}{R_1 + R_2} \right) V_S = \left(\frac{100 \ \Omega}{280 \ \Omega} \right) 15 \text{ V} = 5.357 \text{ V}$$

The meter's resistance in parallel with R_2 is

$$R_2 \| R_M = \left(\frac{R_2 R_M}{R_2 + R_M} \right) = \frac{(100 \ \Omega)(10 \text{ M}\Omega)}{10.0001 \text{ M}\Omega} = 99.999 \ \Omega$$

The voltage actually measured by the meter is

$$V_{R2} = \left(\frac{R_2 \| R_M}{R_1 + R_2 \| R_M} \right) V_S = \left(\frac{99.999 \ \Omega}{279.999 \ \Omega} \right) 15 \text{ V} = 5.357 \text{ V}$$

The voltmeter has no measurable loading effect.

(b) Refer to Figure 6–34(b).

$$V_{R2} = \left(\frac{R_2}{R_1 + R_2} \right) V_S = \left(\frac{100 \text{ k}\Omega}{280 \text{ k}\Omega} \right) 15 \text{ V} = 5.357 \text{ V}$$

$$R_2 \| R_M = \frac{R_2 R_M}{R_2 + R_M} = \frac{(100 \text{ k}\Omega)(10 \text{ M}\Omega)}{10.1 \text{ M}\Omega} = 99.01 \text{ k}\Omega$$

The voltage actually measured by the meter is

$$V_{R2} = \left(\frac{R_2 \| R_M}{R_1 + R_2 \| R_M} \right) V_S = \left(\frac{99.01 \text{ k}\Omega}{279.01 \text{ k}\Omega} \right) 15 \text{ V} = 5.323 \text{ V}$$

The loading effect of the voltmeter reduces the voltage by a very small amount.

(c) Refer to Figure 6–34(c).

$$V_{R2} = \left(\frac{R_2}{R_1 + R_2}\right)V_S = \left(\frac{1.0\ M\Omega}{2.8\ M\Omega}\right)15\ V = 5.357\ V$$

$$R_2 \| R_M = \frac{R_2 R_M}{R_2 + R_M} = \frac{(1.0\ M\Omega)(10\ M\Omega)}{11\ M\Omega} = 909.09\ k\Omega$$

The voltage actually measured is

$$V_{R2} = \left(\frac{R_2 \| R_M}{R_1 + R_2 \| R_M}\right)V_S = \left(\frac{909.09\ k\Omega}{2.709\ M\Omega}\right)15\ V = 5.034\ V$$

The loading effect of the voltmeter reduces the voltage by a noticeable amount. As you can see, the higher the resistance across which a voltage is measured, the more the loading effect.

RELATED PROBLEM

Calculate the voltage across R_2 in Figure 6–34(c) if the meter resistance is 20 MΩ.

SECTION 6–4 CHECKUP

1. Explain why a voltmeter can potentially load a circuit.

2. If a voltmeter with a 10 MΩ internal resistance is measuring the voltage across a 1.0 kΩ resistor, should you normally be concerned about the loading effect?

3. If a voltmeter with a 10 MΩ resistance is measuring the voltage across a 3.3 MΩ resistor, should you be concerned about the loading effect?

4. What is the internal series resistance of a 20,000 Ω/V VOM if it is on the 200 V range?

6–5 THE WHEATSTONE BRIDGE

The Wheatstone bridge circuit can be used to precisely measure resistance. However, the bridge is most commonly used in an automated circuit in conjunction with transducers to measure physical quantities such as strain, temperature, and pressure. Transducers are devices that sense a change in a physical parameter and convert that change into an electrical quantity such as a change in resistance. For example, a strain gage exhibits a change in resistance when it is exposed to mechanical factors such as force, pressure, or displacement. A thermistor exhibits a change in its resistance when it is exposed to a change in temperature. The Wheatstone bridge can be operated in a balanced or an unbalanced condition. The condition of operation depends on the type of application.

After completing this section, you should be able to

• **Analyze and apply a Wheatstone bridge**
  • Determine when a bridge is balanced
  • Determine an unknown resistance with a balanced bridge
  • Determine when a bridge is unbalanced
  • Discuss measurements using an unbalanced bridge

A **Wheatstone bridge** circuit is shown in its most common diamond-shaped configuration in Figure 6–35(a). It consists of four resistors and a dc voltage source connected across the top and bottom points of the diamond. The output voltage is taken across the left and right points of the diamond between *A* and *B*. In part (b), the circuit is drawn in a slightly different way to more clearly show its series-parallel configuration.

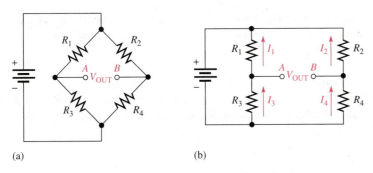

FIGURE 6–35   Wheatstone bridge. Notice that the bridge forms two back-to-back voltage dividers.

The Balanced Wheatstone Bridge

The Wheatstone bridge in Figure 6–35 is in the **balanced bridge** condition when the output voltage (V_{OUT}) between terminals A and B is equal to zero.

$$V_{OUT} = 0 \text{ V}$$

When the bridge is balanced, the voltages across R_1 and R_2 are equal ($V_1 = V_2$) and the voltages across R_3 and R_4 are equal ($V_3 = V_4$). Therefore, the voltage ratios can be written as

$$\frac{V_1}{V_3} = \frac{V_2}{V_4}$$

Substituting IR for V by Ohm's law gives

$$\frac{I_1 R_1}{I_3 R_3} = \frac{I_2 R_2}{I_4 R_4}$$

Since $I_1 = I_3$ and $I_2 = I_4$, all the current terms cancel, leaving the resistor ratios.

$$\frac{R_1}{R_3} = \frac{R_2}{R_4}$$

Solving for R_1 results in the following formula:

$$R_1 = R_3 \left(\frac{R_2}{R_4} \right)$$

This formula allows you to find the value of resistor R_1 in terms of the other resistor values when the bridge is balanced. You can also find the value of any other resistor in a similar way.

USING THE BALANCED WHEATSTONE BRIDGE TO FIND AN UNKNOWN RESISTANCE   Assume that R_1 in Figure 6–35 has an unknown value, which we call R_X. Resistors R_2 and R_4 have fixed values so that their ratio, R_2/R_4, also has a fixed value. Since R_X can be any value, R_3 must be adjusted to make $R_1/R_3 = R_2/R_4$ in order to create a balanced condition. Therefore, R_3 is a variable resistor, which we will call R_V. When R_X is placed in the bridge, R_V is adjusted until the bridge is balanced as indicated by a zero output voltage. Then, the unknown resistance is found as

$$R_X = R_V \left(\frac{R_2}{R_4} \right) \qquad \text{(6–2)}$$

The ratio R_2/R_4 is the scale factor.

An older type of measuring instrument called a *galvanometer* can be connected between the output terminals A and B to detect a balanced condition. The galvanometer is essentially a very sensitive ammeter that senses current in either direction. It differs from a regular ammeter in that the midscale point is zero. Most Wheatstone bridges are now automated and use an amplifier connected across the bridge output to indicate a balanced condition when its output is 0 V. Also, high precision micro-adjustable resistors can be used in demanding applications. The micro-adjustable resistors enable fine adjustment of the bridge resistors during manufacture for applications such as medical sensors, scales, and precision measurements.

From Equation 6–2, the value of R_V at balance multiplied by the scale factor R_2/R_4 is the actual resistance value of R_X. If $R_2/R_4 = 1$, then $R_X = R_V$, if $R_2/R_4 = 0.5$, then $R_X = 0.5 R_V$, and so on. In a practical bridge circuit, the position of the R_V adjustment can be calibrated to indicate the actual value of R_X on a scale or with some other method of display.

EXAMPLE 6–13

Determine the value of R_X in the balanced bridge shown in Figure 6–36. The bridge is balanced ($V_{OUT} = 0$ V) when R_V is set at 1200 Ω.

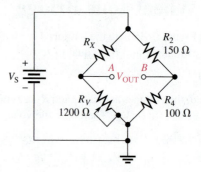

FIGURE 6–36

SOLUTION

The scale factor is

$$\frac{R_2}{R_4} = \frac{150 \ \Omega}{100 \ \Omega} = 1.5$$

The unknown resistance is

$$R_X = R_V\left(\frac{R_2}{R_4}\right) = (1200 \ \Omega)(1.5) = \mathbf{1800 \ \Omega}$$

RELATED PROBLEM

If R_V must be adjusted to 2.2 kΩ to balance the bridge in Figure 6–36, what is R_X?

The Unbalanced Wheatstone Bridge

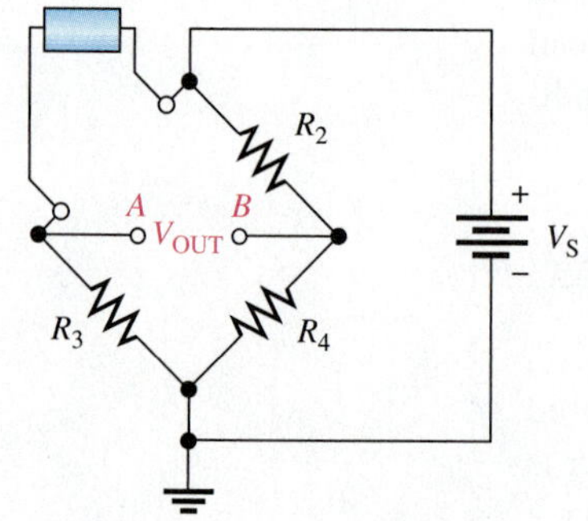

An **unbalanced bridge** condition occurs when V_{OUT} is not equal to zero. The unbalanced bridge is used to measure several types of physical quantities such as mechanical strain, temperature, or pressure. This can be done by connecting a transducer in one leg of the bridge, as shown in Figure 6–37. The resistance of the transducer changes proportionally to the changes in the parameter that it is measuring. If the bridge is balanced at a known point, then the amount of deviation from the balanced condition, as indicated by the output voltage, indicates the amount of change in the parameter being measured. Therefore, the value of the parameter being measured can be determined by the amount that the bridge is unbalanced.

FIGURE 6–37 A bridge circuit for measuring a physical parameter using a transducer.

A BRIDGE CIRCUIT FOR MEASURING TEMPERATURE   If temperature is to be measured, the transducer can be a thermistor, which is a temperature-sensitive resistor. The thermistor resistance changes in a predictable way as the temperature changes. A change in temperature causes a change in thermistor resistance, which causes a corresponding change in the output voltage of the bridge as it becomes unbalanced. The output voltage is proportional to the temperature; therefore, either a voltmeter connected across the output can be calibrated to show the temperature or the output voltage can be amplified and converted to digital form to drive a readout display of the temperature.

A bridge circuit used to measure temperature is designed so that it is balanced at a reference temperature and becomes unbalanced at a measured temperature. For example, let's say the bridge is to be balanced at 25°C. A thermistor will have a known value of resistance at 25°C.

EXAMPLE 6–14

Determine the output voltage of the temperature-measuring bridge circuit in Figure 6–38 if the thermistor is exposed to a temperature of 50°C and its resistance at 25°C is 1.0 kΩ. Assume the resistance of the thermistor decreases to 900 Ω at 50°C.

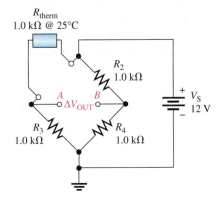

FIGURE 6–38

SOLUTION

Apply the voltage-divider formula to the left side of the bridge at 50°C.

$$V_A = \left(\frac{R_3}{R_3 + R_{therm}}\right)V_S = \left(\frac{1\ k\Omega}{1\ k\Omega + 900\ \Omega}\right)12\ V = 6.32\ V$$

Apply the voltage-divider formula to the right side of the bridge.

$$V_B = \left(\frac{R_4}{R_2 + R_4}\right)V_S = \left(\frac{1\ k\Omega}{2\ k\Omega}\right)12\ V = 6.00\ V$$

The output voltage at 50°C is the difference between V_A and V_B.

$$V_{OUT} = V_A - V_B = 6.32\ V - 6.00\ V = \textbf{0.32 V}$$

Node A is positive with respect to node B.

RELATED PROBLEM

If the temperature is increased to 60°C, causing the thermistor resistance in Figure 6–38 to decrease to 850 Ω, what is V_{OUT}?

STRAIN GAGE APPLICATION OF THE WHEATSTONE BRIDGE   A

Wheatstone bridge with a strain gage can be used to measure forces. A strain gage is a device that exhibits a change in resistance when it is compressed or stretched by the application of an external force. As the resistance of the strain gage changes, the previously balanced bridge becomes unbalanced. This unbalance causes the output voltage to change from zero, and this change can be measured to determine the amount of strain. In strain gages, the resistance change is extremely small. This tiny change unbalances a Wheatstone bridge and can be detected because of its high sensitivity. For example, Wheatstone bridges with strain gages are commonly used in weight scales.

Some resistive transducers have extremely small resistance changes, and these changes are difficult to measure accurately with a direct measurement. In particular, strain gages are one of the most useful resistive transducers that convert the stretching or compression of a fine wire into a change in resistance. When strain causes the wire in the gage to stretch, the resistance increases a small amount and when it compresses, the resistance of the wire decreases.

Strain gages are used in many types of scales, from those that are used for weighing small parts to those for weighing huge trucks. Typically, the gages are mounted on a special block of aluminum that deforms when a weight is on the scale. The strain gages are extremely delicate and must be mounted properly, so the entire assembly is generally prepared as a single unit called a *load cell*. A **load cell** is a transducer that uses strain gages to convert mechanical force into an electrical signal. A wide variety of load cells with different shapes and sizes are available from manufacturers depending on the application. A typical S-shaped load cell for a weighing application that has four strain gages is illustrated in Figure 6–39(a). The gages are mounted so that two of the gages stretch (tension) when a load is placed on the scale and two of the gages compress.

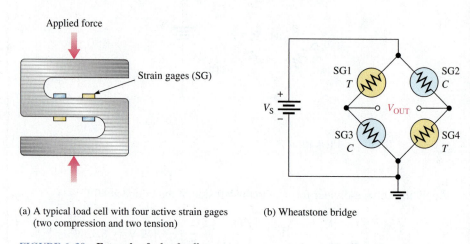

(a) A typical load cell with four active strain gages (two compression and two tension)

(b) Wheatstone bridge

FIGURE 6–39 **Example of a load cell.**

The 4-20 mA Current Loop

Many systems use transducers including strain gages and load cells that have very small signal levels, which are subject to corruption due to interference. Signal conditioning is the process of converting the signal to a usable level that is not likely to pick up unwanted noise. Typically, the signal is converted from an analog signal to a digital signal close to the source.

An older, but widely used, standard for sending information in an industrial environment is the 4-20 mA current loop shown in the block diagram. It is particularly useful when the information needs to be transmitted a long distance (1000 feet or more). The sensor's output is converted to a signal that uses 4 mA to signal the lowest level and 20 mA to signal the highest level. Because the signaling method is current rather than voltage, the line resistance is not a factor, as long as the transmitter can overcome the voltage drop in the line. The major advantage of the method is that it is virtually immune to noise pickup.

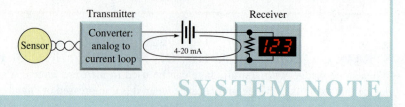

SYSTEM NOTE

Load cells are usually connected to a Wheatstone bridge as shown in Figure 6–39(b) with strain gages (SG) in tension (T) and compression (C) in opposite diagonal legs as shown. The output of the bridge is normally digitized and converted to a reading for a display or sent to a computer for processing. The major advantage of the Wheatstone bridge circuit is that it is capable of accurately measuring very small differences in resistance. The use of four active transducers increases the sensitivity of the measurement and makes the bridge the ideal circuit for instrumentation. The Wheatstone bridge circuit has the added benefit of compensating for temperature variations and wire resistance of connecting wires that would otherwise contribute to inaccuracies.

In addition to scales, strain gages are used with Wheatstone bridges in other types of measurements including pressure measurements, displacement and acceleration measurements to name a few. In pressure measurements, the strain gages are bonded to a flexible diaphragm that stretches when pressure is applied to the transducer. The amount of flexing is related to the pressure, which again converts to a very small resistance change.

SYSTEM EXAMPLE 6–2

A LIQUID LEVEL-SENSING SYSTEM

Load cells were described in the text as force-sensing transducers. Weight is the force gravity exerts on an object, so load cells are widely used in scales. An application that converts the force to another quantity is a liquid-level detection system for a large tank. Cylindrical tanks will usually use three load cells; other shapes may have four cells. A tank containing hazardous materials or subject to wind loading might use more than four load cells.

Typical methods for measuring the liquid-level in a tank include floats, capacitive sensors, and ultrasonic sensors. Certain liquids, like paint, hazardous liquids or those at a high temperature can be a problem for contacting sensors, so weighing systems are an attractive alternative in these cases. Essentially the system described in this example is a weighing system that indicates the liquid level by a basic calculation of net weight. The weight is translated by the computer into a liquid-level indication.

Figure 6–40 shows a liquid-level sensor composed of three load cells arranged so that each load cell has the same load. The tank needs to be initially set level and be able to freely move up and down for this method to work properly. Because the tank needs to move freely in the vertical direction, flexible fittings are used. The load cells must not experience any horizontal motion, and the tank needs to be restrained in the horizontal direction. The tank is fully supported by the load cells, so that the sum of the weights represents the gross weight (tank plus fluid). Each load cell is instrumented with a Wheatstone bridge. A computer calculates the percentage of liquid in the tank and the result is displayed. Figure 6–41 shows a block diagram of the system.

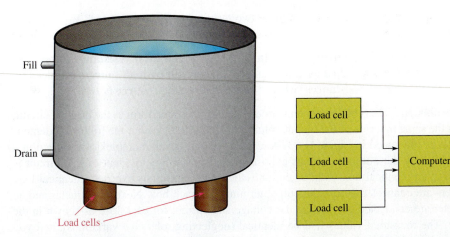

FIGURE 6–40 Tank liquid-level sensing system.   **FIGURE 6–41** Block diagram of system.

SECTION 6–5 CHECKUP

1. Draw a basic Wheatstone bridge circuit.

2. Under what condition is a bridge balanced?

3. What is the unknown resistance in Figure 6–36 when $R_V = 3.3$ kΩ, $R_2 = 10$ kΩ, and $R_4 = 2.2$ kΩ?

4. How is a Wheatstone bridge used in the unbalanced condition?

5. What is a load cell?

6–6 THEVENIN'S THEOREM

Thevenin's theorem provides a method for simplifying a circuit to a standard equivalent form with respect to two output terminals. In many cases, this theorem can be used to simplify the analysis of series-parallel circuits. Another method for simplifying a circuit to an equivalent form is Norton's theorem, which is covered in Appendix C.

After completing this section, you should be able to

- **Apply Thevenin's theorem to simplify a circuit for analysis**
  - **Describe the form of a Thevenin equivalent circuit**
  - **Obtain the Thevenin equivalent voltage source**
  - **Obtain the Thevenin equivalent resistance**
  - **Explain terminal equivalency in the context of Thevenin's theorem**
  - **Thevenize a portion of a circuit**
  - **Thevenize a Wheatstone bridge**

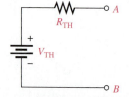

FIGURE 6–42 The general form of a Thevenin equivalent circuit is a voltage source in series with a resistance.

The Thevenin equivalent form of any two-terminal resistive circuit consists of an equivalent voltage source (V_{TH}) and an equivalent resistance (R_{TH}), arranged as shown in Figure 6–42. The values of the equivalent voltage and resistance depend on the values in the original circuit. Any two-terminal resistive circuit can be simplified to a Thevenin equivalent regardless of its complexity.

The equivalent voltage, V_{TH}, is one part of the complete Thevenin equivalent circuit. The other part is R_{TH}.

> **The Thevenin equivalent voltage (V_{TH}) is the open circuit (no-load) voltage between two specified output terminals in a circuit.**

Any component connected between these two terminals effectively "sees" V_{TH} in series with R_{TH}. As defined by **Thevenin's theorem,**

> **The Thevenin equivalent resistance (R_{TH}) is the total resistance appearing between two specified output terminals in a circuit with all sources replaced by their internal resistances (which for an ideal voltage source is zero).**

Although a Thevenin equivalent circuit is not of the same form as the original circuit, it acts the same in terms of the output voltage and current. Consider the following demonstration as illustrated in Figure 6–43. A resistive circuit of any complexity is placed in a box with only the output terminals exposed. The Thevenin equivalent of that circuit is placed in an identical box with, again, only the output terminals exposed. Identical load resistors are connected across the output terminals of each box. Next, a voltmeter and an ammeter are connected to measure the voltage and current for each load as shown in the figure. The measured values will be identical (neglecting tolerance variations), and you will not be able to determine which box contains the original circuit and which contains the Thevenin equivalent of the original circuit. That is, in terms of what you can observe by electrical measurement, both circuits appear to be the same. This condition is sometimes

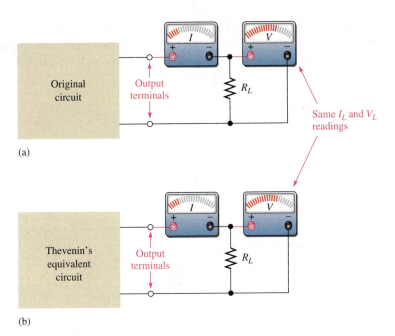

(a)

(b)

FIGURE 6–43   Which box contains the original circuit and which contains the Thevenin equivalent circuit? You cannot tell by observing the meters because the circuits have terminal equivalency.

known as **terminal equivalency** because both circuits look the same from the "viewpoint" of the two output terminals.

To find the Thevenin equivalent of any circuit, you must determine the equivalent voltage, V_{TH}, and the equivalent resistance, R_{TH}. For example, the Thevenin equivalent for the circuit between output terminals A and B is developed in Figure 6–44.

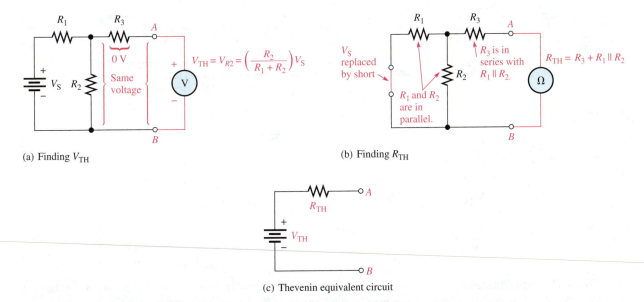

(a) Finding V_{TH}

(b) Finding R_{TH}

(c) Thevenin equivalent circuit

FIGURE 6–44   Example of the simplification of a circuit by Thevenin's theorem.

In Figure 6–44(a), the voltage across the designated terminals A and B is the Thevenin equivalent voltage. In this particular circuit, the voltage from A to B is the same as the voltage across R_2 because there is no current through R_3 and, therefore, no voltage drop across it. V_{TH} is expressed as follows for this particular example:

$$V_{TH} = \left(\frac{R_2}{R_1 + R_2}\right)V_S$$

In Figure 6–44(b), the resistance between terminals A and B with the source replaced by a short to represent its zero internal resistance is the Thevenin equivalent resistance. In this particular circuit, the resistance from A to B is R_3 in series with the parallel combination of R_1 and R_2. Therefore, R_{TH} is expressed as follows:

$$R_{\text{TH}} = R_3 + \frac{R_1 R_2}{R_1 + R_2}$$

The Thevenin equivalent circuit is shown in Figure 6–44(c).

EXAMPLE 6–15

Find the Thevenin equivalent circuit between the output terminals A and B of the circuit in Figure 6–45. If there were a load resistance connected across terminals A and B, it would first have to be removed.

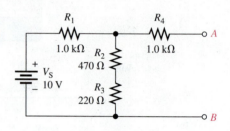

FIGURE 6–45

SOLUTION

Since there is no voltage drop across R_4, V_{AB}, equals the voltage across $R_2 + R_3$ and $V_{\text{TH}} = V_{AB}$, as shown in Figure 6–46(a). Use the voltage-divider principle to find V_{TH}.

$$V_{\text{TH}} = \left(\frac{R_2 + R_3}{R_1 + R_2 + R_3}\right)V_S = \left(\frac{690\ \Omega}{1.69\ \text{k}\Omega}\right)10\ \text{V} = \mathbf{4.08\ V}$$

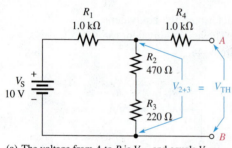

(a) The voltage from A to B is V_{TH} and equals V_{2+3}.

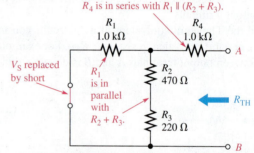

(b) Looking from terminals A and B, R_4 appears in series with the combination of R_1 in parallel with ($R_2 + R_3$).

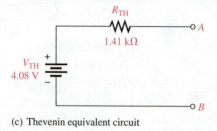

(c) Thevenin equivalent circuit

FIGURE 6–46

To find R_{TH}, first replace the source with a short to simulate a zero internal resistance. Then R_1 appears in parallel with $R_2 + R_3$, and R_4 is in series with the series-parallel combination of R_1, R_2, and R_3 as indicated in Figure 6–46(b).

$$R_{\text{TH}} = R_4 + \frac{R_1(R_2 + R_3)}{R_1 + R_2 + R_3} = 1.0\ \text{k}\Omega + \frac{(1.0\ \Omega)(690\ \Omega)}{1.69\ \text{k}\Omega} = \mathbf{1.41\ k\Omega}$$

The resulting Thevenin equivalent circuit is shown in Figure 6–46(c).

RELATED PROBLEM

Determine V_{TH} and R_{TH} if a 560 Ω resistor is connected in parallel across R_2 and R_3 in Figure 6–45.

Thevenin Equivalency Depends on the Viewpoint

The Thevenin equivalent for any circuit depends on the location of the two output terminals from which the circuit is "viewed." In Figure 6–45, you viewed the circuit from between the two terminals labeled A and B. Any given circuit can have more than one Thevenin equivalent, depending on how the output terminals are designated. For example, if you view the circuit in Figure 6–47 from between terminals A and C, you obtain a completely different result than if you view it from between terminals A and B or from between terminals B and C.

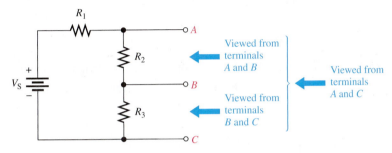

FIGURE 6–47   Thevenin's equivalent depends on the output terminals from which the circuit is viewed.

In Figure 6–48(a), when viewed from between terminals A and C, V_{TH} is the voltage across $R_2 + R_3$ and can be expressed using the voltage-divider formula as

$$V_{TH(AC)} = \left(\frac{R_2 + R_3}{R_1 + R_2 + R_3}\right)V_S$$

Also, as shown in Figure 6–48(b), the resistance between terminals A and C is $R_2 + R_3$ in parallel with R_1 (the source is replaced by a short) and can be expressed as

$$R_{TH(AC)} = R_1 \| (R_2 + R_3) = \frac{R_1(R_2 + R_3)}{R_1 + R_2 + R_3}$$

The resulting Thevenin equivalent circuit is shown in Figure 6–48(c).

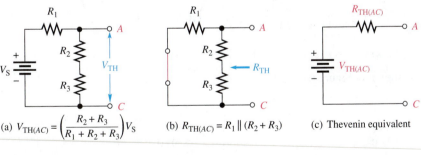

(a) $V_{TH(AC)} = \left(\dfrac{R_2 + R_3}{R_1 + R_2 + R_3}\right)V_S$ 　 (b) $R_{TH(AC)} = R_1 \| (R_2 + R_3)$ 　 (c) Thevenin equivalent

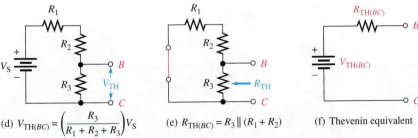

(d) $V_{TH(BC)} = \left(\dfrac{R_3}{R_1 + R_2 + R_3}\right)V_S$ 　 (e) $R_{TH(BC)} = R_3 \| (R_1 + R_2)$ 　 (f) Thevenin equivalent

FIGURE 6–48   Example of a circuit thevenized from two different sets of terminals. Parts (a), (b), and (c) illustrate one set of terminals and parts (d), (e), and (f) illustrate another set of terminals. (The V_{TH} and R_{TH} values are different for each case.)

When viewed from between terminals B and C as indicated in Figure 6–48(d), $V_{TH(BC)}$ is the voltage across R_3 and can be expressed as

$$V_{TH(BC)} = \left(\frac{R_3}{R_1 + R_2 + R_3}\right)V_S$$

As shown in Figure 6–48(e), the resistance between terminals B and C is R_3 in parallel with the series combination of R_1 and R_2.

$$R_{TH(BC)} = R_3 \| (R_1 + R_2) = \frac{R_3(R_1 + R_2)}{R_1 + R_2 + R_3}$$

The resulting Thevenin equivalent is shown in Figure 6–48(f).

EXAMPLE 6–16

(a) Determine the Thevenin equivalent circuit for the circuit in Figure 6–49 viewed from terminals A and C.

(b) Determine the Thevenin equivalent circuit for the circuit in Figure 6–49 viewed from terminals B and C.

FIGURE 6–49

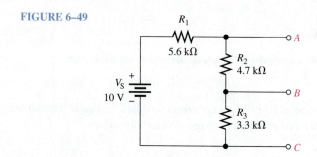

SOLUTION

(a) $V_{TH(AC)} = \left(\dfrac{R_2 + R_3}{R_1 + R_2 + R_3}\right)V_S = \left(\dfrac{4.7\,k\Omega + 3.3\ k\Omega}{5.6\,k\Omega + 4.7\,k\Omega + 3.3\,k\Omega}\right)10\ V = \textbf{5.88 V}$

$R_{TH(AC)} = R_1 \| (R_2 + R_3) = 5.6\,k\Omega \| (4.7\,k\Omega + 3.3\,k\Omega) = \textbf{3.29 k}\boldsymbol{\Omega}$

The Thevenin equivalent circuit is shown in Figure 6–50(a).

(b) $V_{TH(BC)} = \left(\dfrac{R_3}{R_1 + R_2 + R_3}\right)V_S = \left(\dfrac{3.3\,k\Omega}{5.6\,k\Omega + 4.7\,k\Omega + 3.3\,k\Omega}\right)10\ V = \textbf{2.43 V}$

$R_{TH(BC)} = R_3 \| (R_1 + R_2) = 3.3\,k\Omega \| (5.6\,k\Omega + 4.7\,k\Omega) = \textbf{2.5 k}\boldsymbol{\Omega}$

The Thevenin equivalent circuit is shown in Figure 6–50(b).

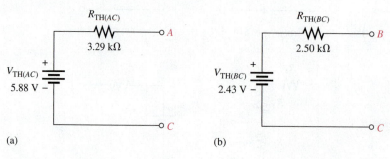

(a)   (b)

FIGURE 6–50

RELATED PROBLEM

Determine the Thevenin equivalent circuit viewed from terminals A and B in Figure 6–49.

Thevenizing a Bridge Circuit

The usefulness of Thevenin's theorem can be illustrated when it is applied to a Wheatstone bridge circuit. For example, consider the case when a load resistor is connected to the output terminals of a Wheatstone bridge, as shown in Figure 6–51. The bridge circuit is difficult to analyze because it is not a straightforward series-parallel arrangement when a load resistance is connected between the output terminals A and B. There are no resistors that are in series or in parallel with another resistor.

Using Thevenin's theorem, you can simplify the bridge circuit to an equivalent circuit viewed from the load resistor as shown step-by-step in Figure 6–52. Study carefully the steps in this figure. Once the equivalent circuit for the bridge is found, the voltage and current for any value of load resistor can easily be determined by Ohm's law.

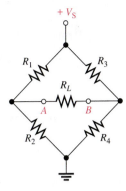

FIGURE 6–51   A Wheatstone bridge with a load resistor connected between the output terminals is not a straightforward series-parallel circuit.

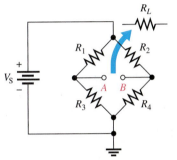

(a) Remove R_L to create an open circuit between the output terminals A and B.

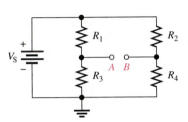

(b) Redraw (if you wish).

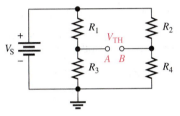

(c) Find V_{TH}:
$$V_{TH} = V_A - V_B = \left(\frac{R_3}{R_1 + R_3}\right)V_S - \left(\frac{R_4}{R_2 + R_4}\right)V_S$$

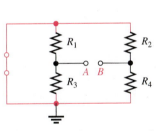

(d) Replace V_S with a short to represent its zero internal resistance. *Note*: The red lines represent the same electrical point as the red lines in Part (e).

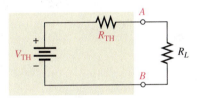

(e) Redraw (if you wish) and find R_{TH}:
$$R_{TH} = R_1 \| R_3 + R_2 \| R_4$$

(f) Thevenin's equivalent circuit (beige block) with R_L reconnected

FIGURE 6–52   Simplifying a Wheatstone bridge with Thevenin's theorem.

EXAMPLE 6–17

Determine the voltage and current for the load resistor, R_L, in the bridge circuit of Figure 6–53.

FIGURE 6–53

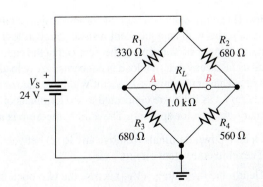

SOLUTION

Step 1: Remove R_L to create an open circuit between A and B.

Step 2: To thevenize the bridge as viewed from between terminals A and B, as was shown in Figure 6–52, first determine V_{TH}.

$$V_{TH} = V_A - V_B = \left(\frac{R_3}{R_1 + R_3}\right)V_S - \left(\frac{R_4}{R_2 + R_4}\right)V_S$$

$$= \left(\frac{680 \ \Omega}{1010 \ \Omega}\right)24 \ V - \left(\frac{560 \ \Omega}{1240 \ \Omega}\right)24 \ V = 16.16 \ V - 10.84 \ V = 5.32 \ V$$

Step 3: Determine R_{TH}.

$$R_{TH} = \frac{R_1 R_3}{R_1 + R_3} + \frac{R_2 R_4}{R_2 + R_4}$$

$$= \frac{(330 \ \Omega)(680 \ \Omega)}{1010 \ \Omega} + \frac{(680 \ \Omega)(560 \ \Omega)}{1240 \ \Omega} = 222 \ \Omega + 307 \ \Omega = 529 \ \Omega$$

Step 4: Place V_{TH} and R_{TH} in series to form the Thevenin equivalent circuit.

Step 5: Connect the load resistor from terminals A to B of the equivalent circuit, and determine the load voltage and current as illustrated in Figure 6–54.

$$V_L = \left(\frac{R_L}{R_L + R_{TH}}\right)V_{TH} = \left(\frac{1.0 \ k\Omega}{1.529 \ k\Omega}\right)5.32 \ V = \mathbf{3.48 \ V}$$

$$I_L = \frac{V_L}{R_L} = \frac{3.48 \ V}{1.0 \ k\Omega} = \mathbf{3.48 \ mA}$$

FIGURE 6–54   Thevenin's equivalent for the Wheatstone bridge

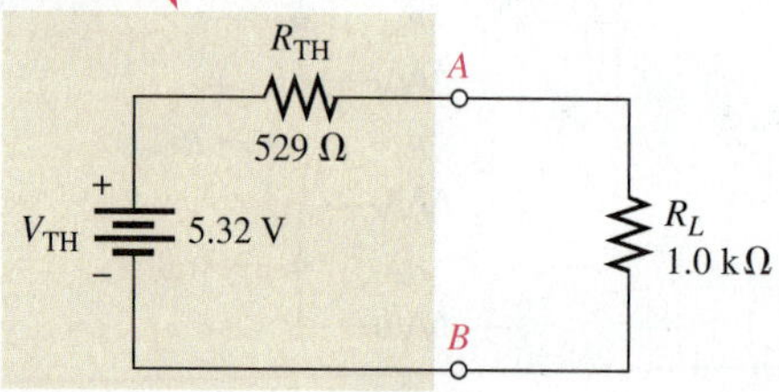

RELATED PROBLEM

Calculate I_L for $R_1 = 2.2 \ k\Omega$, $R_2 = 3.9 \ k\Omega$, $R_3 = 3.3 \ k\Omega$, and $R_4 = 2.7 \ k\Omega$ in Figure 6–53.

MULTISIM

Open Multisim file E06-17. Determine the voltage and current for R_L using a multimeter. Change the resistor values to those specified in the related problem and measure the voltage and current for R_L.

Summary of Thevenin's Theorem

Remember, the Thevenin equivalent circuit for any resistive circuit is *always* an equivalent voltage source in series with an equivalent resistance regardless of the original circuit that it replaces. The significance of Thevenin's theorem is that the equivalent circuit can replace the original circuit as far as any external load is concerned. Any load resistor connected between the terminals of a Thevenin equivalent circuit will have the same current through it and the same voltage across it as if it were connected to the terminals of the original circuit.

A summary of steps for applying Thevenin's theorem is as follows:

Step 1: Open the two terminals (remove any load) between which you want to find the Thevenin equivalent circuit.

Step 2: Determine the voltage (V_{TH}) across the two open terminals.

Step 3: Determine the resistance (R_{TH}) between the two terminals with all sources replaced by their internal resistance. (An ideal voltage source is replaced by a short.)

Step 4: Connect V_{TH} and R_{TH} in series to produce the complete Thevenin equivalent for the original circuit.

Step 5: Replace the load removed in Step 1 across the terminals of the Thevenin equivalent circuit. You can now calculate the load current and load voltage using only Ohm's law, and they have the same value as the load current and load voltage in the original circuit.

Two additional theorems are sometimes used in circuit analysis. One of these is Norton's theorem, which is similar to Thevenin's theorem except that it deals with current sources instead of voltage sources. The other is Millman's theorem, which deals with parallel voltage sources. See Appendix C for a coverage of Norton's theorem and Millman's theorem.

HANDS ON TIP
The Thevenin resistance of a circuit can be measured by connecting a variable resistance on the output of the circuit and adjusting the resistance until the output voltage of the circuit is one-half of its open circuit value. Now, if you remove the variable resistance and measure it, the value equals the Thevenin equivalent resistance of the circuit.

SECTION 6–6 CHECKUP

1. What are the two components of a Thevenin equivalent circuit?

2. Draw the general from of a Thevenin equivalent circuit.

3. Define V_{TH}.

4. Define R_{TH}.

5. For the original circuit in Figure 6–55, determine the Thevenin equivalent circuit as viewed from the output terminals A and B.

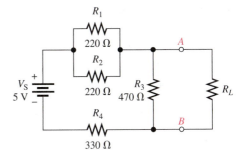

FIGURE 6–55

6–7 THE MAXIMUM POWER TRANSFER THEOREM

The maximum power transfer theorem is important when you need to know the value of the load at which the most power is delivered from the source.

After completing this section, you should be able to

- **Apply the maximum power transfer theorem**
  - **State the theorem**
  - **Determine the value of load resistance for which maximum power is transferred from a given circuit**

The **maximum power transfer** theorem is stated as follows:

> **For a given source voltage, maximum power is transferred from a source to a load when the load resistance is equal to the internal source resistance.**

The source resistance, R_S, of a circuit is the equivalent resistance as viewed from the output terminals using Thevenin's theorem. A Thevenin equivalent circuit with its output resistance and load is shown in Figure 6–56. When $R_L = R_S$, the maximum power possible is transferred from the voltage source to R_L for a given value of V_S.

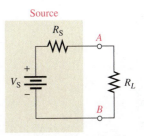

FIGURE 6–56   Maximum power is transferred to the load when $R_L = R_S$.

Practical applications of the maximum power transfer theorem include audio systems such as stereo, radio, and public address. In these systems the resistance of the speaker is the load. The circuit that drives the speaker is a power amplifier. The systems are typically optimized for maximum power to the speakers. Thus, the resistance of the speaker must equal the internal source resistance of the amplifier.

Example 6–18 shows that maximum power occurs when $R_L = R_S$.

EXAMPLE 6–18

The source in Figure 6–57 has an internal resistance of 75 Ω. Determine the load power for each of the following values of the variable load resistance:

(a) 0 Ω **(b)** 25 Ω **(c)** 50 Ω

(d) 75 Ω **(e)** 100 Ω **(f)** 125 Ω

Draw a graph showing the load power versus the load resistance.

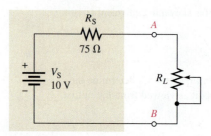

FIGURE 6–57

SOLUTION

Use Ohm's law ($I = V/R$) and the power formula ($P = I^2R$) to find the load power, P_L, for each value of load resistance.

(a) For $R_L = 0\ \Omega$,

$$I = \frac{V_S}{R_S + R_L} = \frac{10V}{75\ \Omega + 0\ \Omega} = 133\ \text{mA}$$

$$P_L = I^2R_L = (133\ \text{mA})^2(0\ \Omega) = \mathbf{0\ mW}$$

(b) For $R_L = 25\ \Omega$,

$$I = \frac{V_S}{R_S + R_L} = \frac{10\ \text{V}}{75\ \Omega + 25\ \Omega} = 100\ \text{mA}$$

$$P_L = I^2R_L = (100\ \text{mA})^2(25\ \Omega) = \mathbf{250\ mW}$$

(c) For $R_L = 50\ \Omega$,

$$I = \frac{V_S}{R_S + R_L} = \frac{10\ \text{V}}{125\ \Omega} = 80\ \text{mA}$$

$$P_L = I^2R_L = (80\ \text{mA})^2(50\ \Omega) = \mathbf{320\ mW}$$

(d) For $R_L = 75\ \Omega$,

$$I = \frac{V_S}{R_S + R_L} = \frac{10\ \text{V}}{150\ \Omega} = 66.7\ \text{mA}$$

$$P_L = I^2R_L = (66.7\ \text{mA})^2(75\ \Omega) = \mathbf{334\ mW}$$

(e) For $R_L = 100\ \Omega$,

$$I = \frac{V_S}{R_S + R_L} = \frac{10\ \text{V}}{175\ \Omega} = 57.1\ \text{mA}$$

$$P_L = I^2 R_L = (57.1\ \text{mA})^2 (100\ \Omega) = \mathbf{326\ mW}$$

(f) For $R_L = 125\ \Omega$,

$$I = \frac{V_S}{R_S + R_L} = \frac{10\ \text{V}}{200\ \Omega} = 50\ \text{mA}$$

$$P_L = I^2 R_L = (50\ \text{mA})^2 (125\ \Omega) = \mathbf{313\ mW}$$

Notice that the load power is greatest when $R_L = R_S = 75\ \Omega$, which is the same as the internal source resistance. When the load resistance is less than or greater than this value, the power drops off, as the curve in Figure 6–58 graphically illustrates.

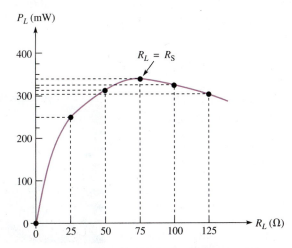

FIGURE 6–58   **Curve showing that the load power is maximum when $R_L = R_S$.**

RELATED PROBLEM

If the source resistance in Figure 6–57 is $600\ \Omega$, what is the maximum power that can be delivered to a load?

SECTION 6–7 CHECKUP

1. State the maximum power transfer theorem.

2. When is maximum power delivered from a source to a load?

3. A given circuit has an internal source resistance of $50\ \Omega$. What will be the value of the load to which the maximum power is delivered?

Some circuits require more than one voltage or current source. For example, most amplifiers operate with two voltage sources: an ac and a dc source. Additionally, some amplifiers require both a positive and a negative dc voltage source for proper operation. When multiple sources are used in a circuit, the superposition theorem provides a method for analysis.

After completing this section, you should be able to

- Apply the superposition theorem to circuit analysis
  - State the superposition theorem
  - List the steps in applying the theorem

The **superposition** theorem is a way to determine currents in a circuit with multiple sources by leaving one source at a time and replacing the other sources by their internal resistances. Recall that the ideal voltage source has a zero internal resistance and the ideal current source has infinite internal resistance. All sources will be treated as ideal in order to simplify the coverage.

The Superposition Theorem in Physics

The superposition theorem is not restricted to electrical and electronic systems. It can be applied to any linear system with two or more sources (stimuli). In physics, it is applied to various types of waves (including sound and water waves), mechanical structures (beams), field theory, quantum mechanics, and even overlapping magnetic or electric fields.

For example, when water, sound, or light waves are combined from independent sources, the result can be predicted within certain limits by the superposition theorem. Where two crests overlap, the wave amplitude increases; when a crest and a trough overlap, they

Anja Kaiser/Fotolia.com

tend to cancel. Another familiar example of superposition is the superimposing of an airplane's forward motion with any wind motion. The vector sum of the two is the net motion of the plane.

SYSTEM NOTE

A general statement of the superposition theorem is as follows:

The current in any given branch of a multiple-source linear circuit can be found by determining the currents in that particular branch produced by each source acting alone, with all other sources replaced by their internal resistances. The total current in the branch is the algebraic sum of the individual source currents in that branch.

The steps in applying the superposition theorem are as follows:

Step 1: Leave one voltage (or current) source at a time in the circuit and replace each of the other voltage (or current) sources with its internal resistance. For ideal sources, a short represents zero internal resistance and an open represents infinite internal resistance.

Step 2: Determine the particular current (or voltage) that you want to find just as if there were only one source in the circuit. This is a component of the total current or voltage that you are looking for.

Step 3: Take the next source in the circuit and repeat Steps 1 and 2 for each source.

Step 4:   To find the actual current in a given branch (with all sources active), algebraically add the results for all sources. Once you find this current, you can determine the voltage using Ohm's law.

The approach to superposition is illustrated in Figure 6–59 for a series-parallel circuit with two ideal voltage sources. Study the steps in this figure.

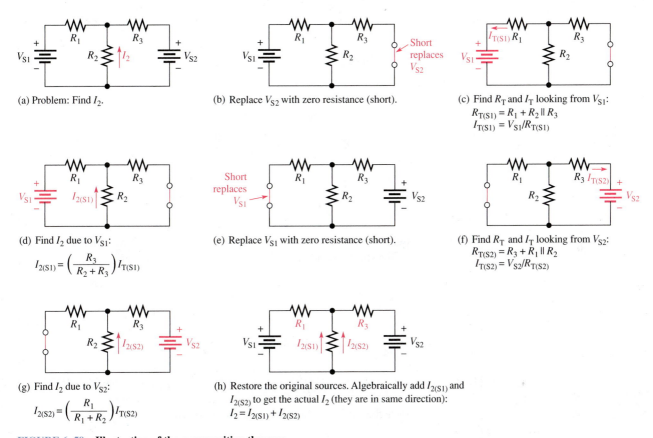

(a) Problem: Find I_2.

(b) Replace V_{S2} with zero resistance (short).

(c) Find R_T and I_T looking from V_{S1}:
$$R_{T(S1)} = R_1 + R_2 \| R_3$$
$$I_{T(S1)} = V_{S1}/R_{T(S1)}$$

(d) Find I_2 due to V_{S1}:
$$I_{2(S1)} = \left(\frac{R_3}{R_2 + R_3}\right)I_{T(S1)}$$

(e) Replace V_{S1} with zero resistance (short).

(f) Find R_T and I_T looking from V_{S2}:
$$R_{T(S2)} = R_3 + R_1 \| R_2$$
$$I_{T(S2)} = V_{S2}/R_{T(S2)}$$

(g) Find I_2 due to V_{S2}:
$$I_{2(S2)} = \left(\frac{R_1}{R_1 + R_2}\right)I_{T(S2)}$$

(h) Restore the original sources. Algebraically add $I_{2(S1)}$ and $I_{2(S2)}$ to get the actual I_2 (they are in same direction):
$$I_2 = I_{2(S1)} + I_{2(S2)}$$

FIGURE 6–59   **Illustration of the superposition theorem.**

EXAMPLE 6–19

Use the superposition theorem to find the current through R_2 and the voltage across it in Figure 6–60.

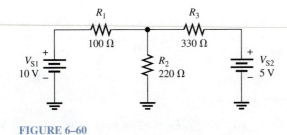

FIGURE 6–60

SOLUTION

Step 1:   Replace V_{S2} with a short to represent its zero internal resistance and find the current through R_2 due to voltage source V_{S1}, as shown in

Figure 6–61. To find I_2, use the current-divider formula. Looking from V_{S1},

$$R_{T(S1)} = R_1 + R_2 \| R_3 = 100\ \Omega + 220\ \Omega \| 330\ \Omega = 232\ \Omega$$

$$I_{T(S1)} = \frac{V_{S1}}{R_{T(S1)}} = \frac{10\ \text{V}}{232\ \Omega} = 43.1\ \text{mA}$$

The component of the total current through R_2 due to V_{S1} is

$$I_{2(S1)} = \left(\frac{R_3}{R_2 + R_3}\right)I_{T(S1)} = \left(\frac{330\ \Omega}{220\ \Omega + 330\ \Omega}\right)43.1\ \text{mA} = 25.9\ \text{mA}$$

Note that this current is upward through R_2.

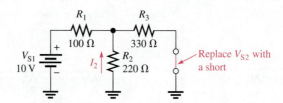

FIGURE 6–61

Step 2:   Find the current through R_2 due to voltage source V_{S2} by replacing V_{S1} with a short, as shown in Figure 6–62. Looking from V_{S2},

$$R_{T(S2)} = R_3 + R_1 \| R_2 = 330\ \Omega + 100\ \Omega \| 220\ \Omega = 399\ \Omega$$

$$I_{T(S2)} = \frac{V_{S2}}{R_{T(S2)}} = \frac{5\ \text{V}}{399\ \Omega} = 12.5\ \text{mA}$$

The component of the total current through R_2 due to V_{S2} is

$$I_{2(S2)} = \left(\frac{R_1}{R_1 + R_2}\right)I_{T(S2)} = \left(\frac{100\ \Omega}{100\ \Omega + 220\ \Omega}\right)12.5\ \text{mA} = 3.90\ \text{mA}$$

Note that this current is upward through R_2.

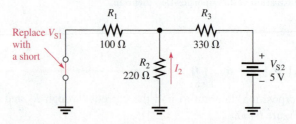

FIGURE 6–62

Step 3:   Both component currents are upward through R_2, so they have the same algebraic sign. Therefore, add the values to get the total current through R_2.

$$I_{2(tot)} = I_{2(S1)} + I_{2(S2)} = 25.9\ \text{mA} + 3.90\ \text{mA} = \textbf{29.8 mA}$$

The voltage across R_2 is

$$V_{R2} = I_{2(tot)}R_2 = (29.8\ \text{mA})(220\ \Omega) = \textbf{6.56 V}$$

RELATED PROBLEM

Determine the total current through R_2 if the polarity of V_{S2} in Figure 6–60 is reversed.

EXAMPLE 6–20

Find the total current through and voltage across R_3 in Figure 6–63.

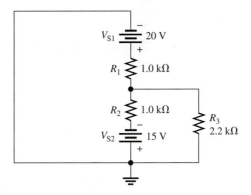

FIGURE 6–63

SOLUTION

Step 1: Find the current through R_3 due to source V_{S1} by replacing source V_{S2} with a short to represent its zero internal resistance, as shown in Figure 6–64. Looking from V_{S1},

$$R_{T(S1)} = R_1 + \frac{R_2 R_3}{R_2 + R_3} = 1.0 \, \text{k}\Omega + \frac{(1.0 \, \text{k}\Omega)(2.2 \, \text{k}\Omega)}{3.2 \, \text{k}\Omega} = 1.69 \, \text{k}\Omega$$

$$I_{T(S1)} = \frac{V_{S1}}{R_{T(S1)}} = \frac{20 \, \text{V}}{1.69 \, \text{k}\Omega} = 11.8 \, \text{mA}$$

Now apply the current-divider formula to get the current through R_3 due to source V_{S1}.

$$I_{3(S1)} = \left(\frac{R_2}{R_2 + R_3} \right) I_{T(S1)} = \left(\frac{1.0 \, \text{k}\Omega}{3.2 \, \text{k}\Omega} \right) 11.8 \, \text{mA} = 3.69 \, \text{mA}$$

Notice that this current is upward through R_3.

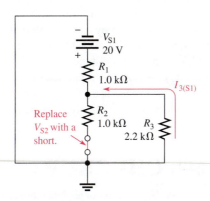

FIGURE 6–64

Step 2: Find the current through R_3 due to source V_{S2} by replacing source V_{S1} with a short, as shown in Figure 6–65. Looking from V_{S2},

$$R_{T(S2)} = R_2 + \frac{R_1 R_3}{R_1 + R_3} = 1.0 \, \text{k}\Omega + \frac{(1.0 \, \text{k}\Omega)(2.2 \, \text{k}\Omega)}{3.2 \, \text{k}\Omega} = 1.69 \, \text{k}\Omega$$

$$I_{T(S2)} = \frac{V_{S2}}{R_{T(S2)}} = \frac{15 \, \text{V}}{1.69 \, \text{k}\Omega} = 8.88 \, \text{mA}$$

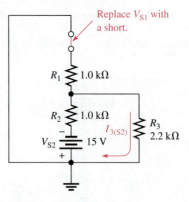

FIGURE 6–65

Now apply the current-divider formula to find the current through R_3 due to source V_{S2}.

$$I_{3(S2)} = \left(\frac{R_1}{R_1 + R_3}\right)I_{T(S2)} = \left(\frac{1.0\,\text{k}\Omega}{3.2\,\text{k}\Omega}\right)8.88\,\text{mA} = 2.78\,\text{mA}$$

Notice that this current is downward through R_3.

Step 3:   Calculate the total current through R_3 and the voltage across it.

$$I_{3(\text{tot})} = I_{3(S1)} - I_{3(S2)} = 3.69\,\text{mA} - 2.78\,\text{mA} = 0.91\,\text{mA} = \mathbf{910\,\mu A}$$
$$V_{R3} = I_{3(\text{tot})}R_3 = (910\,\mu A)(2.2\,\text{k}\Omega) \cong \mathbf{2\,V}$$

The current is upward through R_3.

RELATED PROBLEM

Find the total current through R_3 in Figure 6–63 if V_{S1} is changed to 12 V and its polarity reversed.

Although regulated dc power supplies are close to ideal voltage sources, many ac sources are not. For example, function generators generally have 50 Ω or 600 Ω of internal resistance, which appears as a resistance in series with an ideal source. Also, batteries can look ideal when they are fresh, but as they age, the internal resistance increases. When applying the superposition theorem, it is important to recognize when a source is not ideal and replace it with its equivalent internal resistance.

Current sources are not as common as voltage sources and are also not always ideal. If a current source is not ideal, as in the case of transistors, it should be replaced by its equivalent internal resistance when the superposition theorem is applied.

SECTION 6–8 CHECKUP

1. State the superposition theorem.

2. Why is the superposition theorem useful for analysis of multiple-source linear circuits?

3. Why is an ideal voltage source shorted when the superposition theorem is applied?

4. If, as a result of applying the superposition theorem, two component currents are in opposing directions through a branch of a circuit, in which direction is the net current?

Troubleshooting is the process of identifying and locating a failure or problem in a circuit. Some troubleshooting techniques and the application of logical thought have already been discussed in relation to both series circuits and parallel circuits. A basic premise of troubleshooting is that you must know what to look for before you can successfully troubleshoot a circuit.

After completing this section, you should be able to

- **Troubleshoot series-parallel circuits**
  - **Determine the effects of an open in a circuit**
  - **Determine the effects of a short in a circuit**
  - **Locate opens and shorts**

Opens and shorts are typical problems that occur in electric circuits. As mentioned in Chapter 4, if a resistor burns out, it will normally produce an open. Bad solder connections, broken wires, and poor contacts can also be causes of open paths. Pieces of foreign material, such as solder splashes, broken insulation on wires, and so on, can lead to shorts in a circuit. A short is considered to be a zero resistance path between two points.

In addition to complete opens or shorts, partial opens or partial shorts can develop in a circuit. A partial open would be a much higher than normal resistance, but not infinitely large. A partial short would be a much lower than normal resistance, but not zero.

The following three examples illustrate troubleshooting series-parallel circuits.

EXAMPLE 6–21

From the indicated voltmeter reading in Figure 6–66, determine if there is a fault by applying the APM approach. If there is a fault, identify it as either a short or an open.

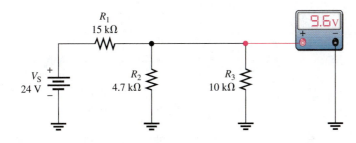

FIGURE 6–66

SOLUTION

Step 1: Analysis

Determine what the voltmeter should be indicating as follows. Since R_2 and R_3 are in parallel, their combined resistance is

$$R_{2\|3} = \frac{R_2 R_3}{R_2 + R_3} = \frac{(4.7\ k\Omega)(10\ k\Omega)}{14.7\ k\Omega} = 3.20\ k\Omega$$

Determine the voltage across the parallel combination by the voltage-divider formula.

$$V_{2\|3} = \left(\frac{R_{2\|3}}{R_1 + R_{2\|3}}\right)V_S = \left(\frac{3.2\ k\Omega}{18.2\ k\Omega}\right)24\ V = 4.22\ V$$

This calculation shows that 4.22 V is the voltage reading that you should get on the meter. However, the meter reads 9.6 V across $R_{2\|3}$. This value is incorrect, and, because it is higher than it should be, either

R_2 or R_3 is probably open. Why? Because if either of these two resistors is open, the resistance across which the meter is connected is larger than expected. A higher resistance will drop a higher voltage in this circuit.

Step 2: Planning

Start trying to find the open resistor by assuming that R_2 is open. If it is, the voltage across R_3 is

$$V_3 = \left(\frac{R_3}{R_1 + R_3}\right)V_S = \left(\frac{10\,\text{k}\Omega}{25\,\text{k}\Omega}\right)24\,\text{V} = 9.6\,\text{V}$$

Since the measured voltage is also 9.6 V, this calculation shows that R_2 is probably open.

Step 3: Measurement

Disconnect power and remove R_2. Measure its resistance to verify it is open. If it is not, inspect the wiring, solder, or connections around R_2, looking for the open.

RELATED PROBLEM

What would be the voltmeter reading if R_3 were open in Figure 6–66? If R_1 were open?

MULTISIM

Open Multisim file E06-21. Determine if a fault exists in the circuit and, if so, isolate the fault to a single component.

EXAMPLE 6–22

Suppose that you measure 24 V with the voltmeter in Figure 6–67. Determine if there is a fault, and, if there is, identify it.

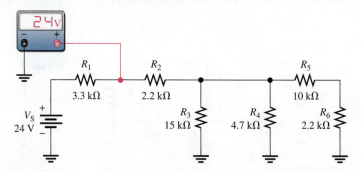

FIGURE 6–67

SOLUTION

Step 1: Analysis

There is no voltage drop across R_1 because both sides of the resistor are at +24 V. Either there is no current through R_1 from the source, which tells you that R_2 is open in the circuit, or R_1 is shorted.

Step 2: Planning

The most probable failure is an open R_2. If it is open, then there will be no current from the source. To verify this, plan to measure across R_2 with the voltmeter. If R_2 is open, the meter will indicate 24 V. The right side of R_2 will be at zero volts because there is no current through any of the other resistors to cause a voltage drop across them.

Step 3: Measurement

The measurement to verify that R_2 is open is shown in Figure 6–68.

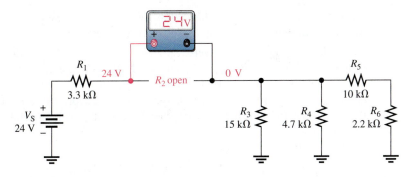

FIGURE 6–68

RELATED PROBLEM

What would be the voltage across an open R_5 in Figure 6–67 assuming no other faults?

EXAMPLE 6–23

The two voltmeters in Figure 6–69 indicate the voltages shown. Apply logical thought and your knowledge of circuit operation to determine if there are any opens or shorts in the circuit and, if so, where they are located.

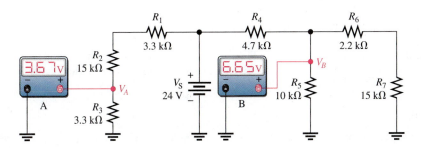

FIGURE 6–69

SOLUTION

Step 1: Determine if the voltmeter readings are correct. R_1, R_2, and R_3 act as a voltage divider. Calculate the voltage (V_A) across R_3 as follows:

$$V_A = \left(\frac{R_3}{R_1 + R_2 + R_3}\right)V_S = \left(\frac{3.3\,\text{k}\Omega}{21.6\,\text{k}\Omega}\right)24\,\text{V} = 3.67\,\text{V}$$

The voltmeter A reading is correct. This indicates that R_1, R_2, and R_3 are connected and are not faulty.

Step 2: See if the voltmeter B reading is correct. $R_6 + R_7$ is in parallel with R_5. The series-parallel combination of R_5, R_6, and R_7 is in series with R_4. Calculate the resistance of the R_5, R_6, and R_7 combination as follows:

$$R_{5\|(6+7)} = \frac{R_5(R_6 + R_7)}{R_5 + R_6 + R_7} = \frac{(10\,\text{k}\Omega)(17.2\,\text{k}\Omega)}{27.2\,\text{k}\Omega} = 6.32\,\text{k}\Omega$$

$R_{5\|(6+7)}$ and R_4 form a voltage divider, and voltmeter B is measuring the voltage across $R_{5\|(6+7)}$. Is it correct? Check as follows:

$$V_B = \left(\frac{R_{5\|(6+7)}}{R_4 + R_{5\|(6+7)}}\right)V_S = \left(\frac{6.32\,\text{k}\Omega}{11\,\text{k}\Omega}\right)24\,\text{V} = 13.8\,\text{V}$$

Thus, the actual measured voltage (6.65 V) at this point is incorrect. Some logical thinking will help to isolate the problem.

Step 3: R_4 is not open, because if it were, the meter would read 0 V. If there were a short across it, the meter would read 24 V. Since the actual voltage is much less than it should be, $R_{5\|(6+7)}$ must be less than the calculated value of 6.32 kΩ. The most likely problem is a short across R_7. If there is a short from the top of R_7 to ground, R_6 is effectively in parallel with R_5. In this case,

$$R_5 \| R_6 = \frac{R_5 R_6}{R_5 + R_6} = \frac{(10\ k\Omega)(2.2\ k\Omega)}{12.2\ k\Omega} = 1.80\ k\Omega$$

Then V_B is

$$V_B = \left(\frac{1.80\ k\Omega}{6.5\ k\Omega}\right)24\ V = 6.65\ V$$

This value for V_B agrees with the voltmeter B reading. So there is a short across R_7. If this were an actual circuit, you would try to find the physical cause of the short.

MULTISIM

Open Multisim file E06-23. Determine if a fault exists in the circuit and, if so, isolate the fault to a single component.

RELATED PROBLEM

If the only fault in Figure 6–69 is that R_2 is shorted, what will voltmeter A read? What will voltmeter B read?

SECTION 6–9 CHECKUP

1. Name two types of common circuit faults.

2. For the following faults in Figure 6–70, determine what voltage would be measured at node A:

(a) no faults

(b) R_1 open

(c) short across R_5

(d) R_3 and R_4 open

(e) R_2 open

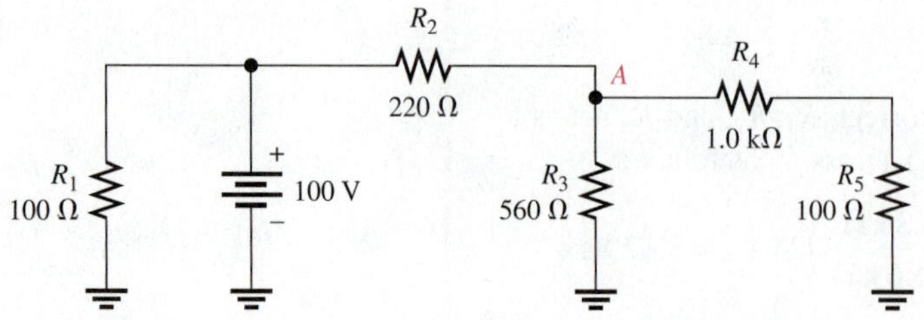

3. In Figure 6–71, one of the resistors in the circuit is open. Based on the meter reading, determine which is the open resistor.

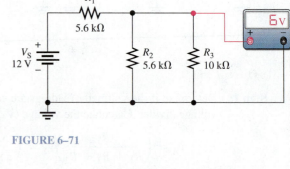

FIGURE 6–71

FIGURE 6–70

SUMMARY

- A series-parallel circuit is a combination of both series current paths and parallel current paths.

- To determine total resistance in a series-parallel circuit, identify the series and parallel relationships, and then apply the formulas for series resistance and parallel resistance from Chapters 4 and 5.

- To find the total current, divide the total voltage by the total resistance.

- To determine branch currents, apply the current-divider formula, Kirchhoff's current law, or Ohm's law. Consider each circuit problem individually to determine the most appropriate method.

- To determine voltage drops across any portion of a series-parallel circuit, use the voltage-divider formula, Kirchhoff's voltage law, or Ohm's law. Consider each circuit problem individually to determine the most appropriate method.

- When a load resistor is connected across a voltage-divider output, the output voltage decreases.

- The load resistor should be large compared to the resistance across which it is connected, in order that the loading effect may be minimized. A *10-times* value is sometimes used as a rule of thumb, but the value depends on the accuracy required for the output voltage.

- To find any current or voltage in a circuit with two or more voltage sources, take the sources one at a time using the superposition theorem.

- A balanced Wheatstone bridge can be used to measure an unknown resistance.

- A bridge is balanced when the output voltage is zero. The balanced condition produces zero current through a load connected across the output terminals of the bridge.

- An unbalanced Wheatstone bridge can be used to measure physical quantities using transducers.

- Any two-terminal resistive circuit, no matter how complex, can be replaced by its Thevenin equivalent.

- The Thevenin equivalent circuit is made up of an equivalent resistance (R_{TH}) in series with an equivalent voltage source (V_{TH}).

- The maximum power transfer theorem states that maximum power is transferred from a source to a load when $R_S = R_L$.

- Opens and shorts in circuits are typical faults.

- Resistors normally open when they fail.

KEY TERMS

Key terms and other bold terms in the chapter are defined in the end-of-book glossary.

Balanced bridge   A bridge circuit that is in the balanced state as indicated by zero volts across the bridge.

Bleeder current   The current left after the total load current is subtracted from the total current into the circuit.

Load current   The output current of a circuit supplied to a load.

Loading   The effect on a circuit when an element that draws current from the circuit is connected across the output terminals.

Maximum power transfer   The condition, when the load resistance equals the source resistance, under which maximum power is transferred from source to load.

Superposition   A method for analyzing circuits with two or more sources by examining the effects of each source by itself and then combining the effects.

Terminal equivalency   A condition that occurs when two circuits produce the same load voltage and load current where the same value of load resistance is connected to either circuit.

Thevenin's theorem   A circuit theorem that provides for reducing any two-terminal resistive circuit to a single equivalent voltage source in series with an equivalent resistance.

Unbalanced bridge   A bridge circuit that is in the unbalanced state as indicated by a voltage across the bridge proportional to the amount of deviation from the balanced state.

Wheatstone bridge   A 4-legged type of bridge circuit with which an unknown resistance can be accurately measured using the balanced state of the bridge. Deviations in resistance can be measured using the unbalanced state.

KEY FORMULAS

(6–1) $I_{BLEEDER} = I_T - I_{RL1} - I_{RL2}$ Bleeder current

(6–2) $R_X = R_V\left(\dfrac{R_2}{R_4}\right)$ Unknown resistance in a Wheatstone bridge

TRUE/FALSE QUIZ

Answers are at the end of the chapter.

1. Parallel resistors are always connected between the same pair of nodes.

2. If one resistor is connected in series with a parallel combination, the series resistor will always have a larger voltage drop than the parallel resistors.

3. In a series-parallel combinational circuit, the same current will always be in parallel resistors.

4. A larger load resistor has a smaller loading effect on a circuit.

5. When measuring dc voltage, a DMM will normally have a small loading effect on a circuit.

6. When measuring dc voltage, the input resistance of a DMM is the same no matter on what scale it is used.

7. When measuring dc voltage, the input resistance of an analog multimeter is the same no matter on what scale it is used.

8. A Thevenin circuit consists of a voltage source with a parallel resistor.

9. The internal resistance of an ideal voltage source is zero.

10. To transfer maximum power to a load, the load resistor should be twice the Thevenin resistance of the source.

SELF-TEST

Answers are at the end of the chapter.

1. Which of the following statements are true concerning Figure 6–72?
    (a) R_1 and R_2 are in series with R_3, R_4, and R_5.
    (b) R_1 and R_2 are in series.
    (c) R_3, R_4, and R_5 are in parallel.
    (d) The series combination of R_1 and R_2 is in parallel with the series combination of R_3, R_4, and R_5.
    (e) answers (b) and (d)

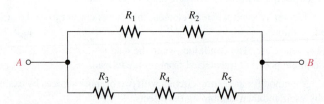

FIGURE 6–72

2. The total resistance of Figure 6–72 can be found with which of the following formulas?
    (a) $R_1 + R_2 + R_3 \| R_4 \| R_5$ (b) $R_1 \| R_2 + R_3 \| R_4 \| R_5$
    (c) $(R_1 + R_2) \| (R_3 + R_4 + R_5)$ (d) none of these answers

3. If all of the resistors in Figure 6–72 have the same value, when voltage is applied across terminals A and B, the current is
    (a) greatest in R_5 (b) greatest in R_3, R_4, and R_5
    (c) greatest in R_1 and R_2 (d) the same in all the resistors

4. Two 1.0 kΩ resistors are in series and this series combination is in parallel with a 2.2 kΩ resistor. The voltage across one of the 1.0 kΩ resistors is 6 V. The voltage across the 2.2 kΩ resistor is
    (a) 6 V   (b) 3 V   (c) 12 V   (d) 13.2 V

5. The parallel combination of a 330 Ω resistor and a 470 Ω resistor is in series with the parallel combination of four 1.0 kΩ resistors. A 10 V source is connected across the circuit. The resistor with the most current has a value of
    (a) 1.0 kΩ   (b) 330 Ω   (c) 470 Ω

6. In the circuit described in Question 5, the resistor(s) with the most voltage has a value of
   (a) 1.0 kΩ   **(b)** 470 Ω   **(c)** 330 Ω

7. In the circuit described in Question 5, the percentage of the total current through any single 1.0 kΩ resistor is
   (a) 100%   **(b)** 25%
   (c) 50%   **(d)** 31.25%

8. The output of a certain voltage divider is 9 V with no load. When a load is connected, the output voltage
   (a) increases
   (b) decreases
   (c) remains the same
   (d) becomes zero

9. A certain voltage divider consists of two 10 kΩ resistors in series. Which of the following load resistors will have the most effect on the output voltage?
   (a) 1.0 MΩ   **(b)** 20 kΩ
   (c) 100 kΩ   **(d)** 10 kΩ

10. When a load resistance is connected to the output of a voltage-divider circuit, the current drawn from the source
    (a) decreases
    (b) increases
    (c) remains the same
    (d) is cut off

11. The output voltage of a balanced Wheatstone bridge is
    (a) equal to the source voltage
    (b) equal to zero
    (c) dependent on all of the resistor values in the bridge
    (d) dependent on the value of the unknown resistor

12. The primary method of analyzing a circuit with two or more voltage sources is usually
    (a) Thevenin's theorem
    (b) Ohm's law
    (c) superposition
    (d) Kirchhoff's law

13. In a certain two-source circuit, one source acting alone produces 10 mA through a given branch. The other source acting alone produces 8 mA in the opposite direction through the same branch. With both sources, the total current through the branch is
    (a) 10 mA
    (b) 8 mA
    (c) 18 mA
    (d) 2 mA

14. A Thevenin equivalent circuit consists of
    (a) a voltage source in series with a resistance
    (b) a voltage source in parallel with a resistance
    (c) a current source in parallel with a resistance
    (d) two voltage sources and a resistance

15. A voltage source with an internal source resistance of 300 Ω transfers maximum power to a
    (a) 150 Ω load
    (b) 50 Ω load
    (c) 300 Ω load
    (d) 600 Ω load

16. You are measuring the voltage at a given point in a circuit that has very high resistance values and the measured voltage is a little lower than it should be. This is possibly because of
    (a) one or more of the resistance values being off
    (b) the loading effect of the voltmeter
    (c) the source voltage is too low
    (d) all of these answers

TROUBLESHOOTING: SYMPTOM AND CAUSE

The purpose of these exercises is to help develop thought processes essential to troubleshooting. Answers are at the end of the chapter.

Determine the cause for each set of symptoms. Refer to Figure 6–73.

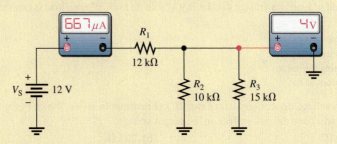

FIGURE 6–73   **The meters indicate the correct readings for this circuit.**

1. *Symptom:* The ammeter reading is too low, and the voltmeter reading is 5.45 V.

   Cause:
   (a) R_1 is open.
   (b) R_2 is open.
   (c) R_3 is open.

2. *Symptom:* The ammeter reading is 1 mA, and the voltmeter reading is 0 V.

   Cause:
   (a) There is a short across R_1.
   (b) There is a short across R_2.
   (c) R_3 is open.

3. *Symptom:* The ammeter reading is near zero, and the voltmeter reading is 12 V.

   Cause:
   (a) R_1 is open.
   (b) R_2 is open.
   (c) Both R_2 and R_3 are open.

4. *Symptom:* The ammeter reading is 444 μA, and the voltmeter reading is 6.67 V.

   Cause:
   (a) R_1 is shorted.
   (b) R_2 is open.
   (c) R_3 is open.

5. *Symptom:* The ammeter reading is 2 mA, and the voltmeter reading is 12 V.

   Cause:
   (a) R_1 is shorted.
   (b) R_2 is shorted.
   (c) Both R_2 and R_3 are open.

PROBLEMS

Answers to odd-numbered problems are at the end of the book.

BASIC PROBLEMS

SECTION 6–1   Identifying Series-Parallel Relationships

1. Identify the series and parallel relationships in Figure 6–74 as seen from the source terminals.

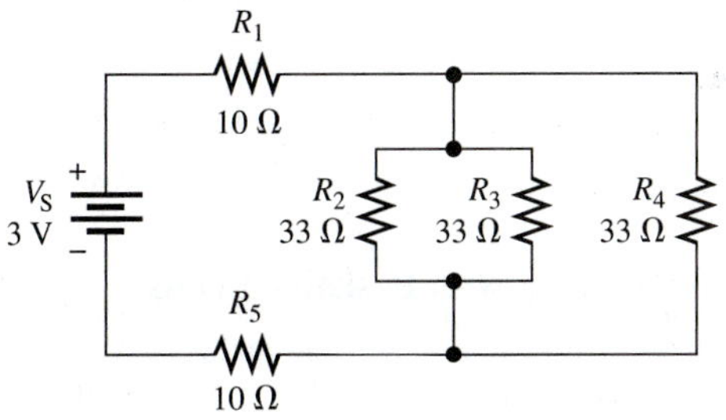

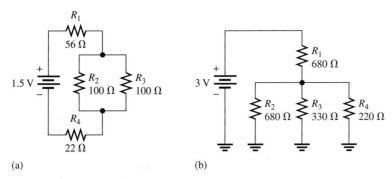

FIGURE 6–74

2. Visualize and draw the following series-parallel combinations:
   (a) R_1 in series with the parallel combination of R_2 and R_3.
   (b) R_1 in parallel with the series combination of R_2 and R_3.
   (c) R_1 in parallel with a branch containing R_2 in series with a parallel combination of four other resistors

3. Visualize and draw the following series-parallel circuits:
   (a) a parallel combination of three branches, each containing two series resistors
   (b) a series combination of three parallel circuits, each containing two parallel resistors

4. In each circuit of Figure 6–75 identify the series and parallel relationships of the resistors viewed from the source.

(a)

(b)

FIGURE 6–75

SECTION 6–2   Analysis of Series-Parallel Resistive Circuits

5. A certain circuit is composed of two parallel resistors. The total resistance is 667 Ω. One of the resistors is 1.0 kΩ. What is the other resistor?

6. For the circuit in Figure 6–76, determine the total resistance between A and B.

7. Determine the total resistance for each circuit in Figure 6–75.

8. Determine the current through each resistor in Figure 6–74; then calculate each voltage drop.

9. Determine the current through each resistor in both circuits of Figure 6–75; then calculate each voltage drop.

10. In Figure 6–77, find the following:
    (a) total resistance between terminals A and B
    (b) total current drawn from a 6 V source connected from A to B

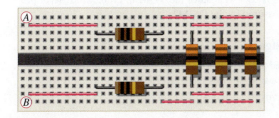

FIGURE 6–76

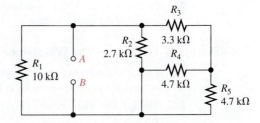

FIGURE 6–77

       (c) current through R_5
       (d) voltage across R_2

11. In Figure 6–77, determine the current through R_2 for $V_{AB} = 6$ V.

12. In Figure 6–77, determine the current through R_4 for $V_{AB} = 6$ V.

SECTION 6–3   Voltage Dividers with Resistive Loads

13. A voltage divider consists of two 56 kΩ resistors and a 15 V source. Calculate the unloaded output voltage taken across one of the 56 kΩ resistors. What will the output voltage be if a load resistor of 1.0 MΩ is connected across the output?

14. A 12 V battery output is divided down to obtain two output voltages. Three 3.3 kΩ resistors are used to provide the two outputs with only one output at a time loaded with 10 kΩ. Determine the output voltages in both cases.

15. Which will cause a smaller decrease in output voltage for a given voltage divider, a 10 kΩ load or a 56 kΩ load?

16. In Figure 6–78, determine the current drain on the battery with no load on the output terminals. With a 10 kΩ load, what is the current from the battery?

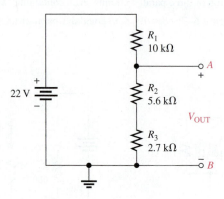

FIGURE 6–78

SECTION 6–4   Loading Effect of a Voltmeter

17. Across which one of the following resistances will a voltmeter with a 10 MΩ internal resistance present the minimum load on a circuit?
    (a) 100 kΩ
    (b) 1.2 MΩ
    (c) 22 kΩ
    (d) 8.2 MΩ

18. A certain voltage divider consists of three 1.0 MΩ resistors connected in series to a 100 V source. Determine the voltage across one of the resistors measured by a 10 MΩ voltmeter.

19. What is the difference between the measured and the actual unloaded voltage in Problem 18?

20. By what percentage does the voltmeter in Problem 18 alter the voltage which it measures?

21. A 10,000 Ω/V VOM is used on the 10 V scale to measure the output of a voltage divider. If the divider consists of two series 100 kΩ resistors, what fraction of the source voltage will be measured across one of the resistors?

22. If a DMM with 10 MΩ input resistance is used instead of the VOM in Problem 21, what percentage of the source voltage will be measured by the DMM?

SECTION 6–5   The Wheatstone Bridge

23. A resistor of unknown value is connected to a Wheatstone bridge circuit. The bridge parameters for a balanced condition are set as follows: $R_V = 18$ kΩ and $R_2/R_4 = 0.02$. What is R_X?

24. A bridge network is shown in Figure 6–79. To what value must R_V be set in order to balance the bridge?

25. Determine the value of R_X in the balanced bridge in Figure 6–80.

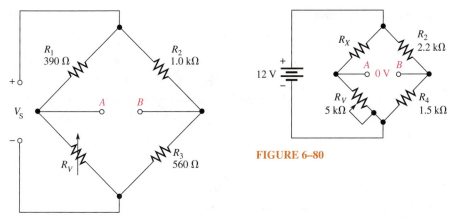

FIGURE 6–79

FIGURE 6–80

26. Determine the output voltage of the unbalanced bridge in Figure 6–81 for a temperature of 65°C. The thermistor has a nominal resistance of 1 kΩ at 25°C and a positive temperature coefficient. Assume that its resistance changes 5 Ω for each C° change in temperature.

FIGURE 6–81

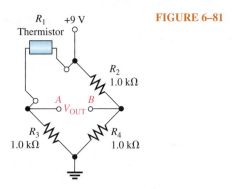

SECTION 6–6   Thevenin's Theorem

27. Reduce the circuit in Figure 6–82 to its Thevenin equivalent as viewed from terminals *A* and *B*.

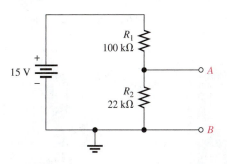

FIGURE 6–82

28. For each circuit in Figure 6–83, determine the Thevenin equivalent as seen from terminals *A* and *B*.

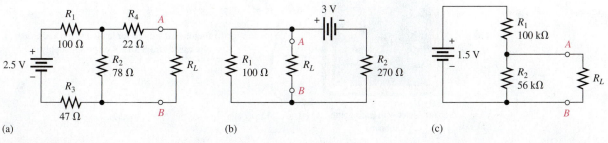

(a)   (b)   (c)

FIGURE 6–83

29. Determine the voltage and current for R_L in Figure 6–84.

FIGURE 6–84

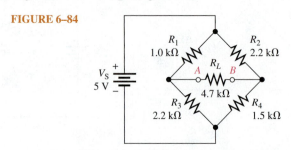

SECTION 6–7   The Maximum Power Transfer Theorem

30. Determine the value of a load resistor connected between terminals A and B in Figure 6–82 for maximum power transfer to the load resistor.

31. A certain Thevenin equivalent circuit has a $V_{TH} = 5.5$ V and an $R_{TH} = 75\ \Omega$. To what value of load resistor will maximum power be transferred?

32. Determine the value of R_L in Figure 6–83(a) for which R_L dissipates maximum power.

SECTION 6–8   The Superposition Theorem

33. In Figure 6–85, use the superposition theorem to find the current in R_3.

34. In Figure 6–85, what is the current through R_2?

FIGURE 6–85

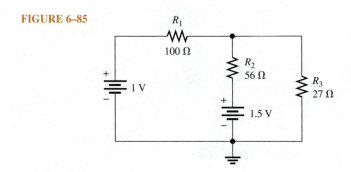

SECTION 6–9   Troubleshooting

35. Is the voltmeter reading in Figure 6–86 correct? If not, what is the problem?

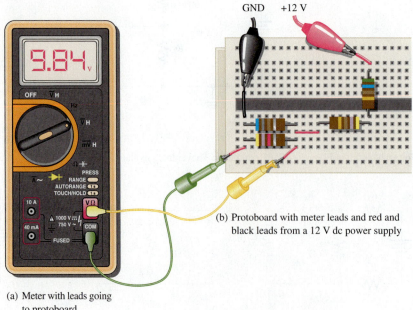

(b) Protoboard with meter leads and red and black leads from a 12 V dc power supply

(a) Meter with leads going to protoboard

FIGURE 6–86

36. If R_2 in Figure 6–87 opens, what voltages will be read at points A, B, and C?

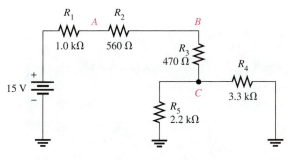

FIGURE 6–87

37. Check the meter readings in Figure 6–88 and locate any fault that may exist.

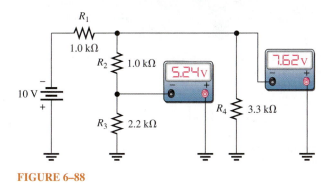

FIGURE 6–88

38. Determine the voltage you would expect to measure across each resistor in Figure 6–87 for each of the following faults. Assume the faults are independent of each other.
   (a) R_1 open   **(b)** R_3 open   **(c)** R_4 open   **(d)** R_5 open
   (e) point C shorted to ground

39. Determine the voltage you would expect to measure across each resistor in Figure 6–88 for each of the following faults:
   (a) R_1 open   **(b)** R_2 open   **(c)** R_3 open   **(d)** a short across R_4

ADVANCED PROBLEMS

40. In each circuit of Figure 6–89, identify the series and parallel relationships of the resistors viewed from the source.

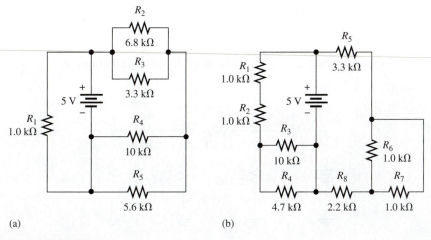

(a)

(b)

FIGURE 6–89

41. Draw the schematic of the PC board layout in Figure 6–90 showing resistor values and identify the series-parallel relationships. Which resistors, if any, can be removed with no effect on R_T?

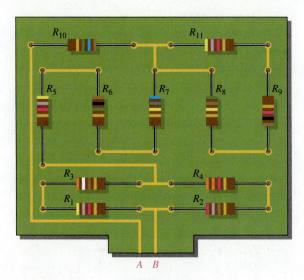

FIGURE 6–90

42. For the circuit shown in Figure 6–91, calculate the following:
   (a) total resistance across the source
   (b) total current from the source
   (c) current through the 910 Ω resistor
   (d) voltage from point A to point B

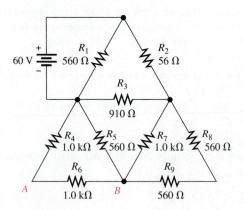

FIGURE 6–91

43. Determine the total resistance and the voltage at points A, B, and C in the circuit of Figure 6–92.

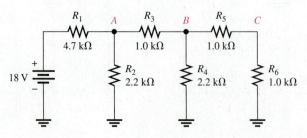

FIGURE 6–92

44. Determine the total resistance between terminals A and B of the circuit in Figure 6–93. Also calculate the current in each branch with 10 V between A and B.

45. What is the voltage across each resistor in Figure 6–93? There are 10 V between A and B.

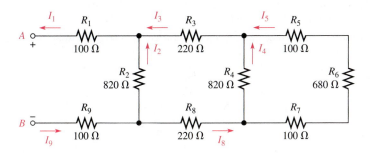

FIGURE 6–93

46. Determine the voltage, V_{AB}, in Figure 6–94.

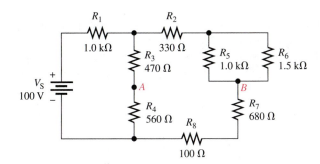

FIGURE 6–94

47. Find the value of R_2 in Figure 6–95.

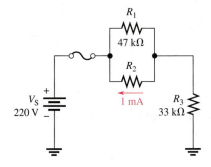

FIGURE 6–95

48. Determine the total resistance and the voltage at points A, B, and C in the circuit of Figure 6–96.

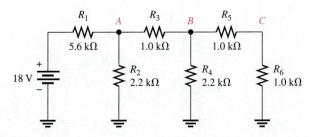

FIGURE 6–96

49. Develop a voltage divider to provide a 6 V output with no load and a minimum of 5.5 V across a 1.0 kΩ load. The source voltage is 24 V, and the unloaded current is not to exceed 100 mA.

50. Determine the resistance values for a voltage divider that must meet the following specifications: The current under an unloaded condition is not to exceed 5 mA. The source voltage is to be 10 V. A 5 V output and a 2.5 V output are required. Draw the circuit. Determine the effect on the output voltages if a 1.0 kΩ load is connected to each output.

51. Using the superposition theorem, calculate the current in the right-most branch of Figure 6–97.

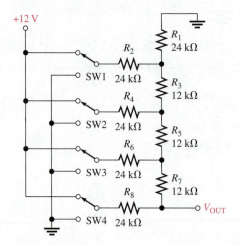

FIGURE 6–97

52. Determine V_{OUT} for the circuit in Figure 6–98 for the following conditions:
    (a) Switch SW2 connected to +12 V and the rest to ground
    (b) Switch SW1 connected to +12 V and the rest to ground

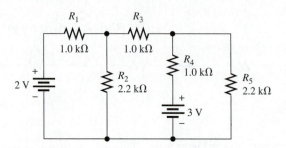

FIGURE 6–98

53. The voltage divider in Figure 6–99 has a switched load. Determine the voltage at each tap (V_1, V_2, and V_3) for each position of the switch.

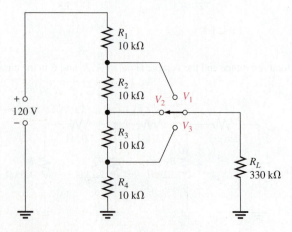

FIGURE 6–99

54. Figure 6–100 shows a dc biasing arrangement for a field-effect transistor amplifier. Biasing is a common method for setting up certain dc voltage levels required for proper amplifier operation. Although you may not be familiar with transistor amplifiers at this point, the dc voltages and currents in the circuit can be determined using methods that you already know.

    (a) Find V_G and V_S with respect to ground

    (b) Determine I_1, I_2, I_D, and I_S

    (c) Find V_{DS} and V_{DG}

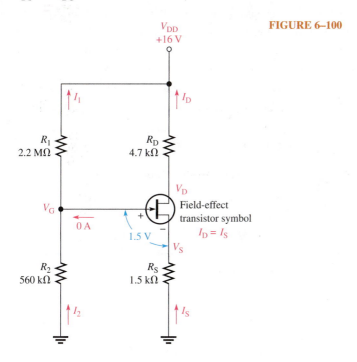

FIGURE 6–100

55. For the circuit in Figure 6–100, what is V_G if R_1 is open?

56. For the circuit in Figure 6–100, what is V_G if R_1 and R_2 are reversed?

57. There is one fault in Figure 6–101. Based on the voltmeter indications, determine what the fault is.

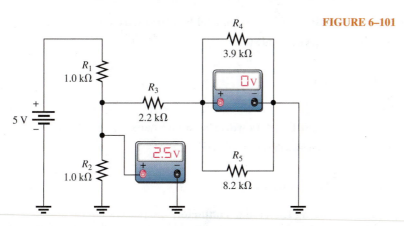

FIGURE 6–101

58. Assume a student decided to increase resistors R_1 and R_2 in Figure 6–30 to ten times the values. The result was a drop in the voltage from 3.7 V to 2.8 V. Explain why this happened.

59. Calculate the base voltage for the circuit in Figure 6–31. Assume there is no loading effect.

60. Repeat problem 59 if the transistors have the same loading effect as a 470 kΩ resistor.

61. Assume a 4-20 mA current loop is terminated with a 150 Ω resistor. What voltage is represented by a 4 mA current? What voltage is represented by a 20 mA current?

62. For the liquid-level sensing system in Figure 6–40, assume the gross weight of the tank is 1700 pounds when full and 300 pounds when empty. What percentage should the display indicate if the load cells indicate a total weight of 700 pounds?

63. For the liquid-level sensing system in Figure 6–40, what is the advantage of using three load cells over four?

MULTISIM TROUBLESHOOTING PROBLEMS

64. Open file P06-64; files are found at www.pearsonhighered.com/floyd. Determine if there is a fault in the circuit. If so, identify the fault.

65. Open file P06-65 and determine if there is a fault in the circuit. If so, identify the fault.

66. Open file P06-66 and determine if there is a fault in the circuit. If so, identify the fault.

67. Open file P06-67 and determine if there is a fault in the circuit. If so, identify the fault.

68. Open file P06-68 and determine if there is a fault in the circuit. If so, identify the fault.

69. Open file P06-69 and determine if there is a fault in the circuit. If so, identify the fault.

70. Open file P06-70 and determine if there is a fault in the circuit. If so, identify the fault.

71. Open file P06-71 and determine if there is a fault in the circuit. If so, identify the fault.

ANSWERS TO SECTION CHECKUPS

SECTION 6–1   Identifying Series-Parallel Relationships

1. See Figure 6–102.

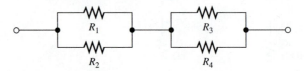

FIGURE 6–102

2. R_1 and R_2 are in series with the parallel combination of R_3 and R_4.

3. All resistors are in parallel.

4. R_1 and R_2 are in parallel; R_3 and R_4 are in parallel.

5. Yes, the two parallel combinations are in series with each other.

SECTION 6–2   Analysis of Series-Parallel Resistive Circuits

1. $R_T = R_1 + R_4 + R_2 \| R_3 = 599\ \Omega$

2. $I_3 = 11.2\ \text{mA}$

3. $V_{R2} = I_2 R_2 = 3.7\ \text{V}$

4. $R_T = 89\ \Omega;\ I_T = 11.2\ \text{mA}$

SECTION 6–3   Voltage Dividers with Resistive Loads

1. The load resistor decreases the output voltage.

2. True

3. $V_{\text{OUT(unloaded)}} = 19.23\ \text{V},\ V_{\text{OUT(loaded)}} = 19.16\ \text{V}$

SECTION 6–4   Loading Effect of a Voltmeter

1. A voltmeter loads a circuit because the internal resistance of the meter appears in parallel with the circuit resistance across which it is connected, reducing the resistance between those two points of the circuit and drawing current from the circuit.

2. No, because the meter resistance is much larger than $1.0\ \text{k}\Omega$.

3. Yes.

4. $4.0\ \text{M}\Omega$

SECTION 6–5   The Wheatstone Bridge

1. See Figure 6–103.

2. A bridge is balanced when the output voltage is zero.

3. $R_X = 15\ \text{k}\Omega$

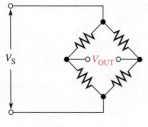

FIGURE 6–103

4. An unbalanced bridge is used to measure transducer-sensed quantities.

5. A load cell is a transducer that uses strain gauges to convert mechanical force into an electrical signal.

SECTION 6–6   Thevenin's Theorem

1. A Thevenin equivalent circuit consists of V_{TH} and R_{TH}.

2. See Figure 6–104.

3. V_{TH} is the open circuit voltage between two points in a circuit.

4. R_{TH} is the resistance as viewed from two terminals in a circuit, with all sources replaced by their internal resistances.

5. See Figure 6–105.

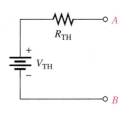

FIGURE 6–104

SECTION 6–7   The Maximum Power Transfer Theorem

1. The maximum power transfer theorem states that maximum power is transferred from a source to a load when the load resistance is equal to the internal source resistance.

2. Maximum power is delivered to a load when $R_L = R_S$.

3. $R_L = R_S = 50 \ \Omega$

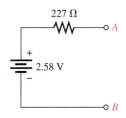

FIGURE 6–105

SECTION 6–8   The Superposition Theorem

1. The total current in any branch of a multiple-source linear circuit is equal to the algebraic sum of the currents due to the individual sources acting alone, with the other sources replaced by their internal resistances.

2. The superposition method allows each source to be treated independently.

3. A short simulates the zero internal resistance of an ideal voltage source.

4. The net current is in the direction of the larger current.

SECTION 6–9   Troubleshooting

1. Opens and short are two common faults.

2. **(a)** 62.8 V   **(b)** 62.8 V   **(c)** 62 V   **(d)** 100 V   **(e)** 0 V

3. The 10 $k\Omega$ resistor is open.

ANSWERS TO RELATED PROBLEMS FOR EXAMPLES

6–1 The added resistor would be in parallel with the combination of R_4 in series with R_2 and R_3 in parallel.

6–2 The added resistor is in parallel with R_5.

6–3 A to gnd: $R_T = R_3 \| (R_1 + R_2) + R_4$

B to gnd: $R_T = R_2 \| (R_1 + R_3) + R_4$

C to gnd: $R_T = R_4$

6–4 None. The new resistor will be shorted by the existing connection between those points.

6–5 55.1 Ω

6–6 128.3 Ω

6–7 $I_1 = 8.93 \ mA; I_3 = 5.85 \ mA; I_T = 18.2 \ mA$

6–8 $V_1 = V_2 = 10.3 \ V; V_3 = 9.70 \ V; V_4 = 3.16 \ V; V_5 = 6.54 \ V$

6–9 $I_1 = 1.42 \ mA, P_1 = 6.67 \ mW; I_2 = 756 \ \mu A, P_2 = 3.55 \ mW;$
$I_3 = 2.18 \ mA, P_3 = 4.75 \ mW; I_4 = 1.13 \ mA, P_4 = 1.28 \ mW;$
$I_5 = 1.06 \ mA, P_5 = 758 \ \mu W; I_6 = 1.06 \ mA, P_6 = 435 \ \mu W$

6–10 3.39 V

6–11 The current will increase because there is a smaller loading effect on the circuit. The new current in R_{L2} is 59 μA.

6–12 5.19 V

6–13 3.3 $k\Omega$

6–14 0.49 V

6–15 2.36 V; 124 Ω

6–16 $V_{TH(AB)} = 3.46$ V; $R_{TH(AB)} = 3.08$ kΩ

6–17 1.17 mA

6–18 41.7 mW

6–19 22.0 mA

6–20 5 mA

6–21 5.73 V, 0 V

6–22 9.46 V

6–23 $V_A = 12$ V; $V_B = 13.8$ V

ANSWERS TO TRUE/FALSE QUIZ

1. T **2.** F **3.** F **4.** T **5.** T **6.** T **7.** F
8. F **9.** T **10.** F

ANSWERS TO SELF-TEST

1. (e) **2.** (c) **3.** (c) **4.** (c) **5.** (b) **6.** (a) **7.** (b)
8. (b) **9.** (d) **10.** (b) **11.** (b) **12.** (c) **13.** (d) **14.** (a)
15. (c) **16.** (d)

ANSWERS TO TROUBLESHOOTING: SYMPTOM AND CAUSE

1. (c) **2.** (b) **3.** (c) **4.** (b) **5.** (a)

MAGNETISM AND ELECTROMAGNETISM

OUTLINE

OBJECTIVES

- Explain the principles of the magnetic field
- Explain the principles of electromagnetism
- Describe the principle of operation for several types of electromagnetic devices
- Explain magnetic hysteresis
- Discuss the principle of electromagnetic induction
- Explain how a dc generator works
- Explain how a dc motor works

KEY TERMS

- Magnetic field
- Lines of force
- Magnetic flux
- Weber (Wb)
- Tesla (T)
- Gauss
- Hall effect
- Electromagnetism
- Electromagnetic field
- Permeability
- Reluctance ($\mathcal{R}$)
- Magnetomotive force (mmf)
- Ampere-turn (At)
- Solenoid
- Relay
- Speaker
- Magnetic field intensity
- Hysteresis
- Retentivity
- Induced voltage (v_{ind})
- Electromagnetic induction
- Induced current (i_{ind})
- Faraday's law
- Lenz's law

INTRODUCTION

This chapter departs from the previous six chapters and introduces two new concepts: magnetism and electromagnetism. The operation of many types of electrical devices is based partially on magnetic or electromagnetic principles. Electromagnetic induction is important in an electrical component called an inductor or coil (covered in Chapter 11).

Two types of magnets are the permanent magnet and the electromagnet. The permanent magnet maintains a constant magnetic field between its two poles with no external excitation. The electromagnet produces a magnetic field only when there is current through it. The electromagnet is basically a coil of wire wound around a magnetic core material. The chapter includes an introduction to dc generators and to dc motors. Motors are widely used in industrial systems, so a basic understanding of motors is important.

VISIT THE WEBSITE
Study aids for this chapter are available at
http://pearsonhighered.com/floyd

7–1 THE MAGNETIC FIELD

A permanent magnet has a magnetic field surrounding it. A **magnetic field** is visualized by **lines of force** that radiate from the north pole (N) to the south pole (S) and back to the north pole through the magnetic material.

After completing this section, you should be able to

- **Explain the principles of the magnetic field**
  - **Define** *magnetic flux*
  - **Define** *magnetic flux density*
  - **Discuss how materials are magnetized**
  - **Explain how a magnetic switch works**

A permanent magnet, such as the bar magnet shown in Figure 7–1, has a magnetic field surrounding it. All magnetic fields have their origin in moving charge, which in solid materials is caused by moving electrons. In certain materials such as iron, atoms can be aligned so that the electron motion is reinforced, creating an observable field that extends in three dimensions. Even some electrical insulators can exhibit this behavior; ceramics make excellent magnets but are electrical insulators.

To explain and illustrate magnetic fields, Michael Faraday drew "lines of force" or flux lines to represent the unseen field. Flux lines are widely used as a description of that field, showing the strength and direction of the field. The flux lines never cross. When lines are close together, the field is more intense; when they are farther apart, the field is weaker. The flux lines are always drawn from the north pole (N) to the south pole (S) of a magnet. Even in small magnets, the number of lines based on the mathematical definition is extremely large, so for clarity, only a few lines are generally shown in drawings of magnetic fields.

When unlike poles of two permanent magnets are placed close together, their magnetic fields produce an attractive force, as indicated in Figure 7–2(a). When two like poles are brought close together, they repel each other, as shown in part (b).

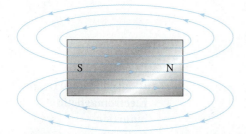

Blue lines represent only a few of the many magnetic lines of force in the magnetic field.

FIGURE 7–1 **Magnetic lines of force around a bar magnet.**

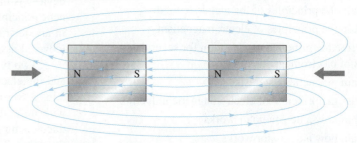

(a) Unlike poles attract.

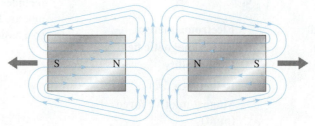

(b) Like poles repel.

FIGURE 7–2 **Magnetic attraction and repulsion.**

When a nonmagnetic material such as paper, glass, wood, or plastic is placed in a magnetic field, the lines of force are unaltered, as shown in Figure 7–3(a). However, when a magnetic material such as iron is placed in the magnetic field, the lines of force tend to change course and pass through the iron rather than through the surrounding air. They do so because the iron provides a magnetic path that is more easily established than that of air. Figure 7–3(b) illustrates this principle. The fact that magnetic lines of force follow a path through iron or other materials is a consideration in the design of shields that prevent stray magnetic fields from affecting sensitive circuits.

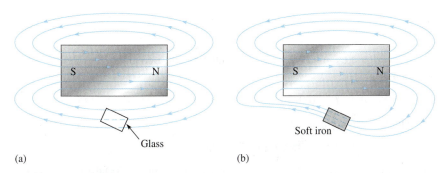

(a)   (b)

FIGURE 7–3   Effect of (a) nonmagnetic and (b) magnetic materials on a magnetic field.

Magnetic Flux (ϕ)

The group of force lines going from the north pole to the south pole of a magnet is called the **magnetic flux**, symbolized by ϕ (the lowercase Greek letter phi). A stronger field is represented by more lines of force. Several factors determine the strength of a magnet, including the material and physical geometry as well as the distance from the magnet. Magnetic field lines tend to be more concentrated at the poles.

The unit of magnetic flux is the **weber (Wb)**. One weber equals 10^8 lines. The weber is a very large unit; thus, in most practical situations, the microweber (μWb) is used. One microweber equals 100 lines of magnetic flux.

Magnetic Flux Density (B)

The **magnetic flux density** is the amount of flux per unit area perpendicular to the magnetic field. Its symbol is B, and its unit is the **tesla (T)**. One tesla equals one weber per square meter (Wb/m^2). The following formula expresses the flux density:

$$B = \frac{\phi}{A}$$ (7–1)

where ϕ is the flux in webers (Wb) and A is the cross-sectional area in square meters (m^2) of the magnetic field.

EXAMPLE 7–1

Compare the flux and the flux density in the two magnetic cores shown in Figure 7–4. The diagram represents the cross section of a magnetized material. Assume that each dot represents 100 lines or 1 μWb.

SOLUTION

The flux is simply the number of lines. In Figure 7–4(a) there are 49 dots. Each represents 1 μWb, so the flux is 49 μWb. In Figure 7–4(b) there are 72 dots, so the flux is 72 μWb.

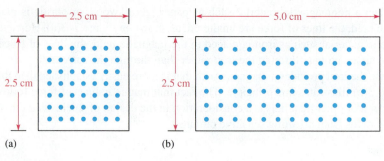

FIGURE 7–4

To calculate the flux density, first calculate the area in m². For Figure 7–4(a) the area is

$$A = l \times w = 0.025 \text{ m} \times 0.025 \text{ m} = 6.25 \times 10^{-4} \text{ m}^2$$

For Figure 7–4(b) the area is

$$A = l \times w = 0.025 \text{ m} \times 0.050 \text{ m} = 1.25 \times 10^{-3} \text{ m}^2$$

Use Equation 7–1 to calculate the flux density. For Figure 7–4(a) the flux density is

$$B = \frac{\phi}{A} = \frac{49 \,\mu\text{Wb}}{6.25 \times 10^{-4} \text{ m}^2} = 78.4 \times 10^{-3} \text{ Wb/m}^2 = 78.4 \times 10^{-3} \text{ T}$$

For Figure 7–4(b) the flux density is

$$B = \frac{\phi}{A} = \frac{72 \,\mu\text{Wb}}{1.25 \times 10^{-3} \text{ m}^2} = 57.6 \times 10^{-3} \text{ Wb/m}^2 = 57.6 \times 10^{-3} \text{ T}$$

The data in Table 7–1 compares the two cores. Note that the core with the largest flux does not necessarily have the highest flux density.

TABLE 7–1

	FLUX (Wb)	AREA (m²)	FLUX DENSITY (T)
Figure 7–4(a):	49 μWb	$6.25 \times 10^{-4} \text{ m}^2$	78.4×10^{-3} T
Figure 7–4(b):	72 μWb	$1.25 \times 10^{-3} \text{ m}^2$	57.6×10^{-3} T

RELATED PROBLEM*

What is the flux density if the same flux shown in Figure 7–4(a) is in a core that is 5.0 cm × 5.0 cm?

*Answers are at the end of the chapter.

EXAMPLE 7–2

If the flux density in a certain magnetic material is 0.23 T and the area of the material is 0.38 in.², what is the flux through the material?

SOLUTION

First, 0.38 in.² must be converted to square meters. 39.37 in. = 1 m; therefore,

$$A = 0.38 \text{ in.}^2 \left[1 \text{ m}^2 / (39.37 \text{ in.})^2 \right] = 245 \times 10^{-6} \text{ m}^2$$

The flux through the material is

$$\phi = BA = (0.23 \text{ T})(245 \times 10^{-6} \text{ m}^2) = \mathbf{56.4 \ \mu Wb}$$

RELATED PROBLEM

Calculate B if $A = 0.05$ in.^{2} and $\phi = 1000 \ \mu$Wb.

THE GAUSS   Although the tesla (T) is the SI unit for flux density, another unit called the **gauss (G)** from the CGS (centimeter-gram-second) system, is used (10^4 G = 1 T). In fact, the instrument used to measure flux density is the gaussmeter. A typical gaussmeter is shown in Figure 7–5. This particular gaussmeter is a portable unit with four ranges that can measure magnetic fields as small as the earth's field (about 0.5 G, but changes depending on location) to strong fields such as in an MRI unit (about 10,000 G). The unit *gauss* is still in widespread use, so you should be familiar with it as well as the *tesla*.

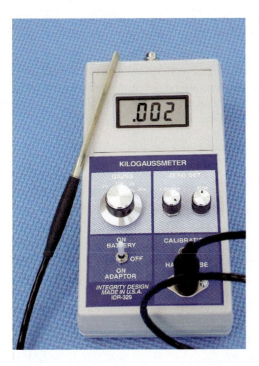

FIGURE 7–5   A dc gaussmeter.
(Integrity Model IDR-329 distributed by Less EMF Inc.)

How Materials Become Magnetized

Ferromagnetic materials such as iron, nickel, and cobalt become magnetized when placed in the magnetic field of a magnet. We have all seen a permanent magnet pick up paper clips, nails, or iron filings. In these cases, the object becomes magnetized (that is, it actually becomes a magnet itself) under the influence of the permanent magnetic field and becomes attracted to the magnet. When removed from the magnetic field, the object tends to lose its magnetism.

The magnetic material affects not only the magnetic flux density at the poles but also how the magnetic flux density falls off as distance from the poles increases. The physical size also affects the flux density. For example, two disk magnets (both made from sintered Alnico) have very similar densities near the pole, but the larger magnet has much higher flux density, as you move away from the pole, as shown in Figure 7–6. Notice that the flux density falls off rapidly as you move away from the pole. This type of plot can illustrate if a given magnet is effective for a specific application that depends on the distance in which the magnet must work.

Ferromagnetic materials have minute magnetic domains created within their atomic structure by the orbital motion and spin of electrons. These domains can be thought of as

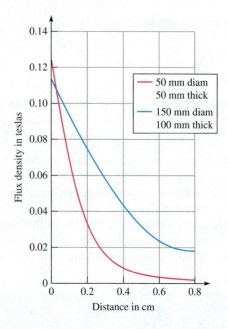

FIGURE 7–6 Example of magnetic flux densities for two disk magnets as a function of distance. The blue curve represents the larger magnet.

very small bar magnets with north and south poles. When the material is not exposed to an external magnetic field, the magnetic domains are randomly oriented, as shown in Figure 7–7(a). When the material is placed in a magnetic field, the domains align themselves, as shown in part (b). Thus, the object itself effectively becomes a magnet.

(a) The magnetic domains (N ◀—— S) are randomly oriented in the unmagnetized material.

(b) The magnetic domains become aligned when the material is magnetized.

FIGURE 7–7 Ferromagnetic domains in (a) an unmagnetized and (b) a magnetized material.

The type of material is an important parameter for the actual flux density of magnets. Table 7–2 lists the flux densities of typical magnetic fields in teslas. For permanent magnets, the numbers given are based on the flux density of the field that is typical if measured close to the pole. As previously discussed, these values can drop significantly as distance from the poles is increased. The strongest field most people will ever experience is about 1 T (10,000 G) if they have an MRI exam. The strongest commercially available permanent

TABLE 7–2 • Flux density of various magnetic fields.	
SOURCE	**TYPICAL FLUX DENSITY IN TESLAS (T)**
Earth's magnetic field	4×10^{-5} (varies with location)
Small "refrigerator" magnets	0.08 to 0.1
Ceramic magnets	0.2 to 0.3
Alnico 5 reed switch magnet	0.1 to 0.2
Neodymium magnets	0.3 to 0.52
Magnetic resonance imaging (MRI)	1
The strongest steady magnetic field ever achieved in a laboratory	45

magnets are neodymium-iron-boron composites (NdFeB). To find the flux density in gauss, multiply the values in teslas by 10^4 (10,000).

Sensitive Magnetic Sensor

A new sensor developed by the National Institute of Standards and Technology (NIST) has the ability to measure the very tiny magnetic field associated with a human heartbeat. The heart's magnetic signature is measured in picoteslas (trillionths of a tesla), which is a million times smaller than Earth's magnetic field. The measurement had to be made in a building with magnetic shielding to block out effects of Earth's magnetic field.

The miniature sensors may eventually find application as manetocardiograms, as a supplement to electrocardiograms.

Courtesy of S. Knappe/NIST.

SYSTEM NOTE

Applications

Permanent magnets are widely used in brushless motors (discussed in Section 7–7), magnetic separators, speakers, microphones, automobiles, and magnetic resonance-imaging devices. They are also commonly used in switches, such as the normally closed switch illustrated in Figure 7–8. When the magnet is near the switch mechanism, as in Figure 7–8 (a), the switch is closed. When the magnet is moved away, as in part (b), the spring pulls the arm open. Magnetic switches are widely used in security systems.

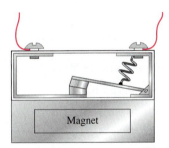

(a) Contact is closed when magnet is near.

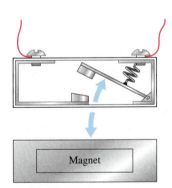

(b) Contact opens when magnet is moved away.

FIGURE 7–8 **Operation of a magnetic switch.**

System Applications for Magnetic Switches

Magnetic switches are widely used in systems. One important application is for proximity detectors, which are useful for counting operations or measuring the speed of movement in rotating machinery. Reed type magnetic switches sense a moving magnet as it moves by the switch. Reed switches can open and close up to 1000 times per second and can have an expected lifetime of three billion cycles.

Magnetic switches are also used in float detectors for monitoring liquid level. The float has a magnet that moves up and down with the fluid. As the magnets pass by the switch, the reeds are pulled together closing the switch.

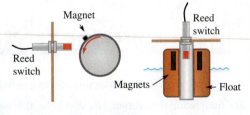

SYSTEM NOTE

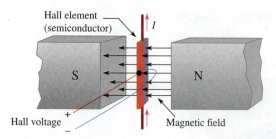

Hall element (semiconductor)

Hall voltage

Magnetic field

FIGURE 7–9   The Hall effect. The Hall voltage is induced across the Hall element—the red side is positive; the blue side is negative.

Another important application of permanent magnets is in sensors that take advantage of an effect known as the Hall effect. The **Hall effect** is the occurrence of a small voltage (a few μV) that is generated on opposite sides of a thin current-carrying conductor or semiconductor (the Hall element) that is in a magnetic field. A voltage called the Hall voltage appears across the Hall element, as illustrated in Figure 7–9. The Hall voltage is due to the forces exerted on the electrons as they traverse the magnetic field, causing an excess of charge on one side of the Hall element. Although the effect was first noticed in a conductor, it is more pronounced in semiconductors, which are normally used in Hall-effect sensors.

Notice that the magnetic field, the electric current, and the Hall voltage are all at right angles to each other. This voltage is amplified and can be used to detect the presence of the magnetic field. The detection of a magnetic field is useful in sensor applications.

Hall-effect sensors are widely used because they are small, inexpensive, and have no moving parts. In addition, they are noncontacting sensors, so they can last for billions of repeated operations, a clear advantage over contacting sensors that can wear out. Hall-effect sensors can detect the nearby presence of a magnet by sensing its magnetic field. For this reason, they can be used for position measurements or for sensing motion. They are used in conjunction with other sensor elements to measure current, temperature, or pressure.

There are many system applications for Hall-effect sensors. In automobiles, Hall-effect sensors are used to measure various parameters such as throttle angle, crankshaft and camshaft positions, distributor position, tachometer, power seat and rear-view mirror positions. A few other applications for Hall-effect sensors include measuring parameters for rotating devices such as drills, fans, flow meter vanes, and disk speed detection. They are also used in dc motors as will be seen in Section 7–7.

SECTION 7–1 CHECKUP*

1. When the north poles of two magnets are placed close together, do they repel or attract each other?

2. What is the difference between magnetic flux and magnetic flux density?

3. What are two units for measuring magnetic flux density?

4. What is the flux density when $\phi = 4.5$ μWb and $A = 5 \times 10^{-3}$ m^2?

*Answers are at the end of the chapter.

7–2 ELECTROMAGNETISM

Electromagnetism is the production of a magnetic field by current in a conductor.

After completing this section, you should be able to

- Explain the principles of electromagnetism
  - Determine the direction of the magnetic lines of force
  - Define *permeability*
  - Define *reluctance*
  - Define *magnetomotive force*
  - Describe a basic electromagnet

Current produces a magnetic field, called an **electromagnetic field**, around a conductor, as illustrated in Figure 7–10. The invisible lines of force of the magnetic field form a concentric circular pattern around the conductor and are continuous along its length. The direction of the lines of force surrounding a conductor for the given current direction is as indicated. The lines of force are in a clockwise direction. When current direction is reversed, the magnetic field lines are in a counterclockwise direction.

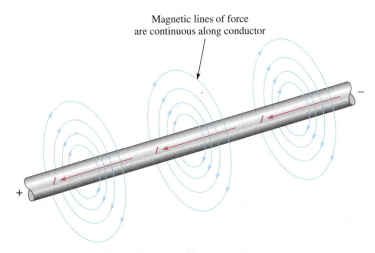

FIGURE 7–10   **Magnetic field around a current-carrying conductor. The red arrows indicate the direction of electron flow current (− to +).**

Although the magnetic field cannot be seen, it is capable of producing visible effects. For example, if a current-carrying wire is inserted through a sheet of paper in a perpendicular direction, iron filings placed on the surface of the paper arrange themselves along the magnetic lines of force in concentric rings, as illustrated in Figure 7–11(a). Part (b) of the figure illustrates that the needle of a compass placed in the electromagnetic field will point in the direction of the lines of force. The field is stronger closer to the conductor and becomes weaker with increasing distance from the conductor.

LEFT-HAND RULE   An aid to remembering the direction of the lines of force is illustrated in Figure 7–12. Imagine that you are grasping the conductor with your left hand, with your thumb pointing in the direction of current. Your fingers indicate the direction of the magnetic lines of force.

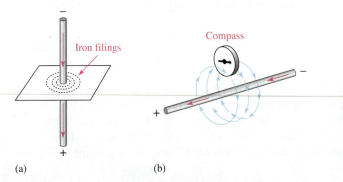

(a)   (b)

FIGURE 7–11   **Visible effects of an electromagnetic field.**

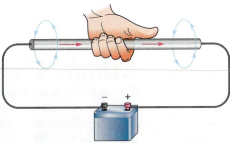

FIGURE 7–12   **Illustration of left-hand rule. The left hand rule is used for electron flow current (− to +).**

Electromagnetic Properties

Several important properties related to electromagnetic fields are now discussed.

PERMEABILITY (μ)   The ease with which a magnetic field can be established in a given material is measured by the **permeability** of that material. The higher the permeability, the more easily a magnetic field can be established. The symbol of permeability is μ (the Greek letter mu).

A material's permeability depends on its type. The permeability of a vacuum (μ_0) is $4\pi \times 10^{-7}$ Wb/At·m (webers/ampere-turn·meter) and is used as a reference. Ferromagnetic materials typically have permeabilities hundreds of times larger than that of a vacuum, indicating that a magnetic field can be set up with relative ease in these materials. Ferromagnetic materials include iron, steel, nickel, cobalt, and their alloys.

The *relative permeability* (μ_r) of a material is the ratio of its absolute permeability (μ) to the permeability of a vacuum (μ_0).

$$\mu_r = \frac{\mu}{\mu_0} \tag{7–2}$$

Because it is a ratio of permeabilities, μ_r has no units. Typical magnetic materials, such as iron, have a relative permeability of a few hundred. Highly permeable materials can have a relative permeability as high as 100,000.

RELUCTANCE ($\mathcal{R}$)   The opposition to the establishment of a magnetic field in a material is called **reluctance ($\mathcal{R}$)**. The value of reluctance is directly proportional to the length (l) of the magnetic path, and inversely proportional to the permeability (μ) and to the cross-sectional area (A) of the material, as expressed by the following equation:

$$\mathcal{R} = \frac{l}{\mu A} \tag{7–3}$$

Reluctance in magnetic circuits is analogous to resistance in electric circuits. The unit of reluctance can be derived using l in meters, A (area) in square meters, and μ in Wb/At·m as follows:

$$\mathcal{R} = \frac{l}{\mu A} = \frac{\cancel{m}}{(\text{Wb/At} \cdot \cancel{m})(\cancel{m}^2)} = \frac{\text{At}}{\text{Wb}}$$

At/Wb is ampere-turns/weber.

Equation 7–3 is similar to Equation 2–6 for determining wire resistance. Recall Equation 2–6 is

$$R = \frac{\rho l}{A}$$

The reciprocal of resistivity (ρ) is conductivity (σ). By substituting $1/\sigma$ for ρ, Equation 2–6 can be written as

$$R = \frac{l}{\sigma A}$$

Compare this last equation for wire resistance with Equation 7–3. The length (l) and the area (A) have the same meaning in both equations. The conductivity (σ) in electrical circuits is analogous to permeability (μ) in magnetic circuits. Also, resistance (R) in electric circuits is analogous to reluctance ($\mathcal{R}$) in magnetic circuits; both are oppositions. Typically, the reluctance of a magnetic circuit is 50,000 At/Wb or more, depending on the size and type of material.

EXAMPLE 7–3

Calculate the reluctance of a torus (a doughnut-shaped core) made of low-carbon steel. The inner radius of the torus is 1.75 cm and the outer radius of the torus is 2.25 cm. Assume the permeability of low-carbon steel is 2×10^{-4} Wb/At·m.

SOLUTION

You must convert centimeters to meters before you calculate the area and length. From the dimensions given, the thickness (diameter) is 0.5 cm = 0.005 m. Thus, the cross-sectional area is

$$A = \pi r^2 = \pi(0.0025)^2 = 1.96 \times 10^{-5}\ \text{m}^2$$

The length is equal to the circumference of the torus measured at the average radius of 2.0 cm or 0.020 m.

$$l = C = 2\pi r = 2\pi(0.020\ \text{m}) = 0.125\ \text{m}$$

Substituting values into Equation 7–3,

$$\mathcal{R} = \frac{l}{\mu A} = \frac{0.125\ \text{m}}{(2 \times 10^{-4}\ \text{Wb/At} \cdot \text{m})(1.96 \times 10^{-5}\ \text{m}^2)} = \mathbf{31.9 \times 10^6\ \text{At/Wb}}$$

RELATED PROBLEM

What happens to the reluctance if cast steel with a permeability of 5×10^{-4} Wb/At · m is substituted for the cast iron core?

EXAMPLE 7–4

Mild steel has a relative permeability of 800. Calculate the reluctance of a mild steel core that has a length of 10 cm, and has a cross section of 1.0 cm × 1.2 cm.

SOLUTION

First, determine the permeability of mild steel.

$$\mu = \mu_0\mu_r = (4\pi \times 10^{-7}\ \text{Wb/At} \cdot \text{m})(800) = 1.00 \times 10^{-3}\ \text{Wb/At} \cdot \text{m}$$

Next, convert the length to meters and the area to square meters.

$$l = 10\ \text{cm} = 0.10\ \text{m}$$
$$A = 0.010\ \text{m} \times 0.012\ \text{m} = 1.2 \times 10^{-4}\ \text{m}^2$$

Substituting values into Equation 7–3,

$$\mathcal{R} = \frac{l}{\mu A} = \frac{0.10\ \text{m}}{(1.00 \times 10^{-3}\ \text{Wb/At} \cdot \text{m})(1.2 \times 10^{-4}\ \text{m}^2)} = \mathbf{8.33 \times 10^5\ \text{At/Wb}}$$

RELATED PROBLEM

What happens to the reluctance if the core is made from 78 Permalloy with a relative permeability of 4000?

MAGNETOMOTIVE FORCE (mmf)   As you have learned, current in a conductor produces a magnetic field. The cause of a magnetic field is called the **magnetomotive force (mmf)**. Magnetomotive force is something of a misnomer because in a physics sense mmf is not really a force, but rather it is a direct result of the movement of charge (current). The unit of mmf, the **ampere-turn (At)**, is established on the basis of the current in a coil of wire. The formula for mmf is

$$F_m = NI \tag{7–4}$$

where F_m is the magnetomotive force, N is the number of turns of wire, and I is the current in amperes.

Figure 7–13 illustrates that a number of turns of wire carrying a current around a magnetic material creates a force that sets up flux lines through the magnetic path. The amount of flux depends on the magnitude of the mmf and on the reluctance of the material, as expressed by the following equation:

$$\phi = \frac{F_m}{\mathcal{R}}$$

(7–5)

Equation 7–5 is known as the *Ohm's law for electromagnetic circuits* because the flux (ϕ) is analogous to current, the mmf (F_m) is analogous to voltage, and the reluctance ($\mathcal{R}$) is analogous to resistance. Like other phenomena in science, the flux is an effect, the mmf is a cause, and the reluctance is an opposition.

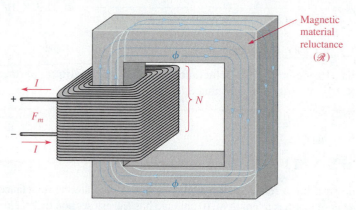

FIGURE 7–13   **A basic electromagnetic circuit.**

One important difference between an electric circuit and a magnetic circuit is that in magnetic circuits, Equation 7–5 is only valid up to a certain point before the magnetic material saturates (flux becomes a maximum). You will see this when you look at magnetization curves in Section 7–4. Another difference that was already noted is that flux does occur in permanent magnets with no source of mmf. In a permanent magnet, the flux is due to internal electron motion rather than to an external current. No equivalent effect occurs in electric circuits.

EXAMPLE 7–5

How much flux is established in the magnetic path of Figure 7–14 if the reluctance of the material is 2.8×10^5 At/Wb?

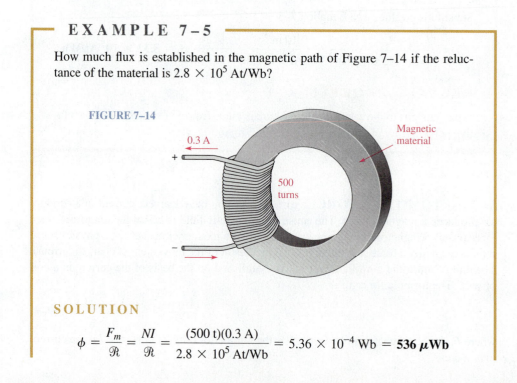

FIGURE 7–14

SOLUTION

$$\phi = \frac{F_m}{\mathcal{R}} = \frac{NI}{\mathcal{R}} = \frac{(500 \text{ t})(0.3 \text{ A})}{2.8 \times 10^5 \text{ At/Wb}} = 5.36 \times 10^{-4} \text{ Wb} = \mathbf{536 \, \mu Wb}$$

RELATED PROBLEM

How much flux is established in the magnetic path of Figure 7–14 if the reluctance is 7.5×10^3 At/Wb, the number of turns is 300, and the current is 0.18 A?

EXAMPLE 7–6

There is 0.1 ampere of current through a coil with 400 turns.

(a) What is the mmf?

(b) What is the reluctance of the circuit if the flux is 250 μWb?

SOLUTION

(a) $N = 400$ and $I = 0.1$ A

$F_m = NI = (400 \text{ t})(0.1 \text{ A}) = \mathbf{40 \text{ At}}$

(b) $\mathcal{R} = \dfrac{F_m}{\phi} = \dfrac{40 \text{ At}}{250 \, \mu\text{Wb}} = \mathbf{1.60 \times 10^5 \text{ At/Wb}}$

RELATED PROBLEM

Rework the example for $I = 85$ mA, $N = 500$, and $\phi = 500 \, \mu$Wb.

In many magnetic circuits, the core is not continuous. For example, if an air gap is cut into the core, it will increase the reluctance of the magnetic circuit. This means that more current is required to establish the same flux as before because an air gap represents a significant opposition to establishing flux. The situation is analogous to a series electrical circuit; the total reluctance of the magnetic circuit is the sum of the reluctance of the core and the reluctance of the air gap.

The Electromagnet

An electromagnet is based on the properties that you have just learned. A basic electromagnet is simply a coil of wire wound around a core material that can be easily magnetized.

The shape of the electromagnet can be designed for various applications. For example, Figure 7–15 shows a U-shaped magnetic core. When the coil of wire is connected to a battery and there is current, as shown in part (a), a magnetic field is established as indicated. If the current direction is reversed, as shown in part (b), the direction of the magnetic field is also reversed. The closer the north and south poles are brought together, the smaller the air gap between them becomes, and the easier it becomes to establish a magnetic field because the reluctance is reduced.

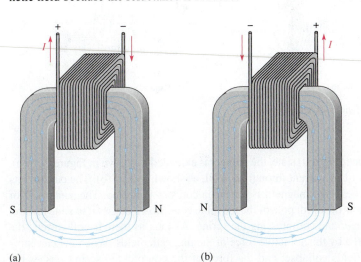

FIGURE 7–15 **Reversing the current in the coil causes the electromagnetic field to reverse.**

(a)   (b)

SECTION 7–2 CHECKUP

1. Explain the difference between magnetism and electromagnetism.

2. What happens to the magnetic field in an electromagnet when the current through the coil is reversed?

3. State Ohm's law for electromagnetic circuits.

4. Compare each quantity in Question 3 to its electrical counterpart.

7–3 ELECTROMAGNETIC DEVICES

Many types of useful devices such as tape recorders, electric motors, speakers, solenoids, and relays are based on electromagnetism. The transformer is another important example and will be covered in Chapter 14.

After completing this section, you should be able to

- **Describe the principle of operation for several types of electromagnetic devices**
  - **Discuss how a solenoid and a solenoid valve work**
  - **Discuss how a relay works**
  - **Discuss how a speaker works**
  - **Discuss the basic analog meter movement**
  - **Explain a magnetic disk and tape read/write operation**
  - **Explain the concept of the magneto-optical disk**

The Solenoid

The **solenoid** is a type of electromagnetic device that has a movable iron core called a *plunger*. The movement of this iron core depends on both an electromagnetic field and a mechanical spring force. The basic structure of a solenoid is shown in Figure 7–16. It consists of a cylindrical coil of wire wound around a nonmagnetic hollow form. A stationary iron core is fixed in position at the end of the shaft, and a sliding iron core is attached to the stationary core with a spring.

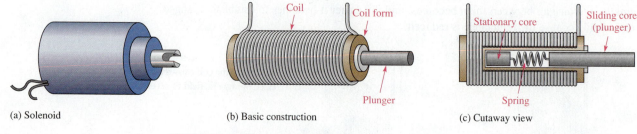

(a) Solenoid

(b) Basic construction

(c) Cutaway view

FIGURE 7–16   **Basic solenoid structure.**

In the at-rest (or unenergized) state, the plunger is extended as shown in Figure 7–17(a). The solenoid is energized by current through the coil, as shown in part (b). The current sets up an electromagnetic field that magnetizes both iron cores as indicated. The south pole of the stationary core attracts the north pole of the movable core, which causes it to slide inward, thus retracting the plunger and compressing the spring. As long as there is coil current, the plunger remains retracted by the attractive force of the magnetic fields. When the current is cut off, the magnetic fields collapse; and the force of the compressed spring pushes the

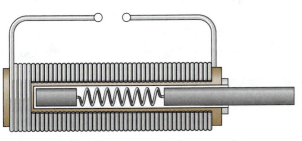

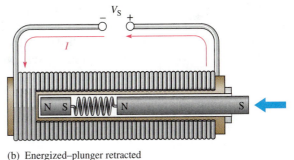

(a) Unenergized (no voltage or current)–plunger extended

(b) Energized–plunger retracted

FIGURE 7–17 Basic solenoid operation.

plunger back out. The solenoid is used for applications such as opening and closing valves and automobile door locks.

SOLENOID VALVES In industrial controls, **solenoid valves** are widely used to control the flow of air, water, steam, oils, refrigerants, and other fluids. Solenoid valves are used in both pneumatic (air) and hydraulic (oil) systems, common in machine controls. Solenoid valves are also common in the aerospace and medical fields. Solenoid valves can either move a plunger to open or close a port or can rotate a blocking flap a fixed amount.

A solenoid valve consists of two functional units: a solenoid coil that provides the magnetic field to provide the required movement to open or close the valve and a valve body, which is isolated from the coil assembly via a leakproof seal and includes a pipe and butterfly valve. Figure 7–18 shows a cutaway of one type of solenoid valve. When the solenoid is energized, the butterfly valve is turned to open a normally closed (NC) valve or to close a normally open (NO) valve.

Solenoid valves are available with a wide variety of configurations including normally open or normally closed valves. They are rated for different types of fluids (for example, gas or water), pressures, number of pathways, sizes, and more. The same valve may control more than one line and may have more than one solenoid to move.

FIGURE 7–18 A basic solenoid valve structure.

The Relay

Relays differ from solenoids in that the electromagnetic action is used to open or close electrical contacts rather than to provide mechanical movement. Figure 7–19 shows the basic operation of an armature-type relay with one normally open (NO) contact and one normally closed (NC) contact (single pole–double throw). When there is no coil current,

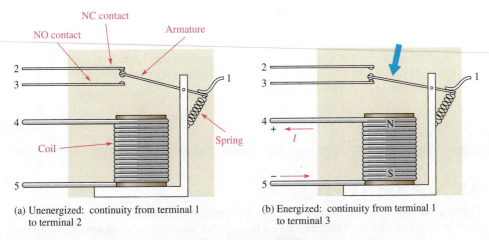

(a) Unenergized: continuity from terminal 1 to terminal 2

(b) Energized: continuity from terminal 1 to terminal 3

FIGURE 7–19 Basic structure of a single-pole–double-throw armature relay.

the armature is held against the upper contact by the spring, thus providing continuity from terminal 1 to terminal 2, as shown in Figure 7–19(a). When energized with coil current, the armature is pulled down by the attractive force of the electromagnetic field and makes connection with the lower contact to provide continuity from terminal 1 to terminal 3, as shown in Figure 7–19(b).

A typical armature relay and its schematic symbol are shown in Figure 7–20.

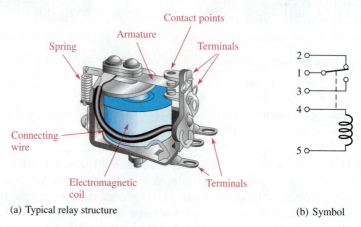

(a) Typical relay structure

(b) Symbol

FIGURE 7–20 **A typical armature relay.**

Another widely used type of relay is the *reed relay,* which is shown in Figure 7–21. The reed relay, like the armature relay, uses an electromagnetic coil. The contacts are thin reeds of magnetic material and are usually located inside the coil. When there is no coil current, the reeds are in the open position as shown in Figure 7–21(b). When there is current through the coil, the reeds make contact because they are magnetized and attract each other as shown in part (c).

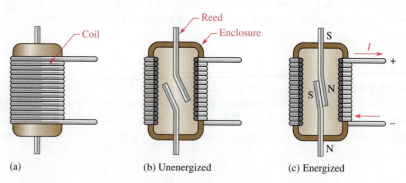

(a)

(b) Unenergized

(c) Energized

FIGURE 7–21 **Basic structure of a reed relay.**

Reed relays are faster, are more reliable, and produce less contact arcing than armature relays. However, they have less current-handling capability than armature relays and are more susceptible to mechanical shock.

SYSTEM EXAMPLE 7–1

AN ALARM SYSTEM

Figure 7–22 shows a simplified intrusion detection system that uses a relay to turn on an audible alarm and lights. The relay uses a battery that allows it to function and sound the audible alarm even if ac power is off. The relay uses 50 mA to close and the audible alarm draws 2 A from the battery. Notice that the contacts for the relay are physically close but drawn separated for clarity of reading the schematic. If any one of the magnetic detection switches closes when the system is *on*, the relay is energized, providing current to the audible alarm and turning on ac power. The system is designed such that the high current for the audible alarm is not in the detection switches; most magnetic switches are low-current devices, so the relay provides a path for the high-current output.

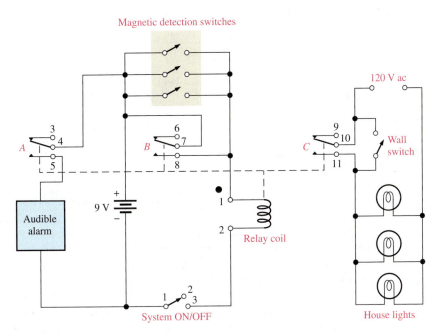

FIGURE 7–22   Simplified security alarm system.

The Speaker

A **speaker** is an electromagnetic device that converts electrical signals into sound. Essentially, it is a linear motor that alternately attracts and repels an electromagnet to and from a permanent magnet called a donut magnet. Figure 7–23 shows the key parts of a speaker. The audio signal is connected using very flexible wires to a cylindrical coil called the voice coil. The voice coil and its movable core form an electromagnet, which is suspended in an accordion-like structure called the spider. The spider acts like an accordion spring, keeping the voice coil in the center and restoring it to the rest position when there is no input signal.

Current from the audio input alternates back and forth and powers the electromagnet; when there is more current, the attraction or repulsion is greater. When the input current reverses direction, the polarity of the electromagnet reverses direction also, faithfully following the input signal. The voice coil and its moving magnet are firmly attached to the cone. The cone is a flexible diaphragm that vibrates to produce sound.

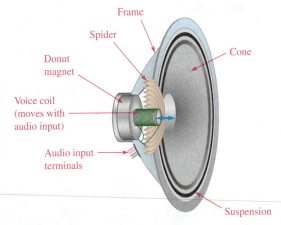

FIGURE 7–23   Key parts of a speaker (cutaway view).

Meter Movement

The d'Arsonval meter movement is the most common type used in analog multimeters. In this type of meter movement, the pointer is deflected in proportion to the amount of current through a coil. Figure 7–24 shows a basic d'Arsonval meter movement. It consists of a coil

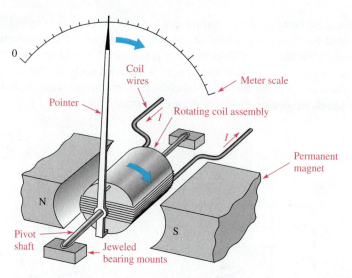

FIGURE 7–24 The basic d'Arsonval meter movement.

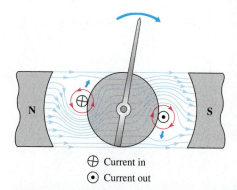

FIGURE 7–25 When the electromagnetic field interacts with the permanent magnetic field, forces are exerted on the rotating coil assembly, causing it to move clockwise and thus deflecting the pointer.

of wire wound on a bearing-mounted assembly that is placed between the poles of a permanent magnet. A pointer is attached to the moving assembly. With no current through the coil, a spring mechanism keeps the pointer at its left-most (zero) position. When there is current through the coil, electromagnetic forces act on the coil, causing a rotation to the right. The amount of rotation depends on the amount of current.

Figure 7–25 illustrates how the interaction of magnetic fields produces rotation of the coil assembly. Current is inward at the "cross" and outward at the "dot" in the single winding shown. The inward current produces a counterclockwise electromagnetic field that reinforces the permanent magnetic field below it. The result is an upward force on the left side of the coil as shown. A downward force is developed on the right side of the coil, where the current is outward. These forces produce a clockwise rotation of the coil assembly and are opposed by a spring mechanism. The indicated forces and the spring force are balanced at the value of the current. When current is removed, the spring force returns the pointer to its zero position.

Magnetic Disk and Tape Read/Write Head

A simplified diagram of a magnetic disk or tape surface read/write operation is shown in Figure 7–26. A data bit (1 or 0) is written on the magnetic surface by the magnetization of a small segment of the surface as it moves by the write head. The direction of the magnetic

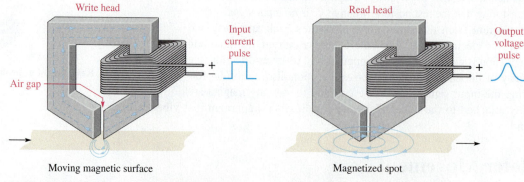

(a) The magnetic flux from the write head follows the low reluctance path through the moving magnetic surface.

(b) When read head passes over magnetized spot, an induced voltage appears at the output.

FIGURE 7–26 Read/write function on a magnetic surface.

flux lines is controlled by the direction of the current pulse in the winding, as shown in Figure 7–26(a) for the case of a positive pulse. At the air gap in the write head, the magnetic flux takes a path through the surface of the storage device. This magnetizes a small spot on the surface in the direction of the field. A magnetized spot of one polarity represents a binary 1, and one of the opposite polarity represents a binary 0. Once a spot on the surface is magnetized, it remains until written over with an opposite magnetic field.

When the magnetic surface passes a read head, the magnetized spots produce magnetic fields in the read head, which induce voltage pulses in the winding. The polarity of these pulses depends on the direction of the magnetized spot and indicates whether the stored bit is a 1 or a 0. This process is illustrated in Figure 7–26(b). Often the read and write heads are combined into a single unit.

The Magneto-Optical Disk

The magneto-optical disk uses an electromagnet and laser beams to read and write (record) data on a magnetic surface. Magneto-optical disks are formatted in tracks and sectors similar to hard disks. However, because of the ability of a laser beam to be precisely directed to an extremely small spot, magneto-optical disks are capable of storing much more data than standard magnetic hard disks.

Figure 7–27(a) illustrates a small cross-sectional area of a disk before recording, with an electromagnet positioned below it. Tiny magnetic particles, represented by the arrows, are all magnetized in the same direction.

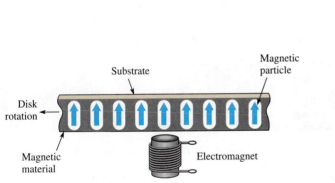

(a) Small cross-section of unrecorded disk

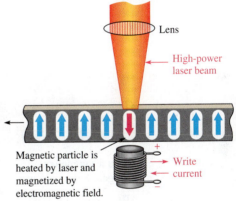

(b) Writing: A high-power laser beam heats the spot, causing the magnetic particle to align with the electromagnetic field.

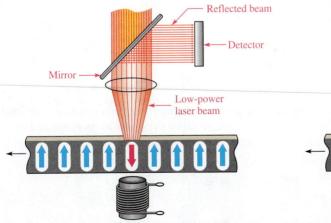

(c) Reading: A low-power laser beam reflects off of the reversed-polarity magnetic particle and its polarization shifts. If the particle is not reversed, the polarization of the reflected beam is unchanged.

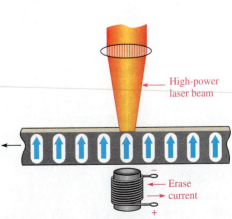

(d) Erasing: The electromagnetic field is reversed as the high-power laser beam heats the spot, causing the magnetic particle to be restored to the original polarity.

FIGURE 7–27   Basic concept of the magneto-optical disk.

Writing (recording) on the disk is accomplished by applying an external magnetic field opposite to the direction of the magnetic particles as indicated in Figure 7–27(b) and then directing a high-power laser beam to heat the disk at a precise spot where a binary 1 is to be stored. The disk material, a magneto-optic alloy, is highly resistant to magnetization at room temperature; but at the spot where the laser beam heats the material, the inherent direction of magnetism is reversed by the external magnetic field produced by the electromagnet. At points where binary 0s are to be stored, the laser beam is not applied and the inherent upward direction of the magnetic particle remains.

As illustrated in Figure 7–27(c), reading data from the disk is accomplished by turning off the external magnetic field and directing a low-power laser beam at a spot where a bit is to be read. Basically, if a binary 1 is stored at the spot (reversed magnetization), the low-power laser beam is reflected and its polarization is shifted; but if a binary 0 is stored, the polarization of the reflected laser beam is unchanged. A detector senses the difference in the polarity of the reflected laser beam to determine if the bit being read is a 1 or a 0.

Figure 7–27(d) shows that the disk is erased by restoring the original magnetic direction of each particle by reversing the external magnetic field and applying the high-power laser beam.

SECTION 7–3 CHECKUP

1. Explain the difference between a solenoid and a relay.

2. What is the movable part of a solenoid called?

3. What is the movable part of a relay called?

4. Upon what basic principle is the d'Arsonval meter movement based?

7–4 MAGNETIC HYSTERESIS

When a magnetizing force is applied to a material, the magnetic flux density in the material changes in a certain way.

After completing this section, you should be able to

- **Explain magnetic hysteresis**
  - **State the formula for magnetic field intensity**
  - **Discuss a hysteresis curve**
  - **Define** *retentivity*

Magnetic Field Intensity (*H*)

The **magnetic field intensity** (also called *magnetizing force*) in a material is defined to be the magnetomotive force (F_m) per unit length (l) of the material, as expressed by the following equation. The unit of magnetic field intensity (H) is ampere-turns per meter (At/m).

$$H = \frac{F_m}{l} \tag{7–6}$$

where $F_m = NI$. Note that the magnetic field intensity (H) depends on the number of turns (N) of the coil of wire, the current (I) through the coil, and the length (l) of the material. It does not depend on the type of material.

Since $\phi = F_m/\mathcal{R}$, as F_m increases, the flux increases. Also, the magnetic field intensity (H) increases. Recall that the flux density (B) is the flux per unit cross-sectional area

$(B = \phi/A)$, so B is also proportional to H. The curve showing how these two quantities (B and H) are related is called the *B-H* curve, or the hysteresis curve. The parameters that influence both B and H are illustrated in Figure 7–28.

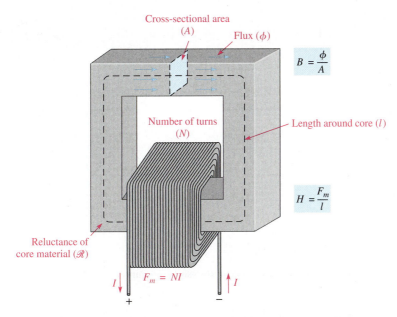

Cross-sectional area
(A)

Flux (ϕ)

$$B = \frac{\phi}{A}$$

Number of turns
(N)

Length around core (l)

$$H = \frac{F_m}{l}$$

Reluctance of
core material $(\mathscr{R})$

I $F_m = NI$ I

$+$ $-$

FIGURE 7–28   **Parameters that determine the magnetic field intensity (H) and the flux density (B).**

The Hysteresis Curve and Retentivity

Hysteresis is a characteristic of a magnetic material whereby a change in magnetization lags the application of the magnetic field intensity. The magnetic field intensity (H) can be readily increased or decreased by varying the current through the coil of wire, and it can be reversed by reversing the voltage polarity across the coil.

Figure 7–29 illustrates the development of the hysteresis curve. Let's start by assuming a magnetic core is unmagnetized so that $B = 0$. As the magnetic field intensity (H) is increased from zero, the flux density (B) increases proportionally as indicated by the curve in Figure 7–29(a). When H reaches a certain value, the value of B begins to level off. As H continues to increase, B reaches a saturation value (B_{sat}) when H reaches a value (H_{sat}), as illustrated in Figure 7–29(b). Once saturation is reached, a further increase in H will not increase B.

Now, if H is decreased to zero, B will fall back along a different path to a residual value (B_R), as shown in Figure 7–29(c). This indicates that the material continues to be magnetized even when the magnetic field intensity is removed ($H = 0$). The ability of a material, once magnetized, to maintain a magnetized state without magnetic field intensity is called **retentivity**. The retentivity of a material is indicated by the ratio of B_R to B_{sat}.

Reversal of the magnetic field intensity is represented by negative values of H on the curve and is achieved by reversing the current in the coil of wire. An increase in H in the negative direction causes saturation to occur at a value ($-H_{sat}$) where the flux density is at its maximum negative value, as indicated in Figure 7–29(d).

When the magnetic field intensity is removed ($H = 0$), the flux density goes to its negative residual value ($-B_R$), as shown in Figure 7–29(e). From the $-B_R$ value, the flux density follows the curve indicated in part (f) back to its maximum positive value when the magnetic field intensity equals H_{sat} in the positive direction.

The complete *B-H* curve is shown in Figure 7–29(g) and is called the *hysteresis curve*. The magnetic field intensity required to make the flux density zero is called the *coercive force, H_C.*

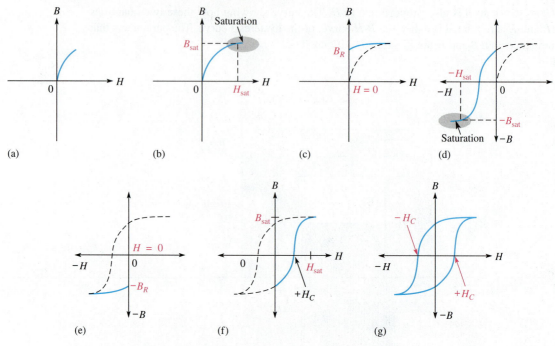

FIGURE 7–29 Development of a magnetic hysteresis (*B-H*) curve.

Materials with a low retentivity do not retain a magnetic field very well while those with high retentivities exhibit values of B_R very close to the saturation value of *B*. Depending on the application, retentivity in a magnetic material can be an advantage or a disadvantage. In permanent magnets and memory cores, for example, high retentivity is required. In ac motors, retentivity is undesirable because the residual magnetic field must be overcome each time the current reverses, thus wasting energy.

SECTION 7–4 CHECKUP

1. For a given wirewound core, how does an increase in current through the coil affect the flux density?

2. Define *retentivity*.

7–5 ELECTROMAGNETIC INDUCTION

In this section, you are introduced to electromagnetic induction. Electromagnetic induction is what makes transformers, electrical generators, electrical motors, and many other devices possible.

After completing this section, you should be able to

- **Discuss the principle of electromagnetic induction**
  - **Explain how voltage is induced in a conductor in a magnetic field**
  - **Determine polarity of an induced voltage**
  - **Discuss forces on a conductor in a magnetic field**
  - **State Faraday's law**
  - **State Lenz's law**
  - **Explain how a crankshaft position sensor works**

Relative Motion

When a straight conductor is moved perpendicular to a magnetic field, there is a relative motion between the conductor and the magnetic field. Likewise, when a magnetic field is moved past a stationary conductor, there is also relative motion. In either case, this relative motion results in an **induced voltage (v_{ind})** across the conductor, as Figure 7–30 indicates. This principle is known as **electromagnetic induction**. The lowercase v stands for instantaneous voltage. The amount of the induced voltage depends on the rate at which the conductor and the magnetic field move with respect to each other: The faster the relative motion, the greater the induced voltage.

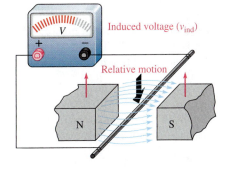

(a) Conductor moving downward   (b) Magnetic field moving upward

FIGURE 7–30   **Relative motion between a straight conductor and a magnetic field.**

Vibrating Wire Strain Gage

A strain gage that works on the principle of magnetic induction is the vibrating wire strain gage. It consists of a strong magnetic wire stretched between two blocks and encased in a tube. A sensor in the tube has a permanent magnet and a plucking coil.

When the wire is plucked by the sensor, it vibrates at a natural frequency. The frequency is picked up by the coil and is sent to instruments that record it. When the assembly is subject to strain, the wire is stretched, changing its frequency. Thus, the frequency recorded is related to strain. The gage is well suited to systems that have wet and damp environments because it is a sealed unit.

SYSTEM NOTE

Polarity of the Induced Voltage

If the conductor in Figure 7–30 is moved first one way and then another in the magnetic field, a reversal of the polarity of the induced voltage will be observed. When the relative motion of the conductor is downward, a voltage is induced with the polarity indicated in Figure 7–31(a). When the relative motion of the conductor is upward, the polarity is as indicated in part (b) of the figure.

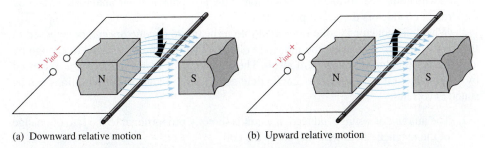

(a) Downward relative motion   (b) Upward relative motion

FIGURE 7–31   **Polarity of induced voltage depends on direction of motion of the conductor relative to the magnetic field.**

When a straight conductor moves perpendicular to a constant magnetic field, the induced voltage is given by

$$v_{ind} = B_\perp lv \qquad (7–7)$$

where v_{ind} is the induced voltage in volts, $B_\perp$ is the component of the magnetic flux density that is perpendicular to the moving conductor (in teslas), l is the length of the conductor that is exposed to the magnetic field, and v is the velocity of the conductor in m/s.

EXAMPLE 7–7

Assume the conductor in Figure 7–31 is 10 cm long and the pole face of the magnet is 5.0 cm wide. The flux density is 0.5 T, and the conductor is moved upward at a velocity of 0.8 m/s. What voltage is induced in the conductor?

SOLUTION

Although the conductor is 10 cm, only 5.0 cm (0.05 m) is in the magnetic field because of the size of the pole faces. Therefore,

$$v_{ind} = B_\perp lv = (0.5\ \text{T})(0.05\ \text{m})(0.8\ \text{m/s}) = \textbf{20 mV}$$

RELATED PROBLEM

What is the induced voltage if the velocity is doubled?

Induced Current

When a load resistor is connected to the conductor in Figure 7–31, the voltage induced by the relative motion between the conductor and the magnetic field will cause a current in the load, as shown in Figure 7–32. This current is called the **induced current** (i_{ind}). The lowercase i stands for instantaneous current.

The action of producing a voltage and a current in a load by moving a conductor across a magnetic field is the basis for electrical generators. A single conductor will have a small induced current, so practical generators use coils with many turns. Also, the existence of a conductor in a moving magnetic field is fundamental to the concept of inductance in an electric circuit.

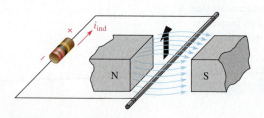

FIGURE 7–32 **Induced current (i_{ind}) in a load as the conductor moves through the magnetic field.**

Faraday's Law

Michael Faraday discovered the principle of electromagnetic induction in 1831. The key idea behind Faraday's law is that a changing magnetic field can induce a voltage in a conductor. Sometimes Faraday's law is stated as Faraday's law of induction. Faraday experimented with coils, and his law is an extension of the principle of electromagnetic induction for straight conductors discussed previously.

When a conductor is coiled into multiple turns, more conductors can be exposed to the magnetic field, increasing the induced voltage. When the flux is caused to change by any means, an induced voltage will result. The change in the magnetic field can be caused by relative motion between the magnetic field and the coil. Faraday's observations can be stated as follows:

1. The amount of voltage induced in a coil is directly proportional to the rate of change of the magnetic field with respect to the coil.

2. The amount of voltage induced in a coil is directly proportional to the number of turns of wire in the coil.

Faraday's first observation is demonstrated in Figure 7–33, where a bar magnet is moved through a coil, thus creating a changing magnetic field. In part (a) of the figure, the magnet is moved at a certain rate, and a certain induced voltage is produced as indicated. In part (b), the magnet is moved at a faster rate through the coil, creating a greater induced voltage.

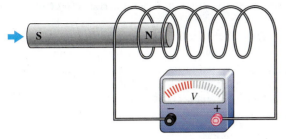

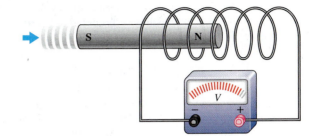

(a) As the magnet moves slowly to the right, its magnetic field is changing with respect to the coil, and a voltage is induced.

(b) As the magnet moves more rapidly to the right, its magnetic field is changing more rapidly with respect to the coil, and a greater voltage is induced.

FIGURE 7–33   A demonstration of Faraday's first observation: The amount of induced voltage is directly proportional to the rate of change of the magnetic field with respect to the coil.

Faraday's second observation is demonstrated in Figure 7–34. In part (a), the magnet is moved through the coil and a voltage is induced as shown. In part (b), the magnet is moved at the same rate through a coil that has a greater number of turns. The greater number of turns creates a greater induced voltage.

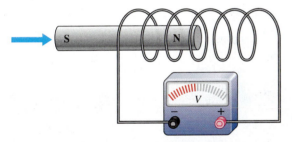

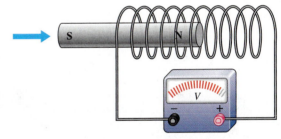

(a) Magnet moves through a coil and induces a voltage.

(b) Magnet moves at same rate through a coil with more turns (loops) and induces a greater voltage.

FIGURE 7–34   A demonstration of Faraday's second observation: The amount of induced voltage is directly proportional to the number of turns in the coil.

Faraday's law is stated as follows:

> **The voltage induced across a coil equals the number of turns in the coil times the rate of change of the magnetic flux.**

Any *relative* motion between the magnetic field and the magnet will produce the changing magnetic field that will induce a voltage in the coil. The changing magnetic field can even be induced by ac applied to an electromagnet, just as if there were motion. This type of changing magnetic field is the basis for transformer action in ac circuits, as you will see in Chapter 14.

Lenz's Law

You have learned that a changing magnetic field induces a voltage in a coil that is directly proportional to the rate of change of the magnetic field and the number of turns in the coil. **Lenz's law** defines the polarity or direction of the induced voltage.

> **When the current through a coil changes, the polarity of the induced voltage created by the changing magnetic field is such that it always opposes the change in current that caused it.**

An Application of Electromagnetic Induction

Automotive control systems include a large number of sensors. For optimum engine performance, it is necessary to know the position and/or speed of the crankshaft to control engine ignition timing, fuel mixture, tachometer, and antilock brakes. The most common method for sensing the crankshaft (or camshaft) position is with a Hall-effect sensor, which was discussed in Section 7–1.

Another widely used method is to detect a change in the magnetic field as a metallic tab moves though the air gap in a magnetic assembly. A basic concept is shown in Figure 7–35. A steel disk with protruding tabs is connected to the end of the crankshaft. As the crankshaft turns, the tabs move through the magnetic field. Steel has a much lower reluctance than does air, so the magnetic flux increases when a tab is in the air gap. This change in the magnetic flux causes an induced voltage to appear across the coil, indicating the position of the crankshaft.

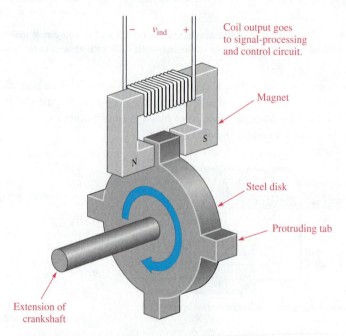

FIGURE 7–35   A crankshaft position sensor that produces a voltage when a tab passes through the air gap of the magnet.

Forces on a Current-Carrying Conductor in a Magnetic Field (Motor Action)

Figure 7–36(a) shows current inward through a wire in a magnetic field. The electromagnetic field set up by the current interacts with the permanent magnetic field; as a result, the permanent lines of force above the wire tend to be deflected down under the wire because

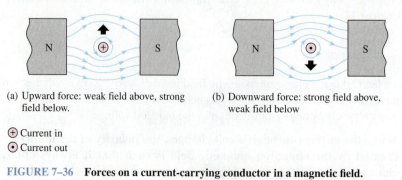

(a) Upward force: weak field above, strong field below.

(b) Downward force: strong field above, weak field below

⊕ Current in
⊙ Current out

FIGURE 7–36   Forces on a current-carrying conductor in a magnetic field.

they are opposite in direction to the electromagnetic lines of force. Therefore, the flux density above is reduced, and the magnetic field is weakened. The flux density below the conductor is increased, and the magnetic field is strengthened. An upward force on the conductor results, and the conductor tends to move toward the weaker magnetic field. Figure 7–36(b) shows the current outward, resulting in a force on the conductor in the downward direction. These upward and downward forces on a conductor are the basis for electric motors.

The force on a current-carrying conductor is given by the equation

$$F = BIl \tag{7-8}$$

where F is the force in newtons, B is the magnetic flux density in teslas, I is the current in amperes, and l is the length of the conductor exposed to the magnetic field in meters.

EXAMPLE 7–8

Assume a magnetic pole face is a square that is 3.0 cm on a side. Find the force on a conductor that has a current of 2 A if the conductor is perpendicular to the field and the flux density is 0.35 T.

SOLUTION

The length of conductor exposed to the magnetic flux is 3.0 cm (0.030 m). Therefore,

$$F = BIl = (0.35 \text{ T})(2.0 \text{ A})(0.03 \text{ m}) = \textbf{0.21 N}$$

RELATED PROBLEM

What is the direction of the force if the field is directed up (along the y axis), and the current (electron flow) is directed inward (along the z axis)?

SECTION 7–5 CHECKUP

1. What is the induced voltage across a stationary conductor in a stationary magnetic field?

2. When the rate at which a conductor is moved through a magnetic field is increased, does the induced voltage increase, decrease, or remain the same?

3. When there is current through a conductor in a magnetic field, what happens?

4. If the steel disk in the crankshaft position sensor has stopped with a tab in the magnet's air gap, what is the induced voltage?

7–6 DC GENERATORS

DC generators produce a voltage that is proportional to the magnetic flux and the rotational speed of the armature.

After completing this section, you should be able to

- **Explain how a dc generator works**
  - **Draw an equivalent circuit for a self-excited shunt-type dc generator**
- **Discuss the parts of a dc generator**

Figure 7–37 shows a greatly simplified dc generator consisting of a single loop of wire that rotates in a magnetic field. Notice that each end of the loop is connected to a split-ring arrangement. This conductive metal ring is called a *commutator*. As the wire loop rotates in the magnetic field, the split commutator ring also rotates. Each half of the split ring rubs against the fixed contacts, called *brushes,* and connects the wire to an external circuit.

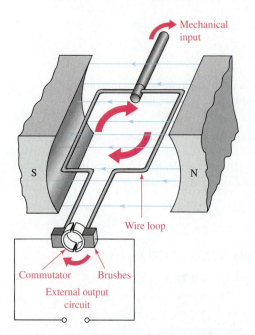

FIGURE 7–37 A simplified dc generator.

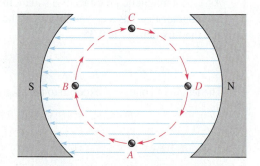

FIGURE 7–38 End view of loop of wire cutting through the magnetic field.

Driven by an external mechanical force, the loop of wire rotates through the magnetic field and cuts through the flux lines at varying angles, as illustrated in Figure 7–38. At position *A* in its rotation, the loop is effectively moving parallel with the magnetic field. Therefore, at this instant, the rate at which it is cutting through the magnetic flux lines is zero. As the loop moves from position *A* to position *B*, it cuts through the flux lines at an increasing rate. At position *B*, it is moving effectively perpendicular to the magnetic field and thus is cutting through a maximum number of lines. As the loop rotates from position *B* to position *C*, the rate at which it cuts the flux lines decreases to minimum (zero) at *C*. From position *C* to position *D*, the rate at which the loop cuts the flux lines increases to a maximum at *D* and then back to a minimum again at *A*.

As you have learned, when a wire moves through a magnetic field, a voltage is induced, and by Faraday's law, the amount of induced voltage is proportional to the number of loops (turns) in the wire and the rate at which it is moving with respect to the magnetic field. The angle at which the wire moves with respect to the magnetic flux lines determines the amount of induced voltage because the rate at which the wire cuts through the flux lines depends on the angle of motion.

Figure 7–39 illustrates how a voltage is induced in the external circuit as a single loop of wire rotates in the magnetic field. Assume that the loop is in its instantaneous horizontal position, so the induced voltage is zero. As the loop continues in its rotation, the induced voltage builds up to a maximum at position *B*, as shown in part (a) of the figure. Then, as the loop continues from *B* to *C*, the voltage decreases to zero at *C*, as shown in part (b).

During the second half of the revolution, shown in Figure 7–39(c) and (d), the brushes switch to opposite commutator sections, so the polarity of the voltage remains the same across the output. Thus, as the loop rotates from position *C* to *D* and then back to *A*, the voltage increases from zero at *C* to a maximum at *D* and back to zero at *A*.

Figure 7–40 shows how the induced voltage varies as a wire loop in a dc generator goes through several rotations (three in this case). This voltage is a dc voltage because its polarity does not change. However, the voltage is pulsating between zero and its maximum value.

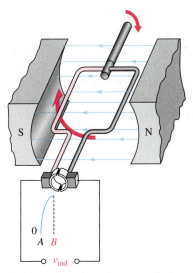

Position *B*: Loop is moving perpendicular to flux lines, and voltage is maximum.

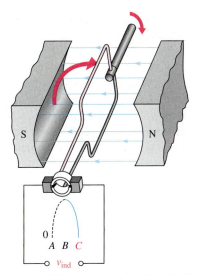

Position *C*: Loop is moving parallel with flux lines, and voltage is zero.

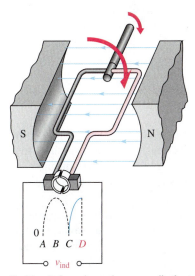

Position *D*: Loop is moving perpendicular to flux lines, and voltage is maximum.

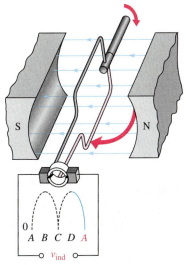

Position *A*: Loop is moving parallel with flux lines, and voltage is zero.

FIGURE 7–39 **Basic operation of a dc generator.**

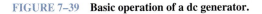

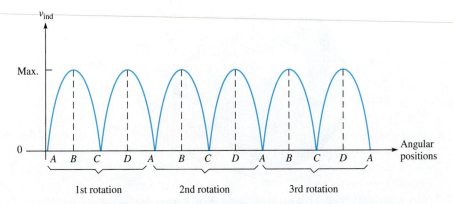

FIGURE 7–40 **Induced voltage over three rotations of a wire loop in a dc generator.**

In practical generators, multiple coils are pressed into slots in a ferromagnetic-core assembly. The entire assembly, called the **rotor**, is connected to bearings and rotates in the magnetic field. Figure 7–41 is a diagram of the rotor core with no wire loops (coils) shown. The commutator is divided into segments, with each pair of segments connected to the end of a coil. With more coils, the voltages from several coils are combined because the brushes can contact more than one of the commutator segments at once. The loops do not reach maximum voltage at the same time, so the pulsating output voltage is much smoother than is the case with only one coil or loop shown previously. The variations can be further smoothed by filters to produce a nearly constant dc output. (Filters are discussed in Chapter 13.)

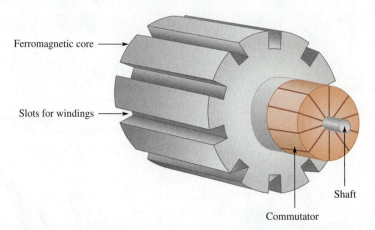

Ferromagnetic core

Slots for windings

Shaft

Commutator

FIGURE 7–41   **A simplified rotor core. The coils are pressed into the slots and connected to the commutator.**

Instead of permanent magnets, most generators use electromagnets to provide the required magnetic field. One advantage to this is that flux density can be controlled, thus controlling the output voltage of the generator. The windings for the electromagnets are appropriately called the field windings. The field windings require current to produce the magnetic field.

The current for the field windings can be provided from a separate voltage source, but this is a disadvantage. A better method is to use the generator itself to provide the current for the electromagnets; this is called a **self-excited generator**. The generator starts because there is normally enough residual magnetism in the field magnets due to hysteresis that causes a small initial field and allows the generator to start producing a voltage. In cases where a generator has not been used for a long time, it may be necessary to provide an external source to the field windings to start it.

The stationary part of a generator (or a motor) includes all of the nonmoving parts and is called the **stator**. Figure 7–42 illustrates a simplified two-pole dc generator showing

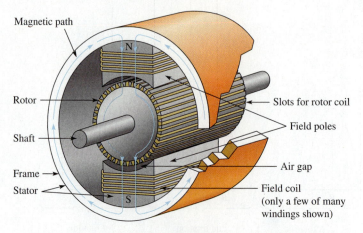

Magnetic path

N

Rotor

Shaft

Frame

Stator

S

Slots for rotor coil

Field poles

Air gap

Field coil
(only a few of many
windings shown)

FIGURE 7–42   **The magnetic structure of a generator (or motor). In this case, the rotor is also the armature because it produces the power.**

the magnetic paths (end caps, bearings, and commutator are not shown). Notice that the frame is part of the magnetic path for the field magnets. To make a generator have high efficiency, the air gap is kept as small as possible. The **armature** is the power-producing component and can be either on the rotor or the stator. In the dc generator described previously, the armature is the rotor because power is produced in the moving conductors and taken from the rotor via the commutator.

Equivalent Circuit for a DC Generator

A self-excited generator can be represented by a basic dc circuit with a coil to produce the magnetic field and a mechanically driven generator, as shown in Figure 7–43. There are other configurations of dc generators, but this represents a common one. In the case shown, the field windings are in parallel with the source; this configuration is called a *shunt-wound generator*. The resistance of the field windings is shown as R_F. For the equivalent circuit, this resistance is shown in series with the field windings. The armature is driven by a mechanical input, causing it to spin; the armature is the generator voltage source, V_G. The armature resistance is shown as the series resistance, R_A. The rheostat, R_{REG}, is in series with the field winding resistance and regulates the output voltage by controlling the current to the field windings and thus the flux density.

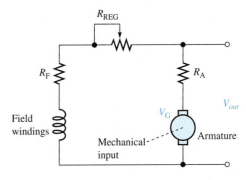

When a load is connected to the output, current in the armature is shared between the load and the field windings. The efficiency of the generator can be calculated as the ratio of the power delivered to the load (P_L) to the total power (P_T), which includes the losses in the armature and the resistance in the field circuit.

FIGURE 7–43   **Equivalent circuit for a self-excited shunt dc generator.**

SECTION 7–6 CHECKUP

1. What is the moving part of a generator called?

2. What is the purpose of a commutator?

3. How does greater resistance in the field windings of a generator affect the output voltage?

4. What is meant by a self-excited generator?

7–7 DC MOTORS

Motors convert electrical energy to mechanical motion by taking advantage of the force produced when a current-carrying conductor is in a magnetic field. A dc motor operates from a dc source and can use either an electromagnet or a permanent magnet to supply the field.

After completing this section, you should be able to

- **Explain how a dc motor works**
  - **Draw an equivalent circuit for a series and a shunt-type dc motor**
  - **Discuss back emf and how it reduces armature current**
  - **Discuss power rating of motors**

Basic Operation

As in the case of generators, motor action is the result of the interaction of magnetic fields. In a dc motor, the rotor field interacts with the magnetic field set up by current in the stator windings. The rotor in all dc motors contains the armature winding, which sets up a magnetic field. The rotor moves because of the attractive force between opposite poles and the repulsive force between like poles, as illustrated in the simplified diagram of Figure 7–44. The rotor moves because of the attraction of its north pole with the south pole of the stator (and vice-versa). As the two poles near each other, the polarity of the rotor current is suddenly switched by the commutator, reversing the magnetic poles of the rotor. The commutator serves as a mechanical switch to reverse the current in the armature just as the unlike poles are near each other, continuing the rotation of the rotor.

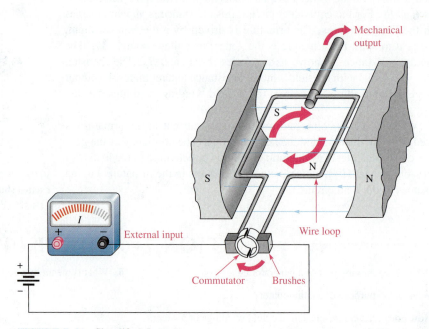

FIGURE 7–44   **Simplified dc motor.**

Motor Maintenance

Industrial systems typically have many motors to perform functions such as pumps, machine operations, conveyers, and robotic control. A motor that is running too hot may be indicating that there is lubrication or bearing issues that need to be addressed. A thermal image can reveal such problems, so in many companies thermal imaging is part of a routine inspection and maintenance program.

Courtesy of Fluke Corporation.

SYSTEM NOTE

Brushless DC Motors

Many dc motors do not use a commutator to reverse the polarity of the current. Instead of supplying current to a moving armature, the magnetic field is rotated in the stator windings using an electronic controller. The direction of current in the field coils is periodically reversed by the controller by producing an ac waveform (or modified ac waveform) from the dc input. This causes the stator field to rotate, and the permanent magnet rotor moves in the same direction to keep up with the rotating field. A common way to sense the position of the rotating magnet is to use a Hall-effect sensor, which provides the controller with position information by sending pulses every time a magnet passes it. Brushless motors have higher reliability than traditional brushed motors because they do not need to have periodic brush replacement, but have the added complexity of the electronic controller. Figure 7–45 shows a cutaway view of a brushless dc motor that includes a pulse width modulation control within the motor housing as well as an optical encoder to indicate the shaft position.

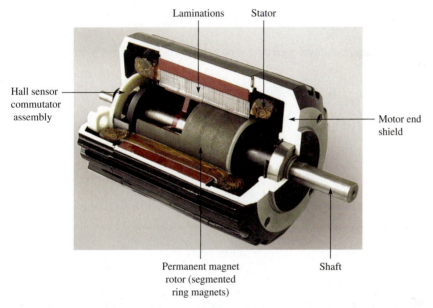

FIGURE 7–45   **Cutaway view of a brushless dc motor.**   (Courtesy of Bodine Electric Company)

Back EMF

When a dc motor is first started, a magnetic field is present from the field windings. Armature current develops another magnetic field that interacts with the one from the field windings and starts the motor turning. The armature windings are now spinning in the presence of the magnetic field, so generator action occurs. In effect, the spinning armature has a voltage developed across it that opposes the original applied voltage in accordance with Lenz's law. This self-generated voltage is called **back emf** (electromotive force). The term *emf* was once common for voltage but is not favored because voltage is not a "force" in the physics sense, but back emf is still applied to the self-generated voltage in motors. Back emf, also called *counter emf,* serves to significantly reduce the armature current when the motor is turning at constant speed.

HANDS ON TIP
One characteristic of dc motors is that if they are allowed to run without a load, the torque can cause the motor to "run away" to a speed beyond the manufacturer's rating. Therefore, dc motors should always be operated with a load to prevent self-destruction.

Motor Ratings

Some motors are rated by the torque they can provide; others are rated by the power they produce. Torque and power are important parameters for any motor. Although torque and power are different physical parameters, if one is known, the other can be obtained.

Torque tends to rotate an object. In a dc motor, the torque is proportional to the amount of flux and to the armature current. Torque, T, in a dc motor can be calculated from the equation

$$T = k\phi I_A \tag{7-9}$$

where T is torque in newton-meters (N-m) k is a constant that depends on physical parameters of the motor, ϕ is magnetic flux in webers (Wb), and I_A is armature current in amperes (A).

Recall that power is defined as the rate of doing work. To calculate power from torque, you must know the speed of the motor in rpm for the torque that you measured. The equation to determine the power, given the torque at a certain speed, is

$$P = 0.105Ts \tag{7-10}$$

where P is power in W, T is torque in N-m, and s is speed of motor in rpm.

EXAMPLE 7–9

What is the power developed by a motor that turns at 350 rpm when the torque is 3.6 N-m?

SOLUTION

Substitute into Equation 7–10.

$$P = 0.105Ts = 0.105(3.6\text{ N-m})(350\text{ rpm}) = \textbf{132 W}$$

RELATED PROBLEM

There are 746 W in one hp. What is the hp of this motor under these conditions?

SYSTEM EXAMPLE 7–2

DC MOTOR REVERSING

DC motors have many applications in systems ranging from small portable drills to large cranes. In many systems that use a dc motor, the motor needs to have the ability to be reversed. For example, a motor that operates a garage door needs to be reversed to open or close the door. To reverse a dc motor, you need to be able to connect separate power leads to the field (stator) and the armature (rotor) and switch the polarity on *either* the rotor or the stator. Only one of the windings is reversed; if both were reversed, the motor would turn in the same direction.

In some cases, the forward or reverse direction may be controlled by a logic operation. In the system described here, a relay is used to interface logic that is used to specify the direction with the motor-reversing switches, which are the relay contacts. A simplified reversing circuit is shown in Figure 7–46. In addition to its role in reversing the motor, the relay provides a second benefit. It completely isolates the low voltage logic circuits from the motor circuit.

The transistor (Q_1) acts as a switch to interface the logic to the relay coil. When the logic is LOW, the transistor is *off*, and the relay is in its unenergized position. When the logic is HIGH, the transistor conducts and activates the relay. The logic circuits are isolated from the motor power by the relay. The purpose of the diode is to avoid noise spikes when the relay is deenergized. The On/Off switch is controlled by the logic to remove power from the motor before changing its direction.

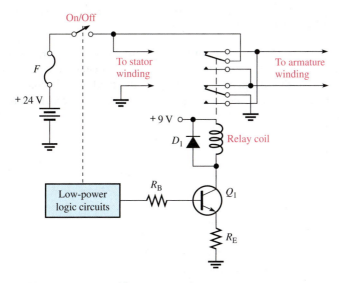

FIGURE 7–46   **DC motor reversing circuit using a relay.**

Series DC Motors

The series dc motor has the field coil windings and the armature coil windings in series. A schematic of this arrangement is shown in Figure 7–47(a). The internal resistance is generally small and consists of field coil resistance, armature winding resistance, and brush resistance. As in the case of generators, dc motors may also contain an interpole winding, as shown, and current limiting for speed control. In a series dc motor, the armature current, field current, and line current are all the same.

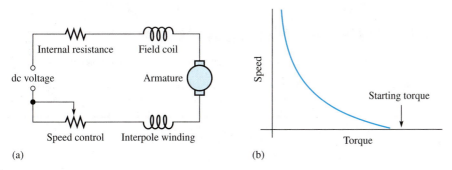

FIGURE 7–47   **Simplified schematic for a series dc motor and torque-speed characteristic.**

As you know, magnetic flux is proportional to the current in a coil. The magnetic flux created by the field windings is proportional to armature current because of the series connection. Thus, when the motor is loaded, armature current rises, and the magnetic flux rises too. Recall that Equation 7–9 showed that the torque in a dc motor is proportional to both the armature current and the magnetic flux. Thus, the series-wound motor will have a very high starting torque when the current is high because flux and armature current are high. For this reason, series dc motors are used when high starting torques are required (such as a starter motor in a car).

The plot of the torque and motor speed for a series dc motor is shown in Figure 7–47(b). The starting torque is at its maximum value. At low speeds, the torque is still very high, but drops off dramatically as the speed increases. As you can see, the speed can be very high if the torque is low; for this reason, the series-wound dc motor always is operated with a load.

Shunt DC Motors

A shunt dc motor has the field coil in parallel with the armature, as shown in the equivalent circuit in Figure 7–48(a). In the shunt motor, the field coil is supplied by a constant voltage source, so the magnetic field set up by the field coils is constant. The armature resistance and the back emf produced by generator action in the armature determine the armature current.

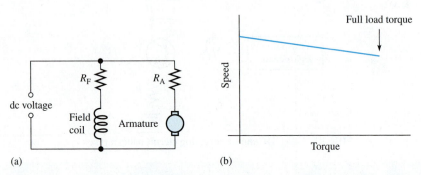

(a)   (b)

FIGURE 7–48   **Simplified schematic for a shunt dc motor and torque-speed characteristic.**

The torque-speed characteristic for a shunt dc motor is quite different than for a series dc motor. When a load is applied, the shunt motor will slow down, causing the back emf to be reduced and the armature current to increase. The increase in armature current tends to compensate for the added load by increasing the torque of the motor. Although the motor has slowed because of the additional load, the torque-speed characteristic is nearly a straight line for the shunt dc motor as shown in Figure 7–48(b). At full load, the shunt dc motor still has high torque.

SECTION 7–7 CHECKUP

1. What creates back emf?

2. How does back emf affect current in a rotating armature as it comes up to speed?

3. What type of dc motor has the highest starting torque?

4. What is the major advantage of a brushless motor over a brushed motor?

SUMMARY

- Unlike magnetic poles attract each other, and like poles repel each other.
- Materials that can be magnetized are called *ferromagnetic*.
- When there is current through a conductor, it produces an electromagnetic field around the conductor.
- You can use the left-hand rule to establish the direction of the electromagnetic lines of force around a conductor.
- An electromagnet is basically a coil of wire around a magnetic core.
- When a conductor moves within a magnetic field, or when a magnetic field moves relative to a conductor, a voltage is induced across the conductor.
- The faster the relative motion between a conductor and a magnetic field, the greater is the induced voltage.
- Table 7–3 summarizes the quantities and SI units used in this chapter.

TABLE 7–3

SYMBOL	QUANTITY	SI UNIT
B	Magnetic flux density	Tesla (T)
ϕ	Flux	Weber (Wb)
μ	Permeability	Webers/ampere-turn · meter (Wb/At · m)
$\mathcal{R}$	Reluctance	Ampere-turns/weber (At/Wb)
F_m	Magnetomotive force (mmf)	Ampere-turn (At)
H	Magnetic field intensity	Ampere-turns/meter (At/m)
F	Force	Newton (N)
T	Torque	Newton-meter (N-m)

- Hall-effect sensors use current to sense the presence of a magnetic field.
- DC generators convert mechanical power to dc electrical power.
- The moving part of a generator or motor is called the rotor; the stationary part is called the stator.
- DC motors convert electrical power to mechanical power.
- Brushless dc motors use a permanent magnet as the rotor, and the stator is the armature.

KEY TERMS

Key terms and other bold terms in the chapter are defined in the end-of-book glossary.

Ampere-turn (At)   The SI unit of magnetomotive force (mmf).

Electromagnetic field   A formation of a group of magnetic lines of force surrounding a conductor created by electrical current in the conductor.

Electromagnetic induction   The phenomenon or process by which a voltage is produced in a conductor when there is relative motion between the conductor and a magnetic or electromagnetic field.

Electromagnetism   The production of a magnetic field by current in a conductor.

Faraday's law   A law stating that the voltage induced across a coil equals the number of turns in the coil times the rate of change of the magnetic flux.

Gauss   A CGS unit of flux density.

Hall effect   A change in current density across a conductor or semiconductor when current in the material is perpendicular to a magnetic field. The change in current density produces a small transverse voltage in the material, called the *Hall voltage*.

Hysteresis   A characteristic of a magnetic material whereby a change in magnetization lags the application of the magnetic field intensity.

Induced current (i_{ind})   A current induced in a conductor as a result of a changing magnetic field.

Induced voltage (v_{ind})   Voltage produced as a result of a changing magnetic field.

Lenz's law   A physical law that states when the current through a coil changes, the polarity of the induced voltage created by the changing magnetic field is such that it always opposes the change in current that caused it. The current cannot change instantaneously.

Lines of force   Magnetic flux lines in a magnetic field radiating from the north pole to the south pole.

Magnetic field   A force field radiating from the north pole to the south pole of a magnet.

Magnetic flux   The lines of force between the north and south poles of a permanent magnet or an electromagnet.

Magnetic field intensity   The amount of mmf per unit length of magnetic material.

Magnetomotive force (mmf)   The cause of a magnetic field, measured in ampere-turns.

Permeability   The measure of ease with which a magnetic field can be established in a material.

Relay   An electromagnetically controlled mechanical device in which electrical contacts are opened or closed by a magnetizing current.

Reluctance ($\mathcal{R}$)   The opposition to the establishment of a magnetic field in a material.

Retentivity   The ability of a material, once magnetized, to maintain a magnetized state without the presence of a magnetizing force.

Solenoid   An electromagnetically controlled device in which the mechanical movement of a shaft or plunger is activated by a magnetizing current.

Speaker   An electromagnetic device that converts electrical signals to sound waves.

Tesla (T)   The SI unit of flux density.

Weber (Wb)   The SI unit of magnetic flux, which represents 10^8 lines.

KEY FORMULAS

(7–1)	$B = \dfrac{\phi}{A}$	Magnetic flux density
(7–2)	$\mu_r = \dfrac{\mu}{\mu_0}$	Relative permeability
(7–3)	$\mathcal{R} = \dfrac{l}{\mu A}$	Reluctance
(7–4)	$F_m = NI$	Magnetomotive force
(7–5)	$\phi = \dfrac{F_m}{\mathcal{R}}$	Magnetic flux
(7–6)	$H = \dfrac{F_m}{l}$	Magnetic field intensity
(7–7)	$v_{\text{ind}} = B_\perp lv$	Induced voltage
(7–8)	$F = BIl$	Force on current-carrying conductor
(7–9)	$T = k\phi I_A$	Torque in a dc motor
(7–10)	$P = 0.105Ts$	Power from torque

TRUE/FALSE QUIZ

Answers are at the end of the chapter.

1. The tesla (T) and the gauss (G) are both units for magnetic flux density.

2. The unit for measuring magnetomotive force (mmf) is the volt.

3. Ohm's law for a magnetic circuit gives the relationship between flux density, magnetomotive force, and reluctance.

4. A solenoid is a form of electromagnetic switch that opens and closes mechanical contacts.

5. A hysteresis curve is a plot of flux density (B) as a function of field intensity (H).

6. To produce an induced voltage in a coil, the magnetic field surrounding it can be changed.

7. The speed of a generator can be controlled with a rheostat in the field windings.

8. A self-excited dc generator will normally have enough residual magnetism in the field magnets to start the generator, producing voltage at the output when it is first turned on.

9. The power developed by a motor is proportional to its torque.

10. In a brushless motor, the magnetic field is supplied by permanent magnets.

SELF-TEST

Answers are at the end of the chapter.

1. When the south poles of two bar magnets are brought close together, there will be
   - (a) a force of attraction
   - (b) a force of repulsion
   - (c) an upward force
   - (d) no force

2. A magnetic field is made up of
   - (a) positive and negative charges
   - (b) magnetic domains
   - (c) flux lines
   - (d) magnetic poles

3. The direction of a magnetic field is from
   - (a) north pole to south pole
   - (b) south pole to north pole
   - (c) inside to outside the magnet
   - (d) front to back

4. Reluctance in a magnetic circuit is analogous to
   - (a) voltage in an electric circuit
   - (b) current in an electric circuit
   - (c) power in an electric circuit
   - (d) resistance in an electric circuit

5. The unit of magnetic flux is the
   - (a) tesla
   - (b) weber
   - (c) gauss
   - (d) ampere-turn

6. The unit of magnetomotive force is the
   - (a) tesla
   - (b) weber
   - (c) ampere-turn
   - (d) electron-volt

7. A unit of magnetic flux density is the
   - (a) tesla
   - (b) weber
   - (c) ampere-turn
   - (d) ampere-turns/meter

8. The electromagnetic activation of a movable shaft is the basis for
   - (a) relays
   - (b) circuit breakers
   - (c) magnetic switches
   - (d) solenoids

9. When there is current through a wire placed in a magnetic field,
   - (a) the wire will overheat
   - (b) the wire will become magnetized
   - (c) a force is exerted on the wire
   - (d) the magnetic field will be cancelled

10. A coil of wire is placed in a changing magnetic field. If the number of turns in the coil is increased, the voltage induced across the coil will
    - (a) remain unchanged
    - (b) decrease
    - (c) increase
    - (d) be excessive

11. If a conductor is moved back and forth at a constant rate in a constant magnetic field, the voltage induced in the conductor will
    - (a) remain constant
    - (b) reverse polarity
    - (c) be reduced
    - (d) be increased

12. In the crankshaft position sensor in Figure 7–35, the induced voltage across the coils is caused by
    - (a) current in the coil
    - (b) rotation of the disk
    - (c) a tab passing through the magnetic field
    - (d) acceleration of the disk's rotational speed

13. The purpose of the commutator in a generator or motor is to
    - (a) change the direction of the current to the rotor as it spins
    - (b) change the direction of the current to the stator windings
    - (c) support the motor or generator shaft
    - (d) provide the magnetic field for the motor or generator

14. In a motor, back emf serves to
    - (a) increase the power from the motor
    - (b) decrease the flux
    - (c) increase the current in the field windings
    - (d) decrease the current in the armature

15. The torque of a motor is proportional to the
    - (a) amount of flux
    - (b) armature current
    - (c) both of the above
    - (d) none of the above

PROBLEMS

Answers to odd-numbered problems are at the end of the book.

BASIC PROBLEMS

SECTION 7–1   The Magnetic Field

1. The cross-sectional area of a magnetic field is increased, but the flux remains the same. Does the flux density increase or decrease?

2. In a certain magnetic field the cross-sectional area is 0.5 m^2 and the flux is 1500 μWb. What is the flux density?

3. What is the flux in a magnetic material when the flux density is 2500×10^{-6} T and the cross-sectional area is 150 cm^2?

4. At a given location, the earth's magnetic field is 0.6 G (gauss). Express this flux density in tesla.

5. A very strong permanent magnet has a field of 100,000 μT. Express this flux density in gauss.

SECTION 7–2   Electromagnetism

6. What happens to the compass needle in Figure 7–11 when the current through the conductor is reversed?

7. What is the relative permeability of a ferromagnetic material whose absolute permeability is 750×10^{-6} Wb/At · m?

8. Determine the reluctance of a material with a length of 0.28 m and a cross-sectional area of 0.08 m^2 if the absolute permeability is 150×10^{-7} Wb/At · m.

9. What is the magnetomotive force in a 500 turn coil of wire with 3 A through it?

SECTION 7–3   Electromagnetic Devices

10. Typically, when a solenoid is activated, is the plunger extended or retracted?
11. (a) What force moves the plunger when a solenoid is activated?
    (b) What force causes the plunger to return to its at-rest position?

12. Explain the sequence of events in the circuit of Figure 7–49 starting when switch 1 (SW1) is closed.

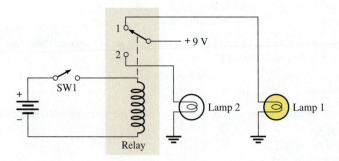

FIGURE 7–49

13. What causes the pointer in a d'Arsonval meter movement to deflect when there is current through the coil?

SECTION 7–4   Magnetic Hysteresis

14. What is the magnetizing force in Problem 9 if the length of the core is 0.2 m?

15. How can the flux density in Figure 7–50 be changed without altering the physical characteristics of the core?

16. In Figure 7–50, determine the following if the winding has 100 turns:
    (a) H (b) ϕ (c) B

17. Determine from the hysteresis curves in Figure 7–51 which material has the most retentivity.

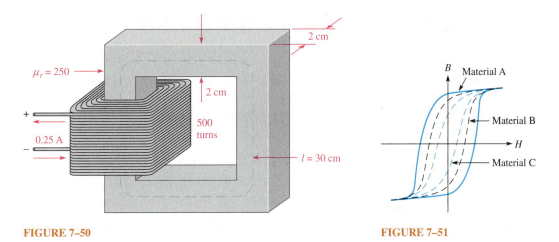

FIGURE 7–50

FIGURE 7–51

SECTION 7–5 Electromagnetic Induction

18. According to Faraday's law, what happens to the induced voltage across a given coil if the rate of change of magnetic flux doubles?

19. The voltage induced across a certain coil is 100 mV. A 100 Ω resistor is connected to the coil terminals. What is the induced current?

20. In Figure 7–35, why is there no induced voltage when the steel disk is not rotating?

21. A 20 cm long conductor is moving upward between the poles of a magnet as shown in Figure 7–52. The pole faces are 8.5 cm on each side and the flux is 1.24 mWb. The motion produces an induced voltage across the conductor of 44 mV. What is the speed of the conductor?

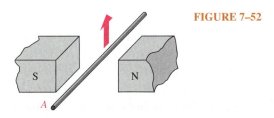

FIGURE 7–52

22. **(a)** For the conductor shown in Figure 7–52, what is the polarity of the end marked with the letter A?
    (b) Assuming a complete path is provided and current is in the direction you indicated, what direction is the induced force on the conductor?

SECTION 7–6 DC Generators

23. If a generator is 80% efficient and delivers 45 W to a load, what is the input power?

24. Assume the self-excited shunt dc generator in Figure 7–43 has a load connected to it that draws 12 A. If the field windings draw 1.0 A, what is the armature current?

25. **(a)** If the output voltage in Problem 24 is 14 V, what power is supplied to the load?
    (b) What power is dissipated in the field resistance?

SECTION 7–7 DC Motors

26. **(a)** What power is developed by a motor that turns at 1200 rpm and has a torque of 3.0 N-m?
    (b) What is the horsepower rating of the motor? (746 W = 1 hp)

27. Assume a motor dissipates 12 W internally when it delivers 50 W to a load. What is the efficiency?

ADVANCED PROBLEMS

28. A basic one-loop dc generator is rotated at 60 rps. How many times each second does the dc output voltage peak (reach a maximum)?

29. Assume that another loop of wire, 90° from the first loop, is added to the dc generator in Problem 28. Describe the output voltage. Let the maximum voltage be 10 V.

30. Explain the purpose of each of the three sets of relay contacts (A, B, and C) in Figure 7–22.

31. For the DC motor reversing circuit in Figure 7–46, assume the polarity to the armature is opposite to the polarity of the stator. Is the relay coil energized or not? Explain your answer.

MULTISIM TROUBLESHOOTING PROBLEMS

32. Open file P07-32; files are found at www.pearsonhighered.com/floyd. Determine if there is a fault in the circuit. If so, identify the fault.

33. Open file P07-33 and determine if there is a fault in the circuit. If so, identify the fault.

ANSWERS TO SECTION CHECKUPS

SECTION 7–1   The Magnetic Field

1. The north poles repel.
2. Magnetic flux is the group of lines of force that make up a magnetic field; magnetic flux density is the flux per area.
3. The gauss and the tesla
4. $B = \phi/A = 900\,\mu\text{T}$

SECTION 7–2   Electromagnetism

1. Electromagnetism is produced by current through a conductor. An electromagnetic field exists only when there is current. A magnetic field exists independently of current.
2. The direction of the magnetic field also reverses when the current is reversed.
3. Flux equals magnetomotive force divided by reluctance.
4. Flux is analogous to current, mmf is analogous to voltage, and reluctance is analogous to resistance.

SECTION 7–3   Electromagnetic Devices

1. A solenoid provides mechanical movement of a shaft. A relay provides an electrical contact closure.
2. The movable part of a solenoid is the plunger.
3. The movable part of a relay is the armature.
4. The d'Arsonval meter movement is based on the interaction of magnetic fields.

SECTION 7–4   Magnetic Hysteresis

1. In a wirewound core, an increase in current increases the flux density.
2. Retentivity is the ability of a material to remain magnetized after removal of the magnetizing force.

SECTION 7–5   Electromagnetic Induction

1. The induced voltage is zero.
2. The induced voltage increases.
3. A force is exerted on the current-carrying conductor in a magnetic field.
4. The induced voltage is zero.

SECTION 7–6   DC Generators

1. The rotor
2. The commutator reverses the current in the rotating coil.
3. Greater resistance will reduce the magnetic flux, causing the output voltage to drop.
4. A generator in which the field windings derive their current from the output

SECTION 7–7   DC Motors

1. Back emf is a voltage developed in a motor because of generator action as the rotor turns. It opposes the original supply voltage.
2. Back emf reduces the armature current.
3. A series-wound motor
4. Higher reliability because there are no brushes to wear out

ANSWERS TO RELATED PROBLEMS FOR EXAMPLES

7–1 19.6×10^{-3} T

7–2 31.0 T

7–3 The reluctance is reduced to 12.8×10^{6} At/Wb.

7–4 Reluctance is 165.7×10^{3} At/Wb

7–5 7.2 mWb

7–6 $F_m = 42.5$ At; $\mathcal{R} = 8.5 \times 10^{4}$ At/Wb

7–7 40 mV

7–8 The direction is along the negative x axis.

7–9 0.18 hp

ANSWERS TO SELF-TEST

1. (b)   **2.** (c)   **3.** (a)   **4.** (d)   **5.** (b)   **6.** (c)   **7.** (a)   **8.** (d)

9. (c)   **10.** (c)   **11.** (b)   **12.** (c)   **13.** (a)   **14.** (d)   **15.** (c)

ANSWERS TO TRUE/FALSE QUIZ

1. T   **2.** F   **3.** F   **4.** F   **5.** T   **6.** T   **7.** F   **8.** T

9. T   **10.** F

CHAPTER 8

INTRODUCTION TO ALTERNATING CURRENT AND VOLTAGE

OUTLINE

OBJECTIVES

- Identify a sinusoidal waveform and measure its characteristics
- Determine the voltage and current values of sine waves
- Describe angular relationships of sine waves
- Mathematically analyze a sinusoidal waveform
- Apply the basic circuit laws to resistive ac circuits
- Describe how an alternator generates electricity
- Explain how ac motors convert electrical energy into rotational motion
- Identify the characteristics of basic nonsinusoidal waveforms
- Use the oscilloscope to measure waveforms
- Describe the types of signal sources and explain the purpose of typical controls

KEY TERMS

Sine wave	Alternator
Waveform	Induction motor
Cycle	Synchronous motor
Period (T)	Squirrel cage
Frequency (f)	Slip
Hertz (Hz)	Pulse
Oscillator	Pulse width (t_W)
Function generator	Rise time (t_r)
Instantaneous value	Fall time (t_f)
Peak value	Periodic
Amplitude	Duty cycle
Peak-to-peak value	Ramp
rms value	Fundamental frequency
Degree	Harmonics
Radian	Oscilloscope
Phase	Modulation

VISIT THE WEBSITE
Study aids for this chapter are available at
http://pearsonhighered.com/floyd

INTRODUCTION

This chapter provides an introduction to alternating current (ac) circuits. Alternating voltages and currents fluctuate with time and periodically change polarity and direction according to certain patterns called *waveforms*. Particular emphasis is given to the sinusoidal waveform (sine wave) because of its basic importance in ac circuits.

Alternators, which generate sine waves, and ac motors are introduced. Other types of waveforms are also introduced, including pulse, triangular, and sawtooth. The use of an oscilloscope for displaying and measuring waveforms is discussed. Frequently, specialized waveforms are needed to test systems. The chapter concludes with a discussion on signal generating.

8–1 THE SINUSOIDAL WAVEFORM

The sinusoidal waveform or sine wave is the fundamental type of alternating current (ac) and alternating voltage. It is also referred to as a sinusoidal wave, or, simply, sinusoid. The electrical service provided by the power companies is in the form of sinusoidal voltage and current. In addition, other types of repetitive waveforms are composites of many individual sine waves called harmonics.

After completing this section, you should be able to

- **Identify a sinusoidal waveform and measure its characteristics**
  - **Define and determine the period**
  - **Define and determine the frequency**
  - **Relate the period and the frequency**
  - **Describe two types of electronic signal generators**

Sine waves, or sinusoids, are produced by two types of sources: rotating electrical machines (ac generators) or electronic oscillator circuits, which are used in instruments commonly known as electronic signal generators. Electronic signal generators are introduced in this section and covered in more detail in Section 8–10. Alternators, which generate ac by electromechanical means, are covered in Section 8–6. Figure 8–1 shows the symbol used to represent a source of sinusoidal voltage.

Figure 8–2 is a graph that shows the general shape of a sine wave, which can be either an alternating current or an alternating voltage. Voltage (or current) is displayed on the vertical axis, and time (*t*) is displayed on the horizontal axis. Notice how the voltage (or current) varies with time. Starting at zero, the voltage (or current) increases to a positive maximum (peak), returns to zero, and then increases to a negative maximum (peak) before returning again to zero, thus completing one full cycle.

FIGURE 8–1   **Symbol for a sinusoidal voltage source.**

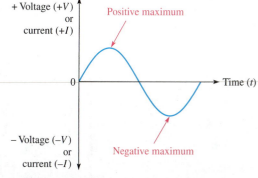

FIGURE 8–2   **Graph of one cycle of a sine wave.**

Polarity of a Sine Wave

As mentioned, a sine wave changes polarity at its zero value; that is, it alternates between positive and negative values. When a sinusoidal voltage source (V_s) is applied to a resistive circuit, as in Figure 8–3, an alternating sinusoidal current results. When the voltage changes polarity, the current correspondingly changes direction as indicated.

During the positive alternation of the source voltage V_s, the current is in the direction shown in Figure 8–3(a). During a negative alternation of the source voltage, the current is in the opposite direction, as shown in Figure 8–3(b). The combined positive and negative alternations make up one cycle of a sine wave.

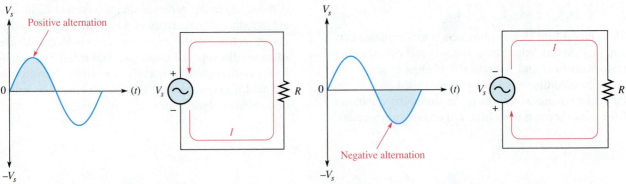

(a) During a positive alternation of the voltage, current is in the direction shown.

(b) During a negative alternation of the voltage, current reverses direction, as shown.

FIGURE 8–3   **Alternating current and voltage.**

Period of a Sine Wave

A sine wave varies with time (t) in a definable manner.

> **The time required for a given sine wave to complete one full cycle is called the period (T).**

Figure 8–4(a) illustrates the period of a sine wave. Typically, a sine wave continues to repeat itself in identical cycles, as shown in Figure 8–4(b). Since all cycles of a repetitive sine wave are the same, the period is always a fixed value for a given sine wave. The period of a sine wave can be measured from a zero crossing to the next corresponding zero crossing, as indicated in Figure 8–4(a). The period can also be measured from any peak in a given cycle to the corresponding peak in the next cycle.

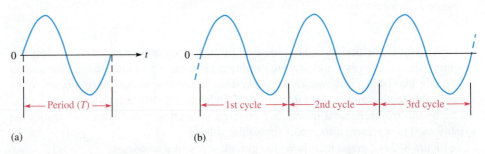

(a)

(b)

FIGURE 8–4   **The period of a given sine wave is the same for each cycle.**

EXAMPLE 8–1

What is the period of the sine wave in Figure 8–5?

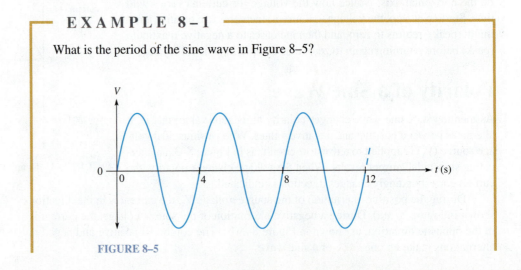

FIGURE 8–5

SOLUTION

As shown in Figure 8–5, it takes twelve seconds (12 s) to complete three cycles. Therefore, to complete one cycle it takes four seconds (4 s), which is the period.

$$T = \mathbf{4\,s}$$

RELATED PROBLEM*

What is the period if a given sine wave goes through five cycles in 12 s?

Answers are at the end of the chapter.

EXAMPLE 8–2

Show three possible ways to measure the period of the sine wave in Figure 8–6. How many cycles are shown?

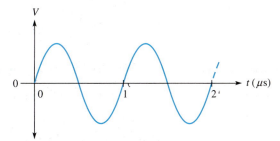

FIGURE 8–6

SOLUTION

Method 1: The period can be measured from one zero crossing to the corresponding zero crossing in the next cycle (the slope must be the same at the corresponding zero crossings).

Method 2: The period can be measured from the positive peak in one cycle to the positive peak in the next cycle.

Method 3: The period can be measured from the negative peak in one cycle to the negative peak in the next cycle.

These measurements are indicated in Figure 8–7, where **two cycles** of the sine wave are shown. Keep in mind that you obtain the same value for the period no matter which corresponding peaks or corresponding zero crossings on the waveform you use.

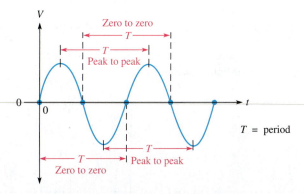

FIGURE 8–7   Measurement of the period of a sine wave.

RELATED PROBLEM

If a positive peak occurs at 1 ms and the next positive peak occurs at 2.5 ms, what is the period?

Frequency of a Sine Wave

Frequency is the number of cycles that a sine wave completes in one second.

The more cycles completed in one second, the higher the frequency. Frequency (f) is measured in units of **hertz (Hz)**. One hertz is equivalent to one cycle per second; for example, 60 Hz is 60 cycles per second. Figure 8–8 shows two sine waves. The sine wave in part (a) completes two full cycles in one second. The one in part (b) completes four cycles in one second. Therefore, the sine wave in part (b) has twice the frequency of the one in part (a).

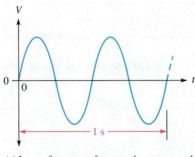
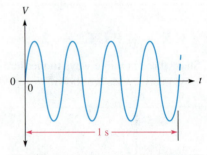

(a) Lower frequency: fewer cycles per second   (b) Higher frequency: more cycles per second

FIGURE 8–8   Illustration of frequency.

Relationship of Frequency and Period

The formulas for the relationship between frequency and period are

$$f = \frac{1}{T} \qquad\qquad (8\text{–}1)$$

$$T = \frac{1}{f} \qquad\qquad (8\text{–}2)$$

There is a reciprocal relationship between f and T. Knowing one, you can calculate the other with the x^{-1} or $1/x$ key on your calculator. (On some calculators, the reciprocal key is a secondary function.) This inverse relationship indicates that a sine wave with a longer period goes through fewer cycles in one second than one with a shorter period.

EXAMPLE 8–3

Which sine wave in Figure 8–9 has the higher frequency? Determine the frequency and the period of both waveforms.

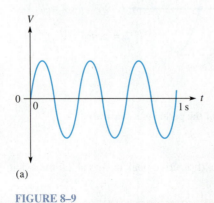
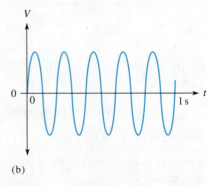

(a)   (b)

FIGURE 8–9

SOLUTION

The sine wave in Figure 8–9(b) has the higher frequency because it completes more cycles in 1 s than does the one in part (a).

In Figure 8–9(a), three cycles are completed in 1 s. Therefore,

$$f = \textbf{3 Hz}$$

One cycle takes 0.333 s (one-third second), so the period is

$$T = 0.333 \text{ s} = \textbf{333 ms}$$

In Figure 8–9(b), five cycles are completed in 1 s. Therefore,

$$f = \textbf{5 Hz}$$

One cycle takes 0.2 s (one-fifth second), so the period is

$$T = 0.2 \text{ s} = \textbf{200 ms}$$

RELATED PROBLEM

If the time interval between consecutive negative peaks of a given sine wave is 50 μs, what is the frequency?

EXAMPLE 8–4

The period of a certain sine wave is 10 ms. What is the frequency?

SOLUTION

Use Equation 8–1.

$$f = \frac{1}{T} = \frac{1}{10 \text{ ms}} = \frac{1}{10 \times 10^{-3} \text{ s}} = \textbf{100 Hz}$$

RELATED PROBLEM

A certain sine wave goes through four cycles in 20 ms. What is the frequency?

EXAMPLE 8–5

The frequency of a sine wave is 60 Hz. What is the period?

SOLUTION

Use Equation 8–2.

$$T = \frac{1}{f} = \frac{1}{60 \text{ Hz}} = \textbf{16.7 ms}$$

RELATED PROBLEM

If $f = 1$ kHz, what is T?

Electronic Signal Generators

The signal generator is an instrument that electronically produces sine waves for use in testing or controlling electronic circuits and systems. There are a variety of signal generators, ranging from special-purpose instruments that produce only one type of waveform in a limited frequency range to programmable instruments that produce a wide range of frequencies

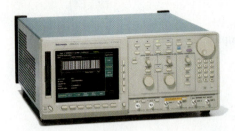

FIGURE 8–10 **A typical arbitrary waveform generator. Copyright © Tektronix, Inc. Reproduced by permission.**

and a variety of waveforms. All signal generators consist basically of an **oscillator**, which is an electronic circuit that produces a repetitive waveform on its output with only the dc supply voltage as an input.

Electronic signal generators are important in testing various types of systems. One type of signal generator is the **function generator**, so called because it can generate a variety of waveforms such as sine, triangle, and square waves. An **arbitrary waveform generator** is another generator that can generate complex signals that simulate different types of inputs. For example, a laboratory that does testing of structures for earthquake safety may need to simulate the complex signal from an earthquake; an arbitrary waveform generator such as shown in Figure 8–10 can easily simulate this. Section 8–10 covers these generators and other types of electronic signal generators in more detail.

SECTION 8–1 CHECKUP*

1. Describe one cycle of a sine wave.
2. At what point does a sine wave change polarity?
3. How many maximum points does a sine wave have during one cycle?
4. How is the period of a sine wave measured?
5. Define *frequency*, and state its unit.
6. Determine f when $T = 5 \, \mu s$.
7. Determine T when $f = 120 \, Hz$.

*Answers are at the end of the chapter.

8–2 VOLTAGE AND CURRENT VALUES OF SINE WAVES

Five ways to express and measure the value of a sine wave in terms of its voltage or its current magnitude are instantaneous, peak, peak-to-peak, rms, and average values.

After completing this section, you should be able to

- **Determine the voltage and current values of sine waves**
  - **Find the instantaneous value at any point**
  - **Find the peak value**
  - **Find the peak-to-peak value**
  - **Define *rms* and find the rms value**
  - **Explain why the average value of an alternating sine wave is always zero over a complete cycle**
  - **Find the half-cycle average value**

Instantaneous Value

Figure 8–11 illustrates that at any point in time on a sine wave, the voltage (or current) has an **instantaneous value**. This instantaneous value is different at different points along the curve. Instantaneous values are positive during the positive alternation and negative during the negative alternation. Instantaneous values of voltage and current are symbolized by lowercase v and i, respectively. The curve in part (a) shows voltage only, but it applies equally for current when v's are replaced with i's. An example of instantaneous values is shown in part (b), where the instantaneous voltage is 3.1 V at 1 μs, 7.07 V at 2.5 μs, 10 V at 5 μs, 0 V at 10 μs, −3.1 V at 11 μs, and so on.

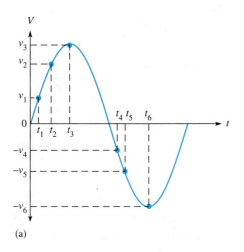

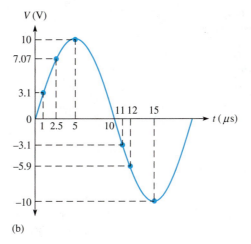

FIGURE 8–11   Example of instantaneous values of a sinusoidal voltage.

Peak Value

The **peak value** of a sine wave is the value of voltage (or current) at the positive or the negative maximum (peaks) with respect to zero. Since positive and negative peak values are equal in **magnitude**, a sine wave is characterized by a single peak value, as is illustrated in Figure 8–12. For a given sine wave, the peak value is constant and is represented by V_p or I_p. The maximum or peak value of a sine wave is also called its **amplitude**. The amplitude is measured from the 0 V line to the peak. In the figure, the peak voltage is 8 V, which is also its amplitude.

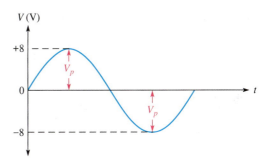

FIGURE 8–12   Peak voltage.

Peak-to-Peak Value

The **peak-to-peak value** of a sine wave, as illustrated in Figure 8–13, is the voltage (or current) from the positive peak to the negative peak. It is always twice the peak value as expressed by the following equations. Peak-to-peak values are represented by V_{pp} or I_{pp}.

$$V_{pp} = 2V_p \qquad (8–3)$$

$$I_{pp} = 2I_p \qquad (8–4)$$

In Figure 8–13, the peak-to-peak voltage is 16 V.

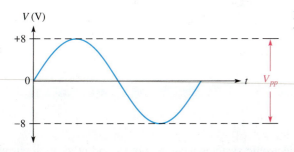

FIGURE 8–13   Peak-to-peak voltage.

RMS Value

The term *rms* stands for *root mean square*. Most ac voltmeters display the rms voltage. The 120 V at your wall outlet is an rms value. The **rms value**, also referred to as the **effective value**, of a sinusoidal voltage is actually a measure of the heating effect of the sine wave. For example, when a resistor is connected across an ac (sinusoidal) voltage source, as shown in Figure 8–14(a), it dissipates a certain amount of heat. Figure 8–14(b) shows the same resistor

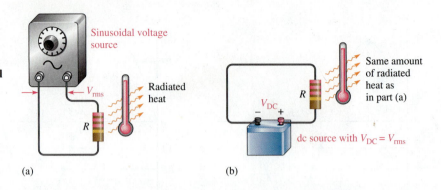

FIGURE 8–14 When the same amount of heat is being produced by the resistor in both setups, the sinusoidal voltage has an rms value equal to the dc voltage.

connected across a dc voltage source. The value of the ac voltage can be adjusted so that the resistor gives off the same amount of heat as it does when connected to the dc source.

The rms value of a sinusoidal voltage is equal to the dc voltage that produces the same amount of heat in a resistance as does the sinusoidal voltage.

You can convert the peak value of a sine wave to the corresponding rms value by using the following relationships for either voltage or current:

$$V_{rms} = 0.707V_p \qquad (8\text{–}5)$$

$$I_{rms} = 0.707I_p \qquad (8\text{–}6)$$

Using these formulas, you can also determine the peak value if you know the rms value.

$$V_p = \frac{V_{rms}}{0.707}$$

$$V_p = 1.414V_{rms} \qquad (8\text{–}7)$$

Similarly,

$$I_p = 1.414I_{rms} \qquad (8\text{–}8)$$

To find the peak-to-peak values, simply double the peak value, which is equivalent to multiplying the rms value by 2.828.

$$V_{pp} = 2.828V_{rms} \qquad (8\text{–}9)$$

and

$$I_{pp} = 2.828I_{rms} \qquad (8\text{–}10)$$

SYSTEM EXAMPLE 8–1

CLAMP METERS

Most industrial electrical and electronic systems have ac motors and other high current loads used for fans, pumps, and conveyors. It is often impractical, time consuming, and a safety hazard to break open these circuits to check current, particularly when a process is ongoing. The traditional clamp meter is a noncontacting meter designed to measure large ac current by sensing the changing magnetic field surrounding the conductor through transformer action (transformers are covered in Chapter 14). The user selects the scale and clamps the jaws or flexible current probe around the conductor. The meter then displays the current based in the selected scale. It is a quick and easy way to check motor current to verify that loading is not excessive or to check for phase balance in three-phase motors.

Originally, clamp meters were designed to only measure ac current; today they also frequently have all of the traditional DMM functions available (resistance, voltage, dc current and sometimes frequency). Meters that can measure direct current (dc) use a Hall-effect sensor. The meter in Figure 8–15 is a multifunction meter that is measuring dc current from a car battery.

Some clamp meters have min/max capability. The min/max function will show the current range. This is useful for reading the currents in a 4-20 mA current loop. (See System Note on page 240.) The min/max capability is also useful for troubleshooting a problem such as determining the maximum current for a circuit breaker that trips intermittently.

FIGURE 8–15   A multifunction clamp meter with detachable jaw.   Courtesy of Fluke Corporation.

Average Value

The average value of a sine wave when taken over one complete cycle is always zero because the positive values (above the zero crossing) offset the negative values (below the zero crossing).

To be useful for comparison purposes and in determining the average value of a rectified voltage such as found in power supplies, the **average value** of a sine wave is defined over a half-cycle rather than over a full cycle. The average value is expressed in terms of the peak value as follows for both voltage and current sine waves:

$$V_{avg} = 0.637V_p \qquad (8\text{–}11)$$

$$I_{avg} = 0.637I_p \qquad (8\text{–}12)$$

The half-cycle average value of a voltage sine wave is illustrated in Figure 8–16.

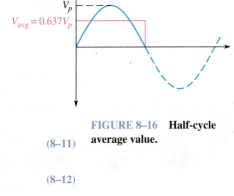

FIGURE 8–16   Half-cycle average value.

EXAMPLE 8–6

Determine V_p, V_{pp}, V_{rms}, and the half-cycle V_{avg} for the sine wave in Figure 8–17.

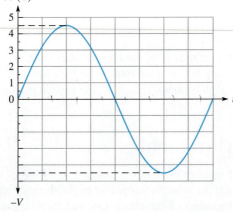

FIGURE 8–17

SOLUTION

As taken directly from the graph, $V_p = $ **4.5 V**. From this value, calculate the other values.

$$V_{pp} = 2V_p = 2(4.5 \text{ V}) = \textbf{9 V}$$

$$V_{rms} = 0.707V_p = 0.707(4.5 \text{ V}) = \textbf{3.18 V}$$

$$V_{avg} = 0.637V_p = 0.637(4.5 \text{ V}) = \textbf{2.87 V}$$

RELATED PROBLEM

If $V_p = 25$ V, determine V_{pp}, V_{rms}, and V_{avg} for a sine wave.

SECTION 8–2 CHECKUP

1. Determine V_{pp} in each case when
     (a) $V_p = 1$ **(b)** $V_{rms} = 1.414$ V   **(c)** $V_{avg} = 3$ V

2. Determine V_{rms} in each case when
     (a) $V_p = 2.5$ V   **(b)** $V_{pp} = 10$ V   **(c)** $V_{avg} = 1.5$ V

3. Determine the half-cycle V_{avg} in each case when
     (a) $V_p = 10$ V   **(b)** $V_{rms} = 2.3$ V   **(c)** $V_{pp} = 60$ V

8–3 ANGULAR MEASUREMENT OF A SINE WAVE

As you have seen, sine waves can be measured along the horizontal axis on a time basis; however, since the time for completion of one full cycle or any portion of a cycle is frequency-dependent, it is often useful to specify points on the sine wave in terms of an angular measurement expressed in degrees or radians. Angular measurement is independent of frequency.

After completing this section, you should be able to

- **Describe angular relationships of sine waves**
  - **Show how to measure a sine wave in terms of angles**
  - **Define** *radian*
  - **Convert radians to degrees**
  - **Determine the phase of a sine wave**

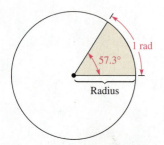

FIGURE 8–18 Angular measurement showing relationship of the radian to degrees.

A sinusoidal voltage can be produced by an alternator, which is an ac generator. There is a direct relationship between the rotation of the rotor in an alternator and the sine wave output. Thus, the angular measurement of the rotor's position is directly related to the angle assigned to the sine wave.

Angular Measurement

A **degree** is an angular measurement corresponding to 1/360 of a circle or a complete revolution. A **radian** (rad) is the angle formed when the distance along the circumference of a circle is equal to the radius of the circle. One radian is equivalent to 57.3°, as illustrated in Figure 8–18. In a 360° revolution, there are 2π radians.

The Greek letter π (pi) represents the ratio of the circumference of any circle to its diameter and has a constant value of approximately 3.1416.

Scientific calculators have a π function so that the actual numerical value does not have to be entered.

Table 8–1 lists several values of degrees and the corresponding radian values. These angular measurements are illustrated in Figure 8–19.

TABLE 8–1	
DEGREES (°)	**RADIANS (RAD)**
0	0
45	$\pi/4$
90	$\pi/2$
135	$3\pi/4$
180	π
225	$5\pi/4$
270	$3\pi/2$
315	$7\pi/4$
360	2π

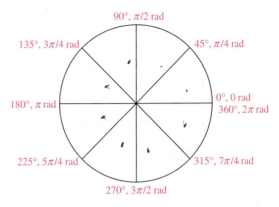

FIGURE 8–19   Angular measurements starting at 0° and going counterclockwise.

Radian/Degree Conversion

Degrees can be converted to radians using Equation 8–13.

$$\textbf{rad} = \left(\frac{\pi \ \textbf{rad}}{180°} \right) \times \textbf{degrees} \qquad (8\text{–}13)$$

Similarly, radians can be converted to degrees with Equation 8–14.

$$\textbf{degrees} = \left(\frac{180°}{\pi \ \textbf{rad}} \right) \times \textbf{rad} \qquad (8\text{–}14)$$

EXAMPLE 8–7

(a) Convert 60° to radians.   **(b)** Convert $\pi/6$ radian to degrees.

SOLUTION

(a) $\text{Rad} = \left(\frac{\pi \ \text{rad}}{180°} \right) 60° = \frac{\pi}{3} \ \textbf{rad}$ **(b)** $\text{Degrees} = \left(\frac{180°}{\pi \ \text{rad}} \right) \left(\frac{\pi}{6} \ \text{rad} \right) = \textbf{30°}$

RELATED PROBLEM

(a) Convert 15° to radians.   **(b)** Convert $5\pi/8$ radians to degrees.

Sine Wave Angles

The angular measurement of a sine wave is based on 360° or 2π rad for a complete cycle. A half-cycle is 180° or π rad; a quarter-cycle is 90° or $\pi/2$ rad; and so on. Figure 8–20(a) shows angles in degrees over a full cycle of a sine wave; part (b) shows the same points in radians.

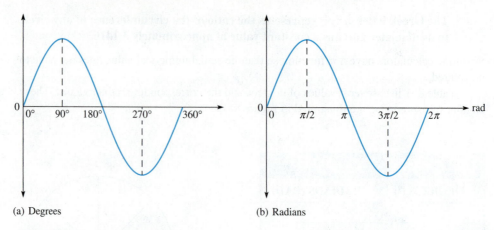

(a) Degrees   (b) Radians

FIGURE 8–20   **Sine wave angles.**

Phase of a Sine Wave

The **phase** of a sine wave is an angular measurement that specifies the position of that sine wave relative to a reference. Figure 8–21 shows one cycle of a sine wave to be used as the reference. Note that the first positive-going crossing of the horizontal axis (zero crossing) is at $0°$ (0 rad), and the positive peak is at $90°$ ($\pi/2$ rad). The negative-going zero crossing is at $180°$ (π rad), and the negative peak is at $270°$ ($3\pi/2$ rad). The cycle is completed at $360°$ (2π rad). When the sine wave is shifted left or right with respect to this reference, there is a phase shift.

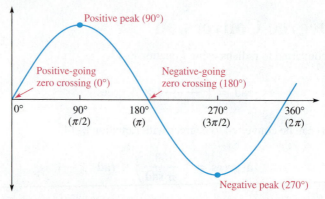

FIGURE 8–21   **Phase reference.**

Figure 8–22 illustrates phase shifts of a sine wave. In part (a), sine wave B is shifted to the right by $90°$ ($\pi/2$ rad) with respect to sine wave A. Thus, there is a phase angle of $90°$ between sine wave A and sine wave B. In terms of time, the positive peak of sine wave B occurs later than the positive peak of sine wave A because time increases to the right

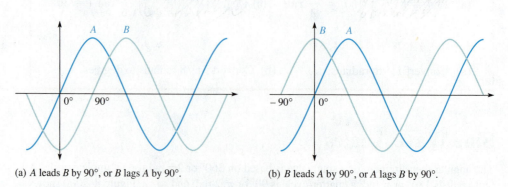

(a) A leads B by $90°$, or B lags A by $90°$.   (b) B leads A by $90°$, or A lags B by $90°$.

FIGURE 8–22   **Illustration of a phase shift.**

along the horizontal axis. In this case, sine wave B is said to **lag** sine wave A by 90° or π/2 radians. In other words, sine wave A leads sine wave B by 90°.

In Figure 8–22(b), sine wave B is shown shifted left by 90° with respect to sine wave A. Thus, again there is a phase angle of 90° between sine wave A and sine wave B. In this case, the positive peak of sine wave B occurs earlier in time than that of sine wave A; therefore, sine wave B is said to **lead** sine wave A by 90°. In both cases there is a 90° phase angle between the two waveforms.

EXAMPLE 8–8

What are the phase angles between sine waves A and B in Figure 8–23(a) and 8–23(b)?

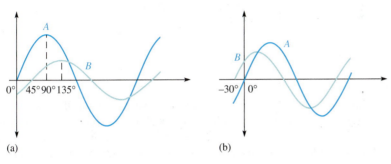

(a)   (b)

FIGURE 8–23

SOLUTION

In Figure 8–23(a), the zero crossing of sine wave A is at 0°, and the corresponding zero crossing of sine wave B is at 45°. There is a **45°** phase angle between the two waveforms with sine wave A leading.

In Figure 8–23(b) the zero crossing of sine wave B is at −30°, and the corresponding zero crossing of sine wave A is at 0°. There is a **30°** phase angle between the two waveforms with sine wave B leading.

RELATED PROBLEM

If the positive-going zero crossing of one sine wave is at 15° relative to the 0° reference and that of the second sine wave is at 23° relative to the 0° reference, what is the phase angle between the sine waves?

When you measure the phase shift between two waveforms on an oscilloscope, you should align the waveforms vertically and make them appear to have the same amplitude. This is done by taking one of the channels out of vertical calibration and adjusting the corresponding waveform until its apparent amplitude equals that of the other waveform. This procedure eliminates the error caused if both waveforms are not measured at their exact center.

Polyphase Power

One important application of phase-shifted sine waves is in electrical power systems. Electrical utilities generate ac with three phases that are separated by 120° as shown in Figure 8–24. The reference is called neutral. Normally, three-phase power is delivered to the user with four lines—three hot lines and neutral. There are important advantages to three-phase power for ac motors. Three-phase motors are more efficient and simpler than an equivalent single-phase motor. Motors are discussed further in Section 8–7.

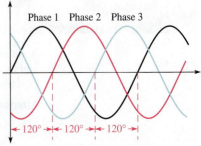

FIGURE 8–24 **Three-phase power waveforms.**

Phase Converters

Many industrial motors and other devices use three-phase motors, which are simpler in construction, more efficient, and more reliable than single-phase motors. Three-phase motors are particularly useful for operations that require frequent starting and stopping or reversal such as used with machine tools.

Because of costs, some smaller industrial plants do not have three-phase power installed. For these plants, phase converters can supply three-phase power from a single-phase source. Several different technologies are available for converting single-phase to three-phase power. The solid-state phase converter shown here uses a digital signal processor (DSP) to monitor the load and generate the required three-phase power. If the load changes, the converter responds to keep constant voltage in all three phases.

Courtesy of Phase Technologies LLC.

SYSTEM NOTE

The three phases can be split up by the utility company to supply three separate single-phase systems. If only one of the three phases plus neutral is supplied, the result is standard 120 V, which is single-phase power. Single-phase power is distributed to residential and small commercial buildings; it consists of two 120 V hot lines that are 180° out of phase with each other and a neutral, which is grounded at the service entrance. The two hot lines allow for connecting to 240 V for high-power appliances (dryers, air conditioners).

SECTION 8–3 CHECKUP

1. When the positive-going zero crossing of a sine wave occurs at 0°, at what angle does each of the following points occur?

   (a) positive peak   **(b)** negative-going zero crossing

   (c) negative peak   **(d)** end of first complete cycle

2. A half cycle is completed in _____ degrees, or _____ radians.

3. A full cycle is completed in _____ degrees, or _____ radians.

4. Determine the phase angle between the sine waves B and C in Figure 8–25.

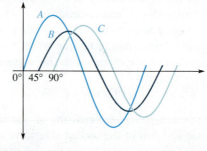

FIGURE 8–25

8–4 THE SINE WAVE FORMULA

A sine wave can be graphically represented by voltage or current values on the vertical axis and by angular measurement (degrees or radians) along the horizontal axis. This graph can be expressed mathematically, as you will see.

After completing this section, you should be able to

- **Mathematically analyze a sinusoidal waveform**
  - **State the sine wave formula**
  - **Find instantaneous values using the sine wave formula**

A generalized graph of one cycle of a sine wave is shown in Figure 8–26. The sine wave amplitude, A, is the maximum value of the voltage or current on the vertical axis; angular values run along the horizontal axis. The variable y is an instantaneous value that can represent either voltage or current at a given angle, θ. The symbol θ is the Greek letter *theta*.

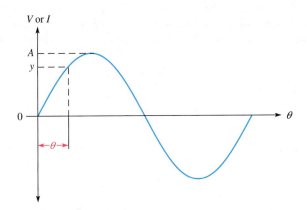

FIGURE 8–26   One cycle of a generic sine wave showing amplitude and phase.

All electrical sine waves follow a specific mathematical formula. The general expression for the sine wave curve in Figure 8–26 is

$$y = A \sin \theta \tag{8–15}$$

This formula states that any point on the sine wave, represented by an instantaneous value y, is equal to the maximum value A times the sine (sin) of the angle θ at that point. For example, a certain voltage sine wave has a peak value of 10 V. The instantaneous voltage at a point 60° along the horizontal axis can be calculated as follows, where $y = v$ and $A = V_p$:

$$v = V_p \sin \theta = (10 \text{ V})\sin 60° = (10 \text{ V})0.866 = 8.66 \text{ V}$$

Figure 8–27 shows this particular instantaneous value on the curve. You can find the sine of any angle on most calculators by first entering the value of the angle and then pressing the SIN key. The calculator must be in the degree mode.

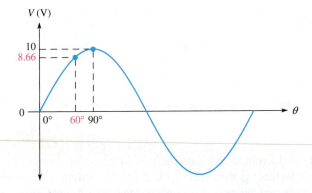

FIGURE 8–27   Illustration of the instantaneous value of a voltage sine wave at $\theta = 60°$.

Derivation of the Sine Wave Formula

As you move along the horizontal axis of a sine wave, the angle increases and the magnitude (height along the y axis) varies. At any given instant, the magnitude of a sine wave can be described by the values of the phase angle and the amplitude (maximum height) and can, therefore, be represented as a **phasor** quantity. A phasor quantity is one that has both magnitude and direction (phase angle). A phasor is represented graphically as an arrow that rotates around a fixed point. The length of a sine wave phasor is the peak value (amplitude), and its angular position as it rotates is the phase angle. One full cycle of a sine wave can be viewed as the rotation of a phasor through 360°.

Figure 8–28 illustrates a phasor rotating counterclockwise through a complete revolution of 360°. If the tip of the phasor is projected over to a graph with the phase angles running along the horizontal axis, a sine wave is "traced out," as shown in the figure. At each angular position of the phasor, there is a corresponding magnitude value. As you can see, at 90° and at 270°, the amplitude of the sine wave is maximum and equal to the length of the phasor. At 0° and at 180°, the sine wave is equal to zero because the phasor lies horizontally at these points.

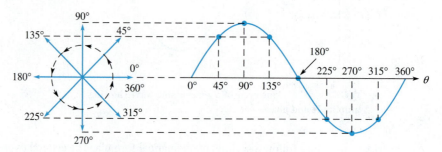

FIGURE 8–28   Sine wave represented by a rotating phasor.

Let's examine a phasor representation at one specific angle. Figure 8–29 shows the voltage phasor at an angular position of 45° and the corresponding point on the sine wave. The instantaneous value, v, of the sine wave at this point is related to both the position (angle) and the length (amplitude) of the phasor. The vertical distance from the phasor tip down to the horizontal axis represents the instantaneous value of the sine wave at that point.

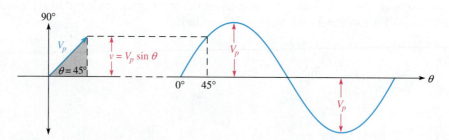

FIGURE 8–29   Right triangle derivation of sine wave formula, $v = V_p \sin \theta$.

Notice that when a vertical line is drawn from the phasor tip down to the horizontal axis, a **right triangle** is formed, as shown in Figure 8–29. The length of the phasor is the **hypotenuse** of the triangle, and the vertical projection is the opposite side. From trigonometry, *the opposite side of a right triangle is equal to the hypotenuse times the sine of the angle θ.* In this case, the length of the phasor is the peak value of the voltage sine wave, V_p. Thus, the opposite side of the triangle, which is the instantaneous value, can be expressed as

$$v = V_p \sin \theta \qquad\qquad (8\text{--}16)$$

This formula also applies to a current sine wave.

$$i = I_p \sin \theta \qquad\qquad (8\text{--}17)$$

Expressions for Phase-Shifted Sine Waves

When a sine wave is shifted to the right of the reference (lags) by a certain angle, ϕ, (Greek letter phi) as illustrated in Figure 8–30(a), the general expression is

$$y = A \sin(\theta - \phi) \qquad \text{(8–18)}$$

where y represents instantaneous voltage or current, and A represents the peak value (amplitude). When a sine wave is shifted to the left of the reference (leads) by a certain angle, ϕ, as shown in Figure 8–30(b), the general expression is

$$y = A \sin(\theta + \phi) \qquad \text{(8–19)}$$

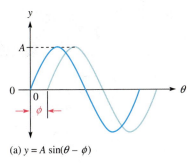
(a) $y = A \sin(\theta - \phi)$

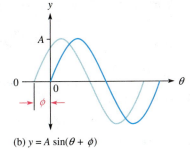
(b) $y = A \sin(\theta + \phi)$

FIGURE 8–30   Shifted sine waves.

EXAMPLE 8–9

Determine the instantaneous value at 90° on the horizontal axis for each voltage sine wave in Figure 8–31.

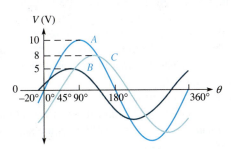

FIGURE 8–31

SOLUTION

Sine wave A is the reference. Sine wave B is shifted left 20° with respect to A, so B leads. Sine wave C is shifted right 45° with respect to A, so C lags.

$$v_A = V_p \sin \theta = (10 \text{ V})\sin 90° = \mathbf{10 \text{ V}}$$
$$v_B = V_p \sin(\theta + \phi_B) = (5 \text{ V})\sin(90° + 20°) = (5 \text{ V})\sin 110° = \mathbf{4.70 \text{ V}}$$
$$v_C = V_p \sin(\theta - \phi_C) = (8 \text{ V})\sin(90° - 45°) = (8 \text{ V})\sin 45° = \mathbf{5.66 \text{ V}}$$

RELATED PROBLEM

A sine wave has a peak value of 20 V. What is the instantaneous value at +65° from the 0° reference?

SECTION 8–4 CHECKUP

1. Determine the sine of the following angles:

    (a) 30°

    (b) 60°

    (c) 90°

2. Calculate the instantaneous value at 120° for the sine wave in Figure 8–27.

3. Determine the instantaneous value at 45° on a voltage sine wave that leads the zero reference by 10° ($V_p = 10$ V).

8–5 ANALYSIS OF AC CIRCUITS

When a time-varying ac voltage such as a sinusoidal voltage is applied to a circuit, the circuit laws and power formulas that you studied earlier still apply. Ohm's law, Kirchhoff's laws and the power formulas apply to ac circuits in the same way that they apply to dc circuits.

After completing this section, you should be able to

- **Apply the basic circuit laws to resistive ac circuits**
  - **Apply Ohm's law to resistive circuits with ac sources**
  - **Apply Kirchhoff's voltage law and current law to resistive circuits with ac sources**
  - **Determine power in resistive ac circuits**
  - **Determine total voltages that have both ac and dc components**

Sine wave generator

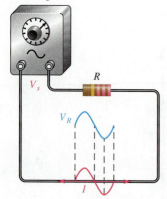

FIGURE 8–32 **A sinusoidal voltage produces a sinusoidal current.**

If a sinusoidal voltage is applied across a resistor, as shown in Figure 8–32, there is a sinusoidal current. The current is zero when the voltage is zero and is maximum when the voltage is maximum. When the voltage changes polarity, the current reverses direction. As a result, the voltage and current are said to be in phase with each other.

When you use Ohm's law in ac circuits, remember that both the voltage and the current must be expressed consistently, that is, both as peak values, both as rms values, both as average values, and so on.

Kirchhoff's voltage and current laws apply to ac circuits as well as to dc circuits. Figure 8–33 illustrates Kirchhoff's voltage law in a resistive circuit that has a sinusoidal voltage source. As you can see, the source voltage is the sum of all the voltage drops across the resistors, just as in a dc circuit.

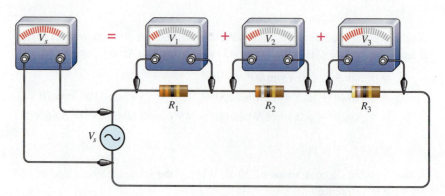

FIGURE 8–33 **Illustration of Kirchhoff's voltage law in an ac circuit.**

Power in resistive ac circuits is determined the same as for dc circuits except that you must use rms values of current and voltage. Recall that the rms value of a sine wave voltage is equivalent to a dc voltage of the same value in terms of its heating effect. The general power formulas are restated for a resistive ac circuit as

$$P = V_{rms}I_{rms}$$

$$P = \frac{V_{rms}^2}{R}$$

$$P = I_{rms}^2 R$$

EXAMPLE 8–10

Determine the rms voltage across each resistor and the rms current in Figure 8–34. The source voltage is given as an rms value. Also, determine the total power.

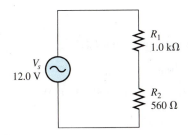

FIGURE 8–34

SOLUTION

The total resistance of the circuit is

$$R_{tot} = R_1 + R_2 = 1.0 \text{ k}\Omega + 560 \text{ }\Omega = 1.56 \text{ k}\Omega$$

Use Ohm's law to find the rms current.

$$I_{rms} = \frac{V_{s(rms)}}{R_{tot}} = \frac{12.0 \text{ V}}{1.56 \text{ k}\Omega} = \textbf{7.69 mA}$$

The rms voltage drop across each resistor is

$$V_{1(rms)} = I_{rms}R_1 = (7.69 \text{ mA})(1.0 \text{ k}\Omega) = \textbf{7.69 V}$$
$$V_{2(rms)} = I_{rms}R_2 = (7.69 \text{ mA})(560 \text{ }\Omega) = \textbf{4.31 V}$$

The total power is

$$P_{tot} = I_{rms}^2 R_{tot} = (7.69 \text{ mA})^2(1.56 \text{ k}\Omega) = \textbf{92.3 mW}$$

RELATED PROBLEM

Repeat this example for a source voltage of 10 V peak.

MULTISIM

Open Multisim file E08-10; files are found at www.pearsonhighered.com/floyd. Measure the rms voltage across each resistor and compare to the calculated values. Change the source voltage to a peak value of 10 V, measure each resistor voltage, and compare to your calculated values.

EXAMPLE 8–11

All values in Figure 8–35 are given in rms.

(a) Find the unknown peak voltage drop in Figure 8–35(a).

(b) Find the total rms current in Figure 8–35(b).

(c) Find the total power in Figure 8–35(b).

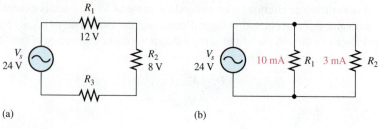

FIGURE 8–35

SOLUTION

(a) Use Kirchhoff's voltage law to find V_3.

$$V_s = V_1 + V_2 + V_3$$
$$V_{3(rms)} = V_{s(rms)} - V_{1(rms)} - V_{2(rms)} = 24\ V - 12\ V - 8\ V = 4\ V$$

Convert rms to peak.

$$V_{3(p)} = 1.414 V_{3(rms)} = 1.414(4\ V) = \mathbf{5.66\ V}$$

(b) Use Kirchhoff's current law to find I_{tot}.

$$I_{tot(rms)} = I_{1(rms)} + I_{2(rms)} = 10\ mA + 3\ mA = \mathbf{13\ mA}$$

(c) $P_{tot} = V_{rms}I_{rms} = (24\ V)(13\ mA) = \mathbf{312\ mW}$

RELATED PROBLEM

A series circuit has the following voltage drops: $V_{1(rms)} = 3.50\ V$, $V_{2(p)} = 4.25\ V$, $V_{3(avg)} = 1.70\ V$. Determine the peak-to-peak source voltage.

Superimposed DC and AC Voltages

In many practical circuits, you will find both dc and ac voltages combined. An example of this is in amplifier circuits where ac signal voltages are superimposed on dc operating voltages. Figure 8–36 shows a dc source and an ac source in series. These two voltages will add algebraically to produce an ac voltage "riding" on a dc level, as measured across the resistor.

If V_{DC} is greater than the peak value of the sinusoidal voltage, the combined ac and dc voltage is a sine wave that never reverses polarity and is therefore nonalternating. That is, the sine wave is riding on a dc level, as shown in Figure 8–37(a). If V_{DC} is less than the peak value of the sine wave, the sine wave will be negative during a portion of its lower half-cycle, as illustrated in Figure 8–37(b), and is therefore alternating. In either case, the sine wave will reach a maximum voltage equal to $V_{DC} + V_p$, and it will reach a minimum voltage equal to $V_{DC} - V_p$.

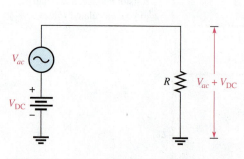

FIGURE 8–36 **Superimposed dc and ac voltages.**

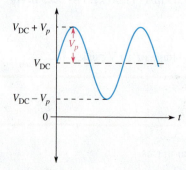

(a) $V_{DC} > V_p$. The sine wave never goes negative.

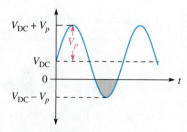

(b) $V_{DC} < V_p$. The sine wave reverses polarity during a portion of its cycle.

FIGURE 8–37 **Sine waves with dc levels.**

EXAMPLE 8–12

Determine the maximum and minimum voltages across the resistor in each circuit of Figure 8–38 and show the resulting waveforms.

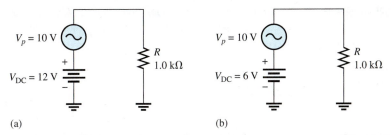

(a)　　　　　　　　　　　(b)

FIGURE 8–38

SOLUTION

In Figure 8–38(a), the maximum voltage across R is

$$V_{max} = V_{DC} + V_p = 12 \text{ V} + 10 \text{ V} = \textbf{22 V}$$

The minimum voltage across R is

$$V_{min} = V_{DC} - V_p = 12 \text{ V} - 10 \text{ V} = \textbf{2 V}$$

Therefore, $V_{R(tot)}$ is a nonalternating sine wave that varies from +22 V to +2 V, as shown in Figure 8–39(a).

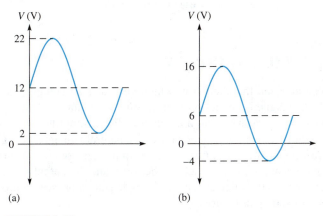

(a)　　　　　　　　　　　(b)

FIGURE 8–39

In Figure 8–38(b), the maximum voltage across R is

$$V_{max} = V_{DC} + V_p = 6 \text{ V} + 10 \text{ V} = \textbf{16 V}$$

The minimum voltage across R is

$$V_{min} = V_{DC} - V_p = \textbf{−4 V}$$

Therefore, $V_{R(tot)}$ is an alternating sine wave that varies from +16 V to −4 V, as shown in Figure 8–39(b).

RELATED PROBLEM

Explain why the waveform in Figure 8–39(a) is nonalternating but the waveform in part (b) is considered to be alternating.

SECTION 8–5 CHECKUP

tage with a half-cycle average value of 12.5 V
rcuit with a resistance of 330 Ω. What is the
the circuit?

age drops in a series resistive circuit are 6.2 V,
d 7.8 V. What is the rms value of the source
voltage?

3. What is the maximum positive value of the resulting total voltage when a sine wave with $V_p = 5$ V is added to a dc voltage of +2.5 V?

4. Will the resulting voltage in Question 3 alternate polarity?

5. If the dc voltage in Question 3 is −2.5 V, what is the maximum positive value of the resulting total voltage?

8–6 ALTERNATORS (AC GENERATORS)

An **alternator** is an ac generator that converts energy of motion into electrical energy. Although it is similar to a dc generator, the alternator is more efficient than the dc generator. Alternators are widely used in vehicles, boats, and other applications even when dc is the final output.

After completing this section, you should be able to

- Describe how an alternator generates electricity
  - Identify the main parts of an alternator, including the rotor, stator, and slip rings
  - Explain why the output of a rotating-field alternator is taken from the stator
  - Describe the purpose of the slip rings
  - Explain how an alternator can be used to produce dc

Simplified Alternator

Both the dc generator and the alternator, which generates ac voltage, are based on the principle of electromagnetic induction that produces a voltage when there is relative motion between a magnetic field and a conductor. For a simplified alternator, a single rotating loop passes permanent magnetic poles. The natural voltage that is generated by a rotating loop is ac. In an alternator, instead of the split rings used in a dc generator, solid rings called slip rings are used to connect to the rotor, and the output is ac. The simplest form of an alternator has the same appearance as a dc generator (see Figure 7–37) except for the slip rings, as shown in Figure 8–40.

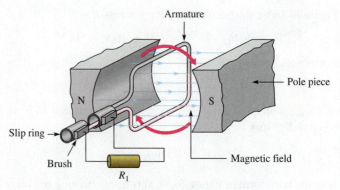

FIGURE 8–40   A simplified alternator.

Frequency

In the simplified alternator in Figure 8–40, each revolution of the loop produces one cycle of a sine wave. The positive and negative peaks occur when the loop cuts the maximum number of flux lines. The rate the loop spins determines the time for one complete cycle and the frequency. If it takes 1/60 of a second to make a revolution, the period of the sine wave is 1/60 of a second and the frequency is 60 Hz. Thus, the faster the loop rotates, the higher the frequency.

Another way of achieving a higher frequency is to use more magnetic poles. When four poles are used instead of two, as shown in Figure 8–41, the conductor passes under a north and a south pole during one-half a revolution, which doubles the frequency. Alternators can have many more poles, depending on the requirements; some have as many as 100. The number of poles and the speed of the rotor determine the frequency in accordance with the following equation:

$$f = \frac{Ns}{120} \hspace{3cm} \text{(8–20)}$$

where f is the frequency in hertz, N is the number of poles, and s is the rotational speed in revolutions per minute.

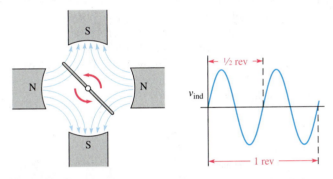

FIGURE 8–41   **Four poles produce twice the frequency of two poles for the same rotational speed.**

EXAMPLE 8–13

Assume a large alternator is turned by a turbine at 300 rpm and has 24 poles. What is the output frequency?

SOLUTION

$$f = \frac{Ns}{120} = \frac{(24)(300 \text{ rpm})}{120} = \textbf{60 Hz}$$

RELATED PROBLEM

At what speed must the rotor move to produce a 50 Hz output?

Practical Alternators

The single loop in our simplified alternator produces only a tiny voltage. In a practical alternator, hundreds of loops are wound on a magnetic core, which forms the rotor. Practical alternators usually have fixed windings surrounding the rotor instead of permanent magnets. Depending on the type of alternator, these fixed windings can either provide the magnetic field (in which case they are called field windings) or act as the fixed conductors that produce the output (in which case they are the armature windings).

ROTATING-ARMATURE ALTERNATORS   In a rotating-armature alternator, the magnetic field is stationary and is supplied by permanent magnets or electromagnets operated from dc. With electromagnets, field windings are used instead of permanent magnets and provide a fixed magnetic field that interacts with the rotor coils. Power is generated in the rotating assembly and supplied to the load through the slip rings.

In a rotating-armature alternator, the rotor is the component from which power is taken. In addition to hundreds of windings, the practical rotating-armature alternator usually has many pole pairs in the stator that alternate as north and south poles, which serve to increase the output frequency.

ROTATING-FIELD ALTERNATOR   The rotating-armature alternator is generally limited to low-power applications because all output current must pass through the slip rings and brushes. To avoid this problem, rotating-field alternators take the output from the stator coils and use a rotating magnet, hence the name. Small alternators may have a permanent magnet for a rotor, but most use an electromagnet formed by a wound rotor. A relatively small amount of dc is supplied to the rotor (through the slip rings) to power the electromagnet. As the rotating magnetic field sweeps by the stator windings, power is generated in the stator. The stator is therefore the armature in this case.

Figure 8–42 shows how a rotating-field alternator can generate three-phase sine waves. (For simplicity, a permanent magnet is shown for the rotor.) AC is generated in each windings as the north pole and the south pole of the rotor alternately sweep by a stator winding. If the north pole generates the positive portion of the sine wave, the south pole will generate the negative portion; thus, one rotation produces a complete sine wave. Each winding has a sine wave output; but because the windings are separated by 120°, the three sine waves are also shifted by 120°. This produces the three-phase output as shown. Most alternators generate three-phase voltage because it is more efficient to produce and is widely used in industry. If the final output is dc, three-phase is easier to convert to dc.

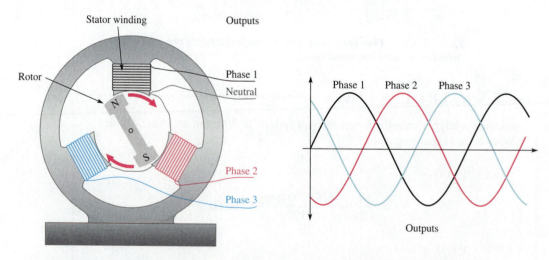

FIGURE 8–42   **The rotor shown is a permanent magnet that produces a strong magnetic field. As it sweeps by each stator winding, a sine wave is produced across that winding. The neutral is the reference.**

Rotor Current

A wound rotor offers important control advantages to alternators. A wound rotor enables control over the strength of the magnetic field by controlling the rotor current and hence the output voltage. For wound rotors, dc must be supplied to the rotor. This current is usually supplied through brushes and slip rings, which are made of a continuous ring of material, (unlike a commutator, which is segmented). Because the brushes need to pass only the magnetizing rotor current, they last longer and are smaller than the brushes in an equiva-lent dc generator, which pass all of the output current.

In wound-rotor alternators, the only current through the brushes and slip rings is the dc that is used to maintain its magnetic field. The dc is usually derived from a small portion of the output current, which is taken from the stator and converted to dc. Large alternators, such as in power stations, may have a separate dc generator, called an **exciter**, to supply current to the field coils. An exciter can respond very fast to changes in output voltage to keep the alternator's output constant, an important consideration in high-power alterna-tors. Some exciters are set up using a stationary field with the armature on the rotating main shaft. The result is a brushless system because the exciter output is on the rotating shaft. A brushless system eliminates the primary maintenance issue with large alternators in cleaning, repairing, and replacing brushes.

An Application

Alternators are used in nearly all modern automobiles, trucks, tractors, and other vehicles. In vehicles, the output is usually three-phase ac taken from stator windings and then con-verted to dc with diodes that are housed inside the alternator case. (Diodes are solid-state devices that allow current in only one direction.) Current to the rotor is controlled by a voltage regulator, which is also internal to the alternator. The voltage regulator keeps the output voltage relatively constant for engine speed changes or changing loads. Alternators have replaced dc generators in automobiles and most other applications because they are more efficient and more reliable.

Important parts of a small alternator, such as you might find in an automobile are shown in Figure 8–43. Like the self-excited generator discussed in Section 7–6, the rotor has a small residual magnetism to begin with, so an ac voltage is generated in the stator as soon as the rotor starts spinning. This ac is converted to dc by a set of rectifier diodes. A portion of the dc is used to provide current to the rotor; the rest is available for the loads. The amount of current required by the rotor is much less than the total current from the alternator, so it can easily provide the required current to the load.

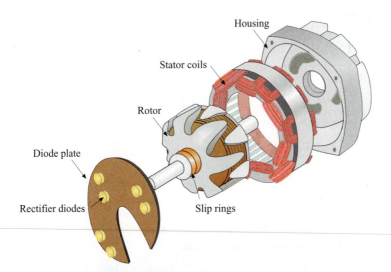

FIGURE 8–43 **Simplified view of a rotor, stator, and diode plate for a small alternator that produces dc.**

In addition to the fact that three-phase voltage is more efficient to produce, it can produce a stable dc output easily by using two diodes in each winding. Since vehicles require dc for the charging system and loads, the output of the alternator is converted to dc internally using a rectifier diode array mounted on a diode plate. Thus, a standard three-phase automotive alternator will normally have six diodes inside to convert the output to dc. (Some alternators have six independent stator coils and twelve diodes.)

SYSTEM EXAMPLE 8–2

AUTOMOTIVE CHARGING SYSTEM

The schematic for a basic automotive charging system is shown in Figure 8–44. The alternator is at the heart of the system. The alternator provides ac which is rectified (changed to dc) by the diode array. The dc is sent to the battery through a heavy wire.

There are normally four terminals on an alternator that are connected to the rest of the charging system. The terminals are labeled B, L, IG, and S. Terminal B is connected to the battery and carries the charging current. Terminal L is connected to the charge warning lamp. The charge lamp is located on the dash panel and is illuminated when there is a difference in voltage across it; if the lamp has equal positive voltages on both sides, it is off. Terminal IG is connected to the run side of the ignition switch, which activates the regulator. If voltage is removed from either side of the lamp, it illuminates to indicate a problem with the charging system. Terminal S is a "sense" connection, which is connected to the battery to provide a battery voltage monitor for the voltage regulator.

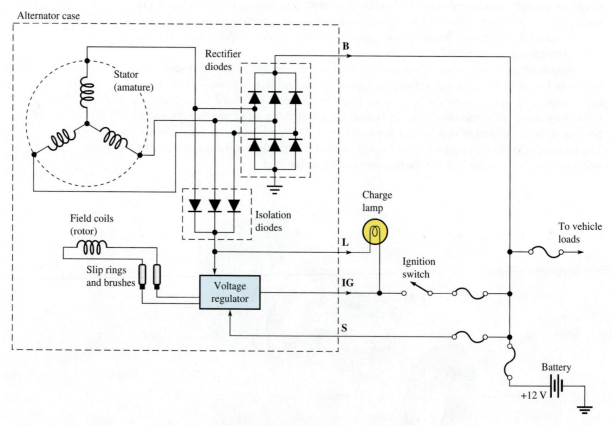

FIGURE 8–44   A basic automotive charging system.

SECTION 8–6 CHECKUP

1. What two factors affect the frequency of an alternator?

2. What is the advantage of taking the output from the stator in a rotating-field alternator?

3. What is an exciter?

4. What is the purpose of the diodes in an automobile alternator?

Motors are electromagnetic devices that represent the most common loads for ac in power applications. AC motors are used to operate household appliances such as heat pumps, refrigerators, washers, dryers, and vacuums. In industry, ac motors are used in many applications to move and process materials as well as refrigeration and heating units, machining operations, pumps, and much more. In this section, the two major types of ac motors, induction motors and synchronous motors, are introduced.

After completing this section, you should be able to

- • Explain how ac motors convert electrical energy into rotational motion
- • Cite the main differences between induction and synchronous motors
- • Explain how the magnetic field rotates in an ac motor
- • Explain how an induction motor develops torque

AC Motor Classification

The two major classifications of ac motors are induction motors and synchronous motors. Several considerations determine which of these types are best for any given application. These considerations include the speed and power requirements, voltage rating, load characteristics (such as starting torque required), efficiency requirements, maintenance requirements, and operating environment (such as underwater operation or temperature).

An **induction motor** is so named because a magnetic field *induces* current in the rotor, creating a magnetic field that interacts with the stator field. Normally, there is no electrical connection to the rotor[1], so there is no need for slip rings or brushes, which tend to wear out. The rotor current is caused by electromagnetic induction, which also occurs in transformers (covered in Chapter 14), so induction motors are said to work by transformer action.

In a **synchronous motor**, the rotor moves in sync (at the same rate) as the rotating field of the stator. Synchronous motors are used in applications where maintaining constant speed is important. Synchronous motors are not self-starting and must receive starting torque from an external source or from built-in starting windings. Like alternators, synchronous motors use slip rings and brushes to provide current to the rotor.

Rotating Stator Field

Both synchronous and induction ac motors have a similar arrangement for the stator windings, which allow the magnetic field of the stator to rotate. The rotating stator field is equivalent to moving a magnet in a circle except that the rotating field is produced electrically, with no moving parts.

How can the magnetic field in the stator rotate if the stator itself does not move? The rotating field is created by the changing ac itself. Let's look at a rotating field with a three-phase stator, as shown in Figure 8–45. Notice that one of the three phases "dominates" at different times. When phase 1 is at 90°, the current in the phase 1 winding is at a maximum and current in the other windings is smaller. Therefore, the stator magnetic field will be oriented toward the phase-1 stator winding. As the phase-1 current declines, the phase-2 current increases, and the field rotates toward the phase-2 winding. The magnetic field will be oriented toward the phase-2 winding when current in it is a maximum. As the phase-2 current declines, the phase-3 current increases, and the field rotates toward the phase-3 winding. The process repeats as the field returns to the phase-1 winding. Thus, the field rotates at a rate determined by the frequency of the applied voltage. With a more detailed analysis, it can be shown that the magnitude of the field is unchanged; only the direction of the field changes.

HANDS ON TIP
New motor technology for high power motors will use high temperature superconductors (HTS). The new technology significantly reduces the size and weight of large motors by hundreds of tons. One application is for ships' propulsion motors, which will reduce fuel consumption and free up space on board. A 49,000 HP HTS motor has passed initial tests for the Navy and will be installed in the near future on Navy ships.

---

[1]An exception is the wound-rotor motor, which is a type of large induction motor that generally has limited application.

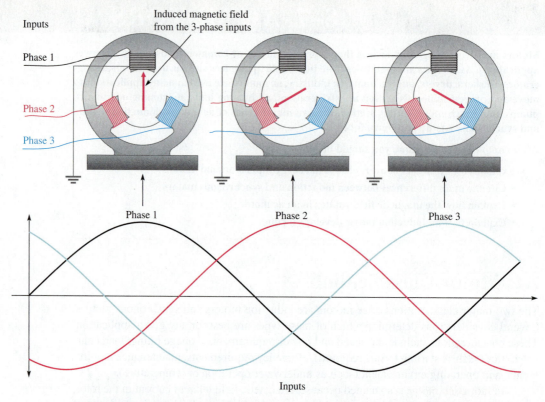

FIGURE 8–45   **The application of three phases to the stator produces a net magnetic field as shown by the red arrow. The rotor (not shown) moves in response to this field.**

As the stator field moves, the rotor moves in sync with it in a synchronous motor but lags behind in an induction motor. The rate the stator field moves is called the *synchronous speed* of the motor.

Induction Motors

The theory of operation is essentially the same for both single-phase and three-phase induction motors. Both types use the rotating field described previously, but the single-phase motor requires starting windings or another method to produce torque for starting the motor, whereas the three-phase motor is self-starting. When starting windings are employed in a single-phase motor, they are removed from the circuit by a mechanical centrifugal switch as the motor speeds up.

The core of the induction motor's rotor consists of an aluminum frame that forms the conductors for the circulating current in the rotor. (Some larger induction motors use copper bars.) The aluminum frame is similar in appearance to the exercise wheel for pet squirrels (common in the early 20th century), so it is aptly called a **squirrel cage**, illustrated in Figure 8–46. The aluminum squirrel cage itself is the *electrical* path; it is embedded within

FIGURE 8–46   **Diagram of a squirrel-cage rotor.**

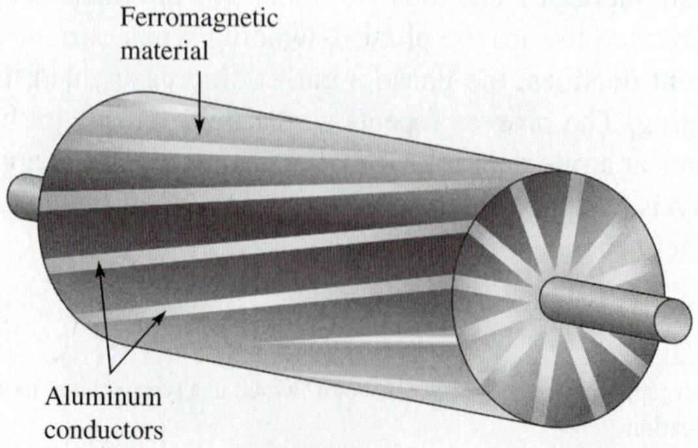

Ferromagnetic material

Aluminum conductors

a ferromagnetic material to provide a low reluctance *magnetic* path through the rotor. In addition, the rotor has cooling fins that may be molded into the same piece of aluminum as the squirrel cage. The entire assembly must be balanced so that it spins easily and without vibrating.

OPERATION OF AN INDUCTION MOTOR   When the magnetic field from the stator moves across the squirrel cage of the inductor, a current is generated in the squirrel cage. This current creates a magnetic field that reacts with the moving field of the stator, causing the rotor to start turning. The rotor will try to "catch-up" with the moving field, but cannot, in a condition known as slip. **Slip** is defined as the difference between the synchronous speed of the stator and the rotor speed. The rotor can never reach the synchronous speed of the stator field because, if it did, it would not cut any field lines and the torque would drop to zero. Without torque, the rotor could not turn itself.

Initially, before the rotor starts moving, there is no back emf, so the stator current is high. As the rotor speeds up, it generates a back emf that opposes the stator current. As the motor speeds up, the torque produced balances the load and the current is just enough to keep the rotor turning. The running current is significantly lower than the initial start-up current because of the back emf. If the load on the motor is then increased, the motor will slow down and generate less back emf. This increases the current to the motor and increases the torque it can apply to the load. Thus, an induction motor can operate over a range of speeds and torque. Maximum torque occurs when the rotor is spinning at about 75% of the synchronous speed.

Synchronous Motors

Recall that an induction motor develops no torque if it runs at the synchronous speed, so it must run slower than the synchronous speed, depending on the load. The synchronous motor will run at the synchronous speed and still develop the required torque for different loads. The only way to change the speed of a synchronous motor is to change the frequency.

The fact that synchronous motors maintain a constant speed for all load conditions is a major advantage in certain industrial operations and in applications where clock or timing requirements are involved (such as a telescope drive motor or a chart recorder). In fact, the first application of synchronous motors was in electric clocks (in 1917).

Another important advantage to large synchronous motors is their efficiency. Although their original cost is higher than a comparable induction motor, the savings in power will often pay for the cost difference in a few years.

OPERATION OF A SYNCHRONOUS MOTOR   Essentially, the rotating stator field of the synchronous motor is identical to that of an induction motor. The primary difference in the two motors is in the rotor. The induction motor has a rotor that is electrically isolated from a supply, and the synchronous motor uses a magnet to follow the rotating stator field. Small synchronous motors use a permanent magnet for the rotor; larger motors use an electromagnet. When an electromagnet is used, dc is supplied from an external source via slip rings as in case of the alternator.

SECTION 8–7 CHECKUP

1. What is the main difference between an induction motor and a synchronous motor?

2. What happens to the magnitude of the rotating stator field as it moves?

3. What is the purpose of a squirrel cage?

4. With reference to motors, what does the term *slip* mean?

Sine waves are important in electronics, but they are not the only type of ac or time-varying waveform. Two other major types of waveforms are the pulse waveform and the triangular waveform.

After completing this section, you should be able to

- **Identify the characteristics of basic nonsinusoidal waveforms**
- **Discuss the properties of a pulse waveform**
- **Define** *duty cycle*
- **Discuss the properties of triangular and sawtooth waveforms**
- **Discuss the harmonic content of a waveform**

Pulse Waveforms

Basically, a **pulse** can be described as a very rapid transition (**leading edge**) from one voltage or current level (**baseline**) to another level; and then, after an interval of time, a very rapid transition (**trailing edge**) back to the original baseline level. The transitions in level are called *steps*. An ideal pulse consists of two opposite-going steps of equal amplitude. When the leading or trailing edge is positive-going, it is called a **rising edge**. When the leading or trailing edge is negative-going, it is called a **falling edge**.

Figure 8–47(a) shows an ideal positive-going pulse consisting of two equal but opposite instantaneous steps separated by an interval of time called the **pulse width**. Figure 8–47(b) shows an ideal negative-going pulse. The height of the pulse measured from the baseline is its voltage (or current) amplitude. Commonly, analysis is simplified by treating all pulses as ideal (composed of instantaneous steps and perfectly rectangular in shape).

FIGURE 8–47   Ideal pulses.

(a) Positive-going pulse

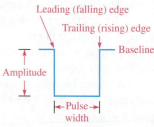

(b) Negative-going pulse

Actual pulses, however, are never ideal. Pulses cannot change from one level to another instantaneously. Time is always required for a transition (step), as illustrated in Figure 8–48(a). As you can see, there is an interval of time during which the pulse is rising from its lower value to its higher value. This interval is called the *rise time, t_r*.

Rise time is the time required for the pulse to go from 10% of its amplitude to 90% of its amplitude.

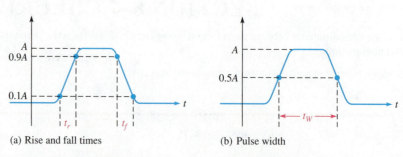

(a) Rise and fall times

(b) Pulse width

FIGURE 8–48   Nonideal pulse.

The interval of time during which the pulse is falling from its higher value to its lower value is called the *fall time*, t_f.

> **Fall time** is the time required for the pulse to go from 90% of its amplitude to 10% of its amplitude.

Pulse width (t_W) also requires a precise definition for the nonideal pulse because the leading and trailing edges are not vertical.

> **Pulse width** is the time between the point on the leading edge where the value is 50% of the amplitude and the point on the trailing edge where the value is 50% of the amplitude.

Pulse width is shown in Figure 8–48(b).

REPETITIVE PULSES   Any waveform that repeats itself at fixed intervals is **periodic**. Figure 8–49 shows some examples of periodic pulse waveforms. Notice that in each case, the pulses repeat at regular intervals. The rate at which the pulses repeat is the **pulse repetition frequency**, which is the fundamental frequency of the waveform. The frequency can be expressed in hertz or in pulses per second. The time from one pulse to the corresponding point on the next pulse is the period (T). The relationship between frequency and period is the same as with the sine wave, $f = 1/T$.

An important characteristic of periodic pulse waveforms is the duty cycle.

> The **duty cycle** is the ratio of the pulse width (t_W) to the period (T) and is usually expressed as a percentage.

$$\text{Percent duty cycle} = \left(\frac{t_W}{T}\right)100\% \tag{8–21}$$

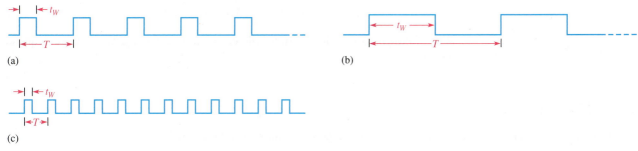

(a)

(b)

(c)

FIGURE 8–49   Repetitive pulse waveforms.

EXAMPLE 8–14

Determine the period, frequency, and duty cycle for the pulse waveform in Figure 8–50.

FIGURE 8–50

SOLUTION

As indicated in Figure 8–50, the period is

$$T = 10 \ \mu s$$

Use Equations 8–1 and 8–21 to determine the frequency and duty cycle.

$$f = \frac{1}{T} = \frac{1}{10\ \mu s} = \textbf{100 kHz}$$

$$\text{Percent duty cycle} = \left(\frac{t_W}{T}\right)100\% = \left(\frac{1\ \mu s}{10\ \mu s}\right)100\% = \textbf{10\%}$$

RELATED PROBLEM

A certain pulse waveform has a frequency of 200 kHz and a pulse width of 0.25 μs. Determine the duty cycle expressed as a percentage.

SQUARE WAVES   A square wave is a pulse waveform with a duty cycle of 50%. Thus, the pulse width is equal to one-half of the period. A square wave is shown in Figure 8–51.

FIGURE 8–51   Square wave.

THE AVERAGE VALUE OF A PULSE WAVEFORM   The **average value** of a pulse waveform is equal to its baseline value plus the product of its duty cycle and its amplitude. The lower level of a positive-going waveform or the upper level of a negative-going waveform is taken as the baseline. The formula for average voltage is as follows:

$$V_{\text{avg}} = \textbf{baseline} + \textbf{(duty cycle)(amplitude)} \tag{8–22}$$

The following example illustrates the calculation of the average voltage of pulse waveforms.

EXAMPLE 8–15

Determine the average voltage of each of the positive-going waveforms in Figure 8–52.

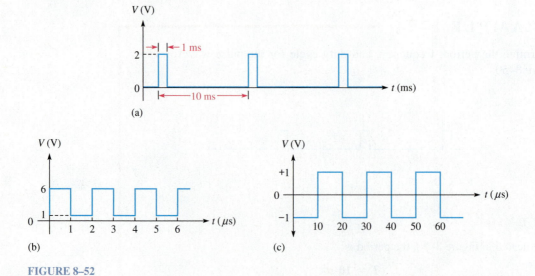

FIGURE 8–52

SOLUTION

In Figure 8–52(a), the baseline is at 0 V, the amplitude is 2 V, and the duty cycle is 10%. The average voltage is

$$V_{avg} = \text{baseline} + (\text{duty cycle})(\text{amplitude})$$
$$= 0\ V + (0.1)(2\ V) = \mathbf{0.2\ V}$$

The waveform in Figure 8–52(b) has a baseline of $+1$ V, an amplitude of 5 V, and a duty cycle of 50%. The average voltage is

$$V_{avg} = \text{baseline} + (\text{duty cycle})(\text{amplitude})$$
$$= 1\ V + (0.5)(5\ V) = 1\ V + 2.5\ V = \mathbf{3.5\ V}$$

The waveform in Figure 8–52(c) is a square wave with a baseline of -1 V and an amplitude of 2 V. The duty cycle is 50%. The average voltage is

$$V_{avg} = \text{baseline} + (\text{duty cycle})(\text{amplitude})$$
$$= -1\ V + (0.5)(2\ V) = -1\ V + 1\ V = \mathbf{0\ V}$$

This is an alternating square wave, and, like an alternating sine wave, it has an average voltage of zero over a full cycle.

RELATED PROBLEM

If the baseline of the waveform in Figure 8–52(a) is shifted to $+1$ V, what is the average voltage?

Triangular and Sawtooth Waveforms

Triangular and sawtooth waveforms are formed by voltage or current ramps. A **ramp** is a linear increase or decrease in the voltage or current. Figure 8–53 shows both positive- and negative-going ramps. In part (a), the ramp has a positive slope; in part (b), the ramp has a negative slope. The slope of a voltage ramp is $\pm V/t$ and the slope of a current ramp is $\pm I/t$.

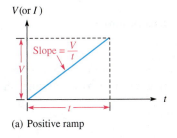
(a) Positive ramp

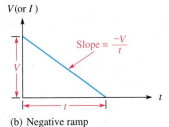
(b) Negative ramp

FIGURE 8–53   **Voltage ramps.**

EXAMPLE 8–16

What are the slopes of the voltage ramps in Figure 8–54?

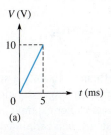
(a)

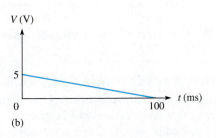
(b)

FIGURE 8–54

SOLUTION

In Figure 8–54(a), the voltage increases from 0 V to +10 V in 5 ms. Thus, $V = 10$ V and $t = 5$ ms. The slope is

$$\frac{V}{t} = \frac{10\text{ V}}{5\text{ ms}} = \textbf{2 V/ms}$$

In Figure 8–54(b), the voltage decreases from +5 V to 0 V in 100 ms. Thus, $V = -5$ V and $t = 100$ ms. The slope is

$$\frac{V}{t} = \frac{-5\text{ V}}{100\text{ ms}} = \textbf{-0.05 V/ms}$$

RELATED PROBLEM

A certain voltage ramp has a slope of $+12$ V/μs. If the ramp starts at zero, what is the voltage at 0.01 ms?

TRIANGULAR WAVEFORMS Figure 8–55 shows that a **triangular waveform** is composed of positive-going and negative-going ramps having equal slopes. The period of this waveform can be measured from one peak to the next corresponding peak, as illustrated. This particular triangular waveform is alternating and has an average value of zero.

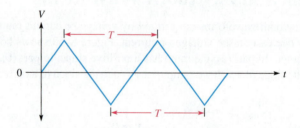

FIGURE 8–55 Alternating triangular waveform.

Figure 8–56 depicts a triangular waveform with a nonzero average value. The frequency for triangular waves is determined in the same way as for sine waves, that is, $f = 1/T$.

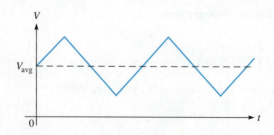

FIGURE 8–56 Nonalternating triangular waveform.

SAWTOOTH WAVEFORMS The **sawtooth waveform** is actually a special case of the triangular waveform consisting of two ramps, one of much longer duration than the other. Sawtooth waveforms are used in many electronic systems. For example, a sawtooth waveform is used in automatic test equipment, control systems, and certain types of displays, including analog oscilloscopes.

Figure 8–57 is an example of a sawtooth waveform. Notice that it consists of a positive-going ramp of relatively long duration, followed by a negative-going ramp of relatively short duration.

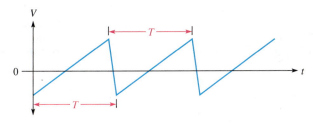

FIGURE 8–57   **Alternating sawtooth waveform.**

Harmonics

A repetitive nonsinusoidal waveform contains sinusoidal waveforms with a fundamental frequency and harmonic frequencies. The **fundamental frequency** is the repetition rate of the waveform, and the **harmonics** are higher-frequency sine waves that are multiples of the fundamental.

ODD HARMONICS   *Odd harmonics* are frequencies that are odd multiples of the fundamental frequency of a waveform. For example, a 1 kHz square wave consists of a fundamental of 1 kHz and odd harmonics of 3 kHz, 5 kHz, 7 kHz, and so on. The 3 kHz frequency in this case is the third harmonic; the 5 kHz frequency is the fifth harmonic; and so on.

EVEN HARMONICS   *Even harmonics* are frequencies that are even multiples of the fundamental frequency. For example, if a certain wave has a fundamental of 200 Hz, the second harmonic is 400 Hz, the fourth harmonic is 800 Hz, the sixth harmonic is 1200 Hz, and so on. These are even harmonics.

COMPOSITE WAVEFORM   Any variation from a pure sine wave produces harmonics. A nonsinusoidal wave is a composite of the fundamental and the harmonics. Some types of waveforms have only odd harmonics, some have only even harmonics, and some contain both. The shape of the wave is determined by its harmonic content. Generally, only the fundamental frequency and the first few harmonics are of significant importance in determining the wave shape.

A square wave is an example of a waveform that consists of a fundamental frequency and only odd harmonics. When the instantaneous values of the fundamental and each odd harmonic are added algebraically at each point, the resulting curve will have the shape of a square wave, as illustrated in Figure 8–58. In part (a), the fundamental and the third harmonic

HANDS ON TIP
The frequency response of an oscilloscope limits the accuracy with which waveforms can be accurately displayed. To view pulse waveforms, the frequency response must be high enough for all significant harmonics of the waveform. For example, a 100 MHz oscilloscope distorts a 100 MHz pulse waveform because the third, fifth, and higher harmonics are greatly attenuated.

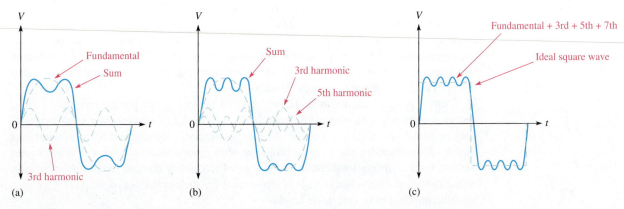

FIGURE 8–58   **Odd harmonics combine to produce a square wave. The plot is for a continuous pulse that theoretically extends from −∞ to +∞.**

produce a wave shape that begins to resemble a square wave. In part (b), the fundamental, third, and fifth harmonics produce a closer resemblance. When the seventh harmonic is included, as in part (c), the resulting wave shape becomes even more like a square wave. As more harmonics are included, a square wave is approached.

SYSTEM EXAMPLE 8–3

ANALYZING SIGNALS IN THE FREQUENCY DOMAIN

As you have learned, signals can be thought of as being composed of various sine waves. All waves can be broken into sine waves that are related to each other. A plot of the amplitude of the sine waves as a function of their frequency is called a frequency domain plot. (The domain refers to the independent variable.) The same wave can be plotted with time as the independent variable, which is referred to as a time domain plot. Figure 8–59 shows the time and frequency domain views of a periodic signal. Notice that the time domain view shows a summation of all of the frequencies in the signal and the frequency domain view breaks the signal into its constituent frequencies.

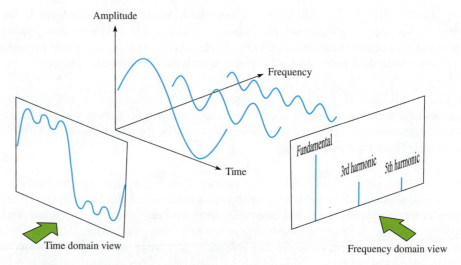

FIGURE 8–59 Time and frequency domain views of a signal.

The time and frequency domains are the two principal domains for describing electronic signals. The relationship between the time and frequency domains was first described mathematically by Joseph Fourier in 1807 in describing heat flow, so the conversion between domains is called Fourier analysis.

The spectrum analyzer is an instrument that is widely used in radio-frequency (rf) and microwave systems. A spectrum analyzer breaks the signal of interest into its sinusoidal components and displays these components to the user. The oscilloscope, which is introduced in Section 8–9, is a time domain instrument (although some specialized oscilloscopes can display the frequency domain).

Many systems use radio-frequency devices including cell phones, wireless LANs, certain remote controls, medical devices, and various tracking devices (for tracking packages or even livestock!). As a result of these applications, the demand for bandwidth has increased significantly and so have problems with interference between various systems. Testing for interfering frequencies is the type of frequency domain measurement ideally suited to the spectrum analyzer. It is also useful for measurements of frequency, power, modulation, distortion, and noise levels.

SECTION 8–8 CHECKUP

1. Define the following parameters:

   (a) rise time   (b) fall time   (c) pulse width

2. In a certain repetitive positive-going pulse waveform, the pulses are 200 μs wide and occur once every millisecond. What is the frequency of this waveform?

3. Determine the duty cycle, amplitude, and average value of the waveform in Figure 8–60(a).

4. What is the period of the triangular wave in Figure 8–60(b)?

5. What is the frequency of the sawtooth wave in Figure 8–60(c)?

6. Define *fundamental frequency.*

7. What is the second harmonic of a fundamental frequency of 1 kHz?

8. What is the fundamental frequency of a square wave having a period of 10 μs?

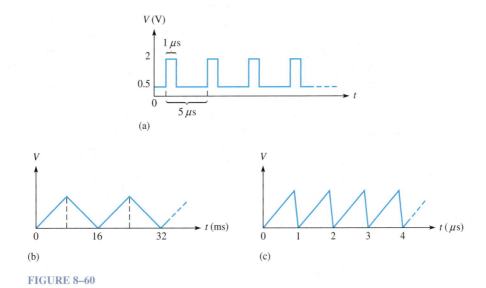

(a)

(b)

(c)

FIGURE 8–60

8–9 THE OSCILLOSCOPE

The oscilloscope, or (scope for short) is a widely used and versatile test instrument for observing and measuring waveforms.

After completing this section, you should be able to

- **Use an oscilloscope to measure waveforms**
  - **Recognize common oscilloscope controls**
- **Measure the amplitude of a waveform**
- **Measure the period and frequency of a waveform**

The oscilloscope is basically a graph-displaying device that traces a graph of a measured electrical signal on its screen. In most applications, the graph shows how signals change over time. The vertical axis of the display screen represents voltage and the horizontal axis represents time. You can measure amplitude, period, and frequency of a signal using an oscilloscope. Also, you can determine the pulse width, duty cycle, rise time, and fall time of a pulse waveform. Most scopes can display at least two signals on the screen at one time, enabling you to observe their time relationship. Two different digital oscilloscopes are shown in Figure 8–61.

FIGURE 8–61   **Digital
oscilloscopes.**   Copyright ©
Tektronix. Reproduced by permission.

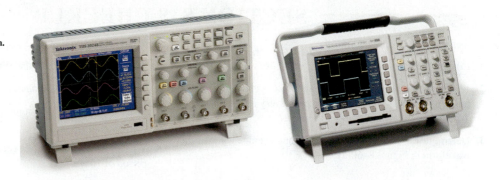

FIGURE 8–61   **Digital
oscilloscopes.**   Copyright ©
Tektronix. Reproduced by permission.

Two basic types of oscilloscopes, analog and digital, can be used to view digital waveforms. The older analog scope works by applying the measured waveform directly to control the up and down motion of the electron beam in the cathode-ray tube (CRT) as it sweeps across the screen. As a result, the beam traces out the waveform pattern on the screen. The digital scope converts the measured waveform to digital information by a sampling process in an analog-to-digital converter (ADC). The digital information is then used to reconstruct the waveform on the screen.

The digital scope is more widely used than the analog scope. However, either type can be used in many applications; each has characteristics that make it more suitable for certain situations. An analog scope displays waveforms as they occur in "real time." Digital scopes are useful for measuring transient pulses that may occur randomly or only once. Also, because information about the measured waveform can be stored in a digital scope, it may be viewed at some later time, printed out, or thoroughly analyzed by a computer or other means.

Basic Operation of Analog Oscilloscopes

To measure a voltage, a probe must be connected from the scope to the point in a circuit at which the voltage is present. Generally, a ×10 probe is used that reduces (attenuates) the signal amplitude by ten. The signal goes through the probe into the vertical circuits where it is either further attenuated (reduced) or amplified depending on the actual amplitude and on where you set the vertical control of the scope. The vertical circuits then drive the vertical deflection plates of the CRT. Also, the signal goes to the trigger circuits that trigger the horizontal circuits to initiate repetitive horizontal sweeps of the electron beam across the screen using a sawtooth waveform. There are many sweeps per second so that the beam appears to form a solid line across the screen in the shape of the waveform. This basic operation is illustrated in Figure 8–62.

FIGURE 8–62   **Block diagram
of an analog oscilloscope.**

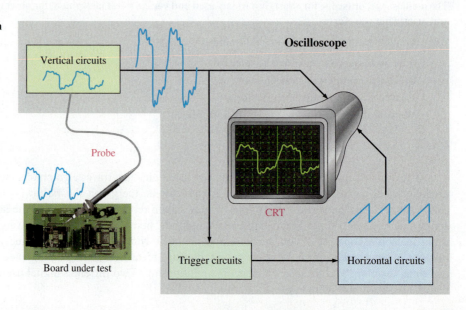

Basic Operation of Digital Oscilloscopes

Some parts of a digital scope are similar to the analog scope. However, the digital scope is more complex than an analog scope and typically has an LCD screen rather than a CRT. Rather than displaying a waveform as it occurs, the digital scope first acquires the measured analog waveform and converts it to a digital format using an analog-to-digital converter (ADC). The digital data is stored and processed. The data then goes to the reconstruction and display circuits for display in its original analog form. Figure 8–63 shows a block diagram for a digital oscilloscope.

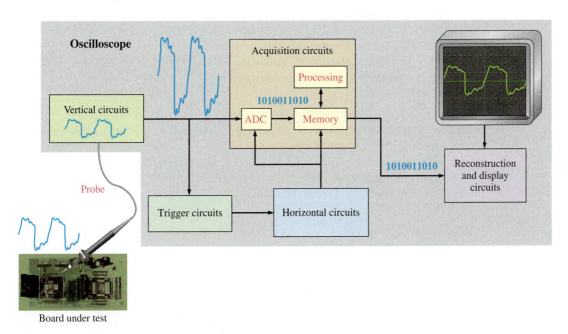

FIGURE 8–63   **Block diagram of a digital oscilloscope.**

Oscilloscope Controls

A front panel view of a typical dual-channel oscilloscope is shown in Figure 8–64. Instruments vary depending on model and manufacturer, but most have certain common features.

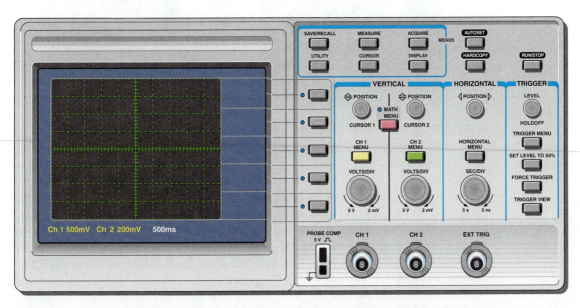

FIGURE 8–64   **A typical dual-channel oscilloscope. Numbers below screen indicate the values for each division on the vertical (voltage) and horizontal (time) scales and can be varied using the vertical and horizontal controls on the scope.**

For example, the two vertical sections contain a Position control, a channel menu button, and a Volts/Div control. The horizontal section contains a Sec/Div control.

Some of the main controls are now discussed. Refer to the user manual for complete details of your particular scope.

VERTICAL CONTROLS   In the Vertical section of the scope in Figure 8–64, there are identical controls for each of the two channels (CH1 and CH2). The position control lets you move a displayed waveform up or down vertically on the screen. The Menu button provides for the selection of several items that appear on the screen, such as the coupling modes (ac, dc, or ground) and coarse or fine adjustment for the Volts/Div. The Volts/Div control adjusts the number of volts represented by each vertical division on the screen. The Volts/Div setting for each channel is displayed on the bottom of the screen. The Math Menu button provides a selection of operations that can be performed on the input waveforms, such as subtraction and addition of signals.

HORIZONTAL CONTROLS   In the Horizontal section, the controls apply to both channels. The Position control lets you move a displayed waveform left to right horizontally on the screen. The Horizontal Menu button provides for the selection of several items that appear on the screen such as the main time base, expanded view of a portion of a waveform, and other parameters. The Sec/Div control adjusts the time represented by each horizontal division or main time base. The Sec/Div setting is displayed at the bottom of the screen.

TRIGGER CONTROLS   In the Trigger section, the Level control determines the point on the triggering waveform where triggering occurs to initiate the sweep to display input waveforms. The Trigger Menu button provides for the selection of several items that appear on the screen including edge or slope triggering, trigger source, trigger mode, and other parameters. There is also an input for an external trigger signal.

Triggering stabilizes a waveform on the screen and properly triggers on a pulse that occurs only one time or randomly. Also it allows you to observe time delays between two waveforms. Figure 8–65 compares a triggered to an untriggered signal. The untriggered signal tends to drift across the screen producing what appears to be multiple waveforms.

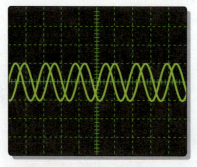
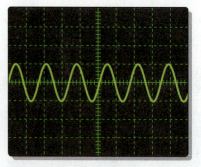

(a) Untriggered waveform display   (b) Triggered waveform display

FIGURE 8–65   **Comparison of an untriggered and a triggered waveform on an oscilloscope.**

COUPLING A SIGNAL INTO THE SCOPE   Coupling is the method used to connect a signal voltage to be measured into the oscilloscope. The DC and AC coupling modes are selected from the Vertical menu. DC coupling allows a waveform including its dc component to be displayed. AC coupling blocks the dc component of a signal so that you see the waveform centered at 0 V. The Ground mode allows you to connect the channel input to ground to see where the 0 V reference is on the screen. Figure 8–66 illustrates the result of DC and AC coupling using a sinusoidal waveform that has a dc component.

The voltage probe, shown in Figure 8–67, is used for connecting a signal to the scope. Since all instruments tend to affect the circuit being measured due to loading, most scope probes provide an attenuation network to minimize loading effects. Probes that attenuate

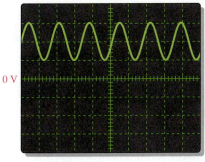

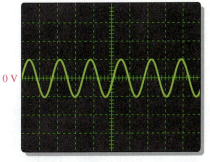

(a) DC coupled waveform

(b) AC coupled waveform

FIGURE 8–66   Displays of the same waveform having a dc component.

the measured signal by a factor of 10 are called ×10 (times ten) probes. Probes with no attenuation are called ×1 (times one) probes. Most oscilloscopes automatically adjust the calibration for the attenuation of the type of probe being used. For most measurements, the ×10 probe should be used. However, if you are measuring very small signals, a ×1 may be the best choice.

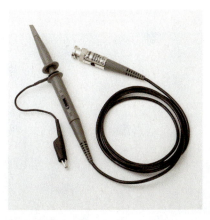

FIGURE 8–67   An oscilloscope voltage probe.   Copyright © Tektronix, Inc. Reproduced by permission.

The probe has an adjustment that allows you to compensate for the input capacitance of the scope. Most scopes have a probe compensation output that provides a calibrated square wave for probe compensation. Before making a measurement, you should make sure that the probe is properly compensated to eliminate any distortion introduced. Typically, there is a screw or other means of adjusting compensation on a probe. Figure 8–68 shows scope waveforms for three probe conditions: properly compensated, undercompensated, and overcompensated. If the waveform appears either over- or undercompensated, adjust the probe until the properly compensated square wave is achieved.

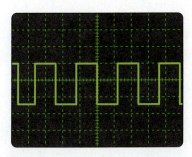

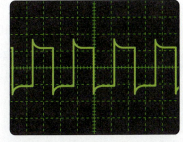

Properly compensated

Undercompensated

Overcompensated

FIGURE 8–68   Probe compensation conditions.

EXAMPLE 8–17

Determine the peak-to-peak value and period of each sine wave in Figure 8–69 from the digital scope screen displays and the settings for Volts/Div and Sec/Div, which are indicated under the screens. Sine waves are centered vertically on the screens.

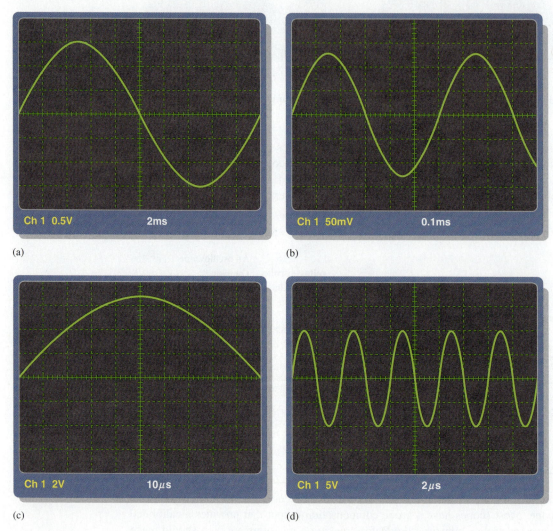

Ch 1 0.5V   2ms   (a)

Ch 1 50mV   0.1ms   (b)

Ch 1 2V   10μs   (c)

Ch 1 5V   2μs   (d)

FIGURE 8–69

SOLUTION

Looking at the vertical scale in Figure 8–69(a),

$$V_{pp} = 6 \text{ divisions} \times 0.5 \text{ V/division} = \textbf{3.0 V}$$

From the horizontal scale (one cycle covers ten divisions),

$$T = 10 \text{ divisions} \times 2 \text{ ms/division} = \textbf{20 ms}$$

Looking at the vertical scale in Figure 8–69(b),

$$V_{pp} = 5 \text{ divisions} \times 50 \text{ mV/division} = \textbf{250 mV}$$

From the horizontal scale (one cycle covers six divisions),

$$T = 6 \text{ divisions} \times 0.1 \text{ ms/division} = 0.6 \text{ ms} = \textbf{600 μs}$$

Looking at the vertical scale in Figure 8–69(c),

$$V_{pp} = 6.8 \text{ divisions} \times 2 \text{ V/division} = \textbf{13.6 V}$$

From the horizontal scale (one-half cycle covers ten divisions),

$$T = 20 \text{ divisions} \times 10 \,\mu\text{s/division} = \textbf{200} \,\boldsymbol{\mu}\textbf{s}$$

Looking at the vertical scale in Figure 8–69(d),

$$V_{pp} = 4 \text{ divisions} \times 5 \text{ V/division} = \textbf{20 V}$$

From the horizontal scale (one cycle covers two divisions),

$$T = 2 \text{ divisions} \times 2 \,\mu\text{s/division} = \textbf{4} \,\boldsymbol{\mu}\textbf{s}$$

RELATED PROBLEM

Determine the rms value and the frequency for each waveform displayed in Figure 8–69.

SECTION 8–9 CHECKUP

1. What is the main difference between a digital and an analog oscilloscope?

2. Is voltage read horizontally or vertically on a scope screen?

3. What does the Volts/Div control on an oscilloscope do?

4. What does the Sec/Div control on an oscilloscope do?

5. When should you use a ×10 probe for making a voltage measurement?

8–10 SIGNAL SOURCES

The function generator was introduced in Section 8–1. In this section, we'll look in more depth at function generators and other signal sources, which are commonly used for testing circuits. The function generator is found at nearly every electronic workbench. Other signal sources include specialized generators and digital pattern generators.

After completing this section, you should be able to

- **Describe the types of signal sources and explain the purpose of typical controls**
  - **Discuss types of signal sources and their application**
  - **Describe important specifications for signal sources**
  - **Describe the function selection, frequency/amplitude adjustment controls, and the dc offset/duty cycle control for a typical function generator**

Types of Signal Sources

Signals can be generated from a variety of low-power electronic instruments, ranging from basic sine wave oscillators to generators that have a selection of waveforms including sine waves, ramps, pulses and even arbitrary waveforms from specialized generators. Except for low-end instruments, the output frequency and amplitude are calibrated and can be varied over a specified range. Instruments for producing waveforms for test are generally categorized as low frequency, radio frequency, and microwave. Traditionally, low-frequency generators extend from dc to about 1 MHz. Radio-frequency sources generally fall in the region from 100 kHz to 1 GHz, and microwave sources are from 1 GHz up. The wide variety of generators on the market has blurred the distinction between these traditional ranges; the range may vary depending on the wave shape.

Some important types of signal sources are function generators, arbitrary function generators (AFGs), signal generators (sometimes referred to as rf generators), swept-frequency oscillators (or sweepers), pulse generators, and digital-pattern generators (DPGs).

Function generators Function generatorsuse a free-running oscillator to provide a selection of sine, square, and triangle waveforms in a single instrument. Function generators may also produce pulses, ramps, and other waveforms.

Arbitrary function generators *(AFGs)* Arbitrary function generators are the most versatile signal generators. They are digital synthesizers that allow the user to create custom waveforms from a mathematical formula, a captured waveform, or simulation software. AFGs create an analog waveform using a controller and a digital-to-analog converter (DAC).

Signal generators Signal generators produce high-frequency sine waves and modulated sine waves. A large variety of signal generators are available with frequencies that go from dc to microwave, and many use frequency synthesizers to generate precise waveforms. A frequency synthesizer generates waveforms digitally from a fixed-reference oscillator.

Swept-frequency oscillators Swept-frequency oscillators generate a sine wave output that varies at a cyclic rate between two selected frequencies. They allow the user to view signals directly in the frequency domain. Some models produce a logarithmic or other nonlinear sweep output.

Pulse generators Pulse generators are specialized instruments for producing pulses with fast rise times over a wide frequency range and with varying duty cycles. They frequently have outputs for specific logic levels. Most pulse generators allow the user to control a variety of pulse parameters, such as amplitude, rise time, offsets, triggering, and polarity. In addition, some units produce pulse pairs and trains.

Digital-pattern generators *(DPGs)* Digital-pattern generators produce a digital sequence for testing digital equipment. Some DPGs can generate a series of standard test patterns for performance checks.

Swept-Frequency Measurements

Many times a system needs to be tested for its frequency response. Swept-frequency oscillators are a type of signal generator that has an output that can be set for a given range depending on the system under test. For example, a radar transmitter may have multiple stages that need to work together to produce a desired output. The frequencies are in the microwave region, so the generator is a specialized microwave oscillator.

A swept-frequency oscillator can generate a small high-frequency signal that is used for tuning purposes. The response is converted to a signal that can be observed on an oscilloscope while the system is tuned in real time.

SYSTEM NOTE

Specifications for Signal Generators

The most important specifications for signal generators are those for mode, frequency range, spectral purity, amplitude range, modulation, and output impedance. These parameters are discussed here:

Mode The mode is a specification of signal types that can be output. Specialized generators may only have one mode, such as a sine wave. Function generators will have several modes (waveform types) to choose from. Typically these are sine, square, triangle, positive and negative pulse, and positive and negative ramps. The duty cycle, which is the on/off time for a pulse, may be adjustable.

Frequency range This is the span of frequencies over which instrument performance is specified. Accuracy and resolution limits are also generally given with the frequency range specification. The accuracy specification depends on the internal oscillator's accuracy as well as the type of dial controls (mechanical or digital). A precise digital control is of little value if the internal oscillator is not accurate, nor is it meaningful to have an accurate oscillator if the control is an imprecise mechanical dial.

Spectral purity Recall that a pure sine wave consists of a single fundamental line in the frequency domain. One type of distortion called *phase noise* can broaden the fundamental line, meaning the signal is not a pure sine wave. Another distortion is called *harmonic distortion*, which means multiples of the fundamental frequency are present in the output. These and other noise sources can create problems with certain system tests, particularly for verifying communication equipment. Even if the equipment is working properly, a signal generator with distortion can make the circuit under test to look worse than it is.

Amplitude range This specification gives the output voltage amplitude range or the maximum and minimum power delivered to a specified load. In addition to the ac voltage amplitude range, output specifications include the dc offset range as well as the accuracy, resolution, and flatness of the output across the frequency range. A related adjustment is the amount of dc offset. The dc offset range is either a positive or negative dc voltage that can be added to the output; it is typically a variable amount up to about 10 V.

Modulation Modulation is the process in which a signal containing information is used to modify the characteristic of another signal, such as amplitude frequency or pulse width, so that the information on the first is also contained in the second. A signal generator with modulation allows the user to change a high-frequency waveform (called a *carrier* in communications systems) by a lower-frequency waveform called the *modulating signal.* The modulation signal can be supplied from an external source or may be generated within the generator; typically, a sine, square, triangle, or ramp can be selected as the modulation signal. The parameter being modulated can be the high-frequency signal's amplitude, frequency, or phase. The various modulation methods are useful for testing the performance of different communication systems.

Output impedance No matter how complicated the internal circuitry, signal generators can be modeled as a Thevenin circuit, which consists of a voltage source in series with a resistance. The output impedance is equivalent to the resistance in series with an ideal source. Generators are normally marked with the output impedance on the output terminals. Typical values are 50 Ω, 75 Ω, or 600 Ω. Keep in mind that connecting a load to a source that has a finite output resistance will change the output voltage amplitude of the generator.

The change in output voltage amplitude can affect the desired measurement to the point of making it meaningless if it is not taken into account. For example, to measure the frequency response of a circuit without having the generator affect the results, the amplitude must be maintained at a constant value from the generator. The output of the circuit under test is monitored while the frequency is changed. Each change in frequency changes the input impedance of the amplifier and causes a different loading effect to occur. To prevent input loading from affecting the output, the amplitude of the generator should be readjusted to the same value each time a new frequency is tested.

Waveform Modes

Depending on the complexity, waveform generators can have from one to several waveform modes from which to choose. Common waveform modes are as follows.

- **Continuous** The output is steady at a specific frequency, amplitude, and offset. This is the basic output of signal generators.
- **Triggered** The output is initiated by an internal or external (user-supplied) trigger. The trigger can be generated by external equipment, a manual pushbutton, or sent over a control bus.
- **Burst** This is the same as the triggered mode, except the output is programmed for a specific number of cycles at a time, which can be repeated as set by a separate oscillator. The burst can be triggered manually or with a command in programmable instruments.

- **Gated** The output is enabled for the duration of an external gate signal.
- **Swept** The frequency of the output changes in some predetermined way. Sweeps can be triggered or programmed (start-and-stop frequency), linear, or logarithmic.
- **Modulation** The output is modulated by an internal or external waveform. Modulated outputs are useful for testing communication systems.

The Basic Function Generator

Function generators are characterized by the various waveforms they produce; they all produce sine, square, and triangle waves and may also include pulses and ramp (sawtooth) waveforms. Sine and square waves are commonly used for general-purpose testing of circuits such as amplifiers. Pulses that have a duty cycle that is not near 50% are useful for digital testing because it is easy to tell if the pulses have been inverted; they don't have the same appearance when inverted as when not. (If a square wave is inverted, it still looks like a square wave.) A versatile function generator is the Agilent 33220A, shown in Figure 8–70. This generator includes more functions than a basic generator, such as the ability to store waveforms in a memory. To set up a standard waveform on the Agilent 33220A, you can set the controls as indicated in the figure.

1. Press the power switch to turn on power.
2. Choose a function switch to select the type of waveform (sinusoidal, pulse, etc.). The function key will be lighted.
3. Use graph mode to view all setup parameters at once.
4. Select a pushbutton to adjust a function (frequency, amplitude, etc.)
5. Adjust the selected function with the keypad or the circular knob.

FIGURE 8–70   **The Agilent 33220A function generator.**   (© Agilent Technologies, Inc. 2012, Reproduced with Permission, Courtesy of Agilent Technologies, Inc.)

An option for function generators is a display. Some models do not include a display, and the user verifies the generator signal using an oscilloscope. Others, such as the Agilent 33220A have a digital readout that indicates the precise frequency (or period), offset, etc. and shows a visual indication of the waveform mode. It also has a graph mode that shows a graphical view of the selected waveform with a soft key menu of available parameters that can be displayed (frequency, period, rise time, etc.).

Function generators often have outputs that are designed for testing digital systems. Digital systems are characterized by two voltage levels, called HIGH and LOW, which are determined by the type of logic to be tested. The pulse output may be set by the manufacturer or may be set by the user depending on the generator. Digital systems use fast signals, so the rise time specification is important. This specification is the time required for the signal to make a 10% to 90% change in the output level.

Another option available for function generators is the ability to control the output with a computer or controller. This is particular useful in automatic test systems. The system can change the output based on the particular requirements for a test.

EXAMPLE 8–18

Assume a function generator is showing the signal shown on the oscilloscope display in Figure 8–71. The oscilloscope controls are set as shown. Determine the type of waveform that is selected on the function generator, the frequency setting, and the peak-to-peak amplitude. Often the peak-to-peak is the simplest measurement with an oscilloscope and, as the name implies, is measured between the maximum excursions of the waveform.

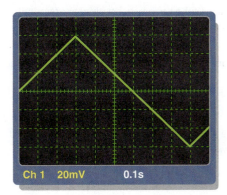

Ch 1   20mV   0.1s

FIGURE 8–71   **The horizontal axis represents 0 V.**

SOLUTION

The waveform that is selected is a **triangle wave**.
    The period is

$$T = (10\,\text{div})(0.1\,\mu\text{s/div}) = 1.0\,\mu\text{s}$$

The frequency setting is

$$f = \frac{1}{T} = \frac{1}{1.0\,\mu\text{s}} = \textbf{1.0 MHz}$$

The peak-to-peak amplitude is

$$V_{pp} = (6.0\,\text{div})(20\,\text{mV/div}) = \textbf{1.20 V}_{\textbf{pp}}$$

RELATED PROBLEM

What setting of dc offset will produce a triangle waveform that goes from 0 V to 1.20 V? (With no dc offset, the waveform is centered on 0 V.)

SECTION 8–10 CHECKUP

1. What type of signals can be generated with an arbitrary function generator?

2. What is a swept-frequency oscillator?

3. Why is spectral purity important for a function generator?

4. What is modulation?

5. What is burst mode for a function generator?

SUMMARY

- The sine wave is a time-varying, periodic waveform.
- The sine wave is the fundamental type of alternating current (ac) and alternating voltage.
- Alternating current changes direction in response to changes in the polarity of the source voltage.
- One cycle of an alternating sine wave consists of a positive alternation and a negative alternation.
- The average value of a sine wave determined over one half-cycle is 0.637 times the peak value. The average value of a sine wave determined over a full cycle is zero.
- A full cycle of a sine wave is 360°, or 2π radians. A half-cycle is 180°, or π radian. A quarter-cycle is 90° or $\pi/2$ radians.
- Phase angle is the difference in degrees (or radians) between two sine waves or between a sine wave and a reference wave.
- The angular position of a phasor represents the angle of a sine wave, and the length of the phasor represents the amplitude.
- Voltages and currents must all be expressed with consistent units when you apply Ohm's or Kirchhoff's laws in ac circuits.
- Power in a resistive ac circuit is determined using rms voltage and/or current values.
- Alternators (ac generators) produce power when there is relative motion between a magnetic field and a conductor.
- Most alternators take the output from the stator. The rotor provides a moving magnetic field.
- Two major types of ac motors are induction motors and synchronous motors.
- Induction motors have a rotor that turns in response to a rotating field from the stator.
- Synchronous motors move at a constant speed in sync with the field of the stator.
- A pulse consists of a transition from a baseline level to an amplitude level, followed by a transition back to the baseline level.
- A triangular or sawtooth wave consists of positive-going and negative-going ramps.
- Harmonic frequencies are odd or even multiples of the repetition rate (or fundamental frequency) of a nonsinusoidal waveform.
- Conversions of sine wave values are summarized in Table 8–2.

TABLE 8–2

TO CHANGE FROM	TO	MULTIPLY BY
Peak	rms	0.707
Peak	Peak-to-peak	2
Peak	Average	0.637
rms	Peak	1.414
Peak-to-peak	Peak	0.5
Average	Peak	1.57

- Signal sources include function generators and a number of specialized instruments.
- Key specifications for signal generators include mode, frequency range, spectral purity, amplitude range, modulation, and output impedance.

KEY TERMS

Key terms and other bold terms in the chapter are defined in the end-of-book glossary.

Alternator   An ac generator, alternators convert mechanical energy to electrical energy.

Amplitude   The maximum value of a voltage or current.

Cycle   One repetition of a periodic waveform.

Degree   The unit of angular measure corresponding to 1/360 of a complete revolution.

Duty cycle   A characteristic of a pulse waveform that indicates the percentage of time that a pulse is present during a cycle; the ratio of pulse width to period expressed as either a fraction or as a percentage.

Fall time (t_f)   The time interval required for a pulse to change from 90% to 10% of its amplitude.

Frequency (f)   A measure of the rate of change of a periodic function; the number of cycles completed in 1 s. The unit of frequency is the hertz.

Function generator   An instrument that produces more than one type of waveform.

Fundamental frequency   The repetition rate of a waveform.

Harmonics   The frequencies contained in a composite waveform, which are integer multiples of the pulse repetition frequency (fundamental).

Hertz (Hz)   The unit of frequency. One hertz equals one cycle per second.

Induction motor   An ac motor that achieves excitation to the rotor by transformer action.

Instantaneous value   The voltage or current value of a waveform at a given instant in time.

Modulation   The process in which a signal containing information is used to modify the characteristics of another signal such as amplitude frequency or pulse width so that the information on the first is also contained in the second.

Oscillator   An electronic circuit that produces a repetitive waveform on its output with only the dc supply voltage as an input.

Oscilloscope   A measurement instrument that displays signal waveforms on a screen.

Peak-to-peak value   The voltage or current value of a waveform measured from its minimum to its maximum points.

Peak value   The voltage or current value of a waveform at its maximum positive or negative points.

Period (T)   The time interval of one complete cycle of a periodic waveform.

Periodic   Characterized by a repetition at fixed-time intervals.

Phase   The relative angular displacement of a time-varying waveform in terms of its occurrence with respect to a reference.

Pulse   A type of waveform that consists of two equal and opposite steps in voltage or current separated by a time interval.

Pulse width (t_W)   The time interval between the opposite steps of an ideal pulse. For a nonideal pulse, the time between the 50% points on the leading and trailing edges.

Radian   A unit of angular measurement. There are 2π radians in one complete 360° revolution. One radian equals 57.3°.

Ramp   A type of waveform characterized by a linear increase or decrease in voltage or current.

Rise time (t_r)   The time interval required for a pulse to change from 10% to 90% of its amplitude.

rms value   The value of a sinusoidal voltage that indicates its heating effect, also known as the effective value. It is equal to 0.707 times the peak value. *rms* stands for root mean square.

Sine wave   A type of waveform that follows a cyclic sinusoidal pattern defined by the formula $v = A \sin \theta$.

Slip   The difference between the synchronous speed of the stator field and the rotor speed in an induction motor.

Squirrel cage   An aluminum frame within the rotor of an induction motor that forms the electrical conductors for a rotating current.

Synchronous motor   An ac motor in which the rotor moves at the same rate as the rotating magnetic field of the stator.

Waveform   The pattern of variations of a voltage or current showing how the quantity changes with time.

KEY FORMULAS

(8–1) $f = \dfrac{1}{T}$ Frequency

(8–2) $T = \dfrac{1}{f}$ Period

(8–3)	$V_{pp} = 2V_p$		Peak-to-peak voltage (sine wave)
(8–4)	$I_{pp} = 2I_p$		Peak to peak current (sine wave)
(8–5)	$V_{rms} = 0.707V_p$		Root-mean-square voltage (sine wave)
(8–6)	$I_{rms} = 0.707I_p$		Root-mean-square current (sine wave)
(8–7)	$V_p = 1.414V_{rms}$		Peak voltage (sine wave)
(8–8)	$I_p = 1.414I_{rms}$		Peak current (sine wave)
(8–9)	$V_{pp} = 2.828V_{rms}$		Peak-to-peak voltage (sine wave)
(8–10)	$I_{pp} = 2.828I_{rms}$		Peak-to-peak current (sine wave)
(8–11)	$V_{avg} = 0.637V_p$		Half-cycle average voltage (sine wave)
(8–12)	$I_{avg} = 0.637I_p$		Half-cycle average current (sine wave)

$$(8\text{–}13) \quad rad = \left(\frac{\pi\ rad}{180°}\right) \times degrees \qquad \text{Degrees to radian conversion}$$

$$(8\text{–}14) \quad degrees = \left(\frac{180°}{\pi\ rad}\right) \times rad \qquad \text{Radian to degrees conversion}$$

(8–15)	$y = A \sin \theta$	General formula for a sine wave
(8–16)	$v = V_p \sin \theta$	Sinusoidal voltage
(8–17)	$i = I_p \sin \theta$	Sinusoidal current
(8–18)	$y = A \sin(\theta - \phi)$	Lagging sine wave
(8–19)	$y = A \sin(\theta + \phi)$	Leading sine wave

$$(8\text{–}20) \quad f = \frac{Ns}{120} \qquad \text{Output frequency of an alternator}$$

$$(8\text{–}21) \quad \text{Percent duty cycle} = \left(\frac{t_W}{T}\right)100\% \qquad \text{Duty cycle}$$

$$(8\text{–}22) \quad V_{avg} = \text{baseline} + (\text{duty cycle})(\text{amplitude}) \qquad \text{Average voltage of pulse waveform}$$

TRUE/FALSE QUIZ

Answers are at the end of the chapter.

1. The period of a 60 Hz sine wave is 16.7 ms.

2. The rms and average value of a sine wave are the same.

3. A sine wave with a peak value of 10 V has the same heating effect as a 10 V dc source.

4. The peak value of a sine wave is the same as its amplitude.

5. The number of radians in 360° is 2π.

6. In a three-phase electrical system, the phases are separated by 60°.

7. The purpose of an exciter is to supply dc rotor current to an alternator.

8. In an automotive alternator, the output current is taken from the rotor through slip rings.

9. A maintenance issue with induction motors is brush replacement.

10. A synchronous motor can be used when constant speed is required.

11. Clamp meters are available that can measure dc.

12. The purpose of the diode array in an automobile alternator is to convert dc to ac.

SELF-TEST

Answers are at the end of the chapter.

1. The difference between alternating current (ac) and direct current (dc) is
   (a) ac changes value and dc does not   (b) ac changes direction and dc does not
   (c) both (a) and (b)   (d) neither (a) nor (b)

2. During each cycle, a sine wave reaches a peak value
   (a) one time   (b) two times   (c) four times
   (d) a number of times depending on the frequency

3. A sine wave with a frequency of 12 kHz is changing at a faster rate than a sine wave with a frequency of
   (a) 20 kHz   (b) 15,000 Hz   (c) 10,000 Hz   (d) 1.25 MHz

4. A sine wave with a period of 2 ms is changing at a faster rate than a sine wave with a period of
   (a) 1 ms   (b) 0.0025 s   (c) 1.5 ms   (d) 1000 μs

5. When a sine wave has a frequency of 60 Hz, in 10 s it goes through
   (a) 6 cycles   (b) 10 cycles   (c) 1/16 cycle   (d) 600 cycles

6. If the peak value of a sine wave is 10 V, the peak-to-peak value is
   (a) 20 V   (b) 5 V   (c) 100 V   (d) none of these

7. If the peak value of a sine wave is 20 V, the rms value is
   (a) 14.14 V   (b) 6.37 V   (c) 7.07 V   (d) 0.707 V

8. The average value of a 10 V peak sine wave over one complete cycle is
   (a) 0 V   (b) 6.37 V   (c) 7.07 V   (d) 5 V

9. The average half-cycle value of a sine wave with a 20 V peak is
   (a) 0 V   (b) 6.37 V   (c) 12.74 V   (d) 14.14 V

10. One sine wave has a positive-going zero crossing at 10° and another sine wave has a positive-going zero crossing at 45°. The phase angle between the two waveforms is
    (a) 55°   (b) 35°   (c) 0°   (d) none of these

11. The instantaneous value of a 15 A peak sine wave at a point 32° from its positive-going zero crossing is
    (a) 7.95 A   (b) 7.5 A   (c) 2.13 A   (d) 7.95 V

12. If the rms current through a 10 kΩ resistor is 5 mA, the rms voltage drop across the resistor is
    (a) 70.7 V   (b) 7.07 V   (c) 5 V   (d) 50 V

13. Two series resistors are connected to an ac source. If there is 6.5 V rms across one resistor and 3.2 V rms across the other, the peak source voltage is
    (a) 9.7 V   (b) 9.19 V   (c) 13.72 V   (d) 4.53 V

14. An advantage of a three-phase induction motor is that it
    (a) maintains constant speed for any load   (b) does not require starting windings
    (c) has a wound rotor   (d) all of the above

15. The difference in the synchronous speed of the stator field and the rotor speed of a motor is called
    (a) differential speed   (b) loading   (c) lag   (d) slip

16. A 10 kHz pulse waveform consists of pulses that are 10 μs wide. Its duty cycle is
    (a) 100%   (b) 10%   (c) 1%   (d) not determinable

17. The duty cycle of a square wave
    (a) varies with the frequency   (b) varies with the pulse width
    (c) both (a) and (b)   (d) is 50%

18. A waveform mode that changes the output between a start-and-stop frequency is called
    (a) triggered   (b) burst
    (c) swept   (d) modulation

TROUBLESHOOTING: SYMPTOM AND CAUSE

The purpose of these exercises is to help develop thought processes essential to troubleshooting. Answers are at the end of the chapter.

Determine the cause for each set of symptoms. Refer to Figure 8–72.

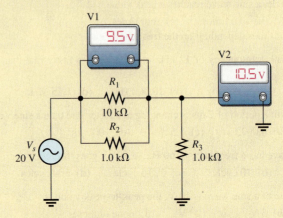

FIGURE 8–72   The ac meters indicate the correct readings for this circuit.

1. *Symptom:* The voltmeter 1 reading is 0 V, and the voltmeter 2 reading is 20 V.
   Cause:
   (a) R_1 is open.   **(b)** R_2 is open.   **(c)** R_3 is open.

2. *Symptom:* The voltmeter 1 reading is 20 V, and the voltmeter 2 reading is 0 V.
   Cause:
   (a) R_1 is open.   **(b)** R_2 is shorted.   **(c)** R_3 is shorted.

3. *Symptom:* The voltmeter 1 reading is 18.2 V, and the voltmeter 2 reading is 1.8 V.
   Cause:
   (a) R_1 is open.   **(b)** R_2 is open.   **(c)** R_1 is shorted.

4. *Symptom:* Both voltmeter readings are 10 V.
   Cause:
   (a) R_1 is open.   **(b)** R_1 is shorted.   **(c)** R_2 is open.

5. *Symptom:* The voltmeter 1 reading is 16.7 V, and the voltmeter 2 reading is 3.3 V.
   Cause:
   (a) R_1 is shorted.   **(b)** R_2 is 10 kΩ instead of 1 kΩ.   **(c)** R_3 is 10 kΩ instead of 1 kΩ.

PROBLEMS

Answers to odd-numbered problems are at the end of the book.

BASIC PROBLEMS

SECTION 8–1   The Sinusoidal Waveform

1. Calculate the frequency for each of the following periods:
   (a) 1 s   **(b)** 0.2 s   **(c)** 50 ms   **(d)** 1 ms   **(e)** 500 μs   **(f)** 10 μs

2. Calculate the period for each of the following frequencies:
   (a) 1 Hz   **(b)** 60 Hz   **(c)** 500 Hz   **(d)** 1 kHz   **(e)** 200 kHz   **(f)** 5 MHz

3. A sine wave goes through 5 cycles in 10 μs. What is its period?

4. A sine wave has a frequency of 50 kHz. How many cycles does it complete in 10 ms?

5. How long does it take a 10 kHz sine wave to complete 100 cycles?

SECTION 8–2   Voltage and Current Values of Sine Waves

6. A sine wave has a peak value of 12 V. Determine the following voltage values:
 (a) rms   **(b)** peak-to-peak   **(c)** half-cycle average

7. A sinusoidal current has an rms value of 5 mA. Determine the following current values:
 (a) peak   **(b)** half-cycle average   **(c)** peak-to-peak

8. For the sine wave in Figure 8–73, determine the peak, peak-to-peak, rms, and half-cycle average values.

9. If each horizontal division in Figure 8–73 is 1 ms, determine the instantaneous voltage value at
 (a) 1 ms   **(b)** 2 ms   **(c)** 4 ms   **(d)** 7 ms

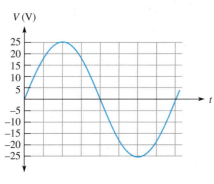

FIGURE 8–73

SECTION 8–3   Angular Measurement of a Sine Wave

10. In Figure 8–73, what is the instantaneous voltage at
 (a) 45°   **(b)** 90°   **(c)** 180°

11. Sine wave *A* has a positive-going zero crossing at 30° from a reference. Sine wave *B* has a positive-going zero crossing at 45° from the same reference. Determine the phase angle between the two signals. Which signal leads?

12. One sine wave has a positive peak at 75° and another has a positive peak at 100°. How much is each sine wave shifted in phase from the 0° reference? What is the phase angle between them?

13. Draw two sine waves as follows: Sine wave *A* is the reference, and sine wave *B* lags *A* by 90°. Both have equal amplitudes.

14. Convert the following angular values from degrees to radians:
 (a) 30°   **(b)** 45°   **(c)** 78°   **(d)** 135°   **(e)** 200°   **(f)** 300°

15. Convert the following angular values from radians to degrees:
 (a) $\pi/8$ **(b)** $\pi/3$ **(c)** $\pi/2$ **(d)** $3\pi/5$ **(e)** $6\pi/5$ **(f)** 1.8π

SECTION 8–4   The Sine Wave Formula

16. A certain sine wave has a positive-going zero crossing at 0° and an rms value of 20 V. Calculate its instantaneous value at each of the following angles:
 (a) 15°   **(b)** 33°   **(c)** 50°   **(d)** 110°
 (e) 70°   **(f)** 145°   **(g)** 250°   **(h)** 325°

17. For a particular 0° reference sinusoidal current, the peak value is 100 mA. Determine the instantaneous value at each of the following points:
 (a) 35°   **(b)** 95°   **(c)** 190°   **(d)** 215°   **(e)** 275°   **(f)** 360°

18. For a 0° reference sine wave with an rms value of 6.37 V, determine its instantaneous value at each of the following points:
 (a) $\pi/8$ rad   **(b)** $\pi/4$ rad   **(c)** $\pi/2$ rad   **(d)** $3\pi/4$ rad
 (e) π rad   **(f)** $3\pi/2$ rad   **(g)** 2π rad

19. Sine wave *A* lags sine wave *B* by 30°. Both have peak values of 15 V. Sine wave *A* is the reference with a positive-going crossing at 0°. Determine the instantaneous value of sine wave *B* at 30°, 45°, 90°, 180°, 200°, and 300°.

20. Repeat Problem 19 for the case when sine wave *A* leads sine wave *B* by 30°.

SECTION 8–5 **Analysis of AC Circuits**

21. A sinusoidal voltage is applied to the resistive circuit in Figure 8–74. Determine the following:
   (a) I_{rms} (b) I_{avg} (c) I_p (d) I_{pp} (e) i at the positive peak

22. Find the half-cycle average values of the voltages across R_1 and R_2 in Figure 8–75. All values shown are rms.

23. Determine the rms voltage across R_3 in Figure 8–76.

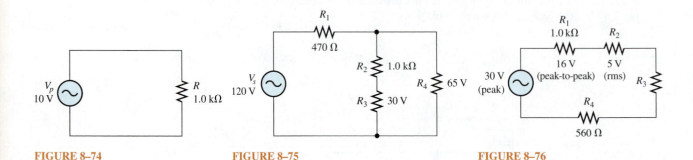

FIGURE 8–74   **FIGURE 8–75**   **FIGURE 8–76**

24. A sine wave with an rms value of 10.6 V is riding on a dc level of 24 V. What are the maximum and minimum values of the resulting waveform?

25. How much dc voltage must be added to a 3 V rms sine wave in order to make the resulting voltage nonalternating (no negative values)?

26. A 6 V peak sine wave is riding on a dc voltage of 8 V. If the dc voltage is lowered to 5 V, how far negative will the sine wave go?

SECTION 8–6 **Alternators (AC Generators)**

27. The conductive wire loop on the rotor of a simple two-pole, single-phase generator rotates at a rate of 250 rps. What is the frequency of the induced output voltage?

28. A certain four-pole generator has a speed of rotation of 3600 rpm. What is the frequency of the voltage produced by this generator?

29. At what speed of rotation must a four-pole generator be operated to produce a 400 Hz sinusoidal voltage?

30. A common frequency for alternators on aircraft is 400 Hz. How many poles does a 400 Hz alternator have if the speed of rotation is 3000 rpm?

SECTION 8–7 **AC Motors**

31. What is the main difference between a one-phase induction motor and a three-phase induction motor?

32. Explain how the field in a three-phase motor rotates if there are no moving parts to the field coils.

SECTION 8–8 **Nonsinusoidal Waveforms**

33. From the graph in Figure 8–77, determine the approximate values of t_r, t_f, t_W, and amplitude.

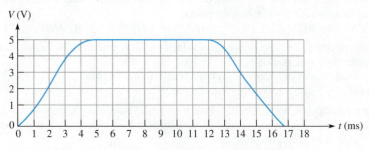

FIGURE 8–77

34. Determine the duty cycle for each pulse waveform in Figure 8–78.

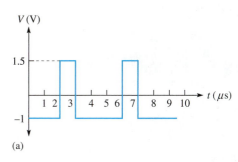

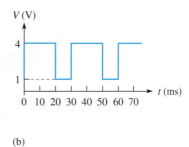

(a)

(b)

FIGURE 8–78

35. Find the average value of each positive-going pulse waveform in Figure 8–78.
36. What is the frequency of each waveform in Figure 8–78.
37. What is the frequency of each sawtooth waveform in Figure 8–79?

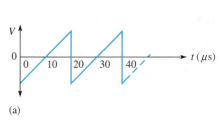

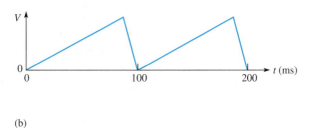

(a)

(b)

FIGURE 8–79

38. A square wave has a period of 40 μs. List the first six odd harmonics.
39. What is the fundamental frequency of the square wave mentioned in Problem 38?

SECTION 8–9   The Oscilloscope

40. Determine the peak value and the period of the sine wave displayed on the scope screen in Figure 8–80. The horizontal axis is 0 V.
41. Determine the rms value and the frequency of the sine wave displayed on the scope screen in Figure 8–80.
42. Determine the rms value and the frequency of the sine wave displayed on the scope screen in Figure 8–81. The horizontal axis is 0 V.

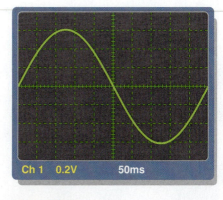

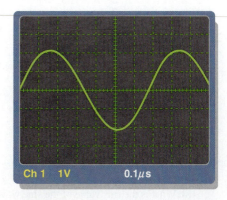

FIGURE 8–80

FIGURE 8–81

43. Find the amplitude, pulse width, and duty cycle for the pulse waveform displayed on the scope screen in Figure 8–82. The horizontal axis is 0 V.

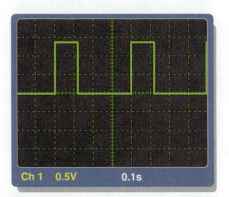

FIGURE 8–82

SECTION 8–10   Signal Sources

44. Why is spectral purity important for a sinusoidal signal generator?

45. What is a modulating signal and how is it applied to an analog signal generator?

46. Assume you are testing a circuit with different frequencies to measure its response. Why is it important to check the signal level from the generator each time the frequency is changed?

47. What is the difference between burst mode and gated mode for a generator?

ADVANCED PROBLEMS

48. A certain sine wave has a frequency of 2.2 kHz and an rms value of 25 V. Assuming a given cycle begins (zero crossing) at $t = 0$ s, what is the change in voltage from 0.12 ms to 0.2 ms?

49. Figure 8–83 shows a sinusoidal voltage source in series with a dc source. Effectively, the two voltages are superimposed. Draw the total voltage across R_L. Determine the maximum current through R_L and the average voltage across R_L.

50. A nonsinusoidal waveform called a *stairstep* is shown in Figure 8–84. Determine its average value.

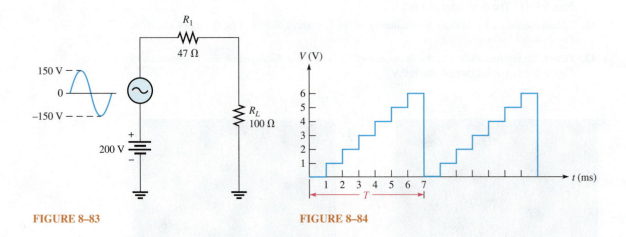

FIGURE 8–83   **FIGURE 8–84**

51. Refer to the oscilloscope screen in Figure 8–85.
    (a) How many cycles are displayed?
    (b) What is the rms value of the sine wave?
    (c) What is the frequency of the sine wave?

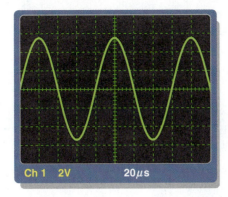

FIGURE 8–85

52. Accurately draw on a grid representing the scope screen in Figure 8–85 how the sine wave will appear if the Volts/Div control is moved to a setting of 5 V.

53. Accurately draw on a grid representing the scope screen in Figure 8–85 how the sine wave will appear if the Sec/Div control is moved to a setting of 10 μs.

54. Based on the instrument settings and an examination of the scope display and the circuit board in Figure 8–86, determine the frequency and peak value of the input signal and output signal. The waveform shown is channel 1. Draw the channel 2 waveform as it would appear on the scope with the indicated settings.

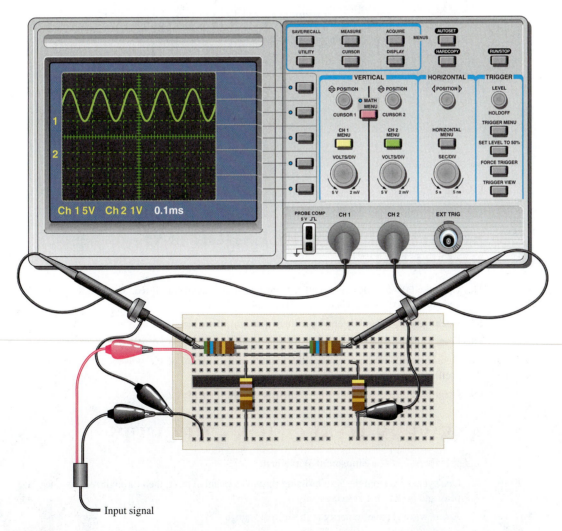

FIGURE 8–86

55. Examine the circuit board and the oscilloscope display in Figure 8–87 and determine the peak value and the frequency of the unknown input signal.

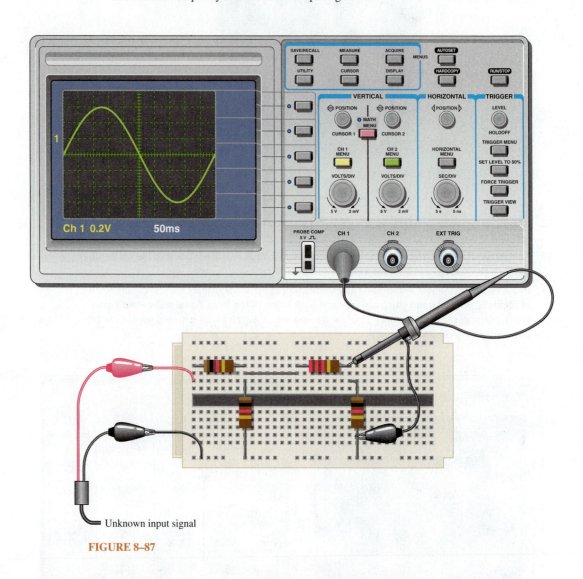

Unknown input signal

FIGURE 8–87

MULTISIM

MULTISIM TROUBLESHOOTING PROBLEMS

56. Open file P08–56; files are found at www.pearsonhighered.com/floyd. Measure the peak value and period of the voltage sine wave using the oscilloscope.

57. Open file P08–57 and determine if there is a fault. If so, identify it.

58. Open file P08–58 and determine if there is a fault. If so, identify it.

59. Open file P08–59 and measure the amplitude and period of the pulse waveform using the oscilloscope.

60. Open file P08–60 and determine if there is a fault. If so, identify it.

ANSWERS TO SECTION CHECKUPS

SECTION 8–1   The Sinusoidal Waveform

1. One cycle goes from the zero crossing through a positive peak, then through zero to a negative peak and back to the zero crossing.

2. A sine wave changes polarity at the zero crossings.

3. A sine wave has two maximum points (peaks) in one cycle.

4. The period is measured from one zero crossing to the next corresponding zero crossing, or from one peak to the next corresponding peak.

5. Frequency is the number of cycles completed in one second and its unit is the hertz.

6. $f = 1/5 \, \mu s = 200 \, \text{kHz}$

7. $T = 1/120 \, \text{Hz} = 8.33 \, \text{ms}$

SECTION 8–2 Voltage and Current Values of Sine Waves

1. (a) $V_{pp} = 2(1 \, \text{V}) = 2 \, \text{V}$
   (b) $V_{pp} = 2(1.414)(1.414 \, \text{V}) = 4 \, \text{V}$
   (c) $V_{pp} = 2(1.57)(3 \, \text{V}) = 9.42 \, \text{V}$

2. (a) $V_{rms} = (0.707)(2.5 \, \text{V}) = 1.77 \, \text{V}$
   (b) $V_{rms} = (0.5)(0.707)(10 \, \text{V}) = 3.54 \, \text{V}$
   (c) $V_{rms} = (0.707)(1.57)(1.5 \, \text{V}) = 1.66 \, \text{V}$

3. (a) $V_{avg} = (0.637)(10 \, \text{V}) = 6.37 \, \text{V}$
   (b) $V_{avg} = (0.637)(1.414)(2.3 \, \text{V}) = 2.07 \, \text{V}$
   (c) $V_{avg} = (0.637)(0.5)(60 \, \text{V}) = 19.1 \, \text{V}$

SECTION 8–3 Angular Measurement of a Sine Wave

1. (a) The positive peak is at 90°.
   (b) The negative-going zero crossing is at 180°.
   (c) The negative peak is at 270°.
   (d) The cycle ends at 360°.

2. A half-cycle is completed in 180° or π radians.

3. A full cycle is completed in 360° or 2π radians.

4. $\theta = 90° - 45° = 45°$

SECTION 8–4 The Sine Wave Formula

1. (a) $\sin 30° = 0.5$
   (b) $\sin 60° = 0.866$
   (c) $\sin 90° = 1$

2. $v = 10 \sin 120° = 8.66 \, \text{V}$

3. $v = 10 \sin(45° + 10°) = 8.19 \, \text{V}$

SECTION 8–5 Analysis of AC Circuits

1. $I_p = V_p/R = (1.57)(12.5 \, \text{V})/330 \, \Omega = 59.5 \, \text{mA}$

2. $V_{s(rms)} = (0.707)(25.3 \, \text{V}) = 17.9 \, \text{V}$

3. $+V_{max} = 5 \, \text{V} + 2.5 \, \text{V} = 7.5 \, \text{V}$

4. Yes, it will alternate.

5. $+V_{max} = 5 \, \text{V} - 2.5 \, \text{V} = 2.5 \, \text{V}$

SECTION 8–6 Alternators (AC Generators)

1. The number of poles and the rotor speed.

2. The brushes do not have to handle the output current.

3. A dc generator that supplies rotor current to larger alternators.

4. The diodes convert the ac from the stator to dc for the final output.

SECTION 8–7 AC Motors

1. The difference is the rotors. In an induction motor, the rotor obtains current by transformer action; in a synchronous motor, the rotor is a permanent magnet or an electromagnet that is supplied current from an external source through slip rings and brushes.

2. The magnitude is constant.

3. The squirrel cage is composed of the electrical conductors that generate current in the rotor.

4. Slip is the difference between the synchronous speed of the stator field and the rotor speed.

SECTION 8–8   Nonsinusoidal Waveforms

1. **(a)** Rise time is the time interval from 10% to 90% of the amplitude;
   (b) Fall time is the time interval from 90% to 10% of the amplitude;
   (c) Pulse width is the time interval from 50% of the leading pulse edge to 50% of the trailing pulse edge.
2. $f = 1/1 \text{ ms} = 1 \text{ kHz}$
3. $d = (1/5)100\% = 20\%$; Ampl. 1.5 V; $V_{avg} = 0.5 \text{ V} + 0.2(1.5 \text{ V}) = 0.8 \text{ V}$
4. $T = 16 \text{ ms}$
5. $f = 1/T = 1/1 \, \mu\text{s} = 1 \text{ MHz}$
6. Fundamental frequency is the repetition rate of the waveform.
7. 2nd harmonic: 2 kHz
8. $f = 1/10 \, \mu\text{s} = 100 \text{ kHz}$

SECTION 8–9   The Oscilloscope

1. Digital: Signal is converted to digital for processing and then reconstructed for display.
   Analog: Signal drives display directly.
2. Voltage is measured vertically.
3. The Volts/Div control sets the amount of volts represented by each division on the vertical scale.
4. The Sec/Div control sets the amount of time represented by each division on the horizontal scale.
5. Use a ×10 probe most of the time except when measuring very small voltages.

SECTION 8–10   Signal Sources

1. Any waveform within the limits (amplitude and frequency) of the generator. Waves can be specified by various methods.
2. A generator that has a sine wave output that is varied at a cyclic rate between two selected frequencies.
3. Certain measurements of a unit under test can be distorted if the generator does not have good spectral purity.
4. The process in which a signal containing information is used to modify the characteristic of another signal.
5. A mode where the output is triggered for a specific number of cycles at a time.

ANSWERS TO RELATED PROBLEMS FOR EXAMPLES

8–1   2.4 s

8–2   1.5 ms

8–3   20 kHz

8–4   200 Hz

8–5   1 ms

8–6 $V_{pp} = 50 \text{ V}$; $V_{rms} = 17.7 \text{ V}$; $V_{avg} = 15.9 \text{ V}$

8–7   **(a)** $\pi/12$ rad
         (b) 112.5°

8–8   8°

8–9   18.1 V

8–10 $I_{rms} = 4.53 \text{ mA}$; $V_{1(rms)} = 4.53 \text{ V}$; $V_{2(rms)} = 2.54 \text{ V}$; $P_{tot} = 32 \text{ mV}$

8–11 23.7 V

8–12 The waveform in part (a) never goes negative. The waveform in part (b) goes negative for a portion of its cycle.

8–13 250 rpm

8–14 5%

8–15 1.2 V

8–16 120 V

8–17 **(a)** $V_{rms} = 1.06$ V; $f = 50$ Hz
(b) $V_{rms} = 88.4$ mV; $f = 1.67$ kHz
(c) $V_{rms} = 4.81$ V; $f = 5$ kHz
(d) $V_{rms} = 7.07$ V; $f = 250$ kHz

8–18 0.6 V

ANSWERS TO TRUE/FALSE QUIZ

1. T   **2.** F   **3.** F   **4.** T   **5.** T   **6.** F

7. T   **8.** F   **9.** F   **10.** T   **11.** T   **12.** F

ANSWERS TO SELF-TEST

1. (b)   **2.** (b)   **3.** (c)   **4.** (b)   **5.** (d)   **6.** (a)

7. (a)   **8.** (a)   **9.** (c)   **10.** (b)   **11.** (a)   **12.** (d)

13. (c)   **14.** (b)   **15.** (d)   **16.** (b)   **17.** (d)   **18.** (c)

ANSWERS TO TROUBLESHOOTING: SYMPTOM AND CAUSE

1. (c)

2. (c)

3. (b)

4. (a)

5. (b)

CHAPTER 9

CAPACITORS

OUTLINE

OBJECTIVES

- Describe the basic structure and characteristics of a capacitor
- Discuss various types of capacitors
- Analyze series capacitors
- Analyze parallel capacitors
- Describe how a capacitor operates in a dc switching circuit
- Describe how a capacitor operates in an ac circuit
- Discuss some capacitor applications

KEY TERMS

Capacitor
Dielectric
Capacitance
Farad (F)
Dielectric strength
Temperature coefficient

Dielectric constant
Charging
RC time constant
Exponential
Transient time
Capacitive reactance
Instantaneous power
True power
Reactive power
VAR (volt-ampere reactive)
Filter
Ripple voltage
Coupling
Decoupling
Bypass

INTRODUCTION

The *capacitor* is a device that can store electrical charge, thereby creating an electric field which in turn stores energy. The measure of the charge-storing ability of a capacitor is its *capacitance*.

In this chapter, the basic capacitor is introduced and its characteristics are studied. The physical construction and electrical properties of various types of capacitors are discussed. Series and parallel combinations are analyzed, and the basic behavior of capacitors in both dc and ac circuits is studied. Representative circuits that are found in many systems also are discussed. One example, found in nearly all systems that are not battery powered, is the power supply that is used to convert ac utility voltage to dc. Capacitors are important elements of power supplies, and they are discussed in this chapter.

VISIT THE WEBSITE
Study aids for this chapter are available at
http://pearsonhighered.com/floyd

A **capacitor** is a passive electrical component that stores electrical charge and has the property of capacitance.

After completing this section, you should be able to

- Describe the basic structure and characteristics of a capacitor
  - Explain how a capacitor stores charge
  - Define *capacitance* and state its unit
  - Explain how a capacitor stores energy
  - Discuss voltage rating and temperature coefficient
  - Explain capacitor leakage
  - Specify how the physical characteristics of a capacitor affect the capacitance

Basic Construction

In its simplest form, a basic capacitor is an electrical device constructed of two parallel conductive plates separated by an insulating material called the **dielectric**. Connecting leads are attached to the parallel plates. A basic capacitor is shown in Figure 9–1(a), and the schematic symbol for this type of capacitor is shown in part (b).

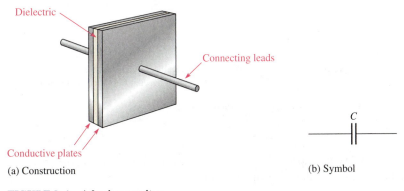

C

(a) Construction

(b) Symbol

FIGURE 9–1   A basic capacitor.

How a Capacitor Stores Charge

In the neutral state, both plates of a capacitor have an equal number of free electrons, as indicated in Figure 9–2(a). When the capacitor is connected to a dc voltage source through a resistor, as shown in part (b), electrons (negative charge) are removed from plate A, and an equal number are deposited on plate B. As plate A loses electrons and plate B gains electrons, plate A becomes positive with respect to plate B. During this charging process, electrons flow only through the connecting leads and the source. No electrons flow through the dielectric of the capacitor because it is an insulator. The movement of electrons ceases when the voltage across the capacitor equals the source voltage, as indicated in Figure 9–2(c). If the capacitor is disconnected from the source, it retains the stored charge for a long period of time (the length of time depends on the type of capacitor) and still has the voltage across it, as shown in Figure 9–2(d). A charged capacitor can act as a temporary battery.

Capacitance

The amount of charge that a capacitor can store per unit of voltage across its plates is its capacitance, designated C. That is, **capacitance** is a measure of a capacitor's ability to store

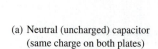

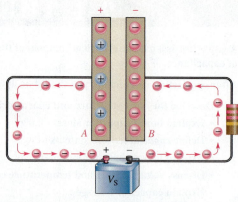

(a) Neutral (uncharged) capacitor (same charge on both plates)

(b) When connected to a voltage source, electrons flow from plate A to plate B as the capacitor charges.

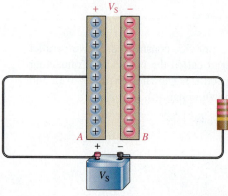

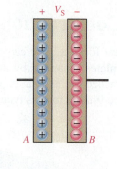

(c) After the capacitor charges to V_S, no electrons flow.

(d) Ideally, the capacitor retains charge when disconnected from the voltage source.

FIGURE 9–2   Illustration of a capacitor storing charge.

charge. The more charge per unit of voltage that a capacitor can store, the greater its capacitance, as expressed by the following formula:

$$C = \frac{Q}{V} \tag{9–1}$$

where C is capacitance, Q is charge, and V is voltage.

By rearranging the terms in Equation 9–1, you can obtain two other formulas.

$$Q = CV \tag{9–2}$$

$$V = \frac{Q}{C} \tag{9–3}$$

THE UNIT OF CAPACITANCE   The **farad (F)** is the basic unit of capacitance. Recall that the coulomb (C) is the unit of electrical charge.

> **One farad is the amount of capacitance when one coulomb of charge is stored with one volt across the plates.**

Most capacitors that are used in electronics work have capacitance values in microfarads (μF) and picofarads (pF). Occasionally, a schematic will show a capacitance in nanofarads (nF). A microfarad is one-millionth of a farad ($1\,\mu$F $= 1 \times 10^{-6}$ F); and a picofarad is one-trillionth of a farad (1 pF $= 1 \times 10^{-12}$ F). Conversions for farads, microfarads, and picofarads are given in Table 9–1.

TABLE 9–1 • Conversions for farads, microfarads, and picofarads.

TO CONVERT FROM	TO	MOVE THE DECIMAL POINT
Farads	Microfarads	right 6 places ($\times 10^6$)
Farads	Picofarads	right 12 places ($\times 10^{12}$)
Microfarads	Farads	left 6 places ($\times 10^{-6}$)
Microfarads	Picofarads	right 6 places ($\times 10^6$)
Nanofarads	Farads	left 9 places ($\times 10^{-9}$)
Nanofarads	Microfarads	left 3 places ($\times 10^{-3}$)
Picofarads	Farads	left 12 places ($\times 10^{-12}$)
Picofarads	Microfarads	left 6 places ($\times 10^{-6}$)

EXAMPLE 9–1

(a) A certain capacitor stores 50 microcoulombs (50 μC) when 10 V are applied across its plates. What is its capacitance?

(b) A 2.2 μF capacitor has 100 V across its plates. How much charge does it store?

(c) Determine the voltage across a 100 pF capacitor that is storing 2 μC of charge.

SOLUTION

(a) $C = \dfrac{Q}{V} = \dfrac{50\ \mu C}{10\ V} = \textbf{5}\ \boldsymbol{\mu}\textbf{F}$

(b) $Q = CV = (2.2\ \mu F)(100\ V) = \textbf{220}\ \boldsymbol{\mu}\textbf{C}$

(c) $V = \dfrac{Q}{C} = \dfrac{2\ \mu C}{100\ pF} = \textbf{20\ kV}$

RELATED PROBLEM*

Determine V if $C = 1000$ pF and $Q = 10\ \mu$C.

*Answers are at the end of the chapter.

EXAMPLE 9–2

Convert the following values to microfarads:

(a) 0.00001 F   **(b)** 0.0047 F   **(c)** 1000 pF   **(d)** 220 pF

SOLUTION

(a) 0.00001 F $\times 10^6\ \mu$F/F $= \textbf{10}\ \boldsymbol{\mu}\textbf{F}$

(b) 0.0047 F $\times 10^6\ \mu$F/F $= \textbf{4700}\ \boldsymbol{\mu}\textbf{F}$

(c) 1000 pF $\times 10^{-6}\ \mu$F/pF $= \textbf{0.001}\ \boldsymbol{\mu}\textbf{F}$

(d) 220 pF $\times 10^{-6}\ \mu$F/pF $= \textbf{0.00022}\ \boldsymbol{\mu}\textbf{F}$

RELATED PROBLEM

Convert 47,000 pF to microfarads.

E X A M P L E   9 – 3

Convert the following values to picofarads:

(a) 0.1×10^{-8} F   **(b)** 0.000027 F   **(c)** 0.01 μF   **(d)** 0.0047 μF

SOLUTION

(a) 0.1×10^{-8} F $\times 10^{12}$ pF/F = **1000 pF**

(b) 0.000027 F $\times 10^{12}$ pF/F = **27 $\times 10^6$ pF**

(c) 0.01 μF $\times 10^6$ pF/μF = **10,000 pF**

(d) 0.0047 μF $\times 10^6$ pF/μF = **4700 pF**

RELATED PROBLEM

Convert 100 μF to picofarads.

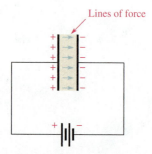

Lines of force

FIGURE 9–3 **The electric field stores energy in a capacitor. The beige area indicates the dielectric.**

How a Capacitor Stores Energy

A capacitor stores energy in an electric field that is established by the opposite charges stored on the two plates. The electric field is represented by lines of force between the positive and negative charges and is concentrated within the dielectric, as shown in Figure 9–3.

The plates in Figure 9–3 have acquired a charge because they are connected to a battery. This creates an electric field between the plates, which stores energy. The energy stored in the electric field is directly related to the size of the capacitor and to the square of the voltage as given by the following equation for the energy stored:

$$W = \frac{1}{2}CV^2 \qquad (9\text{–}4)$$

When capacitance (C) is in farads and voltage (V) is in volts, the energy (W) is in joules.

Voltage Rating

Every capacitor has a limit on the amount of voltage that it can withstand across its plates. The voltage rating specifies the maximum dc voltage that can be applied without risk of damage to the device. If this maximum voltage, commonly called the *breakdown voltage* or *working voltage*, is exceeded, permanent damage to the capacitor can result.

You must consider both the capacitance and the voltage rating before you use a capacitor in a circuit application. The choice of capacitance value is based on particular circuit requirements. The voltage rating should always be above the maximum voltage expected in a particular application.

DIELECTRIC STRENGTH   The breakdown voltage of a capacitor is determined by the **dielectric strength** of the dielectric material used. The dielectric strength is expressed in V/mil (1 mil = 0.001 in. = 2.54×10^{-5} meter). Table 9–2 lists typical values for several materials. Exact values vary depending on the specific composition of the material.

The dielectric strength of a capacitor can best be explained by an example. Assume that a certain capacitor has a plate separation of 1 mil and that the dielectric material is ceramic. This particular capacitor can withstand a maximum voltage of 1000 V because its dielectric strength is 1000 V/mil. If the maximum voltage is exceeded, the dielectric may break down and conduct current, causing permanent damage to the capacitor. If the ceramic capacitor has a plate separation of 2 mils, its breakdown voltage is 2000 V.

TABLE 9–2 • Some common dielectric materials and their typical dielectric strengths.	
MATERIAL	**DIELECTRIC STRENGTH (V/MIL)**
Air	80
Oil	375
Ceramic	1000
Paper (paraffined)	1200
Teflon®	1500
Mica	1500
Glass	2000

Temperature Coefficient

The **temperature coefficient** indicates the amount and direction of a change in capacitance value with temperature. A positive temperature coefficient means that the capacitance increases with an increase in temperature or decreases with a decrease in temperature. A negative coefficient means that the capacitance decreases with an increase in temperature or increases with a decrease in temperature.

Temperature coefficients typically are specified in parts per million per Celsius degree (ppm/°C). For example, a negative temperature coefficient of 150 ppm/°C for a 1 μF capacitor means that for every degree Celsius rise in temperature, the capacitance decreases by 150 pF (there are one million picofarads in one microfarad).

Leakage

No insulating material is perfect. The dielectric of any capacitor will conduct some very small amount of current. Thus, the charge on a capacitor will eventually leak off. Some types of capacitors have higher leakages than others. An equivalent circuit for a nonideal capacitor is shown in Figure 9–4. The parallel resistor R_{leak} represents the extremely high resistance (several hundred kΩ or more) of the dielectric material through which there is leakage current.

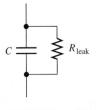

FIGURE 9–4 Equivalent circuit for a nonideal capacitor.

Capacitive Sensors

Capacitive sensors are common in systems that measure quantities like pressure, proximity, humidity, liquid level, and more. A capacitive sensor is a variable capacitor that responds to a physical stimulus. In the case of a pressure sensor, a conductive, flexible diaphragm forms one plate of the capacitor. The other plate is fixed. Pressure causes the diaphragm to move, thus changing the capacitance. The change in capacitance can be sensed by an external circuit.

In the case of a proximity sensor, one of the plates can actually be a moving object (such as the cam on a rotating shaft). Other sensors can change the dielectric as in the case of a liquid-level sensor.

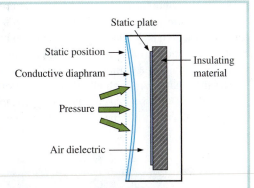

 SYSTEM NOTE

Physical Characteristics of a Capacitor

The following parameters are important in establishing the capacitance and the voltage rating of a capacitor: plate area, plate separation, and dielectric constant.

PLATE AREA   *Capacitance is directly proportional to the physical size of the plates as determined by the plate area, A.* A large plate area produces more capacitance, and a smaller plate area produces less capacitance. Figure 9–5(a) shows that the plate area of a parallel plate capacitor is the area of one of the plates. If the plates are moved in relation to each other, as shown in Figure 9–5(b), the overlapping area determines the effective plate area. This variation in effective plate area is the basis for a certain type of variable capacitor.

PLATE SEPARATION   *Capacitance is inversely proportional to the distance between the plates.* The plate separation is designated *d*, as shown in Figure 9–6. A greater separation of the plates produces a lesser capacitance, as illustrated in the figure. The breakdown voltage is directly proportional to the plate separation. The further the plates are separated, the greater the breakdown voltage.

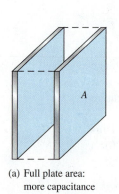 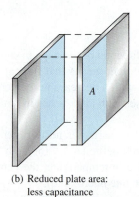

(a) Full plate area:
    more capacitance

(b) Reduced plate area:
    less capacitance

FIGURE 9–5   Capacitance is directly proportional to plate area (A).

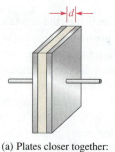 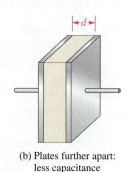

(a) Plates closer together:
    more capacitance

(b) Plates further apart:
    less capacitance

FIGURE 9–6   Capacitance is inversely proportional to the distance *d* between the plates.

DIELECTRIC CONSTANT   As you know, the insulating material between the plates of a capacitor is called the *dielectric*. Every dielectric material has the ability to concentrate the lines of force of the electric field existing between the oppositely charged plates of a capacitor and thus increase the capacity for energy storage. The measure of a material's ability to establish an electric field is called the **dielectric constant** or *relative permittivity,* symbolized by ε_r (the Greek letter epsilon).

Capacitance is directly proportional to the dielectric constant. The dielectric constant of a vacuum is defined as 1, and that of air is very close to 1. These values are used as a reference, and all other materials have values of ε_r specified with respect to that of a vacuum or air. For example, a material with $\varepsilon_r = 5$ can have a capacitance five times greater than that of air, with all other factors being equal.

Table 9–3 lists several common dielectric materials and a typical dielectric constant for each. The values can vary because it depends on the specific composition of the material.

TABLE 9–3 • Some common dielectric materials and their typical dielectric constants.	
MATERIAL	**TYPICAL ε_r VALUE**
Air (vacuum)	1.0
Teflon®	2.0
Paper (paraffined)	2.5
Oil	4.0
Mica	5.0
Glass	7.5
Ceramic	1200

The dielectric constant (relative permittivity) is dimensionless because it is a relative measure. It is a ratio of the absolute permittivity of a material, ε, to the absolute permittivity of a vacuum, ε_0, as expressed by the following formula:

$$\varepsilon_r = \frac{\varepsilon}{\varepsilon_0}$$

(9–5)

The value of ε_0 is 8.85×10^{-12} F/m (farads per meter).

FORMULA FOR CAPACITANCE   You have seen how capacitance is directly related to plate area, A, and the dielectric constant, ε_r, and inversely related to plate separation, d. An exact formula for calculating the capacitance in terms of these three quantities is

$$C = \frac{A\varepsilon_r(8.85 \times 10^{-12}\ \text{F/m})}{d}$$

(9–6)

where A is in square meters (m^2), d is in meters (m), and C is in farads (F).

EXAMPLE 9–4

Determine the capacitance in μF of a parallel plate capacitor having a plate area of 0.01 m^2 and a plate separation of 0.5 mil (1.27×10^{-5} m). The dielectric is mica, which has a dielectric constant of 5.0.

SOLUTION

Use Equation 9–6.

$$C = \frac{A\varepsilon_r(8.85 \times 10^{-12}\ \text{F/m})}{d} = \frac{(0.01\ \text{m}^2)\,(5.0)\,(8.85 \times 10^{-12}\ \text{F/m})}{1.27 \times 10^{-5}\ \text{m}} = \mathbf{0.035\ \mu F}$$

RELATED PROBLEM

Determine C in μF where $A = 3.6 \times 10^{-5}\ m^2$, $d = 1$ mil (2.54×10^{-5} m), and ceramic is the dielectric.

SECTION 9–1 CHECKUP*

1. Define *capacitance*.

2. (a) How many microfarads in one farad?

   (b) How many picofarads in one farad?

   (c) How many picofarads in one microfarad?

3. Convert 0.0015 μF to picofarads. To farads.

4. How much energy in joules is stored by a 0.01 μF capacitor with 15 V across its plates?

5. (a) When the plate area of a capacitor is increased, does the capacitance increase or decrease?

   (b) When the distance between the plates is increased, does the capacitance increase or decrease?

6. The plates of a ceramic capacitor are separated by 2 mils. What is the typical breakdown voltage?

*Answers are at the end of the chapter.

9–2 TYPES OF CAPACITORS

Capacitors normally are classified according to the type of dielectric material. The most common types of dielectric materials are mica, ceramic, plastic-film, and electrolytic (aluminum oxide and tantalum oxide).

After completing this section, you should be able to

- **Discuss various types of capacitors**
  - Describe the characteristics of mica, ceramic, plastic-film, and electrolytic capacitors
  - Describe types of variable capacitors
  - Identify capacitor labeling
  - Discuss capacitance measurement

Fixed Capacitors

MICA CAPACITORS   Two types of mica capacitors are stacked-foil and silver-mica. The basic construction of the stacked-foil type is shown in Figure 9–7. It consists of alternate layers of metal foil and thin sheets of mica. The metal foil forms the plate, with alternate foil sheets connected together to increase the plate area. More layers are used to increase the plate area, thus increasing the capacitance. The mica/foil stack is encapsulated in an insulating material such as Bakelite®, as shown in Figure 9–7(b). The silver-mica capacitor is formed in a similar way by stacking mica sheets with silver electrode material screened on them.

Mica capacitors are generally available with capacitance values ranging from 1 pF to 0.1 μF and voltage ratings from 100 V dc to 2500 V dc and higher. Mica has a typical dielectric constant of 5.

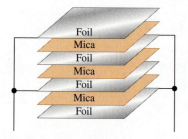

(a) Stacked layer arrangement

(b) Layers are pressed together and encapsulated.

FIGURE 9–7   **Construction of a typical radial-lead mica capacitor.**

CERAMIC CAPACITORS   Ceramic dielectrics provide very high dielectric constants (1200 is typical). As a result, comparatively high capacitance values can be achieved in a small physical size. Ceramic capacitors are commonly available in a ceramic disk form, as shown in Figure 9–8; in a multilayer radial-lead configuration, as shown in Figure 9–9; or in a leadless ceramic chip, as shown in Figure 9–10.

Ceramic capacitors typically are available in capacitance values ranging from 1 pF to 100 μF with voltage ratings up to 6 kV.

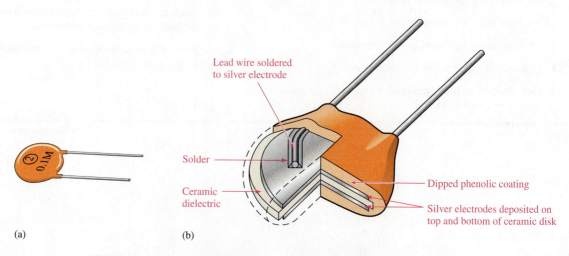

Lead wire soldered to silver electrode

Solder

Ceramic dielectric

Dipped phenolic coating

Silver electrodes deposited on top and bottom of ceramic disk

(a)

(b)

FIGURE 9–8   **A ceramic disk capacitor and its basic construction.**

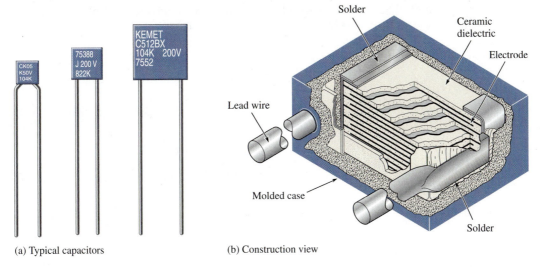

(a) Typical capacitors

(b) Construction view

FIGURE 9–9   **Examples of ceramic capacitors.**

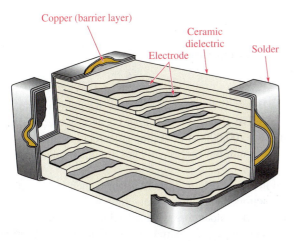

FIGURE 9–10   **Construction view of a typical ceramic chip capacitor used for surface mounting on printed circuit boards.**

PLASTIC-FILM CAPACITORS   Common dielectric materials used in plastic-film capacitors include polycarbonate, propylene, polyester, polystyrene, polypropylene, and mylar. Some of these types have capacitance values up to 100 μF but most are less than 1 μF.

Figure 9–11 shows a basic construction used in many plastic-film capacitors. A thin strip of plastic-film dielectric is sandwiched between two thin metal strips that act as plates. One lead is connected to the inner plate and one is connected to the outer plate as indicated. The strips are then rolled in a spiral configuration and encapsulated in a molded case. Thus, a large plate area can be packaged in a relatively small physical size, thereby achieving large capacitance values. Another method uses metal deposited directly on the film dielectric to form the plates.

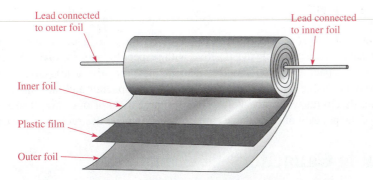

FIGURE 9–11   **Basic construction of axial-lead tubular plastic-film dielectric capacitors.**

Figure 9–12(a) shows typical plastic-film capacitors. Figure 9–12(b) shows a construction view for one type of an axial-lead plastic-film capacitor.

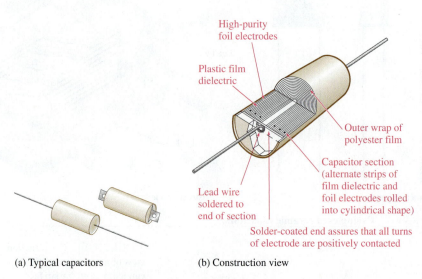

(a) Typical capacitors   (b) Construction view

FIGURE 9–12   Examples of plastic-film capacitors.

ELECTROLYTIC CAPACITORS   Electrolytic capacitors are polarized so that one plate is positive and the other negative. These capacitors are generally used for high capacitance values from $1\ \mu F$ up to over $200,000\ \mu F$, but they have relatively low break-down voltages (350 V is a typical maximum but higher voltages are occasionally found) and high amounts of leakage.

Manufacturers have developed new electrolytic capacitors with much larger capacitance values; however, these new capacitors have lower voltage ratings than smaller-value capacitors and tend to be expensive. Super capacitors with capacitances of hundreds of farads are available. These capacitors are useful for battery backup and for applications like small motor starters that require a very large capacitance.

Electrolytic capacitors offer much higher capacitance values than mica or ceramic capacitors, but their voltage ratings are typically lower. While other capacitors use two similar plates, the electrolytic capacitor consists of one plate made of aluminum foil and another plate made of a conducting electrolyte applied to a material such as plastic film. These two "plates" are separated by a layer of aluminum oxide that forms on the surface of the aluminum plate. Figure 9–13(a) illustrates the construction of a typical aluminum electrolytic capacitor with axial leads. Other electrolytic capacitors with radial leads are shown in Figure 9–13(b); the symbol for an electrolytic capacitor is shown in part (c).

Tantalum electrolytic capacitors can be in either a tubular configuration similar to Figure 9–13 or "tear drop" shape as shown in Figure 9–14. In the tear drop configuration, the positive plate is actually a pellet of tantalum powder rather than a sheet of foil. Tantalum pentoxide forms the dielectric, and manganese dioxide forms the negative plate.

Because of the process used for the insulating pentoxide dielectric, the metallic (aluminum or tantalum) plate is always positive with respect to the electrolyte plate, and thus all electrolytic capacitors are polarized. The metal plate (positive lead) is usually indicated by a plus sign or some other obvious marking on the capacitor and must always be connected in a dc circuit where the voltage across the capacitor does not change polarity. Reversal of the polarity of the voltage can result in complete destruction of the capacitor.

Variable Capacitors

Variable capacitors are used in a circuit when there is a need to adjust the capacitance value either manually or automatically. These capacitors are generally less than 300 pF but

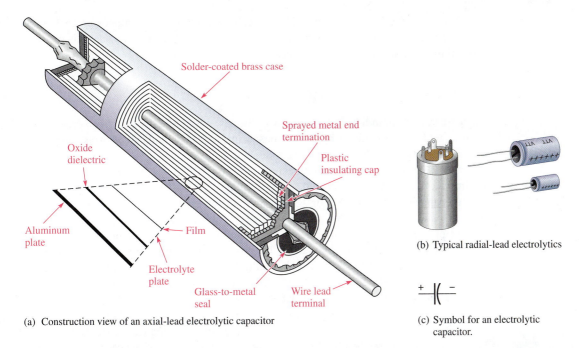

Solder-coated brass case

Sprayed metal end termination

Plastic insulating cap

Oxide dielectric

Aluminum plate

Film

Electrolyte plate

Glass-to-metal seal

Wire lead terminal

(a) Construction view of an axial-lead electrolytic capacitor

(b) Typical radial-lead electrolytics

(c) Symbol for an electrolytic capacitor.

FIGURE 9–13 **Examples of electrolytic capacitors.**

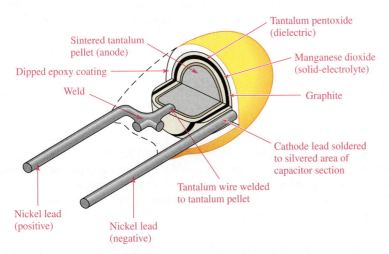

Tantalum pentoxide (dielectric)

Sintered tantalum pellet (anode)

Manganese dioxide (solid-electrolyte)

Dipped epoxy coating

Weld

Graphite

Cathode lead soldered to silvered area of capacitor section

Tantalum wire welded to tantalum pellet

Nickel lead (positive)

Nickel lead (negative)

FIGURE 9–14 **Construction view of a typical "tear drop" shaped tantalum electrolytic capacitor.**

are available in larger values for specialized application. The schematic symbol for a variable capacitor is shown in Figure 9–15.

Adjustable capacitors that normally have slotted screw-type adjustments and are used for very fine adjustments in a circuit are called **trimmers**. Ceramic or mica is a common dielectric in these types of capacitors, and the capacitance usually is changed by adjusting the plate separation. Generally, trimmer capacitors have values less than 100 pF. Figure 9–16 shows some typical devices.

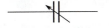

FIGURE 9–15 **Schematic symbol for a variable capacitor.**

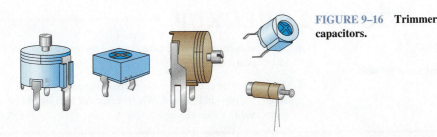

FIGURE 9–16 **Trimmer capacitors.**

The **varactor** is a semiconductive device that exhibits a capacitance characteristic that is varied by changing the voltage across its terminals. This device usually is covered in detail in a course on electronic devices.

Capacitor Labeling

Capacitor values are indicated on the body of the capacitor either by numerical or alphanumerical labels and sometimes by color codes. Capacitor labels indicate various parameters such as capacitance, voltage rating, and tolerance.

Some capacitors carry no unit designation for capacitance. In these cases, the units are implied by the value indicated and are recognized by experience. For example, a ceramic capacitor marked .001 or .01 has units of microfarads because picofarad values that small are not available. As another example, a ceramic capacitor labeled 50 or 330 has units of picofarads because microfarad units that large are not available in this type. In some cases a 3-digit designation is used. The first two digits are the first two digits of the capacitance value. The third digit is the multiplier or number of zeros after the second digit. For example, 103 means 10,000 pF. In some instances, the units are labeled as pF or μF; sometimes the microfarad unit is labeled as MF or MFD.

A voltage rating appears on some types of capacitors with WV or WVDC and is omitted on others. When it is omitted, the voltage rating can be determined from information supplied by the manufacturer. The tolerance of the capacitor is usually labeled as a percentage, such as $\pm 10\%$. The temperature coefficient is indicated by a *parts per million* marking. This type of label consists of a P or N followed by a number. For example, N750 means a negative temperature coefficient of 750 ppm/°C, and P30 means a positive temperature coefficient of 30 ppm/°C. An NP0 designation means that the positive and negative coefficients are zero; thus the capacitance does not change with temperature. Certain types of capacitors are color coded. Refer to Appendix B for additional capacitor labeling and color code information.

Capacitance Measurement

A capacitance meter, such as the one shown in Figure 9–17, can be used to check the value of a capacitor. Also, many DMMs provide a capacitance measurement feature. All capacitors change value over a period of time, some more than others. Ceramic capacitors, for example, often exhibit a 10% to 15% change in the value during the first year. Electrolytic capacitors are particularly subject to value change due to drying of the electrolytic solution. In other cases, capacitors may be labeled incorrectly or the wrong value may have been installed in the circuit. Although a value change represents less than 25% of defective capacitors, a value check should be made to quickly eliminate this as a source of trouble when troubleshooting a circuit. Values from 200 pF to 50,000 μF can be measured by simply connecting the capacitor, setting the switch, and reading the value on the display.

Some capacitance meters can also be used to check for leakage current in capacitors. In order to check for leakage, a sufficient voltage must be applied across the capacitor to simulate operating conditions. This is automatically done by the test instrument. Over 40% of all defective capacitors have excessive leakage current and electrolytic capacitors are particularly susceptible to this problem.

FIGURE 9–17   A typical capacitance meter.   (Photo courtesy of B&K Precision Corp.)

SECTION 9–2 CHECKUP

1. How are capacitors commonly classified?

2. What is the difference between a fixed and a variable capacitor?

3. What type of capacitor normally is polarized?

4. What precautions must be taken when a polarized capacitor is installed in a circuit?

5. An electrolytic capacitor is connected between a negative supply voltage and ground. Which capacitor lead should be connected to ground?

The total capacitance of a series connection of capacitors is less than the individual capacitance of any of the capacitors. Capacitors in series divide voltage across them inversely proportional to their capacitance. Thus, a large capacitor in series with a smaller one will have less voltage than a smaller one.

After completing this section, you should be able to

- **Analyze series capacitors**
  - **Determine total capacitance**
  - **Determine capacitor voltages**

When capacitors are connected in series, the total capacitance is less than the smallest capacitance value because the effective plate separation increases. The calculation of total series capacitance is analogous to the calculation of total resistance of parallel resistors (Chapter 5).

Two capacitors can be used in series to show how the total capacitance is determined. Figure 9–18 shows two capacitors, which initially are uncharged, connected in series with a dc voltage source. When the switch is closed, as shown in part (a), current begins.

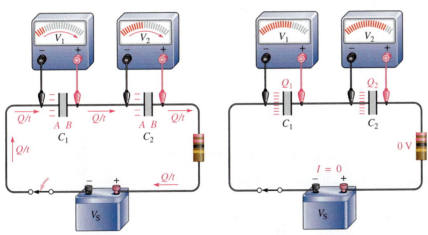

(a) While charging, $I = Q/t$ is the same at all points. Capacitor voltages are increasing.

(b) Both capacitors store the same amount of charge ($Q_T = Q_1 = Q_2$).

FIGURE 9–18   **Capacitors in series produce a total capacitance that is less than the smallest value.**

Recall that current is the same at all points in a series circuit and that current is defined as the rate of flow of charge ($I = Q/t$). In a certain period of time, a certain amount of charge moves through the circuit. Since current is the same everywhere in the circuit of Figure 9–18(a), the same amount of charge is moved from the negative side of the source to plate A of C_1, and from plate B of C_1 to plate A of C_2, and from plate B of C_2 to the positive side of the source. As a result, of course, the same amount of charge is deposited on the plates of both capacitors in a given period of time, and the total charge (Q_T) moved through the circuit in that period of time equals the charge stored by C_1 and also equals the charge stored by C_2.

$$Q_T = Q_1 = Q_2$$

As the capacitors charge, the voltage across each one increases as indicated.

Figure 9–18(b) shows the capacitors after they have been completely charged and the current has ceased. Both capacitors store an equal amount of charge (Q), and the voltage across each one depends on its capacitance value ($V = Q/C$). By Kirchhoff's voltage law, which applies to capacitive circuits as well as to resistive circuits, the sum of the capacitor voltages equals the source voltage.

$$V_S = V_1 + V_2$$

Using the formula $V = Q/C$, you can substitute into the formula for Kirchhoff's law and get the following relationship (where $Q = Q_T = Q_1 = Q_2$):

$$\frac{Q}{C_T} = \frac{Q}{C_1} + \frac{Q}{C_2}$$

The Q can be factored out of the right side of the equation and canceled with the Q on the left side as follows:

$$\frac{\cancel{Q}}{C_T} = \cancel{Q}\left(\frac{1}{C_1} + \frac{1}{C_2}\right)$$

Thus, you have the following relationship for two capacitors in series:

$$\frac{1}{C_T} = \frac{1}{C_1} + \frac{1}{C_2}$$

Taking the reciprocal of both sides of this equation gives the formula for the total capacitance for two capacitors in series.

$$C_T = \frac{1}{\dfrac{1}{C_1} + \dfrac{1}{C_2}}$$

This equation can also be expressed equivalently as

$$C_T = \frac{C_1 C_2}{C_1 + C_2} \qquad (9\text{--}7)$$

EXAMPLE 9–5

Find the total capacitance C_T in Figure 9–19.

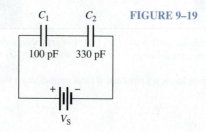

FIGURE 9–19

SOLUTION

$$C_T = \frac{C_1 C_2}{C_1 + C_2} = \frac{(100\,\text{pF})(330\,\text{pF})}{100\,\text{pF} + 330\,\text{pF}} = \mathbf{76.7\ pF}$$

RELATED PROBLEM

Determine C_T if $C_1 = 470$ pF and $C_2 = 680$ pF in Figure 9–19.

GENERAL FORMULA   Although more than two capacitors in series is rare in practical circuits, you should be aware of the general formula, should the situation arise. The equation for two series capacitors can be extended to any number of capacitors in series, as shown in Figure 9–20.

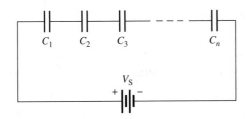

FIGURE 9–20   **General circuit with *n* capacitors in series.**

The formula for total capacitance for any number of capacitors in series is developed as follows; the subscript *n* can be any number.

$$\frac{1}{C_T} = \frac{1}{C_1} + \frac{1}{C_2} + \frac{1}{C_3} + \cdots + \frac{1}{C_n}$$

$$C_T = \cfrac{1}{\cfrac{1}{C_1} + \cfrac{1}{C_2} + \cfrac{1}{C_3} + \cdots + \cfrac{1}{C_n}} \qquad (9\text{–}8)$$

Remember,

The total series capacitance is always less than the smallest capacitance.

EXAMPLE 9–6

Determine the total capacitance in Figure 9–21.

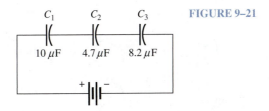

FIGURE 9–21

SOLUTION

$$C_T = \cfrac{1}{\cfrac{1}{C_1} + \cfrac{1}{C_2} + \cfrac{1}{C_3}} = \cfrac{1}{\cfrac{1}{10\ \mu F} + \cfrac{1}{4.7\ \mu F} + \cfrac{1}{8.2\ \mu F}} = \textbf{2.30}\ \mu\textbf{F}$$

RELATED PROBLEM

If another 4.7 μF capacitor is connected in series with the three existing capacitors in Figure 9–21, what is the value of C_T?

Capacitor Voltages

The voltage across each capacitor in a series connection depends on its capacitance value according to the formula $V = Q/C$. You can determine the voltage across any individual capacitor in series with the following formula:

$$V_x = \left(\frac{C_T}{C_x}\right) V_S \qquad (9\text{–}9)$$

where C_x is any capacitor in series, such as C_1, C_2, and C_3, and so on, and V_x is the voltage across C_x.

The largest-value capacitor in a series connection will have the smallest voltage across it. The smallest-value capacitor will have the largest voltage across it.

EXAMPLE 9–7

Find the voltage across each capacitor in Figure 9–22.

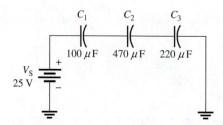

FIGURE 9–22

SOLUTION

Calculate the total capacitance.

$$C_T = \frac{1}{\frac{1}{C_1} + \frac{1}{C_2} + \frac{1}{C_3}} = \frac{1}{\frac{1}{100\,\mu F} + \frac{1}{470\,\mu F} + \frac{1}{220\,\mu F}} = 60\,\mu F$$

The voltages are as follows:

$$V_1 = \left(\frac{C_T}{C_1}\right)V_S = \left(\frac{60\,\mu F}{100\,\mu F}\right)25\,V = \textbf{15.0 V}$$

$$V_2 = \left(\frac{C_T}{C_2}\right)V_S = \left(\frac{60\,\mu F}{470\,\mu F}\right)25\,V = \textbf{3.19 V}$$

$$V_3 = \left(\frac{C_T}{C_3}\right)V_S = \left(\frac{60\,\mu F}{220\,\mu F}\right)25\,V = \textbf{6.82 V}$$

RELATED PROBLEM

Another 470 μF capacitor is connected in series with the existing capacitors in Figure 9–22. Determine the voltage across the new capacitor, assuming all the capacitors are initially uncharged.

SYSTEM EXAMPLE 9–1

FEEDBACK IN A COLPITTS OSCILLATOR

Oscillators are circuits that generate a repetitive wave. Sinusoidal oscillators are used in many systems and for test signals. The circuit shown in Figure 9–23 is a sinusoidal oscillator designed for producing 550 kHz sine waves, which are used in AM radio receivers to mix with the incoming frequency from the antenna. This circuit is an example of one that uses a feedback network; a portion of the output is returned by the feedback network to the input to be amplified and reinforce the output. The feedback network is shown in the shaded box. Our focus is on the capacitors in the feedback network. Although the circuit is an ac circuit, the equations for capacitance and capacitor voltages developed with dc sources can be applied.

Capacitors C_1 and C_2 are in series and form a resonant circuit with L (resonant circuits are covered in Chapter 13). The total capacitance of C_1 in series with C_2 is found by Equation 9–7:

$$C_T = \frac{C_1 C_2}{C_1 + C_2} = \frac{(2.2 \text{ nF})(220 \text{ pF})}{2.2 \text{ nF} + 220 \text{ pF}} = 200 \text{ pF}$$

The amplifier's signals are referenced to ground, so the feedback signal, V_f, is the fraction of the output developed across C_1. Notice that V_f is returned to the input of the amplifier. You can find the fraction of output voltage that is fed back (V_f/V_{out}) by rearranging Equation 9–9 and substituting, so V_{out} for V_S and V_f for V_x.

$$\frac{V_x}{V_S} = \frac{V_f}{V_{out}} = \frac{C_T}{C_1} = \left(\frac{200 \text{ pF}}{2.2 \text{ nF}}\right) = 0.0909 = 9.1\%$$

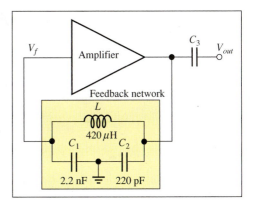

FIGURE 9–23 A Colpitts oscillator showing the feedback network.

As you can see from this result, the larger capacitor has the smaller fraction of the output voltage. The gain of the amplifier has to be high enough to keep oscillations going and restore the output to 100%. The feedback network determines the required gain for the amplifier portion of the Colpitts oscillator. The feedback has returned only 9.1% of the output, so the gain needs to be 11 (9.1% $\times$ 11 = 100%).

SECTION 9–3 CHECKUP

1. Is the total capacitance of series capacitors less than or greater than the value of the smallest capacitor?

2. The following capacitors are in series: 100 pF, 220 pF, and 560 pF. What is the total capacitance?

3. A 0.01 μF and a 0.015 μF capacitor are in series. Determine the total capacitance.

4. Determine the voltage across the 0.01 μF capacitor in Question 3 if 10 V are connected across the two series capacitors.

9–4 PARALLEL CAPACITORS

Capacitances add when they are connected in parallel. As in all parallel circuits, the voltage across each capacitor is the same as the source voltage.

After completing this section, you should be able to

- **Analyze parallel capacitors**
  - **Determine total capacitance**

When capacitors are connected in parallel, the total capacitance is the sum of the individual capacitances because the effective plate area increases. The calculation of total parallel capacitance is analogous to the calculation of total series resistance (Chapter 4).

Figure 9–24 shows two parallel capacitors connected to a dc voltage source. When the switch is closed, as shown in part (a), current begins. A total amount of charge (Q_T) moves through the circuit in a certain period of time. Part of the total charge is stored by C_1 and part by C_2. The portion of the total charge that is stored by each capacitor depends on its capacitance value according to the relationship $Q = CV$.

Figure 9–24(b) shows the capacitors after they have been completely charged and the current has stopped. Since the voltage across both capacitors is the same, the larger capacitor

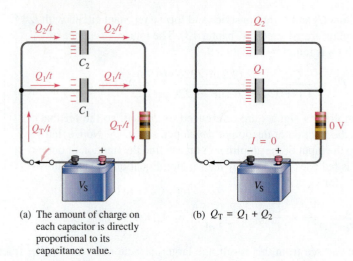

(a) The amount of charge on each capacitor is directly proportional to its capacitance value.

(b) $Q_T = Q_1 + Q_2$

FIGURE 9–24 Capacitors in parallel produce a total capacitance that is the sum of the individual capacitances.

stores more charge. If the capacitors are equal in value, they store an equal amount of charge. The charge stored by both of the capacitors together equals the total charge that was delivered from the source.

$$Q_T = Q_1 + Q_2$$

Since $Q = CV$, Equation 9–2, you can use substitution to get the following relationship:

$$C_T V_S = C_1 V_S + C_2 V_S$$

Because all the V_S terms are equal, they can be canceled. Therefore, the total capacitance for two capacitors in parallel is

$$C_T = C_1 + C_2 \qquad\qquad (9\text{–}10)$$

EXAMPLE 9–8

What is the total capacitance in Figure 9–25? What is the voltage across each capacitor?

FIGURE 9–25

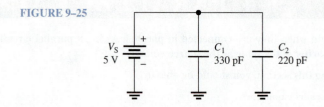

V_S 5 V C_1 330 pF C_2 220 pF

SOLUTION

The total capacitance is

$$C_T = C_1 + C_2 = 330 \text{ pF} + 220 \text{ pF} = \mathbf{550 \text{ pF}}$$

The voltage across each capacitor in parallel is equal to the source voltage.

$$V_S = V_1 = V_2 = \mathbf{5 \text{ V}}$$

RELATED PROBLEM

What is C_T if a 100 pF capacitor is connected in parallel with C_1 and C_2 in Figure 9–25?

GENERAL FORMULA   There are a few cases in practical circuits where you may find more than two parallel capacitors. In some industrial plants, parallel capacitors can be switched into the circuit depending on the load. For this reason, you should be aware of the general formula. Equation 9–10 can be extended to any number of capacitors in parallel, as shown in Figure 9–26. The expanded formula is as follows; the subscript n can be any number.

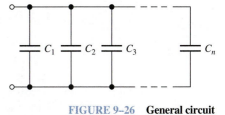

$$C_T = C_1 + C_2 + C_3 + \cdots + C_n \qquad (9\text{–}11)$$

FIGURE 9–26   **General circuit with n capacitors in parallel.**

EXAMPLE 9–9

Determine C_T in Figure 9–27.

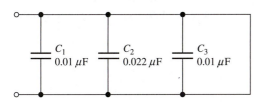

FIGURE 9–27

SOLUTION

$$C_T = C_1 + C_2 + C_3 = 0.01\ \mu F + 0.022\ \mu F + 0.01\ \mu F = \mathbf{0.042\ \mu F}$$

RELATED PROBLEM

If one more $0.01\ \mu F$ capacitor is connected in parallel in Figure 9–27, what is C_T?

SECTION 9–4 CHECKUP

1. How is total parallel capacitance determined?

2. In a certain application, you need $0.05\ \mu F$. The only capacitor value available is $0.01\ \mu F$, and this capacitor is available in large quantities. How can you get the total capacitance that you need?

3. The following capacitors are in parallel: 10 pF, 56 pF, 33 pF, and 68 pF. What is C_T?

9–5 CAPACITORS IN DC CIRCUITS

A capacitor will charge up when it is connected to a dc voltage source. The buildup of charge across the plates occurs in a predictable manner that is dependent on the capacitance and the resistance in a circuit.

After completing this section, you should be able to

- **Describe how a capacitor operates in a dc switching circuit**
  - **Describe the charging and discharging of a capacitor**
  - **Define *RC time constant***
  - **Relate the time constant to charging and discharging**
  - **Write equations for the charging and discharging curves**
  - **Explain why a capacitor blocks constant dc**

Charging a Capacitor

A capacitor will charge when it is connected to a dc voltage source, as shown in Figure 9–28. The capacitor in part (a) is uncharged; that is, plate *A* and plate *B* have equal numbers of free electrons. When the switch is closed, as shown in part (b), the source moves electrons away from plate *A* through the circuit to plate *B* as the arrows indicate. As plate *A* loses electrons and plate *B* gains electrons, plate *A* becomes positive with respect to plate *B*. As this **charging** process continues, the voltage across the plates builds up rapidly until it is equal to the source voltage, V_S, but opposite in polarity, as shown in part (c). When the capacitor is fully charged, there is no current.

A capacitor blocks constant dc.

When the charged capacitor is disconnected from the source, as shown in Figure 9–28(d), it remains charged for long periods of time, depending on its leakage resistance. The charge on an electrolytic capacitor generally leaks off more rapidly than in other types of capacitors.

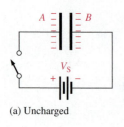
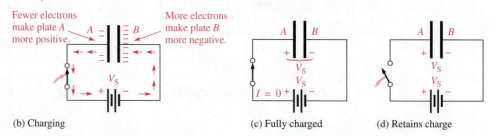

(a) Uncharged   (b) Charging   (c) Fully charged   (d) Retains charge

FIGURE 9–28   **Charging a capacitor.**

Discharging a Capacitor

When a conductor is connected across a charged capacitor, as shown in Figure 9–29, the capacitor will discharge. In this particular case, a very low resistance path (the conductor) is connected across the capacitor with a switch. Before the switch is closed, the capacitor is charged to 50 V, as indicated in part (a). When the switch is closed, as shown in part (b), the excess electrons on plate *B* move through the circuit to plate *A* (indicated by the arrows); as a result of the current through the low resistance of the conductor, the energy stored by the capacitor is dissipated in the resistance of the conductor. The charge is neutralized when the numbers of free electrons on both plates are again equal. At this time, the voltage across the capacitor is zero, and the capacitor is completely discharged, as shown in part (c).

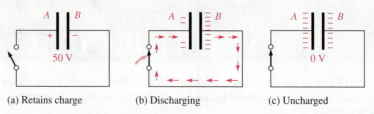

(a) Retains charge   (b) Discharging   (c) Uncharged

FIGURE 9–29   **Discharging a capacitor.**

Current and Voltage During Charging and Discharging

Notice in Figures 9–28 and 9–29 that the direction of electron flow during discharge is opposite to that during charging. It is important to understand that, ideally, *there is no current through the dielectric of the capacitor during charging or discharging because the dielectric is an insulating material.* There is current from one plate to the other only through the external circuit.

Figure 9–30(a) shows a capacitor connected in series with a resistor and a switch to a dc voltage source. Initially, the switch is open and the capacitor is uncharged with zero volts across its plates. At the instant the switch is closed, the current jumps to its maximum value and the capacitor begins to charge. The current is maximum initially because the capacitor has zero volts across it and, therefore, effectively acts as a short; thus, the current is limited only by the resistance. As time passes and the capacitor charges, the current decreases and the voltage across the capacitor (V_C) increases. The resistor voltage is proportional to the current during this charging period.

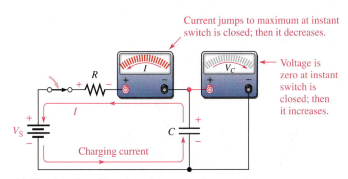

(a) Charging: Capacitor voltage increases as the current and resistor voltage decrease.

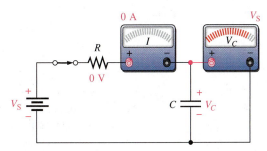

(b) Fully charged: Capacitor voltage equals source voltage. The current is zero.

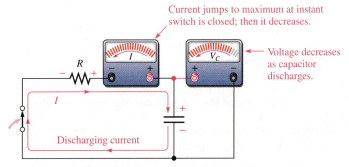

(c) Discharging: Capacitor voltage, resistor voltage, and the current decrease from their initial maximum values. Note that the discharge current is opposite to the charge current.

FIGURE 9–30   **Current and voltage in a charging and discharging capacitor.**

After a certain period of time, the capacitor reaches full charge. At this point, the current is zero and the capacitor voltage is equal to the dc source voltage, as shown in Figure 9–30(b). If the switch were opened now, the capacitor would retain its full charge (neglecting any leakage).

In Figure 9–30(c), the voltage source has been removed. When the switch is closed, the capacitor begins to discharge. Initially, the current jumps to a maximum but in a direction opposite to its direction during charging. As time passes, the current and capacitor voltage decrease. The resistor voltage is always proportional to the current. When the capacitor has fully discharged, the current and the capacitor voltage are zero.

Remember these two rules about capacitors in dc circuits:

1. A capacitor appears as an *open* to constant voltage.

2. A capacitor appears as a *short* to an instantaneous change in voltage.

Now let's examine in more detail how the voltage and current change with time in a capacitive circuit.

Surge Resistor

A typical small power supply for electronic systems has a bridge rectifier to change ac to pulsating dc, a filter capacitor to smooth the pulses, and a regulator to provide a constant dc output. Many times, you will see a small-value resistor between the rectifier and the regulator. This resistor limits inrush current to the capacitor when the supply is first turned on. The capacitor is typically a large value and has the highest current when it is initially uncharged. By limiting charging current, the surge resistor prevents blowing a fuse or tripping a circuit breaker.

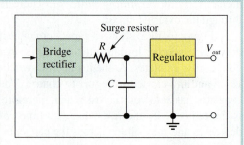

SYSTEM NOTE

The *RC* Time Constant

In a practical situation, there cannot be capacitance without some resistance in a circuit. It may simply be the small resistance of a wire, or it may be a designed-in resistance. Because of this, the charging and discharging characteristics of a capacitor must always be considered with the associated series resistance included. The resistance introduces the element of *time* in the charging and discharging of a capacitor.

When a capacitor charges or discharges through a resistance, a certain time is required for the capacitor to charge fully or discharge fully. The voltage across a capacitor cannot change instantaneously because a finite time is required to move charge from one point to another. The time constant of a series *RC* circuit determines the rate at which the capacitor charges or discharges.

> The *RC* time constant is a fixed time interval that equals the product of the resistance and the capacitance in a series *RC* circuit.

The time constant is expressed in units of seconds when resistance is in ohms and capacitance is in farads. It is symbolized by τ (Greek letter tau), and the formula is

$$\tau = RC \tag{9--12}$$

Recall that $I = Q/t$. The current depends on the amount of charge moved in a given time. When the resistance is increased, the charging current is reduced, thus increasing the charging time of the capacitor. When the capacitance is increased, the amount of charge increases; thus, for the same current, more time is required to charge the capacitor.

EXAMPLE 9–10

A series *RC* circuit has a resistance of 1.0 MΩ and a capacitance of 4.7 μF. What is the time constant?

SOLUTION

$$\tau = RC = (1.0 \times 10^6\,\Omega)(4.7 \times 10^{-6}\,\text{F}) = \textbf{4.7 s}$$

RELATED PROBLEM

A series *RC* circuit has a 270 kΩ resistor and a 3300 pF capacitor. What is the time constant in μs?

When a capacitor is charging or discharging between two voltage levels, the charge on the capacitor changes by approximately 63% of the difference in the levels in one time constant. An uncharged capacitor charges to approximately 63% of its fully charged voltage in one time constant. When a fully charged capacitor is discharging, its voltage drops to approximately 100% − 63% = 37% of its initial voltage in one time constant. This change also corresponds to a 63% change.

The Charging and Discharging Curves

A capacitor charges and discharges following a nonlinear curve, as shown in Figure 9–31. In these graphs, the approximate percentage of full charge is shown at each time-constant interval. This type of curve follows a precise mathematical formula and is called an **exponential** curve. The charging curve is an increasing exponential, and the discharging curve is a decreasing exponential. It takes five time constants to reach 99% (considered 100%) of the final voltage. This five time-constant interval is generally accepted as the time to fully charge or discharge a capacitor and is called the **transient time**.

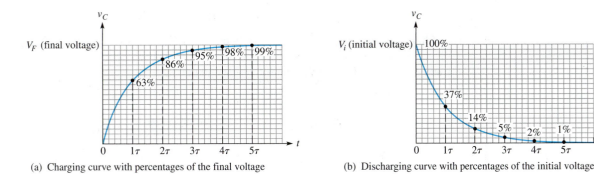

(a) Charging curve with percentages of the final voltage   (b) Discharging curve with percentages of the initial voltage

FIGURE 9–31 **Exponential voltage curves for the charging and discharging of a capacitor in an *RC* circuit.**

GENERAL FORMULA   The general expressions for either increasing or decreasing exponential curves are given in the following equations for both instantaneous voltage and current:

$$v = V_F + (V_i - V_F)e^{-t/\tau} \tag{9–13}$$

$$i = I_F + (I_i - I_F)e^{-t/\tau} \tag{9–14}$$

where V_F and I_F are the final values of voltage and current and V_i and I_i are the initial values of voltage and current. The lowercase italic letters v and i are the instantaneous values of the capacitor voltage and current at time t, and e is the base of natural logarithms. The e^x key on a calculator makes it easy to work with this exponential term.

CHARGING FROM ZERO   The formula for the special case in which an increasing exponential voltage curve begins at zero ($V_i = 0$), as shown in Figure 9–31(a), is given in Equation 9–15. It is developed as follows, starting with the general formula, Equation 9–13.

$$v = V_F + (V_i - V_F)e^{-t/\tau} = V_F + (0 - V_F)e^{-t/RC} = V_F - V_F e^{-t/RC}$$

Factoring out V_F, you have

$$v = V_F(1 - e^{-t/RC}) \tag{9–15}$$

Using Equation 9–15, you can calculate the value of the charging voltage of a capacitor at any instant of time if it is initially uncharged. You can calculate an increasing current by substituting i for v and I_F for V_F in Equation 9–15.

EXAMPLE 9–11

In Figure 9–32, determine the capacitor voltage 50 μs after the switch is closed if the capacitor initially is uncharged. Draw the charging curve.

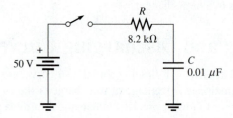

FIGURE 9–32

SOLUTION

The time constant is

$$\tau = RC = (8.2\text{ k}\Omega)(0.01\text{ }\mu\text{F}) = 82\text{ }\mu\text{s}$$

The voltage to which the capacitor will fully charge is 50 V (this is V_F). The initial voltage is zero. Notice that 50 μs is less than one time constant; so the capacitor will charge less than 63% of the full voltage in that time.

$$v_C = V_F(1 - e^{-t/RC}) = (50\text{ V})(1 - e^{-50\mu s/82\mu s}) = \textbf{22.8 V}$$

The charging curve for the capacitor is shown in Figure 9–33.

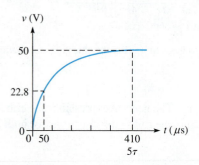

FIGURE 9–33

RELATED PROBLEM

Determine the capacitor voltage 15 μs after the switch is closed in Figure 9–32.

DISCHARGING TO ZERO   The formula for the special case in which a decreasing exponential voltage curve ends at zero ($V_F = 0$), as shown in Figure 9–31(b), is derived from the general formula as follows:

$$v = V_F + (V_i - V_F)e^{-t/\tau} = 0 + (V_i - 0)e^{-t/RC}$$

This reduces to

$$v = V_i e^{-t/RC} \tag{9–16}$$

where V_i is the voltage at the beginning of the discharge. You can use this formula to calculate the discharging voltage at any instant. The exponent $-t/RC$ can also be written as $-t/\tau$.

EXAMPLE 9–12

Determine the capacitor voltage in Figure 9–34 at a point in time 6 ms after the switch is closed. Draw the discharging curve.

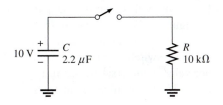

FIGURE 9–34

SOLUTION

The discharge time constant is

$$\tau = RC = (10 \text{ k}\Omega)(2.2 \text{ }\mu\text{F}) = 22 \text{ ms}$$

The initial capacitor voltage is 10 V. Notice that 6 ms is less than one time constant, so the capacitor will discharge less than 63%. Therefore, it will have a voltage greater than 37% of the initial voltage at 6 ms.

$$v_C = V_i e^{-t/RC} = (10 \text{ V})e^{-6\text{ms}/22\text{ms}} = \textbf{7.61 V}$$

The discharging curve for the capacitor is shown in Figure 9–35.

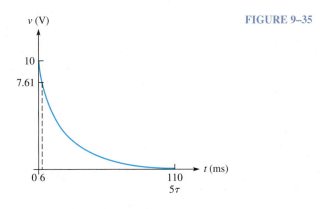

FIGURE 9–35

RELATED PROBLEM

In Figure 9–34, change R to 2.2 kΩ and determine the capacitor voltage 1 ms after the switch is closed.

GRAPHICAL METHOD USING UNIVERSAL EXPONENTIAL CURVES

The universal curves in Figure 9–36 provide a graphic solution of the charge and discharge of capacitors. Example 9–13 illustrates this graphical method.

FIGURE 9–36   Normalized universal exponential curves.

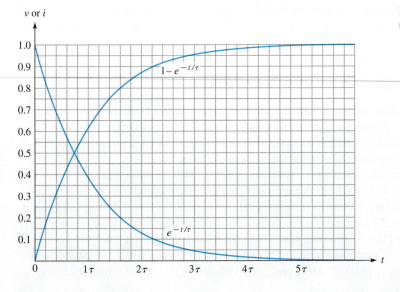

EXAMPLE 9–13

How long will it take the capacitor in Figure 9–37 to charge to 75 V? What is the capacitor voltage 2 ms after the switch is closed? Use the normalized universal curves in Figure 9–36 to determine the answers.

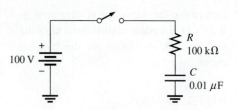

FIGURE 9–37

SOLUTION

The final voltage is 100 V, which is the 100% level (1.0) on the graph. The value 75 V is 75% of the maximum, or 0.75 on the graph. You can see that this value occurs at 1.4 time constants. In this circuit one time constant is $RC = (100\text{ k}\Omega)(0.01\text{ }\mu\text{F}) = 1$ ms. Therefore, the capacitor voltage reaches 75 V at 1.4 ms after the switch is closed.

The capacitor is at approximately 86 V (0.86 on the vertical axis) in 2 ms. These graphic solutions are shown in Figure 9–38.

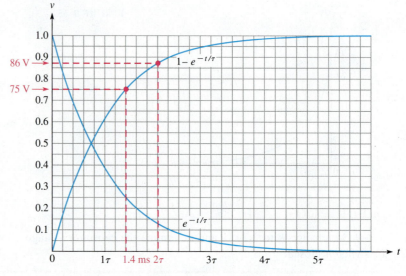

FIGURE 9–38

RELATED PROBLEM

Using the normalized universal exponential curves, determine how long it will take the capacitor in Figure 9–37 to charge to 50 V from zero. What is the capacitor voltage 3 ms after switch closure?

Response to a Square Wave

A common case that illustrates the rising and falling exponential occurs when an *RC* circuit is driven with a square wave that has a long period compared to the time constant. The square wave provides on and off action but, unlike a single switch, it provides a discharge path back through the generator when the wave drops back to zero.

When the square wave rises, the voltage across the capacitor rises exponentially toward the maximum value of the square wave in a time that depends on the time constant. When the square wave returns to the zero level, the capacitor voltage decreases exponentially, again depending on the time constant. The internal resistance of the square wave generator is part of the *RC* time constant; however, it can be ignored if it is small compared to *R*. The following example shows the waveforms for the case where the period is long compared to the time constant; other cases will be covered in detail in Chapter 15.

EXAMPLE 9–14

For the circuit in Figure 9–39(a), calculate the voltage across the capacitor every 0.1 ms for one complete period of the input waveform shown in Figure 9–39(b). Then sketch the capacitor waveform. Assume the internal resistance of the generator is negligible.

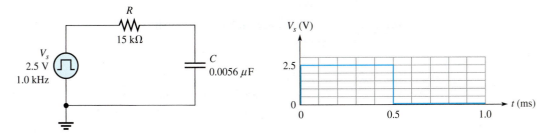

FIGURE 9–39

SOLUTION

$$\tau = RC = (15 \text{ k}\Omega)(0.0056 \,\mu\text{F}) = 0.084 \text{ ms}$$

The period of the square wave is 1 ms, which is approximately 12τ. This means that 6τ will elapse after each change of the pulse, allowing the capacitor to fully charge and fully discharge.

For the increasing exponential,

$$v = V_F(1 - e^{-t/RC}) = V_F(1 - e^{-t/\tau})$$

At 0.1 ms: $v = 2.5 \text{ V}(1 - e^{-0.1\text{ms}/0.084\text{ms}}) = 1.74 \text{ V}$
At 0.2 ms: $v = 2.5 \text{ V}(1 - e^{-0.2\text{ms}/0.084\text{ms}}) = 2.27 \text{ V}$
At 0.3 ms: $v = 2.5 \text{ V}(1 - e^{-0.3\text{ms}/0.084\text{ms}}) = 2.43 \text{ V}$
At 0.4 ms: $v = 2.5 \text{ V}(1 - e^{-0.4\text{ms}/0.084\text{ms}}) = 2.48 \text{ V}$
At 0.5 ms: $v = 2.5 \text{ V}(1 - e^{-0.5\text{ms}/0.084\text{ms}}) = 2.49 \text{ V}$

For the decreasing exponential,

$$v = V_i(e^{-t/RC}) = V_i(e^{-t/\tau})$$

In the equation, time is shown from the point when the change occurs (subtracting 0.5 ms from the actual time). For example, at 0.6 ms, $t = 0.6 \text{ ms} - 0.5 \text{ ms} = 0.1 \text{ ms}$.

At 0.6 ms: $v = 2.5 \text{ V}(e^{-0.1\text{ms}/0.084\text{ms}}) = 0.76 \text{ V}$
At 0.7 ms: $v = 2.5 \text{ V}(e^{-0.2\text{ms}/0.084\text{ms}}) = 0.23 \text{ V}$
At 0.8 ms: $v = 2.5 \text{ V}(e^{-0.3\text{ms}/0.084\text{ms}}) = 0.07 \text{ V}$
At 0.9 ms: $v = 2.5 \text{ V}(e^{-0.4\text{ms}/0.084\text{ms}}) = 0.02 \text{ V}$
At 1.0 ms: $v = 2.5 \text{ V}(e^{-0.5\text{ms}/0.084\text{ms}}) = 0.01 \text{ V}$

Figure 9–40 is a graph of these results.

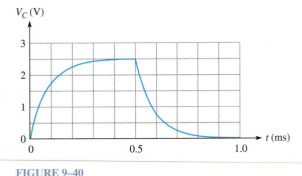

RELATED PROBLEM

What is the capacitor voltage at 0.65 ms?

FIGURE 9–40

SECTION 9–5 CHECKUP

1. Determine the time constant when $R = 1.2 \text{ k}\Omega$ and $C = 1000 \text{ pF}$.

2. If the circuit in Question 1 is charged with a 5 V source, how long will it take the initially uncharged capacitor to reach full charge? At full charge, what is the capacitor voltage?

3. A certain circuit has a time constant of 1 ms. If it is charged with a 10 V battery, what will the capacitor voltage be at each of the following times: 2 ms, 3 ms, 4 ms, and 5 ms? The capacitor is initially uncharged.

4. A capacitor is charged to 100 V. If it is discharged through a resistor, what is the capacitor voltage at one time constant?

9–6 CAPACITORS IN AC CIRCUITS

As you know, a capacitor blocks constant dc. A capacitor passes ac with an amount of opposition, called capacitive reactance, that depends on the frequency of the ac.

After completing this section, you should be able to

- **Describe how a capacitor operates in an ac circuit**
  - **Define *capacitive reactance***
  - **Determine the value of capacitive reactance in a given circuit**
  - **Calculate the capacitance for series and parallel capacitors**
  - **Explain why a capacitor causes a phase shift between voltage and current**
  - **Discuss instantaneous, true, and reactive power in a capacitor**

Capacitive Reactance, X_C

In Figure 9–41, a capacitor is shown connected to a sinusoidal voltage source. When the source voltage is held at a constant amplitude value and its frequency is increased, the amplitude of the current increases. Also, when the frequency of the source is decreased, the current amplitude decreases.

Fixed amplitude ac voltage source
f
I
C
f
I
C

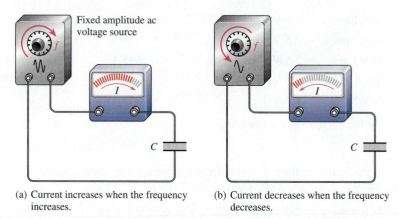

(a) Current increases when the frequency increases.

(b) Current decreases when the frequency decreases.

FIGURE 9–41 The current in a capacitive circuit varies directly with the frequency of the source voltage.

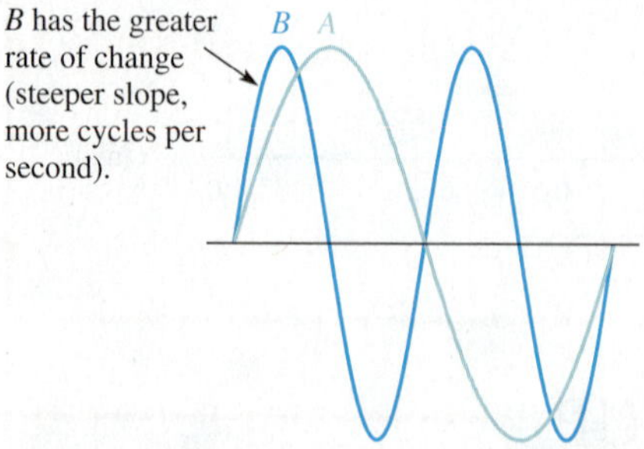

B has the greater rate of change (steeper slope, more cycles per second).
B
A

FIGURE 9–42 Rate of change of a sine wave increases when frequency increases.

When the frequency of the voltage increases, its rate of change also increases. This relationship is illustrated in Figure 9–42, where the frequency is doubled. Now, if the rate at which the voltage is changing increases, the amount of charge moving through the circuit in a given period of time must also increase. More charge in a given period of time means more current. For example, a tenfold increase in frequency means that the capacitor is charging and discharging 10 times more often in a given time interval. The rate of charge movement has increased 10 times. This means the current has increased by 10 because $I = Q/t$.

An increase in the amount of current with a fixed amount of voltage indicates that opposition to the current has decreased. Therefore, the capacitor offers opposition to current, and that opposition varies *inversely* with frequency.

The opposition to sinusoidal current in a capacitor is called capacitive reactance.

The symbol for capacitive reactance is X_C, and its unit is the ohm (Ω).

You have just seen how frequency affects the opposition to current (capacitive reactance) in a capacitor. Now let's see how the capacitance (C) itself affects the reactance.

Figure 9–43(a) shows that when a sinusoidal voltage with a fixed amplitude and fixed frequency is applied to a 1 μF capacitor, there is a certain amount of alternating current. When the capacitance value is increased to 2 μF, the current increases, as shown in Figure 9–43(b). Thus, when the capacitance is increased, the opposition to current (capacitive reactance) decreases. Therefore, not only is the capacitive reactance *inversely* proportional to frequency, but it is also *inversely* proportional to capacitance. This relationship can be stated as follows:

X_C **is proportional to** $\dfrac{1}{fC}$.

It can be proven that the constant of proportionality that relates X_C to $1/fC$ is $1/2\pi$. Therefore, the formula for capacitive reactance (X_C) is

$$X_C = \frac{1}{2\pi fC} \tag{9–17}$$

X_C is in ohms when f is in hertz and C is in farads. The 2π term comes from the fact that, as you learned in Chapter 8, a sine wave can be described in terms of rotational motion, and one revolution contains 2π radians.

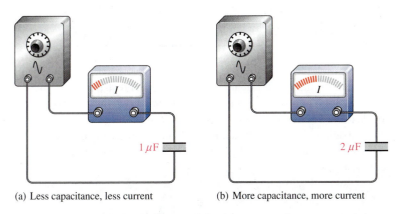

(a) Less capacitance, less current   (b) More capacitance, more current

FIGURE 9–43   **For a fixed voltage and fixed frequency, the current varies directly with the capacitance value.**

EXAMPLE 9–15

A sinusoidal voltage is applied to a capacitor, as shown in Figure 9–44. The frequency of the sine wave is 1.0 kHz. Determine the capacitive reactance.

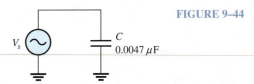

FIGURE 9–44

V_s C $0.0047\,\mu$F

SOLUTION

$$X_C = \frac{1}{2\pi fC} = \frac{1}{2\pi(1.0 \times 10^3\ \text{Hz})(0.0047 \times 10^{-6}\ \text{F})} = \textbf{33.9 k}\boldsymbol{\Omega}$$

RELATED PROBLEM

Determine the frequency required to make the capacitive reactance in Figure 9–44 equal to 10 kΩ.

Reactance for Series Capacitors

When capacitors are in series in an ac circuit, the total capacitance is smaller than the smallest individual capacitance. Because the total capacitance is smaller, the total capacitive reactance (opposition to current) must be larger than any individual capacitive reactance. With series capacitors, the total capacitive reactance ($X_{C(tot)}$) is the sum of the individual reactances.

$$X_{C(tot)} = X_{C1} + X_{C2} + X_{C3} + \cdots + X_{Cn} \qquad (9\text{--}18)$$

Compare this formula with Equation 4–1 for finding the total resistance of series resistors. In both cases, you simply add the individual oppositions.

Reactance for Parallel Capacitors

In ac circuits with parallel capacitors, the total capacitance is the sum of the capacitances. Recall that the capacitive reactance is inversely proportional to the capacitance. Because the total parallel capacitance is larger than any individual capacitances, the total capacitive reactance must be smaller than the reactance of any individual capacitor. With parallel capacitors, the total reactance is found by

$$X_{C(tot)} = \cfrac{1}{\cfrac{1}{X_{C1}} + \cfrac{1}{X_{C2}} + \cfrac{1}{X_{C3}} + \cdots + \cfrac{1}{X_{Cn}}} \qquad (9\text{--}19)$$

Compare this formula with Equation 5–1 for parallel resistors. As in the case of parallel resistors, the total opposition (resistance or reactance) is the reciprocal of the sum of the reciprocals of the individual oppositions.

For two capacitors in parallel, Equation 9–19 can be reduced to the product-over-sum form. This is useful because, for most practical circuits, more than two capacitors in parallel is not common.

$$X_{C(tot)} = \frac{X_{C1}X_{C2}}{X_{C1} + X_{C2}}$$

EXAMPLE 9–16

What is the total capacitive reactance in each circuit in Figure 9–45?

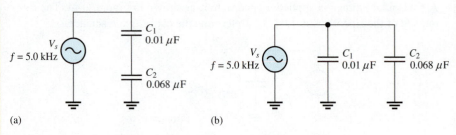

(a)   (b)

FIGURE 9–45

SOLUTION

The reactances of the individual capacitors are the same in both circuits.

$$X_{C1} = \frac{1}{2\pi f C_1} = \frac{1}{2\pi(5.0 \text{ kHz})(0.01 \ \mu\text{F})} = 3.18 \text{ k}\Omega$$

$$X_{C2} = \frac{1}{2\pi f C_2} = \frac{1}{2\pi(5.0 \text{ kHz})(0.068 \ \mu\text{F})} = 468 \ \Omega$$

Series Circuit: For the capacitors in series in Figure 9–45(a), the total reactance is the sum of X_{C1} and X_{C2}, as given in Equation 9–18.

$$X_{C(tot)} = X_{C1} + X_{C2} = 3.18 \text{ k}\Omega + 468 \text{ }\Omega = \mathbf{3.65 \text{ k}\Omega}$$

Alternatively, you can obtain the total series reactance by first finding the total capacitance using Equation 9–7. Then substitute that value into Equation 9–17 to calculate the total reactance.

$$C_{tot} = \frac{C_1 C_2}{C_1 + C_2} = \frac{(0.01 \text{ }\mu\text{F})(0.068 \text{ }\mu\text{F})}{0.01 \text{ }\mu\text{F} + 0.068 \text{ }\mu\text{F}} = 0.0087 \text{ }\mu\text{F}$$

$$X_{C(tot)} = \frac{1}{2\pi f C_{tot}} = \frac{1}{2\pi(5.0 \text{ kHz})(0.0087 \text{ }\mu\text{F})} = \mathbf{3.65 \text{ k}\Omega}$$

Parallel Circuit: For the capacitors in parallel in Figure 9–45(b), determine the total reactance from the product-over-sum rule using X_{C1} and X_{C2}.

$$X_{C(tot)} = \frac{X_{C1} X_{C2}}{X_{C1} + X_{C2}} = \frac{(3.18 \text{ k}\Omega)(468 \text{ }\Omega)}{3.18 \text{ k}\Omega + 468 \text{ }\Omega} = 408 \text{ }\Omega$$

RELATED PROBLEM

Determine the total parallel capacitive reactance by first finding the total capacitance.

OHM'S LAW The reactance of a capacitor is analogous to the resistance of a resistor. In fact, both are expressed in ohms. Since both R and X_C are forms of opposition to current, Ohm's law applies to capacitive circuits as well as to resistive circuits.

$$I = \frac{V}{X_C} \tag{9–20}$$

When applying Ohm's law in ac circuits, you must express both the current and the voltage in the same way, that is, both in rms, both in peak, and so on.

EXAMPLE 9–17

Determine the rms current in Figure 9–46.

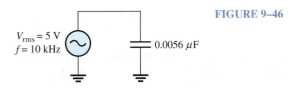

FIGURE 9–46

$V_{rms} = 5$ V
$f = 10$ kHz

$0.0056 \text{ }\mu\text{F}$

SOLUTION

First, find X_C.

$$X_C = \frac{1}{2\pi f C} = \frac{1}{2\pi(10 \times 10^3 \text{ Hz})(0.0056 \times 10^{-6} \text{ F})} = 2.84 \text{ k}\Omega$$

Then, apply Ohm's law.

$$I_{rms} = \frac{V_{rms}}{X_C} = \frac{5 \text{ V}}{2.84 \text{ k}\Omega} = \mathbf{1.76 \text{ mA}}$$

RELATED PROBLEM

Change the frequency in Figure 9–46 to 25 kHz and determine the rms value of the current.

MULTISIM

Open Multisim file E09-17. Measure the rms current and compare with the calculated value. Change the frequency of the source voltage to 25 kHz and measure the current.

Capacitive Voltage Divider

In ac circuits, capacitors can be used in applications that require a voltage divider. The voltage across a series capacitor was given as Equation 9–9, which is repeated here.

$$V_x = \left(\frac{C_{tot}}{C_x}\right)V_s$$

A resistive voltage divider is expressed in terms of a resistance ratio, which is a ratio of oppositions. You can think of the capacitive voltage divider by applying this idea from a resistive divider, but using reactance in place of resistance. The equation for the voltage across a capacitor in a capacitive voltage divider can be written as

$$V_x = \left(\frac{X_{Cx}}{X_{C(tot)}}\right)V_s \qquad\qquad (9\text{–}21)$$

where X_{Cx} is the reactance of capacitor C_x, $X_{C(tot)}$ is the total capacitive reactance, and V_x is the voltage across capacitor C_x. Compare this equation with Equation 4–5. Either Equation 9–9 or Equation 9–21 can be used to find the voltage from a divider as illustrated in the following example.

EXAMPLE 9–18

What is the voltage across C_2 in the circuit of Figure 9–47?

FIGURE 9–47

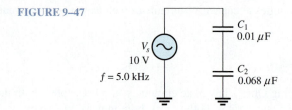

SOLUTION

The reactance of the individual capacitors and the total reactance were determined in Example 9–16. Substituting into Equation 9–21,

$$V_2 = \left(\frac{X_{C2}}{X_{C(tot)}}\right)V_s = \left(\frac{468\ \Omega}{3.65\ \text{k}\Omega}\right)10\ \text{V} = \mathbf{1.28\ V}$$

Notice that the voltage across the larger capacitor is the smaller fraction of the total. You can obtain the same result from Equation 9–9:

$$V_2 = \left(\frac{C_{tot}}{C_2}\right)V_s = \left(\frac{0.0087\ \mu\text{F}}{0.068\ \mu\text{F}}\right)10\ \text{V} = \mathbf{1.28\ V}$$

RELATED PROBLEM

Use Equation 9–21 to determine the voltage across C_1.

Current Leads Capacitor Voltage by 90°

A sinusoidal voltage is shown in Figure 9–48. Notice that the rate at which the voltage is changing varies along the sine wave curve, as indicated by the "steepness" of the curve. At the zero crossings, the curve is changing at a faster rate than anywhere else along the curve. At the peaks, the curve has a zero rate of change because it has just reached its maximum and is at the point of changing direction.

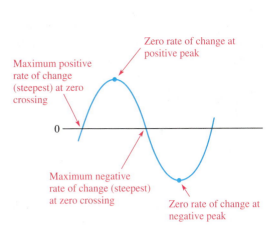

FIGURE 9–48   **The rates of change of a sine wave.**

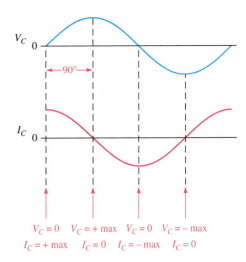

FIGURE 9–49   **Current is always leading the capacitor voltage by 90°.**

The amount of charge stored by a capacitor determines the voltage across it. Therefore, the rate at which the charge is moved ($Q/t = I$) from one plate to the other determines the rate at which the voltage changes. When the current is changing at its maximum rate (at the zero crossings), the voltage is at its maximum value (peak). When the current is changing at its minimum rate (zero at the peaks), the voltage is at its minimum value (zero). This phase relationship is illustrated in Figure 9–49. As you can see, the current peaks occur a quarter of a cycle before the voltage peaks. Thus, the current leads the voltage by 90°.

Power in a Capacitor

As discussed earlier in this chapter, a charged capacitor stores energy in the electric field within the dielectric. An ideal capacitor does not dissipate energy; it only stores energy temporarily. When an ac voltage is applied to a capacitor, energy is stored by the capacitor during one-quarter of the voltage cycle. Then the stored energy is returned to the source during another quarter of the cycle. There is no net energy loss. Figure 9–50 shows the power curve that results from one cycle of capacitor voltage and current.

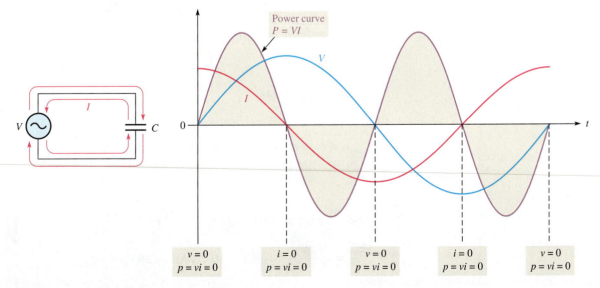

FIGURE 9–50   **Power curve for a capacitor.**

INSTANTANEOUS POWER (p)   The product of instantaneous voltage, v, and instantaneous current, i, gives **instantaneous power**, p. At points where v or i is zero, p is also zero. When both v and i are positive, p is also positive. When either v or i is positive and the other negative, p is negative. When both v and i are negative, p is positive. As you can see, the power follows a sinusoidal-shaped curve. Positive values of power indicate that energy is stored by the capacitor. Negative values of power indicate that energy is returned from the capacitor to the source. Note that the power fluctuates at a frequency twice that of the voltage or current, as energy is alternately stored and returned to the source.

TRUE POWER (P_{true})   Ideally, all of the energy stored by a capacitor during the positive portion of the power cycle is returned to the source during the negative portion. No net energy is lost due to conversion to heat in the capacitor, so the **true power** is zero. Actually, because of leakage and foil resistance in a practical capacitor, a small percentage of the total power is dissipated in the form of true power.

REACTIVE POWER (P_r)   The rate at which a capacitor stores or returns energy is called its **reactive power**, P_r. The reactive power is a nonzero quantity, because at any instant in time, the capacitor is actually taking energy from the source or returning energy to it. Reactive power does not represent an energy loss. The following formulas apply:

$$P_r = V_{rms}I_{rms} \tag{9-22}$$

$$P_r = \frac{V_{rms}^2}{X_C} \tag{9-23}$$

$$P_r = I_{rms}^2 X_C \tag{9-24}$$

Notice that these equations are of the same form as those introduced in Chapter 3 for true power in a resistor. The voltage and current are expressed as rms values. The unit of reactive power is **VAR (volt-ampere reactive)**.

EXAMPLE 9–19

Determine the true power and the reactive power in Figure 9–51.

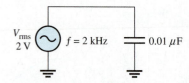

FIGURE 9–51

SOLUTION

The true power P_{true} is always **zero** for an ideal capacitor. Determine the reactive power by first finding the value for the capacitive reactance and then using Equation 9–23.

$$X_C = \frac{1}{2\pi fC} = \frac{1}{2\pi(2 \times 10^3 \text{ Hz})(0.01 \times 10^{-6} \text{ F})} = 7.96 \text{ k}\Omega$$

$$P_r = \frac{V_{rms}^2}{X_C} = \frac{(2 \text{ V})^2}{7.96 \text{ k}\Omega} = 503 \times 10^{-6} \text{ VAR} = \textbf{503 } \mu\textbf{VAR}$$

RELATED PROBLEM

If the frequency is doubled in Figure 9–51, what are the true power and the reactive power?

SECTION 9–6 CHECKUP

1. Calculate X_C for $f = 5$ kHz and $C = 47$ pF.

2. At what frequency is the reactance of a 0.1 μF capacitor equal to 2 kΩ?

3. Calculate the rms current in Figure 9–52.

4. State the phase relationship between current and voltage in a capacitor.

5. A 1 μF capacitor is connected to an ac voltage source of 12 V rms. What is the true power?

6. In Question 5, determine the reactive power at a frequency of 500 Hz.

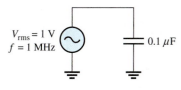

$V_{rms} = 1$ V
$f = 1$ MHz

0.1 μF

FIGURE 9–52

9–7 CAPACITOR APPLICATIONS

Capacitors are widely used in electrical and electronic systems.

After completing this section, you should be able to

- **Discuss some capacitor applications**
  - **Describe a power supply filter**
  - **Explain the purpose of coupling and bypass capacitors**
  - **Discuss the basics of capacitors applied to tuned circuits, timing circuits, and computer memories**

If you pick up any circuit board, open any power supply, or look inside any piece of electronic equipment, chances are you will find capacitors of one type or another. These components are used for a variety of purposes in both dc and ac systems.

Electrical Storage

One of the most basic applications of a capacitor is as a backup voltage source for low-power circuits such as certain types of semiconductor memories in computers. This particular application requires a very high capacitance value and negligible leakage.

A storage capacitor is connected between the dc power supply input to the circuit and ground. When the circuit is operating from its normal power supply, the capacitor remains fully charged to the dc power supply voltage. If the normal power source is disrupted, effectively removing the power supply from the circuit, the storage capacitor temporarily becomes the power source for the circuit.

A capacitor provides voltage and current to a circuit as long as its charge remains sufficient. As current is drawn by the circuit, charge is removed from the capacitor and the voltage decreases. For this reason, a storage capacitor can only be used as a temporary power source. The length of time that a capacitor can provide sufficient power to a circuit depends on the capacitance and the amount of current drawn by the circuit. The smaller the current and the higher the capacitance, the longer the time a capacitor can provide power to a circuit.

Power Supply Filtering

A basic dc power supply consists of a circuit known as a **rectifier** followed by a filter. The rectifier converts the 120 V, 60 Hz sinusoidal voltage available at a standard outlet to a pulsating dc voltage that can be either a half-wave rectified voltage or a full-wave rectified voltage, depending on the type of rectifier circuit. As shown in Figure 9–53(a), a half-wave rectifier removes each negative half-cycle of the sinusoidal voltage. As shown in Figure 9–53(b), a full-wave rectifier actually reverses the polarity of the negative portion of each cycle. Both half-wave and full-wave rectified voltages are dc because, even though they are changing, they do not alternate polarity.

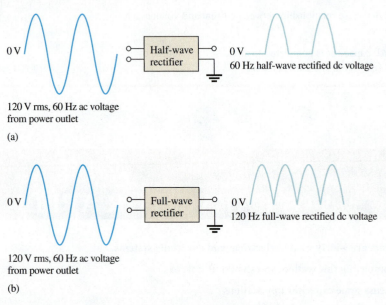

FIGURE 9–53   **Half-wave and full-wave rectifier operation.**

To be useful for powering electronic circuits, the rectified voltage must be changed to constant dc voltage because all circuits require constant power. When connected to the rectifier output, the **filter** nearly eliminates the fluctuations in the rectified voltage and provides a smooth constant-value dc voltage to the load, which is the electronic circuit, as indicated in Figure 9–54.

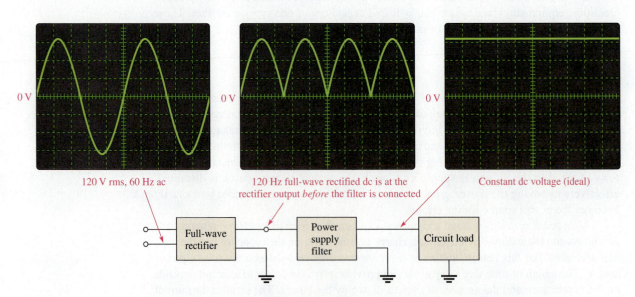

FIGURE 9–54   **Basic block diagram and operation of a dc power supply.**

THE CAPACITOR AS A POWER SUPPLY FILTER   Capacitors are used as filters in dc power supplies because of their ability to store electrical charge. Figure 9–55(a) shows a dc power supply with a full-wave rectifier and a capacitor filter. The operation can be described from a charging and discharging point of view as follows. Assume the capacitor is initially

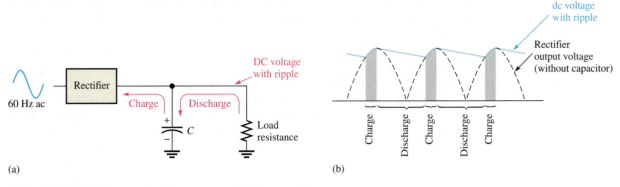

FIGURE 9–55   Basic operation of a power supply filter capacitor.

uncharged. When the power supply is first turned on and the first cycle of the rectified voltage occurs, the capacitor will quickly charge through the low resistance of the rectifier. The capacitor voltage will follow the rectified voltage curve up to the peak of the rectified voltage. As the rectified voltage passes the peak and begins to decrease, the capacitor will begin to discharge very slowly through the high resistance of the load circuit, as indicated in Figure 9–55(b). The amount of discharge is typically very small and is exaggerated in the figure for purposes of illustration. The next cycle of the rectified voltage will recharge the capacitor back to the peak value by replenishing the small amount of charge lost since the previous peak. This pattern of a small amount of charging and discharging continues as long as the power is on.

A rectifier is designed so that it allows current only in the direction to charge the capacitor. The capacitor will not discharge back through the rectifier but will only discharge a small amount through the relatively high resistance of the load. The small fluctuation in voltage due to the charging and discharging of the capacitor is called the **ripple voltage**. A good dc power supply has a very small amount of ripple on its dc output. The discharge time constant of a power supply filter capacitor depends on its capacitance and the resistance of the load; consequently, the higher the capacitance value, the longer the discharge time and, therefore, the smaller the ripple voltage.

DC Blocking and AC Coupling

Capacitors are commonly used to block the constant dc voltage in one part of a circuit from getting to another part. As an example of this, a capacitor is connected between two stages of an amplifier to prevent the dc voltage at the output of stage 1 from affecting the dc voltage at the input of stage 2, as illustrated in Figure 9–56. Assume that, for proper

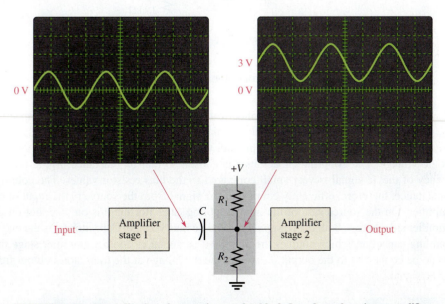

FIGURE 9–56   An application of a capacitor used to block dc and couple ac in an amplifier.

operation, the output of stage 1 has a zero dc voltage and the input to stage 2 has a 3 V dc voltage. The capacitor prevents the 3 V dc at stage 2 from getting to the stage 1 output and affecting its zero value, and vice versa.

If a sinusoidal signal voltage is applied to the input to stage 1, the signal voltage is increased (amplified) and appears on the output of stage 1, as shown in Figure 9–56. The amplified signal voltage is then coupled through the capacitor to the input of stage 2 where it is superimposed on the 3 V dc level and then again amplified by stage 2. In order for the signal voltage to be passed through the capacitor without being reduced, the capacitor must be large enough so that its reactance at the frequency of the signal voltage is negligible. In this type of application, the capacitor is known as a **coupling** capacitor, which ideally appears as an open to dc and as a short to ac. As the signal frequency is reduced, the capacitive reactance increases and, at some point, the capacitive reactance becomes large enough to cause a significant reduction in ac voltage between stage 1 and stage 2.

SYSTEM EXAMPLE 9–2

TRANSISTOR AMPLIFIER

Amplifiers form the backbone of many electronic systems. All transistor amplifiers require dc voltages to establish proper operating conditions for amplifying ac signals. These dc voltages are referred to as bias voltages and are superimposed on the ac signals. Figure 9–57 shows a discrete transistor amplifier with two coupling capacitors to connect the ac signal to and from the amplifier without affecting the bias voltages. The input bias voltage is set up by the voltage divider formed by R_1 and R_2, which divide the 24 V dc supply voltage.

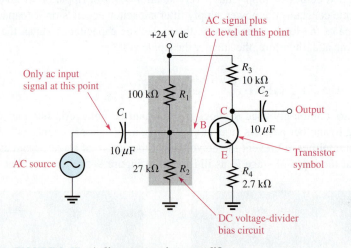

FIGURE 9–57   A discrete transistor amplifier.

When an ac signal voltage is applied to the amplifier, the input coupling capacitor, C_1, prevents ac source from affecting the dc bias voltage. Without the capacitor, the internal source resistance would appear in parallel with R_2 and drastically change the value of the dc voltage. The coupling capacitance is chosen so that its reactance, X_C, at the frequency of the ac signal is very small compared to the bias resistor values. The coupling capacitance, therefore, efficiently couples the ac signal from the source to the input of the amplifier. On the source side of the input coupling capacitor there is only ac, but on the amplifier side there is both the ac signal plus the dc bias voltage. Capacitor C_2 is the output coupling capacitor, which couples the amplified ac signal to another amplifier stage that would be connected to the output. It isolates the dc voltage at the transistor's output from the ac signal.

Power Line Decoupling

Capacitors connected from the dc supply voltage line to ground are used on circuit boards to decouple unwanted voltage transients or spikes that occur on the dc supply voltage because of fast switching digital circuits. A voltage **transient** contains higher frequencies that may affect the operation of the circuits. These transients are shorted to ground through the very low reactance of the decoupling capacitors. **Decoupling** capacitors are often used at various points along the supply voltage line on a circuit board, particularly near integrated circuits (ICs).

Bypassing

Bypass capacitors are used to bypass an ac voltage around a resistor in a circuit without affecting the dc voltage across the resistor. In amplifier circuits, for example, dc voltages called *bias voltages* are required at various points. For the amplifier to operate properly, certain bias voltages must remain constant and, therefore, any ac voltages must be removed. A sufficiently large capacitor connected from a bias point to ground provides a low reactance path to ground for ac voltages, leaving the constant dc bias voltage at the given point. This bypass application is illustrated in Figure 9–58. As frequencies decrease, the bypass capacitor becomes less effective because of its increased reactance.

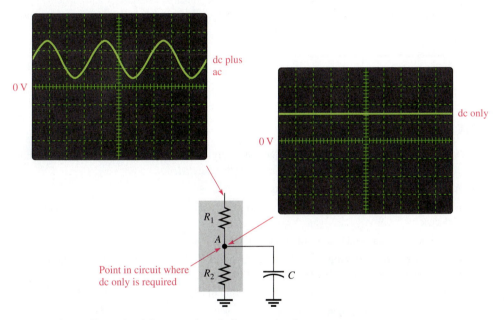

FIGURE 9–58 **Example of the operation of a bypass capacitor.**

Signal Filters

Filters are used for selecting one ac signal with a certain specified frequency from a wide range of signals with many different frequencies or for selecting a certain band of frequencies and eliminating all others. A common example of this application is in radio and television receivers where it is necessary to select the signal transmitted from a given station and eliminate or filter out the signals transmitted from all the other stations in the area.

When you tune your radio or TV, you are actually changing the capacitance in the tuner circuit (which is a type of filter) so that only the signal from the station or channel you want passes through to the receiver circuitry. Capacitors are used in conjunction with resistors, inductors (covered in Chapter 11), and other components in these types of filters.

The main characteristic of a filter is its frequency selectivity, which is based on the fact that the reactance of a capacitor depends on frequency ($X_C = 1/2\pi fC$).

Timing Circuits

Another important area in which capacitors are used is in timing circuits that generate specified time delays or produce waveforms with specific characteristics. Recall that the time constant of a circuit with resistance and capacitance can be controlled by selecting appropriate values for R and C. The charging time of a capacitor can be used as a basic time delay in various types of circuits. An example is the circuit that controls the turn indicators on your car where the light flashes on and off at regular intervals.

Computer Memories

Dynamic computer memories use capacitors as the basic storage element for binary information, which consists of two digits, 1 and 0. A charged capacitor can represent a stored 1 and a discharged capacitor can represent a stored 0. Patterns of 1s and 0s that make up binary data are stored in a memory that consists of an array of capacitors with associated circuitry. You will study this topic in a computer or digital fundamentals course.

SECTION 9–7 CHECKUP

1. Explain how half-wave or full-wave rectified dc voltages are smoothed out by a filter capacitor.

2. Explain the purpose of a coupling capacitor.

3. How large must a coupling capacitor be?

4. Explain the purpose of a decoupling capacitor.

5. Discuss how the relationship of frequency and capacitive reactance is important in frequency-selective circuits such as signal filters.

SUMMARY

- A capacitor is an electrical device that has the ability to store charge. It consists of one or more conductors separated by an insulating material called the *dielectric*.
- A capacitor stores electrical charge on its plates.
- Energy is stored in a capacitor by the electric field created between the charged plates in the dielectric.
- Capacitance is measured in units of farads (F).
- Capacitance is directly proportional to the plate area and the dielectric constant and inversely proportional to the distance between the plates (the dielectric thickness).
- The dielectric constant is an indication of the ability of a material to establish an electric field.
- The dielectric strength is one factor that determines the breakdown voltage of a capacitor.
- Capacitors are commonly classified according to the dielectric material. Typical materials are mica, ceramic, plastic-film, and electrolytic (aluminum oxide and tantalum oxide).
- The total capacitance of series capacitors is less than the smallest capacitance.
- The total capacitance of parallel capacitors is the sum of all the capacitances.
- A capacitor blocks constant dc.
- The time constant determines the charging and discharging time of a capacitor with resistance in series.
- In an *RC* circuit, the voltage and current during charging and discharging make an approximate 63% change during each time-constant interval.
- Five time constants are required for a capacitor to fully charge or discharge. This is called the *transient time*.
- The approximate percentage of final charge after each charging time-constant interval is given in Table 9–4.
- The approximate percentage of initial charge after each discharging time-constant interval is given in Table 9–5.

TABLE 9–4	
NUMBER OF TIME CONSTANTS	**APPROXIMATE % OF FINAL CHARGE**
1	63
2	86
3	95
4	98
5	99 (considered 100%)

TABLE 9–5	
NUMBER OF TIME CONSTANTS	**APPROXIMATE % OF INITIAL CHARGE**
1	37
2	14
3	5
4	2
5	1 (considered 0)

- Alternating current in a capacitor leads the voltage by 90°.
- A capacitor passes ac to an extent that depends on its reactance and the resistance in the rest of the circuit.
- Capacitive reactance is the opposition to ac, expressed in ohms.
- Capacitive reactance (X_C) is inversely proportional to the frequency and to the capacitance value.
- The total capacitive reactance of series capacitors is the sum of the individual reactances.
- The total capacitive reactance of parallel capacitors is the reciprocal of the sum of the reciprocals of the individual reactances.
- Ideally, there is no energy loss in a capacitor and, thus, the true power (watts) is zero. However, in most capacitors there is some small energy loss due to leakage resistance.

KEY TERMS

Key terms and other bold terms in the chapter are defined in the end-of-book glossary.

Bypass   A capacitor connected from a point to ground to remove the ac signal without affecting the dc voltage. A special case of decoupling.

Capacitance   The ability of a capacitor to store electrical charge.

Capacitive reactance   The opposition of a capacitor to sinusoidal current. The unit is the ohm.

Capacitor   An electrical device that has the ability to store charge. It consists of one or more conductors separated by an insulating material called the *dielectric*.

Charging   The process in which a current removes charge from one plate of a capacitor and deposits it on the other plate, making one plate more positive than the other.

Coupling   The method of connecting a capacitor from between two points in a circuit to allow ac to pass from one point to the other while blocking dc.

Decoupling   The method of connecting a capacitor from one point, usually the dc power supply line, to ground to short ac to ground without affecting the dc voltage.

Dielectric   The insulating material between the plates of a capacitor.

Dielectric constant   A measure of the ability of a dielectric material to establish an electric field.

Dielectric strength   A measure of the ability of a dielectric material to withstand voltage without breaking down.

Exponential   A mathematical function described by a natural logarithm (base). The charging and discharging of a capacitor are described by an exponential function.

Farad (F)   The unit of capacitance.

Filter   A type of circuit that passes certain frequencies and rejects all others.

Instantaneous power (p)   The value of power in a circuit at any given instant of time.

RC time constant   A fixed time interval set by R and C values that determines the time response of a series RC circuit. It equals the product of the resistance and the capacitance.

Reactive power (P_r)   The rate at which energy is alternately stored and returned to the source by a capacitor. The unit is the VAR.

Ripple voltage   The small fluctuation in voltage due to the charging and discharging of a capacitor.

Temperature coefficient   A constant specifying the amount of change in the value of a quantity for a given change in temperature.

Transient time   An interval equal to approximately five time constants.
True power (P_{true})   The power that is dissipated in a circuit, usually in the form of heat.
VAR (volt-ampere reactive)   The unit of reactive power.

KEY FORMULAS

(9–1) $C = \dfrac{Q}{V}$ Capacitance in terms of charge and voltage

(9–2) $Q = CV$ Charge in terms of capacitance and voltage

(9–3) $V = \dfrac{Q}{C}$ Voltage in terms of charge and capacitance

(9–4) $W = \dfrac{1}{2}CV^2$ Energy stored by a capacitor

(9–5) $\varepsilon_r = \dfrac{\varepsilon}{\varepsilon_0}$ Dielectric constant (relative permittivity)

(9–6) $C = \dfrac{A\varepsilon_r(8.85 \times 10^{-12}\ \text{F/m})}{d}$ Capacitance in terms of physical parameters

(9–7) $C_T = \dfrac{C_1 C_2}{C_1 + C_2}$ Total series capacitance (two capacitors)

(9–8) $C_T = \dfrac{1}{\dfrac{1}{C_1} + \dfrac{1}{C_2} + \dfrac{1}{C_3} + \cdots + \dfrac{1}{C_n}}$ Total series capacitance (general)

(9–9) $V_x = \left(\dfrac{C_T}{C_x}\right)V_S$ Voltage across series capacitor

(9–10) $C_T = C_1 + C_2$ Two capacitors in parallel

(9–11) $C_T = C_1 + C_2 + C_3 + \cdots + C_n$ n capacitors in parallel

(9–12) $\tau = RC$ RC time constant

(9–13) $v = V_F + (V_i - V_F)e^{-t/\tau}$ Exponential voltage (general)

(9–14) $i = I_F + (I_i - I_F)e^{-t/\tau}$ Exponential current (general)

(9–15) $v = V_F(1 - e^{-t/RC})$ Increasing exponential voltage beginning at zero

(9–16) $v = V_i e^{-t/RC}$ Decreasing exponential voltage ending at zero

(9–17) $X_C = \dfrac{1}{2\pi f C}$ Capacitive reactance

(9–18) $X_{C(tot)} = X_{C1} + X_{C2} + X_{C3} + \cdots + X_{Cn}$ Capacitive reactance for series capacitors

(9–19) $X_{C(tot)} = \dfrac{1}{\dfrac{1}{X_{C1}} + \dfrac{1}{X_{C2}} + \dfrac{1}{X_{C3}} + \cdots + \dfrac{1}{X_{Cn}}}$ Capacitive reactance for parallel capacitors

(9–20) $I = \dfrac{V}{X_C}$ Ohm's law for a capacitor

(9–21) $V_x = \left(\dfrac{X_{Cx}}{X_{C(tot)}}\right)V_s$ Capacitive voltage divider

(9–22) $P_r = V_{rms}I_{rms}$ Reactive power in a capacitor

$$(9-23) \qquad P_r = \frac{V_{rms}^2}{X_C} \qquad\qquad \text{Reactive power in a capacitor}$$

$$(9-24) \qquad P_r = I_{rms}^2 X_C \qquad\qquad \text{Reactive power in a capacitor}$$

TRUE/FALSE QUIZ

Answers are at the end of the chapter.

1. The area of the plates of a capacitor is proportional to the capacitance.

2. A capacitance of 1200 pF is the same as 1.2 μF.

3. When two capacitors are in series with a voltage source, the smaller capacitor will have the larger voltage.

4. When two capacitors are in parallel with a voltage source, the smaller capacitor will have the larger voltage.

5. A capacitor appears as an open to a constant dc.

6. When a capacitor is charging or discharging between two levels, the charge on the capacitor changes by 63% of the difference in one time constant.

7. Capacitive reactance is proportional to the applied frequency.

8. The total reactance of series capacitors is the product-over-sum of the individual reactances.

9. Voltage leads current in a capacitor.

10. The unit of reactive power is the VAR.

SELF-TEST

Answers are at the end of the chapter.

1. Which of the following accurately describes a capacitor?
   (a) The plates are conductive.
   (b) The dielectric is an insulator between the plates.
   (c) There is constant direct current (dc) through a fully charged capacitor.
   (d) A capacitor stores charge indefinitely when disconnected from the source.
   (e) none of the above answers
   (f) all the above answers
   (g) only answers (a) and (b)

2. Which one of the following statements is true?
   (a) There is current through the dielectric of a charging capacitor.
   (b) When a capacitor is connected to a dc voltage source, it will charge to the value of the source.
   (c) A capacitor can be discharged by disconnecting it from the voltage source.

3. A capacitance of 0.01 μF is larger than
   (a) 0.00001 F   (b) 100,000 pF   (c) 1000 pF   (d) all of these answers

4. A capacitance of 1000 pF is smaller than
   (a) 0.01 μF   (b) 0.001 μF   (c) 0.00000001 F   (d) answers (a) and (c)

5. When the voltage across a capacitor is increased, the stored charge
   (a) increases   (b) decreases   (c) remains constant   (d) fluctuates

6. When the voltage across a capacitor is doubled, the stored charge
   (a) stays the same   (b) is halved   (c) increases by four   (d) doubles

7. The voltage rating of a capacitor is increased by
   (a) increasing the plate separation   (b) decreasing the plate separation
   (c) increasing the plate area   (d) answers (b) and (c)

8. The capacitance value is increased by

    (a) decreasing plate area   **(b)** increasing plate separation

    (c) decreasing plate separation   **(d)** increasing plate area

    (e) answers (a) and (b)   **(f)** answers (c) and (d)

9. A 1 μF, a 2.2 μF, and a 0.047 μF capacitor are connected in series. The total capacitance is less than

    (a) 1 μF   **(b)** 2.2 μF   **(c)** 0.047 μF   **(d)** 0.001 μF

10. Four 0.022 μF capacitors are in parallel. The total capacitance is

    (a) 0.022 μF   **(b)** 0.088 μF   **(c)** 0.011 μF   **(d)** 0.044 μF

11. An uncharged capacitor and a resistor are connected in series with a switch and a 12 V battery. At the instant the switch is closed, the voltage across the capacitor is

    (a) 12 V   **(b)** 6 V   **(c)** 24 V   **(d)** 0 V

12. In Question 11, the voltage across the capacitor when it is fully charged is

    (a) 12 V   **(b)** 6 V   **(c)** 24 V   **(d)** -6 V

13. In Question 11, the capacitor will reach full charge in a time equal to approximately

    (a) RC **(b)** $5RC$ **(c)** $12RC$ **(d)** cannot be predicted

14. A sinusoidal voltage is applied across a capacitor. When the frequency of the voltage is increased, the current

    (a) increases   **(b)** decreases   **(c)** remains constant   **(d)** ceases

15. A capacitor and a resistor are connected in series to a sine wave generator. The frequency is set so that the capacitive reactance is equal to the resistance and, thus, an equal amount of voltage appears across each component. If the frequency is decreased,

    (a) $V_R > V_C$ **(b)** $V_C > V_R$ **(c)** $V_R = V_C$ **(d)** $V_C < V_R$

TROUBLESHOOTING: SYMPTOM AND CAUSE

The purpose of these exercises is to help develop thought processes essential to troubleshooting. Answers are at the end of the chapter.

Determine the cause for each set of symptoms. Refer to Figure 9–59.

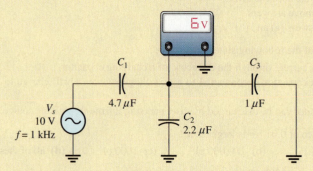

FIGURE 9–59   **The ac meter indicates the correct reading for this circuit.**

1. *Symptom:* The voltmeter reading is 0 V.

    Cause:

    (a) C_1 is shorted.   **(b)** C_2 is shorted.   **(c)** C_3 is open.

2. *Symptom:* The voltmeter reading is 10 V.

    Cause:

    (a) C_1 is shorted.   **(b)** C_2 is open.   **(c)** C_3 is open.

3. *Symptom:* The voltmeter reading is 6.86 V.
   Cause:
   (a) C_1 is open.   (b) C_2 is open.   (c) C_3 is open.

4. *Symptom:* The voltmeter reading is 0 V.
   Cause:
   (a) C_1 is open.   (b) C_2 is open.   (c) C_3 is open.

5. *Symptom:* The voltmeter reading is 8.28 V.
   Cause:
   (a) C_1 is shorted.   (b) C_2 is open.   (c) C_3 is open.

PROBLEMS

Answers to odd-numbered problems are at the end of the book.

BASIC PROBLEMS

SECTION 9–1   The Basic Capacitor

1. (a) Find the capacitance when $Q = 50\ \mu C$ and $V = 10$ V.
   (b) Find the charge when $C = 0.001\ \mu F$ and $V = 1$ kV.
   (c) Find the voltage when $Q = 2$ mC and $C = 200\ \mu F$.

2. Convert the following values from microfarads to picofarads:
   (a) $0.1\ \mu F$ (b) $0.0025\ \mu F$ (c) $5\ \mu F$

3. Convert the following values from picofarads to microfarads:
   (a) 1000 pF   (b) 3500 pF   (c) 250 pF

4. Convert the following values from farads to microfarads:
   (a) 0.0000001 F   (b) 0.0022 F   (c) 0.0000000015 F

5. What size capacitor is capable of storing 10 mJ of energy with 100 V across its plates?

6. A mica capacitor has a plate area of 20 cm^2 and a dielectric thickness of 2.5 mils. What is its capacitance?

7. An air capacitor has plates with an area of 0.1 m^2. The plates are separated by 0.01 m. Calculate the capacitance.

8. A student wants to construct a 1 F capacitor out of two square plates for a science fair project. He plans to use a paper dielectric ($\varepsilon_r = 2.5$) that is 8×10^{-5} m thick. The science fair is to be held in the Astrodome. Will his capacitor fit in the Astrodome? What would be the size of the plates if it could be constructed?

9. A student decides to construct a capacitor using two conducting plates 30 cm on a side. He separates the plates with a paper dielectric ($\varepsilon_r = 2.5$) that is 8×10^{-5} m thick. What is the capacitance of his capacitor?

10. At ambient temperature (25°C), a certain capacitor is specified to be 1000 pF. It has a negative temperature coefficient of 200 ppm/°C. What is its capacitance at 75°C?

11. A 0.001 μF capacitor has a positive temperature coefficient of 500 ppm/°C. How much change in capacitance will a 25°C increase in temperature cause?

SECTION 9–2   Types of Capacitors

12. In the construction of a stacked-foil mica capacitor, how is the plate area increased?

13. What type of capacitor has the higher dielectric constant, mica or ceramic?

14. Show how to connect an electrolytic capacitor across points A and B in Figure 9–60.

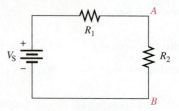

FIGURE 9–60

15. Determine the value of the typographically labeled ceramic disk capacitors in Figure 9–61.
16. Name two types of electrolytic capacitors. How do electrolytics differ from other capacitors?
17. Identify the parts of the ceramic disk capacitor shown in the cutaway view of Figure 9–62 by referring to Figure 9–8(b).

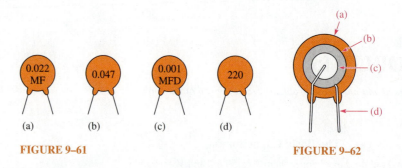

FIGURE 9–61   FIGURE 9–62

SECTION 9–3   Series Capacitors

18. Five 1000 pF capacitors are in series. What is the total capacitance?
19. Find the total capacitance for each circuit in Figure 9–63.

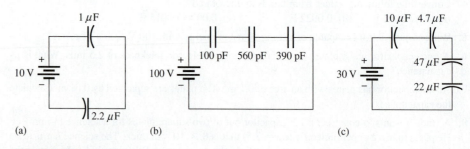

FIGURE 9–63

20. For each circuit in Figure 9–63, determine the voltage across each capacitor.
21. The total charge stored by the series capacitors in Figure 9–64 is 10 μC. Determine the voltage across each of the capacitors.

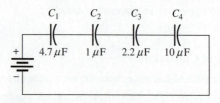

FIGURE 9–64

SECTION 9–4   Parallel Capacitors

22. Determine C_T for each circuit in Figure 9–65.

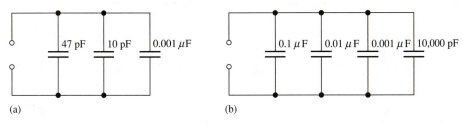

(a)   (b)

FIGURE 9–65

23. Determine the total capacitance and total charge on the capacitors in Figure 9–66.

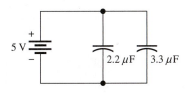

FIGURE 9–66

24. Assume you need a total capacitance of 2.1 μF in a certain timing application, but all that is available are 0.22 μF and 0.47 μF capacitors (in large quantities). How would you get the total capacitance that you need?

SECTION 9–5   Capacitors in DC Circuits

25. Determine the time constant for each of the following series RC combinations:
    (a) $R = 100 \, \Omega, C = 1 \, \mu$F   (b) $R = 10 \, \text{M}\Omega, C = 56$ pF
    (c) $R = 4.7 \, \text{k}\Omega, C = 0.0047 \, \mu$F   (d) $R = 1.5 \, \text{M}\Omega, C = 0.01 \, \mu$F

26. Determine how long it takes the capacitor to reach full charge for each of the following combinations:
    (a) $R = 47 \, \Omega, C = 47 \, \mu$F   (b) $R = 3300 \, \Omega, C = 0.015 \, \mu$F
    (c) $R = 22 \, \text{k}\Omega, C = 100$ pF   (d $R = 4.7 \, \text{M}\Omega, C = 10$ pF

27. In the circuit of Figure 9–67, the capacitor initially is uncharged. Determine the capacitor voltage at the following times after the switch is closed:
    (a) 10 μs   (b) 20 μs   (c) 30 μs   (d) 40 μs   (e) 50 μs

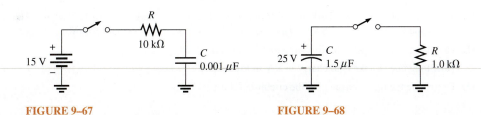

FIGURE 9–67   **FIGURE 9–68**

28. In Figure 9–68, the capacitor is charged to 25 V. Find the capacitor voltage at the following times when the switch is closed:
    (a) 1.5 ms   (b) 4.5 ms   (c) 6 ms   (d) 7.5 ms

29. Repeat Problem 27 for the following time intervals:
    (a) 2 μs   (b) 5 μs   (c) 15 μs

30. Repeat Problem 28 for the following time intervals:
    (a) 0.5 ms   (b) 1 ms   (c) 2 ms

SECTION 9–6   Capacitors in AC Circuits

31. Determine X_C for a 0.047 μF capacitor at each of the following frequencies:
   (a) 10 Hz   **(b)** 250 Hz   **(c)** 5 kHz   **(d)** 100 kHz

32. What is the value of the total capacitive reactance in each circuit in Figure 9–69?

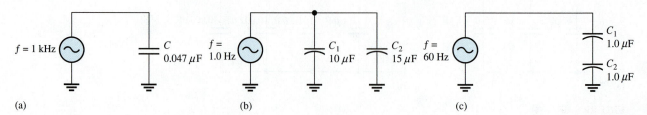

FIGURE 9–69

33. For the circuit in Figure 9–70, find the reactance of each capacitor, the total reactance, and the voltage across each capacitor.

FIGURE 9–70

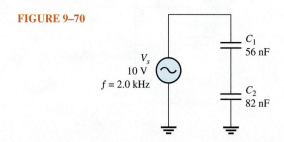

34. In each circuit of Figure 9–69, what frequency is required to produce an $X_{C(tot)}$ of 100 Ω? An $X_{C(tot)}$ of 1 kΩ?

35. A sinusoidal voltage of 20 V rms produces an rms current of 100 mA when connected to a certain capacitor. What is the reactance?

36. A 10 kHz voltage is applied to a 0.0047 μF capacitor, and 1 mA of rms current is measured. What is the rms value of the voltage?

37. Determine the true power and the reactive power in Problem 36.

SECTION 9–7   Capacitor Applications

38. If another capacitor is connected in parallel with the existing capacitor in the power supply filter of Figure 9–52, how is the ripple voltage affected?

39. Ideally, what should the reactance of a bypass capacitor be in order to eliminate a 10 kHz ac voltage at a given point in an amplifier circuit?

ADVANCED PROBLEMS

40. Two series capacitors (one 1 μF, the other of unknown value) are charged from a 12 V source. The 1 μF capacitor is charged to 8 V, and the other to 4 V. What is the value of the unknown capacitor?

41. How long does it take C to discharge to 3 V in Figure 9–68?

42. How long does it take C to charge to 8 V in Figure 9–67?

43. Determine the time constant for the circuit in Figure 9–71.

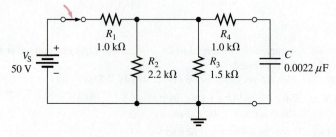

FIGURE 9–71

44. In Figure 9–72, the capacitor initially is uncharged. At $t = 10 \ \mu s$ after the switch is closed, the instantaneous capacitor voltage is 7.2 V. Determine the value of R.

45. (a) The capacitor in Figure 9–73 is uncharged when the switch is thrown into position 1. The switch remains in position 1 for 10 ms and then is thrown into position 2, where it remains indefinitely. Draw the complete waveform for the capacitor voltage.

   (b) If the switch is thrown back to position 1 after 5 ms in position 2, and then is left in position 1, how will the waveform appear?

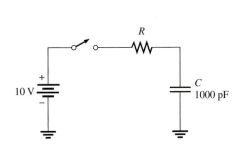

FIGURE 9–72

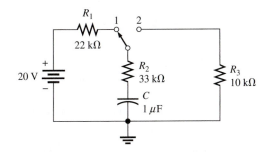

FIGURE 9–73

46. For the Colpitts oscillator in Figure 9–23, assume C_1 is changed to 3.3 nF.
   (a) What is the new total capacitance?
   (b) What is the new feedback fraction?

47. For the Colpitts oscillator in Figure 9–23, assume you need to change C_1 to obtain a feedback fraction of 15%. What value would you select?

48. Assume the discrete amplifier in Figure 9–57 has a 5 kHz, 1 V rms input. Sketch the complete signal as it would appear at the point in the circuit labeled B (the base of the transistor). Assume there is no loading effect of the transistor on the input voltage divider.

49. For the discrete amplifier in Figure 9–57, at what frequency will the reactance of $C_1 = 1.0 \ k\Omega$?

MULTISIM TROUBLESHOOTING PROBLEMS

MULTISIM

50. Open file P09-50; files are found at www.pearsonhighered.com/floyd. Determine if there is a fault. If so, identify it.

51. Open file P09-51 and determine if there is a fault. If so, identify it.

52. Open file P09-52 and determine if there is a fault. If so, identify it.

53. Open file P09-53 and determine if there is a fault. If so, identify it.

54. Open file P09-54 and determine if there is a fault. If so, identify it.

ANSWERS TO SECTION CHECKUPS

SECTION 9–1   The Basic Capacitor

1. Capacitance is the ability (capacity) to store electrical charge.

2. (a) There are 1,000,000 microfarads in one farad.
   (b) There are 1×10^{12} picofarads in one farad.
   (c) There are 1,000,000 picofarads in one microfarad.

3. $0.0015 \ \mu F \times 10^6 \ pF/\mu F = 1500 \ pF; 0.0015 \ \mu F \times 10^{-6} \ F/\mu F = 0.0000000015 \ F$

4. $W = \frac{1}{2}CV^2 = \frac{1}{2}(0.01 \ \mu F)(15 \ V)^2 = 1.125 \ \mu J$

5. (a) Capacitance increases when plate area is increased.
   (b) Capacitance decreases when plate separation is increased.

6. (1000 V/mil)(2 mils) = 2 kV

SECTION 9–2   Types of Capacitors

1. Capacitors are commonly classified by the dielectric material.
2. A fixed capacitance cannot be changed; a variable capacitor can.
3. Electrolytic capacitors are polarized.
4. Be sure that the voltage rating is sufficient and connect the positive end to the positive side of the circuit when installing a polarized capacitor.
5. The positive lead should be connected to ground.

SECTION 9–3   Series Capacitors

1. C_T of series capacitors is less than the smallest value.
2. $C_T = 61.2$ pF
3. $C_T = 0.006$ μF
4. $V = (0.006\ \mu\text{F}/0.01\ \mu\text{F})10\ \text{V} = 6$ V

SECTION 9–4   Parallel Capacitors

1. The individual parallel capacitors are added to get C_T.
2. Use five 0.01 μF capacitors in parallel to get 0.05 μF.
3. $C_T = 167$ pF

SECTION 9–5   Capacitors in DC Circuits

1. $\tau = RC = 1.2\ \mu$s
2. $5\tau = 6\ \mu$s; V_C is approximately 5 V.
3. $v_{2\text{ms}} = (0.86)10\ \text{V} = 8.6\ \text{V}$; $v_{3\text{ms}} = (0.95)10\ \text{V} = 9.5\ \text{V}$;
   $v_{4\text{ms}} = (0.98)10\ \text{V} = 9.8\ \text{V}$; $v_{5\text{ms}} = (0.99)10\ \text{V} = 9.9\ \text{V}$
4. $v_C = (0.37)(100\ \text{V}) = 37$ V

SECTION 9–6   Capacitors in AC Circuits

1. $X_C = 1/2\pi fC = 677$ kΩ
2. $f = 1/2\pi C X_C = 796$ Hz
3. $I_{\text{rms}} = 1\ \text{V}/1.59\ \Omega = 629$ mA
4. Current leads voltage by 90°.
5. $P_{\text{true}} = 0$ W
6. $P_r = (12\ \text{V})^2/318\ \Omega = 0.453$ VAR

SECTION 9–7   Capacitor Applications

1. Once the capacitor charges to the peak voltage, it discharges very little before the next peak thus smoothing the rectified voltage.
2. A coupling capacitor allows ac to pass from one point to another, but blocks constant dc.
3. A coupling capacitor must be large enough to have a negligible reactance at the frequency that is to be passed without opposition.
4. A decoupling capacitor shorts power line voltage transients to ground.
5. X_C is inversely proportional to frequency and so is the filter's ability to pass ac signals.

ANSWERS TO RELATED PROBLEMS FOR EXAMPLES

9–1   10 kV
9–2   0.047 μF
9–3   100,000,000 pF
9–4   62.7 pF
9–5   278 pF

9–6 1.54 μF

9–7 2.83 V

9–8 650 pF

9–9 0.052 μF

9–10 891 μs

9–11 8.36 V

9–12 8.13 V

9–13 0.7 ms; 95 V

9–14 0.42 V

9–15 3.39 kHz

9–16 **(a)** 1.83 kΩ
      (b) 408 Ω

9–17 4.40 mA

9–18 8.72 V

9–19 0 W; 1.01 mVAR

ANSWERS TO TRUE/FALSE QUIZ

1. T   **2.** F   **3.** T   **4.** F   **5.** T   **6.** T   **7.** F   **8.** F   **9.** F   **10.** T

ANSWERS TO SELF-TEST

1. (g)   **2.** (b)   **3.** (c)   **4.** (d)   **5.** (a)   **6.** (d)   **7.** (a)   **8.** (f)

9. (c)   **10.** (b)   **11.** (d)   **12.** (a)   **13.** (b)   **14.** (a)   **15.** (b)

ANSWERS TO TROUBLESHOOTING: SYMPTOM AND CAUSE

1. (b)

2. (a)

3. (c)

4. (a)

5. (b)

CHAPTER 10

RC CIRCUITS

OUTLINE

OBJECTIVES

- Describe the relationship between current and voltage in a series *RC* circuit
- Determine impedance and phase angle in a series *RC* circuit
- Analyze a series *RC* circuit
- Determine impedance and phase angle in a parallel *RC* circuit
- Analyze a parallel *RC* circuit
- Analyze series-parallel *RC* circuits
- Determine power in *RC* circuits
- Discuss some basic *RC* applications
- Troubleshoot *RC* circuits

KEY TERMS

- Impedance (*Z*)
- Phase angle
- *RC* lag circuit
- *RC* lead circuit
- Capacitive susceptance (*B_C*)
- Admittance (*Y*)
- Apparent power (*P_a*)
- Power factor
- Frequency response
- Cutoff frequency
- Bandwidth

INTRODUCTION

An *RC* circuit, which contains both resistance and capacitance, is one of the basic types of reactive circuits. In this chapter, series and parallel *RC* circuits and their responses to sinusoidal voltages are covered. Series-parallel combinations are also examined. Power considerations in *RC* circuits are introduced, and practical aspects of power ratings are discussed. Three *RC* circuit applications are presented to give you an idea of how simple combinations of resistors and capacitors can be applied. Troubleshooting common faults in *RC* circuits is also covered.

The methods for analyzing reactive circuits are similar to those you studied in dc circuits. Reactive circuit problems can be solved at only one frequency at a time, and phasor math must be used.

VISIT THE WEBSITE
Study aids for this chapter are available at
http://pearsonhighered.com/floyd

10–1 SINUSOIDAL RESPONSE OF SERIES *RC* CIRCUITS

When a sinusoidal voltage is applied to a series *RC* circuit, each resulting voltage drop and the current in the circuit(s) are also sinusoidal and have the same frequency as the source voltage. The capacitance causes a phase shift between the voltage and current that depends on the relative values of the resistance and the capacitive reactance.

After completing this section, you should be able to

- **Describe the relationship between current and voltage in a series *RC* circuit**
  - **Discuss voltage and current waveforms**
  - **Discuss phase shift**

As shown in Figure 10–1 for the case of a series *RC* circuit, the resistor voltage (V_R), the capacitor voltage (V_C), and the current (I) are all sine waves with the frequency of the source. Phase shifts are introduced because of the capacitance. As you will learn, the resistor voltage and current are in phase with each other and lead the source voltage in phase. The capacitor voltage lags the source voltage. The phase angle between the current and the capacitor voltage is always 90°. These generalized phase relationships are indicated in Figure 10–1.

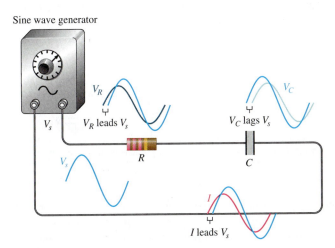

Sine wave generator

FIGURE 10–1   Illustration of sinusoidal response with general phase relationships of V_R, V_C, and I relative to the source voltage. V_R and I are in phase, and V_R and V_C are 90° out of phase.

The amplitudes and the phase relationships of the voltages and current depend on the values of the resistance and the capacitive reactance. When a circuit is purely resistive, the phase angle between the source voltage and the total current is zero. When a circuit is purely capacitive, the phase angle between the source voltage and the total current is 90°, with the current leading the voltage. When there is a combination of both resistance and capacitive reactance in a circuit, the phase angle between the source voltage and the total current is somewhere between zero and 90°, depending on the relative values of the resistance and the capacitive reactance.

SECTION 10–1 CHECKUP*

1. A 60 Hz sinusoidal voltage is applied to an *RC* circuit. What is the frequency of the capacitor voltage? What is the frequency of the current?

2. What determines the phase shift between V_s and I in a series *RC* circuit?

3. When the resistance in a series *RC* circuit is greater than the capacitive reactance, is the phase angle between the source voltage and the total current closer to 0° or to 90°?

10–2 IMPEDANCE AND PHASE ANGLE OF SERIES *RC* CIRCUITS

In circuits where there is no reactance, the opposition to current is strictly resistance. In circuits having both resistance and reactance, the opposition to current is more complex because of the reactance and the resulting phase shift. Impedance, which is the total opposition to ac and includes the effect of phase shift, is introduced in this section.

After completing this section, you should be able to

- **Determine impedance and phase angle in a series *RC* circuit**
  - **Define** *impedance*
  - **Define** *phase angle*
  - **Draw an impedance triangle**
  - **Calculate the total impedance magnitude**
  - **Calculate the phase angle**

The **impedance** of a series *RC* circuit consists of resistance and capacitive reactance and is the total opposition to sinusoidal current. Its unit is the ohm. The **phase angle** is the phase difference between the total current and the source voltage.

In a purely resistive circuit, the impedance is simply equal to the total resistance. In a purely capacitive circuit, the impedance is the total capacitive reactance. The impedance of a series *RC* circuit is determined by both the resistance (R) and the capacitive reactance (X_C). These cases are illustrated in Figure 10–2. The magnitude of the impedance is symbolized by Z.

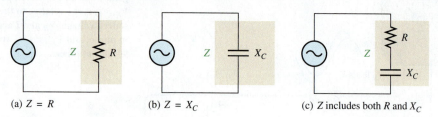

(a) $Z = R$ (b) $Z = X_C$ (c) Z includes both R and X_C

FIGURE 10–2   **Three cases of impedance.**

In ac analysis, both R and X_C are treated as phasor quantities, as shown in the phasor diagram of Figure 10–3(a), with X_C appearing at a $-90°$ angle with respect to R. This relationship comes from the fact that the capacitor voltage in a series *RC* circuit lags the current, and thus the resistor voltage, by 90°. Since Z is the phasor sum of R and X_C, its phasor representation is as shown in Figure 10–3(b). A repositioning of the phasors, as shown in Figure 10–3(c), forms a right triangle, which is called the *impedance triangle*. The length of each phasor represents the magnitude in ohms, and the angle θ (the Greek letter theta) is the phase angle of the *RC* circuit and represents the phase difference between the source voltage and the current.

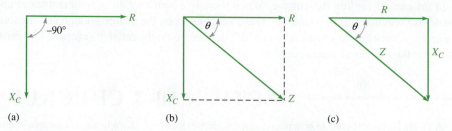

(a)   (b)   (c)

FIGURE 10–3   **Development of the impedance triangle for a series *RC* circuit.**

From right-triangle trigonometry (Pythagorean theorem), the magnitude (length) of the impedance can be expressed in terms of the resistance and capacitive reactance.

$$Z = \sqrt{R^2 + X_C^2}$$

$$(10–1)$$

The magnitude of the impedance (Z), as shown in the *RC* circuit in Figure 10–4, is expressed in ohms.

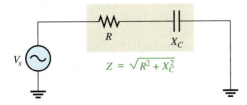

FIGURE 10–4   **Impedance in a series *RC* circuit.**

The value of the phase angle, θ, is expressed as

$$\theta = \tan^{-1}\left(\frac{X_C}{R}\right)$$ (10–2)

The symbol $\tan^{-1}$ stands for *inverse tangent* and can be found on most calculators by pressing the 2nd and TAN^{-1} keys. Another term for inverse tangent is *arctangent* (arctan).

EXAMPLE 10–1

Determine the impedance and the phase angle of the *RC* circuit in Figure 10–5. Draw the impedance triangle.

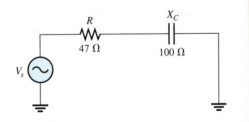

FIGURE 10–5

SOLUTION

The impedance is

$$Z = \sqrt{R^2 + X_C^2} = \sqrt{(47\ \Omega)^2 + (100\ \Omega)^2} = \mathbf{110\ \Omega}$$

The phase angle is

$$\theta = \tan^{-1}\left(\frac{X_C}{R}\right) = \tan^{-1}\left(\frac{100\ \Omega}{47\ \Omega}\right) = \tan^{-1}(2.13) = \mathbf{64.8°}$$

The source voltage lags the current by 64.8°.
   The impedance triangle is shown in Figure 10–6.

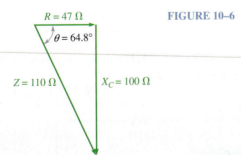

FIGURE 10–6

RELATED PROBLEM*

Find Z and θ for $R = 1.0\ \text{k}\Omega$ and $X_C = 2.2\ \text{k}\Omega$ in Figure 10–5.

Answers are at the end of the chapter.

SECTION 10–2 CHECKUP

1. Define *impedance*.

2. Does the source voltage lead or lag the current in a series *RC* circuit?

3. What causes the phase angle in an *RC* circuit?

4. A series *RC* circuit has a resistance of 33 kΩ and a capacitive reactance of 50 kΩ. What is the value of the impedance? What is the phase angle?

10–3 ANALYSIS OF SERIES *RC* CIRCUITS

Ohm's law and Kirchhoff's voltage law are used in the analysis of series *RC* circuits to determine voltage, current, and impedance. Also, in this section *RC* lead and lag circuits are examined.

After completing this section, you should be able to

- **Analyze a series *RC* circuit**
  - **Apply Ohm's law and Kirchhoff's voltage law to series *RC* circuits**
  - **Determine the phase relationships of the voltages and current**
  - **Show how impedance and phase angle vary with frequency**
  - **Analyze the *RC* lag circuit**
  - **Analyze the *RC* lead circuit**

Ohm's Law

The application of Ohm's law to series *RC* circuits involves the use of the quantities of *Z*, *V*, and *I*. The three equivalent forms of Ohm's law are as follows:

$$V = IZ \tag{10–3}$$

$$I = \frac{V}{Z} \tag{10–4}$$

$$Z = \frac{V}{I} \tag{10–5}$$

The following two examples illustrate the use of Ohm's law.

EXAMPLE 10–2

The current in Figure 10–7 is 0.2 mA. Determine the source voltage and the phase angle. Draw the impedance triangle.

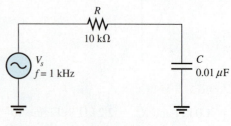

FIGURE 10–7

SOLUTION

The capacitive reactance is

$$X_C = \frac{1}{2\pi fC} = \frac{1}{2\pi(1000 \text{ Hz})(0.01 \text{ } \mu\text{F})} = 15.9 \text{ k}\Omega$$

The impedance is

$$Z = \sqrt{R^2 + X_C^2} = \sqrt{(10 \text{ k}\Omega)^2 + (15.9 \text{ k}\Omega)^2} = 18.8 \text{ k}\Omega$$

Applying Ohm's law yields

$$V_s = IZ = (0.2 \text{ mA})(18.8 \text{ k}\Omega) = \textbf{3.76 V}$$

The phase angle is

$$\theta = \tan^{-1}\left(\frac{X_C}{R}\right) = \tan^{-1}\left(\frac{15.9 \text{ k}\Omega}{10 \text{ k}\Omega}\right) = \textbf{57.8°}$$

The source voltage has a magnitude of 3.76 V and lags the current by 57.8°.
The impedance triangle is shown in Figure 10–8.

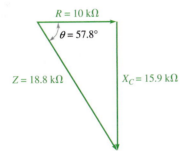

$R = 10 \text{ k}\Omega$

$\theta = 57.8°$

$Z = 18.8 \text{ k}\Omega$

$X_C = 15.9 \text{ k}\Omega$

FIGURE 10–8

RELATED PROBLEM

Determine V_s in Figure 10–7 if $f = 2$ kHz and $I = 200 \text{ } \mu\text{A}$.

EXAMPLE 10–3

Determine the current in the *RC* circuit of Figure 10–9.

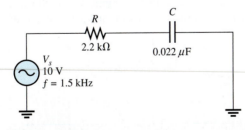

R
2.2 kΩ

C
0.022 μF

V_s
10 V
$f = 1.5$ kHz

FIGURE 10–9

SOLUTION

The capacitive reactance is

$$X_C = \frac{1}{2\pi fC} = \frac{1}{2\pi(1.5 \text{ kHz})(0.022 \text{ } \mu\text{F})} = 4.82 \text{ k}\Omega$$

The impedance is

$$Z = \sqrt{R^2 + X_C^2} = \sqrt{(2.2\text{ k}\Omega)^2 + (4.82\text{ k}\Omega)^2} = 5.30\text{ k}\Omega$$

Applying Ohm's law yields

$$I = \frac{V}{Z} = \frac{10\text{ V}}{5.30\text{ k}\Omega} = \textbf{1.89 mA}$$

RELATED PROBLEM

Determine the phase angle between V_s and I in Figure 10–9.

Phase Relationships of the Current and Voltages

In a series *RC* circuit, the current is the same through both the resistor and the capacitor. Thus, the resistor voltage is in phase with the current, and the capacitor voltage lags the current by 90°. Therefore, there is a phase difference of 90° between the resistor voltage, V_R, and the capacitor voltage, V_C, as shown in the waveform diagram of Figure 10–10.

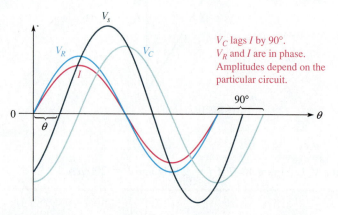

FIGURE 10–10 **Phase relation of the voltages and current in a series *RC* circuit.**

You know from Kirchhoff's voltage law that the sum of the voltage drops must equal the source voltage. However, since V_R and V_C are 90° out of phase with each other, they must be added as phasor quantities, with V_C lagging V_R, as shown in Figure 10–11(a). As shown in Figure 10–11(b), V_s is the phasor sum of V_R and V_C, as expressed in the following equation:

$$V_s = \sqrt{V_R^2 + V_C^2} \tag{10–6}$$

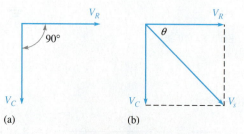

FIGURE 10–11 **Voltage phasor diagram for the waveforms in Figure 10–10.**

The value of the phase angle between the resistor voltage and the source voltage can be expressed as

$$\theta = \tan^{-1}\left(\frac{V_C}{V_R}\right) \qquad (10\text{--}7)$$

Since the resistor voltage and the current are in phase, θ in Equation 10–7 also represents the phase angle between the source voltage and the current and is equivalent to $\tan^{-1}(X_C/R)$.

Figure 10–12 shows the voltage and current phasor diagram that represents the waveform diagram of Figure 10–10.

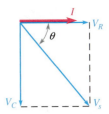

FIGURE 10–12 **Voltage and current phasor diagram for the waveforms in Figure 10–10.**

EXAMPLE 10–4

Determine the source voltage and the phase angle in Figure 10–13. Draw the voltage phasor diagram.

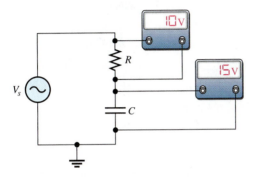

FIGURE 10–13

SOLUTION

Since V_R and V_C are 90° out of phase, you cannot add them directly. The source voltage is the phasor sum of V_R and V_C.

$$V_s = \sqrt{V_R^2 + V_C^2} = \sqrt{(10\text{ V})^2 + (15\text{ V})^2} = \textbf{18 V}$$

The phase angle between the resistor voltage and the source voltage is

$$\theta = \tan^{-1}\left(\frac{V_C}{V_R}\right) = \tan^{-1}\left(\frac{15\text{ V}}{10\text{ V}}\right) = \textbf{56.3°}$$

The voltage phasor diagram is shown in Figure 10–14.

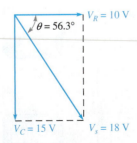

FIGURE 10–14

RELATED PROBLEM

In a certain series *RC* circuit, $V_s = 10$ V and $V_R = 7$ V. Find V_C.

Variation of Impedance and Phase Angle with Frequency

As you know, capacitive reactance varies inversely with frequency. Since $Z = \sqrt{R^2 + X_C^2}$, you can see that when X_C increases, the entire term under the square root sign increases and thus the total impedance also increases; and when X_C decreases, the total impedance also decreases. Therefore, *in a series RC circuit, Z is inversely dependent on frequency.*

Figure 10–15 illustrates how the voltages and current in a series *RC* circuit vary as the frequency increases or decreases, with the source voltage held at a constant value. Part (a) shows that as the frequency is increased, X_C decreases; so the voltage across the capacitor decreases. Also, Z decreases as X_C decreases, causing the current to increase. An increase in the current causes more voltage across R.

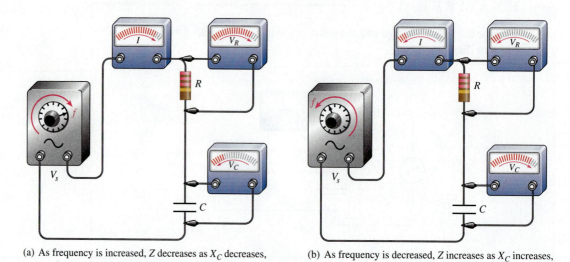

(a) As frequency is increased, Z decreases as X_C decreases, causing I and V_R to increase and V_C to decrease.

(b) As frequency is decreased, Z increases as X_C increases, causing I and V_R to decrease and V_C to increase.

FIGURE 10–15   An illustration of how the variation of impedance affects the voltages and current as the source frequency is varied. The source voltage is held at a constant amplitude.

Figure 10–15(b) shows that as the frequency is decreased, X_C increases; so more voltage is dropped across the capacitor. Also, Z increases as X_C increases, causing the current to decrease. A decrease in the current causes less voltage across R.

Changes in Z and X_C can be observed as shown in Figure 10–16. As the frequency increases, the voltage across Z remains constant because V_s is constant ($V_s = V_Z$). Also, the voltage across C decreases. The increasing current indicates that Z is decreasing. It

FIGURE 10–16   An illustration of how Z and X_C change with frequency.

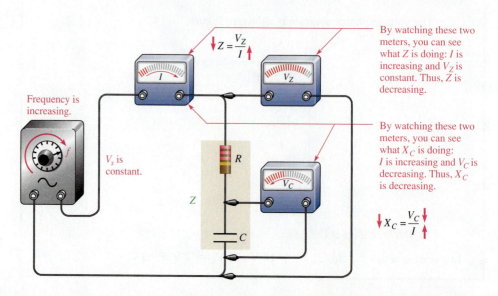

$$Z = \frac{V_Z}{I}$$

Frequency is increasing.

V_s is constant.

Z

By watching these two meters, you can see what Z is doing: I is increasing and V_Z is constant. Thus, Z is decreasing.

By watching these two meters, you can see what X_C is doing: I is increasing and V_C is decreasing. Thus, X_C is decreasing.

$$X_C = \frac{V_C}{I}$$

does so because of the inverse relationship stated in Ohm's law ($Z = V_Z/I$). The increasing current also indicates that X_C is decreasing ($X_C = V_C/I$). The decrease in V_C corresponds to the decrease in X_C.

Since X_C is the factor that introduces the phase angle in a series *RC* circuit, a change in X_C produces a change in the phase angle. As the frequency is increased, X_C becomes smaller, and thus the phase angle decreases. As the frequency is decreased, X_C becomes larger, and thus the phase angle increases. The angle between V_s and V_R is the phase angle of the circuit because I is in phase with V_R.

Figure 10–17 uses the impedance triangle to illustrate the variations in X_C, Z, and θ as the frequency changes. Of course, *R* remains constant. The key point is that because X_C varies inversely with the frequency, the magnitude of the total impedance and the phase angle also vary inversely with the frequency. Example 10–5 illustrates this.

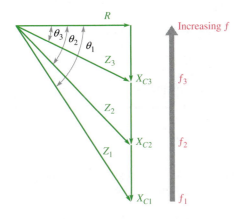

FIGURE 10–17 **As the frequency increases, X_C decreases, Z decreases, and θ decreases. Each value of frequency can be visualized as forming a different impedance triangle.**

EXAMPLE 10–5

For the series *RC* circuit in Figure 10–18, determine the impedance and phase angle for each of the following values of frequency:

(a) 10 kHz   **(b)** 20 kHz   **(c)** 30 kHz

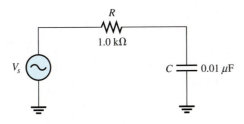

FIGURE 10–18

SOLUTION

(a) For $f = 10$ kHz, calculate the impedance as follows:

$$X_C = \frac{1}{2\pi f C} = \frac{1}{2\pi(10\text{ kHz})(0.01\ \mu\text{F})} = 1.59\text{ k}\Omega$$

$$Z = \sqrt{R^2 + X_C^2} = \sqrt{(1.0\text{ k}\Omega)^2 + (1.59\text{ k}\Omega)^2} = \mathbf{1.88\text{ k}\Omega}$$

The phase angle is

$$\theta = \tan^{-1}\left(\frac{X_C}{R}\right) = \tan^{-1}\left(\frac{1.59\text{ k}\Omega}{1.0\text{ k}\Omega}\right) = \mathbf{57.8°}$$

(b) For $f = 20$ kHz,

$$X_C = \frac{1}{2\pi(20\text{ kHz})(0.01\ \mu\text{F})} = 796\ \Omega$$

$$Z = \sqrt{(1.0\text{ k}\Omega)^2 + (796\ \Omega)^2} = \mathbf{1.28\ k\Omega}$$

$$\theta = \tan^{-1}\left(\frac{796\ \Omega}{1.0\text{ k}\Omega}\right) = \mathbf{38.5°}$$

(c) For $f = 30$ kHz,

$$X_C = \frac{1}{2\pi(30\text{ kHz})(0.01\ \mu\text{F})} = 531\ \Omega$$

$$Z = \sqrt{(1.0\text{ k}\Omega)^2 + (531\ \Omega)^2} = \mathbf{1.13\ k\Omega}$$

$$\theta = \tan^{-1}\left(\frac{531\ \Omega}{1.0\text{ k}\Omega}\right) = \mathbf{28.0°}$$

Notice that as the frequency increases, X_C, Z, and θ decrease.

RELATED PROBLEM

Find the total impedance and phase angle in Figure 10–18 for $f = 1$ kHz.

RC Lag Circuit

An *RC* **lag circuit** is a phase shift circuit in which the output voltage lags the input voltage by a specified angle, ϕ. Phase shift circuits are commonly used in electronic communication systems and for other applications.

A basic series *RC* lag circuit is shown in Figure 10–19(a). Keep in mind that phase angle θ is measured between the source (input) voltage and the current. In terms of voltages, this is equivalent to saying the phase angle is measured between V_{in} and V_R because current and voltage are in phase in a resistor. The output is taken across the capacitor; therefore, the phase lag between V_{in} and V_C is $90° - \theta$. This angle is designated as ϕ and represents the phase difference between the input (V_{in}) and the output (V_{out}), as shown in Figure 10–19(b).

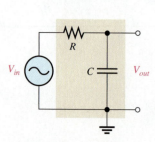

(a) A basic *RC* lag circuit

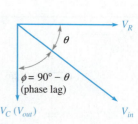

(b) Phasor voltage diagram showing the phase lag between V_{in} and V_{out}

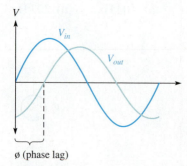

(c) Input and output voltage waveforms

FIGURE 10–19   The *RC* lag circuit ($V_{out} = V_C$).

Since $\theta = \tan^{-1}(X_C/R)$, the value of the phase lag, ϕ, can be expressed as

$$\phi = 90° - \tan^{-1}\left(\frac{X_C}{R}\right) \tag{10–8}$$

The input and output voltage waveforms of the lag circuit are shown in Figure 10–19(c). The exact amount of phase lag between the input and the output depends on the values of the resistance and the capacitive reactance. The magnitude of the output voltage depends on these values also.

EXAMPLE 10–6

Determine the amount of phase lag from input to output in the lag circuit in Figure 10–20.

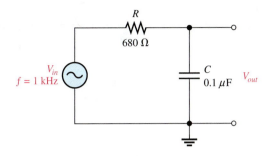

FIGURE 10–20

SOLUTION

First determine the capacitive reactance.

$$X_C = \frac{1}{2\pi fC} = \frac{1}{2\pi(1\text{ kHz})(0.1\ \mu\text{F})} = 1.59\text{ k}\Omega$$

The phase lag between the output voltage and the input voltage is

$$\phi = 90° - \tan^{-1}\left(\frac{X_C}{R}\right) = 90° - \tan^{-1}\left(\frac{1.59\text{ k}\Omega}{680\ \Omega}\right) = \mathbf{23.2°}$$

The output voltage lags the input voltage by 23.2°.

RELATED PROBLEM

In a lag circuit, what happens to the phase lag if the frequency increases?

The phase-lag circuit can be considered as a voltage divider with a portion of the input voltage dropped across *R* and a portion across *C*. The output voltage can be determined with the following formula:

$$V_{out} = \left(\frac{X_C}{\sqrt{R^2 + X_C^2}}\right)V_{in} \qquad (10\text{–}9)$$

EXAMPLE 10–7

For the lag circuit in Figure 10–20 of Example 10–6, determine the output voltage when the input voltage has an rms value of 10 V. Draw the input and output waveforms showing the proper relationships. The values for X_C (1.59 kΩ) and ϕ (23.2°) were found in Example 10–6.

SOLUTION

Use Equation 10–9 to determine the output voltage for the lag circuit in Figure 10–20.

$$V_{out} = \left(\frac{X_C}{\sqrt{R^2 + X_C^2}}\right)V_{in} = \left(\frac{1.59\text{ k}\Omega}{\sqrt{(680\ \Omega)^2 + (1.59\text{ k}\Omega)^2}}\right)10\text{ V} = \mathbf{9.2\text{ V rms}}$$

The waveforms are shown in Figure 10–21. (Note that the rms value is converted to peak value on the plot.)

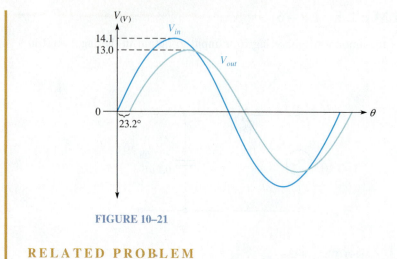

FIGURE 10–21

RELATED PROBLEM

In a lag circuit, what happens to the output voltage if the frequency increases?

EFFECTS OF FREQUENCY ON THE LAG CIRCUIT   Since the circuit phase angle, θ, decreases as frequency increases, the phase lag ϕ between the input and the output voltages increases. You can see this relationship by examining Equation 10–8. Also, the magnitude of V_{out} decreases as the frequency increases because X_C becomes smaller and less of the total input voltage is dropped across the capacitor.

RC Lead Circuit

An *RC* **lead circuit** is a phase shift circuit in which the output voltage leads the input voltage by a specified angle, ϕ. A basic *RC* lead circuit is shown in Figure 10–22(a). Notice how it differs from the lag circuit. Here the output voltage is taken across the resistor. The relationship of the voltages is given in the phasor diagram in Figure 10–22(b). The output voltage, V_{out}, leads V_{in} by an angle that is the same as the circuit phase angle because V_R and I are in phase with each other.

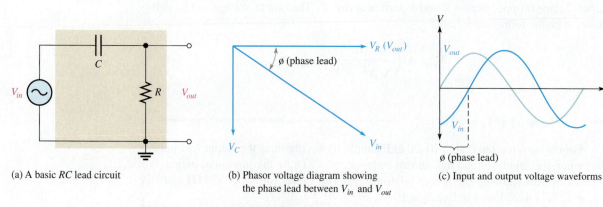

(a) A basic *RC* lead circuit

(b) Phasor voltage diagram showing the phase lead between V_{in} and V_{out}

(c) Input and output voltage waveforms

FIGURE 10–22   **The *RC* lead circuit ($V_{out} = V_R$).**

When the input and output waveforms are displayed on an oscilloscope, a relationship similar to that in Figure 10–22(c) is observed. Of course, the exact amount of phase lead and the output voltage magnitude depend on the values of R and X_C. The value of the phase lead, ϕ, is expressed as

$$\phi = \tan^{-1}\left(\frac{X_C}{R}\right)$$

(10–10)

The output voltage is expressed as

$$V_{out} = \left(\frac{R}{\sqrt{R^2 + X_C^2}}\right)V_{in} \qquad (10\text{–}11)$$

EXAMPLE 10–8

Calculate the phase lead and the output voltage for the circuit in Figure 10–23.

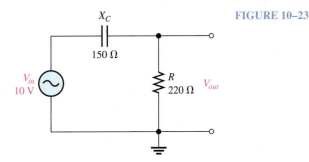

FIGURE 10–23

SOLUTION

The phase lead is

$$\phi = \tan^{-1}\left(\frac{X_C}{R}\right) = \tan^{-1}\left(\frac{150\ \Omega}{220\ \Omega}\right) = \mathbf{34.3°}$$

The output leads the input by 34.3°.
Use Equation 10–11 to determine the output voltage.

$$V_{out} = \left(\frac{R}{\sqrt{R^2 + X_C^2}}\right)V_{in} = \left(\frac{220\ \Omega}{\sqrt{(220\ \Omega)^2 + (150\ \Omega)^2}}\right)10\ \text{V} = \mathbf{8.26\ V}$$

RELATED PROBLEM

How does an increase in *R* affect the phase lead and the output voltage in Figure 10–23?

EFFECTS OF FREQUENCY ON THE LEAD CIRCUIT   Since the phase lead is the same as the circuit phase angle θ, it decreases as frequency increases. The output voltage increases with frequency because as X_C becomes smaller, more of the input voltage is dropped across the resistor.

SYSTEM EXAMPLE 10–1

A BUFFER AMPLIFIER FOR A PREAMP

Often an analog system has a low-level input signal that needs to be amplified to avoid transmission problems before sending it to a controller or processor. For low signal levels, a field-effect transistor (FET) is often used as a buffer amplifier followed by a gain stage, forming a simple preamp system. A buffer amplifier for voltage is one that has a gain of 1 (output voltage equals input voltage) but presents a high impedance to the input signal thus preventing loading of a low-impedance source. Figure 10–24(a) shows a buffer amplifier, which then drives other gain stages. The equivalent input circuit for this amplifier is just the coupling capacitor, C_1, and the bias resistor, R_G, as shown in part (b). For calculating any effect on the driving circuit, this is the only part of the amplifier that needs to be considered because the transistor looks like an open to the input signal.

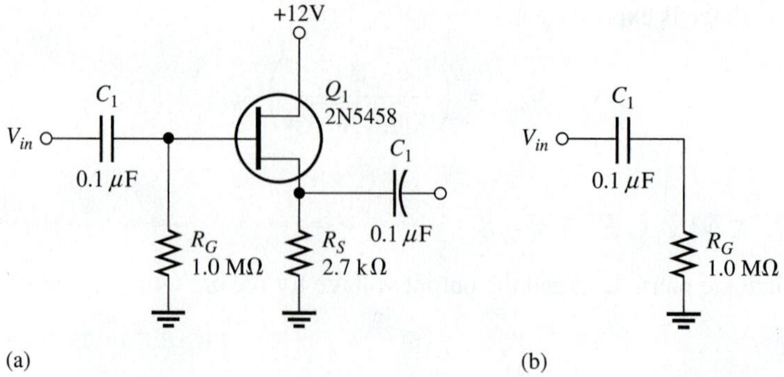

FIGURE 10–24 **A buffer amplifier.**

The input of the amplifier contributes to the lower cutoff frequency of the entire preamp. It can be simplified to the basic *RC* circuit that you studied in this section. In fact, the input to *any* capacitively coupled amplifier can be reduced to this same basic circuit. This simplification allows you to apply basic dc/ac theory to a more-complicated circuit to calculate a response. For example, you can easily determine the frequency at which the voltage across the capacitor is equal to the voltage across the resistor for the buffer amplifier. (This frequency is called the cutoff frequency, f_c, which is discussed in Section 10–8.) If the voltage across the capacitor is equal to the voltage across the resistor, then $X_C = R$. Then,

$$X_C = \frac{1}{2\pi f_c C} = R$$

$$f_c = \frac{1}{2\pi RC} = \frac{1}{2\pi(1.0 \text{ M}\Omega)(0.1 \ \mu\text{F})} = 1.59 \text{ Hz}$$

SECTION 10–3 CHECKUP

1. In a certain series *RC* circuit, $V_R = 4$ V and $V_C = 6$ V. What is the magnitude of the source voltage?

2. In Question 1, what is the phase angle?

3. What is the phase difference between the capacitor voltage and the resistor voltage in a series *RC* circuit?

4. When the frequency of the source voltage in a series *RC* circuit is increased, what happens to each of the following?

   (a) the capacitive reactance   (b) the impedance
   (c) the phase angle

5. A certain *RC* lag circuit consists of a 4.7 kΩ resistor and a 0.022 μF capacitor. Determine the phase lag between the input and output voltages at a frequency of 3 kHz.

6. An *RC* lead circuit has the same component values as the lag circuit in Question 5. What is the magnitude of the output voltage at 3 kHz when the input is 10 V rms?

10–4 IMPEDANCE AND PHASE ANGLE OF PARALLEL *RC* CIRCUITS

In this section, you will learn how to determine the impedance and phase angle of a parallel *RC* circuit. Also conductance (*G*), capacitive susceptance (*B$_C$*), and total admittance (*Y$_{tot}$*) are discussed because of their usefulness in parallel circuit analysis.

After completing this section, you should be able to

- **Determine impedance and phase angle in a parallel *RC* circuit**
  - **Express total impedance in a product-over-sum form**
  - **Express the phase angle in terms of *R* and *X$_C$***
  - **Determine the conductance, capacitive susceptance, and admittance**
  - **Convert admittance to impedance**

Figure 10–25 shows a basic parallel *RC* circuit.

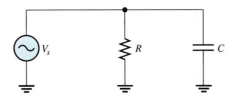

FIGURE 10–25 **Parallel *RC* circuit.**

The expression for the impedance in Equation 10–12 is given in a product-over-sum form similar to the way two resistors in parallel can be expressed. In this case, the denominator is the phasor sum of *R* and *X$_C$*.

$$Z = \frac{RX_C}{\sqrt{R^2 + X_C^2}} \tag{10–12}$$

The value of the phase angle between the source voltage and the total current can be expressed in terms of *R* and *X$_C$* as shown in Equation 10–13.

$$\theta = \tan^{-1}\left(\frac{R}{X_C}\right) \tag{10–13}$$

This formula is derived from an equivalent formula using branch currents (Eq. 10–22), which is introduced in Section 10–5.

EXAMPLE 10–9

For each circuit in Figure 10–26, determine the impedance and the phase angle.

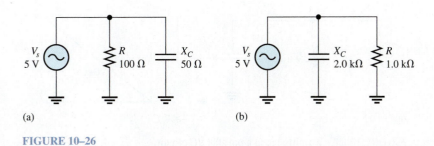

(a)   (b)

FIGURE 10–26

SOLUTION

For the circuit in Figure 10–26(a), the impedance and phase angle are

$$Z = \frac{RX_C}{\sqrt{R^2 + X_C^2}} = \frac{(100\ \Omega)(50\ \Omega)}{\sqrt{(100\ \Omega)^2 + (50\ \Omega)^2}} = \mathbf{44.7\ \Omega}$$

$$\theta = \tan^{-1}\left(\frac{R}{X_C}\right) = \tan^{-1}\left(\frac{100\ \Omega}{50\ \Omega}\right) = \mathbf{63.4°}$$

For the circuit in Figure 10–26(b),

$$Z = \frac{(1.0\ \text{k}\Omega)(2.0\ \text{k}\Omega)}{\sqrt{(1.0\ \text{k}\Omega)^2 + (2.0\ \text{k}\Omega)^2}} = \mathbf{894\ \Omega}$$

$$\theta = \tan^{-1}\left(\frac{1.0\ \text{k}\Omega}{2.0\ \text{k}\Omega}\right) = \mathbf{26.6°}$$

RELATED PROBLEM

Determine Z in Figure 10–26(a) if the frequency is doubled.

Conductance, Capacitive Susceptance, and Admittance

Recall that **conductance (*G*)** is the reciprocal of resistance and is expressed as

$$G = \frac{1}{R} \qquad\qquad (10\text{–}14)$$

Two new terms are now introduced for use in parallel *RC* circuits. Susceptance is the reciprocal of reactance; therefore, **capacitive susceptance (*B_C*)** is the reciprocal of capacitive reactance and is a measure of the ability of a capacitor to permit current. It is expressed as

$$B_C = \frac{1}{X_C} \qquad\qquad (10\text{–}15)$$

Admittance (*Y*) is the reciprocal of impedance and is expressed as

$$Y = \frac{1}{Z} \qquad\qquad (10\text{–}16)$$

The unit of each of these three quantities is the siemens (S), which is the reciprocal of the ohm.

In working with parallel circuits, it is often easier to use conductance (*G*), capacitive susceptance (*B_C*), and admittance (*Y*) rather than resistance (*R*), capacitive reactance (*X_C*) and impedance (*Z*). In a parallel *RC* circuit, as shown in Figure 10–27(a), the total admittance is the phasor sum of the conductance and the capacitive susceptance, as shown in Figure 10–27(b).

$$Y_{tot} = \sqrt{G^2 + B_C^2} \qquad\qquad (10\text{–}17)$$

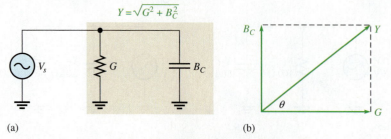

(a)   (b)

FIGURE 10–27   **Admittance in a parallel *RC* circuit.**

EXAMPLE 10–10

Determine the total admittance in Figure 10–28, and then convert it to impedance.

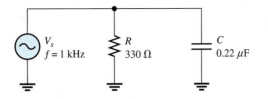

SOLUTION

To determine Y, first calculate the values for G and B_C. Since $R = 330\ \Omega$,

FIGURE 10–28

$$G = \frac{1}{R} = \frac{1}{330\ \Omega} = 3.03\ \text{mS}$$

The capacitive reactance is

$$X_C = \frac{1}{2\pi fC} = \frac{1}{2\pi(1000\ \text{Hz})(0.22\ \mu\text{F})} = 723\ \Omega$$

The capacitive susceptance is

$$B_C = \frac{1}{X_C} = \frac{1}{723\ \Omega} = 1.38\ \text{mS}$$

Therefore, the total admittance is

$$Y_{tot} = \sqrt{G^2 + B_C^2} = \sqrt{(3.03\ \text{mS})^2 + (1.38\ \text{mS})^2} = \textbf{3.33 mS}$$

Convert to impedance.

$$Z = \frac{1}{Y_{tot}} = \frac{1}{3.33\ \text{mS}} = \textbf{300}\ \boldsymbol{\Omega}$$

RELATED PROBLEM

Calculate the admittance in Figure 10–28 if f is increased to 2.5 kHz.

SECTION 10–4 CHECKUP

1. Determine Z if a 1.0 kΩ resistance is in parallel with a 650 Ω capacitive reactance.

2. Define *conductance, capacitive susceptance,* and *admittance.*

3. If $Z = 100\ \Omega$, what is the value of Y?

4. In a certain parallel *RC* circuit, $R = 50\ \Omega$ and $X_C = 75\ \Omega$. Determine Y.

10–5 ANALYSIS OF PARALLEL *RC* CIRCUITS

Ohm's law and Kirchhoff's current law are used in the analysis of *RC* circuits. Current and voltage relationships in a parallel *RC* circuit are examined.

After completing this section, you should be able to

- **Analyze a parallel *RC* circuit**
  - **Apply Ohm's law and Kirchhoff's current law to parallel *RC* circuits**
  - **Show how impedance and phase angle vary with frequency**
  - **Convert from a parallel circuit to an equivalent series circuit**

For convenience in the analysis of parallel circuits, the Ohm's law formulas using impedance—Equations 10–3, 10–4, and 10–5—can be rewritten for admittance using the relation $Y = 1/Z$.

$$V = \frac{I}{Y} \qquad (10\text{–}18)$$

$$I = VY \qquad (10\text{–}19)$$

$$Y = \frac{I}{V} \qquad (10\text{–}20)$$

EXAMPLE 10–11

Determine the total current and the phase angle in Figure 10–29.

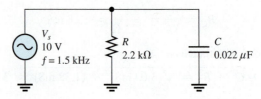

FIGURE 10–29

SOLUTION

First, determine the total admittance. The capacitive reactance is

$$X_C = \frac{1}{2\pi f C} = \frac{1}{2\pi(1.5\ \text{kHz})(0.022\ \mu\text{F})} = 4.82\ \text{k}\Omega$$

The conductance is

$$G = \frac{1}{R} = \frac{1}{2.2\ \text{k}\Omega} = 455\ \mu\text{S}$$

The capacitive susceptance is

$$B_C = \frac{1}{X_C} = \frac{1}{4.82\ \text{k}\Omega} = 207\ \mu\text{S}$$

Therefore, the total admittance is

$$Y_{tot} = \sqrt{G^2 + B_C^2} = \sqrt{(455\ \mu\text{S})^2 + (207\ \mu\text{S})^2} = 500\ \mu\text{S}$$

Next, use Ohm's law to calculate the total current.

$$I_{tot} = VY_{tot} = (10\ \text{V})(500\ \mu\text{S}) = \textbf{5.00 mA}$$

The phase angle is

$$\theta = \tan^{-1}\left(\frac{R}{X_C}\right) = \tan^{-1}\left(\frac{2.2\ \text{k}\Omega}{4.82\ \text{k}\Omega}\right) = \textbf{24.5°}$$

The total current is 5.00 mA, and it leads the source voltage by 24.5°.

RELATED PROBLEM

What is the total current if the frequency is doubled?

MULTISIM

Open Multisim file E10-11. Verify the value of the total current that was calculated. Then measure each branch current. Double the frequency to 3 kHz and measure the total current.

Phase Relationships of the Currents and Voltages

Figure 10–30(a) shows all the currents and voltages in a basic parallel *RC* circuit. As you can see, the source voltage, V_s, appears across both the resistive and the capacitive branches, so V_s, V_R, and V_C are all in phase and of the same magnitude. The total current, I_{tot}, divides at the junction into the two branch currents, I_R and I_C.

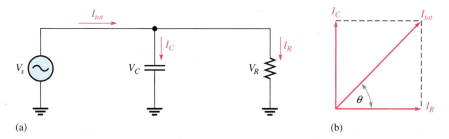

(a)   (b)

FIGURE 10–30   **Currents and voltages in a parallel *RC* circuit. The current directions shown in (a) are instantaneous and, of course, reverse when the source voltage reverses. The current phasors in (b) rotate once per cycle.**

Stray Capacitance

Stray capacitance is the unseen capacitance that is in all electronic circuits and systems. It occurs anytime conductors are separated by an insulator as in the case of a basic transistor. The effect of stray capacitance is to reduce the high-frequency response of analog circuits. In high-speed digital circuits, it limits switching speed. The stray capacitance for a transistor can be drawn as an equivalent parallel *RC* circuit, enabling high-frequency analysis of a transistor circuit by applying a basic circuit concept.

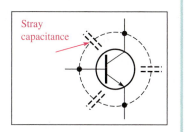

SYSTEM NOTE

The current through the resistor is in phase with the voltage. The current through the capacitor leads the voltage, and thus the resistive current, by 90°. By Kirchhoff's current law, the total current is the phasor sum of the two branch currents, as shown by the phasor diagram in Figure 10–30(b). The total current is expressed as

$$I_{tot} = \sqrt{I_R^2 + I_C^2} \qquad (10\text{–}21)$$

The value of the phase angle between the resistor current and the total current is

$$\theta = \tan^{-1}\left(\frac{I_C}{I_R}\right) \qquad (10\text{–}22)$$

Equation 10–22 is equivalent to Equation 10–13, $\theta = \tan^{-1}(R/X_C)$.

Figure 10–31 shows a complete current and voltage phasor diagram. Notice that I_C leads I_R by 90° and that I_R is in phase with the voltage ($V_s = V_R = V_C$).

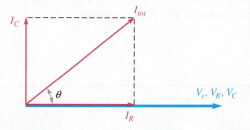

FIGURE 10–31   Current and voltage phasor diagram for a parallel *RC* circuit (amplitudes depend on the particular circuit).

EXAMPLE 10-12

Determine the value of each current in Figure 10–32, and describe the phase relationship of I_R, I_C, and I_{tot}. Draw the current phasor diagram.

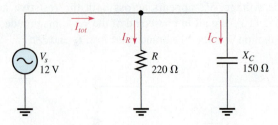

FIGURE 10–32

SOLUTION

The resistor current, the capacitor current, and the total current are expressed as follows:

$$I_R = \frac{V_s}{R} = \frac{12\text{ V}}{220\text{ }\Omega} = \textbf{54.5 mA}$$

$$I_C = \frac{V_s}{X_C} = \frac{12\text{ V}}{150\text{ }\Omega} = \textbf{80 mA}$$

$$I_{tot} = \sqrt{I_R^2 + I_C^2} = \sqrt{(54.5\text{ mA})^2 + (80\text{ mA})^2} = \textbf{96.8 mA}$$

The phase angle is

$$\theta = \tan^{-1}\left(\frac{I_C}{I_R}\right) = \tan^{-1}\left(\frac{80\text{ mA}}{54.5\text{ mA}}\right) = 55.7°$$

I_R is in phase with the source voltage, I_C leads the source voltage by 90°, and I_{tot} leads the source voltage by 55.7°. The current phasor diagram is shown in Figure 10–33.

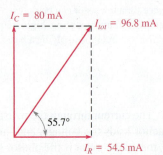

RELATED PROBLEM

In a certain parallel circuit, $I_R = 100$ mA and $I_C = 60$ mA. Determine the total current and the phase angle.

FIGURE 10–33

Note that a parallel *RC* circuit becomes less reactive when X_C is increased. That is, the circuit phase angle becomes smaller. The reason for this effect is that when X_C is increased relative to R, less current is through the capacitive branch, and although the in-phase or resistive current does not increase, it becomes a greater percentage of the total current.

SECTION 10–5 CHECKUP

1. The admittance of a parallel *RC* circuit is 3.5 mS, and the source voltage is 6 V. What is the total current?

2. In a certain parallel *RC* circuit, the resistor current is 10 mA, and the capacitor current is 15 mA. Determine the phase angle and the total current.

3. What is the phase angle between the capacitor current and the source voltage in a parallel *RC* circuit?

10–6 ANALYSIS OF SERIES-PARALLEL *RC* CIRCUITS

The concepts studied in the previous sections are used to analyze circuits with combinations of both series and parallel *R* and *C* components.

After completing this section, you should be able to

- **Analyze series-parallel *RC* circuits**
  - **Determine total impedance**
  - **Calculate currents and voltages**
  - **Measure impedance and phase angle**

As in the case of dc circuits, combinational ac circuits can be solved by combining series or parallel elements into equivalent circuits. The following example demonstrates the analysis of a series-parallel reactive circuit.

EXAMPLE 10–13

In the series-parallel *RC* circuit of Figure 10–34, determine the following:

(a) total impedance   **(b)** total current   **(c)** phase angle by which I_{tot} leads V_s

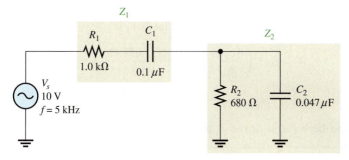

FIGURE 10–34

SOLUTION

(a) First, calculate the magnitudes of the capacitive reactances.

$$X_{C1} = \frac{1}{2\pi(5\ \text{kHz})(0.1\ \mu\text{F})} = 318\ \Omega$$

$$X_{C2} = \frac{1}{2\pi(5\ \text{kHz})(0.047\ \mu\text{F})} = 677\ \Omega$$

One approach is to find the series equivalent resistance and capacitive reactance for the parallel portion of the circuit; then add the resistances $(R_1 + R_{eq})$ to get total resistance and add the reactances $(X_{C1} + X_{C(eq)})$ to get total reactance. From these totals, you can determine the total impedance.
Find the impedance of the parallel portion (Z_2) by first finding the admittance.

$$G_2 = \frac{1}{R_2} = \frac{1}{680\ \Omega} = 1.47\ \text{mS}$$

$$B_{C2} = \frac{1}{X_{C2}} = \frac{1}{677\ \Omega} = 1.48\ \text{mS}$$

$$Y_2 = \sqrt{G_2^2 + B_{C2}^2} = \sqrt{(1.47 \text{ mS})^2 + (1.48 \text{ mS})^2} = 2.09 \text{ mS}$$

$$Z_2 = \frac{1}{Y_2} = \frac{1}{2.09 \text{ mS}} = 478 \text{ }\Omega$$

The phase angle associated with the parallel portion of the circuit is

$$\theta_p = \tan^{-1}\left(\frac{R_2}{X_{C2}}\right) = \tan^{-1}\left(\frac{680 \text{ }\Omega}{677 \text{ }\Omega}\right) = 45.1°$$

The series equivalent values for the parallel portion are

$$R_{eq} = Z_2 \cos \theta_p = (478 \text{ }\Omega)\cos(45.1°) = 337 \text{ }\Omega$$

$$X_{C(eq)} = Z_2 \sin \theta_p = (478 \text{ }\Omega)\sin(45.1°) = 339 \text{ }\Omega$$

The total circuit resistance is

$$R_{tot} = R_1 + R_{eq} = 1000 \text{ }\Omega + 337 \text{ }\Omega = 1.34 \text{ k}\Omega$$

The total circuit reactance is

$$X_{C(tot)} = X_{C1} + X_{C(eq)} = 318 \text{ }\Omega + 339 \text{ }\Omega = 657 \text{ }\Omega$$

The total circuit impedance is

$$Z_{tot} = \sqrt{R_{tot}^2 + X_{C(tot)}^2} = \sqrt{(1.34 \text{ k}\Omega)^2 + (657 \text{ }\Omega)^2} = \mathbf{1.49 \text{ k}\Omega}$$

(b) Use Ohm's law to find the total current.

$$I_{tot} = \frac{V_s}{Z_{tot}} = \frac{10 \text{ V}}{1.49 \text{ k}\Omega} = \mathbf{6.71 \text{ mA}}$$

(c) To find the phase angle, view the circuit as a series combination of R_{tot} and $X_{C(tot)}$. The phase angle by which I_{tot} leads V_s is

$$\theta = \tan^{-1}\left(\frac{X_{C(tot)}}{R_{tot}}\right) = \tan^{-1}\left(\frac{657 \text{ }\Omega}{1.34 \text{ k}\Omega}\right) = \mathbf{26.1°}$$

RELATED PROBLEM

Determine the voltages across Z_1 and Z_2 in Figure 10–34.

MULTISIM

Open Multisim file E10-13. Verify the calculated value of the total current. Measure the current through R_2 and the current through C_2. Measure voltages across Z_1 and Z_2.

SYSTEM EXAMPLE 10–2

PASSIVE OSCILLOSCOPE PROBES

Although there are many types of oscilloscope probes, the most common type is the ×10 passive probe. The ×10 refers to the fact that the probe attenuates the signal by a factor of 10. The purpose of the probe is to transmit as accurately as possible the signal while avoiding noise pickup or affecting the circuit under test (loading).

A typical oscilloscope input has about 20 pF of internal capacitance due to the input amplifier, cables, and stray capacitance. These are lumped into C_{in} in the circuit shown in Figure 10–35. The dotted lines are a reminder that this is not a physical capacitor but has all of the attributes of one. There is also about 1.0 MΩ of internal resistance for the input amplifier in the oscilloscope, shown as R_{in}. To complete the circuit, there is typically a series resistor in the probe itself, R_{probe}, and a small variable capacitor, C_{probe}, that is adjusted to optimize the overall probe/scope response.

At low frequencies, the small capacitors have very high reactance compared to the resistance. Ignoring the capacitors, the resistors form a 10:1 resistive divider. At high frequencies, the capacitive reactance of C_{probe} and C_{in} are both small compared to the resistors. Ignoring the resistances, C_{probe} forms a 10:1 capacitive divider with C_{in}, which keeps

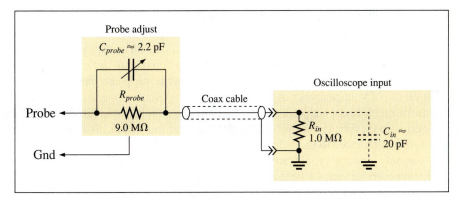

FIGURE 10–35 A passive oscilloscope probe.

the frequency response flat. The net effect of the probe circuit for an oscilloscope is to reduce the input capacitance by the probe's ×10 ratio and increase the input impedance by the same amount, thus reducing circuit loading.

Circuit Measurements

DETERMINING Z_{tot}   Now, let's see how the value of Z_{tot} for the circuit in Example 10–13 can be determined by measurement. First, the total impedance is measured as outlined in the following steps and as illustrated in Figure 10–36 (other ways are also possible).

Step 1: Using a sine wave generator, set the source voltage to a known value (10 V) and the frequency to 5 kHz. Check the voltage with an ac voltmeter and the frequency with a frequency counter or an oscilloscope rather than relying on the marked values on the generator controls.

Step 2: Connect an ac ammeter as shown in Figure 10–36, and measure the total current. Alternately, you can measure the voltage across R_1 and calculate the current.

Step 3: Calculate the total impedance by using Ohm's law.

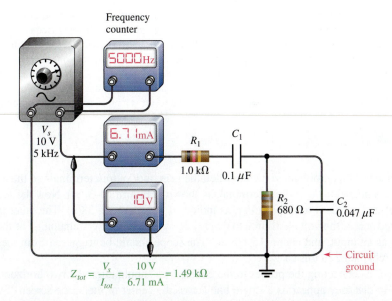

$$Z_{tot} = \frac{V_s}{I_{tot}} = \frac{10\ \text{V}}{6.71\ \text{mA}} = 1.49\ \text{k}\Omega$$

FIGURE 10–36 Determining Z_{tot} by measurement of V_s and I_{tot}.

DETERMINING θ   To measure the phase angle, the source voltage and the total current must be displayed on an oscilloscope screen in the proper time relationship. Two basic types of scope probes are available to measure the quantities with an oscilloscope: the voltage probe and the current probe. The current probe is a convenient device, but it is often not as readily available as a voltage probe. We will confine our phase measurement technique to the use of voltage probes in conjunction with the oscilloscope. Although there are special isolation methods, a typical passive oscilloscope voltage probe has two points that are connected to the circuit: the probe tip and the ground lead. Thus, all voltage measurements must be referenced to ground.

Since only voltage probes are to be used, the total current cannot be measured directly. However, for phase measurement, the voltage across R_1 is in phase with the total current and can be used to establish the phase angle of the current.

Before proceeding with the actual phase measurement, there is a problem with displaying V_{R1}. If the scope probe is connected across the resistor, as indicated in Figure 10–37(a), the ground lead of the scope will short point B to ground, thus bypassing the rest of the components and effectively removing them from the circuit electrically, as illustrated in Figure 10–37(b) (assuming that the scope is not isolated from power line ground).

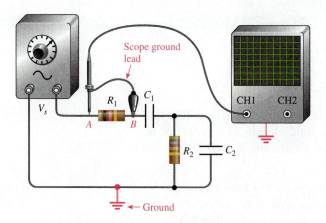

(a) Ground lead on scope probe grounds point B.

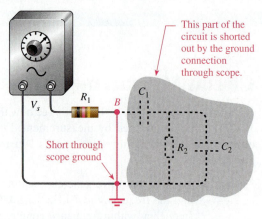

(b) The effect of grounding point B is to short out the rest of the circuit.

FIGURE 10–37   **Effects of measuring directly across a component when the instrument and the circuit are grounded. The scope ground has shorted out part of the circuit.**

High-Frequency Differential Probing

Most oscilloscopes will allow you to combine two single-ended probes in order to make a measurement of an ungrounded component. This is a good method for low frequencies. For high-frequency work, the separate paths for each channel can lead to delay differences between the two signals, resulting in amplitude and timing errors. A better way to probe high-frequency and fast signals is to use a differential probe. A differential probe uses a differential amplifier at the probe tip to subtract the two signals, allowing one channel of the oscilloscope to make a measurement across an ungrounded component. The result is a more accurate measurement.

SYSTEM NOTE

To avoid this problem, you can switch the generator output terminals so that one end of R_1 is connected to the ground terminal, as shown in Figure 10–38(a). Now the scope can be connected across it to display V_{R1}, as indicated in Figure 10–38(b). The other probe is connected across the voltage source to display V_s as indicated. Now channel 1 of the scope has V_{R1} as an input, and channel 2 has V_s. The scope should be triggered from the source voltage (channel 2 in this case).

Before connecting the probes to the circuit, you should align the two horizontal lines (traces) so that they appear as a single line across the center of the scope screen. To do so, ground the probe tips and adjust the vertical position knobs to move the traces toward the

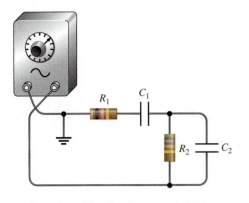

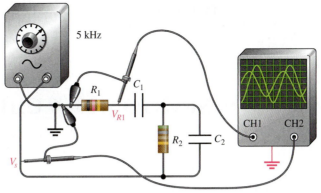

(a) Ground repositioned so that one end of R_1 is grounded.

(b) The scope displays V_{R1} and V_s. V_{R1} represents the phase of the total current.

FIGURE 10–38 **Repositioning ground so that a direct voltage measurement can be made with respect to ground without shorting out part of the circuit.**

center line of the screen until they are superimposed. This procedure ensures that both waveforms have the same zero crossing so that an accurate phase measurement can be made.

Once you have stabilized the waveforms on the scope screen, you can measure the period of the source voltage. Next, use the Volts/Div controls to adjust the amplitudes of the waveforms until they both appear to have the same amplitude. Now, spread the waveforms horizontally by using the Sec/Div control to expand the distance between them. This horizontal distance represents the time between the two waveforms. The number of divisions between the waveforms along any horizontal lines times the Sec/Div setting is equal to the time between them, Δt. Also, you can use the cursors to determine Δt if your oscilloscope has this feature.

Once you have determined the period, T, and the time between the waveforms, Δt, you can calculate the phase angle with the following equation:

$$\theta = \left(\frac{\Delta t}{T}\right)360°$$ (10–23)

Figure 10–39 shows simulated screen displays for an oscilloscope in Multisim. In Figure 10–39(a), the waveforms are aligned and set to the same apparent amplitude by adjusting the fine Volts/Div control. The period of these waveforms is 200 μs. The Sec/Div control is adjusted to spread the waveforms out to read Δt more accurately. As shown in part (b), there are 3.0 divisions between the crossings on the center line. The Sec/Div control is set to 5.0 μs and there are 3.0 divisions between the waveforms.

$$\Delta t = 3.0 \text{ divisions} \times 5.0\ \mu s/\text{division} = 15\ \mu s$$

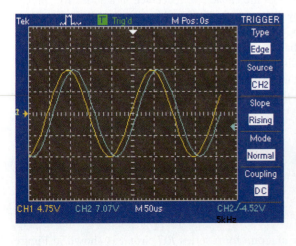

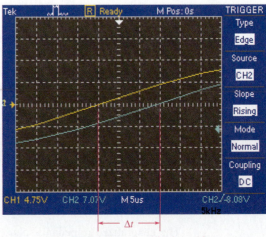

$\Delta t = 3.0$ divisions $\times 5\ \mu$s/division $= 15\ \mu$s

(a)   (b)

FIGURE 10–39 **Determining the phase angle using an oscilloscope.**

The phase angle is

$$\theta = \left(\frac{\Delta t}{T}\right)360° = \left(\frac{15 \ \mu s}{200 \ \mu s}\right)360° = 27°$$

SECTION 10–6 CHECKUP

1. Explain why a scope probe ground should be connected to the circuit ground.

2. What is the voltage across R_1 in Figure 10–34?

3. To what value should S_{probe} in Figure 10–35 be adjusted if $C_{in} = 15$ pF?

10–7 POWER IN *RC* CIRCUITS

In a purely resistive ac circuit, all of the energy delivered by the source is dissipated in the form of heat by the resistance. In a purely capacitive ac circuit, all of the energy delivered by the source is stored by the capacitor during a portion of the voltage cycle and then returned to the source during another portion of the cycle so that there is no net energy conversion to heat. When there is both resistance and capacitance, some of the energy is alternately stored and returned by the capacitance and some is dissipated by the resistance. The amount of energy converted to heat is determined by the relative values of the resistance and the capacitive reactance.

After completing this section, you should be able to

- **Determine power in *RC* circuits**
  - **Explain true and reactive power**
  - **Draw the power triangle**
  - **Define *power factor***
  - **Explain apparent power**
  - **Calculate power in an *RC* circuit**

When the resistance is greater than the capacitive reactance in a series *RC* circuit, more of the total energy delivered by the source is dissipated by the resistor than is stored by the capacitor. Likewise, when the reactance is greater than the resistance, more of the total energy is stored and returned than is converted to heat.

The formulas for the power dissipated in a resistor, sometimes called *true power* (P_{true}), and the power in a capacitor, called *reactive power* (P_r), are restated here. The unit of true power is the watt (W), and the unit of reactive power is the VAR (volt-ampere reactive).

$$P_{\text{true}} = I_{tot}^2 R \tag{10–24}$$

$$P_r = I_{tot}^2 X_C \tag{10–25}$$

Power Triangle for *RC* Circuits

The generalized impedance phasor diagram for a series *RC* circuit is shown in Figure 10–40(a). A relationship for power can also be represented by a similar diagram because the respective magnitudes of the powers, P_{true} and P_r, differ from R and X_C by a factor of I_{tot}^2, as shown in Figure 10–40(b).

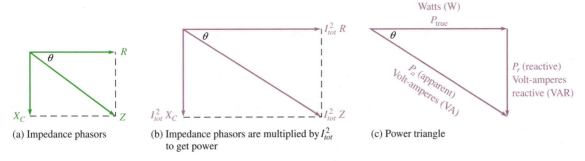

(a) Impedance phasors   (b) Impedance phasors are multiplied by I_{tot}^2 to get power   (c) Power triangle

FIGURE 10–40   **Development of the power triangle for a series *RC* circuit.**

The resultant power, $I_{tot}^2 Z$, represents the **apparent power**, P_a. At any instant in time, P_a is the total power that appears to be transferred between the source and the *RC* circuit. Part of the apparent power is true power, and part of it is reactive power. The unit of apparent power is the volt-ampere (VA). The expression for apparent power is

$$P_a = I_{tot}^2 Z \qquad (10\text{–}26)$$

The diagram in Figure 10–40(b) can be rearranged in the form of a right triangle as shown in Figure 10–40(c), which is called the *power triangle*. Using the rules of trigonometry, P_{true} can be expressed as

$$P_{\text{true}} = P_a \cos \theta$$

Since P_a equals $I_{tot}^2 Z$ or $V_s I_{tot}$, the equation for true power can be written as

$$\boldsymbol{P_{\text{true}} = V_s I_{tot} \cos \theta} \qquad (10\text{–}27)$$

where V_s is the source voltage and I_{tot} is the total current.

For the case of a purely resistive circuit, $\theta = 0°$ and $\cos 0° = 1$, so P_{true} equals $V_s I_{tot}$. For the case of a purely capacitive circuit, $\theta = 90°$ and $\cos 90° = 0$, so P_{true} is zero. As you already know, there is no power dissipated in an ideal capacitor.

Power Factor

The term $\cos \theta$ is called the **power factor** and is stated as

$$\boldsymbol{PF = \cos \theta} \qquad (10\text{–}28)$$

As the phase angle between the source voltage and the total current increases, the power factor decreases, indicating an increasingly reactive circuit. The smaller the power factor, the smaller the power dissipation.

The power factor can vary from 0 for a purely reactive circuit to 1 for a purely resistive circuit. In an *RC* circuit, the power factor is referred to as a leading power factor because the current leads the voltage.

EXAMPLE 10–14

Determine the power factor and the true power in the *RC* circuit of Figure 10–41.

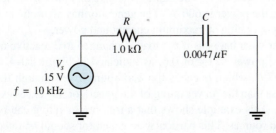

FIGURE 10–41

SOLUTION

Calculate the capacitive reactance and phase angle.

$$X_C = \frac{1}{2\pi fC} = \frac{1}{2\pi(10\text{ kHz})(0.0047\ \mu\text{F})} = 3.39\text{ k}\Omega$$

$$\theta = \tan^{-1}\left(\frac{X_C}{R}\right) = \tan^{-1}\left(\frac{3.39\text{ k}\Omega}{1.0\text{ k}\Omega}\right) = 73.6°$$

The power factor is

$$PF = \cos\theta = \cos(73.6°) = \mathbf{0.282}$$

The impedance is

$$Z = \sqrt{R^2 + X_C^2} = \sqrt{(1.0\text{ k}\Omega)^2 + (3.39\text{ k}\Omega)^2} = 3.53\text{ k}\Omega$$

Therefore, the current is

$$I = \frac{V_s}{Z} = \frac{15\text{ V}}{3.53\text{ k}\Omega} = 4.25\text{ mA}$$

The true power is

$$P_{\text{true}} = V_sI\cos\theta = (15\text{ V})(4.25\text{ mA})(0.282) = \mathbf{18.0\text{ mW}}$$

RELATED PROBLEM

What is the power factor if *f* is reduced by half in Figure 10–41?

MULTISIM

Open Multisim file E10–14. Measure the current and compare to the calculated value. Measure the voltages across *R* and *C* at 10 kHz, 5 kHz, and 20 kHz. Explain your observations.

Significance of Apparent Power

Apparent power is the power that *appears* to be transferred between the source and the load, and it consists of two components: a true power component and a reactive power component.

In all electrical and electronic systems, it is the true power that does the work. The reactive power is simply shuttled back and forth between the source and the load. Ideally, in terms of performing useful work, all of the power transferred to the load should be true power and none of it reactive power. However, in most practical situations the load has some reactance associated with it, and therefore you must deal with both power components.

For any reactive load, there are two components of the total current: the resistive component and the reactive component. If you consider only the true power (watts) in a load, you are dealing with only a portion of the total current that the load demands from a source. In order to have a realistic picture of the actual current that a load will draw, you must consider apparent power (in VA).

A source such as an ac generator can provide current to a load up to some maximum value. *If the load draws more than this maximum value, the source can be damaged.* Figure 10–42(a) shows a 120 V generator that can deliver a maximum current of 5 A to a load. Assume that the generator is rated at 600 W and is connected to a purely resistive load of 24 Ω (power factor of 1). The ammeter shows that the current is 5 A, and the wattmeter indicates that the power is 600 W. The generator has no problem under these conditions, although it is operating at maximum current and power.

Now, consider what happens if the load is changed to a reactive one with an impedance of 18 Ω and a power factor of 0.6, as indicated in Figure 10–42(b). The current is 120 V/18 Ω = 6.67 A, which exceeds the maximum. Even though the wattmeter reads 480 W, which is less than the power rating of the generator, the excessive current probably will cause damage. This example shows that a true power rating can be deceiving and is inappropriate for ac sources. This particular ac generator should be rated at 600 VA, a rating the manufacturer would use, rather than 600 W.

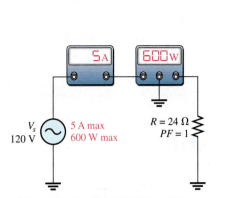

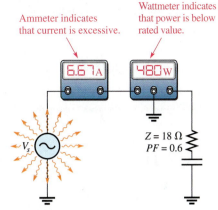

Ammeter indicates that current is excessive.

Wattmeter indicates that power is below rated value.

(a) Generator operating at its limits with a resistive load.

(b) Generator is in danger of internal damage due to excess current, even though the wattmeter indicates that the power is below the maximum wattage rating.

FIGURE 10–42   **The wattage rating of a source is inappropriate when the load is reactive. The rating should be in VA rather than in watts.**

EXAMPLE 10–15

For the circuit in Figure 10–43, find the true power, the reactive power, and the apparent power. X_C has been determined to be 2.0 kΩ.

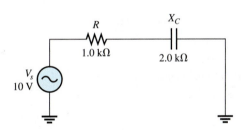

FIGURE 10–43

SOLUTION

First find the total impedance so that you can calculate the current.

$$Z_{tot} = \sqrt{R^2 + X_C^2} = \sqrt{(1.0\,\text{k}\Omega)^2 + (2.0\,\text{k}\Omega)^2} = 2.24\,\text{k}\Omega$$

$$I = \frac{V_s}{Z} = \frac{10\,\text{V}}{2.24\,\text{k}\Omega} = 4.46\,\text{mA}$$

The phase angle, θ, is

$$\theta = \tan^{-1}\left(\frac{X_C}{R}\right) = \tan^{-1}\left(\frac{2.0\,\text{k}\Omega}{1.0\,\text{k}\Omega}\right) = 63.4°$$

The true power is

$$P_{\text{true}} = V_s I \cos\theta = (10\,\text{V})(4.46\,\text{mA})\cos(63.4°) = \textbf{20 mW}$$

Note that the same result is realized if you use the formula $P_{\text{true}} = I^2 R$.

The reactive power is

$$P_r = I^2 X_C = (4.46\,\text{mA})^2(2.0\,\text{k}\Omega) = \textbf{39.8 mVAR}$$

The apparent power is

$$P_a = I^2 Z = (4.46\,\text{mA})^2(2.24\,\text{k}\Omega) = \textbf{44.6 mVA}$$

The apparent power is also the phasor sum of P_{true} and P_r.

$$P_a = \sqrt{P_{true}^2 + P_r^2} = 44.6 \text{ mVA}$$

RELATED PROBLEM

What is the true power in Figure 10–43 if $X_C = 10 \text{ k}\Omega$?

SECTION 10–7 CHECKUP

1. To which component in an *RC* circuit is the power dissipation due?

2. If the phase angle, θ, is 45°, what is the power factor?

3. A certain series *RC* circuit has the following parameter values: $R = 330 \ \Omega$, $X_C = 460 \ \Omega$, and $I = 2 \ A$. Determine the true power, the reactive power, and the apparent power.

10–8 BASIC APPLICATIONS

RC **circuits are found in a variety of applications, often as part of a more complex circuit. Three applications are phase shift circuits applied to an oscillator, frequency-selective circuits (filters), and ac coupling.**

After completing this section, you should be able to

- **Describe some basic *RC* applications**
  - **Discuss how the *RC* circuit is used in an oscillator**
  - **Discuss how the *RC* circuit operates as a filter**
  - **Discuss ac coupling**

The Phase Shift Oscillator

As you know, a series *RC* circuit will shift the phase of the output voltage by an amount that depends on the values of *R* and *C* and the frequency of the signal. This ability to shift phase depending on frequency is vital in certain feedback oscillator circuits. An oscillator is a circuit that generates a periodic waveform and is an important circuit for many electronic systems. You will study oscillators in a devices course, so the focus here is on the application of *RC* circuits for shifting phase. The requirement is that a fraction of the output of the oscillator is returned to the input (called *feedback*) in the proper phase to reinforce the input and sustain oscillations. Generally, the requirement is to feed back the signal with a total of 180° of phase shift.

A single *RC* circuit is limited to phase shifts that are smaller than 90°. The basic *RC* lag circuit discussed in Section 10–3 can be "stacked" to form a complex *RC* network, as shown in Figure 10–44, which shows a specific circuit called a *phase shift oscillator*. The phase shift oscillator typically uses three equal-component *RC* circuits that produce the required 180° phase shift at a certain frequency, which will be the frequency at which the oscillator works. The output of the amplifier is phase shifted by the *RC* network and returned to the input of the amplifier, which provides sufficient gain to maintain oscillations.

The process of putting several *RC* circuits together results in a loading effect, so the overall phase shift is not the same as simply adding the phase shifts of the individual *RC* circuits. The detailed calculation for this circuit involves a lot of tedious phasor math, but

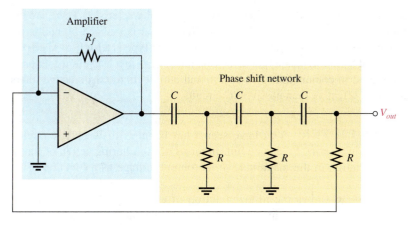

FIGURE 10–44   **Phase shift oscillator.**

the result is simple. With equal components, the frequency at which a 180° phase shift occurs is given by the following equation:

$$f_r = \frac{1}{2\pi \sqrt{6} RC}$$

It also turns out that the *RC* network attenuates (reduces) the signal from the amplifier by a factor of 29; the amplifier must make up for this attenuation by having a gain of −29 (the minus sign takes into account the phase shift).

EXAMPLE 10–16

In Figure 10–45, calculate the output frequency.

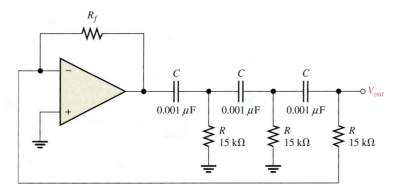

FIGURE 10–45

SOLUTION

$$f_r = \frac{1}{2\pi \sqrt{6} RC} = \frac{1}{2\pi \sqrt{6}(15 \text{ k}\Omega)(0.001 \text{ }\mu\text{F})} = 4.33 \text{ kHz}$$

RELATED PROBLEM

If all of the capacitors are changed to 0.0027 μF, what is the oscillator frequency?

The *RC* Circuit as a Filter

Frequency-selective circuits (filters) permit signals of certain frequencies to pass from the input to the output while blocking all others. That is, ideally all frequencies but the selected ones are filtered out. Frequency-selective circuits are important in many systems, particularly communication systems.

Series *RC* circuits exhibit a frequency-selective characteristic. There are two types. The first one that we examine is a **low-pass circuit** that is realized by taking the output across the capacitor, just as in a lag circuit. The second type is a **high-pass circuit**, which is implemented by taking the output across the resistor, as in a lead circuit. In practice, *RC* circuits are used in conjunction with operational amplifiers to create active filters, which are much more effective than passive *RC* circuits.

LOW-PASS FILTER   You have seen what happens to the phase angle and the output voltage in a series *RC* lag circuit. In terms of the filtering action of the series *RC* circuit, the variation in the magnitude of the output voltage as a function of frequency is important.

Figure 10–46 shows the filtering action of a series *RC* circuit using a specific series of measurements in which the frequency starts at 100 Hz and is increased in increments up to 20 kHz. At each value of frequency, the output voltage is measured. As you can see, the capacitive reactance decreases as frequency increases, thus dropping less voltage across the capacitor while the input voltage is held at a constant 10 V throughout each step. Table 10–1 summarizes the variation of the circuit parameters with frequency.

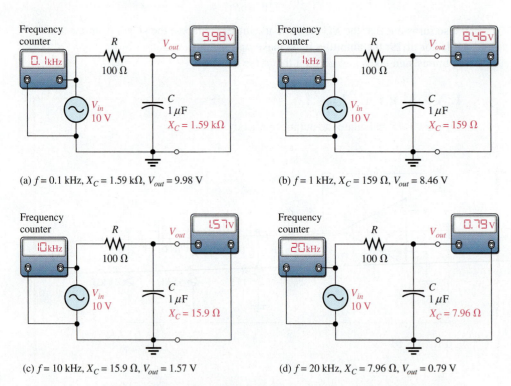

(a) $f = 0.1$ kHz, $X_C = 1.59$ kΩ, $V_{out} = 9.98$ V

(b) $f = 1$ kHz, $X_C = 159$ Ω, $V_{out} = 8.46$ V

(c) $f = 10$ kHz, $X_C = 15.9$ Ω, $V_{out} = 1.57$ V

(d) $f = 20$ kHz, $X_C = 7.96$ Ω, $V_{out} = 0.79$ V

FIGURE 10–46   Example of low-pass filtering action. As frequency increases, V_{out} decreases.

TABLE 10–1				
f (kHz)	$X_C(\Omega)$	$Z_{tot}(\Omega)$	I (mA)	V_{out} (V)
0.1	1,590	1,590	6.29	9.98
1	159	188	53.2	8.46
10	15.9	101	99.0	1.57
20	7.96	100	100	0.79

The **frequency response** of the low-pass *RC* circuit in Figure 10–46 is shown in Figure 10–47, where the measured values are plotted on a graph of V_{out} versus f, and a smooth curve is drawn connecting the points. This graph, called a *response curve,* shows that the output voltage is greater at the lower frequencies and decreases as the frequency increases. The frequency scale is logarithmic.

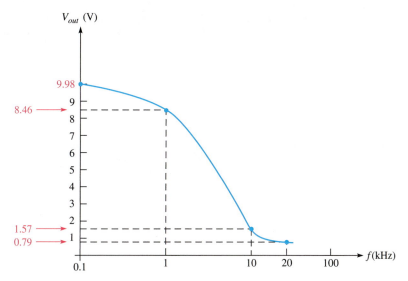

FIGURE 10–47 Frequency response curve for the low-pass *RC* circuit in Figure 10–46.

HIGH-PASS FILTER   To illustrate *RC* high-pass filtering action, Figure 10–48 shows a series of specific measurements. The frequency starts at 10 Hz and is increased in increments up to 10 kHz. As you can see, the capacitive reactance decreases as the frequency increases, thus causing more of the total input voltage to be dropped across the resistor. Table 10–2 summarizes the variation of circuit parameters with frequency.

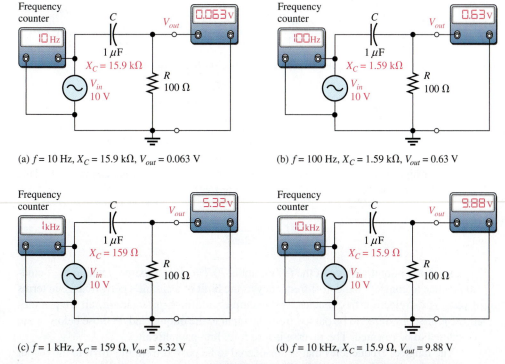

(a) $f = 10$ Hz, $X_C = 15.9$ kΩ, $V_{out} = 0.063$ V

(b) $f = 100$ Hz, $X_C = 1.59$ kΩ, $V_{out} = 0.63$ V

(c) $f = 1$ kHz, $X_C = 159$ Ω, $V_{out} = 5.32$ V

(d) $f = 10$ kHz, $X_C = 15.9$ Ω, $V_{out} = 9.88$ V

FIGURE 10–48   Example of high-pass filtering action. As frequency increases, V_{out} increases.

TABLE 10–2

f (kHz)	$X_C(\Omega)$	$Z_{tot}\ (\Omega)$	I (mA)	V_{out} (V)
.01	15,900	15,900	0.629	0.063
.1	1590	1593	6.28	0.63
1	159	188	53.2	5.32
10	15.9	101	98.8	9.88

In Figure 10–49, the measured values for the high-pass *RC* circuit shown in Figure 10–48 have been plotted to produce a frequency response curve for this circuit. As you can see, the output voltage is greater at the higher frequencies and decreases as the frequency is reduced. The frequency scale is logarithmic.

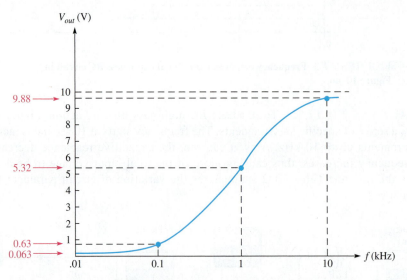

FIGURE 10–49   Frequency response curve for the high-pass *RC* circuit in Figure 10–48.

THE CUTOFF FREQUENCY AND THE BANDWIDTH OF AN *RC* CIRCUIT   The frequency at which the capacitive reactance equals the resistance in a low-pass or high-pass *RC* circuit is called the **cutoff frequency** and is designated f_c. This condition is expressed as $1/(2\pi f_c C) = R$. Solving for f_c results in the following formula:

$$f_c = \frac{1}{2\pi RC} \qquad (10\text{–}29)$$

At f_c, the output voltage of the *RC* circuit is 70.7% of its maximum value. It is standard practice to consider the cutoff frequency as the limit of a circuit's performance in terms of passing or rejecting frequencies. For example, in a high-pass circuit, all frequencies above f_c are considered to be passed from the input to the output, and all those below f_c are considered to be rejected. The reverse is true for a low-pass circuit.

The range of frequencies that is considered to be passed from the input to the output of a circuit is called the **bandwidth**. Figure 10–50 illustrates the bandwidth and the cutoff frequency for a low-pass circuit.

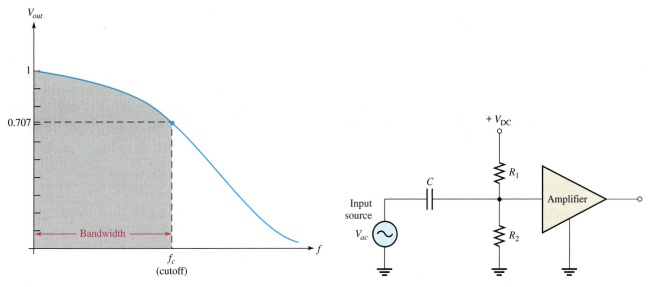

FIGURE 10–50   Normalized general response curve of a low-pass *RC* circuit showing the cutoff frequency and the bandwidth.

FIGURE 10–51   Amplifier bias and signal-coupling circuit.

Coupling an AC Signal into a DC Bias Circuit

Figure 10–51 shows an *RC* circuit that is used to create a dc voltage level with an ac voltage superimposed on it. This type of circuit is commonly found in amplifiers in which the dc voltage is required to **bias** the amplifier to the proper operating point and the signal voltage to be amplified is coupled through a capacitor and superimposed on the dc level. The capacitor prevents the low internal resistance of the signal source from affecting the dc bias voltage. This circuit is similar to the one in System Example 10–1 but uses a voltage-divider to set up the required bias voltage.

In this type of application, a relatively large value of capacitance is selected so that for the frequencies to be amplified, the reactance is very small compared to the resistance of the bias network. When the reactance is very small (ideally zero), there is practically no phase shift or signal voltage dropped across the capacitor. Therefore, all of the signal voltage passes from the source to the input to the amplifier.

Figure 10–52 illustrates the application of the superposition principle to the circuit in Figure 10–51. In part (a), the ac source has been effectively removed from the circuit and

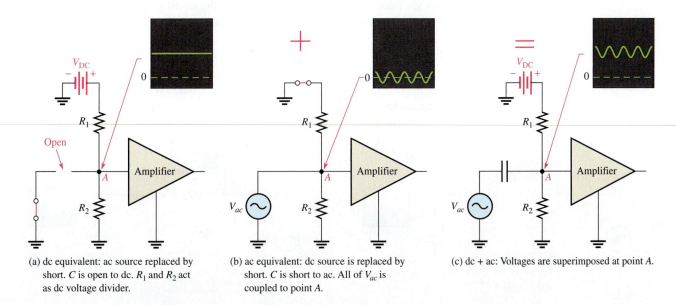

(a) dc equivalent: ac source replaced by short. *C* is open to dc. R_1 and R_2 act as dc voltage divider.

(b) ac equivalent: dc source is replaced by short. *C* is short to ac. All of V_{ac} is coupled to point *A*.

(c) dc + ac: Voltages are superimposed at point *A*.

FIGURE 10–52   The superposition of dc and ac voltages in an *RC* bias and coupling circuit.

replaced with a short to represent its ideal internal resistance. Since C is open to dc, the voltage at point A is determined by the voltage-divider action of R_1 and R_2 and the dc voltage source.

In Figure 10–52(b), the dc source has been effectively removed from the circuit and replaced with a short to represent its ideal internal resistance. Since C appears as a short at the frequency of the ac, the signal voltage is coupled directly to point A and appears across the parallel combination of R_1 and R_2. Figure 10–52(c) illustrates that the combined effect of the superposition of the dc and the ac voltages results in the signal voltage "riding" on the dc level.

SECTION 10–8 CHECKUP

1. In a phase shift oscillator, the *RC* circuit must provide a total phase shift of how many degrees?

2. To achieve a low-pass characteristic with a series *RC* circuit, across which component is the output taken?

10–9 TROUBLESHOOTING

Typical component failures or degradation have an effect on the response of basic *RC* circuits. The APM (analysis, planning, and measurement) method of troubleshooting can be used to locate problems in a circuit.

After completing this section, you should be able to

- **Troubleshoot *RC* circuits**
  - **Find an open resistor**
  - **Find an open capacitor**
  - **Find a shorted capacitor**
  - **Find a leaky capacitor**

EFFECT OF AN OPEN RESISTOR It is easy to see how an open resistor affects the operation of a basic series *RC* circuit, as shown in Figure 10–53. Obviously, there is no path for current, so the capacitor voltage remains at zero and the total voltage, V_s, appears across the open resistor.

EFFECT OF AN OPEN CAPACITOR When the capacitor is open, there is no current; thus, the resistor voltage drop is zero. The total source voltage appears across the open capacitor, as shown in Figure 10–54.

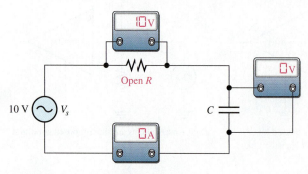

FIGURE 10–53 **Effect of an open resistor.**

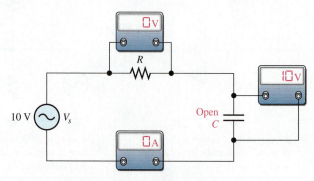

FIGURE 10–54 **Effect of an open capacitor.**

EFFECT OF A SHORTED CAPACITOR   When a capacitor shorts out, the voltage across it is zero, the current equals V_s/R, and the total voltage appears across the resistor, as shown in Figure 10–55.

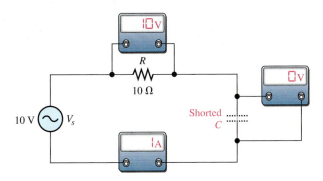

FIGURE 10–55   Effect of a shorted capacitor.

EFFECT OF A LEAKY CAPACITOR   When a capacitor exhibits a high leakage current, the leakage resistance effectively appears in parallel with the capacitor, as shown in Figure 10–56(a). When the leakage resistance is comparable in value to the circuit resistance, R, the circuit response is drastically affected. The circuit, looking from the capacitor toward the source, can be thevenized, as shown in Figure 10–56(b). The Thevenin equivalent resistance is R in parallel with R_{leak} (the source appears as a short), and the Thevenin equivalent voltage is determined by the voltage-divider action of R and R_{leak}.

$$R_{th} = R \| R_{leak} = \frac{RR_{leak}}{R + R_{leak}}$$

$$V_{th} = \left(\frac{R_{leak}}{R + R_{leak}}\right)V_{in}$$

As you can see, the voltage to which the capacitor will charge is reduced since $V_{th} < V_{in}$. Also, the current is increased. The Thevenin equivalent circuit is shown in Figure 10–56(c).

> **HANDS ON TIP**
> Some multimeters have a relatively low frequency response of 1 kHz or less, while others can measure voltages or currents with frequencies up to about 2 MHz. Always check to make sure that your meter is capable of making accurate measurements at the particular frequency at which you are working.

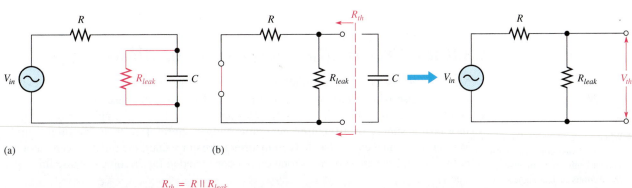

(a)

(b)

(c)

FIGURE 10–56   Effect of a leaky capacitor.

EXAMPLE 10–17

Assume that the capacitor in Figure 10–57 is degraded to a point where its leakage resistance is $10\,\text{k}\Omega$. Determine the phase shift from input to output and the output voltage under the degraded condition.

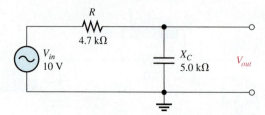

FIGURE 10–57

SOLUTION

The effective circuit resistance is

$$R_{th} = \frac{RR_{leak}}{R + R_{leak}} = \frac{(4.7\,\text{k}\Omega)(10\,\text{k}\Omega)}{14.7\,\text{k}\Omega} = 3.2\,\text{k}\Omega$$

The phase lag is

$$\phi = 90° - \tan^{-1}\left(\frac{X_C}{R_{th}}\right) = 90° - \tan^{-1}\left(\frac{5.0\,\text{k}\Omega}{3.2\,\text{k}\Omega}\right) = \textbf{32.6°}$$

To determine the output voltage, first calculate the Thevenin equivalent voltage.

$$V_{th} = \left(\frac{R_{leak}}{R + R_{leak}}\right)V_{in} = \left(\frac{10\,\text{k}\Omega}{14.7\,\text{k}\Omega}\right)10\,\text{V} = 6.80\,\text{V}$$

$$V_{out} = \left(\frac{X_C}{\sqrt{R_{th}^2 + X_C^2}}\right)V_{th} = \left(\frac{5.0\,\text{k}\Omega}{\sqrt{(3.2\,\text{k}\Omega)^2 + (5.0\,\text{k}\Omega)^2}}\right)6.8\,\text{V} = \textbf{5.73\,V}$$

RELATED PROBLEM

What would the output voltage be if the capacitor were not leaky?

Other Troubleshooting Considerations

Many times, the failure of a circuit to work properly is not the result of a faulty component. A loose wire, a bad contact, or a poor solder joint can cause an open circuit. A short can be caused by a wire clipping or solder splash. Things as simple as not plugging in a power supply or a function generator happen more often than you might think. Wrong values in a circuit (such as an incorrect resistor value), the function generator set at the wrong frequency, or the wrong output connected to the circuit can cause improper operation.

When you have problems with a circuit, always check to make sure that the instruments are properly connected to the circuits and to a power outlet. Also, look for obvious things such as a broken or loose contact, a connector that is not completely plugged in, or a piece of wire or a solder bridge that could be shorting something out.

The point is that you should consider all possibilities, not just faulty components, when a circuit is not working properly. The following example illustrates this approach with a simple circuit using the APM (analysis, planning, and measurement) method.

HANDS ON TIP

When connecting protoboard circuits, always consistently use standard colors for universal connections such as signals, power supply voltage, and ground. For example, you can use green wires for signals, red wires for power supply voltage, and black wires for ground connections. This helps you identify wiring during connecting and troubleshooting.

EXAMPLE 10–18

The circuit represented by the schematic in Figure 10–58 has no output voltage, which is the voltage across the capacitor. You expect to see about 7.4 V at the output. The circuit is physically constructed on a protoboard. Use your troubleshooting skills to find the problem.

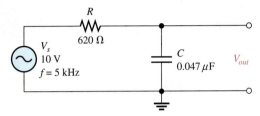

FIGURE 10–58

SOLUTION

Apply the APM method to this troubleshooting problem.

Analysis: First think of the possible causes for the circuit to have no output voltage.

1. There is no source voltage or the frequency is so high that the capacitive reactance is almost zero.
2. There is a short between the output terminals. Either the capacitor could be internally shorted, or there could be some physical short.
3. There is an open between the source and the output. This would prevent current and thus cause the output voltage to be zero. The resistor could be open, or the conductive path could be open due to a broken or loose connecting wire or a bad protoboard contact.
4. There is an incorrect component value. The resistor could be so large that the current and, therefore, the output voltage are negligible. The capacitor could be so large that its reactance at the input frequency is near zero.

Planning: You decide to make some visual checks for problems such as the function generator power cord not plugged in or the frequency set at an incorrect value. Also, broken leads, shorted leads, as well as an incorrect resistor color code or capacitor label often can be found visually. If nothing is discovered after a visual check, then you will make voltage measurements to track down the cause of the problem. You decide to use a digital oscilloscope and a DMM to make the measurements.

Measurement: Assume that you find that the function generator is plugged in and the frequency setting appears to be correct. Also, you find no visible opens or shorts during your visual check, and the component values are correct.

The first step in the measurement process is to check the voltage from the source with the scope. Assume a 10 V rms sine wave with a frequency of 5 kHz is observed at the circuit input as shown in Figure 10–59(a). The correct voltage is present, so *the first possible cause has been eliminated.*

Next, check for a shorted capacitor by disconnecting the source and placing a DMM (set on the ohmmeter function) across the capacitor. If the capacitor is good, an open will be indicated by an OL (overload) in the meter display after a short charging time. Assume the capacitor checks okay, as shown in Figure 10–59(b). *The second possible cause has been eliminated.*

Since the voltage has been "lost" somewhere between the input and the output, you must now look for the voltage. You reconnect the source and measure the voltage across the resistor with the DMM (set on the voltmeter function) from one resistor lead to the other. The voltage across the resistor is zero. This means there is no current, which indicates an open somewhere in the circuit.

Now, you begin tracing the circuit back toward the source looking for the voltage (you could also start from the source and work forward). You can use either the scope or the DMM, but decide to use the multimeter with one lead connected to ground and the other used to probe the circuit. As shown in Figure 10–59(c), the voltage on the right lead of the resistor, point ①, reads zero. Since you already have measured zero voltage across the resistor, the voltage on the left resistor lead at point ② must be zero as the meter indicates. Next, moving the meter probe to point ③, you read 10 V. You have found the voltage! Since there is zero volts on the left resistor lead, and there is 10 V at point ③, one of the two contacts in the protoboard hole into which the wire leads are inserted is bad. It could be that the small contacts were pushed in too far and were bent or broken so that the circuit lead does not make contact.

Move either or both the resistor lead and the wire to another hole in the same row. Assume that when the resistor lead is moved to the hole just above, you have voltage at the output of the circuit (across the capacitor).

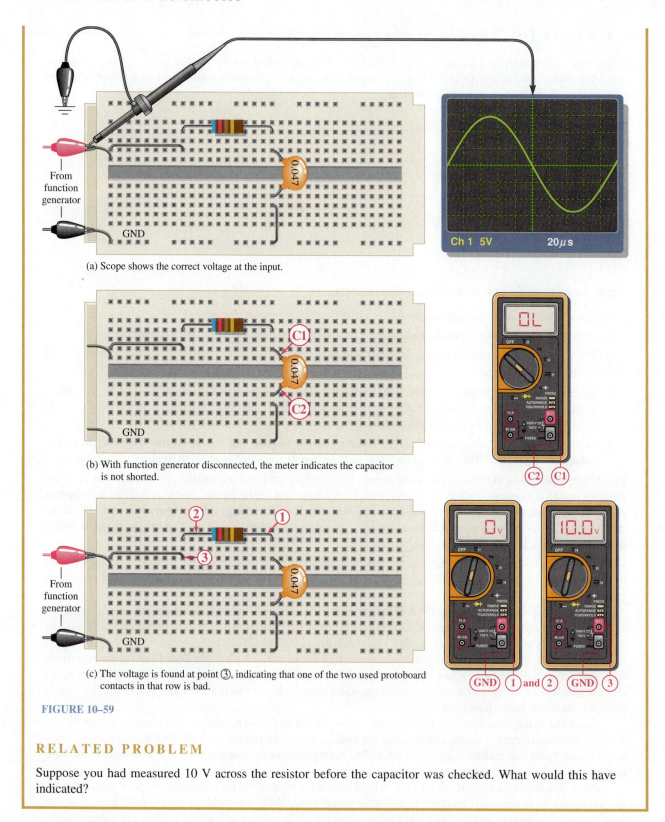

(a) Scope shows the correct voltage at the input.

(b) With function generator disconnected, the meter indicates the capacitor is not shorted.

(c) The voltage is found at point ③, indicating that one of the two used protoboard contacts in that row is bad.

FIGURE 10–59

RELATED PROBLEM

Suppose you had measured 10 V across the resistor before the capacitor was checked. What would this have indicated?

SECTION 10–9 CHECKUP

1. Describe the effect of a leaky capacitor on the dc response of an *RC* circuit.

2. In a series *RC* circuit, if all the applied voltage appears across the capacitor, what is the problem?

3. What faults can cause 0 V across the capacitor in a series *RC* circuit?

SUMMARY

- When a sinusoidal voltage is applied to an *RC* circuit, the current and all the voltage drops are also sine waves.
- Total current in a series or parallel *RC* circuit always leads the source voltage.
- The resistor voltage is always in phase with the current.
- The capacitor voltage always lags the current by 90°.
- In an *RC* circuit, the impedance is determined by both the resistance and the capacitive reactance combined.
- Impedance is expressed in units of ohms.
- The circuit phase angle is the angle between the total current and the source voltage.
- The impedance of a series *RC* circuit varies inversely with frequency.
- The phase angle (θ) of a series *RC* circuit varies inversely with frequency.
- In an *RC* lag circuit, the output voltage lags the input voltage.
- In an *RC* lead circuit, the output voltage leads the input voltage.
- The impedance of a circuit can be determined by measuring the source voltage and the total current and then applying Ohm's law.
- In an *RC* circuit, part of the power is resistive and part reactive.
- The phasor combination of resistive power (true power) and reactive power is called *apparent power*.
- Apparent power is expressed in volt-amperes (VA).
- The power factor (*PF*) indicates how much of the apparent power is true power.
- A power factor of 1 indicates a purely resistive circuit, and a power factor of 0 indicates a purely reactive circuit.
- In a circuit that exhibits frequency selectivity, certain frequencies are passed to the output while others are rejected.

KEY TERMS

Key terms and other bold terms in the chapter are defined in the end-of-book glossary.

Admittance (*Y*)   A measure of the ability of a reactive circuit to permit current; the reciprocal of impedance. The unit is the siemens (S).

Apparent power (*P_a*)   The phasor combination of resistive power (true power) and reactive power. The unit is the volt-ampere (VA).

Bandwidth   The range of frequencies that is considered to be passed from the input to the output of a circuit.

Capacitive susceptance (*B_C*)   The ability of a capacitor to permit current: the reciprocal of capacitive reactance. The unit is the siemens (S).

Cutoff frequency   The frequency at which the output voltage of a filter is 70.7% of the maximum output voltage.

Frequency response   In electric circuits, the variation in the output voltage (or current) over a specified range of frequencies.

Impedance (*Z*)   The total opposition to sinusoidal current expressed in ohms.

Phase angle   The angle between the source voltage and the total current in a reactive circuit.

Power factor   The relationship between volt-amperes and true power or watts. Volt-amperes multiplied by the power factor equals true power.

RC lag circuit   A phase shift circuit in which the output voltage, taken across the capacitor, lags the input voltage by a specified angle.

RC lead circuit   A phase shift circuit in which the output voltage, taken across the resistor, leads the input voltage by a specified angle.

KEY FORMULAS

(10–1) $Z = \sqrt{R^2 + X_C^2}$ Series *RC* impedance

(10–2) $\theta = \tan^{-1}\left(\dfrac{X_C}{R}\right)$ Series *RC* phase angle

(10–3) $V = IZ$ Ohm's law

(10–4) $I = \dfrac{V}{Z}$ Ohm's law

(10–5) $Z = \dfrac{V}{I}$ Ohm's law

(10–6) $V_s = \sqrt{V_R^2 + V_C^2}$ Total voltage in series *RC* circuits

(10–7) $\theta = \tan^{-1}\left(\dfrac{V_C}{V_R}\right)$ Series *RC* phase angle

(10–8) $\phi = 90° - \tan^{-1}\left(\dfrac{X_C}{R}\right)$ Phase angle of lag circuit

(10–9) $V_{out} = \left(\dfrac{X_C}{\sqrt{R^2 + X_C^2}}\right)V_{in}$ Output voltage of lag circuit

(10–10) $\phi = \tan^{-1}\left(\dfrac{X_C}{R}\right)$ Phase angle of lead circuit

(10–11) $V_{out} = \left(\dfrac{R}{\sqrt{R^2 + X_C^2}}\right)V_{in}$ Output voltage of lead circuit

(10–12) $Z = \dfrac{RX_C}{\sqrt{R^2 + X_C^2}}$ Parallel *RC* impedance

(10–13) $\theta = \tan^{-1}\left(\dfrac{R}{X_C}\right)$ Parallel *RC* phase angle

(10–14) $G = \dfrac{1}{R}$ Conductance

(10–15) $B_C = \dfrac{1}{X_C}$ Capacitive susceptance

(10–16) $Y = \dfrac{1}{Z}$ Admittance

(10–17) $Y_{tot} = \sqrt{G^2 + B_C^2}$ Total admittance

(10–18) $V = \dfrac{I}{Y}$ Ohm's law

(10–19) $I = VY$ Ohm's law

(10–20) $Y = \dfrac{I}{V}$ Ohm's law

(10–21) $I_{tot} = \sqrt{I_R^2 + I_C^2}$ Total current in parallel *RC* circuits

(10–22) $\theta = \tan^{-1}\left(\dfrac{I_C}{I_R}\right)$ Parallel *RC* phase angle

(10–23) $\theta = \left(\dfrac{\Delta t}{T}\right)360°$ Phase angle using time measurements

(10–24) $P_{true} = I_{tot}^2 R$ True power (W)

(10–25) $P_r = I_{tot}^2 X_C$ Reactive power (VAR)

(10–26) $P_a = I_{tot}^2 Z$ Apparent power (VA)

(10–27) $P_{true} = V_s I_{tot} \cos\theta$ True power

(10–28) $PF = \cos\theta$ Power factor

(10–29) $f_c = \dfrac{1}{2\pi RC}$ Cutoff frequency of an *RC* circuit

TRUE/FALSE QUIZ

Answers are at the end of the chapter.

1. In a series *RC* circuit, the impedance increases when frequency increases.

2. In a series *RC* lag circuit, the output voltage is taken across the resistor.

3. Admittance is the reciprocal of susceptance.

4. In a parallel *RC* circuit, as frequency is increased, the conductance is unchanged.

5. The phase angle of an *RC* circuit is measured between the source voltage and current.

6. The impedance of a parallel *RC* circuit can be found by applying phasor arithmetic to the product-over-sum rule.

7. If $X_C = R$, the phase shift in a series *RC* circuit is 45° with current leading voltage.

8. The power factor is equal to the tangent of the phase angle.

9. A purely resistive circuit has a power factor of 0.

10. Apparent power is measured in watts.

SELF-TEST

Answers are at the end of the chapter.

1. In a series *RC* circuit, the voltage across the resistance is
   - (a) in phase with the source voltage
   - (b) lagging the source voltage by 90°
   - (c) in phase with the current
   - (d) lagging the current by 90°

2. In a series *RC* circuit, the voltage across the capacitor is
   - (a) in phase with the source voltage
   - (b) lagging the resistor voltage by 90°
   - (c) in phase with the current
   - (d) lagging the source voltage by 90°

3. When the frequency of the voltage applied to a series *RC* circuit is increased, the impedance
   - (a) increases   (b) decreases   (c) remains the same   (d) doubles

4. When the frequency of the voltage applied to a series *RC* circuit is decreased, the phase angle
   - (a) increases   (b) decreases   (c) remains the same   (d) becomes erratic

5. In a series *RC* circuit when the frequency and the resistance are doubled, the impedance
   - (a) doubles   (b) is halved
   - (c) is quadrupled   (d) cannot be determined without values

6. In a series *RC* circuit, 10 V rms is measured across the resistor and 10 V rms is also measured across the capacitor. The rms source voltage is
   - (a) 20 V   (b) 14.14 V   (c) 28.28 V   (d) 10 V

7. The voltages in Problem 6 are measured at a certain frequency. To make the resistor voltage greater than the capacitor voltage, the frequency
   (a) must be increased
   (b) must be decreased
   (c) is held constant
   (d) has no effect

8. When $R = X_C$, the phase angle is
   (a) 0°   (b) +90°   (c) −90°   (d) 45°

9. To decrease the phase angle below 45°, the following condition must exist:
   (a) $R = X_C$ (b) $R < X_C$ (c) $R > X_C$ (d) $R = 10X_C$

10. When the frequency of the source voltage is increased, the impedance of a parallel *RC* circuit
    (a) increases
    (b) decreases
    (c) does not change

11. In a parallel *RC* circuit, there is 1 A rms through the resistive branch and 1 A rms through the capacitive branch. The total rms current is
    (a) 1 A   (b) 2 A   (c) 2.28 A   (d) 1.414 A

12. A power factor of 1 indicates that the circuit phase angle is
    (a) 90°   (b) 45°   (c) 180°   (d) 0°

13. For a certain load, the true power is 100 W and the reactive power is 100 VAR. The apparent power is
    (a) 200 VA   (b) 100 VA   (c) 141.4 VA   (d) 141.4 W

14. Energy sources are normally rated in
    (a) watts
    (b) volt-amperes
    (c) volt-amperes reactive
    (d) none of these

15. If the bandwidth of a certain low-pass filter is 1 kHz, the cutoff frequency is
    (a) 0 Hz   (b) 500 Hz   (c) 2 kHz   (d) 1000 Hz

TROUBLESHOOTING: SYMPTOM AND CAUSE

The purpose of these exercises is to help develop thought processes essential to troubleshooting. Answers are at the end of the chapter.

Determine the cause for each set of symptoms. Refer to Figure 10–60.

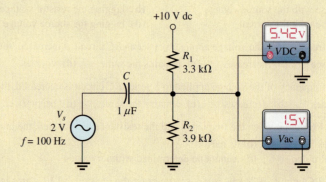

FIGURE 10–60   **The meters indicate the correct readings for this circuit.**

1. *Symptom:* The dc voltmeter reading is 0 V, and the ac voltmeter reading is 1.85 V.

   Cause:

   (a) C is shorted.
   (b) R_1 is open.
   (c) R_2 is open.

2. *Symptom:* The dc voltmeter reading is 5.42 V, and the ac voltmeter reading is 0 V.

   Cause:

   (a) C is shorted.
   (b) C is open.
   (c) A resistor is open.

3. *Symptom:* The dc voltmeter reading is approximately 0 V, and the ac voltmeter reading is 2 V.

   Cause:

   (a) C is shorted.
   (b) C is open.
   (c) R_1 is shorted.

4. *Symptom:* The dc voltmeter reading is 10 V, and the ac voltmeter reading is 0 V.

   Cause:

   (a) C is open.
   (b) C is shorted.
   (c) R_1 is shorted.

5. *Symptom:* The dc voltmeter reading is 10 V, and the ac voltmeter reading is 1.8 V.

   Cause:

   (a) R_1 is shorted.
   (b) R_2 is open.
   (c) C is shorted.

PROBLEMS

Answers to odd-numbered problems are at the end of the book.

BASIC PROBLEMS

SECTION 10–1   Sinusoidal Response of Series *RC* Circuits

1. An 8 kHz sinusoidal voltage is applied to a series *RC* circuit. What is the frequency of the voltage across the resistor? Across the capacitor?

2. What is the waveshape of the current in the circuit of Problem 1?

SECTION 10–2   Impedance and Phase Angle of Series *RC* Circuits

3. Find the impedance of each circuit in Figure 10–61.

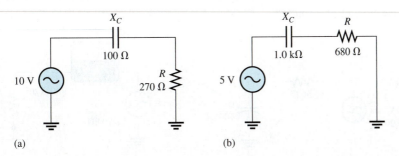

FIGURE 10–61

4. Determine the impedance and the phase angle in each circuit in Figure 10–62.

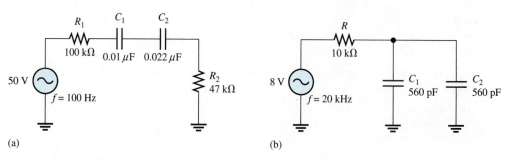

(a)

(b)

FIGURE 10–62

5. For the circuit of Figure 10–63, determine the impedance for each of the following frequencies:
   (a) 100 Hz   **(b)** 500 Hz   **(c)** 1.0 kHz   **(d)** 2.5 kHz

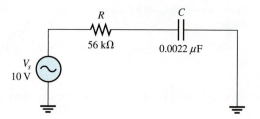

FIGURE 10–63

6. Repeat Problem 5 for $C = 0.0047 \, \mu\text{F}$.

SECTION 10–3   Analysis of Series *RC* Circuits

7. Calculate the total current in each circuit of Figure 10–61.

8. Repeat Problem 7 for the circuits in Figure 10–62.

9. For the circuit in Figure 10–64, draw the phasor diagram showing all voltages and the total current. Indicate the phase angles.

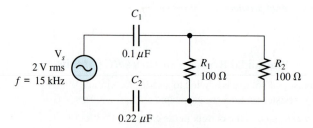

FIGURE 10–64

10. For the circuit in Figure 10–65, determine the following:
   (a) Z **(b)** I **(c)** V_R **(d)** V_C

11. To what value must the rheostat be set in Figure 10–66 to make the total current 10 mA? What is the resulting phase angle?

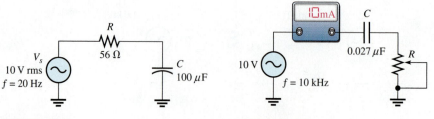

FIGURE 10–65

FIGURE 10–66

12. For the lag circuit in Figure 10–67, determine the phase lag between the input voltage and the output voltage for each of the following frequencies:
    (a) 1 Hz   (b) 100 Hz   (c) 1.0 kHz   (d) 10 kHz

13. Repeat Problem 12 for the lead circuit in Figure 10–68.

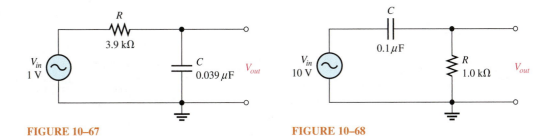

FIGURE 10–67   **FIGURE 10–68**

SECTION 10–4   Impedance and Phase Angle of Parallel *RC* Circuits

14. Determine the impedance for the circuit in Figure 10–69.

15. Determine the impedance and the phase angle in Figure 10–70.

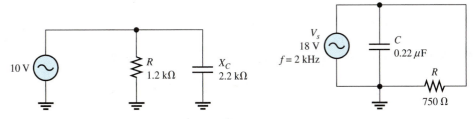

FIGURE 10–69   **FIGURE 10–70**

16. Repeat Problem 15 for the following frequencies:
    (a) 1.5 kHz   (b) 3.0 kHz   (c) 5.0 kHz   (d) 10 kHz

17. Determine the impedance and phase angle in Figure 10–71.

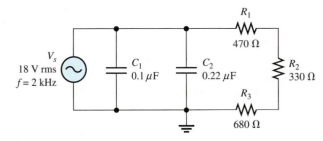

FIGURE 10–71

SECTION 10–5   Analysis of Parallel *RC* Circuits

18. For the circuit in Figure 10–72, find all the currents and voltages.

19. For the parallel circuit in Figure 10–73, find each branch current and the total current. What is the phase angle between the source voltage and the total current?

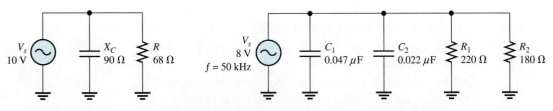

FIGURE 10–72   **FIGURE 10–73**

20. For the circuit in Figure 10–74, determine the following:
    (a) Z **(b)** I_R **(c)** I_C **(d)** I_{tot} **(e)** θ

21. Repeat Problem 20 for $R = 4.7\ \text{k}\Omega$, $C = 0.047\ \mu F$, and $f = 500\ \text{Hz}$.

22. Find the total current for the circuit in Figure 10–75.

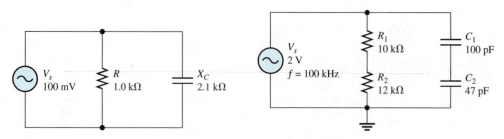

FIGURE 10–74   **FIGURE 10–75**

SECTION 10–6   Analysis of Series-Parallel *RC* Circuits

23. The circuit in Figure 10–76 is a complex circuit consisting of five components. With trigonometry it can be shown that R_1, R_2, C_2, and C_3 are equivalent to a resistance of 3.08 Ω in series with a capacitive reactance of 39.5 Ω. From this given information, determine the voltages across each component in the original circuit.

24. Is the circuit in Figure 10–76 predominantly resistive or predominantly capacitive?

25. Find the current through each branch and the total current in Figure 10–76.

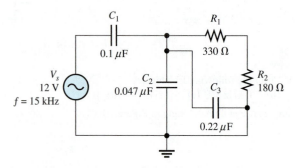

FIGURE 10–76

26. For the circuit in Figure 10–77, determine the following:
    (a) I_{tot} **(b)** θ **(c)** V_{R1} **(d)** V_{R2} **(e)** V_{R3} **(f)** V_C

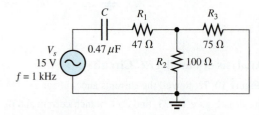

FIGURE 10–77

SECTION 10–7   Power in *RC* Circuits

27. In a certain series *RC* circuit, the true power is 2 W, and the reactive power is 3.5 VAR. Determine the apparent power.

28. In Figure 10–65, what is the true power and the reactive power?

29. What is the power factor for the circuit of Figure 10–75?

30. Determine P_{true}, P_r, P_a, and PF for the circuit in Figure 10–77. Draw the power triangle.

SECTION 10–8   Basic Applications

31. The lag circuit in Figure 10–67 also acts as a low-pass filter. Draw a response curve for this circuit by plotting the output voltage versus frequency for 0 Hz to 10 kHz in 1 kHz increments.

32. Plot the frequency response curve for the circuit in Figure 10–68 for a frequency range of 0 Hz to 10 kHz in 1 kHz increments.

33. Draw the voltage phasor diagram for each circuit in Figures 10–67 and 10–68 for a frequency of 5 kHz with $V_{in} = 1$ V rms.

34. The rms value of the signal voltage out of amplifier A in Figure 10–78 is 50 mV. If the input resistance to amplifier B is 10 kΩ, how much of the signal is lost due to the coupling capacitor (C_c) when the frequency is 3 kHz?

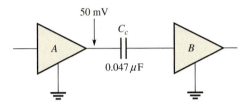

FIGURE 10–78

35. Determine the cutoff frequency for each circuit in Figures 10–67 and 10–68.

36. Determine the bandwidth of the circuit in Figure 10–68.

SECTION 10–9   Troubleshooting

37. Assume that the capacitor in Figure 10–79 is excessively leaky. Show how this degradation affects the output voltage and phase angle, assuming that the leakage resistance is 5 kΩ and the frequency is 10 Hz.

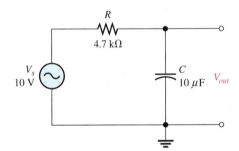

FIGURE 10–79

38. Each of the capacitors in Figure 10–80 has developed a leakage resistance of 2 kΩ. Determine the output voltages under this condition for each circuit.

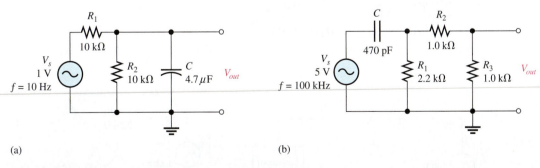

(a)   (b)

FIGURE 10–80

39. Determine the output voltage for the circuit in Figure 10–80(a) for each of the following failure modes, and compare it to the correct output:
    (a) R_1 open   (b) R_2 open   (c) C open   (d) C shorted

40. Determine the output voltage for the circuit in Figure 10–80(b) for each of the following failure modes, and compare it to the correct output:
    (a) *C* open   (b) *C* shorted   (c) R_1 open   (d) R_2 open   (e) R_3 open

ADVANCED PROBLEMS

41. A single 240 V, 60 Hz source drives two loads. Load *A* has an impedance of 50 Ω and a power factor of 0.85. Load *B* has an impedance of 72 Ω and a power factor of 0.95.
    (a) How much current does each load draw?
    (b) What is the reactive power in each load?
    (c) What is the true power in each load?
    (d) What is the apparent power in each load?

42. What value of coupling capacitor is required in Figure 10–81 so that the signal voltage at the input of amplifier 2 is at least 70.7% of the signal voltage at the output of amplifier 1 when the frequency is 20 Hz? Disregard the input resistance of the amplifier.

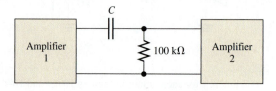

FIGURE 10–81

43. Determine the value of R_1 required to get a phase angle of 30° between the source voltage and the total current in Figure 10–82.

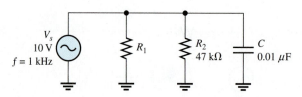

FIGURE 10–82

44. Draw the impedance phasor diagram for the input circuit of the amplifier in System Example 10–1 (Figure 10–24) at the cutoff frequency of 1.59 Hz.

45. A certain load dissipates 1.5 kW of power with an impedance of 12 Ω and a power factor of 0.75. What is its reactive power? What is its apparent power?

46. Determine the series element or elements that are in the block of Figure 10–83 to meet the following overall circuit requirements:
    (a) $P_{true} = 400$ W
    (b) leading power factor (I_{tot} leads V_s)

47. Determine the value of C_2 in Figure 10–84 when $V_A = V_B$.

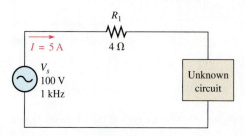

FIGURE 10–83

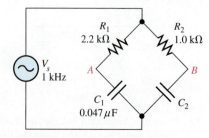

FIGURE 10–84

48. Draw the schematic for the circuit in Figure 10–85 and determine if the waveform on the scope is correct. If there is a fault in the circuit, identify it.

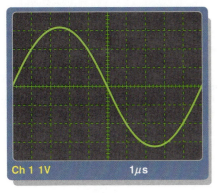

Ch 1 1V   1μs

(a) Oscilloscope display

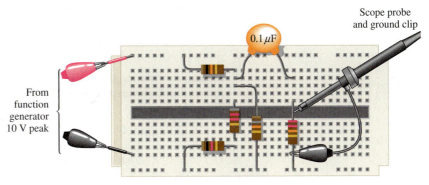

Scope probe
and ground clip

0.1 μF

From
function
generator
10 V peak

(b) Circuit with leads connected

FIGURE 10–85

MULTISIM TROUBLESHOOTING PROBLEMS

MULTISIM

49. Open file P10-49; files are found at www.pearsonhighered.com/floyd. Determine if there is a fault and, if so, identify it.

50. Open file P10-50. Determine if there is a fault and, if so, identify it.

51. Open file P10-51. Determine if there is a fault and, if so, identify it.

52. Open file P10-52. Determine if there is a fault and, if so, identify it.

53. Open file P10-53. Determine if there is a fault and, if so, identify it.

54. Open file P10-54. Determine if there is a fault and, if so, identify it.

55. Open the Multisim file P10-55. Modify the circuit so that you can measure the phase shift as shown in Figure 10–38 by moving ground and using a Tektronix simulated scope as shown in Figure 10–39. You will need to trigger the scope with Normal mode triggering. You should be able to set up the display as shown in Figure 10–39 to measure the phase shift. Then simulate a shorted capacitor C_1 and determine the effect on phase shift. (You will need to reset the amplitudes to make an accurate check.)

ANSWERS TO SECTION CHECKUPS

SECTION 10–1   Sinusoidal Response of Series *RC* Circuits

1. V_C frequency is 60 Hz, *I* frequency is 60 Hz.

2. The capacitive reactance and resistance

3. θ is closer to 0° when $R > X_C$.

SECTION 10–2   Impedance and Phase Angle of Series *RC* Circuits

1. Impedance is the total opposition to sinusoidal current expressed in ohms.

2. V_s lags *I*.

3. The capacitive reactance produces the phase angle.

4. $Z = \sqrt{R^2 + X_C^2} = 59.9 \text{ k}\Omega; \theta = \tan^{-1}(X_C/R) = 56.6°$

SECTION 10–3   Analysis of Series *RC* Circuits

1. $V_s = \sqrt{V_R^2 + V_C^2} = 7.2 \text{ V}$

2. $\theta = \tan^{-1}(V_C/V_R) = 56.3° \text{ s}$

3. $\theta = 90°$

4. **(a)** X_C decreases when f increases.
   (b) Z decreases when f increases.
   (c) θ decreases when f increases.

5. $\phi = 90° - \tan^{-1}(X_C/R) = 62.8°$

6. $V_{out} = (R/\sqrt{R^2 + X_C^2})V_{in} = 8.9$ V rms

SECTION 10–4   Impedance and Phase Angle of Parallel *RC* Circuits

1. $Z = RX_C/\sqrt{R^2 + X_C^2} = 545 \ \Omega$

2. *Conductance* is the reciprocal of resistance; *capacitive susceptance* is the reciprocal of capacitive reactance; and *admittance* is the reciprocal of impedance.

3. $Y = 1/Z = 10$ mS

4. $Y = \sqrt{G^2 + B_C^2} = 24$ mS

SECTION 10–5   Analysis of Parallel *RC* Circuits

1. $I_{tot} = V_s Y = 21$ mA

2. $\theta = \tan^{-1}(I_C/I_R) = 56.3°; I_{tot} = \sqrt{I_R^2 + I_C^2} = 18$ mA

3. $\theta = 90°$

SECTION 10–6   Analysis of Series-Parallel *RC* Circuits

1. All grounds should be connected to an electrical common point to avoid shorting components.

2. $V_1 = I_{tot}R_1 = 6.71$ V

3. 1.67 pF

SECTION 10–7   Power in *RC* Circuits

1. Power dissipation is due to the resistance.

2. $PF = \cos 45° = 0.707$

3. $P_{\text{true}} = I_{tot}^2 R = 1.32$ kW; $P_r = I_{tot}^2 X_C = 1.84$ kVAR; $P_a = I_{tot}^2 Z = 2.26$ kVA

SECTION 10–8   Basic Applications

1. 180°

2. Output is across the capacitor.

SECTION 10–9   Troubleshooting

1. The leakage resistance acts in parallel with *C*, which alters the circuit time constant.

2. The capacitor is open.

3. A shorted capacitor, open resistor, no source voltage, or open contact can cause 0 V across the capacitor.

ANSWERS TO RELATED PROBLEMS FOR EXAMPLES

10–1 $2.42 \ k\Omega; 65.6°$

10–2   2.56 V

10–3   65.5°

10–4   7.14 V

10–5 $15.9 \ k\Omega; 86.4°$

10–6   Phase lag ϕ increases.

10–7   Output voltage decreases.

10–8   Phase lead ϕ decreases; output voltage increases.

10–9 $24.3 \ \Omega$

10–10   4.60 mS

10–11   6.16 mA

10–12 117 mA; 31°

10–13 $V_1 = 7.04$ V; $V_2 = 3.21$ V

10–14 0.146

10–15 990 μW

10–16 1.60 kHz

10–17 7.29 V

10–18 Resistor open

ANSWERS TO TRUE/FALSE QUIZ

1. F   **2.** F   **3.** F   **4.** T   **5.** T   **6.** T   **7.** T   **8.** F   **9.** F   **10.** F

ANSWERS TO SELF-TEST

1. (c)   **2.** (b)   **3.** (b)   **4.** (a)   **5.** (d)   **6.** (b)   **7.** (a)   **8.** (d)

9. (c)   **10.** (b)   **11.** (d)   **12.** (d)   **13.** (c)   **14.** (b)   **15.** (d)

ANSWERS TO TROUBLESHOOTING: SYMPTOM AND CAUSE

1. (b)

2. (b)

3. (a)

4. (c)

5. (b)

CHAPTER 11

INDUCTORS

OUTLINE

OBJECTIVES

- Describe the basic structure and characteristics of an inductor
- Discuss various types of inductors
- Analyze series and parallel inductors
- Analyze inductive dc switching circuits
- Analyze inductive ac circuits
- Discuss some inductor applications

KEY TERMS

Inductor	Henry (H)
Winding	Winding resistance
Coil	RL time constant
Induced voltage	Inductive reactance
Inductance (L)	Quality factor (Q)

INTRODUCTION

Inductance is the property of a coil of wire that opposes a change in current. The basis for inductance is the electromagnetic field that surrounds any conductor when there is current through it. The electrical component designed to have the property of inductance is called an *inductor, coil* or in some applications *choke*. These terms refer to the same type of device. The term *choke* is usually associated with inductors that are used to block high frequencies which are widely used in communication systems and in radio-frequency circuits.

In this chapter, you will study the basic inductor and its characteristics. Various types of inductors are covered in terms of their physical construction and their electrical properties. The basic behavior of inductors in both dc and ac circuits is discussed, and series and parallel combinations are analyzed.

VISIT THE WEBSITE
Study aids for this chapter are available at
http://pearsonhighered.com/floyd

An **inductor** is a passive electrical component, formed by a coil of wire, that exhibits the property of inductance.

After completing this section, you should be able to

- **Describe the basic structure and characteristics of an inductor**
  - **Define** *inductance* **and state its unit**
  - **Discuss induced voltage**
  - **Explain how an inductor stores energy**
  - **Specify how the physical characteristics affect inductance**
  - **Discuss winding resistance and winding capacitance**
  - **Discuss how Faraday's law and Lenz's law explain the voltage induced across an inductor**

When a length of wire is formed into a coil, as shown in Figure 11–1, it becomes an inductor. Current through the coil produces an electromagnetic field. The magnetic lines of force around each loop (turn) in the **winding** of the coil effectively add to the lines of force around the adjoining loops, forming a stronger magnetic field within and around the coil, as shown. The net direction of the total magnetic field creates a north and a south pole. A schematic symbol for the inductor is shown in Figure 11–2.

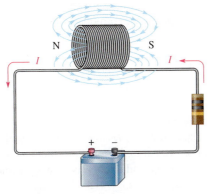

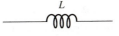

FIGURE 11–1 Current through a coil creates a three-dimensional electromagnetic field. The resistor limits the current.

FIGURE 11–2 Symbol for an inductor.

Inductance

When there is current through an inductor, an electromagnetic field is established. When the current changes, the electromagnetic field also changes. An increase in current expands the field, and a decrease in current reduces it. Therefore, a changing current produces a changing electromagnetic field around the inductor (also known as **coil** and in some applications, **choke**). In turn, the changing electromagnetic field causes an **induced voltage** across the coil in a direction to oppose the change in current. This property is called *self-inductance* but is usually referred to as simply *inductance*, symbolized by *L*.

> **Inductance** **is a measure of a coil's ability to establish an induced voltage as a result of a change in its current, and that induced voltage is in a direction to oppose that change in current.**

THE UNIT OF INDUCTANCE   The **henry**, symbolized by H, is the basic unit of inductance. By definition, the inductance of a coil is one henry when current through the coil, changing at the rate of one ampere per second, induces one volt across the coil. The henry is a large unit, so in practical applications, millihenries (mH), and microhenries (μH) are the most common units. Nanohenries (nH) are also used for very small inductors.

ENERGY STORAGE   An inductor stores energy in the magnetic field created by the current. The energy stored is expressed as

$$W = \frac{1}{2}LI^2 \tag{11–1}$$

As you can see, the energy stored is proportional to the inductance and the square of the current. When current (I) is in amperes and inductance (L) is in henries, the energy (W) is in joules.

Physical Characteristics of an Inductor

The following characteristics are important in establishing the inductance of a coil: the permeability of the core material, the number of turns of wire, the core length, and the cross-sectional area of the core.

CORE MATERIAL   As discussed earlier, an inductor is basically a coil of wire. The material around which the coil is formed is called the **core**. Coils are wound on either non-magnetic or magnetic materials. Examples of nonmagnetic materials are air, wood, copper, plastic, and glass. The permeabilities of these materials are the same as for a vacuum. Examples of magnetic materials are iron, nickel, steel, cobalt, or alloys. These materials have permeabilities that are hundreds or thousands of times greater than that of a vacuum and are classified as *ferromagnetic*. A ferromagnetic core provides a lower reluctance path for the magnetic lines of force and thus permits a stronger magnetic field.

As you learned in Chapter 7, the permeability (μ) of the core material determines how easily a magnetic field can be established. The inductance is directly proportional to the permeability of the core material.

PHYSICAL PARAMETERS   As indicated in Figure 11–3, the number of turns of wire, the length, and the cross-sectional area of the core are factors in setting the value of inductance. The inductance is inversely proportional to the length of the core and directly proportional to the cross-sectional area. Also, the inductance is directly related to the number of turns squared. This relationship is as follows:

$$L = \frac{N^2\mu A}{l} \tag{11–2}$$

where L is the inductance in henries (H), N is the number of turns, μ is the permeability in henries per meter (H/m), A is the cross-sectional area in meters squared (m^2), and l is the core length in meters (m).

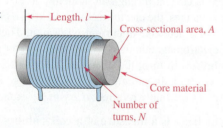

FIGURE 11–3 **Factors that determine the inductance of a coil.**

Length, l

Cross-sectional area, A

Core material

Number of turns, N

E X A M P L E   1 1 – 1

Determine the inductance of the coil in Figure 11–4. The permeability of the core is 0.25×10^{-3} H/m.

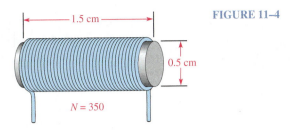

FIGURE 11–4

SOLUTION

1 cm = 0.01 m. Therefore, 1.5 cm = 0.015 m and 0.5 cm = 0.005 m.

$$A = \pi r^2 = \pi (0.25 \times 10^{-2}\,\text{m})^2 = 1.96 \times 10^{-5}\,\text{m}^2$$

$$L = \frac{N^2 \mu A}{l} = \frac{(350)^2 (0.25 \times 10^{-3}\,\text{H/m})(1.96 \times 10^{-5}\,\text{m}^2)}{0.015\,\text{m}} = \textbf{40 mH}$$

RELATED PROBLEM*

Determine the inductance of a coil with 400 turns on a core that is 2 cm long and has a diameter of 1 cm. The permeability is 0.25×10^{-3} H/m.

*Answers are at the end of the chapter.

Winding Resistance

When a coil is made of a certain material (for example, insulated copper wire), that wire has a certain resistance per unit of length. When many turns of wire are used to construct a coil, the total resistance may be significant. This inherent resistance is called the *dc resistance* or the **winding resistance (R_W)**.

Although this resistance is distributed along the length of the wire, as shown in Figure 11–5(a), it is sometimes indicated in a schematic as resistance appearing in series with the inductance of the coil, as shown in Figure 11–5(b). In many applications, the winding resistance can be ignored and the coil can be considered an ideal inductor. In other cases, the resistance must be considered.

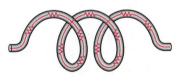

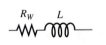

FIGURE 11–5   Winding resistance of a coil.

(a) The wire has resistance distributed along its length.

(b) Equivalent circuit

Winding Capacitance

When two conductors are placed side by side, there is always some capacitance between them. Thus, when many turns of wire are placed close together in a coil, a certain amount of stray capacitance, called *winding capacitance* (C_W), is a natural side effect. In many applications, this winding capacitance is very small and has no significant effect. In other cases, particularly at high frequencies, it may become quite important.

The equivalent circuit for an inductor with both its winding resistance (R_W) and its winding capacitance (C_W) is shown in Figure 11–6. The capacitance effectively acts in parallel. The total of the stray capacitances between each loop of the winding is indicated in a schematic as a capacitance appearing in parallel with the coil and its winding resistance, as shown in Figure 11–6(b).

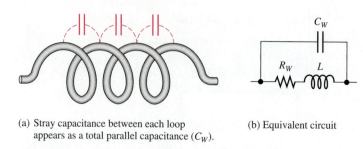

(a) Stray capacitance between each loop
appears as a total parallel capacitance (C_W).

(b) Equivalent circuit

FIGURE 11–6   **Winding capacitance of a coil.**

Review of Faraday's Law

Faraday's law was introduced in Chapter 7 and is reviewed here because of its importance in the study of inductors. Faraday found that by moving a magnet through a coil of wire, a voltage was induced across the coil, and that when a complete path was provided, the induced voltage caused an induced current. He observed that

> **The amount of voltage induced in a coil is directly proportional to the rate of change of the magnetic field with respect to the coil.**

This principle is illustrated in Figure 11–7, where a bar magnet is moved through a coil of wire. An induced voltage is indicated by the voltmeter connected across the coil. The faster the magnet is moved, the greater the induced voltage.

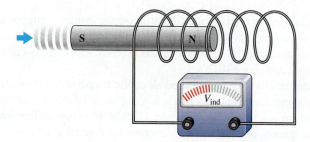

FIGURE 11–7   **Induced voltage is created by a changing magnetic field.**

When a wire is formed into a certain number of loops or turns and is exposed to a changing magnetic field, a voltage is induced across the coil. The induced voltage is proportional to the number of turns, N, of the wire in the coil and to the rate at which the magnetic field changes.

Lenz's Law

Lenz's law, also introduced in Chapter 7, adds to Faraday's law by defining the direction of induced voltage.

> **When the current through a coil changes and an induced voltage is created as a result of the changing magnetic field, the direction of the induced voltage is such that it always opposes the change in current.**

Figure 11–8 illustrates Lenz's law. In part (a), the current is constant and is limited by R_1. There is no induced voltage because the magnetic field is unchanging. In part (b), the switch suddenly is closed, placing R_2 in parallel with R_1 and thus reducing the resistance. Naturally, the current tries to increase and the magnetic field begins to expand, but the induced voltage opposes this attempted increase in current for an instant.

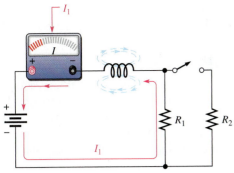

(a) Switch open: Constant current and constant magnetic field; no induced voltage.

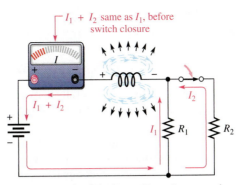

(b) At instant of switch closure: Expanding magnetic field induces voltage, which opposes an increase in total current. The total current remains the same at this instant.

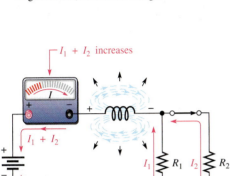

(c) Right after switch closure: The rate of expansion of the magnetic field decreases, allowing the current to increase exponentially as induced voltage decreases.

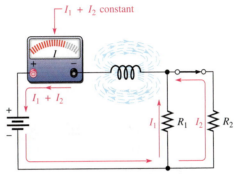

(d) Switch remains closed: Current and magnetic field reach constant value, no induced voltage.

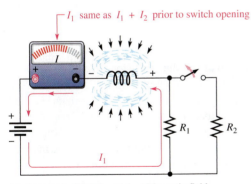

(e) At instant of switch opening: Magnetic field begins to collapse, creating an induced voltage, which opposes a decrease in current.

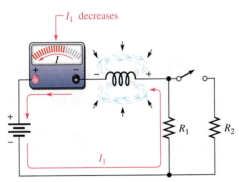

(f) After switch opening: Rate of collapse of magnetic field decreases, allowing current to decrease exponentially back to original value.

FIGURE 11–8   **Demonstration of Lenz's law in an inductive circuit: When the current tries to change suddenly, the electromagnetic field changes and induces a voltage in a direction that opposes that change in current.**

In Figure 11–8(c), the induced voltage gradually decreases, allowing the current to increase. In part (d), the current has reached a constant value as determined by the parallel resistors, and the induced voltage is zero. In part (e), the switch has been suddenly opened, and, for an instant, the induced voltage prevents any decrease in current. In part (f), the induced voltage gradually decreases, allowing the current to decrease back to a value determined by R_1. Notice that the induced voltage has a polarity that opposes any current change. The polarity of the induced voltage is opposite that of the battery voltage for an increase in current and aids the battery voltage for a decrease in current.

SYSTEM EXAMPLE 11–1

SWITCH-MODE POWER SUPPLY

Power supplies generally convert ac to dc; they are found in virtually all electronics equipment. Switch-mode power supplies (SMPS) became prevalent in the 1970s after the development of very fast switching transistors and IC pulse width modulators (PWM). The Apple II computer, released in 1977, became one of the first computer systems to use a switch-mode power supply, resulting in a much more efficient and cooler running supply. Today, nearly all computers use switch-mode power supplies.

A simplified schematic of a step-down SMPS is shown in Figure 11–9. The switch, SW, represents a switching MOSFET transistor that rapidly alternates between the open and closed position (this happens about 20,000 times per second). V represents an unregulated dc input voltage. The pulse-width modulator is the controller; it compares the output to a reference voltage and sets the pulse width to keep the output constant.

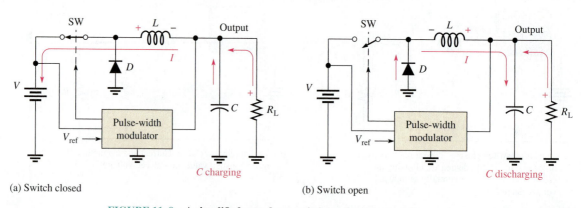

(a) Switch closed   (b) Switch open

FIGURE 11–9   A simplified step-down switch-mode power supply.

Our focus is on the inductor, L, a key component in a SMPS. Recall from Lenz's law that an inductor opposes a change in current. In part (a), the transistor is *on* as represented by the closed switch; the major current path to the load is shown with the red arrows. The induced voltage across the inductor is opposite to that of the dc input as the magnetic field builds and tends to keep the current constant in the load. In part (b), the switch is open (transistor *off*), and current is supplied through the diode and from the capacitor. Notice the inductor has the same polarity as the source when its magnetic field collapses. The induced voltage is opposite to what it was. The result is a nearly constant current in the load.

SECTION 11–1 CHECKUP*

1. List the parameters that contribute to the inductance of a coil.

2. Describe what happens to L when

   (a) N is increased

   (b) the core length is increased

   (c) the cross-sectional area of the core is decreased

   (d) a ferromagnetic core is replaced by an air core

3. Explain why inductors have some winding resistance.

4. Explain why inductors have some winding capacitance.

*Answers are at the end of the chapter.

11–2 TYPES OF INDUCTORS

Inductors normally are classified according to the type of core material.

After completing this section, you should be able to

- **Discuss various types of inductors**
  - **Describe the basic types of fixed inductors**
  - **Distinguish between fixed and variable inductors**

Inductors are made in a variety of shapes and sizes. Basically, they fall into two general categories: fixed and variable. The standard schematic symbols are shown in Figure 11–10.

Both fixed and variable inductors can be classified according to the type of core material. Three common types are the air core, the iron core, and the ferrite core. Each has a unique symbol, as shown in Figure 11–11. Large-value iron-core inductors are rarely used.

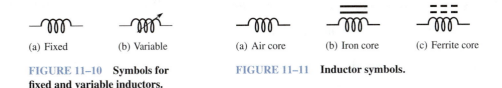

(a) Fixed   (b) Variable   (a) Air core   (b) Iron core   (c) Ferrite core

FIGURE 11–10 **Symbols for fixed and variable inductors.**

FIGURE 11–11 **Inductor symbols.**

Adjustable (variable) inductors usually have a screw-type adjustment that moves a sliding core in and out, thus changing the inductance. A wide variety of inductors exists, and some are shown in Figure 11–12. Small fixed inductors are usually encapsulated in an insulating material that protects the fine wire in the coil. Encapsulated inductors have an appearance similar to a small resistor. Small-value inductors are available for high-frequency applications as surface-mount types and include some with ferrite-core construction.

FIGURE 11–12 **Typical inductors.**

Shielded Inductors

Inductors can radiate electromagnetic energy in certain cases, causing interference in nearby circuits. To avoid this, engineers will sometimes specify shielded inductors. Shielded inductors are available in small surface-mount packages that are soldered in place. Shielded surface-mount inductors are available from 1 μH to 1 mH.

SYSTEM NOTE

SECTION 11–2 CHECKUP

1. Name two general categories of inductors.

2. Identify the inductor symbols in Figure 11–13.

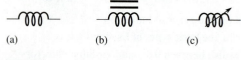

(a)   (b)   (c)

FIGURE 11–13

11–3 SERIES AND PARALLEL INDUCTORS

When inductors are connected in series, the total inductance increases. When inductors are connected in parallel, the total inductance decreases.

After completing this section, you should be able to

- **Analyze series and parallel inductors**
  - **Determine total series inductance**
  - **Determine total parallel inductance**

Total Series Inductance

When inductors are connected in series, as in Figure 11–14, the total inductance, L_T, is the sum of the individual inductances. The formula for L_T is expressed in the following equation for the general case of n inductors in series:

$$L_T = L_1 + L_2 + L_3 + \cdots + L_n \qquad (11\text{–}3)$$

Notice that the calculation of total inductance in series is analogous to the calculations of total resistance in series (Chapter 4) and total capacitance in parallel (Chapter 9).

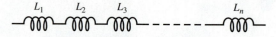

FIGURE 11–14   **Induction in series.**

EXAMPLE 11–2

Determine the total inductance for each of the series connections in Figure 11–15.

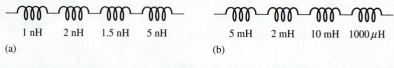

1 nH   2 nH   1.5 nH   5 nH   5 mH   2 mH   10 mH   1000 μH

(a)   (b)

FIGURE 11–15

SOLUTION

In Figure 11–15(a),

$$L_T = 1\,\text{nH} + 2\,\text{nH} + 1.5\,\text{nH} + 5\,\text{nH} = 9.5\,\text{nH}$$

In Figure 11–15(b),

$$L_T = 5\,\text{mH} + 2\,\text{mH} + 10\,\text{mH} + 1\,\text{mH} = \mathbf{18\,mH}$$

Note: $1000\,\mu\text{H} = 1\,\text{mH}$

RELATED PROBLEM

What is the total inductance of three $50\,\mu\text{H}$ inductors in series?

SYSTEM EXAMPLE 11–2

HARTLEY OSCILLATORS

Recall from System Example 9–1 that an oscillator is a circuit that generates a repetitive wave. The circuit shown in Figure 11–16 is a Hartley oscillator that uses inductors in series. As in the case of the Colpitts oscillator, this circuit uses positive feedback—a portion of the output is returned by the feedback network to the input to be amplified and reinforce the output. The feedback is across L_1 and is referenced to ground. Although the circuit is an ac circuit, the equations for inductors and inductive voltages developed with dc sources can be applied.

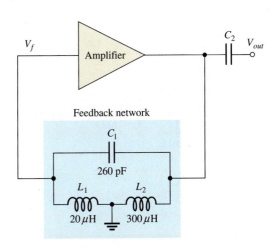

FIGURE 11–16   A Hartley oscillator.

Inductors L_1 and L_2 are in series and form part of a resonant circuit (resonant circuits are covered in Chapter 13). Find the total inductance by simply adding the inductors.

$$L_T = L_1 + L_2 = 20\,\mu\text{H} + 300\,\mu\text{H} = 320\,\mu\text{H}$$

The amplifier's signals are referenced to ground, so the feedback signal, V_f, is the fraction of the output developed across L_1. Notice that V_f is returned to the input of the amplifier. You can find the fraction of output voltage that is fed back (V_f/V_{out}) by applying the voltage divider idea to the series inductors.

$$\frac{V_f}{V_{out}} = \frac{L_1}{L_T} = \left(\frac{20\,\mu\text{H}}{320\,\mu\text{H}}\right) = 0.0625 = 6.2\%$$

Unlike capacitors, the smaller inductor has the smaller fraction of the output voltage.

Total Parallel Inductance

When inductors are connected in parallel, as in Figure 11–17, the total inductance is less than the smallest inductance. The general formula states that the reciprocal of the total inductance is equal to the sum of the reciprocals of the individual inductances.

$$\frac{1}{L_T} = \frac{1}{L_1} + \frac{1}{L_2} + \frac{1}{L_3} + \cdots + \frac{1}{L_n}$$

You can calculate total inductance, L_T, by taking the reciprocals of both sides of the previous equation.

$$L_T = \frac{1}{\dfrac{1}{L_1} + \dfrac{1}{L_2} + \dfrac{1}{L_3} + \cdots + \dfrac{1}{L_n}} \qquad (11\text{–}4)$$

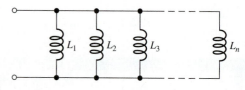

FIGURE 11–17 Inductors in parallel.

This calculation for total inductance in parallel is analogous to the calculations of total parallel resistance (Chapter 5) and total *series* capacitance (Chapter 9). For series-parallel combinations of inductors, determine the total inductance in the same way as total resistance in series-parallel resistive circuits (Chapter 6).

EXAMPLE 11–3

Determine L_T in Figure 11–18.

FIGURE 11–18

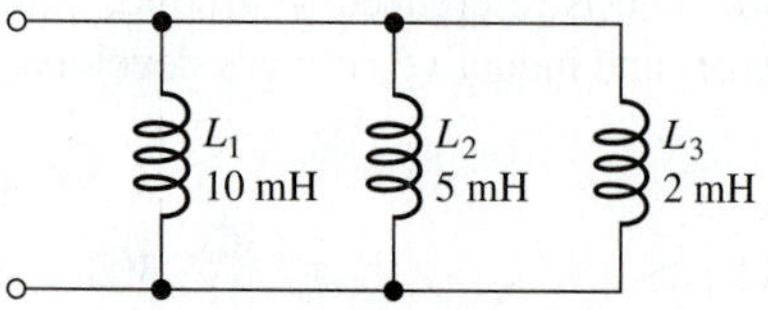

L_1 10 mH L_2 5 mH L_3 2 mH

SOLUTION

$$L_T = \frac{1}{\dfrac{1}{L_1} + \dfrac{1}{L_2} + \dfrac{1}{L_3}} = \frac{1}{\dfrac{1}{10 \text{ mH}} + \dfrac{1}{5 \text{ mH}} + \dfrac{1}{2 \text{ mH}}} = \frac{1}{0.8 \text{ mH}} = \textbf{1.25 mH}$$

RELATED PROBLEM

Determine L_T for the following inductors in parallel: 50 μH, 80 μH, 100 μH, and 150 μH.

SECTION 11–3 CHECKUP

1. State the rule for combining inductors in series.

2. What is L_T for a series connection of 100 μH, 500 μH, and 2 mH?

3. Five 100 mH coils are connected in series. What is the total inductance?

4. Compare the total inductance in parallel with the smallest-value individual inductor.

5. The calculation of total parallel inductance is analogous to that for total parallel resistance.   (True or False)

6. Determine L_T for each parallel combination:

   (a) 100 mH, 50 mH, and 10 mH

   (b) 40 μH and 60 μH

Energy is stored in the electromagnetic field of an inductor when it is connected to a dc voltage source. The buildup of current through the inductor occurs in a predictable manner, which is dependent on the time constant of the circuit. The time constant is determined by the inductance and resistance in a circuit.

After completing this section, you should be able to

- **Analyze inductive dc switching circuits**
  - **Define *RL time constant***
  - **Describe the increase and decrease of current in an inductor**
  - **Relate the time constant to the energizing and deenergizing of an inductor**
  - **Describe induced voltage**
  - **Write the exponential equations for current in an inductor**

When there is constant direct current in an inductor, there is no induced voltage. There is, however, a voltage drop due to the winding resistance of the coil. The inductance itself appears as a short to dc. Energy is stored in the magnetic field according to the formula $W = \frac{1}{2}LI^2$. The only energy conversion to heat occurs in the winding resistance ($P = I^2R_W$). This condition is illustrated in Figure 11–19.

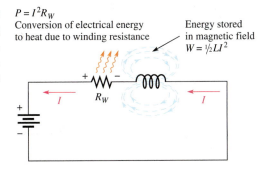

$P = I^2R_W$
Conversion of electrical energy to heat due to winding resistance

Energy stored in magnetic field $W = \frac{1}{2}LI^2$

FIGURE 11–19   Energy storage and conversion to heat in an inductor.

The *RL* Time Constant

Because the inductor's basic action is to develop a voltage that opposes a change in its current, it follows that current cannot change instantaneously in an inductor. A certain time is required for the current to make a change from one value to another. The rate at which the current changes is determined by the *RL* time constant.

> The ***RL* time constant** is a fixed time interval that equals the ratio of the inductance to the resistance.

The formula is

$$\tau = \frac{L}{R} \qquad\qquad (11\text{–}5)$$

where τ is in seconds when inductance (L) is in henries and resistance (R) is in ohms.

EXAMPLE 11–4

A series *RL* circuit has a resistance of 1.0 kΩ and an inductance of 1.0 mH. What is the time constant?

SOLUTION

$$\tau = \frac{L}{R} = \frac{1.0\,\text{mH}}{1.0\,\text{k}\Omega} = \frac{1.0 \times 10^{-3}\,\text{H}}{1.0 \times 10^{3}\,\Omega} = 1.0 \times 10^{-6}\,\text{s} = \mathbf{1.0\ \mu s}$$

RELATED PROBLEM

Find the time constant for $R = 2.2$ kΩ and $L = 500$ μH.

Current in an Inductor

INCREASING CURRENT

In a series *RL* circuit, the current will increase to approximately 63% of its full value in one time-constant interval after voltage is applied. This buildup of current (analogous to the buildup of capacitor voltage during the charging in an *RC* circuit) follows an exponential curve and reaches the approximate percentage of final current as indicated in Table 11–1 and as illustrated in Figure 11–20.

TABLE 11–1 • Percentage of final current after each time-constant interval during current buildup.	
NUMBER OF TIME CONSTANTS	**APPROXIMATE % OF FINAL CURRENT**
1	63
2	86
3	95
4	98
5	99 (considered 100%)

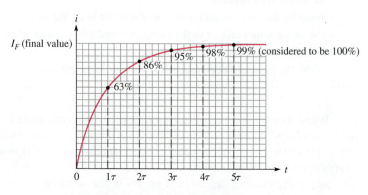

FIGURE 11–20 Increasing current in an inductor.

The change in current over five time-constant intervals is illustrated in Figure 11–21. When the current reaches its final value at approximately 5τ, it ceases to change. At this time, the inductor acts as a short (except for winding resistance) to the constant current. The final value of the current is

$$I_F = \frac{V_S}{R} = \frac{10 \text{ V}}{1.0 \text{ k}\Omega} = 10 \text{ mA}$$

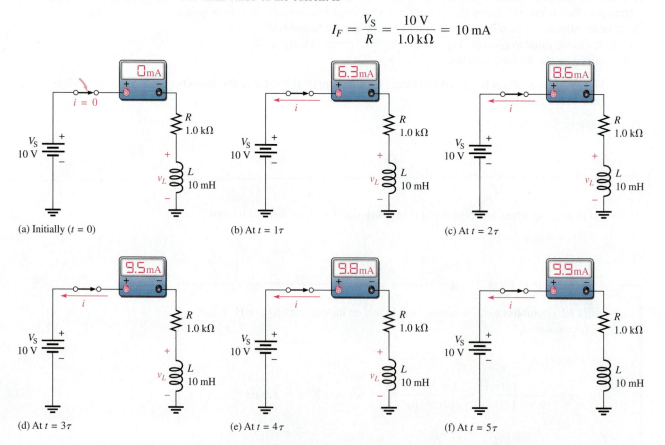

FIGURE 11–21 Illustration of the exponential buildup of current in an inductor. The current increases approximately 63% during each time-constant interval after the switch is closed. A voltage (v_L) is induced in the coil that tends to oppose the increase in current.

EXAMPLE 11–5

Calculate the *RL* time constant for Figure 11–22. Then determine the current and the time at each time-constant interval, measured from the instant the switch is closed.

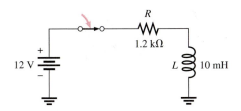

FIGURE 11–22

SOLUTION

The *RL* time constant is

$$\tau = \frac{L}{R} = \frac{10 \text{ mH}}{1.2 \text{ k}\Omega} = \textbf{8.33 } \boldsymbol{\mu}\textbf{s}$$

The final current is

$$I_F = \frac{V_S}{R} = \frac{12 \text{ V}}{1.2 \text{ k}\Omega} = 10 \text{ mA}$$

Use the time-constant percentage values from Table 11–1.

At 1τ: $i = 0.63(10 \text{ mA}) = \textbf{6.3 mA}$; $t = \textbf{8.33 } \boldsymbol{\mu}\textbf{s}$

At 2τ: $i = 0.86(10 \text{ mA}) = \textbf{8.6 mA}$; $t = \textbf{16.7 } \boldsymbol{\mu}\textbf{s}$

At 3τ: $i = 0.95(10 \text{ mA}) = \textbf{9.5 mA}$; $t = \textbf{25.0 } \boldsymbol{\mu}\textbf{s}$

At 4τ: $i = 0.98(10 \text{ mA}) = \textbf{9.8 mA}$; $t = \textbf{33.3 } \boldsymbol{\mu}\textbf{s}$

At 5τ: $i = 0.99(10 \text{ mA}) = 9.9 \text{ mA} \cong \textbf{10 mA}$; $t = \textbf{41.7 } \boldsymbol{\mu}\textbf{s}$

RELATED PROBLEM

Repeat the calculations if $R = 680 \text{ }\Omega$ and $L = 100 \text{ }\mu\text{H}$.

DECREASING CURRENT   Current in an inductor decreases exponentially according to the approximate percentage values shown in Table 11–2 and in Figure 11–23.

TABLE 11–2 • Percentage of initial current after each time-constant interval while current is decreasing.

NUMBER OF TIME CONSTANTS	APPROXIMATE % OF INITIAL CURRENT
1	37
2	14
3	5
4	2
5	1 (considered 0)

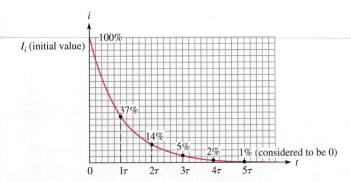

FIGURE 11–23 **Decreasing current in an inductor.**

The change in current over five time-constant intervals is illustrated in Figure 11–24. When the current reaches its final value of approximately 0 A, it ceases to change. Before the switch is opened, the current through L is at a constant value of 10 mA, which is determined by R_1 because L acts ideally as a short. When the switch is opened, the induced inductor voltage initially provides 10 mA through R_2. The current then decreases by 63% during each time constant interval.

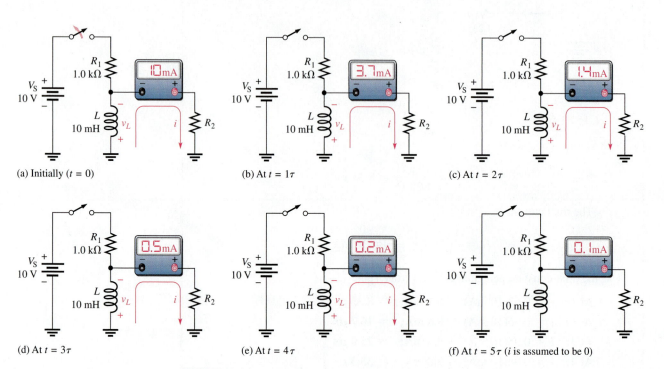

(a) Initially ($t = 0$)

(b) At $t = 1\tau$

(c) At $t = 2\tau$

(d) At $t = 3\tau$

(e) At $t = 4\tau$

(f) At $t = 5\tau$ (i is assumed to be 0)

FIGURE 11–24

Response to a Square Wave

A good way to demonstrate both increasing and decreasing current in an RL circuit is to use a square wave voltage as the input. The square wave is a useful signal for observing the dc response of a circuit because it provides on and off action similar to a switch. Pulse response will be covered further in Chapter 15. When the square wave goes from its low level to its high level, the current in the circuit responds by exponentially increasing to its final value. When the square wave returns to the zero level, the current in the circuit responds by exponentially decreasing to its zero value. Figure 11–25 shows input voltage and current waveforms. The current will have the same shape as the voltage across the resistor, so a two-channel difference measurement can show the waveform.

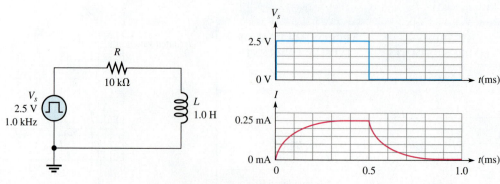

FIGURE 11–25

EXAMPLE 11–6

For the circuit in Figure 11–25, what is the current at 0.1 ms and at 0.6 ms?

SOLUTION

The *RL* time constant for the circuit is

$$\tau = \frac{L}{R} = \frac{1.0 \, \text{H}}{10 \, \text{k}\Omega} = 0.1 \, \text{ms}$$

If the square wave generator period is long enough for the current to reach its maximum value in 5τ, the current will increase exponentially and during each time constant interval will have a value equal to the percentage of the final current given in Table 11–1. The final current is

$$I_F = \frac{V_s}{R} = \frac{2.5 \, \text{V}}{10 \, \text{k}\Omega} = 0.25 \, \text{mA}$$

The current at 0.1 ms is

$$i = 0.63(0.25 \, \text{mA}) = \mathbf{0.158 \, mA}$$

At 0.6 ms, the square wave input has been at the 0 V level for 0.1 ms, or 1τ; and, the current decreases from the maximum value toward its final value of 0 mA by 63%. Therefore,

$$i = 0.25 \, \text{mA} - 0.63(0.25 \, \text{mA}) = \mathbf{0.092 \, mA}$$

RELATED PROBLEM

What is the current at 0.2 ms and 0.8 ms?

Voltages in the Series *RL* Circuit

As you know, when current changes in an inductor, a voltage is induced. Let's examine what happens to the induced voltage across the inductor in the series circuit in Figure 11–26 during one complete cycle of a square wave input. Keep in mind that the generator produces a level that is like switching a dc source on and then puts an "automatic" low resistance (ideally zero) path across the source when it returns to its zero level.

An ammeter placed in the circuit shows the current in the circuit at any instant in time. The V_L waveform is the voltage across the inductor. In Figure 11–26(a), the square wave has just transitioned from zero to its maximum value of 2.5 V. In accordance with Lenz's law, a voltage is induced across the inductor that opposes this *change* as the magnetic field surrounding the inductor builds up. There is no current in the circuit due to the equal but opposing voltages.

As the magnetic field builds up, the induced voltage across the inductor decreases, and current is in the circuit. After 1τ, the induced voltage across the inductor has decreased by 63%, which causes the current to increase by 63% to 0.158 mA. This is shown in Figure 11–26(b) at the end of one time constant (0.1 ms).

The voltage on the inductor continues to exponentially decrease to zero, at which point the current is limited only by the circuit resistance. Then the square wave goes back to zero (at $t = 0.5$ ms) as shown in Figure 11–26(c). Again a voltage is induced across the inductor opposing this change. This time, the polarity of the inductor voltage is reversed due to the collapsing magnetic field. Although the source voltage is 0, the collapsing magnetic field maintains current in the same direction until the current decreases to zero, as shown in Figure 11–26(d).

The voltage across the resistor, V_R, in Figure 11–26(d) is found by subtracting the voltage across the inductor, V_L, from the source voltage, V_s, according to Kirchhoff's voltage law. The shape of the V_R waveform is the same as the current waveform in Figure 11–25.

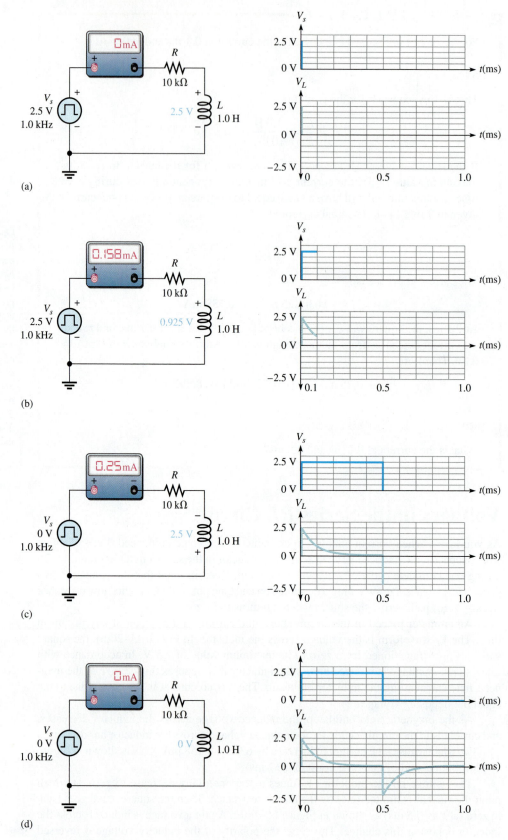

FIGURE 11–26

EXAMPLE 11–7

(a) The circuit in Figure 11–27 has a square wave input. What is the highest frequency that can be used and still observe the complete waveform across the inductor?

(b) Assume the generator is set to the frequency determined in (a). Describe the voltage waveform across the resistor?

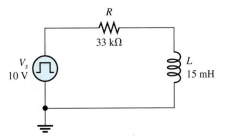

FIGURE 11–27

SOLUTION

(a) $\tau = \dfrac{L}{R} = \dfrac{15\text{ mH}}{33\text{ k}\Omega} = 0.454\ \mu s$

The period needs to be ten times longer than τ to observe the entire wave.

$$T = 10\tau = 4.54\ \mu s$$

$$f = \frac{1}{T} = \frac{1}{4.54\ \mu s} = \textbf{220 kHz}$$

(b) The voltage across the resistor has the same shape as the current waveform. The general shape was shown in Figure 11–26 and has a maximum value of 10 V (the same as V_s, assuming no winding resistance).

MULTISIM

Open Multisim file E11-07. Set the frequency to the calculated value and observe the voltage waveform across the inductor.

RELATED PROBLEM

What is the maximum voltage across the resistor for $f = 220$ kHz?

The Exponential Formulas

The formulas for the exponential current and voltage in an *RL* circuit are similar to those used in Chapter 9 for the *RC* circuit, and the universal exponential curves in Figure 9–36 apply to inductors as well as capacitors. The general formulas for *RL* circuits are stated as follows:

$$v = V_F + (V_i - V_F)e^{-Rt/L} \tag{11–6}$$

$$i = I_F + (I_i - I_F)e^{-Rt/L} \tag{11–7}$$

where V_F and I_F are the final values, V_i and I_i are the initial values, and v and i are the instantaneous values of the inductor voltage or current at time t.

INCREASING CURRENT   The formula for the special case in which an increasing exponential current curve begins at zero ($I_i = 0$) is

$$i = I_F (1 - e^{-Rt/L}) \tag{11–8}$$

Using Equation 11–8, you can calculate the value of the increasing inductor current at any instant of time. You can calculate voltage by substituting v for i and V_F for I_F in Equation 11–8. The exponent $-Rt/L$ can also be written as $-t/(L/R) = -t/\tau$.

EXAMPLE 11–8

In Figure 11–28, determine the inductor current 30 μs after the switch is closed.

FIGURE 11–28

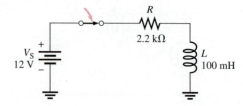

SOLUTION

The RL time constant is

$$\tau = \frac{L}{R} = \frac{100 \text{ mH}}{2.2 \text{ k}\Omega} = 45.5 \ \mu s$$

The final current is

$$I_F = \frac{V_S}{R} = \frac{12 \text{ V}}{2.2 \text{ k}\Omega} = 5.45 \text{ mA}$$

The initial current is zero. Notice that 30 μs is less than one time constant, so the current will reach less than 63% of its final value in that time.

$$i_L = I_F(1 - e^{-Rt/L}) = 5.45 \text{ mA}(1 - e^{-0.66}) = \textbf{2.63 mA}$$

RELATED PROBLEM

In Figure 11–28, determine the inductor current 55 μs after the switch is closed.

DECREASING CURRENT The formula for the special case in which a decreasing exponential current has a final value of zero is

$$i = I_i e^{-Rt/L} \tag{11–9}$$

This formula can be used to calculate the decreasing inductor current at any instant.

EXAMPLE 11–9

In Figure 11–29, what is the current at each microsecond interval for one complete cycle of the input square wave as shown? After calculating the current at each time, sketch the current waveform.

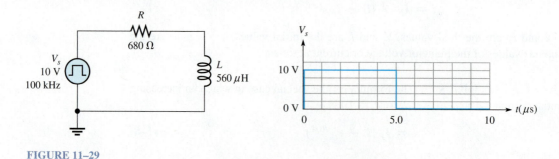

FIGURE 11–29

SOLUTION

The *RL* time constant is

$$\tau = \frac{L}{R} = \frac{560 \ \mu\text{H}}{680 \ \Omega} = 0.824 \ \mu\text{s}$$

When the pulse goes from 0 V to 10 V at $t = 0$, the current increases exponentially. The final current is

$$I_F = \frac{V_s}{R} = \frac{10 \ \text{V}}{680 \ \Omega} = 14.7 \ \text{mA}$$

For the increasing current, $i = I_F(1 - e^{-Rt/L}) = I_F(1 - e^{-t/\tau})$.

At 1 μs: $i = 14.7 \ \text{mA}(1 - e^{-1\mu\text{s}/0.824\mu\text{s}}) = $ **10.3 mA**

At 2 μs: $i = 14.7 \ \text{mA}(1 - e^{-2\mu\text{s}/0.824\mu\text{s}}) = $ **13.4 mA**

At 3 μs: $i = 14.7 \ \text{mA}(1 - e^{-3\mu\text{s}/0.824\mu\text{s}}) = $ **14.3 mA**

At 4 μs: $i = 14.7 \ \text{mA}(1 - e^{-4\mu\text{s}/0.824\mu\text{s}}) = $ **14.6 mA**

At 5 μs: $i = 14.7 \ \text{mA}(1 - e^{-5\mu\text{s}/0.824\mu\text{s}}) = $ **14.7 mA**

When the pulse goes from 10 V to 0 V at $t = 5 \ \mu$s, the current decreases exponentially.

For the decreasing current,

$$i = I_i(e^{-Rt/L}) = I_i(e^{-t/\tau})$$

The initial current is the value at 5 μs, which is 14.7 mA.

At 6 μs: $i = 14.7 \ \text{mA}(e^{-1\mu\text{s}/0.824\mu\text{s}}) = $ **4.37 mA**

At 7 μs: $i = 14.7 \ \text{mA}(e^{-2\mu\text{s}/0.824\mu\text{s}}) = $ **1.30 mA**

At 8 μs: $i = 14.7 \ \text{mA}(e^{-3\mu\text{s}/0.824\mu\text{s}}) = $ **0.38 mA**

At 9 μs: $i = 14.7 \ \text{mA}(e^{-4\mu\text{s}/0.824\mu\text{s}}) = $ **0.11 mA**

At 10 μs: $i = 14.7 \ \text{mA}(e^{-5\mu\text{s}/0.824\mu\text{s}}) = $ **0.03 mA**

Figure 11–30 is a graph of these results.

FIGURE 11–30

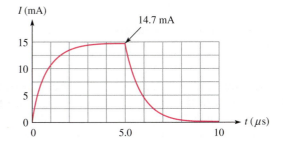

MULTISIM

Open Multisim file E11-09. Measure the current by placing a very small-value series resistor between the inductor and ground and observing the voltage across it.

RELATED PROBLEM

What is the current at 0.5 μs?

SECTION 11–4 CHECKUP

1. A 15 mH inductor with a winding resistance of 10 Ω has a constant direct current of 10 mA through it. What is the voltage drop across the inductor?

2. A 20 V dc source is connected to a series *RL* circuit with a switch. At the instant of switch closure, what are the values of i and v_L?

3. In the same circuit as in Question 2, after a time interval equal to 5τ from switch closure, what is v_L?

4. In a series *RL* circuit where $R = 1.0 \ \text{k}\Omega$ and $L = 500 \ \mu$H, what is the time constant? Determine the current 0.25 μs after a switch connects 10 V across the circuit.

An inductor passes ac but with an amount of opposition that depends on the frequency of the ac.

After completing this section, you should be able to

- **Analyze inductive ac circuits**
  - **Define** *inductive reactance*
  - **Determine the value of inductive reactance in a given circuit**
  - **Discuss instantaneous, true, and reactive power in an inductor**

Inductive Reactance, X_L

In Figure 11–31, an inductor is shown connected to a sinusoidal voltage source. When the source voltage is held at a constant amplitude value and its frequency is increased, the amplitude of the current decreases. Also, when the frequency of the source is decreased, the current amplitude increases.

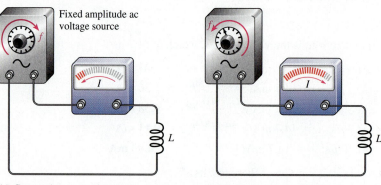

(a) Current decreases when the frequency increases.

(b) Current increases when the frequency decreases.

FIGURE 11–31   **The current in an inductive circuit varies inversely with the frequency of the source voltage.**

When the frequency of the source voltage increases, its rate of change also increases, as you already know. Now, if the frequency of the source voltage is increased, the frequency of the current also increases. According to Faraday's and Lenz's laws, this increase in frequency induces more voltage across the inductor in a direction to oppose the current and causes it to decrease in amplitude. Similarly, a decrease in frequency will cause an increase in current.

A decrease in the amount of current with an increase in frequency for a fixed amount of voltage indicates that opposition to the current has increased. Thus, the inductor offers opposition to current, and that opposition varies *directly* with frequency.

Inductive reactance is the opposition to sinusoidal current in an inductor.

The symbol for inductive reactance is X_L, and its unit is the ohm (Ω).

You have just seen how frequency affects the opposition to current (inductive reactance) in an inductor. Now let's see how the inductance, L, affects the reactance. Figure 11–32(a) shows that when a sinusoidal voltage with a fixed amplitude and fixed frequency is applied to a 1 mH inductor, there is a certain amount of alternating current. When the inductance value is increased to 2 mH, the current decreases, as shown in part (b). Thus, when the inductance is increased, the opposition to current (inductive reactance) increases. So not only is the inductive reactance *directly* proportional to frequency, but it is also *directly* proportional to inductance. This relationship can be stated as follows:

X_L is proportional to fL.

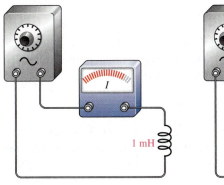
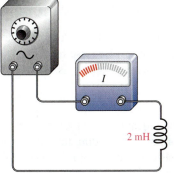

(a) Less inductance, more current.

(b) More inductance, less current.

FIGURE 11–32 **For a fixed voltage and fixed frequency, the current varies inversely with the inductance value.**

Gyrators

Inductors tend to be nonideal components and larger inductors do not lend themselves to integrated circuits because of physical size. A circuit that mimics an inductor is useful for active filter applications and network synthesis. The circuit uses resistors and capacitors and two operational amplifiers to form an equivalent circuit for an inductor. A gyrator can be included within an IC.

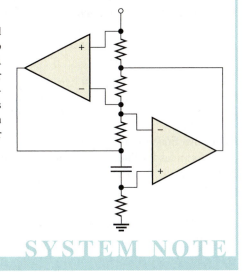

SYSTEM NOTE

It can be proven that the constant of proportionality is 2π, so the formula for inductive reactance (X_L) is

$$X_L = 2\pi f L \qquad (11\text{--}10)$$

X_L is in ohms when f is in hertz and L is in henries. As with capacitive reactance, the 2π term comes from the relationship of the sine wave to rotational motion.

EXAMPLE 11–10

A sinusoidal voltage is applied to the circuit in Figure 11–33. The frequency is 10 kHz. Determine the inductive reactance.

SOLUTION

Convert 10 kHz to 10×10^3 Hz and 5 mH to 5×10^{-3} H. Then, the inductive reactance is

FIGURE 11–33

$$X_L = 2\pi f L = 2\pi(10 \times 10^3 \text{ Hz})(5 \times 10^{-3} \text{ H}) = \textbf{314 } \boldsymbol{\Omega}$$

RELATED PROBLEM

What is X_L in Figure 11–33 if the frequency is increased to 35 kHz?

Reactance for Series Inductors

As given in Equation 11–3, the total inductance of series inductors is the sum of the individual inductances. Because reactance is directly proportional to the inductance, the total reactance of series inductors is the sum of the individual reactances.

$$X_{L(tot)} = X_{L1} + X_{L2} + X_{L3} + \cdots + X_{Ln} \qquad (11\text{--}11)$$

Notice that Equation 11–11 has the same form as Equation 11–3. It also has the same form as the formulas for finding the total opposition to current such as the total resistance of series resistors or the total reactance of series capacitors. When combining the resistance or reactance of the same type of component in series (resistors, inductors, or capacitors), you simply add the individual oppositions to obtain the total.

Reactance for Parallel Inductors

In an ac circuit with parallel inductors, Equation 11–4 was given to find the total inductance. It stated that the total inductance is the reciprocal of the sum of the reciprocals of the inductors. Likewise, the total inductive reactance is the reciprocal of the sum of the reciprocals of the individual reactances.

$$X_{L(tot)} = \cfrac{1}{\cfrac{1}{X_{L1}} + \cfrac{1}{X_{L2}} + \cfrac{1}{X_{L3}} + \cdots + \cfrac{1}{X_{Ln}}} \qquad (11\text{--}12)$$

Notice that Equation 11–12 has the same form as Equation 11–4. It also has the same form as the formulas for finding the total resistance of parallel resistors or the total reactance of parallel capacitors. When combining the resistance or reactance of the same type of component in parallel (resistors, inductors, or capacitors), you take the reciprocal of the sum of the reciprocals to obtain the total opposition.

For two inductors in parallel, Equation 11–12 can be reduced to the product-over-sum form.

$$X_{L(tot)} = \frac{X_{L1}X_{L2}}{X_{L1} + X_{L2}}$$

EXAMPLE 11–11

What is the total inductive reactance of each circuit of Figure 11–34?

FIGURE 11–34

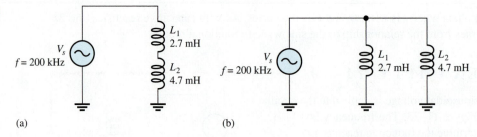

(a) (b)

SOLUTION

The reactances of the individual inductors are the same in both circuits.

$$X_{L1} = 2\pi f L_1 = 2\pi(200\text{ kHz})(2.7\text{ mH}) = 3.39\text{ k}\Omega$$

$$X_{L2} = 2\pi f L_2 = 2\pi(200\text{ kHz})(4.7\text{ mH}) = 5.91\text{ k}\Omega$$

For the series inductors in Figure 11–34(a), the total reactance is the sum of X_{L1} and X_{L2}, as given in Equation 11–11.

$$X_{L(tot)} = X_{L1} + X_{L2} = 3.39\text{ k}\Omega + 5.91\text{ k}\Omega = \mathbf{9.30\text{ k}\Omega}$$

For the inductors in parallel in Figure 11–34(b), determine the total reactance from the product-over-sum rule using X_{L1} and X_{L2}.

$$X_{L(tot)} = \frac{X_{L1}X_{L2}}{X_{L1} + X_{L2}} = \frac{(3.39 \text{ k}\Omega)(5.91 \text{ k}\Omega)}{3.39 \text{ k}\Omega + 5.91 \text{ k}\Omega} = \textbf{2.15 k}\boldsymbol{\Omega}$$

You can also obtain the total reactance for either series or parallel inductors by first finding the total inductance and then substituting in Equation 11–10 to find the total reactance.

For the series inductors,

$$L_T = L_1 + L_2 = 2.7 \text{ mH} + 4.7 \text{ mH} = 7.4 \text{ mH}$$

$$X_{L(tot)} = 2\pi f L_T = 2\pi(200 \text{ kHz})(7.4 \text{ mH}) = \textbf{9.30 k}\boldsymbol{\Omega}$$

For the parallel inductors,

$$L_T = \frac{L_1 L_2}{L_1 + L_2} = \frac{(2.7 \text{ mH})(4.7 \text{ mH})}{2.7 \text{ mH} + 4.7 \text{ mH}} = 1.71 \text{ mH}$$

$$X_{L(tot)} = 2\pi f L_T = 2\pi(200 \text{ kHz})(1.71 \text{ mH}) = \textbf{2.15 k}\boldsymbol{\Omega}$$

RELATED PROBLEM

What is the total inductive reactance for each circuit in Figure 11–34 if $L_1 = 1$ mH and L_2 is unchanged?

OHM'S LAW   The reactance of an inductor is analogous to the resistance of a resistor. In fact, X_L, just like X_C and R, is expressed in ohms. Since inductive reactance is a form of opposition to current, Ohm's law applies to inductive circuits as well as to resistive circuits and capacitive circuits; and it is stated as follows:

$$I = \frac{V}{X_L} \qquad (11\text{–}13)$$

When applying Ohm's law in ac circuits, you must express both the current and the voltage in the same way, that is, both in rms, both in peak, and so on.

EXAMPLE 11–12

Determine the rms current in Figure 11–35.

FIGURE 11–35

V_s
5 V rms
$f = 10$ kHz

L
100 mH

SOLUTION

Convert 10 kHz to 10×10^3 Hz and 100 mH to 100×10^{-3} H. Then calculate X_L.

$$X_L = 2\pi f L = 2\pi(10 \times 10^3 \text{ Hz})(100 \times 10^{-3} \text{ H}) = 6283 \ \Omega$$

Apply Ohm's law.

$$I_{rms} = \frac{V_{rms}}{X_L} = \frac{5 \text{ V}}{6283 \ \Omega} = \textbf{796 } \boldsymbol{\mu}\textbf{A}$$

RELATED PROBLEM

Determine the rms current in Figure 11–35 for the following values: $V_{rms} = 12$ V, $f = 4.9$ kHz, and $L = 680 \ \mu$H.

MULTISIM

Open Multisim file E11-12. Measure the rms current and compare to the calculated value. Change the circuit values to those given in the related problem and measure the rms current.

Current Lags Inductor Voltage by 90°

As you know, a sinusoidal voltage has a maximum rate of change at its zero crossings and a zero rate of change at the peaks. From Faraday's law you know that the amount of voltage induced across a coil is directly proportional to the rate at which the current is changing. Therefore, the coil voltage is maximum at the zero crossings of the current where the rate of change of the current is the greatest. Also, the amount of voltage is zero at the peaks of the current where the rate of change is zero. This phase relationship is illustrated in Figure 11–36. As you can see, the current peaks occur a quarter cycle after the voltage peaks. Thus, the current *lags* the voltage by 90°. Recall that in a capacitor, the current *leads* the voltage by 90°.

FIGURE 11–36   Current is always lagging the inductor voltage by 90°.

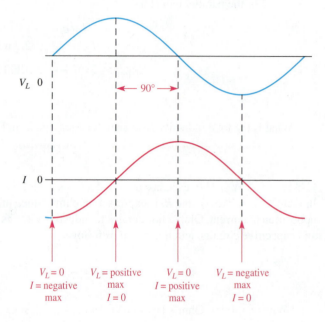

Power in an Inductor

As discussed earlier, an inductor stores energy in its magnetic field when there is current through it. An ideal inductor (assuming no winding resistance) does not dissipate energy; it only stores it. When an ac voltage is applied to an inductor, energy is stored by the inductor during a portion of the cycle; then the stored energy is returned to the source during another portion of the cycle. No net energy is lost in an ideal inductor due to conversion to heat. Figure 11–37 shows the power curve that results from one cycle of inductor current and voltage. Compare the power curve for an inductor with that of a capacitor in Figure 9–50. Notice that main difference is that voltage and current are interchanged between the plots.

FIGURE 11–37   Power curve for an inductor.

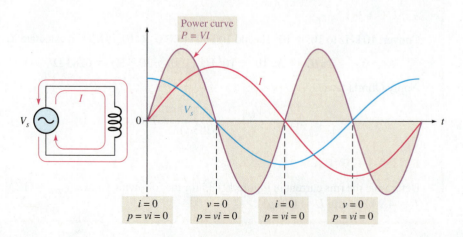

INSTANTANEOUS POWER (p)   The product of instantaneous voltage, v, and instantaneous current, i, gives instantaneous power, p. At points where v or i is zero, p is also zero. When both v and i are positive, p is also positive. When either v or i is positive and the other negative, p is negative. When both v and i are negative, p is positive. As you can see in Figure 11–37, the power follows a sinusoidal-shaped curve. Positive values of power indicate that energy is stored by the inductor. Negative values of power indicate that energy is returned from the inductor to the source. Note that the power fluctuates at a frequency twice that of the voltage or current, as energy is alternately stored and returned to the source.

TRUE POWER (P_{true})   Ideally, all of the energy stored by an inductor during the positive portion of the power cycle is returned to the source during the negative portion. No net energy is lost due to conversion to heat in the inductor, so the power is zero. Actually, because of winding resistance in a practical inductor, some power is always dissipated, and there is a very small amount of true power, which can normally be neglected.

$$P_{true} = I_{rms}^2 R_W \qquad (11\text{–}14)$$

REACTIVE POWER (P_r)   The rate at which an inductor stores or returns energy is called its **reactive power**, P_r, with a unit of VAR (volt-ampere reactive). The reactive power is a nonzero quantity because at any instant in time the inductor is actually taking energy from the source or returning energy to it. Reactive power does not represent an energy loss due to conversion to heat. The following formulas apply:

$$P_r = V_{rms} I_{rms} \qquad (11\text{–}15)$$

$$P_r = \frac{V_{rms}^2}{X_L} \qquad (11\text{–}16)$$

$$P_r = I_{rms}^2 X_L \qquad (11\text{–}17)$$

EXAMPLE 11–13

A 10 V rms signal with a frequency of 10 kHz is applied to a 10 mH coil with a winding resistance of 40 Ω. Determine the reactive power (P_r) and the true power (P_{true}).

SOLUTION

First, calculate the inductive reactance and current values.

$$X_L = 2\pi f L = 2\pi(10\text{ kHz})(10\text{ mH}) = 628\ \Omega$$

$$I = \frac{V_s}{X_L} = \frac{10\text{ V}}{628\ \Omega} = 15.9\text{ mA}$$

Then using Equation 11–17,

$$P_r = I^2 X_L = (15.9\text{ mA})^2(628\ \Omega) = \mathbf{159\ mVAR}$$

The true power is

$$P_{true} = I^2 R_W = (15.9\text{ mA})^2(40\ \Omega) = \mathbf{10.1\ mW}$$

RELATED PROBLEM

What happens to the reactive power if the frequency increases?

The Quality Factor (Q)

The **quality factor (Q)** is the ratio of the reactive power in the inductor to the true power in the winding resistance of the coil. It is a ratio of the power in L to the power in R_W. The quality factor is important in resonant circuits, which are studied in Chapter 13. A formula for Q is developed as follows:

$$Q = \frac{\text{reactive power}}{\text{true power}} = \frac{P_r}{P_{\text{true}}} = \frac{I^2 X_L}{I^2 R_W}$$

The I^2 terms cancel, leaving

$$Q = \frac{X_L}{R_W} \qquad\qquad (11\text{–}18)$$

Note that Q is a ratio of like units and, therefore, has no unit itself. The quality factor is also known as unloaded Q because it is defined with no load across the coil. Notice also that Q depends on the frequency because X_L is dependent on frequency.

SECTION 11–5 CHECKUP

1. State the phase relationship between current and voltage in an inductor.

2. Calculate X_L for $f = 500$ kHz and $L = 1.0$ mH

3. At what frequency is the reactance of a 50 μH inductor equal to 800 Ω?

4. Calculate the rms current in Figure 11–38.

5. An ideal 50 mH inductor is connected to a 12 V rms source. What is the true power? What is the reactive power at a frequency of 1 kHz?

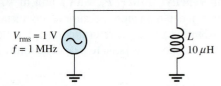

$V_{\text{rms}} = 1$ V
$f = 1$ MHz

L
$10\,\mu$H

FIGURE 11–38

11–6 INDUCTOR APPLICATIONS

Inductors are not as versatile as capacitors and tend to be more limited in their application due, in part, to size, cost factors, and nonideal behavior (internal resistance, etc.). One of the most common applications for inductors is noise reduction applications.

After completing this section, you should be able to

- **Discuss some inductor applications**
  - **Discuss two ways in which noise enters a circuit**
  - **Describe the suppression of electromagnetic interference (EMI)**
  - **Explain how a ferrite bead is used**
  - **Discuss the basics of tuned circuits**

Noise Suppression

One of the most important applications of inductors has to do with suppressing unwanted electrical noise. The inductors used in these applications are generally wound on a closed core to avoid having the inductor become a source of radiated noise itself. Two types of noise are conductive noise and radiated noise.

CONDUCTIVE NOISE   Many systems have common conductive paths connecting different parts of the system, which can conduct high frequency noise from one part of the system to another. Consider the case of two circuits connected with common lines as shown in Figure 11–39(a). A path for high frequency noise exists though the common grounds, creating a condition known as a *ground loop*. Ground loops are particularly a problem in instrumentation systems, where a transducer may be located a distance from the recording system and noise current in the ground can affect the signal.

If the signal of interest changes slowly, a special inductor, called a *longitudinal choke,* can be installed in the signal line as shown in Figure 11–39(b). The longitudinal choke is a form of transformer (covered in Chapter 14) that acts as an inductor in each signal line. The ground loop sees a high impedance path, thus reducing the noise, while the low-frequency signal is coupled through the low impedance of the choke.

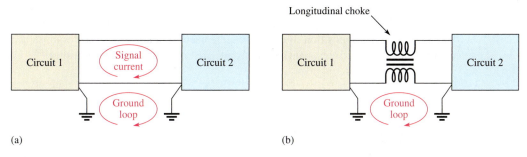

(a)   (b)

FIGURE 11–39

Switching circuits also tend to generate high-frequency noise (above 10 MHz) by virtue of the high-frequency components present. (Recall from Section 8–5 that a fast pulse contains many high-frequency harmonics.) Certain types of power supplies use high-speed switching circuits, which are a source of both conductive and radiated noise. Because an inductor's impedance increases with frequency, inductors are good for blocking electrical noise from these supplies, which should carry only dc. Inductors are frequently installed in the power supply lines to suppress this conductive noise so that one circuit does not adversely affect another circuit. One or more capacitors may also be used in conjunction with the inductor to improve filtering action.

RADIATED NOISE   Noise can also enter a circuit by way of an electromagnetic field. The noise source can be an adjacent circuit or a nearby power supply. There are several approaches to reducing the effects of radiated noise. Usually, the first step is to determine the cause of the noise and isolate it using shielding or filtering. Inductors are widely used in filters that are used to suppress radio-frequency noise. The inductor used for noise suppression must be carefully selected so as not to become a source of radiated noise itself. For high frequencies (>20 MHz), inductors wound on highly permeable toroidal cores are widely used, as they tend to keep the magnetic flux restricted to the core.

HANDS ON TIP
When breadboarding circuits that include small inductors, it is best to use encapsulated inductors for structural strength. Inductors generally have extremely fine coil wire that is connected to a much larger size of lead wires.

In unencapsulated inductors, these contact points are very vulnerable to breaking if the inductor is frequently inserted and removed from a protoboard.

RF Chokes

Inductors used for the purpose of blocking very high frequencies are called *radio frequency (RF) chokes*. RF chokes are used for conductive or radiated noise. They are special inductors designed to block high frequencies from getting into or leaving parts of a system by providing a high impedance path for high frequencies. In general, the choke is placed in series with the line for which RF suppression is required. Depending on the frequency of the interference, different types of chokes are required. A common type of electromagnetic interference (EMI) filter wraps the signal line on a toroidal core several times. The toroidal configuration is desired because it contains the magnetic field so that the choke does not become a source of noise itself.

Another common type of RF choke is a ferrite bead. All wires have inductance, and the ferrite bead is a small ferromagnetic material that is strung onto the wire to increase its inductance. The impedance presented by the bead is a function of both the material and the frequency, as well as the size of the bead. It is an effective and inexpensive "choke" for high frequencies. Ferrite beads are common in high-frequency communication systems. Sometimes several are strung together in series to increase the effective inductance.

Tuned Circuits

Inductors are used in conjunction with capacitors to provide frequency selection in communications systems. These tuned circuits allow a narrow band of frequencies to be selected while all other frequencies are rejected. The tuners in your TV and radio receivers are based on this principle and permit you to select one channel or station out of the many that are available.

Frequency selectivity is based on the fact that the reactances of both capacitors and inductors depend on the frequency and on the interaction of these two components when connected in series or parallel. Since the capacitor and the inductor produce opposite phase shifts, their combined opposition to current can be used to obtain a desired response at a selected frequency. Tuned (resonant) *RLC* circuits are covered in detail in Chapter 13.

SECTION 11–6 CHECKUP

1. Name two types of unwanted noise.

2. What do the letters EMI stand for?

3. How is a ferrite bead used?

SUMMARY

- Inductance is a measure of a coil's ability to establish an induced voltage as a result of a change in its current.
- An inductor opposes a change in its own current.
- Faraday's law states that relative motion between a magnetic field and a coil induces a voltage across the coil.
- Lenz's law states that the polarity of induced voltage is such that the resulting induced current is in a direction that opposes the change in the magnetic field that produced it.
- Energy is stored by an inductor in its magnetic field.
- One henry is the amount of inductance when current, changing at the rate of one ampere per second, induces one volt across the inductor.
- Inductance is directly proportional to the square of the turns, the permeability, and the cross-sectional area of the core. It is inversely proportional to the length of the core.
- The permeability of a core material is an indication of the ability of the material to establish a magnetic field.
- Inductors add in series.
- Total parallel inductance is less than that of the smallest inductor in parallel.
- The time constant for a series *RL* circuit is the inductance divided by the resistance.
- In an *RL* circuit, the increasing or decreasing voltage and current in an inductor make an approximately 63% change during each time-constant interval.
- Increasing and decreasing current and voltage follow exponential curves.
- Voltage leads current by 90° in an inductor.
- Inductive reactance (X_L) is directly proportional to frequency and inductance.
- The true power in an inductor is zero; that is, there is no energy conversion to heat in an ideal inductor, only in its winding resistance.

KEY TERMS

Key terms and other bold terms in the chapter are defined in the end-of-book glossary.

Coil   A common term for an inductor.

Henry (H)   The unit of inductance.

Induced voltage   Voltage produced as a result of a changing magnetic field.

Inductance   The property of an inductor whereby a change in current causes the inductor to produce a voltage that opposes the change in current.

Inductive reactance   The opposition of an inductor to sinusoidal current. The unit is the ohm.

Inductor   A passive electrical component, formed by a coil of wire, that exhibits the property of inductance.

Quality factor (Q)   The ratio of reactive power to true power.

RL time constant   A fixed time interval, set by the L and R values, that determines the time response of a circuit.

Winding   The loops or turns of wire in an inductor.

Winding resistance   The resistance of the length of wire that makes up a coil.

KEY FORMULAS

(11–1) $W = \dfrac{1}{2}LI^2$ Energy stored by an inductor

(11–2) $L = \dfrac{N^2 \mu A}{l}$ Inductance in terms of physical parameters

(11–3) $L_T = L_1 + L_2 + L_3 + \cdots + L_n$ Series inductance

(11–3) $L_T = \dfrac{1}{\dfrac{1}{L_1} + \dfrac{1}{L_2} + \dfrac{1}{L_3} + \cdots + \dfrac{1}{L_n}}$ Total parallel inductance

(11–5) $\tau = \dfrac{L}{R}$ RL time constant

(11–6) $v = V_F + (V_i - V_F)e^{-Rt/L}$ Exponential voltage (general)

(11–7) $i = I_F + (I_i - I_F)e^{-Rt/L}$ Exponential current (general)

(11–8) $i = I_F(1 - e^{-Rt/L})$ Increasing exponential current beginning at zero

(11–9) $i = I_i e^{-Rt/L}$ Decreasing exponential current ending at zero

(11–10) $X_L = 2\pi f L$ Inductive reactance

(11–11) $X_{L(tot)} = X_{L1} + X_{L2} + X_{L3} + \cdots + X_{Ln}$ Series inductive reactance

(11–12) $X_{L(tot)} = \dfrac{1}{\dfrac{1}{X_{L1}} + \dfrac{1}{X_{L2}} + \dfrac{1}{X_{L3}} + \cdots + \dfrac{1}{X_{Ln}}}$ Parallel inductive reactance

(11–13) $I = \dfrac{V}{X_L}$ Ohm's law

(11–14) $P_{\text{true}} = I_{\text{rms}}^2 R_W$ True power

(11–15) $P_r = V_{\text{rms}} I_{\text{rms}}$ Reactive power

$$(11\text{–}16) \quad P_r = \frac{V_{rms}^2}{X_L} \qquad\qquad \text{Reactive power}$$

$$(11\text{–}17) \quad P_r = I_{rms}^2 X_L \qquad\qquad \text{Reactive power}$$

$$(11\text{–}18) \quad Q = \frac{X_L}{R_W} \qquad\qquad \text{Quality factor}$$

TRUE/FALSE QUIZ

Answers are at the end of the chapter.

1. Lenz's law states that the amount of voltage induced in a coil is proportional to the rate of change of the magnetic field with respect to the coil.

2. An ideal inductor has no winding resistance.

3. The total inductance of two parallel inductors is equal to the product-over-sum of the individual inductors.

4. The total inductance of parallel inductors is always less than that of the smallest inductor.

5. The time constant of an *RL* circuit is given by the formula $\tau = R/L$.

6. If a series *RL* circuit is connected to a dc source, the maximum current is limited by the total inductance.

7. Kirchhoff's voltage law does not apply to inductive circuits.

8. Inductive reactance is directly proportional to the frequency.

9. In an ac inductive circuit, the current lags the voltage in the inductor.

10. The frequency of the power curve for an inductive circuit is equal to the frequency of the applied voltage.

SELF-TEST

Answers are at the end of the chapter.

1. An inductance of 0.05 μH is larger than
   (a) 0.0000005 H   (b) 0.000005 H   (c) 0.000000008 H   (d) 0.00005 mH

2. An inductance of 0.33 mH is smaller than
   (a) 33 μH   (b) 330 μH   (c) 0.05 mH   (d) 0.0005 H

3. When the current through an inductor increases, the amount of energy stored in the electromagnetic field
   (a) decreases   (b) remains constant   (c) increases   (d) doubles

4. When the current through an inductor doubles, the stored energy
   (a) doubles   (b) quadruples   (c) is halved   (d) does not change

5. The winding resistance of a coil can be decreased by
   (a) reducing the number of turns   (b) using a larger wire
   (c) changing the core material   (d) either answer (a) or (b)

6. The inductance of an iron-core coil increases if
   (a) the number of turns is increased   (b) the iron core is removed
   (c) the length of the core is increased   (d) larger wire is used

7. Four 10 mH inductors are in series. The total inductance is
   (a) 40 mH   (b) 2.5 mH   (c) 40,000 μH   (d) answers (a) and (c)

8. A 1 mH, a 3.3 mH, and a 0.1 mH inductor are connected in parallel. The total inductance is
   (a) 4.4 mH   (b) greater than 3.3 mH
   (c) less than 0.1 mH   (d) answers (a) and (b)

9. An inductor, a resistor, and a switch are connected in series to a 12 V battery. At the instant the switch is closed, the inductor voltage is

   (a) 0 V   (b) 12 V   (c) 6 V   (d) 4 V

10. A sinusoidal voltage is applied across an inductor. When the frequency of the voltage is increased, the current

   (a) decreases   (b) increases   (c) does not change   (d) momentarily goes to zero

11. An inductor and a resistor are in series with a sinusoidal voltage source. The frequency is set so that the inductive reactance is equal to the resistance. If the frequency is increased, then

   (a) $V_R > V_L$ (b) $V_L < V_R$ (c) $V_L = V_R$ (d) $V_L > V_R$

TROUBLESHOOTING: SYMPTOM AND CAUSE

The purpose of these exercises is to help develop thought processes essential to troubleshooting. Answers are at the end of the chapter.

Determine the cause for each set of symptoms. Refer to Figure 11–40.

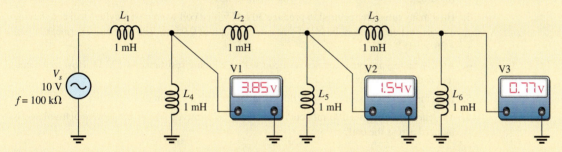

FIGURE 11–40   **The ac meters indicate the correct readings for this circuit.**

1. *Symptom:* All of the voltmeter readings are 0 V.

   Cause:

   (a) The source is off or faulty.
   (b) L_1 is open.
   (c) either (a) or (b)

2. *Symptom:* All of the voltmeter readings are 0 V.

   Cause:

   (a) L_4 is completely shorted.
   (b) L_5 is completely shorted.
   (c) L_6 is completely shorted.

3. *Symptom:* The voltmeter 1 reading is 5 V, and the voltmeter 2 and voltmeter 3 readings are 0 V.

   Cause:

   (a) L_4 is open.
   (b) L_2 is open.
   (c) L_5 is shorted.

4. *Symptom:* The voltmeter 1 reading is 4 V, the voltmeter 2 reading is 2 V, and the voltmeter 3 reading is 0 V.

   Cause:

   (a) L_3 is open.
   (b) L_6 is shorted.
   (c) either (a) or (b)

5. *Symptom:* The voltmeter 1 reading is 4 V, the voltmeter 2 reading is 2 V, and the voltmeter 3 reading is 2 V.

   Cause:

   (a) L_3 is shorted.
   (b) L_6 is open.
   (c) either (a) or (b)

PROBLEMS

Answers to odd-numbered problems are at the end of the book.

BASIC PROBLEMS

SECTION 11–1   The Basic Inductor

1. Convert the following to millihenries:
   (a) 1 H   **(b)** 250 μH
   (c) 10 μH   **(d)** 0.0005 H

2. Convert the following to microhenries:
   (a) 300 mH   **(b)** 0.08 H
   (c) 5 mH   **(d)** 0.00045 mH

3. How many turns are required to produce 30 mH with a coil wound on a cylindrical core having a cross-sectional area of 10×10^{-5} m^2 and a length of 0.05 m? The core has a permeability of 1.26×10^{-6}.

4. A 12 V battery is connected across a coil with a winding resistance of 120 Ω. How much current is there in the coil?

5. How much energy is stored by a 100 mH inductor with a current of 1 A?

6. The current through a 100 mH coil is changing at a rate of 200 mA/s. How much voltage is induced across the coil?

SECTION 11–3   Series and Parallel Inductors

7. Five inductors are connected in series. The lowest value is 5 μH. If the value of each inductor is twice that of the preceding one, and if the inductors are connected in order of ascending values, what is the total inductance?

8. Suppose that you require a total inductance of 50 mH. You have available a 10 mH coil and a 22 mH coil. How much additional inductance do you need?

9. Determine the total parallel inductance for the following coils in parallel: 75 μH, 50 μH, 25 μH, and 15 μH.

10. You have a 12 mH inductor, and it is your smallest value. You need an inductance of 8 mH. What value can you use in parallel with the 12 mH to obtain 8 mH?

11. Determine the total inductance of each circuit in Figure 11–41.

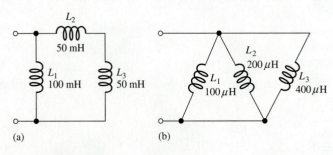

(a)   (b)

FIGURE 11–41

12. Determine the total inductance of each circuit in Figure 11–42.

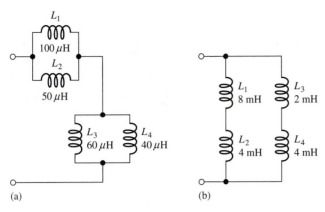

(a)

(b)

FIGURE 11–42

SECTION 11–4 Inductors in DC Circuits

13. Determine the time constant for each of the following series *RL* combinations:
   (a) $R = 100\ \Omega,\ L = 100\ \mu\text{H}$ **(b)** $R = 4.7\ \text{k}\Omega,\ L = 10\ \text{mH}$
   (c) $R = 1.5\ \text{k}\Omega,\ L = 3\ \text{mH}$

14. In a series *RL* circuit, determine how long it takes the current to build up to its full value for each of the following:
   (a) $R = 56\ \Omega,\ L = 50\ \mu\text{H}$ **(b)** $R = 3300\ \Omega,\ L = 15\ \text{mH}$
   (c) $R = 22\ \text{k}\Omega,\ L = 100\ \text{mH}$

15. In the circuit of Figure 11–43, there is initially no current. Determine the inductor voltage at the following times after the switch is closed:
   (a) $10\ \mu\text{s}$ **(b)** $20\ \mu\text{s}$ **(c)** $30\ \mu\text{s}$ **(d)** $40\ \mu\text{s}$ **(e)** $50\ \mu\text{s}$

FIGURE 11–43

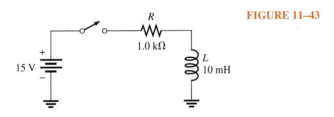

16. In Figure 11–44, calculate the current at each of the following times. Assume an ideal inductor and voltage source.
   (a) $10\ \mu\text{s}$ **(b)** $20\ \mu\text{s}$ **(c)** $30\ \mu\text{s}$

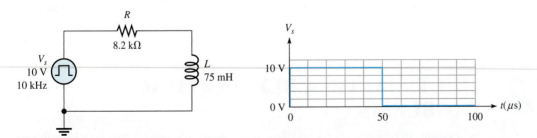

FIGURE 11–44

SECTION 11–6 Inductors in AC Circuits

17. Find the total reactance for each circuit in Figure 11–41 when a voltage with a frequency of 500 kHz is applied across the terminals.

18. Find the total reactance for each circuit in Figure 11–42 when a 400 kHz signal is applied.

19. Determine the total rms current in Figure 11–45. What are the currents through L_2 and L_3?

FIGURE 11–45

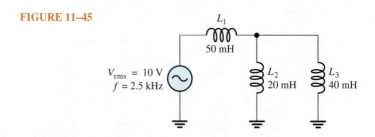

20. What frequency will produce a total rms current of 500 mA in each circuit of Figure 11–42 with an rms input voltage of 10 V?

21. Determine the reactive power in Figure 11–45, neglecting the winding resistance.

ADVANCED PROBLEMS

22. Determine the time constant for the circuit in Figure 11–46.

FIGURE 11–46

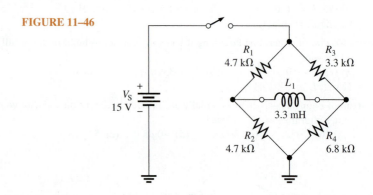

23. What is the voltage across the inductor in Figure 11–44 at each of the following times?
    (a) 60 μs   (b) 70 μs   (c) 80 μs

24. What is the voltage across the resistor in Figure 11–44 at a time of 60 μs?

25. (a) What is the current in the inductor 1.0 μs after the switch closes in Figure 11–46.
    (b) What is the current after 5τ have elapsed?

26. For the circuit in Figure 11–46, assume the switch has been closed for more than 5τ and is opened. What is the current in the inductor 1.0 μs after the switch is opened?

27. In System Example 11–2, assume $f = 550$ kHz. What is the reactance of each inductor? What is the total reactance of the two series inductors?

28. For System Example 11–2, assume you need to change L_1 to provide 8% feedback. What is the new value?

MULTISIM

MULTISIM TROUBLESHOOTING PROBLEMS

29. Open file P11-29; files are found at www.pearsonhighered.com/floyd. Test the circuit. If there is a fault, identify it.

30. Open file P11-30 and test the circuit. If there is a fault, identify it.

31. Determine if there is a fault in the circuit in file P11-31. If so, identify it.

32. Find and specify any faulty component in the circuit in file P11-32.

33. Is there a short or an open in the circuit in file P11-33? If so, identify the component that is faulty.

ANSWERS TO SECTION CHECKUPS

SECTION 11–1   The Basic Inductor

1. Parameters that determine inductance are turns of wire, permeability, cross-sectional area, and core length.

2. **(a)** When N increases, L increases.

   (b) When the core length increases, L decreases.

   (c) When the cross-sectional area decreases, L decreases.

   (d) For an air core, L decreases.

3. All wire has some resistance, and because inductors are made from turns of wire, there is always resistance in the winding.

4. Adjacent turns of wire in a coil act as plates of a capacitor and produce a small capacitance.

SECTION 11–2   Types of Inductors

1. Two categories of inductors are fixed and variable.

2. **(a)** air core   **(b)** iron core   **(c)** variable

SECTION 11–3   Series and Parallel Inductors

1. Inductances are added in series.

2. $L_T = 2600 \, \mu\text{H}$

3. $L_T = 5 \times 100 \, \text{mH} = 500 \, \text{mH}$

4. The total parallel inductance is smaller than that of the smallest-value individual inductor in parallel.

5. True

6. **(a)** $L_T = 7.69 \, \text{mH}$ **(b)** $L_T = 24 \, \mu\text{H}$

SECTION 11–4   Inductors in DC Circuits

1. $V_L = (10 \, \text{mA})(10 \, \Omega) = 100 \, \text{mV}$

2. Initially, $i = 0 \, \text{V}$, $v_L = 20 \, \text{V}$

3. After 5τ, $v_L = 0 \, \text{V}$

4. $\tau = 500 \, \mu\text{H}/1.0 \, \text{k}\Omega = 500 \, \text{ns}$, $i_L = 3.93 \, \text{mA}$

SECTION 11–5   Inductors in AC Circuits

1. Voltage leads current by $90°$ in an inductor.

2. $X_L = 2\pi f L = 3.14 \, \text{k}\Omega$

3. $f = X_L/2\pi L = 2.55 \, \text{MHz}$

4. $I_{\text{rms}} = 15.9 \, \text{mA}$

5. $P_{\text{true}} = 0 \, \text{W}$, $P_r = 458 \, \text{mVAR}$

SECTION 11–6   Inductor Applications

1. Conductive and radiated

2. Electromagnetic interference

3. It is placed on a wire to increase its inductance, creating an RF choke.

ANSWERS TO RELATED PROBLEMS FOR EXAMPLES

11–1.   157 mH

11–2.   150 μH

11–3.   20.3 μH

11–4.   227 ns

11–5. $I_F = 17.6$ mA, $\tau = 147$ ns

At 1τ: $i = 11.1$ mA; $t = 147$ ns

At 2τ: $i = 15.1$ mA; $t = 294$ ns

At 3τ: $i = 16.7$ mA; $t = 441$ ns

At 4τ: $i = 17.2$ mA; $t = 588$ ns

At 5τ: $i = 17.4$ mA; $t = 735$ ns

11–6. At 0.2 ms: $i = 0.215$ mA

At 0.8 ms: $i = 0.0125$ mA

11–7. 10 V if R_W is neglected.

11–8. 3.83 mA

11–9. 6.7 mA

11–10. 1100 Ω

11–11. (a) 7.17 kΩ; (b) 1.04 kΩ

11–12. 573 mA

11–13. P_r decreases.

ANSWERS TO TRUE/FALSE QUIZ

1. F   **2.** T   **3.** T   **4.** T   **5.** F   **6.** F   **7.** F

8. F   **9.** T   **10.** F

ANSWERS TO SELF-TEST

1. (c)   **2.** (d)   **3.** (c)   **4.** (b)   **5.** (d)   **6.** (a)   **7.** (d)

8. (c)   **9.** (b)   **10.** (a)   **11.** (d)

ANSWERS TO TROUBLESHOOTING: SYMPTOM AND CAUSE

1. (c)   **2.** (a)   **3.** (b)   **4.** (a)   **5.** (b)

CHAPTER 12

RL CIRCUITS

OUTLINE

OBJECTIVES

- Describe the relationship between current and voltage in an *RL* circuit
- Determine impedance and phase angle in a series *RL* circuit
- Analyze a series *RL* circuit
- Determine impedance and phase angle in a parallel *RL* circuit
- Analyze a parallel *RL* circuit
- Analyze series-parallel *RL* circuits
- Determine power in *RL* circuits
- Discuss how the *RL* circuit operates as a filter
- Troubleshoot *RL* circuits

KEY TERMS

RL lag circuit
RL lead circuit
Inductive susceptance (B_L)

INTRODUCTION

An *RL* circuit contains both resistance and inductance. In this chapter, series and parallel *RL* circuits and their responses to sinusoidal voltages are covered. In addition, series-parallel combinations are examined. Power considerations in *RL* circuits are introduced, and practical aspects of the power factor are discussed. A method of improving the power factor is presented. Troubleshooting common faults in *RL* circuits is also covered.

The methods for analyzing reactive circuits are similar to those you studied in dc circuits. Reactive circuit problems can be solved at only one frequency at a time, and phasor math must be used. Frequently in systems, the frequency of interest is the utility power frequency (50 Hz or 60 Hz). The inductance is not an ordinary component but may be the winding for a motor, solenoid, or transformer. You can still apply basic *RL* theory to these cases.

As you study this chapter, note both the differences and the similarities in the response of *RL* circuits compared to *RC* circuits.

VISIT THE WEBSITE
Study aids for this chapter are available at
http://pearsonhighered.com/floyd

12–1 SINUSOIDAL RESPONSE OF *RL* CIRCUITS

As with an *RC* circuit, all currents and voltages in any type of *RL* circuit are sinusoidal when the input voltage is sinusoidal. The inductance causes a phase shift between the voltage and the current that depends on the relative values of the resistance and the inductive reactance. Because of its winding resistance, the inductor is generally not as "ideal" as a resistor or capacitor. However, it will usually be treated as ideal for purposes of illustration.

After completing this section, you should be able to

- **Describe the relationship between current and voltage in an *RL* circuit**
  - **Discuss voltage and current waveforms**
  - **Discuss phase shift**

In an *RL* circuit, the resistor voltage and the current lag the source voltage. The inductor voltage leads the source voltage. Ideally, the phase angle between the current and the inductor voltage is always 90°. These generalized phase relationships are indicated in Figure 12–1. Notice how they differ from those of the *RC* circuit that was discussed in Chapter 10.

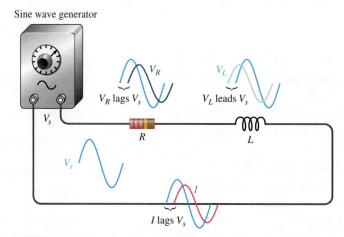

FIGURE 12–1 Illustration of sinusoidal response with general phase relationships of V_R, V_L, and *I* relative to the source voltage. V_R and *I* are in phase; V_R and V_L are 90° out of phase with each other.

The amplitudes and the phase relationships of the voltages and current depend on the values of the resistance and the inductive reactance. When a circuit is purely inductive, the phase angle between the source voltage and the total current is 90°, with the current lagging the voltage. When there is a combination of both resistance and inductive reactance in a circuit, the phase angle is somewhere between zero and 90°, depending on the relative values of the resistance and the inductive reactance. Because all inductors have winding resistance, ideal conditions may be approached but never reached in practice.

SECTION 12–1 CHECKUP*

1. A 1 kHz sinusoidal voltage is applied to an *RL* circuit. What is the frequency of the resulting current?

2. When the resistance in an *RL* circuit is greater than the inductive reactance, is the phase angle between the source voltage and the total current closer to 0° or to 90°?

12–2 IMPEDANCE AND PHASE ANGLE OF SERIES *RL* CIRCUITS

Impedance was introduced in Section 10–2 for *RC* circuits and represents the total opposition to sinusoidal current. As in the case with *RC* circuits, impedance is the combination of the resistive and reactive quantities, which can be represented by phasors. Because of phase differences, the total impedance must be treated as a phasor quantity.

After completing this section, you should be able to

- **Determine impedance and phase angle in a series *RL* circuit**
  - **Draw the impedance triangle**
  - **Calculate impedance magnitude**
  - **Calculate the phase angle**

The **impedance** of a series *RL* circuit is the total opposition to sinusoidal current and its unit is the ohm. The **phase angle** is the phase difference between the total current and the source voltage. Impedance (Z) is determined by the resistance (R) and the inductive reactance (X_L), as indicated in Figure 12–2.

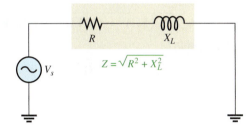

FIGURE 12–2   Impedance of a series *RL* circuit.

In ac analysis, both R and X_L are treated as phasor quantities, as shown in the phasor diagram of Figure 12–3(a), with X_L appearing at a $+90°$ angle with respect to R. This relationship comes from the fact that the inductor voltage leads the current, and thus the resistor voltage, by 90°. Since Z is the phasor sum of R and X_L, its phasor representation is as shown in Figure 12–3(b). A repositioning of the phasors, as shown in part (c), forms a right triangle. This formation, as you have learned, is called the *impedance triangle*. The length of each phasor represents the magnitude of the quantity, and θ is the phase angle between the source voltage and the current in the *RL* circuit.

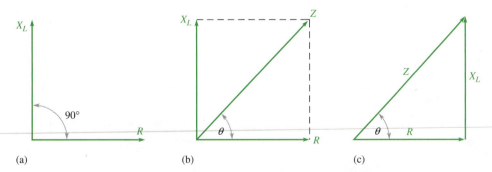

FIGURE 12–3   Development of the impedance triangle for a series *RL* circuit.

The magnitude of the impedance, Z, of the series *RL* circuit can be expressed in terms of the resistance and reactance as

$$Z = \sqrt{R^2 + X_L^2} \qquad (12\text{–}1)$$

where Z is expressed in ohms.

The phase angle, θ, is expressed as

$$\theta = \tan^{-1}\left(\frac{X_L}{R}\right)$$ (12–2)

EXAMPLE 12–1

Determine the impedance and phase angle of the circuit in Figure 12–4. Draw the impedance triangle.

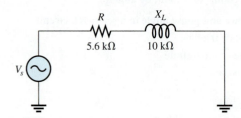

FIGURE 12–4

SOLUTION

The impedance is

$$Z = \sqrt{R^2 + X_L^2} = \sqrt{(5.6\,\text{k}\Omega)^2 + (10\,\text{k}\Omega)^2} = \mathbf{11.5\,k\Omega}$$

The value of the phase angle is

$$\theta = \tan^{-1}\left(\frac{X_L}{R}\right) = \tan^{-1}\left(\frac{10\,\text{k}\Omega}{5.6\,\text{k}\Omega}\right) = \mathbf{60.8°}$$

The source voltage leads the current by 60.8°. The impedance triangle is shown in Figure 12–5.

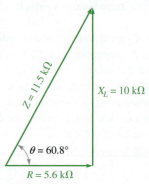

FIGURE 12–5

RELATED PROBLEM*

In a series *RL* circuit, $R = 1.8\,\text{k}\Omega$ and $X_L = 950\,\Omega$. Determine the impedance and phase angle.

*Answers are at the end of the chapter.

SECTION 12–2 CHECKUP

1. Does the source voltage lead or lag the current in a series *RL* circuit?

2. What is the relationship between X_L and R when the phase angle is 45°?

3. How does the phase angle in an *RL* circuit differ from the phase angle in an *RC* circuit?

4. A series *RL* circuit has a resistance of 33 kΩ and an inductive reactance of 50 kΩ. Determine Z and θ.

Ohm's law and Kirchhoff's voltage law are used in the analysis of series *RL* circuits to determine voltage, current, and impedance. Also, in this section *RL* lead and lag circuits are examined.

After completing this section, you should be able to

- **Analyze a series *RL* circuit**
  - Apply Ohm's law and Kirchhoff's voltage law to series *RL* circuits
  - Determine the phase relationships of the voltages and current
  - Show how impedance and phase angle vary with frequency
  - Discuss and analyze the *RL* lag circuit
  - Discuss and analyze the *RL* lead circuit

Ohm's Law

The application of Ohm's law to series *RL* circuits involves the use of the quantities of Z, V, and I. The three equivalent forms of Ohm's law were stated in Chapter 10 for *RC* circuits. They apply also to *RL* circuits and are restated here:

$$V = IZ \qquad I = \frac{V}{Z} \qquad Z = \frac{V}{I}$$

The following example illustrates the use of Ohm's law.

EXAMPLE 12–2

The current in Figure 12–6 is 200 μA. Determine the source voltage.

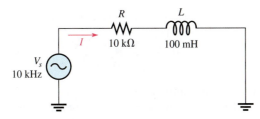

FIGURE 12–6

SOLUTION

From Equation 11–10, the inductive reactance is

$$X_L = 2\pi fL = 2\pi(10 \text{ kHz})(100 \text{ mH}) = 6.28 \text{ k}\Omega$$

The impedance is

$$Z = \sqrt{R^2 + X_L^2} = \sqrt{(10 \text{ k}\Omega)^2 + (6.28 \text{ k}\Omega)^2} = 11.8 \text{ k}\Omega$$

Applying Ohm's law yields

$$V_s = IZ = (200 \text{ }\mu\text{A})(11.8 \text{ k}\Omega) = \mathbf{2.36 \text{ V}}$$

RELATED PROBLEM

If the source voltage in Figure 12–6 is 5 V, what would be the current?

MULTISIM

Open Multisim file E12-02; files are found at www.pearsonhighered.com/floyd. Measure the current at 10 kHz, 5 kHz, and 20 kHz. Explain the results of your measurement.

Phase Relationships of the Current and Voltages

In a series *RL* circuit, the current is the same through both the resistor and the inductor. Thus, the resistor voltage is in phase with the current, and current lags the inductor voltage by 90°. Therefore, there is a phase difference of 90° between the resistor voltage, V_R, and the inductor voltage, V_L, as shown in the waveform diagram of Figure 12–7.

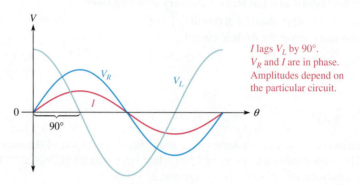

FIGURE 12–7 **Phase relation of current and voltages in a series *RL* circuit.**

From Kirchhoff's voltage law, the sum of the voltage drops must equal the source voltage. However, since V_R and V_L are not in phase with each other, they must be added as phasor quantities with V_L leading V_R by 90°, as shown in Figure 12–8(a). As shown in part (b), V_s is the phasor sum of V_R and V_L. This equation can be expressed as

$$V_s = \sqrt{V_R^2 + V_L^2} \qquad (12\text{--}3)$$

The phase angle between the resistor voltage and the source voltage can be expressed as

$$\theta = \tan^{-1}\left(\frac{V_L}{V_R}\right) \qquad (12\text{--}4)$$

Since the resistor voltage and the current are in phase, θ in Equation 12–4 also represents the phase angle between the source voltage and the current and is equivalent to $\tan^{-1}(X_L/R)$.

Figure 12–9 shows a voltage and current phasor diagram that represents the waveform diagram of Figure 12–7.

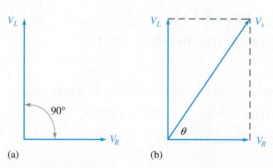

(a)   (b)

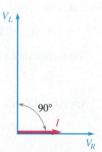

FIGURE 12–8 **Voltage phasor diagram for the waveforms in Figure 12–7.**

FIGURE 12–9 **Voltage and current phasor diagram for the waveforms in Figure 12–7.**

EXAMPLE 12–3

Determine the source voltage and the phase angle in Figure 12–10. Draw the voltage phasor diagram.

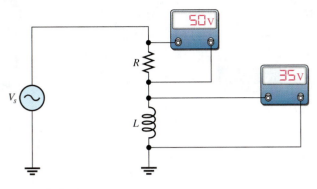

FIGURE 12–10

SOLUTION

Since V_R and V_L are 90° out of phase, you cannot add them directly. The source voltage is the phasor sum of V_R and V_L.

$$V_s = \sqrt{V_R^2 + V_L^2} = \sqrt{(50 \text{ V})^2 + (35 \text{ V})^2} = \mathbf{61 \text{ V}}$$

The phase angle between the resistor voltage and the source voltage is

$$\theta = \tan^{-1}\left(\frac{V_L}{V_R}\right) = \tan^{-1}\left(\frac{35 \text{ V}}{50 \text{ V}}\right) = \mathbf{35°}$$

The voltage phasor diagram is shown in Figure 12–11.

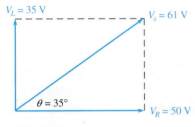

FIGURE 12–11

RELATED PROBLEM

With the information given, can you determine the current in Figure 12–10?

Variation of Impedance and Phase Angle with Frequency

As you know, inductive reactance varies directly with frequency. When X_L increases, the total impedance also increases; and when X_L decreases, the total impedance decreases. Thus, *in an RL circuit, Z is directly dependent on frequency.*

Figure 12–12 illustrates how the voltages and current in a series *RL* circuit vary as the frequency increases or decreases, with the source voltage held at a constant value. Part (a) shows that as frequency is increased, the increase in X_L will cause more voltage to appear across the inductor and less current will be in the circuit. Thus, V_L increases and V_R decreases. Also, Z is larger because there is less total current with the same total voltage across the circuit.

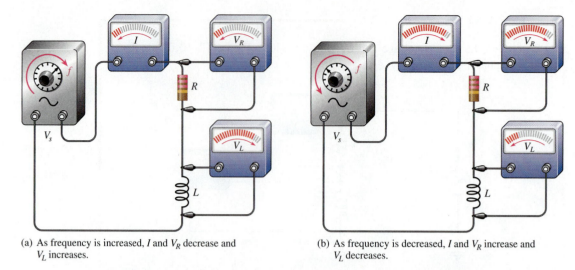

(a) As frequency is increased, *I* and V_R decrease and V_L increases.

(b) As frequency is decreased, *I* and V_R increase and V_L decreases.

FIGURE 12–12   **An illustration of how the variation of impedance affects the voltages and current as the source frequency is varied. The source voltage is held at a constant amplitude.**

Figure 12–12(b) shows that as frequency decreases, the voltage across the inductor decreases because X_L decreases. Also, *Z* decreases, causing the current to increase. An increase in current causes more voltage across the resistor.

Changes in *Z* and X_L can be observed as shown in Figure 12–13. As the frequency increases, the voltage across *Z* remains constant because V_s is constant ($V_s = V_Z$), but the voltage across *L* increases. The decreasing current indicates that *Z* is increasing. It does so because of the inverse relationship stated in Ohm's law ($Z = V_Z/I$). The decreasing current also indicates that X_L is increasing. The increase in V_L corresponds to the increase in X_L.

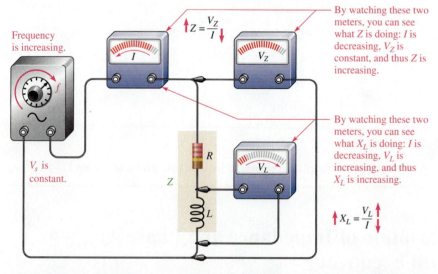

FIGURE 12–13   **Observing changes in *Z* and X_L with frequency by watching the meters and recalling Ohm's law.**

HANDS ON TIP
As you know, some multimeters have a relatively low frequency response. There are other things of which you should be aware. One is that most ac meters are accurate only if the waveform being measured is sinusoidal. Another is that accuracy when measuring small ac voltage is usually less than for dc measurements. Finally, loading can affect the accuracy of meter readings.

Since X_L is the factor that introduces the phase angle in a series *RL* circuit, a change in X_L produces a change in the phase angle. As the frequency is increased, X_L becomes greater, and thus the phase angle increases. As the frequency is decreased, X_L becomes smaller, and thus the phase angle decreases. The angle between V_s and V_R is the phase angle of the circuit because *I* is in phase with V_R. The variations of phase angle with frequency are illustrated with the impedance triangle as shown in Figure 12–14.

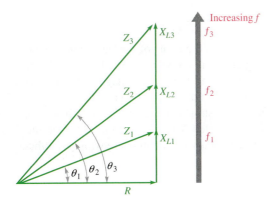

FIGURE 12–14   **As the frequency increases, the phase angle θ increases.**

EXAMPLE 12–4

For the series *RL* circuit in Figure 12–15, determine the impedance and the phase angle for each of the following frequencies:

(a) 10 kHz   **(b)** 20 kHz   **(c)** 30 kHz

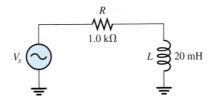

FIGURE 12–15

SOLUTION

(a) For $f = 10$ kHz, calculate the impedance as follows:

$$X_L = 2\pi fL = 2\pi(10 \text{ kHz})(20 \text{ mH}) = 1.26 \text{ k}\Omega$$

$$Z = \sqrt{R^2 + X_L^2} = \sqrt{(1.0 \text{ k}\Omega)^2 + (1.26 \text{ k}\Omega)^2} = \mathbf{1.61 \text{ k}\Omega}$$

The phase angle is

$$\theta = \tan^{-1}\left(\frac{X_L}{R}\right) = \tan^{-1}\left(\frac{1.26 \text{ k}\Omega}{1.0 \text{ k}\Omega}\right) = \mathbf{51.6°}$$

(b) For $f = 20$ kHz,

$$X_L = 2\pi(20 \text{ kHz})(20 \text{ mH}) = 2.51 \text{ k}\Omega$$

$$Z = \sqrt{(1.0 \text{ k}\Omega)^2 + (2.51 \text{ k}\Omega)^2} = \mathbf{2.70 \text{ k}\Omega}$$

$$\theta = \tan^{-1}\left(\frac{2.51 \text{ k}\Omega}{1.0 \text{ k}\Omega}\right) = \mathbf{68.3°}$$

(c) For $f = 30$ kHz,

$$X_L = 2\pi(30 \text{ kHz})(20 \text{ mH}) = 3.77 \text{ k}\Omega$$

$$Z = \sqrt{(1.0 \text{ k}\Omega)^2 + (3.77 \text{ k}\Omega)^2} = \mathbf{3.90 \text{ k}\Omega}$$

$$\theta = \tan^{-1}\left(\frac{3.77 \text{ k}\Omega}{1.0 \text{ k}\Omega}\right) = \mathbf{75.1°}$$

Notice that as the frequency increases, X_L, Z, and θ also increase.

RELATED PROBLEM

Determine Z and θ in Figure 12–15 if $f = 100$ kHz.

RL Lag Circuit

The ***RL* lag circuit** is a phase shift circuit in which the output voltage lags the input voltage by a specified angle, ϕ. A basic series *RL* lag circuit is shown in Figure 12–16(a). Notice that the output voltage is taken across the resistor and the input voltage is the total voltage applied across the circuit. The relationship of the voltages is shown in the phasor diagram in Figure 12–16(b), and a waveform diagram is shown in Figure 12–16(c). Notice that the output voltage, V_{out}, lags V_{in} by an angle, designated ϕ, that is the same as the circuit phase angle. The angles are equal, of course, because V_R and I are in phase with each other.

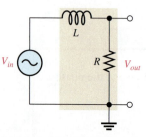

(a) A basic *RL* lag circuit

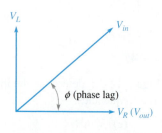

(b) Phasor voltage diagram showing phase lag between V_{in} and V_{out}

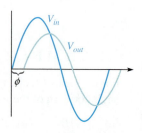

(c) Input and output waveforms

FIGURE 12–16   The *RL* lag circuit ($V_{out} = V_R$).

The phase lag, ϕ, can be expressed as

$$\phi = \tan^{-1}\left(\frac{X_L}{R}\right) \tag{12–5}$$

EXAMPLE 12–5

Calculate the phase lag for each circuit in Figure 12–17.

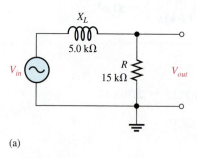

(a)

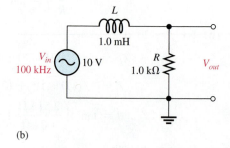

(b)

FIGURE 12–17

SOLUTION

For the circuit in Figure 12–17(a),

$$\phi = \tan^{-1}\left(\frac{X_L}{R}\right) = \tan^{-1}\left(\frac{5.0\,\text{k}\Omega}{15\,\text{k}\Omega}\right) = \mathbf{18.4°}$$

The output lags the input by 18.4°.

For the circuit in Figure 12–17(b), first determine the inductive reactance.

$$X_L = 2\pi fL = 2\pi(100\,\text{kHz})(1.0\,\text{mH}) = 628\,\Omega$$

The phase lag is

$$\phi = \tan^{-1}\left(\frac{X_L}{R}\right) = \tan^{-1}\left(\frac{628\ \Omega}{1.0\ k\Omega}\right) = \mathbf{32.1°}$$

The output lags the input by 32.1°.

RELATED PROBLEM

In a certain lag circuit, $R = 5.6\ k\Omega$ and $X_L = 3.5\ k\Omega$. Determine the phase lag between input and output.

The phase-lag circuit can be considered as a voltage divider with a portion of the input voltage dropped across L and a portion across R. The output voltage can be determined with the following formula:

$$V_{out} = \left(\frac{R}{\sqrt{R^2 + X_L^2}}\right)V_{in} \qquad\qquad (12\text{–}6)$$

EXAMPLE 12–6

The input voltage in Figure 12–17(b) of Example 12–5 has an rms value of 10 V. Determine the output voltage for the lag circuit shown in Figure 12–17(b). Draw the waveform relationships for the input and output voltages. The phase lag (32.1°) and X_L (628 Ω) were found in Example 12–5.

SOLUTION

Use Equation 12–6 to determine the output voltage for the lag circuit in Figure 12–17(b).

$$V_{out} = \left(\frac{R}{\sqrt{R^2 + X_L^2}}\right)V_{in} = \left(\frac{1.0\ k\Omega}{1.18\ k\Omega}\right)10\ V = \mathbf{8.47\ V\ rms}$$

The waveforms are shown in Figure 12–18.

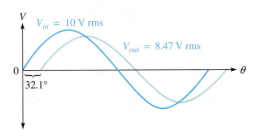

FIGURE 12–18

RELATED PROBLEM

In a lag circuit, $R = 4.7\ k\Omega$ and $X_L = 6\ k\Omega$. If the rms input voltage is 20 V, what is the output voltage?

MULTISIM

Open Multisim file E12-06. Measure the output voltage and compare to the calculated value.

EFFECTS OF FREQUENCY ON THE LAG CIRCUIT   Since the circuit phase angle and the phase lag are the same, an increase in frequency causes an increase in phase lag. Also, an increase in frequency causes a decrease in the magnitude of the output voltage because X_L becomes greater and more of the total voltage is dropped across the inductor and less across the resistor.

RL Lead Circuit

The ***RL* lead circuit** is a phase shift circuit in which the output voltage leads the input voltage by a specified angle, ϕ. A basic series *RL* lead circuit is shown in Figure 12–19(a). Notice how this circuit differs from the lag circuit. Here the output voltage is taken across the inductor rather than across the resistor. The relationship of the voltages is shown in the phasor diagram of Figure 12–19(b) and in the waveform plot of Figure 12–19(c). Notice that the output voltage, V_{out}, leads V_{in} by an angle (phase lead) that is the difference between 90° and the circuit phase angle θ.

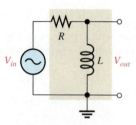

(a) A basic *RL* lead circuit

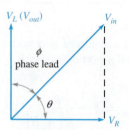

(b) Phasor voltage diagram showing V_{out} leading V_{in}

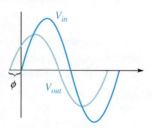

(c) Input and output voltage waveforms

FIGURE 12–19 The *RL* lead circuit ($V_{out} = V_L$).

Since $\theta = \tan^{-1}(X_L/R)$, the phase lead, ϕ, can be expressed as

$$\phi = 90° - \tan^{-1}\left(\frac{X_L}{R}\right) \tag{12–7}$$

Equivalently,

$$\phi = \tan^{-1}\left(\frac{R}{X_L}\right)$$

Again, the phase-lead circuit can be considered as a voltage divider with the voltage across L being the output. The expression for the output voltage is

$$V_{out} = \left(\frac{X_L}{\sqrt{R^2 + X_L^2}}\right)V_{in} \tag{12–8}$$

EXAMPLE 12–7

Determine the amount of phase lead and output voltage in the lead circuit in Figure 12–20.

FIGURE 12–20

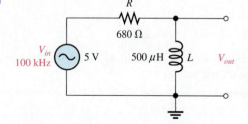

SOLUTION

First, determine the inductive reactance.

$$X_L = 2\pi fL = 2\pi(100\text{ kHz})(500\ \mu\text{H}) = 314\ \Omega$$

The phase lead is

$$\phi = 90° - \tan^{-1}\left(\frac{X_L}{R}\right) = 90° - \tan^{-1}\left(\frac{314 \ \Omega}{680 \ \Omega}\right) = \mathbf{65.2°}$$

The output leads the input by 65.2°.
    The output voltage is

$$V_{out} = \left(\frac{X_L}{\sqrt{R^2 + X_L^2}}\right)V_{in} = \left(\frac{314 \ \Omega}{\sqrt{(680 \ \Omega)^2 + (314 \ \Omega)^2}}\right)5 \ V = \mathbf{2.1 \ V}$$

MULTISIM

Open Multisim file E12-07.
Measure the output voltage and
compare to the calculated
value.

RELATED PROBLEM

In a certain lead circuit, $R = 2.2 \ k\Omega$ and $X_L = 1 \ k\Omega$. What is the phase lead?

EFFECTS OF FREQUENCY ON THE LEAD CIRCUIT   Since the circuit phase angle, θ, increases as frequency increases, the phase lead between the input and the output voltages decreases. Also, the amplitude of the output voltage increases as the frequency increases because X_L becomes greater and more of the total input voltage is dropped across the inductor.

SECTION 12–3 CHECKUP

1. In a certain series *RL* circuit, $V_R = 2 \ V$ and $V_L = 3 \ V$. What is the magnitude of the total voltage?

2. In Question 1, what is the phase angle?

3. When the frequency of the source voltage in a series *RL* circuit is increased, what happens to the inductive reactance? What happens to the impedance? What happens to the phase angle?

4. A certain *RL* lead circuit consists of a 3.3 kΩ resistor and a 15 mH inductor. Determine the phase lead between input and output at a frequency of 5 kHz.

5. An *RL* lag circuit has the same component values as the lead circuit in Question 4. What is the magnitude of the output voltage at 5 kHz when the input is 10 V rms?

12–4 IMPEDANCE AND PHASE ANGLE OF PARALLEL *RL* CIRCUITS

In this section, you will learn how to determine the impedance and phase angle of a parallel *RL* circuit. Also, inductive susceptance and admittance of a parallel *RL* circuit are introduced.

After completing this section, you should be able to

- **Determine impedance and phase angle in a parallel *RL* circuit**
  - **Express total impedance in a product-over-sum form**
  - **Express the phase angle in terms of *R* and X_L**
  - **Determine inductive susceptance and admittance**
  - **Convert admittance to impedance**

A basic parallel *RL* circuit is shown in Figure 12–21. The expression for the impedance, using the product-over-sum rule, is

$$Z = \frac{RX_L}{\sqrt{R^2 + X_L^2}} \tag{12–9}$$

The phase angle between the source voltage and the total current can be expressed in terms of R and X_L as

$$\theta = \tan^{-1}\left(\frac{R}{X_L}\right)$$

(12–10)

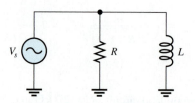

FIGURE 12–21   Parallel RL circuit.

EXAMPLE 12–8

For each circuit in Figure 12–22, determine the impedance and the phase angle.

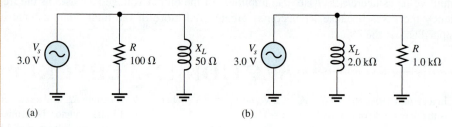

FIGURE 12–22

SOLUTION

For the circuit in Figure 12–22(a), the impedance and phase angle are

$$Z = \frac{RX_L}{\sqrt{R^2 + X_L^2}} = \frac{(100\ \Omega)(50\ \Omega)}{\sqrt{(100\ \Omega)^2 + (50\ \Omega)^2}} = \textbf{44.7}\ \mathbf{\Omega}$$

$$\theta = \tan^{-1}\left(\frac{R}{X_L}\right) = \tan^{-1}\left(\frac{100\ \Omega}{50\ \Omega}\right) = \textbf{63.4°}$$

For the circuit in Figure 12–22(b),

$$Z = \frac{(1.0\ \text{k}\Omega)(2.0\ \text{k}\Omega)}{\sqrt{(1.0\ \text{k}\Omega)^2 + (2.0\ \text{k}\Omega)^2}} = \textbf{894}\ \mathbf{\Omega}$$

$$\theta = \tan^{-1}\left(\frac{1.0\ \text{k}\Omega}{2.0\ \text{k}\Omega}\right) = \textbf{26.6°}$$

The voltage leads the current, compared to the parallel RC case where the voltage lags the current.

RELATED PROBLEM

In a parallel RL circuit $R = 10\ \text{k}\Omega$ and $X_L = 14\ \text{k}\Omega$. Find Z and θ.

Conductance, Susceptance, and Admittance

As you know from Section 10–4, conductance (G) is the reciprocal of resistance, susceptance (B) is the reciprocal of reactance, and admittance (Y) is the reciprocal of impedance.

For parallel *RL* circuits, **conductance** (*G*) is expressed as

$$G = \frac{1}{R} \qquad\qquad (12\text{–}11)$$

Inductive susceptance (*B_L*) is expressed as

$$B_L = \frac{1}{X_L} \qquad\qquad (12\text{–}12)$$

Admittance (*Y*) is expressed as

$$Y = \frac{1}{Z} \qquad\qquad (12\text{–}13)$$

As with the *RC* circuit, the unit for *G*, *B_L*, and *Y* is the siemens (S).

In the basic parallel *RL* circuit shown in Figure 12–23(a), the total admittance is the phasor sum of the conductance and the inductive susceptance, as shown in part (b).

$$Y_{tot} = \sqrt{G^2 + B_L^2} \qquad\qquad (12\text{–}14)$$

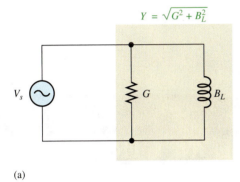
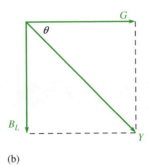

(a)   (b)

FIGURE 12–23   **Admittance in a parallel *RL* circuit.**

EXAMPLE 12–9

Determine the total admittance for the circuit in Figure 12–24; then convert it to impedance.

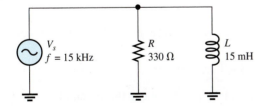

FIGURE 12–24

SOLUTION

To determine *Y*, first calculate the values for *G* and *B_L*. Since *R* = 330 Ω,

$$G = \frac{1}{R} = \frac{1}{330\ \Omega} = 3.03\ \text{mS}$$

The inductive reactance is

$$X_L = 2\pi f L = 2\pi(15\ \text{kHz})(15\ \text{mH}) = 1.41\ \text{k}\Omega$$

The inductive susceptance is

$$B_L = \frac{1}{X_L} = \frac{1}{1.41\ \text{k}\Omega} = 0.707\ \text{mS}$$

Therefore, the total admittance is

$$Y_{tot} = \sqrt{G^2 + B_L^2} = \sqrt{(3.03 \text{ mS})^2 + (0.707 \text{ mS})^2} = \textbf{3.11 mS}$$

Convert to impedance.

$$Z = \frac{1}{Y_{tot}} = \frac{1}{3.11 \text{ mS}} = \textbf{321 } \boldsymbol{\Omega}$$

RELATED PROBLEM

What is the total admittance of the circuit in Figure 12–24 if *f* is decreased to 5 kHz?

SECTION 12–4 CHECKUP

1. If $Y = 50$ mS, what is the value of Z?

2. In a certain parallel *RL* circuit, $R = 470 \text{ } \Omega$ and $X_L = 750 \text{ } \Omega$. Determine the admittance.

3. In the circuit of Question 2, does the total current lead or lag the source voltage and by what phase angle?

12–5 ANALYSIS OF PARALLEL *RL* CIRCUITS

Ohm's law and Kirchhoff's current law are used in the analysis of parallel *RL* circuits. Current and voltage relationships in a parallel *RL* circuit are examined.

After completing this section, you should be able to

- **Analyze a parallel *RL* circuit**
  - **Apply Ohm's law and Kirchhoff's current law to parallel *RL* circuits**
  - **Determine total current and phase angle**

The following example applies Ohm's law to the analysis of a parallel *RL* circuit.

EXAMPLE 12–10

Determine the total current and the phase angle in the circuit of Figure 12–25.

FIGURE 12–25

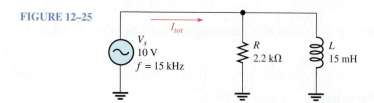

SOLUTION

First, determine the total admittance. The inductive reactance is

$$X_L = 2\pi fL = 2\pi(15 \text{ kHz})(15 \text{ mH}) = 1.41 \text{ k}\Omega$$

The conductance is

$$G = \frac{1}{R} = \frac{1}{2.2 \text{ k}\Omega} = 455 \text{ }\mu\text{S}$$

The inductive susceptance is

$$B_L = \frac{1}{X_L} = \frac{1}{1.41 \text{ k}\Omega} = 709 \text{ }\mu\text{S}$$

Therefore, the total admittance is

$$Y_{tot} = \sqrt{G^2 + B_L^2} = \sqrt{(455 \text{ }\mu\text{S})^2 + (709 \text{ }\mu\text{S})^2} = 842 \text{ }\mu\text{S}$$

Next, use Ohm's law to calculate the total current.

$$I_{tot} = VY_{tot} = (10 \text{ V})(842 \text{ }\mu\text{S}) = \textbf{8.42 mA}$$

The phase angle is

$$\theta = \tan^{-1}\left(\frac{R}{X_L}\right) = \tan^{-1}\left(\frac{2.2 \text{ k}\Omega}{1.41 \text{ k}\Omega}\right) = \textbf{57.3°}$$

The total current is 8.42 mA, and it lags the source voltage by 57.3°.

RELATED PROBLEM

Determine the total current and the phase angle if the frequency is reduced to 8.0 kHz in Figure 12–25.

Phase Relationships of the Currents and Voltages

Figure 12–26(a) shows all the currents and voltages in a basic parallel *RL* circuit. As you can see, the source voltage, V_s, appears across both the resistive and the inductive branches, so V_s, V_R, and V_L are all in phase and of the same magnitude. The total current, I_{tot}, divides at the junction into the two branch currents, I_R and I_L. The current and voltage phasor diagram is shown in Figure 12–26(b).

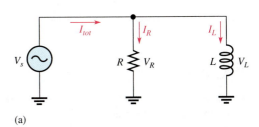

(a)

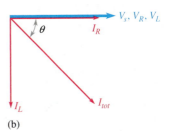

(b)

FIGURE 12–26   **Currents and voltages in a parallel *RL* circuit. The current directions shown in part (a) are instantaneous and, of course, reverse when the source voltage reverses during each cycle.**

The current through the resistor is in phase with the voltage. The current through the inductor lags the voltage and the resistor current by 90°. By Kirchhoff's current law, the total current is the phasor sum of the two branch currents. The total current is expressed as

$$I_{tot} = \sqrt{I_R^2 + I_L^2} \tag{12–15}$$

The phase angle between the resistor current and the total current is

$$\theta = \tan^{-1}\left(\frac{I_L}{I_R}\right) \tag{12–16}$$

EXAMPLE 12–11

For the circuit in Figure 12–25, draw the current phasor diagram to show the phase relationships between the currents and the source voltage.

SOLUTION

The conductance, susceptance, admittance, total current, and phase angle were calculated in Example 12–10.

$$G = 455 \ \mu S$$

$$B_L = 709 \ \mu S$$

$$Y_{tot} = 842 \ \mu S$$

$$I_{tot} = 8.42 \ \text{mA}$$

$$\theta = 57.3°$$

The branch currents are

$$I_R = GV = (455 \ \mu S)(10 \ V) = 4.55 \ \text{mA (in phase with } V_s)$$

$$I_L = B_L V = (709 \ \mu S)(10 \ V) = 7.09 \ \text{mA (lags } V_s \text{ by 90°)}$$

The current phasor diagram is shown in Figure 12–27. The blue phasor represents the voltage in the circuit.

FIGURE 12–27

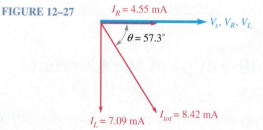

$I_R = 4.55 \ \text{mA}$
V_s, V_R, V_L
$\theta = 57.3°$
$I_L = 7.09 \ \text{mA}$
$I_{tot} = 8.42 \ \text{mA}$

RELATED PROBLEM

Find I_{tot} for the circuit in Figure 12–25 if a second 15 mH inductor is in parallel with the first one.

SECTION 12–5 CHECKUP

1. The admittance of a parallel *RL* circuit is 4 mS, and the source voltage is 8 V. What is the total current?

2. In a certain parallel *RL* circuit, the resistor current is 12 mA, and the inductor current is 20 mA. Determine the phase angle and the total current.

3. What is the phase angle between the inductor current and the source voltage in a parallel *RL* circuit?

The concepts studied in the previous sections are used to analyze circuits with combinations of both series and parallel R and L components.

After completing this section, you should be able to

- **Analyze series-parallel RL circuits**
  - **Determine total impedance and phase angle**
  - **Calculate currents and voltages**

The following two examples illustrate the procedures used to analyze a series-parallel reactive circuit.

EXAMPLE 12–12

In the series-parallel RL circuit of Figure 12–28, determine the values of the following:

(a) Z_{tot} **(b)** I_{tot} **(c)** θ

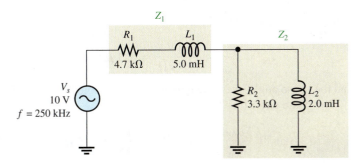

FIGURE 12–28

SOLUTION

(a) First, calculate the magnitudes of the inductive reactances.

$$X_{L1} = 2\pi fL_1 = 2\pi(250 \text{ kHz})(5.0 \text{ mH}) = 7.85 \text{ k}\Omega$$

$$X_{L2} = 2\pi fL_2 = 2\pi(250 \text{ kHz})(2.0 \text{ mH}) = 3.14 \text{ k}\Omega$$

One approach is to find the series equivalent resistance and inductive reactance for the parallel portion of the circuit; then add the resistances $(R_1 + R_{eq})$ to get the total resistance and add the reactances $(X_{L1} + X_{L(eq)})$ to get the total reactance. From these totals, you can determine the total impedance.
Determine the impedance of the parallel portion (Z_2) as follows:

$$G_2 = \frac{1}{R_2} = \frac{1}{3.3 \text{ k}\Omega} = 303 \text{ } \mu\text{S}$$

$$B_{L2} = \frac{1}{X_{L2}} = \frac{1}{3.14 \text{ k}\Omega} = 318 \text{ } \mu\text{S}$$

$$Y_2 = \sqrt{G_2^2 + B_L^2} = \sqrt{(303 \text{ } \mu\text{S})^2 + (318 \text{ } \mu\text{S})^2} = 439 \text{ } \mu\text{S}$$

Then

$$Z_2 = \frac{1}{Y_2} = \frac{1}{439 \ \mu S} = 2.28 \ k\Omega$$

The phase angle associated with the parallel portion of the circuit is

$$\theta_p = \tan^{-1}\left(\frac{R_2}{X_{L2}}\right) = \tan^{-1}\left(\frac{3.3 \ k\Omega}{3.14 \ k\Omega}\right) = 46.4°$$

The series equivalent values for the parallel portion are found by breaking the impedance into an equivalent resistance in series with an equivalent inductive reactance as follows:

$$R_{eq} = Z_2\cos\theta_p = (2.28 \ k\Omega)\cos(46.4°) = 1.57 \ k\Omega$$

$$X_{L(eq)} = Z_2\sin\theta_p = (2.28 \ k\Omega)\sin(46.4°) = 1.65 \ k\Omega$$

The total circuit resistance is

$$R_{tot} = R_1 + R_{eq} = 4.7 \ k\Omega + 1.57 \ k\Omega = 6.27 \ k\Omega$$

The total circuit reactance is

$$X_{L(tot)} = X_{L1} + X_{L(eq)} = 7.85 \ k\Omega + 1.65 \ k\Omega = 9.50 \ k\Omega$$

The total circuit impedance is

$$Z_{tot} = \sqrt{R_{tot}^2 + X_{L(tot)}^2} = \sqrt{(6.27 \ k\Omega)^2 + (9.50 \ k\Omega)^2} = \mathbf{11.4 \ k\Omega}$$

(b) Use Ohm's law to find the total current.

$$I_{tot} = \frac{V_s}{Z_{tot}} = \frac{10 \ V}{11.4 \ k\Omega} = \mathbf{877 \ \mu A}$$

(c) To find the phase angle, view the circuit as a series combination of R_{tot} and $X_{L(tot)}$. The phase angle by which I_{tot} lags V_s is

$$\theta = \tan^{-1}\left(\frac{X_{L(tot)}}{R_{tot}}\right) = \tan^{-1}\left(\frac{9.50 \ k\Omega}{6.27 \ k\Omega}\right) = \mathbf{56.6°}$$

MULTISIM

Open Multisim file E12-12. Measure the current through each component. Measure the voltage across Z_1 and Z_2.

RELATED PROBLEM

(a) Determine the voltage across the series part of the circuit in Figure 12–28.

(b) Determine the voltage across the parallel part of the circuit.

EXAMPLE 12–13

Determine the voltage across each component in Figure 12–29. Draw a current and voltage phasor diagram.

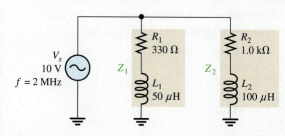

FIGURE 12–29

SOLUTION

First, calculate X_{L1} and X_{L2}.

$$X_{L1} = 2\pi fL_1 = 2\pi(2 \text{ MHz})(50 \text{ }\mu\text{H}) = 628 \text{ }\Omega$$

$$X_{L2} = 2\pi fL_2 = 2\pi(2 \text{ MHz})(100 \text{ }\mu\text{H}) = 1.26 \text{ k}\Omega$$

Now, determine the impedance of each branch.

$$Z_1 = \sqrt{R_1^2 + X_{L1}^2} = \sqrt{(330 \text{ }\Omega)^2 + (628 \text{ }\Omega)^2} = 709 \text{ }\Omega$$

$$Z_2 = \sqrt{R_2^2 + X_{L2}^2} = \sqrt{(1.0 \text{ k}\Omega)^2 + (1.26 \text{ k}\Omega)^2} = 1.61 \text{ k}\Omega$$

Calculate each branch current.

$$I_1 = \frac{V_s}{Z_1} = \frac{10 \text{ V}}{709 \text{ }\Omega} = 14.1 \text{ mA}$$

$$I_2 = \frac{V_s}{Z_2} = \frac{10 \text{ V}}{1.61 \text{ k}\Omega} = 6.21 \text{ mA}$$

Now, use Ohm's law to find the voltage across each component.

$$V_{R1} = I_1R_1 = (14.1 \text{ mA})(330 \text{ }\Omega) = \textbf{4.65 V}$$

$$V_{L1} = I_1X_{L1} = (14.1 \text{ mA})(628 \text{ }\Omega) = \textbf{8.85 V}$$

$$V_{R2} = I_2R_2 = (6.21 \text{ mA})(1.0 \text{ k}\Omega) = \textbf{6.21 V}$$

$$V_{L2} = I_2X_{L2} = (6.21 \text{ mA})(1.26 \text{ k}\Omega) = \textbf{7.82 V}$$

Now determine the values of the angles associated with each branch current.

$$\theta_1 = \tan^{-1}\left(\frac{X_{L1}}{R_1}\right) = \tan^{-1}\left(\frac{628 \text{ }\Omega}{330 \text{ }\Omega}\right) = 62.3°$$

$$\theta_2 = \tan^{-1}\left(\frac{X_{L2}}{R_2}\right) = \tan^{-1}\left(\frac{1.26 \text{ k}\Omega}{1.0 \text{ k}\Omega}\right) = 51.6°$$

Thus, I_1 lags V_s by 62.3°, and I_2 lags V_s by 51.6°, as indicated in Figure 12–30(a).

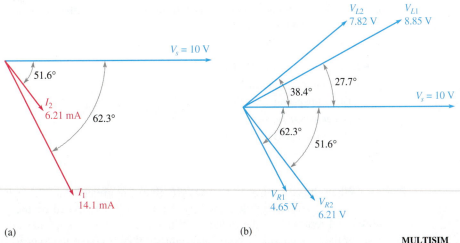

(a)

(b)

FIGURE 12–30

The phase relationships of the voltages are determined as follows:

- V_{R1} is in phase with I_1 and therefore lags V_s by 62.3°.
- V_{L1} leads I_1 by 90°, so its angle is 90° − 62.3° = 27.7°.

MULTISIM

Open Multisim file E12-13. Measure the voltage across each component and compare to the calculated values.

- V_{R2} is in phase with I_2 and therefore lags V_s by 51.6°.
- V_{L2} leads I_2 by 90°, so its angle is 90° − 51.6° = 38.4°.

These phase relationships are shown in Figure 12–30(b).

RELATED PROBLEM

What effect does an increase in frequency have on the total current in Figure 12–30?

SECTION 12–6 CHECKUP

1. Determine the total current for the circuit in Figure 12–29. *Hint:* Find the sum of the horizontal components of I_1 and I_2 and the sum of the vertical components of I_1 and I_2. Then apply the Pythagorean theorem to get I_{tot}.

2. What is the total impedance of the circuit in Figure 12–29?

12–7 POWER IN *RL* CIRCUITS

In a purely resistive ac circuit, all of the energy delivered by the source is dissipated in the form of heat by the resistance. In a purely inductive ac circuit, all of the energy delivered by the source is stored by the inductor in its magnetic field during a portion of the voltage cycle and then returned to the source during another portion of the cycle so that there is no net energy conversion to heat. When there is both resistance and inductance, some of the energy is alternately stored and returned by the inductance and some is dissipated by the resistance. The amount of energy converted to heat is determined by the relative values of the resistance and the inductive reactance.

After completing this section, you should be able to

- **Determine power in *RL* circuits**
  - **Explain true and reactive power**
  - **Draw the power triangle**
  - **Define** *power factor*
  - **Explain power factor correction**

When the resistance in a series *RL* circuit is greater than the inductive reactance, more of the total energy delivered by the source is dissipated by the resistance than is stored and returned by the inductor. When the reactance is greater than the resistance, more of the total energy is stored and returned than is converted to heat.

As you know, the power dissipated in a resistor is called the true power. The power in an inductor is **reactive power** and is expressed as

$$P_r = I^2 X_L \qquad (12\text{–}17)$$

The generalized power triangle for a series *RL* circuit is shown in Figure 12–31. The **apparent power, P_a,** is the resultant of the true power, P_{true}, and the reactive power, P_r.

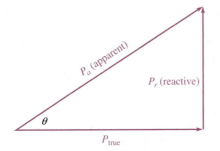

FIGURE 12–31   **Generalized power triangle for an *RL* circuit.**

EXAMPLE 12–14

Determine the power factor, the true power, the reactive power, and the apparent power in Figure 12–32.

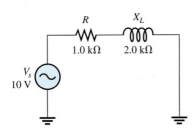

FIGURE 12–32

SOLUTION

The impedance of the circuit is

$$Z = \sqrt{R^2 + X_L^2} = \sqrt{(1.0\,\text{k}\Omega)^2 + (2.0\,\text{k}\Omega)^2} = 2.24\,\text{k}\Omega$$

The current is

$$I = \frac{V_s}{Z} = \frac{10\,\text{V}}{2.24\,\text{k}\Omega} = 4.46\,\text{mA}$$

The phase angle is

$$\theta = \tan^{-1}\left(\frac{X_L}{R}\right) = \tan^{-1}\left(\frac{2.0\,\text{k}\Omega}{1.0\,\text{k}\Omega}\right) = 63.4°$$

Power factor was defined in Equation 10–28. The power factor is

$$PF = \cos\theta = \cos(63.4°) = \mathbf{0.448}$$

The true power is

$$P_{\text{true}} = V_s I \cos\theta = (10\,\text{V})(4.46\,\text{mA})(0.448) = \mathbf{20\,mW}$$

The reactive power is

$$P_r = I^2 X_L = (4.46\,\text{mA})^2(2.0\,\text{k}\Omega) = \mathbf{39.8\,mVAR}$$

The apparent power is

$$P_a = I^2 Z = (4.46\,\text{mA})^2(2.24\,\text{k}\Omega) = \mathbf{44.6\,mVA}$$

RELATED PROBLEM

If the frequency in Figure 12–32 is increased, what happens to P_{true}, P_r, and P_a?

Significance of the Power Factor

Recall that the **power factor** equals the cosine of θ ($PF = \cos \theta$). As the phase angle between the source voltage and the total current increases, the power factor decreases, indicating an increasingly reactive circuit. A smaller power factor indicates less true power and more reactive power. The power factor of inductive loads is called a *lagging power factor* because the current lags the source voltage.

As you learned in Chapter 10, the power factor (*PF*) is important in determining how much useful power (true power) is transferred to a load. The highest power factor is 1, which indicates that all of the current to a load is in phase with the voltage (resistive). When the power factor is 0, all of the current to a load is 90° out of phase with the voltage (reactive).

Motor Phase Shift

Large ac induction motors in industry can have a current that is five or six times the maximum rated current for a brief time. The starting impedance is purely inductive, meaning that while starting, the current lags the terminal voltage by 90° and the power factor is high. When the motor is up to speed, the current drops off to its rated current (or less), and the load looks much more resistive; thus, the phase shift and the power factor are reduced. Most large induction motors use a motor starter that inserts current-limiting resistors in series during startup to reduce the very high starting currents.

yukosourov/Fotolia.com

SYSTEM NOTE

Generally, a power factor as close to 1 as possible is desirable because most of the power transferred from the source to the load is useful or true power. True power goes only one way—from source to load—and performs work on the load in terms of energy dissipation. Reactive power simply goes back and forth between the source and the load with no net work being done. Energy must be used in order for work to be done.

Many practical loads have inductance as a result of their particular function, and it is essential for their proper operation. Examples are transformers, electric motors, and speakers, to name a few. Therefore, inductive (and capacitive) loads are important considerations.

To see the effect of the power factor on system requirements, refer to Figure 12–33. This figure shows a representation of a typical inductive load consisting effectively of

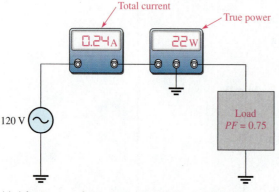

(a) A lower power factor means more total current for a given power dissipation (watts). A larger source is required to deliver the true power (watts).

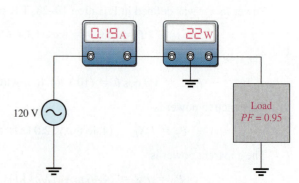

(b) A higher power factor means less total current for a given power dissipation. A smaller source can deliver the same true power (watts).

FIGURE 12–33 **Illustration of the effect of the power factor on system requirements such as source rating (VA) and conductor size. Although the loads are different, they dissipate the same true power.**

inductance and resistance in parallel. Part (a) shows a load with a relatively low power factor (0.75), and part (b) shows a load with a relatively high power factor (0.95). Although both loads, as indicated by the wattmeters, dissipate the same amount of true power, the low power factor load in Figure 12–33(a) draws more current from the source than does the high power factor load in Figure 12–33(b), as indicated by the ammeters. Therefore, the source in part (a) must have a higher VA rating than the one in part (b). Also, the lines connecting the source to the load must be a larger wire gauge than those in part (b), a condition that becomes significant when very long transmission lines are required, such as in power distribution.

Figure 12–33 has demonstrated that a higher power factor is an advantage in delivering power more efficiently to a load. Also, because the electric company charges for apparent power, it is less expensive to have a high power factor.

POWER FACTOR CORRECTION   The power factor of an inductive load can be increased by the addition of a capacitor in parallel, as shown in Figure 12–34. The capacitor compensates for the phase lag of the total current by creating a capacitive component of current that is 180° out of phase with the inductive component. This has a canceling effect and reduces the phase angle and the total current, as illustrated in the figure. This causes the power factor to increase.

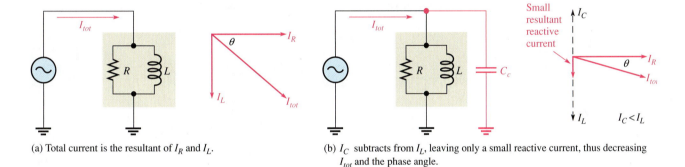

(a) Total current is the resultant of I_R and I_L.   (b) I_C subtracts from I_L, leaving only a small reactive current, thus decreasing I_{tot} and the phase angle.

FIGURE 12–34   **Example of how the power factor can be increased by the addition of a compensating capacitor (C_c). As θ decreases, *PF* increases.**

In industrial environments, three-phase motors, arc welders, and similar loads are common and tend to look like inductive loads, which cause the current to lag behind the voltage in each phase. The generalized power triangle, shown in Figure 12–31, can be applied to three-phase circuits by considering each phase as having true power and apparent power components. Ideally, the current waveform must track the voltage waveform for each phase (implying *PF* = 1), and the power drawn from each phase must be the same (this is referred to a *load balancing*.) With a balanced load, the capacitor-correction method can be used by connecting a compensating capacitor across each line, as was shown for the single-phase case.

In some cases, the load on each phase is unbalanced; different loads may cause distortion of the current waveform. Distortion is caused by nonlinear loads such as switched-mode power supplies. Loads that are intermittently on (such as arc welders) can also change the required power factor correction. These cases require more complicated solutions than just a compensating capacitor in each line; active circuits are available that can vary the phase angle.

SECTION 12–7 CHECKUP

1. To which component in an *RL* circuit is the power dissipation due?

2. Calculate the power factor when $\theta = 50°$.

3. A certain series *RL* circuit consists of a 470 Ω resistor and an inductive reactance of 620 Ω at the operating frequency. Determine P_{true}, P_r, and P_a when $I = 100$ mA.

Like *RC* circuits, series *RL* circuits also exhibit a frequency-selective characteristic. An application of *RL* circuits is a basic frequency selective (filter) circuit.

After completing this section, you should be able to

- Discuss how the *RL* circuit operates as a filter

Low-Pass Characteristic

You have seen what happens to the phase angle and the output voltage in a series *RC* lag circuit. In terms of the filtering action of the series *RL* circuit, the variation in the magnitude of the output voltage as a function of frequency is important.

Figure 12–35 shows the filtering action of a series *RL* circuit using a specific series of measurements in which the frequency starts at 100 Hz and is increased in increments up to 20 kHz. At each value of frequency, the output voltage is measured. As you can see, the inductive reactance increases as frequency increases, thus causing less voltage to be dropped across the resistor while the input voltage is held at a constant 10 V throughout each step. The **frequency response** curve for these particular values would appear similar to the response curve in Figure 10–47 for the low-pass *RC* circuit.

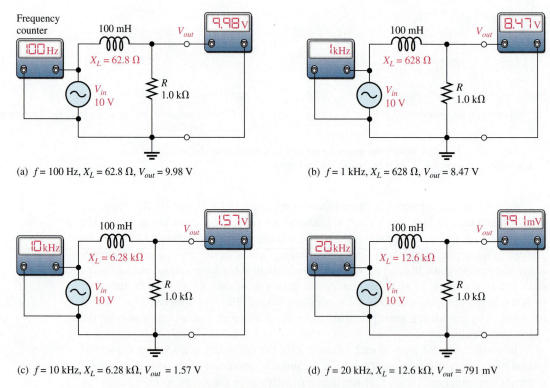

(a) $f = 100$ Hz, $X_L = 62.8\ \Omega$, $V_{out} = 9.98$ V

(b) $f = 1$ kHz, $X_L = 628\ \Omega$, $V_{out} = 8.47$ V

(c) $f = 10$ kHz, $X_L = 6.28$ kΩ, $V_{out} = 1.57$ V

(d) $f = 20$ kHz, $X_L = 12.6$ kΩ, $V_{out} = 791$ mV

FIGURE 12–35   **Example of low-pass filtering action. Winding resistance has been neglected. As the input frequency increases, the output voltage decreases.**

High-Pass Characteristic

To illustrate *RL* high-pass filtering action, Figure 12–36 shows a series of specific measurements. The frequency starts at 10 Hz and is increased in increments up to 10 kHz. As you can see, the inductive reactance increases as the frequency increases, thus causing more voltage to be dropped across the inductor. Again, when the values are plotted, the response curve is similar to the one for the high-pass *RC* circuit that was shown in Figure 10–49.

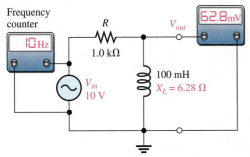

(a) $f = 10$ Hz, $X_L = 6.28\ \Omega$, $V_{out} = 62.8$ mV

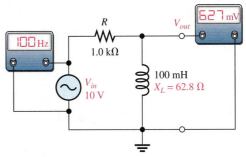

(b) $f = 100$ Hz, $X_L = 62.8\ \Omega$, $V_{out} = 627$ mV

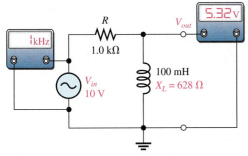

(c) $f = 1$ kHz, $X_L = 628\ \Omega$, $V_{out} = 5.32$ V

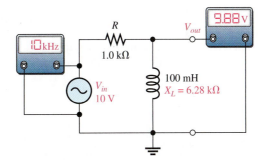

(d) $f = 10$ kHz, $X_L = 6.28$ kΩ, $V_{out} = 9.88$ V

FIGURE 12–36   Example of high-pass filtering action. Winding resistance has been neglected. As the input frequency increases, the output voltage increases.

The Cutoff Frequency of an *RL* Filter

The frequency at which the inductive reactance equals the resistance in a low-pass or high-pass *RL* filter is called the **cutoff frequency** and is designated f_c. This condition is expressed as $2\pi f_c L = R$. Solving for f_c results in the following formula:

$$f_c = \frac{R}{2\pi L} \qquad (12\text{–}18)$$

As with the *RC* filter, the output voltage is 70.7% of its maximum value at f_c. In a high-pass circuit, all frequencies above f_c are considered to be passed and all those below f_c are considered to be rejected. The reverse, of course, is true for a low-pass circuit. **Bandwidth**, defined in Chapter 10, applies to both *RC* and *RL* series circuits.

EXAMPLE 12–15

Calculate the cutoff frequency for the filters in Figure 12–37. Which is a high-pass filter and which is a low-pass filter?

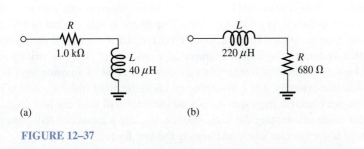

(a)   (b)

FIGURE 12–37

SOLUTION

For Figure 12–32(a), the cutoff frequency is

$$f_c = \frac{R}{2\pi L} = \frac{1.0\,\text{k}\Omega}{2\pi\,(40\,\mu\text{H})} = \textbf{3.98 MHz}$$

This is a **high-pass filter**.

For Figure 12–32(b), the cutoff frequency is

$$f_c = \frac{R}{2\pi L} = \frac{680\,\Omega}{2\pi(220\,\mu\text{H})} = \textbf{492 kHz}$$

This is a **low-pass filter**.

RELATED PROBLEM

Assume you want to change the cutoff frequency of the high-pass filter to 10 MHz by changing the resistor. What value of resistance would you specify?

SYSTEM EXAMPLE 12–1

EMI FILTERS

An important application for inductors is for filters that select certain frequencies and reject others. A pervasive problem facing electronic engineers is the conducted or radiated interference called electromagnetic interference (EMI) that can create problems in other circuits and systems. In many systems, power supply inputs will contain an electromagnetic interference (EMI) input filter to limit potential interference from other circuits that are common to the supply through conduction.

The design of EMI filters starts by characterizing the problem, including the harmonic content of the interference, the type of noise, and the path (conductive or radiated). The interference can be any frequency but is typically in the radio frequency domain. The design of EMI filters can be quite complicated, depending on the sensitivity of the circuit being protected and the particular interference. Computers are often used as a design aid.

Figure 12–38 shows an actual EMI circuit that was designed for a dc-to-dc converter. Other EMI filters may look entirely different than this one but still have the goal of reducing interference.

FIGURE 12–38 An EMI filter.

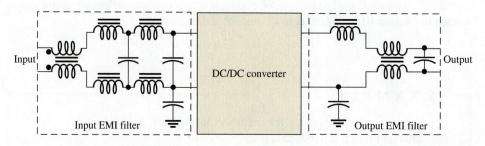

While the design of an EMI filter is beyond the scope of this course, it illustrates the application of inductors in complex filters. The filters in this section of text are basic *RL* filters, which are simplified frequency-selective circuits compared to the complicated EMI filter in this system example. An example of a basic filter (but not strictly *RL* or *RC*) is given in Figure 12–39, which represents a cross-over network to route high frequencies to one speaker (the "tweeter") and low frequencies to the other (the "woofer"). The speakers themselves are complex impedances, but an inductor will pass the low frequencies on to the woofer while attenuating the high frequencies, and a capacitor will pass the high frequencies on to the tweeter while attenuating the low frequencies.

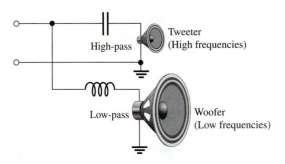

FIGURE 12–39   **A cross-over network.**

Tweeter
(High frequencies)

High-pass

Low-pass

Woofer
(Low frequencies)

SECTION 12–8 CHECKUP

1. To achieve a low-pass characteristic with a series *RL* circuit, across which component is the output taken?

2. What is EMI?

3. What is the purpose of an EMI filter?

12–9   TROUBLESHOOTING

Typical component failures have an effect on the frequency response of basic *RL* circuits.

After completing this section, you should be able to

- **Troubleshoot *RL* circuits**
  - **Find an open inductor**
  - **Find an open resistor**
  - **Find an open in a parallel circuit**

EFFECT OF AN OPEN INDUCTOR   The most common failure mode for inductors occurs when the winding opens as a result of excessive current or a mechanical contact failure. For this reason, inductors are often encapsulated; this helps avoid breaking connections. It is easy to see how an open coil affects the operation of a basic series *RL* circuit, as shown in Figure 12–40. Obviously, there is no path for current, so the resistor voltage is zero and the total source voltage appears across the open inductor.

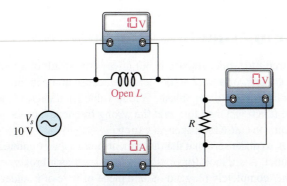

V_s
10 V

Open L

R

FIGURE 12–40   **Effect of an open coil.**

EFFECT OF AN OPEN RESISTOR   When the resistor is open, there is no current, and the inductor voltage is zero. The total input voltage appears across the open resistor, as shown in Figure 12–41.

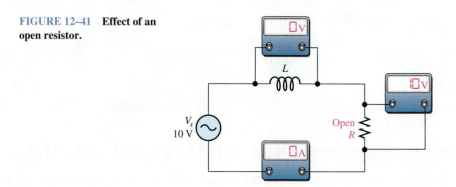

FIGURE 12–41   Effect of an open resistor.

OPEN COMPONENTS IN PARALLEL CIRCUITS   In a parallel *RL* circuit, an open resistor or inductor will cause the total current to decrease because the total impedance will increase. Obviously, the branch with the open component will have zero current. Figure 12–42 illustrates these conditions.

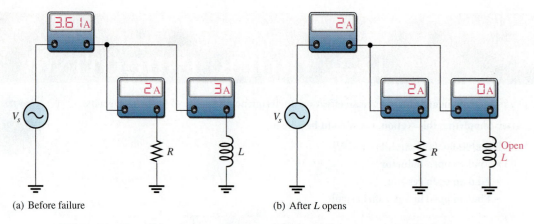

(a) Before failure   (b) After *L* opens

FIGURE 12–42   Effect of an open component in a parallel circuit with V_s constant.

EFFECT OF AN INDUCTOR WITH SHORTED WINDINGS   Although a rare occurrence, it is possible for some of the windings of coils to short together as a result of damaged insulation. This failure mode is much less likely than the open coil. A sufficient number of shorted windings may result in a reduction in inductance because the inductance of a coil is proportional to the square of the number of turns.

Other Troubleshooting Considerations

The failure of a circuit to work properly is not always the result of a faulty component. A loose wire, a bad contact, or a poor solder joint can cause an open circuit. A short can be caused by a wire clipping or solder splash. Wrong values in a circuit (such as an incorrect resistor value), the function generator set at the wrong frequency, or the wrong output connected to the circuit can cause improper operation.

Always check to make sure that the instruments are properly connected to the circuits and to a power outlet. Also, look for obvious things such as a broken or loose contact, a connector that is not completely plugged in, or a piece of wire or a solder bridge that could be shorting something out.

SECTION 12–9 CHECKUP

1. Why are inductors frequently encapsulated?

2. Describe the effect of an inductor with shorted windings on the response of a series *RL* circuit.

3. In the circuit of Figure 12–43, indicate whether I_{tot}, V_{R1}, and V_{R2} increase or decrease as a result of *L* opening.

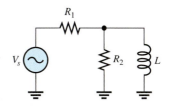

FIGURE 12–43

SUMMARY

- When a sinusoidal voltage is applied to an *RL* circuit, the current and all the voltage drops are also sine waves.
- Total current in an *RL* circuit always lags the source voltage.
- The resistor voltage is always in phase with the current.
- In an ideal inductor, the voltage always leads the current by 90°.
- In an *RL* lag circuit, the output voltage lags the input voltage in phase.
- In an *RL* lead circuit, the output voltage leads the input voltage in phase.
- In an *RL* circuit, the impedance is determined by both the resistance and the inductive reactance combined.
- Impedance is expressed in units of ohms.
- The impedance of an *RL* circuit varies directly with frequency.
- The phase angle (θ) of a series *RL* circuit varies directly with frequency.
- You can determine the impedance of a circuit by measuring the source voltage and the total current and then applying Ohm's law.
- In an *RL* circuit, part of the power is resistive and part reactive.
- The phasor combination of resistive power (true power) and reactive power is called *apparent power.*
- The power factor indicates how much of the apparent power is true power.
- A power factor of 1 indicates a purely resistive circuit, and a power factor of 0 indicates a purely reactive circuit.
- In a circuit that exhibits frequency selectivity, certain frequencies are passed to the output while others are rejected.

KEY TERMS

Key terms and other bold terms in the chapter are defined in the end-of-book glossary.

Inductive susceptance (B_L)   The reciprocal of inductive reactance. The unit is the siemens.

RL lag circuit   A phase shift circuit in which the output voltage, taken across the resistor, lags the input voltage by a specified angle.

RL lead circuit   A phase shift circuit in which the output voltage, taken across the inductor, leads the input voltage by a specified angle.

KEY FORMULAS

12–1 $Z = \sqrt{R^2 + X_L^2}$ Series *RL* impedance

12–2 $\theta = \tan^{-1}\left(\dfrac{X_L}{R}\right)$ Series *RL* phase angle

12–3 $\qquad V_s = \sqrt{V_R^2 + V_L^2}$ $\qquad\qquad\qquad$ Total voltage in a series *RL* circuit

12–4 $\qquad \theta = \tan^{-1}\left(\dfrac{V_L}{V_R}\right)$ $\qquad\qquad$ Series *RL* phase angle

12–5 $\qquad \phi = \tan^{-1}\left(\dfrac{X_L}{R}\right)$ $\qquad\qquad$ Phase angle of an *RL* lag circuit

12–6 $\qquad V_{out} = \left(\dfrac{R}{\sqrt{R^2 + X_L^2}}\right)V_{in}$ $\qquad$ Output voltage of lag circuit

12–7 $\qquad \phi = 90° - \tan^{-1}\left(\dfrac{X_L}{R}\right)$ $\qquad$ Phase angle of an *RL* lead circuit

12–8 $\qquad V_{out} = \left(\dfrac{X_L}{\sqrt{R^2 + X_L^2}}\right)V_{in}$ $\qquad$ Output voltage of an *RL* lead circuit

12–9 $\qquad Z = \dfrac{RX_L}{\sqrt{R^2 + X_L^2}}$ $\qquad\qquad$ Parallel *RL* impedance

12–10 $\qquad \theta = \tan^{-1}\left(\dfrac{R}{X_L}\right)$ $\qquad\qquad$ Parallel *RL* phase angle

12–11 $\qquad G = \dfrac{1}{R}$ $\qquad\qquad\qquad$ Conductance

12–12 $\qquad B_L = \dfrac{1}{X_L}$ $\qquad\qquad\qquad$ Inductive susceptance

12–13 $\qquad Y = \dfrac{1}{Z}$ $\qquad\qquad\qquad$ Admittance

12–14 $\qquad Y_{tot} = \sqrt{G^2 + B_L^2}$ $\qquad\qquad$ Total admittance

12–15 $\qquad I_{tot} = \sqrt{I_R^2 + I_L^2}$ $\qquad\qquad$ Total current in parallel *RL* circuit

12–16 $\qquad \theta = \tan^{-1}\left(\dfrac{I_L}{I_R}\right)$ $\qquad\qquad$ Parallel *RL* phase angle

12–17 $\qquad P_r = I^2 X_L$ $\qquad\qquad\qquad$ Reactive power

12–18 $\qquad f_c = \dfrac{R}{2\pi L}$ $\qquad\qquad\qquad$ Cutoff frequency of *RL* filter

TRUE/FALSE QUIZ

Answers are at the end of the chapter.

1. In an ac circuit where $R = X_L$, the phase angle is 45°.

2. In an ac series *RL* circuit where 3.0 V is dropped across the resistor and 4.0 V is dropped across the inductor, it follows that the source voltage is 5.0 V.

3. In an ac series *RL* circuit, the current and voltage are in phase for the inductor.

4. In an ac parallel *RL* circuit, the inductive susceptance is always less than the admittance.

5. In an ac parallel *RL* circuit, the voltage across the inductor is out of phase with the voltage across the resistor.

6. The unit *siemens* is used to measure both susceptance and admittance.

7. The reciprocal of impedance is susceptance.

8. If the power factor of a circuit is 0.5, the reactive power and true power are equal.

9. A purely resistive circuit has a power factor of 0.

10. A high-pass series *RL* filter has the output taken from across the resistor.

SELF-TEST

Answers are at the end of the chapter.

1. In a series *RL* circuit, the resistor voltage
   - **(a)** leads the source voltage
   - **(b)** lags the source voltage
   - **(c)** is in phase with the source voltage
   - **(d)** is in phase with the current
   - **(e)** answers (a) and (d)
   - **(f)** answers (b) and (d)

2. When the frequency of the voltage applied to a series *RL* circuit is increased, the impedance
   - **(a)** decreases   **(b)** increases   **(c)** does not change

3. When the frequency of the voltage applied to a series *RL* circuit is decreased, the phase angle
   - **(a)** decreases   **(b)** increases   **(c)** does not change

4. If the frequency is doubled and the resistance is halved, the impedance of a series *RL* circuit
   - **(a)** doubles
   - **(b)** halves
   - **(c)** remains constant
   - **(d)** cannot be determined without values

5. To reduce the current in a series *RL* circuit, the frequency should be
   - **(a)** increased   **(b)** decreased   **(c)** constant

6. In a series *RL* circuit, 10 V rms is measured across the resistor, and 10 V rms is measured across the inductor. The peak value of the source voltage is
   - **(a)** 14.14 V   **(b)** 28.28 V
   - **(c)** 10 V   **(d)** 20 V

7. The voltages in Question 6 are measured at a certain frequency. To make the resistor voltage greater than the inductor voltage, the frequency is
   - **(a)** increased   **(b)** decreased
   - **(c)** doubled   **(d)** not a factor

8. When the resistor voltage in a series *RL* circuit becomes greater than the inductor voltage, the phase angle
   - **(a)** increases   **(b)** decreases   **(c)** is not affected

9. When the frequency is increased, the impedance of a parallel *RL* circuit
   - **(a)** increases   **(b)** decreases   **(c)** remains constant

10. In a parallel *RL* circuit, there are 2 A rms in the resistive branch and 2 A rms in the inductive branch. The total rms current is
    - **(a)** 4 A   **(b)** 5.656 A   **(c)** 2 A   **(d)** 2.828 A

11. You are observing two voltage waveforms on an oscilloscope. The time base (sec/div) of the scope is set to 10 μs. One half-cycle of the waveforms covers the ten horizontal divisions. The positive-going zero crossing of one waveform is at the leftmost division, and the positive-going zero crossing of the other is three divisions to the right. The phase angle between these two waveforms is
    - **(a)** 18°   **(b)** 36°   **(c)** 54°   **(d)** 180°

12. Which of the following power factors results in the least energy conversion to heat in an *RL* circuit?
    - **(a)** 1   **(b)** 0.9   **(c)** 0.5   **(d)** 0.1

13. If a load is purely inductive and the reactive power is 10 VAR, the apparent power is
    - **(a)** 0 VA   **(b)** 10 VA
    - **(c)** 14.14 VA   **(d)** 3.16 VA

14. For a certain load, the true power is 10 W and the reactive power is 10 VAR. The apparent power is
    - **(a)** 5 VA   **(b)** 20 VA
    - **(c)** 14.14 VA   **(d)** 100 VA

15. The cutoff frequency of a certain low-pass *RL* filter is 20 kHz. The bandwidth is
    - **(a)** 20 kHz   **(b)** 40 kHz
    - **(c)** 0 kHz   **(d)** unknown

TROUBLESHOOTING: SYMPTOM AND CAUSE

The purpose of these exercises is to help develop thought processes essential to troubleshooting. Answers are at the end of the chapter.

Determine the cause for each set of symptoms. Refer to Figure 12–44.

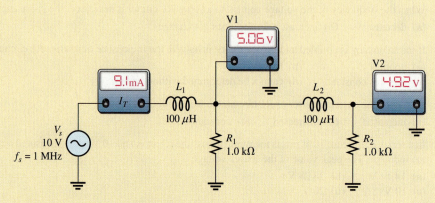

FIGURE 12–44 The ac meters indicate the correct readings for this circuit.

1. *Symptom:* The ammeter reading is 15.9 mA, and the voltmeter 1 and voltmeter 2 readings are zero.

   Cause:

   (a) L_1 is open.
   (b) L_2 is open.
   (c) R_1 is shorted.

2. *Symptom:* The ammeter reading is 8.47 mA, the voltmeter 1 reading is 8.47 V, and the voltmeter 2 reading is 0 V.

   Cause:

   (a) L_2 is open.
   (b) R_2 is open.
   (c) R_2 is shorted.

3. *Symptom:* The ammeter reading is slightly less than 20 mA, and both voltmeter readings are slightly less than 10 V.

   Cause:

   (a) L_1 is shorted.
   (b) R_1 is open.
   (c) The source frequency is incorrectly set very low.

4. *Symptom:* The ammeter reading is 4.55 mA, voltmeter 1 reads 2.53 V, and voltmeter 2 reads 2.15 V.

   Cause:

   (a) The source frequency is incorrectly set at 500 kHz.
   (b) The source voltage is incorrectly set at 5 V.
   (c) The source frequency is incorrectly set at 2 MHz.

5. *Symptom:* All meter readings are zero.

   Cause:

   (a) The voltage source is faulty or off.
   (b) L_1 is open.
   (c) Either (a) or (b)

PROBLEMS

Answers to odd-numbered problems are at the end of the book.

BASIC PROBLEMS

SECTION 12–1   Sinusoidal Response of *RL* Circuits

1. A 15 kHz sinusoidal voltage is applied to a series *RL* circuit. Determine the frequency of *I*, V_R, and V_L.

2. What are the waveshapes of *I*, V_R, and V_L in Problem 1?

SECTION 12–2   Impedance and Phase Angle of Series *RL* Circuits

3. Find the impedance of each circuit in Figure 12–45.

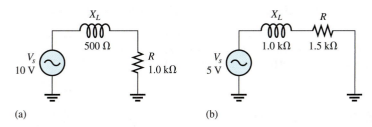

(a)   **(b)**

FIGURE 12–45

4. Determine the impedance and phase angle for the circuit in Figure 12–46.

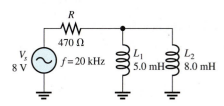

FIGURE 12–46

5. In Figure 12–47, determine the impedance at each of the following frequencies:
   (a) 100 Hz   **(b)** 500 Hz   **(c)** 1 kHz   **(d)** 2 kHz

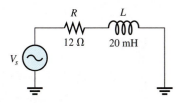

FIGURE 12–47

6. Determine the values of *R* and X_L in a series *RL* circuit for the following values of impedance and phase angle:
   (a) $Z = 20\ \Omega, \theta = 45°$ **(b)** $Z = 500\ \Omega, \theta = 35°$
   (c) $Z = 2.5\ k\Omega, \theta = 72.5°$ **(d)** $Z = 998\ \Omega, \theta = 45°$

SECTION 12–3   Analysis of Series *RL* Circuits

7. Assume that the inductance in Figure 12–45(a) is 796 μH. What is the source frequency?

8. Determine the voltage across the total resistance and across the total inductance in Figure 12–46.

9. Find the current for each circuit of Figure 12–45.

10. Calculate the total current for the circuit of Figure 12–46.

11. Determine θ for the circuit in Figure 12–48.

12. If the inductance in Figure 12–48 is doubled, does θ increase or decrease, and by how many degrees?

13. Draw the waveforms for V_s, V_R, and V_L in Figure 12–48. Show the proper phase relationships.

14. For the circuit in Figure 12–49, find V_R and V_L for each of the following frequencies:
    (a) 60 Hz   **(b)** 200 Hz   **(c)** 500 Hz   **(d)** 1 kHz

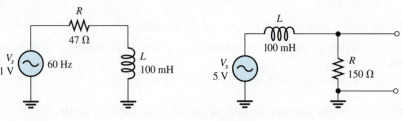

      FIGURE 12–48   **FIGURE 12–49**

SECTION 12–4   Impedance and Phase Angle of Parallel *RL* Circuits

15. What is the impedance for the circuit in Figure 12–50?

16. Repeat Problem 15 for the following frequencies:
    (a) 1.5 kHz   **(b)** 3 kHz   **(c)** 5 kHz   **(d)** 10 kHz

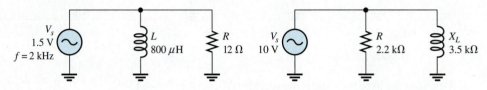

FIGURE 12–50   **FIGURE 12–51**

17. What is the admittance of the circuit in Figure 12–51?

18. What is the admittance of the circuit in Figure 15–51 if the frequency is doubled?

19. At what frequency does X_L equal R in Figure 12–50?

SECTION 12–5   Analysis of Parallel *RL* Circuits

20. Find the total current and each branch current in Figure 12–51.

21. Determine the following quantities in Figure 12–52:
    (a) Z **(b)** I_R **(c)** I_L **(d)** I_{tot} **(e)** θ

22. Convert the circuit in Figure 12–53 to an equivalent series form.

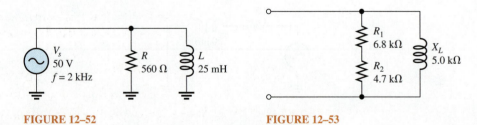

    FIGURE 12–52   **FIGURE 12–53**

SECTION 12–6   Analysis of Series-Parallel *RL* Circuits

23. Determine the voltage across each element in Figure 12–54.

24. Is the circuit in Figure 12–54 predominantly resistive or predominantly inductive?

25. Find the current in each branch and the total current in Figure 12–54.

      FIGURE 12–54

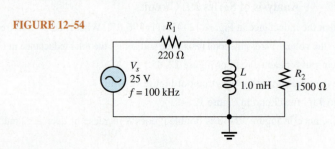

SECTION 12–7   Power in *RL* Circuits

26. In a certain *RL* circuit, the true power is 100 mW, and the reactive power is 340 mVAR. What is the apparent power?

27. Determine the true power and the reactive power in Figure 12–48.

28. What is the power factor in Figure 12–51?

29. Determine P_{true}, P_r, P_a, and *PF* for the circuit in Figure 12–54. Draw the power triangle.

SECTION 12–8   *RL* Filters

30. Plot the response curve for the circuit in Figure 12–49. Show the output voltage versus frequency in 200 Hz increments from 0 to 1200 Hz.

31. Using the same procedure as in Problem 30, plot the response curve for Figure 12–55.

32. Draw the voltage phasor diagram for the circuit in Figure 12–55 for a frequency of 500 Hz.

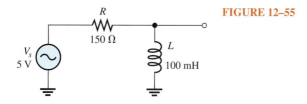

FIGURE 12–55

SECTION 12–9   Troubleshooting

33. Determine the voltage across each component in Figure 12–56 if L_1 were open.

34. Determine if the voltage reading on the DMM in Figure 12–57 is correct or not. Explain your answer.

FIGURE 12–56

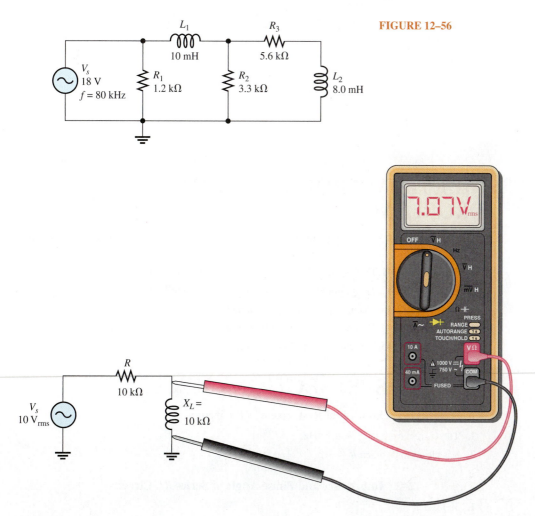

FIGURE 12–57

ADVANCED PROBLEMS

35. Determine the voltage across the inductors in Figure 12–58.

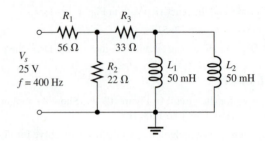

FIGURE 12–58

36. Is the circuit in Figure 12–58 predominantly resistive or predominantly inductive?

37. Find the total current in Figure 12–58.

38. For the circuit in Figure 12–59, determine the following:
    (a) Z_{tot} **(b)** I_{tot} **(c)** θ **(d)** V_L **(e)** V_{R3}

39. For the circuit in Figure 12–60, determine the following:
    (a) I_{R1} **(b)** I_{L1} **(c)** I_{L2} **(d)** I_{R2}

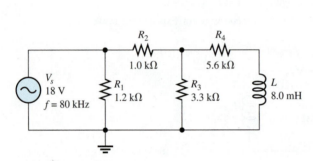

FIGURE 12–59

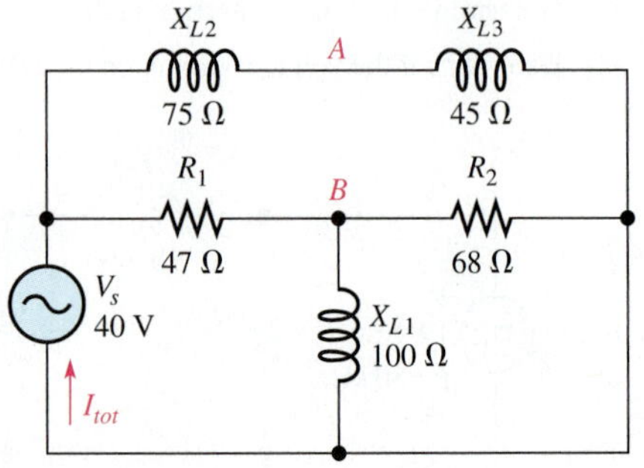

FIGURE 12–60

MULTISIM TROUBLESHOOTING PROBLEMS

40. Open file P12-40; files are found at www.pearsonhighered.com/floyd. Determine if there is a fault and, if so, identify it.

41. Open file P12-41. Determine if there is a fault and, if so, identify it.

42. Open file P12-42. Determine if there is a fault and, if so, identify it.

43. Open file P12-43. Determine if there is a fault and, if so, identify it.

44. Open file P12-44. Determine if there is a fault and, if so, identify it.

45. Open file P12-45. Determine if there is a fault and, if so, identify it.

ANSWERS TO SECTION CHECKUPS

SECTION 12–1   Sinusoidal Response of *RL* Circuits

1. The current frequency is at 1 kHz.

2. θ is closer to 0° when $R > X_L$.

SECTION 12–2   Impedance and Phase Angle of Series *RL* Circuits

1. V_s leads I.

2. $X_L = R$

3. In an *RL* circuit, current lags voltage; in an *RC* circuit, current leads voltage.

4. $Z = \sqrt{R^2 + X_L^2} = 59.9 \text{ k}\Omega; \theta = \tan^{-1}(X_L/R) = 56.6°$

SECTION 12–3   Analysis of Series *RL* Circuits

1. $V_s = \sqrt{V_R^2 + V_L^2} = 3.61 \text{ V}$

2. $\theta = \tan^{-1}(X_L/V_R) = 56.3°$

3. When *f* increases, X_L increases, *Z* increases, and θ increases.

4. $\phi = 90° - \tan^{-1}(X_L/R) = 81.9°$

5. $V_{out} = \left(\dfrac{R}{\sqrt{R^2 + X_L^2}} \right) V_{in} = 9.90 \text{ V}$

SECTION 12–4   Impedance and Phase Angle of Parallel *RL* Circuits

1. $Z = 1/Y = 20 \ \Omega$

2. $Y = \sqrt{G^2 + B_L^2} = 2.5 \text{ mS}$

3. I_{tot} lags V_s by 32.1°.

SECTION 12–5   Analysis of Parallel *RL* Circuits

1. $I_{tot} = V_s Y = 32 \text{ mA}$

2. $\theta = \tan^{-1}(I_L/I_R) = 59.0°; I_{tot} = \sqrt{I_R^2 + I_L^2} = 23.3 \text{ mA}$

3. $\theta = 90°$

SECTION 12–6   Analysis of Series-Parallel *RL* Circuits

1. $I_{tot} = \sqrt{(I_1\cos\theta_1 + I_2\cos\theta_2)^2 + (I_1\sin\theta_1 + I_2\sin\theta_2)^2} = 20.2 \text{ mA}$

2. $Z = V_s/I_{tot} = 494 \ \Omega$

SECTION 12–7   Power in *RL* Circuits

1. Power dissipation is in the resistor.

2. $PF = \cos 50° = 0.643$

3. $P_{true} = I^2 R = 4.7 \text{ W}; P_r = I^2 X_L = 6.2 \text{ VAR}; P_a = \sqrt{P_{true}^2 + P_r^2} = 7.78 \text{ VA}$

SECTION 12–8   *RL* Filters

1. Output is across the resistor.

2. Electromagnetic interference, caused when a circuit conducts or radiates electrical noise into another circuit

3. To reduce interference

SECTION 12–9   Troubleshooting

1. Encapsulation helps avoid open circuits due to stressing the leads on an inductor.

2. Shorted windings reduce *L* and thereby reduce X_L at any given frequency.

3. I_{tot} decreases, V_{R1} decreases, and V_{R2} increases when *L* opens.

ANSWERS TO RELATED PROBLEMS FOR EXAMPLES

12–1   2.04 kΩ; 27.8°

12–2   423 μA

12–3   No

12–4   12.6 kΩ; 85.5°

12–5   32°

12–6   12.3 V rms

12–7   65.6°

12–8 8.14 kΩ; 35.5°

12–9 3.70 mS

12–10 14.0 mA; 71.1°

12–11 14.9 mA

12–12 **(a)** 8.04 V
(b) 2.00 V

12–13 Current decreases.

12–14 P_{true}, P_r, and P_a decrease.

12–15 2.5 kΩ

ANSWERS TO TRUE/FALSE QUIZ

1. T   **2.** T   **3.** F   **4.** T   **5.** F   **6.** T   **7.** F
8. T   **9.** F   **10.** F

ANSWERS TO SELF-TEST

1. (f)   **2.** (b)   **3.** (a)   **4.** (d)   **5.** (a)   **6.** (d)   **7.** (b)   **8.** (b)
9. (a)   **10.** (d)   **11.** (c)   **12.** (d)   **13.** (b)   **14.** (c)   **15.** (a)

ANSWERS TO TROUBLESHOOTING: SYMPTOM AND CAUSE

1. (c)   **2.** (a)   **3.** (c)   **4.** (b)   **5.** (c)

CHAPTER 13

RLC CIRCUITS AND RESONANCE

OUTLINE

OBJECTIVES

- Determine the impedance and phase angle of a series *RLC* circuit
- Analyze series *RLC* circuits
- Analyze a circuit for series resonance
- Analyze series resonant filters
- Analyze parallel *RLC* circuits
- Analyze a circuit for parallel resonance
- Analyze the operation of parallel resonant filters
- Discuss some applications of resonant circuits

KEY TERMS

Series resonance	Decibel (dB)
Resonant frequency	Selectivity
Band-pass filter	Band-stop filter
Cutoff frequency	Parallel resonance
Half-power frequency	

INTRODUCTION

In this chapter, you will learn about frequency response of circuits with combinations of resistance, capacitance, and inductance (*RLC*). Both series and parallel *RLC* circuits, including the concepts of series and parallel resonance will be discussed.

Because it is the basis for frequency selectivity, resonance in electrical circuits is important to the operation of many types of electronic systems, particularly in the area of communications. For example, the ability of a radio or television receiver to select a certain frequency transmitted by a particular station and to eliminate frequencies from other stations is based on the principle of resonance.

The operation of both band-pass and band-stop filters is based on resonance of circuits containing inductance and capacitance, and these filters are discussed in this chapter. Both types of filters are widely used in communication systems.

VISIT THE WEBSITE

Study aids for this chapter are available at
http://pearsonhighered.com/floyd

A series *RLC* circuit contains resistance, inductance, and capacitance. Since inductive reactance and capacitive reactance have opposite effects on the circuit phase angle, the total reactance is less than either individual reactance.

After completing this section, you should be able to

- **Determine the impedance and phase angle of a series *RLC* circuit**
  - **Calculate total reactance**
  - **Determine whether a circuit is predominately inductive or capacitive**

A series *RLC* circuit is shown in Figure 13–1. It contains resistance, inductance, and capacitance.

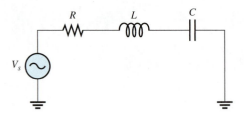

FIGURE 13–1 Series *RLC* circuit.

As you know, inductive reactance (X_L) causes the total current to lag the source voltage. Capacitive reactance (X_C) has the opposite effect: It causes the current to lead the voltage. Thus, X_L and X_C tend to offset each other. When they are equal, they cancel, and the total reactance is zero. In any case, the total reactance in a series circuit is

$$X_{tot} = |X_L - X_C| \qquad (13\text{–}1)$$

The term $|X_L - X_C|$ means the absolute value of the difference of the two reactances. That is, the sign of the result is considered positive no matter which reactance is greater. For example, $3 - 7 = -4$, but the absolute value is

$$|3 - 7| = 4$$

When $X_L > X_C$, the circuit is predominantly inductive; and when $X_C > X_L$, the circuit is predominantly capacitive.

The total impedance for a series *RLC* circuit is given by

$$Z_{tot} = \sqrt{R^2 + X_{tot}^2} \qquad (13\text{–}2)$$

and the value of the phase angle between V_s and I is

$$\theta = \tan^{-1}\left(\frac{X_{tot}}{R}\right) \qquad (13\text{–}3)$$

The phase angle is positive (current leading voltage) when the circuit is inductive and negative (current lagging voltage) when the circuit is capacitive.

EXAMPLE 13–1

Determine the total impedance and the phase angle for the series *RLC* circuit in Figure 13–2.

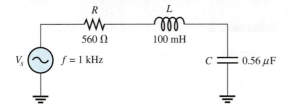

FIGURE 13–2

SOLUTION

First, find X_C and X_L.

$$X_C = \frac{1}{2\pi fC} = \frac{1}{2\pi(1\text{ kHz})(0.56\ \mu\text{F})} = 284\ \Omega$$

$$X_L = 2\pi fL = 2\pi(1\text{ kHz})(100\text{ mH}) = 628\ \Omega$$

In this case, X_L is greater than X_C, so the circuit is more inductive than capacitive. The magnitude of the total reactance is

$$X_{tot} = |X_L - X_C| = |628\ \Omega - 284\ \Omega| = 344\ \Omega \qquad \text{(inductive)}$$

The total circuit impedance is

$$Z_{tot} = \sqrt{R^2 + X_{tot}^2} = \sqrt{(560\ \Omega)^2 + (344\ \Omega)^2} = \mathbf{657\ \Omega}$$

The phase angle (between I and V_s) is

$$\theta = \tan^{-1}\left(\frac{X_{tot}}{R}\right) = \tan^{-1}\left(\frac{344\ \Omega}{560\ \Omega}\right) = \mathbf{31.6°} \qquad \text{(current lagging } V_s\text{)}$$

RELATED PROBLEM*

Increase the frequency in Figure 13–2 to 2000 Hz and determine Z and θ.

*Answers are at the end of the chapter.

As you have seen, when the inductive reactance is greater than the capacitive reactance, the circuit appears to be inductive, and the current lags the source voltage. When the capacitive reactance is greater, the circuit appears to be capacitive, and the current leads the source voltage.

SECTION 13–1 CHECKUP*

1. Explain how you can determine if a series *RLC* circuit is inductive or capacitive.

2. In a given series *RLC* circuit, X_C is 150 Ω and X_L is 80 Ω. What is the total reactance in ohms? Is the circuit inductive or capacitive?

3. Determine the impedance for the circuit in Question 2 when $R = 45\ \Omega$. What is the phase angle? Is the current leading or lagging the source voltage?

*Answers are at the end of the chapter.

Recall that capacitive reactance varies inversely with frequency and that inductive reactance varies directly with frequency. In this section, the combined effects of the reactances as a function of frequency are examined.

After completing this section, you should be able to

- **Analyze series *RLC* circuits**
  - **Determine current in a series *RLC* circuit**
  - **Determine the voltages in a series *RLC* circuit**
  - **Determine the phase angle**

Figure 13–3 shows that for a typical series *RLC* circuit the total reactance behaves as follows: Starting at a very low frequency, X_C is high, X_L is low, and the circuit is predominantly capacitive. As the frequency is increased, X_C decreases and X_L increases until a value is reached where $X_C = X_L$ and the two reactances cancel, making the circuit purely resistive. This condition is **series resonance** and will be studied in Section 13–3. As the frequency is increased further, X_L becomes greater than X_C, and the circuit is predominantly inductive. Example 13–2 illustrates how the impedance and phase angle change as the source frequency is varied.

FIGURE 13–3 **How X_C and X_L vary with frequency.**

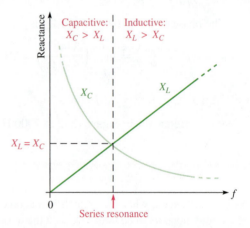

The graph of X_L is a straight line, and the graph of X_C is curved, as shown in Figure 13–3. The general equation for a straight line is $y = mx + b$, where m is the slope of the line and b is the y-axis intercept point. The formula $X_L = 2\pi fL$ fits this general straight-line formula, where $y = X_L$ (a variable), $m = 2\pi L$ (a constant), $x = f$ (a variable), and $b = 0$ as follows: $X_L = 2\pi Lf + 0$.

The X_C curve is called a *hyperbola,* and the general equation of a hyperbola is $xy = k$. The equation for capacitive reactance, $X_C = 1/2\pi fC$, can be rearranged as $X_C f = 1/2\pi C$ where $x = X_C$ (a variable), $y = f$ (a variable), and $k = 1/2\pi C$ (a constant).

EXAMPLE 13–2

For each of the following frequencies of the source voltage, find the impedance and the phase angle for the circuit in Figure 13–4. Note the change in the impedance and the phase angle with frequency.

(a) $f = 1$ kHz

(b) $f = 3.5$ kHz

(c) $f = 5$ kHz

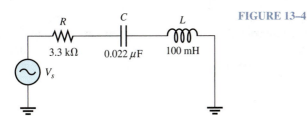

FIGURE 13–4

SOLUTION

(a) At $f = 1$ kHz,

$$X_C = \frac{1}{2\pi fC} = \frac{1}{2\pi(1 \text{ kHz})(0.022 \ \mu\text{F})} = 7.23 \text{ k}\Omega$$

$$X_L = 2\pi fL = 2\pi(1 \text{ kHz})(100 \text{ mH}) = 628 \ \Omega$$

The circuit is highly capacitive because X_C is much larger than X_L. The magnitude of the total reactance is

$$X_{tot} = |X_L - X_C| = |628 \ \Omega - 7.23 \text{ k}\Omega| = 6.6 \text{ k}\Omega$$

The impedance is

$$Z = \sqrt{R^2 + X_{tot}^2} = \sqrt{(3.3 \text{ k}\Omega)^2 + (6.6 \text{ k}\Omega)^2} = \mathbf{7.38 \text{ k}\Omega}$$

The phase angle is

$$\theta = \tan^{-1}\left(\frac{X_{tot}}{R}\right) = \tan^{-1}\left(\frac{6.60 \text{ k}\Omega}{3.3 \text{ k}\Omega}\right) = \mathbf{63.4°}$$

I leads V_s by 63.4°.

(b) At $f = 3.5$ kHz,

$$X_C = \frac{1}{2\pi(3.5 \text{ kHz})(0.022 \ \mu\text{F})} = 2.07 \text{ k}\Omega$$

$$X_L = 2\pi(3.5 \text{ kHz})(100 \text{ mH}) = 2.20 \text{ k}\Omega$$

The circuit is very close to being purely resistive but is slightly inductive because X_L is slightly larger than X_C. The total reactance, impedance, and phase angle are

$$X_{tot} = |2.20 \text{ k}\Omega - 2.07 \text{ k}\Omega| = 130 \ \Omega$$

$$Z = \sqrt{(3.3 \text{ k}\Omega)^2 + (130 \ \Omega)^2} = \mathbf{3.30 \text{ k}\Omega}$$

$$\theta = \tan^{-1}\left(\frac{130 \ \Omega}{3.3 \text{ k}\Omega}\right) = \mathbf{2.26°}$$

I lags V_s by 2.26°.

(c) At $f = 5$ kHz,

$$X_C = \frac{1}{2\pi(5 \text{ kHz})(0.022 \ \mu\text{F})} = 1.45 \text{ k}\Omega$$

$$X_L = 2\pi(5 \text{ kHz})(100 \text{ mH}) = 3.14 \text{ k}\Omega$$

The circuit is now predominantly inductive because $X_L > X_C$. The total reactance, impedance, and phase angle are

$$X_{tot} = |3.14 \text{ k}\Omega - 1.45 \text{ k}\Omega| = 1.69 \text{ k}\Omega$$

$$Z = \sqrt{(3.3 \text{ k}\Omega)^2 + (1.69 \text{ k}\Omega)^2} = \mathbf{3.71 \text{ k}\Omega}$$

$$\theta = \tan^{-1}\left(\frac{1.69 \text{ k}\Omega}{3.3 \text{ k}\Omega}\right) = \mathbf{27.1°}$$

I lags V_s by 27.1°.

Notice how the circuit changed from capacitive to inductive as the frequency increased. The phase condition changed from the current leading to the current lagging. It is important to note that both the impedance and the phase angle decreased to a minimum and then began increasing again as the frequency went up.

RELATED PROBLEM

Determine *Z* for $f = 7$ kHz in Figure 13–4.

In a series *RLC* circuit, the capacitor voltage and the inductor voltage are always 180° out of phase with each other. For this reason, V_C and V_L subtract from each other, and thus the voltage across *L* and *C* combined is always less than the larger individual voltage across either component, as illustrated in Figure 13–5 and in the waveform diagram of Figure 13–6.

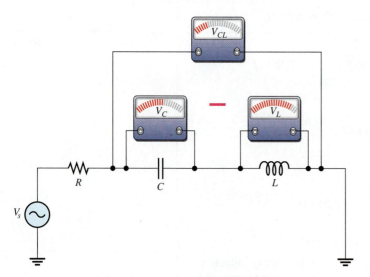

FIGURE 13–5   The voltage across the series combination of *C* and *L* is always less than the larger individual voltage across either *C* or *L*.

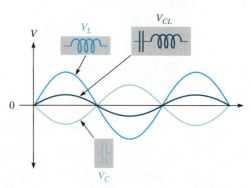

FIGURE 13–6   Inductor voltage and capacitor voltage effectively subtract because they are out of phase.

In the next example, Ohm's law is used to find the current and voltages in a series *RLC* circuit.

EXAMPLE 13–3

Find the voltage across each component in Figure 13–7, and draw a complete voltage phasor diagram. Also find the voltage across *L* and *C* combined.

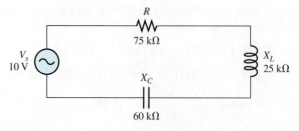

FIGURE 13–7

SOLUTION

First, find the total reactance.

$$X_{tot} = |X_L - X_C| = |25\,\text{k}\Omega - 60\,\text{k}\Omega| = 35\,\text{k}\Omega$$

The total impedance is

$$Z_{tot} = \sqrt{R^2 + X_{tot}^2} = \sqrt{(75\,\text{k}\Omega)^2 + (35\,\text{k}\Omega)^2} = 82.8\,\text{k}\Omega$$

Apply Ohm's law to find the current.

$$I = \frac{V_s}{Z_{tot}} = \frac{10\,\text{V}}{82.8\,\text{k}\Omega} = 121\,\mu\text{A}$$

Now, apply Ohm's law to find the voltages across *R*, *L*, and *C*.

$$V_R = IR = (121\,\mu\text{A})(75\,\text{k}\Omega) = \textbf{9.08 V}$$

$$V_L = IX_L = (121\,\mu\text{A})(25\,\text{k}\Omega) = \textbf{3.03 V}$$

$$V_C = IX_C = (121\,\mu\text{A})(60\,\text{k}\Omega) = \textbf{7.26 V}$$

The voltage across *L* and *C* combined is

$$V_{CL} = V_C - V_L = 7.26\,\text{V} - 3.03\,\text{V} = \textbf{4.23 V}$$

The circuit phase angle is

$$\theta = \tan^{-1}\!\left(\frac{X_{tot}}{R}\right) = \tan^{-1}\!\left(\frac{35\,\text{k}\Omega}{75\,\text{k}\Omega}\right) = 25°$$

Since the circuit is capacitive ($X_C > X_L$), the current leads the source voltage by 25°.

The phasor diagram is shown in Figure 13–8. Notice that V_L is leading V_R by 90°, and V_C is lagging V_R by 90°. Also, there is a 180° phase difference between V_L and V_C. If the current phasor were shown, it would be at the same angle as V_R.

FIGURE 13–8

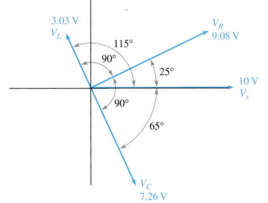

RELATED PROBLEM

What will happen to the current as the frequency of the source voltage in Figure 13–7 is increased?

SECTION 13–2 CHECKUP

1. The following voltages occur in a certain series *RLC* circuit: $V_R = 24\,\text{V}$, $V_L = 15\,\text{V}$, $V_C = 45\,\text{V}$. Determine the source voltage.

2. When $R = 10\,\text{k}\Omega$, $X_C = 18\,\text{k}\Omega$, and $X_L = 12\,\text{k}\Omega$, does the current lead or lag the source voltage? Why?

3. Determine the total reactance in Question 2.

13-3 SERIES RESONANCE

In a series *RLC* circuit, series resonance occurs when $X_L = X_C$. The frequency at which resonance occurs is called the **resonant frequency** and is designated f_r.

After completing this section, you should be able to

- **Analyze a circuit for series resonance**
  - Define *resonant frequency*
  - Explain why the reactances cancel at resonance
  - Determine the series resonant frequency
  - Calculate the current and voltages at resonance
  - Determine the impedance at resonance

Figure 13–9 illustrates the series resonant condition. Since the reactances are equal and effectively cancel, the impedance is purely resistive. These resonant conditions are stated as

$$X_L = X_C$$

$$Z_r = R$$

FIGURE 13–9 At the resonant frequency (f_r), the reactances are equal in magnitude and effectively cancel, leaving $Z_r = R$.

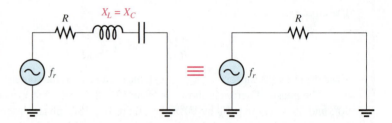

EXAMPLE 13–4

Assume the series *RLC* circuit in Figure 13–10 is at resonance. Determine X_C and Z.

FIGURE 13–10

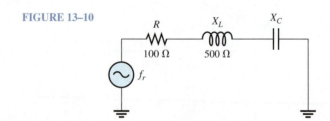

SOLUTION

X_L equals X_C at resonance; therefore,

$$X_C = X_L = \textbf{500 } \boldsymbol{\Omega}$$

Since the reactances cancel each other, the impedance equals the resistance.

$$Z_r = R = \textbf{100 } \boldsymbol{\Omega}$$

RELATED PROBLEM

Just below the resonant frequency, is the circuit more inductive or more capacitive?

At the series resonant frequency, the voltages across *C* and *L* are equal in magnitude because the reactances are equal. The same current is through both since they are in series ($IX_C = IX_L$). Also, V_L and V_C are always 180° out of phase with each other.

During any given cycle of the input voltage, the polarities of the voltages across C and L are opposite, as shown in parts (a) and (b) of Figure 13–11. The equal and opposite voltages across C and L cancel, leaving zero volts from point A to point B as shown in the figure. Since there is no voltage drop from A to B but there is still current, the total reactance must be zero, as indicated in Figure 13–11(c). Also, the voltage phasor diagram in part (d) shows that V_C and V_L are equal in magnitude and 180° out of phase with each other.

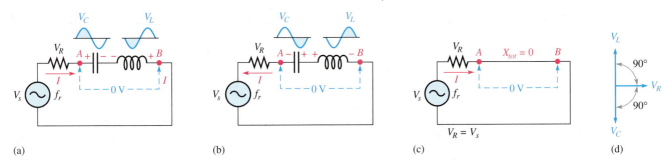

(a)   (b)   (c)   (d)

FIGURE 13–11   At the resonant frequency, f_r, the voltages across C and L are equal in magnitude. Since they are 180° out of phase with each other, they cancel, leaving 0 V across the CL combination (point A to point B). The section of the circuit from A to B effectively looks like a short at resonance (neglecting winding resistance).

Series Resonant Frequency

For a given series RLC circuit, resonance occurs at only one specific frequency. A formula for this resonant frequency for the ideal case is developed as follows:

$$X_L = X_C$$

Substitute the reactance formulas, and solve for the resonant frequency (f_r).

$$2\pi f_r L = \frac{1}{2\pi f_r C}$$

$$(2\pi f_r L)(2\pi f_r C) = 4\pi^2 f_r^2 LC = 1$$

$$f_r^2 = \frac{1}{4\pi^2 LC}$$

Take the square root of both sides. The formula for resonant frequency is

$$f_r = \frac{1}{2\pi\sqrt{LC}} \qquad\qquad (13\text{–}4)$$

EXAMPLE 13–5

Find the series resonant frequency for the circuit in Figure 13–12.

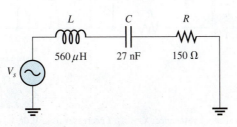

FIGURE 13–12

SOLUTION

The resonant frequency is

$$f_r = \frac{1}{2\pi\sqrt{LC}} = \frac{1}{2\pi\sqrt{(560\ \mu H)(27\ nF)}} = \textbf{40.9 kHz}$$

RELATED PROBLEM

If $C = 0.01\ \mu F$ in Figure 13–12, what is the resonant frequency?

Voltage and Current Amplitudes in a Series *RLC* Circuit

The current and voltage amplitudes in a series *RLC* circuit vary as the frequency is increased from below the resonant frequency, through resonance, and then above resonance. Figure 13–13 illustrates the general response of a circuit in terms of changes in the

FIGURE 13–13 An illustration of how the voltage and current amplitudes respond in a series *RLC* circuit as the frequency is increased from below to above its resonant value. The source voltage is held at a constant amplitude.

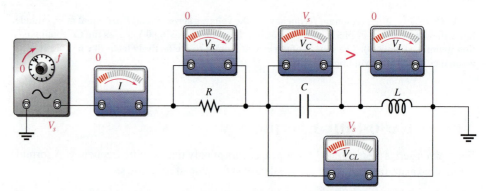

(a) As frequency is increased below resonance from 0: $X_C > X_L$, I increases from 0, V_R increases from 0, V_C increases from V_s, V_L increases from 0, and V_{CL} decreases from V_s.

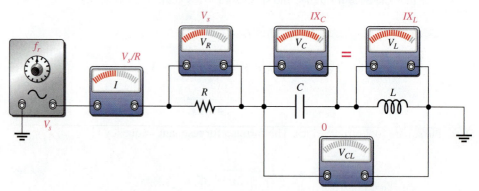

(b) At the resonant frequency, $X_C = X_L$.

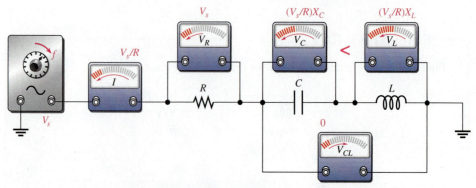

(c) As frequency is increased above resonance: $X_C < X_L$, I decreases from V_s/R, V_R decreases from V_s, V_C decreases from $(V_s/R)X_C$, V_L decreases from $(V_s/R)X_L$, and V_{CL} increases from 0.

current and in the voltage drops. The Q (quality factor) of the circuit is assumed to be sufficiently high so it has no effect on the response. The Q is the ratio of reactive power to true power and is discussed in Section 13–4.

BELOW THE RESONANT FREQUENCY   Figure 13–13(a) indicates the response as the source frequency is increased from zero toward f_r. At $f = 0$ Hz (dc), the capacitor appears open and blocks the current. Thus, there is no voltage across R or L, and the entire source voltage appears across C. The impedance of the circuit is infinitely large at 0 Hz because X_C is infinite (C is open). As the frequency begins to increase, X_C decreases and X_L increases, causing the total reactance, $X_C - X_L$, to decrease. As a result, the impedance decreases and the current increases. As the current increases, V_R also increases, and both V_C and V_L increase. The voltage across C and L combined decreases from its maximum value of V_s because as V_C and V_L approach the same value, their difference becomes less.

AT THE RESONANT FREQUENCY   When the frequency reaches its resonant value, f_r, as shown in Figure 13–13(b), V_C and V_L can each be much larger than the source voltage. V_C and V_L cancel, leaving 0 V across the combination of C and L because the voltages are equal in magnitude but opposite in phase. At this point the total impedance is equal to R and is at its minimum value because the total reactance is zero. Thus, the current is at its maximum value, V_s/R, and V_R is at its maximum value, which is equal to the source voltage.

ABOVE THE RESONANT FREQUENCY   As the frequency is increased above resonance, as indicated in Figure 13–13(c), X_L continues to increase and X_C continues to decrease, causing the total reactance, $X_L - X_C$, to increase. As a result, there is an increase in impedance and a decrease in current. As the current decreases, V_R also decreases and V_C and V_L decrease. As V_C and V_L decrease, their difference becomes greater, so V_{CL} increases. As the frequency becomes very high, the current approaches zero, both V_R and V_C approach zero, and V_L approaches V_s.

Figure 13–14(a) and (b) illustrate the responses of current and voltage to increasing frequency, respectively. As frequency is increased, the current increases below resonance, reaches a peak at the resonant frequency, and then decreases above resonance. The resistor voltage responds in the same way as the current.

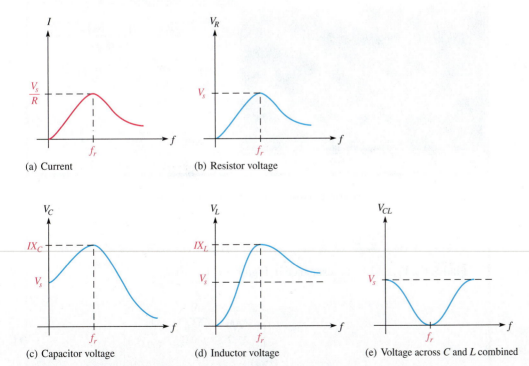

(a) Current

(b) Resistor voltage

(c) Capacitor voltage

(d) Inductor voltage

(e) Voltage across C and L combined

FIGURE 13–14   **Generalized current and voltage magnitudes as a function of frequency in a series *RLC* circuit. V_C and V_L can be much larger than the source voltage. The shapes of the graphs depend on particular circuit values.**

Capacitive and Inductive Sensors

Sensors are widely used by many systems to measure certain parameters such as temperature, pressure, or other variable. A common method of indirectly measuring the capacitance or inductance of a sensor is to measure the resonant frequency of an oscillator as the sensor variable changes. One unusual example is to infer the progress of a shock wave as it crushes a cable and changes its capacitance.

eplosione/Fotolia.com

SYSTEM NOTE

The general slopes of the V_C and V_L curves are indicated in Figure 13–14(c) and (d). The voltages are maximum at resonance but drop off above and below f_r. The voltages across L and C at resonance are exactly equal in magnitude but 180° out of phase, so they cancel. Thus, the total voltage across both L and C is zero, and $V_R = V_s$ at resonance. Individually, V_L and V_C can be much greater than the source voltage. Keep in mind that V_L and V_C are always opposite in polarity regardless of the frequency, but only at resonance are their magnitudes equal. The voltage across the C and L combination decreases as the frequency increases below resonance, reaching a minimum of zero at the resonant frequency; then it increases above resonance, as shown in Figure 13–14(e).

The curves in Figure 13–14 are generalized curves, which depend on the particular components in the circuit. The shape will depend greatly on Q. You can plot these curves for any circuit by constructing the circuit in Multisim and using the Bode plotter. The **Bode plotter** in Multisim is a fictitious instrument that plots the frequency response based on where the probes are placed. The response across the resistor in Example 13–5 is shown in Figure 13–15.

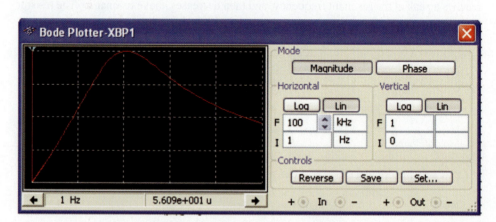

FIGURE 13–15   **Multisim response across the resistor in Example 13–5.**

E X A M P L E   1 3 – 6

Find I, V_R, V_L, and V_C at resonance in Figure 13–16.

FIGURE 13–16

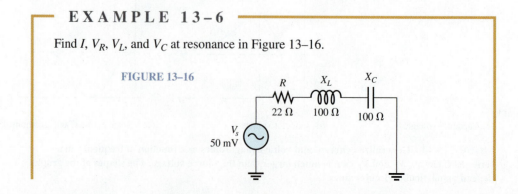

SOLUTION

At resonance, I is maximum and equal to V_s/R.

$$I = \frac{V_s}{R} = \frac{50\text{ mV}}{22\text{ }\Omega} = \textbf{2.27 mA}$$

Apply Ohm's law to obtain the voltages.

$$V_R = IR = (2.27\text{ mA})(22\text{ }\Omega) = \textbf{50 mV}$$
$$V_L = IX_L = (2.27\text{ mA})(100\text{ }\Omega) = \textbf{227 mV}$$
$$V_C = IX_C = (2.27\text{ mA})(100\text{ }\Omega) = \textbf{227 mV}$$

Notice that all of the source voltage is dropped across the resistor. Also, of course, V_L and V_C are equal in magnitude but opposite in phase. This causes the voltages to cancel, making the total reactive voltage zero.

RELATED PROBLEM

What is the current at resonance in Figure 13–16 if $X_L = X_C = 1\text{ k}\Omega$?

The Impedance of a Series *RLC* Circuit

Figure 13–17 shows a general graph of impedance versus frequency superimposed on the curves for X_C and X_L. At zero frequency, both X_C and Z are infinitely large and X_L is zero because the capacitor looks like an open at 0 Hz and the inductor looks like a short. As the frequency increases, X_C decreases and X_L increases. Since X_C is larger than X_L at frequencies below f_r, Z decreases along with X_C. At f_r, $X_C = X_L$ and $Z = R$. At frequencies above f_r, X_L becomes increasingly larger than X_C, causing Z to increase.

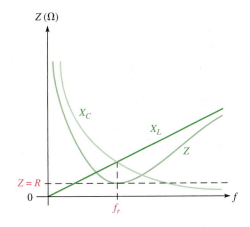

FIGURE 13–17   Series *RLC* impedance as a function of frequency.

EXAMPLE 13–7

For the circuit in Figure 13–18, determine the impedance at the following frequencies:

(a) f_r

(b) 1 kHz below f_r

(c) 1 kHz above f_r

FIGURE 13–18

(circuit diagram: V_s source, R = 100 Ω, L = 100 mH, C = 0.01 μF)

SOLUTION

(a) At f_r, the impedance is equal to R.

$$Z = R = \mathbf{100\ \Omega}$$

To determine the impedance above and below f_r, first calculate the resonant frequency.

$$f_r = \frac{1}{2\pi\sqrt{LC}} = \frac{1}{2\pi\sqrt{(100\ \text{mH})(0.01\ \mu\text{F})}} = 5.03\ \text{kHz}$$

(b) At 1 kHz below f_r, the frequency and reactances are as follows:

$$f = f_r - 1\ \text{kHz} = 5.03\ \text{kHz} - 1\ \text{kHz} = 4.03\ \text{kHz}$$

$$X_C = \frac{1}{2\pi fC} = \frac{1}{2\pi(4.03\ \text{kHz})(0.01\ \mu\text{F})} = 3.95\ \text{k}\Omega$$

$$X_L = 2\pi fL = 2\pi(4.03\ \text{kHz})(100\ \text{mH}) = 2.53\ \text{k}\Omega$$

Therefore, the total reactance and the impedance at $f_r - 1$ kHz are

$$X_{tot} = |X_L - X_C| = |2.53\ \text{k}\Omega - 3.95\ \text{k}\Omega| = 1.42\ \text{k}\Omega$$

$$Z = \sqrt{R^2 + X_{tot}^2} = \sqrt{(100\ \Omega)^2 + (1.42\ \text{k}\Omega)^2} = \mathbf{1.42\ k\Omega}$$

(c) At 1 kHz above f_r,

$$f = 5.03\ \text{kHz} + 1\ \text{kHz} = 6.03\ \text{kHz}$$

$$X_C = \frac{1}{2\pi(6.03\ \text{kHz})(0.01\ \mu\text{F})} = 2.64\ \text{k}\Omega$$

$$X_L = 2\pi(6.03\ \text{kHz})(100\ \text{mH}) = 3.79\ \text{k}\Omega$$

Therefore, the total reactance and the impedance at $f_r + 1$ kHz are

$$X_{tot} = |3.79\ \text{k}\Omega - 2.64\ \text{k}\Omega| = 1.15\ \text{k}\Omega$$

$$Z = \sqrt{(100\ \Omega)^2 + (1.15\ \text{k}\Omega)^2} = \mathbf{1.15\ k\Omega}$$

In part (b), Z is capacitive; in part (c), Z is inductive.

MULTISIM

Open Multisim file E13-07. Measure the current and the voltage across each element at the resonant frequency, 1 kHz below resonance, and 1 kHz above resonance.

RELATED PROBLEM

What happens to the impedance if f is decreased below 4.03 kHz? Above 6.03 kHz?

The Phase Angle of a Series *RLC* Circuit

At frequencies below resonance, $X_C > X_L$, and the current leads the source voltage, as indicated in Figure 13–19(a). The phase angle decreases as the frequency approaches the resonant value and is 0° at resonance, as indicated in part (b). At frequencies above resonance, $X_L > X_C$, and the current lags the source voltage, as indicated in part (c). As the frequency goes higher, the phase angle approaches 90°. A plot of phase angle versus frequency is shown in Figure 13–19(d).

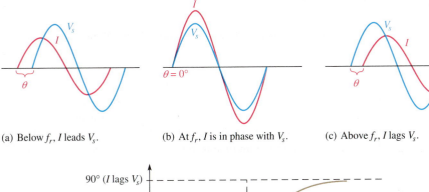

(a) Below f_r, I leads V_s.   (b) At f_r, I is in phase with V_s.   (c) Above f_r, I lags V_s.

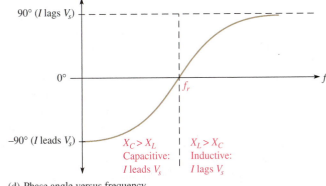

(d) Phase angle versus frequency.

FIGURE 13–19   **The phase angle as a function of frequency in a series *RLC* circuit.**

SECTION 13–3 CHECKUP

1. What is the condition for series resonance?

2. Why is the current maximum at the resonant frequency?

3. Calculate the resonant frequency for $C = 1000$ pF and $L = 1000$ μH.

4. In Question 3, is the circuit inductive, capacitive, or resistive at 50 kHz?

13–4 SERIES RESONANT FILTERS

A common use of series *RLC* circuits is in filter applications. In this section, you will learn the basic configurations for passive band-pass and band-stop filters and several important filter characteristics.

After completing this section, you should be able to

- **Analyze series resonant filters**
  - Identify a basic series resonant band-pass filter
  - Define and determine *bandwidth*
  - Define the *half-power frequency*
  - Discuss dB measurement
  - Define *selectivity*
  - Discuss filter quality factor (*Q*)
  - Identify a series resonant band-stop filter

The Band-Pass Filter

A basic series resonant band-pass filter is shown in Figure 13–20. Notice that the series *LC* portion is placed between the input and the output and that the output is taken across the resistor.

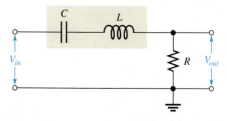

FIGURE 13–20   A basic series resonant band-pass filter.

A **band-pass filter** allows signals at the resonant frequency and at frequencies within a certain band (or range) extending below and above the resonant value to pass from input to output without a significant reduction in amplitude. Signals at frequencies lying outside this specified band (called the **passband**) are reduced in amplitude to below a certain level and are considered to be rejected by the filter.

The filtering action is the result of the impedance characteristic of the filter. As you learned in Section 13–3, the impedance is minimum at resonance and has increasingly higher values below and above the resonant frequency. At very low frequencies, the impedance is very high and tends to block the current. As the frequency increases, the impedance drops, allowing more current and thus more voltage across the output resistor. At the resonant frequency, the impedance is very low and equal to the total resistance of the circuit (winding resistance plus *R*). At this point there is maximum current and the resulting output voltage is maximum. As the frequency goes above resonance, the impedance again increases, causing the current and the resulting output voltage to drop. Figure 13–21 illustrates the general frequency response of a series resonant band-pass filter.

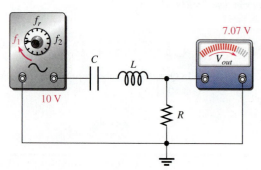

(a) As the frequency increases to f_1, V_{out} increases to 7.07 V.

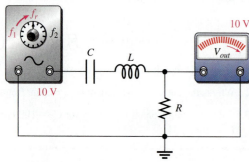

(b) As the frequency increases from f_1 to f_r, V_{out} increases from 7.07 V to 10 V.

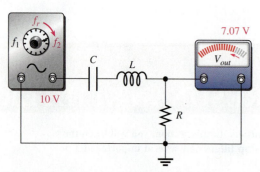

(c) As the frequency increases from f_r to f_2, V_{out} decreases from 10 V to 7.07 V.

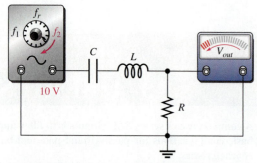

(d) As the frequency increases above f_2, V_{out} decreases below 7.07 V.

FIGURE 13–21   Example of the frequency response of a series resonant band-pass filter with the input voltage at a constant 10 V rms. The winding resistance of the coil is neglected.

Bandwidth of the Passband

The bandwidth (*BW*) of a band-pass filter is the range of frequencies for which the current (or output voltage) is equal to or greater than 70.7% of its value at the resonant frequency. Figure 13–22 shows the bandwidth on the response curve for a band-pass filter.

The frequencies at which the output of a filter is 70.7% of its maximum are the **cutoff frequencies**. Notice in Figure 13–22 that frequency f_1, which is below f_r, is the frequency at which I (or V_{out}) is 70.7% of the resonant value (I_{max}); f_1 is commonly called the *lower cutoff frequency*. At frequency f_2, above f_r, the current (or V_{out}) is again 70.7% of its maximum; f_2 is called the *upper cutoff frequency*. Other names for f_1 and f_2 are *−3 dB frequencies, critical frequencies, band frequencies,* and *half-power frequencies.* (The term *decibel,* abbreviated dB, is defined later in this section.)

The formula for calculating the bandwidth is

$$BW = f_2 - f_1 \qquad (13\text{--}5)$$

The unit of bandwidth is the hertz (Hz), the same as for frequency.

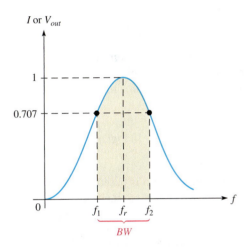

FIGURE 13–22 **Generalized response curve of a series resonant band-pass filter.**

EXAMPLE 13–8

A certain series resonant band-pass filter has a maximum current of 100 mA at the resonant frequency. What is the value of the current at the cutoff frequencies?

SOLUTION

Current at the cutoff frequencies is 70.7% of maximum.

$$I_{f1} = I_{f2} = 0.707I_{max} = 0.707(100 \text{ mA}) = \textbf{70.7 mA}$$

RELATED PROBLEM

Will a change in the cutoff frequencies affect the current at the new cutoff frequencies if the maximum current remains 100 mA?

EXAMPLE 13–9

A resonant circuit has a lower cutoff frequency of 8 kHz and an upper cutoff frequency of 12 kHz. Determine the bandwidth.

SOLUTION

$$BW = f_2 - f_1 = 12 \text{ kHz} - 8 \text{ kHz} = \textbf{4 kHz}$$

RELATED PROBLEM

What is the bandwidth when $f_1 = 1.0 \text{ MHz}$ and $f_2 = 1.2 \text{ MHz}$?

Half-Power Points of the Filter Response

As previously mentioned, the upper and lower cutoff frequencies are sometimes called the **half-power frequencies**. This term is derived from the fact that the true power from the source at these frequencies is one-half the power delivered at the resonant frequency. The following steps show that this relationship is true for a series resonant circuit.

At resonance,

$$P_{max} = I_{max}^2 R$$

The power at f_1 (or f_2) is

$$P_{f1} = I_{f1}^2 R = (0.707 I_{max})^2 R = (0.707)^2 I_{max}^2 R = 0.5 I_{max}^2 R = 0.5 P_{max}$$

Decibel (dB) Measurement

Another common term for the upper and lower cutoff frequencies is *−3 dB frequencies*. The **decibel (dB)** is ten times the logarithmic ratio of two powers, which can be used to express the input-to-output relationship of a filter. The following equation expresses a power ratio in decibels:

$$\text{dB} = 10 \log\left(\frac{P_{out}}{P_{in}}\right) \tag{13–6}$$

The decibel formula using a voltage ratio is based on the previous formula for the power ratio with voltages measured across equal resistances.

$$\text{dB} = 20 \log\left(\frac{V_{out}}{V_{in}}\right) \tag{13–7}$$

EXAMPLE 13–10

At a certain frequency, the output power of a filter is 5 W and the input power is 10 W. Express the power ratio in decibels.

SOLUTION

$$10 \log\left(\frac{P_{out}}{P_{in}}\right) = 10 \log\left(\frac{5\text{ W}}{10\text{ W}}\right) = 10 \log(0.5) = \textbf{−3.01 dB}$$

RELATED PROBLEM

Express in decibels the ratio $V_{out}/V_{in} = 0.2$.

THE −3 dB FREQUENCIES The output of a filter is said to be down 3 dB at the cutoff frequencies. As you know, this frequency is the point at which the output voltage is 70.7% of the maximum voltage at resonance. We can show that the 70.7% point is the same as 3 dB below maximum (or −3 dB) as follows. The maximum voltage is the 0 dB reference.

$$20 \log\left(\frac{0.707 V_{max}}{V_{max}}\right) = 20 \log(0.707) = -3\text{ dB}$$

Selectivity of a Band-Pass Filter

The response curve in Figure 13–22 is also called a *selectivity curve*. **Selectivity** defines how well a resonant circuit responds to a certain frequency and discriminates against all others. *The narrower the bandwidth, the greater the selectivity.*

In the ideal response, we assume that a resonant circuit accepts frequencies within its bandwidth and completely eliminates frequencies outside the bandwidth. Such is not actually the case, however, because signals with frequencies outside the bandwidth are not completely eliminated. Their magnitudes, however, are greatly reduced. The further the frequencies are from the cutoff frequencies, the greater the reduction, as illustrated in Figure 13–23(a). An ideal selectivity curve is shown in Figure 13–23(b).

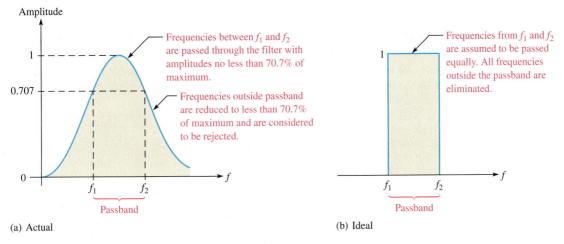

(a) Actual

(b) Ideal

FIGURE 13–23   **Generalized selectivity curve of a band-pass filter.**

The factor that determines the selectivity of a practical band-pass filter is the steepness of the response curve, as illustrated in Figure 13–24. A more selective filter attenuates (reduces) frequencies outside the passband more quickly than a less selective filter. In communication systems, a highly selective filter will differentiate between a desired signal (such as from a radio station) and a nearby one.

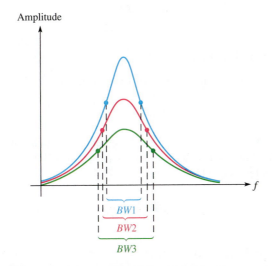

FIGURE 13–24   **Comparative selectivity curves.**
The blue curve represents the greatest selectivity.

The Quality Factor (Q) of a Resonant Circuit

Recall that the quality factor (Q) for a coil was defined in Section 11–5 as the ratio of the reactive power (at a specified frequency) to the true power, which is dissipated in the winding resistance of the coil (L). In a series resonant circuit, the Q of the circuit includes any other resistance in series with the coil, so the circuit Q may be lower than the Q of the coil alone because R includes more than just the coil resistance. The Q of a series resonant circuit is the ratio of the reactive power in *L* to the true power in *R*. The quality factor is important in resonant circuits. A formula for *Q* is developed as follows:

$$Q = \frac{\text{reactive power}}{\text{true power}} = \frac{I^2 X_L}{I^2 R}$$

The I^2 terms cancel, leaving

$$Q = \frac{X_L}{R} \qquad (13\text{–}8)$$

Since *Q* varies with frequency because X_L varies, we are interested mainly in *Q* at resonance. Note that *Q* is a ratio of like units (ohms) which cancel and, therefore, *Q* has no unit itself. The quality factor is also known as *unloaded Q* because it is defined with no load across the coil.

EXAMPLE 13–11

Determine *Q* at resonance for the circuit in Figure 13–25.

FIGURE 13–25

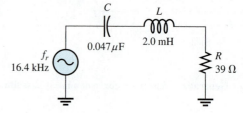

SOLUTION

Determine the inductive reactance.

$$X_L = 2\pi f_r L = 2\pi(16.4 \text{ kHz})(2.0 \text{ mH}) = 206 \ \Omega$$

The quality factor is

$$Q = \frac{X_L}{R} = \frac{206 \ \Omega}{39 \ \Omega} = \mathbf{5.3}$$

RELATED PROBLEM

Calculate *Q* at resonance if *C* in Figure 13–25 is halved. The resonant frequency will increase.

HOW *Q* AFFECTS BANDWIDTH   A higher value of circuit *Q* results in a smaller bandwidth. A lower value of *Q* causes a larger bandwidth. A formula for the bandwidth of a resonant circuit in terms of *Q* is

$$BW = \frac{f_r}{Q} \qquad (13\text{–}9)$$

EXAMPLE 13–12

What is the bandwidth of the filter in Figure 13–26?

FIGURE 13–26

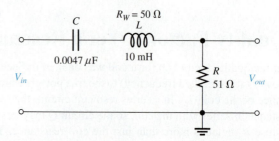

SOLUTION

The total resistance is

$$R_{tot} = R + R_W = 51\ \Omega + 50\ \Omega = 101\ \Omega$$

Determine the bandwidth as follows:

$$f_r = \frac{1}{2\pi\sqrt{LC}} = \frac{1}{2\pi\sqrt{(10\ \text{mH})(0.0047\ \mu\text{F})}} = 23.2\ \text{kHz}$$

$$X_L = 2\pi f_r L = 2\pi(23.2\ \text{kHz})(10\ \text{mH}) = 1.46\ \text{k}\Omega$$

$$Q = \frac{X_L}{R_{tot}} = \frac{1.46\ \text{k}\Omega}{101\ \Omega} = 14.5$$

$$BW = \frac{f_r}{Q} = \frac{23.2\ \text{kHz}}{14.5} = \mathbf{1.60\ kHz}$$

RELATED PROBLEM

Determine the bandwidth if L is changed to 50 mH with the some winding resistance in Figure 13–26.

MULTISIM

Open Multisim file E13-12. Determine the bandwidth by measurement. How closely does this result compare to the calculated value?

The Band-Stop Filter

The basic series resonant band-stop filter is shown in Figure 13–27. Notice that the output voltage is taken across the *LC* portion of the circuit. This filter is still a series *RLC* circuit, just as the band-pass filter is. The difference is that in this case, the output voltage is taken across the combination of *L* and *C* rather than across *R*.

The **band-stop filter** rejects signals with frequencies between the lower and upper cutoff frequencies and passes those signals with frequencies below and above the cutoff values, as shown in the response curve of Figure 13–28. The range of frequencies between the lower and upper cutoff points is called the **stopband**. This type of filter is also referred to as a *band-elimination filter, band-reject filter,* or a *notch filter.*

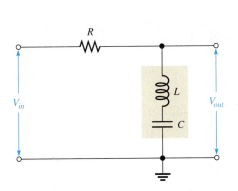

FIGURE 13–27 A basic series resonant band-stop filter.

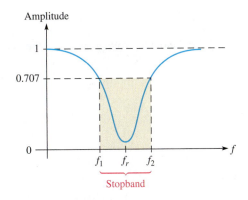

FIGURE 13–28 Generalized response curve for a band-stop filter.

All the characteristics that have been discussed in relation to the band-pass filter apply equally to the band-stop filter, with the exception that the response curve of the output voltage is opposite. For the band-pass filter, V_{out} is maximum at resonance. For the band-stop filter, V_{out} is minimum at resonance.

At very low frequencies, the *LC* combination appears as a near open due to the high X_C, thus allowing most of the input voltage to pass through to the output. As the frequency increases, the impedance of the *LC* combination decreases until, at resonance, it is zero (ideally). Thus, the input signal is shorted to ground, and there is very little output voltage. As the frequency goes above its resonant value, the *LC* impedance increases, allowing an increasing amount of voltage to be dropped across it. The general frequency response of a series resonant band-stop filter is illustrated in Figure 13–29.

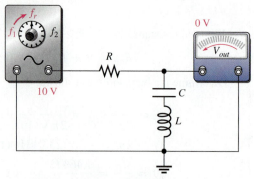

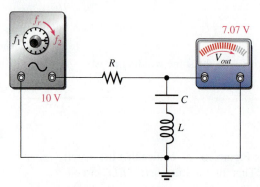

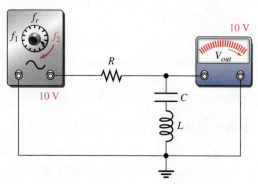

(a) As frequency increases to f_1, V_{out} decreases from 10 V to 7.07 V.

(b) As frequency increases from f_1 to f_r, V_{out} decreases from 7.07 V to 0 V.

(c) As frequency increases from f_r to f_2, V_{out} increases from 0 V to 7.07 V.

(d) As frequency increases above f_2, V_{out} increases toward 10 V.

FIGURE 13–29 **Example of the frequency response of a series resonant band-stop filter with V_{in} at a constant 10 V rms. The winding resistance is neglected.**

EXAMPLE 13–13

Find the output voltage at f_r and the bandwidth in Figure 13–30.

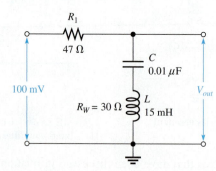

FIGURE 13–30

SOLUTION

Since $X_L = X_C$ at resonance, we can apply the voltage-divider formula to find V_{out}.

$$V_{out} = \left(\frac{R_W}{R_1 + R_W}\right)V_{in} = \left(\frac{30\ \Omega}{77\ \Omega}\right)100\ \text{mV} = \mathbf{39.0\ mV}$$

Determine the bandwidth as follows:

$$f_r = \frac{1}{2\pi\sqrt{LC}} = \frac{1}{2\pi\sqrt{(15\text{ mH})(0.01\ \mu\text{F})}} = 13.0\text{ kHz}$$

$$X_L = 2\pi f_r L = 2\pi(13.0\text{ kHz})(15\text{ mH}) = 1.22\text{ k}\Omega$$

$$Q = \frac{X_L}{R} = \frac{X_L}{R_1 + R_W} = \frac{1.22\text{ k}\Omega}{77\ \Omega} = 15.9$$

$$BW = \frac{f_r}{Q} = \frac{1.22\text{ kHz}}{15.9} = \mathbf{810\text{ Hz}}$$

RELATED PROBLEM

What happens to V_{out} if the frequency is increased above resonance? Below resonance?

SECTION 13–4 CHECKUP

1. The output voltage of a certain band-pass filter is 15 V at the resonant frequency. What is its value at the cutoff frequencies?

2. For a certain band-pass filter, $f_r = 120$ kHz and $Q = 12$. What is the bandwidth of the filter?

3. In a band-stop filter, is the current minimum or maximum at resonance? Is the output voltage minimum or maximum at resonance?

13–5 PARALLEL *RLC* CIRCUITS

In this section, you will learn how to determine the impedance and phase angle of a parallel *RLC* circuit. Current relationships and series-parallel to parallel conversions are discussed.

After completing this section, you should be able to

- **Analyze parallel *RLC* circuits**
  - **Calculate the impedance**
  - **Calculate the phase angle**
  - **Determine all currents**
  - **Convert from a series-parallel *RLC* circuit to an equivalent parallel form**

Impedance and Phase Angle

The circuit in Figure 13–31 consists of the parallel combination of R, L, and C. To find the admittance, add the conductance (G) and the total susceptance (B_{tot}) as phasor quantities. B_{tot} is the absolute value of the difference of the inductive susceptance and the capacitive susceptance.

$$B_{tot} = |B_L - B_C|$$

Therefore, the formula for admittance is

$$Y = \sqrt{G^2 + B_{tot}^2} \qquad (13\text{–}10)$$

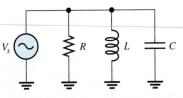

FIGURE 13–31   Parallel *RLC* circuit.

The total impedance is the reciprocal of the admittance.

$$Z_{tot} = \frac{1}{Y}$$

The phase angle of the circuit is given by the following formula:

$$\boldsymbol{\theta} = \tan^{-1}\left(\frac{\boldsymbol{B_{tot}}}{\boldsymbol{G}}\right) \tag{13–11}$$

When the frequency is above its resonant value ($X_C < X_L$), the impedance of the circuit in Figure 13–31 is predominantly capacitive because the capacitive current is greater, and the total current leads the source voltage. When the frequency is below its resonant value ($X_L < X_C$), the impedance of the circuit is predominantly inductive, and the total current lags the source voltage.

EXAMPLE 13–14

Determine the total impedance and the phase angle for the parallel *RLC* circuit in Figure 13–32.

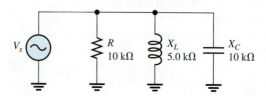

FIGURE 13–32

SOLUTION

First, determine the admittance as follows:

$$G = \frac{1}{R} = \frac{1}{10\,k\Omega} = 100\,\mu S$$

$$B_C = \frac{1}{X_C} = \frac{1}{10\,k\Omega} = 100\,\mu S$$

$$B_L = \frac{1}{X_L} = \frac{1}{5.0\,k\Omega} = 200\,\mu S$$

$$B_{tot} = |B_L - B_C| = 100\,\mu S$$

$$Y = \sqrt{G^2 + B_{tot}^2} = \sqrt{(100\,\mu S)^2 + (100\,\mu S)^2} = 141.4\,\mu S$$

From *Y*, you can get Z_{tot}.

$$Z_{tot} = \frac{1}{Y} = \frac{1}{141.4\,\mu S} = \textbf{7.07 k}\boldsymbol{\Omega}$$

The phase angle is

$$\theta = \tan^{-1}\left(\frac{B_{tot}}{G}\right) = \tan^{-1}\left(\frac{100\,\mu S}{100\,\mu S}\right) = \textbf{45}\boldsymbol{°}$$

RELATED PROBLEM

Is the impedance of the circuit in Figure 13–32 predominantly inductive or predominantly capacitive?

Current Relationships

In a parallel *RLC* circuit, the current in the capacitive branch and the current in the inductive branch are always 180° out of phase with each other (neglecting any coil resistance). For this reason, I_C and I_L subtract from each other, and thus the total current into the parallel branches of *L* and *C* is always less than the largest individual branch current, as illustrated in Figure 13–33 and in the waveform diagram of Figure 13–34. Of course, the current in the resistive branch is always 90° out of phase with both reactive currents, as shown in the current phasor diagram of Figure 13–35. Notice that I_C is plotted on the positive *y*-axis and that I_L is plotted on the negative *y*-axis. The total current can be expressed as

$$I_{tot} = \sqrt{I_R^2 + I_{CL}^2} \qquad (13\text{–}12)$$

Coaxial Cable

Coaxial cable (or "coax" for short) is shielded cable used for transmitting high frequencies and fast-pulsed waveforms. The shielding helps keep external signals from affecting the signal in the cable and forms part of the electrical circuit. In addition to the shielding, coax is characterized by its impedance, attenuation, and capacitance.

Albert Lozano-Nieto/Fotolia

SYSTEM NOTE

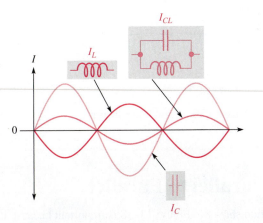

FIGURE 13–33 **The total current into the parallel combination of *C* and *L* is the absolute value of the difference of the two branch currents** $(I_{CL} = |I_C - I_L|)$.

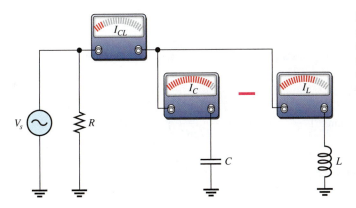

FIGURE 13–34 I_C **and** I_L **effectively subtract from each other.**

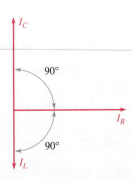

FIGURE 13–35 **Current phasor diagram for a parallel *RLC* circuit.**

where I_{CL} is the absolute value of the difference between the two currents, $|I_C - I_L|$, and is the total current into the *L* and *C* branches.

The phase angle can also be expressed in terms of the branch currents as

$$\theta = \tan^{-1}\left(\frac{I_{CL}}{I_R}\right) \tag{13–13}$$

EXAMPLE 13–15

Find each branch current and the total current in Figure 13–36. Draw a diagram of their relationship.

FIGURE 13–36

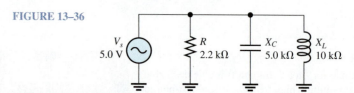

SOLUTION

Apply Ohm's law to find each branch current.

$$I_R = \frac{V_s}{R} = \frac{5.0 \text{ V}}{2.2 \text{ k}\Omega} = \textbf{2.27 mA}$$

$$I_C = \frac{V_s}{X_C} = \frac{5.0 \text{ V}}{5.0 \text{ k}\Omega} = \textbf{1.0 mA}$$

$$I_L = \frac{V_s}{X_L} = \frac{5.0 \text{ V}}{10 \text{ k}\Omega} = \textbf{0.5 mA}$$

The total current is the phasor sum of the branch currents.

$$I_{CL} = |I_C - I_L| = 0.5 \text{ mA}$$
$$I_{tot} = \sqrt{I_R^2 + I_{CL}^2} = \sqrt{(2.27 \text{ mA})^2 + (0.5 \text{ mA})^2} = \textbf{2.32 mA}$$

The phase angle is

$$\theta = \tan^{-1}\left(\frac{I_{CL}}{I_R}\right) = \tan^{-1}\left(\frac{0.5 \text{ mA}}{2.27 \text{ mA}}\right) = 12.4°$$

The total current is 2.32 mA, leading V_s by 12.4°. Figure 13–37 is the current phasor diagram for the circuit.

FIGURE 13–37

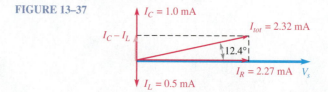

RELATED PROBLEM

Will the total current increase or decrease if the frequency in Figure 13–36 is increased? Why?

Conversion of Series-Parallel to Parallel

The particular series-parallel configuration shown in Figure 13–38 is important because it represents a circuit having parallel *L* and *C* branches, with the winding resistance of the coil taken into account as a series resistance in the *L* branch.

The series-parallel circuit in Figure 13–38 can be viewed in an equivalent purely parallel form, as indicated in Figure 13–39. This equivalent form will simplify the analysis of parallel resonant characteristics that will be discussed in Section 13–6.

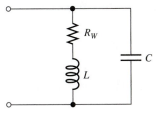

FIGURE 13–38   **A series-parallel *RLC* circuit $(Q = X_L/R_W)$.**

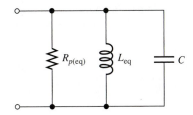

FIGURE 13–39   **Parallel equivalent form of the circuit in Figure 13–38.**

The equivalent inductance, L_{eq}, and the equivalent parallel resistance, $R_{p(eq)}$, are given by the following formulas:

$$L_{eq} = L\left(\frac{Q^2 + 1}{Q^2}\right) \qquad (13\text{–}14)$$

$$R_{p(eq)} = R_W(Q^2 + 1) \qquad (13\text{–}15)$$

where Q is the quality factor of the coil, X_L/R_W. Derivations of these formulas are quite involved and thus are not given here. Notice in the equations that for a $Q \geq 10$, the value of L_{eq} is approximately the same as the original value of L. For example, if $L = 10\,\text{mH}$ and $Q = 10$, then

$$L_{eq} = 10\,\text{mH}\left(\frac{10^2 + 1}{10^2}\right) = 10\,\text{mH}(1.01) = 10.1\,\text{mH}$$

The equivalency of the two circuits means that at a given frequency, when the same value of voltage is applied to both circuits, the same total current is in both circuits and the phase angles are the same. Basically, an equivalent circuit simply makes circuit analysis more convenient.

EXAMPLE 13–16

Convert the series-parallel circuit in Figure 13–40 to an equivalent parallel form at the given frequency.

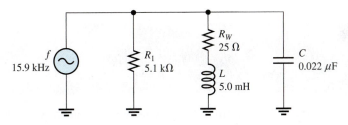

FIGURE 13–40

SOLUTION

Determine the inductive reactance.

$$X_L = 2\pi f L = 2\pi(15.9\,\text{kHz})(5.0\,\text{mH}) = 500\,\Omega$$

The Q of the coil is

$$Q = \frac{X_L}{R_W} = \frac{500\,\Omega}{25\,\Omega} = 20$$

Since $Q > 10$, then $L_{eq} \cong L = 5.0$ mH.
   The equivalent parallel resistence is

$$R_{p(eq)} = R_W(Q^2 + 1) = (25 \, \Omega)(20^2 + 1) = 10.0 \, k\Omega$$

This equivalent resistance ($R_{p(eq)}$) appears in parallel with R_1 as shown in Figure 13–41(a). When combined, they give a total parallel resistance ($R_{p(tot)}$) of 3.38 kΩ, as indicated in Figure 13–41(b).

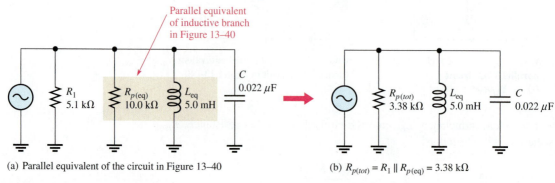

(a) Parallel equivalent of the circuit in Figure 13–40

(b) $R_{p(tot)} = R_1 \| R_{p(eq)} = 3.38$ kΩ

FIGURE 13–41

RELATED PROBLEM

Find the equivalent parallel circuit if $R_W = 10 \, \Omega$ in Figure 13–40.

SECTION 13–5 CHECKUP

1. In a three-branch parallel circuit, $R = 1500 \, \Omega$, $X_C = 1000 \, \Omega$, and $X_L = 500 \, \Omega$ at a certain frequency. Determine the current in each branch when $V_s = 12$ V.

2. Is the circuit in Question 1 capacitive or inductive? Why?

3. Find the equivalent parallel inductance and resistance for a 20 mH coil with a winding resistance of 10 Ω at a frequency of 1 kHz.

13–6 PARALLEL RESONANCE

In this section, we will first look at the resonant condition in an ideal parallel *LC* circuit. Then, we will examine the more realistic case where the resistance of the coil is taken into account.

After completing this section, you should be able to

• **Analyze a circuit for parallel resonance**
   • Describe parallel resonance in an ideal circuit
   • Describe parallel resonance in a nonideal circuit
   • Explain how impedance varies with frequency
   • Determine current and phase angle at resonance
   • Determine parallel resonant frequency
   • Discuss the effects of loading a parallel resonant circuit

Condition for Ideal Parallel Resonance

Ideally, **parallel resonance** occurs when $X_L = X_C$. The frequency at which resonance occurs is called the *resonant frequency,* just as in the series resonant circuit. When $X_L = X_C$, the two branch currents, I_C and I_L, are equal in magnitude, and, of course, they are always 180° out of phase with each other. Thus, the two currents cancel and the total current is zero, as presented in Figure 13–42. In this ideal case, the winding resistance of the coil is assumed to be zero.

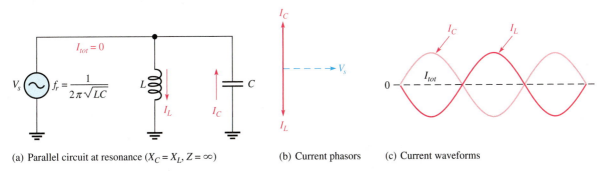

(a) Parallel circuit at resonance ($X_C = X_L, Z = \infty$)   (b) Current phasors   (c) Current waveforms

FIGURE 13–42   **An ideal parallel *LC* circuit at resonance.**

Since the total current is zero, the impedance of the parallel *LC* circuit is infinitely large (∞). These ideal resonant conditions are stated as follows:

$$X_L = X_C$$
$$Z_r = \infty$$

Parallel Resonant Frequency

For an ideal parallel resonant circuit, the frequency at which resonance occurs is determined by the same formula as in series resonant circuits.

$$f_r = \frac{1}{2\pi\sqrt{LC}} \qquad\qquad (13\text{–}16)$$

Currents in a Parallel Resonant Circuit

The currents in a parallel *LC* circuit vary as the frequency is increased from below the resonant value, through resonance, and then above the resonant value. Figure 13–43 illustrates the general response of an ideal circuit in terms of changes in the currents.

BELOW THE RESONANT FREQUENCY   Figure 13–43(a) indicates the response as the source frequency is increased from zero toward f_r. At very low frequencies, X_C is very high and X_L is very low, so most of the current is through L. As the frequency increases, the current through L decreases and the current through C increases, causing the total current to decrease. At all times, I_L and I_C are 180° out of phase with each other, and thus the total current is the difference of the two branch currents. During this time, the impedance is increasing, as indicated by the decrease in total current.

AT THE RESONANT FREQUENCY   When the frequency reaches its resonant value, f_r, as shown in Figure 13–43(b), X_C and X_L are equal, so I_C and I_L cancel because they are equal in magnitude but opposite in phase. At this point the total current is at its minimum value of zero. Since I_{tot} is zero, Z is infinite. Thus, the ideal parallel *LC* circuit appears as an open at f_r.

HANDS ON TIP
When you check the resonant frequency of an *LC* circuit using a function generator, the Thevenin resistance of the generator increases the Q of the circuit. A generator with an internal resistance of 600 Ω can significantly broaden the bandwidth of the resonant circuit. To reduce the Q and obtain a narrower bandwidth, a small parallel resistor can be placed across the generator output. The drawback of this method is that the signal amplitude from the generator is reduced.

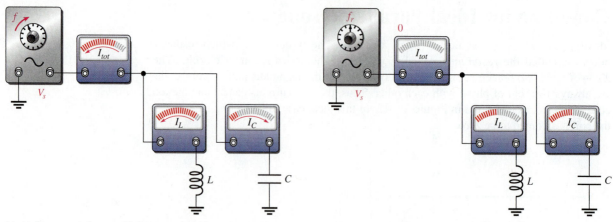

(a) As frequency is increased below resonance: $X_C > X_L$, $I_L > I_C$, I_L decreases, I_C increases, and I_{tot} decreases.

(b) At the resonant frequency: $X_C = X_L$, $I_L = I_C$, and $I_{tot} = 0$.

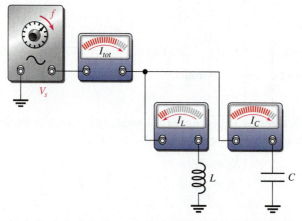

(c) As frequency is increased above resonance: $X_L > X_C$, $I_L < I_C$, I_L decreases, I_C increases, and I_{tot} increases.

FIGURE 13–43   An illustration of how the currents respond in a parallel *LC* circuit as the frequency is varied from below resonance to above resonance. The source voltage amplitude is constant.

ABOVE THE RESONANT FREQUENCY   As the frequency is increased above resonance, as indicated in Figure 13–43(c), X_C continues to decrease and X_L continues to increase, causing the branch currents again to be unequal, with I_C being larger. As a result, total current increases and impedance decreases. As the frequency becomes very high, the impedance becomes very small due to the dominance of a very small X_C in parallel with a very large X_L.

In summary, the current dips to a minimum as the impedance peaks to a maximum at parallel resonance. The expression for the total current into the *L* and *C* branches is

$$I_{tot} = |I_L - I_C| \qquad (13\text{–}17)$$

EXAMPLE 13–17

A resonant circuit in a communication receiver has a 680 μH inductor in parallel with a 180 pF capacitor as shown in Figure 13–44.

(a) What is the resonant frequency?

(b) If 2.0 V is across the parallel combination, what is the current in each component and the total current?

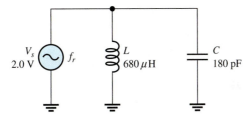

FIGURE 13–44

SOLUTION

(a) The resonant frequency is

$$f_r = \frac{1}{2\pi\sqrt{LC}} = \frac{1}{2\pi\sqrt{(680\ \mu H)(180\ pF)}} = \textbf{455 kHz}$$

(b) $X_L = 2\pi f L = 2\pi(455\ \text{kHz})(680\ \mu H) = 1.94\ \text{k}\Omega$

$X_C = X_L = 1.94\ \text{k}\Omega$

$I_L = \dfrac{V_s}{X_L} = \dfrac{2.0\ \text{V}}{1.94\ \text{k}\Omega} = \textbf{1.03 mA}$

$I_C = I_L = \textbf{1.03 mA}$

The total current at resonance is

$$I_{tot} = |I_L - I_C| = \textbf{0 A}$$

RELATED PROBLEM

Find the current in the inductor and capacitor if the circuit is tuned to 470 kHz.

MULTISIM

Open Multisim file E13-17. Determine the resonant frequency by measurement. Determine the currents through *L* and *C* at the resonant frequency.

Tank Circuit

The parallel resonant *LC* circuit is often called a **tank circuit**. The term *tank circuit* refers to the fact that the parallel resonant circuit stores energy in the magnetic field of the coil and in the electric field of the capacitor. The stored energy is transferred back and forth between the capacitor and the coil on alternate half-cycles as the current goes first one way and then the other when the inductor deenergizes and the capacitor charges, and vice versa. This concept is illustrated in Figure 13–45.

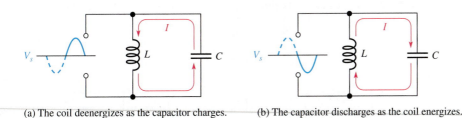

(a) The coil deenergizes as the capacitor charges.   (b) The capacitor discharges as the coil energizes.

FIGURE 13–45   **Energy storage in an ideal parallel resonant tank circuit.**

Parallel Resonant Conditions in a Nonideal Circuit

So far, the resonance of an ideal parallel *LC* circuit has been examined. Now let's consider resonance in a tank circuit with the winding resistance taken into account. Figure 13–46 shows a nonideal tank circuit and its parallel *RLC* equivalent.

FIGURE 13–46 A practical treatment of parallel resonant circuits includes the winding resistance.

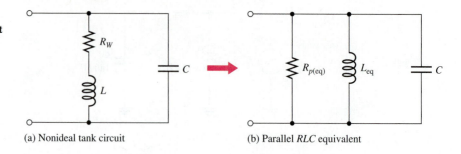

(a) Nonideal tank circuit   (b) Parallel *RLC* equivalent

If the winding resistance is the only resistance in the circuit, the quality factor, Q, of the circuit at resonance is simply the Q of the coil.

$$Q = \frac{X_L}{R_W}$$

In terms of circuit component values, the Q can also be expressed as

$$Q = \frac{1}{R_W}\sqrt{\frac{L}{C}}$$

These equations for Q include only the winding resistance, R_W, for the coil and neglect other resistance that may be present. The internal resistance of the voltage source and any load resistance can lower the Q. One common way to measure the actual Q is to insert a relatively small resistor, called a *sense resistor* into the circuit. The sense resistor introduces minimal loading but develops a small voltage that can be used to determine the current. Experiment 27 in the lab manual[1] illustrates measuring the Q of a resonant circuit using a sense resistor.

The expressions for the equivalent inductance and the equivalent parallel resistance were given in Equations 13–14 and 13–15 as

$$L_{eq} = L\left(\frac{Q^2 + 1}{Q^2}\right)$$

$$R_{p(eq)} = R_W(Q^2 + 1)$$

Recall that for $Q \geq 10$, $L_{eq} \cong L$.

At parallel resonance,

$$X_{L(eq)} = X_C$$

In the parallel equivalent circuit, $R_{p(eq)}$ is in parallel with an ideal coil and a capacitor, so the L and C branches act as an ideal tank circuit that has an infinite impedance at resonance, as shown in Figure 13–47. Therefore, the total impedance of the nonideal tank circuit at resonance can be expressed as simply the equivalent parallel resistance.

$$Z_r = R_W(Q^2 + 1)$$

FIGURE 13–47 At resonance, the parallel *LC* portion appears open and the source sees only $R_{p(eq)}$, which equals $R_W(Q^2 + 1)$.

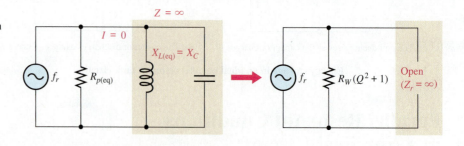

[1] *Experiments for DC/AC Fundamentals: A System Approach* by David M. Buchla.

EXAMPLE 13–18

Determine the impedance of the circuit in Figure 13–48 at the resonant frequency ($f_r \cong 17{,}794$ Hz).

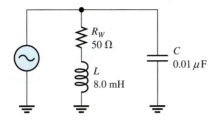

FIGURE 13–48

SOLUTION

Before you can calculate the impedance, you must find the quality factor. To get Q, first find the inductive reactance.

$$X_L = 2\pi f_r L = 2\pi(17{,}794 \text{ Hz})(8.0 \text{ mH}) = 894 \ \Omega$$

$$Q = \frac{X_L}{R_W} = \frac{894 \ \Omega}{50 \ \Omega} = 17.9$$

$$Z_r = R_W(Q^2 + 1) = 50 \ \Omega(17.9^2 + 1) = \mathbf{16.0 \ k\Omega}$$

RELATED PROBLEM

Determine Z_r in Figure 13–48 if the winding resistance is 10 Ω.

Variation of the Impedance with Frequency

The impedance of a parallel resonant circuit is maximum at the resonant frequency and decreases at lower and higher frequencies, as indicated by the curve in Figure 13–49.

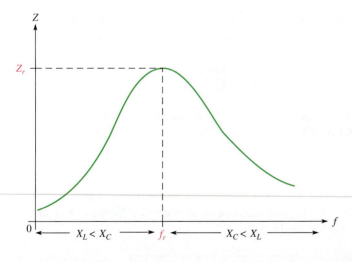

FIGURE 13–49 **Generalized impedance curve for a parallel resonant circuit. The circuit is inductive below f_r, resistive at f_r, and capacitive above f_r.**

At very low frequencies, X_L is very small and X_C is very high, so the total impedance is essentially equal to that of the inductive branch. As the frequency goes up, the impedance also increases, and the inductive reactance dominates (because it is less than X_C) until the resonant frequency is reached. At this point, of course, $X_L \cong X_C$ (for $Q > 10$) and the impedance is at its maximum. As the frequency goes above resonance, the capacitive reactance dominates (because it is less than X_L) and the impedance decreases.

Current and Phase Angle at Resonance

In the ideal tank circuit, the total current from the source at resonance is zero because the impedance is infinite. In the nonideal case, there is some total current at the resonant frequency, and it is determined by the impedance at resonance.

$$I_{tot} = \frac{V_s}{Z_r}$$

(13–18)

The phase angle of the parallel resonant circuit is 0° because the impedance is purely resistive at the resonant frequency.

Parallel Resonant Frequency in a Nonideal Circuit

As you know, when the winding resistance is considered, the resonant condition is

$$X_{L(eq)} = X_C$$

which can be expressed as

$$2\pi f_r L\left(\frac{Q^2 + 1}{Q^2}\right) = \frac{1}{2\pi f_r C}$$

Solving for f_r in terms of Q yields

$$f_r = \frac{1}{2\pi\sqrt{LC}}\sqrt{\frac{Q^2}{Q^2 + 1}}$$

When $Q \geq 10$, the term with the Q factors is approximately 1.

$$\sqrt{\frac{Q^2}{Q^2 + 1}} = \sqrt{\frac{100}{101}} = 0.995 \cong 1$$

Therefore, the parallel resonant frequency is approximately the same as the series resonant frequency as long as Q is equal to or greater than 10.

$$f_r \cong \frac{1}{2\pi\sqrt{LC}} \qquad \text{for } Q \geq 10$$

For the case where the R_W of the coil is the only resistance in the circuit, a precise expression for f_r in terms of the circuit component values is

$$f_r = \frac{\sqrt{1 - (R_W^2 C/L)}}{2\pi\sqrt{LC}}$$

(13–19)

This precise formula is seldom necessary and for most practical situations the simpler equation $f_r = 1/(2\pi\sqrt{LC})$ is sufficient. However, the next example illustrates the use of Equation 13–19.

EXAMPLE 13–19

Find the frequency, impedance, and total current at resonance for the circuit in Figure 13–50 using Equation 13–19.

FIGURE 13–50

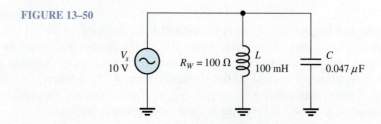

V_s = 10 V, $R_W = 100\ \Omega$, L = 100 mH, C = 0.047 μF

SOLUTION

The precise resonant frequency is

$$f_r = \frac{\sqrt{1 - (R_W^2 C/L)}}{2\pi\sqrt{LC}} = \frac{\sqrt{1 - [(100\ \Omega)^2(0.047\ \mu F)/100\ mH]}}{2\pi\sqrt{(0.047\ \mu F)(100\ mH)}} = \mathbf{2.32\ kHz}$$

Calculate impedance as follows:

$$X_L = 2\pi f_r L = 2\pi(2.32\ kHz)(100\ mH) = 1.46\ k\Omega$$

$$Q = \frac{X_L}{R_W} = \frac{1.46\ k\Omega}{100\ \Omega} = 14.6$$

$$Z_r = R_W(Q^2 + 1) = 100\ \Omega(14.6^2 + 1) = \mathbf{21.4\ k\Omega}$$

The total current is

$$I_{tot} = \frac{V_s}{Z_r} = \frac{10\ V}{21.4\ k\Omega} = \mathbf{467\ \mu A}$$

RELATED PROBLEM

Repeat this example using the formula $f_r = 1/(2\pi\sqrt{LC})$ and compare the results.

MULTISIM

Open Multisim file E13-19. Determine the resonant frequency by measurement. Determine the total current and the currents through L and C at the resonant frequency. How do these results compare with the calculated values?

An External Load Resistance Affects a Tank Circuit

In a basic parallel resonant circuit with no load resistor, the Q of the circuit is determined only by the Q of the coil and some loading due to the source resistance. However, the source loading effect is complicated by frequency dependency. At resonance, the source resistance has a small current, which is due to the nonideal characteristic of the tank circuit. As the circuit moves away from resonance, this current increases due to phase shifts. To simplify the discussion, we will ignore the effect of source loading.

In most practical circuits, as illustrated in Figure 13–51(a), an external load (R_L) dissipates most of the power delivered by the source. In this case, the load resistor will lower the Q of the circuit—the smaller the load resistor, the lower the Q. The load resistor, R_L, effectively is in parallel with the equivalent parallel resistance of the coil, $R_{p(eq)}$; both are combined to determine the total parallel resistance, $R_{p(tot)}$, as indicated in Figure 13–51(b) and given as

$$R_{p(tot)} = R_L \parallel R_{p(eq)} \tag{13–20}$$

In the case of an unloaded tank circuit, $R_{p(tot)}$ reduces to $R_{p(eq)}$.

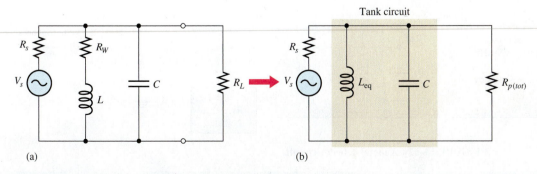

FIGURE 13–51 **Tank circuit with a load resistor and its equivalent circuit.**

The overall Q, designated Q_O, is given by the approximate equation

$$Q_O \cong \frac{R_{p(tot)}}{X_L}$$

(13–21)

Notice that this equation is reversed from the Q of a series circuit or the Q of an inductor alone. However, the unloaded case for a parallel circuit will give the same result as the Q of the inductor alone, so either equation can be used with no load.

SYSTEM EXAMPLE 13–1

A BASIC METAL DETECTOR

Resonant circuits are used in many oscillators. Figure 13–52 shows a circuit that uses two oscillators to make a basic metal detector. The resonant circuit of oscillator-1 is the focus of this example and is shown in the yellow highlighted box. The oscillator is a Colpitts type first introduced in System Example 9–1. This particular oscillator will change frequencies when it is near metal because of the change in inductance of the sensing coil (L_1).

To calculate the oscillation frequency of oscillator-1, we apply

$$f_r = \frac{1}{2\pi\sqrt{LC}}$$

Capacitor C_1 has almost no effect on the resonant frequency and can be ignored; however, it is in the circuit to block dc. C_2 and C_3 are in series; together with L_1 they form an equivalent tank circuit (or parallel resonant circuit).

Oscillator-2 is designed to oscillate at a *slightly* different frequency from oscillator-1. It develops a fixed reference frequency that interacts with the frequency generated by oscillator-1. The output of each oscillator is mixed in a nonlinear mixer. (Recall that the superposition theorem allows you to add sources *linearly*, so it does not apply in this case.) By mixing the signals in a nonlinear circuit, two new frequencies appear at the output: the sum of oscillator-1 plus oscillator-2 and the difference in their frequencies; this lower difference frequency is known as a beat frequency. A low-pass filter, consisting of L_2 and R_4, allows only the beat frequency through, which is an audio tone. It is amplified and sent to headphones.

The frequency of oscillator-1 will change when it is near metal due to the magnetic coupling, causing reduced inductance. As a result, the beat frequency also changes and is heard in the headphones as a changing squeal.

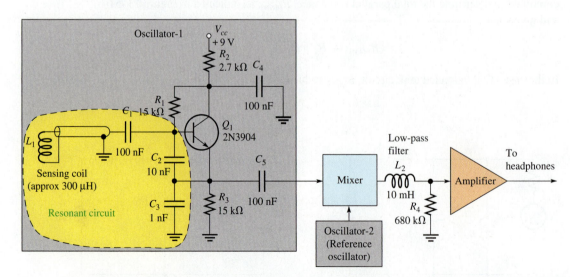

FIGURE 13–52 A metal detector circuit.

13–7 PARALLEL RESONANT FILTERS

Parallel resonant circuits are commonly applied to band-pass and band-stop filters.

After completing this section, you should be able to

- **Analyze the operation of parallel resonant filters**
  - **Show how a band-pass filter is implemented**
  - **Define *bandwidth***
  - **Explain how loading affects selectivity**
  - **Show how a band-stop filter is implemented**
  - **Determine the resonant frequency, bandwidth, and output voltage of band-pass and band-stop parallel resonant filters**

The Band-Pass Filter

A basic parallel resonant band-pass filter is shown in Figure 13–53. Notice that the output is taken across the tank circuit in this application.

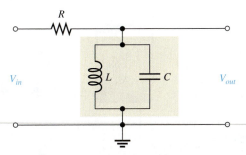

FIGURE 13–53   **A basic parallel resonant band-pass filter.**

The bandwidth and cutoff frequencies for a parallel resonant band-pass filter are defined in the same way as for the series resonant circuit, and the formulas given in Section 13–4 still apply. General band-pass frequency response curves showing both V_{out} and I_{tot} versus frequency are given in Figures 13–54(a) and 13–54(b), respectively.

FIGURE 13–54 **Generalized frequency response curves for a parallel resonant band-pass filter.**

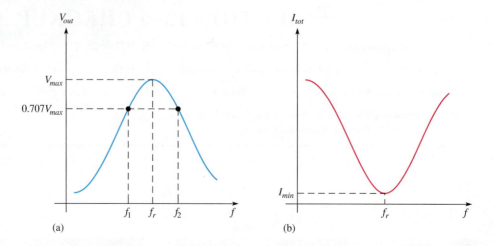

(a)

(b)

The filtering action is as follows: At very low frequencies, the impedance of the tank circuit is very low, and therefore only a small amount of the input voltage is dropped across it; the rest is dropped across *R*. As the frequency increases, the impedance of the tank circuit increases, and, as a result, the output voltage increases. When the frequency reaches its resonant value, the impedance is at its maximum and most of the input voltage is developed across the tank circuit. As the frequency goes above resonance, the impedance begins to decrease, causing the output voltage to decrease. The general response of a parallel resonant band-pass filter is illustrated in Figure 13–55. This illustration shows how the current and the output voltage change with frequency.

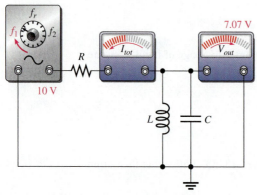

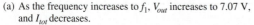

(a) As the frequency increases to f_1, V_{out} increases to 7.07 V, and I_{tot} decreases.

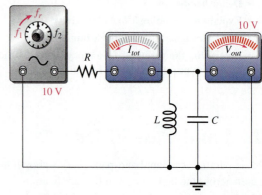

(b) As the frequency increases from f_1 to f_r, V_{out} increases from 7.07 V to 10 V, and I_{tot} decreases to its minimum value.

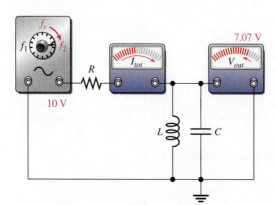

(c) As the frequency increases from f_r to f_2, V_{out} decreases from 10 V to 7.07 V, and I_{tot} increases from its minimum.

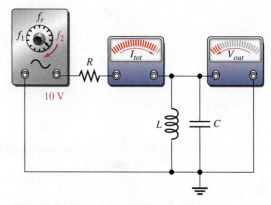

(d) As the frequency increases above f_2, V_{out} decreases below 7.07 V, and I_{tot} continues to increase.

FIGURE 13–55 **Example of the response of a parallel resonant band-pass filter with the input voltage at a constant 10 V rms.**

EXAMPLE 13–20

A certain parallel resonant band-pass filter has a maximum output voltage of 4 V at f_r. What is the value of V_{out} at the cutoff frequencies?

SOLUTION

V_{out} at the cutoff frequencies is 70.7% of maximum.

$$V_{out(1)} = V_{out(2)} = 0.707 V_{out(max)} = 0.707(4 \text{ V}) = \textbf{2.83 V}$$

RELATED PROBLEM

What is V_{out} at the cutoff frequencies if V_{out} at the resonant frequency is 10 V?

EXAMPLE 13–21

A parallel resonant circuit has a lower cutoff frequency of 3.5 kHz and an upper cutoff frequency of 6 kHz. What is the bandwidth?

SOLUTION

$$BW = f_2 - f_1 = 6 \text{ kHz} - 3.5 \text{ kHz} = \textbf{2.5 kHz}$$

RELATED PROBLEM

A filter has a lower cutoff frequency of 520 kHz and a bandwidth of 10 kHz. What is the upper cutoff frequency?

EXAMPLE 13–22

A certain parallel resonant band-pass filter has a resonant frequency of 12 kHz and a Q of 10. What is its bandwidth?

SOLUTION

$$BW = \frac{f_r}{Q} = \frac{12 \text{ kHz}}{10} = \textbf{1.2 kHz}$$

RELATED PROBLEM

A certain parallel resonant band-pass filter has a resonant frequency of 100 MHz and a bandwidth of 4 MHz. What is its Q?

LOADING EFFECTS   When a resistive load is connected across the output of a filter as shown in Figure 13–56(a), the Q of the filter is reduced. Since $BW = f_r/Q$, the bandwidth is increased, thus reducing the selectivity. Also, the impedance of the filter at resonance is decreased because R_L effectively appears in parallel with $R_{p(eq)}$. Thus, the maximum output voltage is reduced by the voltage-divider effect of $R_{p(tot)}$ and the internal source resistance R_s, as illustrated in Figure 13–56(b). Figure 13–56(c) shows the effect of a load on the specific resonant circuit described in Example 13–23. The effect of a load drops the peak and increases the bandwidth as shown.

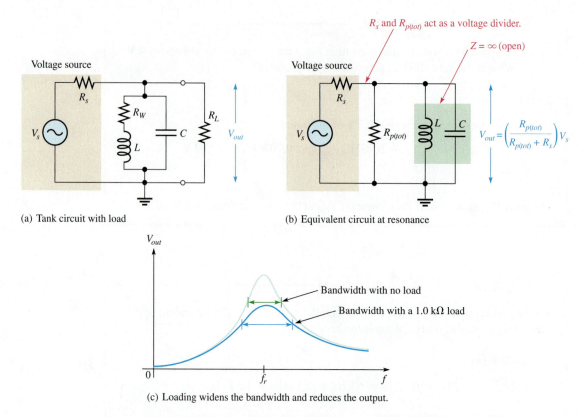

R_s and $R_{p(tot)}$ act as a voltage divider.

$Z = \infty$ (open)

(a) Tank circuit with load

(b) Equivalent circuit at resonance

$V_{out} = \left(\dfrac{R_{p(tot)}}{R_{p(tot)} + R_s}\right)V_s$

Bandwidth with no load

Bandwidth with a 1.0 kΩ load

(c) Loading widens the bandwidth and reduces the output.

FIGURE 13–56   **Effects of loading on a parallel resonant band-pass filter. The resistor, *R*, in Figure 13–53 is now represented by the internal source resistance R_s.**

E X A M P L E   1 3 – 2 3

(a) Determine f_r, V_{out}, and *BW* at resonance for the filter in Figure 13–57(a). The inductor has a winding resistance of 10.5 Ω. The internal source resistance is 50 Ω.

(b) Repeat part (a) when the filter is loaded with a 1.0 kΩ resistance and compare the results.

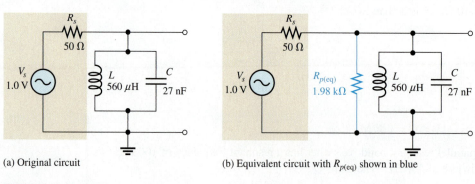

(a) Original circuit

(b) Equivalent circuit with $R_{p(eq)}$ shown in blue

FIGURE 13–57

SOLUTION

(a) $f_r = \dfrac{1}{2\pi\sqrt{LC}} = \dfrac{1}{2\pi\sqrt{(560\ \mu H)(27\ nF)}} = \textbf{40.9 kHz}$

The circuit is unloaded. To calculate the output voltage, begin by finding the *Q* of the coil and $R_{p(eq)}$.

$$X_L = 2\pi fL = 2\pi(40.9\ \text{kHz})(560\ \mu H) = 144\ \Omega$$

$$Q = \frac{X_L}{R_W} = \frac{144\ \Omega}{10.5\ \Omega} = 13.7$$

Apply Equation 13–15 to calculate $R_{p(eq)}$ (parallel equivalent resistance of the coil).

$$R_{p(eq)} = R_W(Q^2 + 1) = 10.5\,\Omega(13.7 + 1) = 1.98\,k\Omega$$

Because the circuit is unloaded, $R_{p(tot)} = R_{p(eq)} = 1.98\,k\Omega$, as shown in Figure 13–57(b). At resonance, the tank circuit looks like an open, so apply the voltage-divider formula to $R_{p(tot)}$ and R_s to find V_{out}.

$$V_{out} = \left(\frac{R_{p(tot)}}{R_{p(tot)} + R_s}\right)V_s = \left(\frac{1.98\,k\Omega}{1.98\,k\Omega + 50\,\Omega}\right)1.0\,V = \textbf{975 mV}$$

To calculate the BW, apply Equation 13–21.

$$Q_O \cong \frac{R_{p(tot)}}{X_L} \cong \frac{1.98\,k\Omega}{144\,\Omega} = 13.8$$

Notice that this is the same Q (within round off error) as the inductor alone, because the circuit is unloaded.

$$BW = \frac{f_r}{Q_O} = \frac{40.9\,kHz}{13.8} = \textbf{2.96 kHz}$$

(b) The resonant frequency is not significantly affected by the load and is equal to

$$f_r = \textbf{40.9 kHz}$$

The load drops the output voltage because it appears to be in parallel with $R_{p(eq)}$. Apply Equation 13–20.

$$R_{p(tot)} = R_L \| R_{p(eq)} = 1.0\,k\Omega \| 2.07\,k\Omega = 674\,\Omega$$

Applying the voltage-divider formula,

$$V_{out} = \left(\frac{R_{p(tot)}}{R_{p(tot)} + R_s}\right)V_s = \left(\frac{674\,\Omega}{674\,\Omega + 50\,\Omega}\right)1.0\,V = \textbf{930 mV}$$

To calculate the BW, apply Equation 13–21.

$$Q_O = \frac{R_{p(tot)}}{X_L} = \frac{674\,\Omega}{144\,\Omega} = 4.68$$

$$BW = \frac{f_r}{Q_O} = \frac{40.9\,kHz}{4.68} = \textbf{8.74 kHz}$$

The load decreased the output voltage and caused the bandwidth to increase.

RELATED PROBLEM

How is Q_O affected by a larger-value load resistor?

The Band-Stop Filter

A basic parallel resonant band-stop filter is shown in Figure 13–58. The output is taken across a load resistor in series with the tank circuit. The load resistor has only a small effect on Q in this configuration, and generally the effect can be ignored.

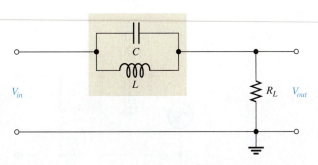

FIGURE 13–58   A basic parallel resonant band-stop filter.

The variation of the tank circuit impedance with frequency produces the familiar current response that has been previously discussed; that is, the current is minimum at resonance and increases on both sides of resonance. Since the output voltage is across the series load resistor, the output voltage follows the current, thus creating the band-stop response characteristic, as indicated in Figure 13–59.

FIGURE 13–59 Band-stop filter response.

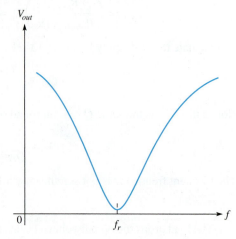

Actually, the band-stop filter in Figure 13–58 can be viewed as a voltage divider created by Z_r of the tank and the load resistance. Thus, the output voltage at f_r is

$$V_{out} = \left(\frac{R_L}{R_L + Z_r} \right) V_{in}$$

Wave Traps

A *wave trap* is a resonant circuit designed to significantly reduce interference from a specific source. For example, a wave trap to prevent interference from a citizen band radio at 27 MHz is a parallel resonant circuit tuned to that frequency. Other frequencies can pass through the filter with little effect. The circuit shown above is a wave trap designed to be inserted in 300 Ω twin lead, which is a commonly used radio frequency transmission line. The wave trap is installed in a shielded enclosure with the shield connected to ground.

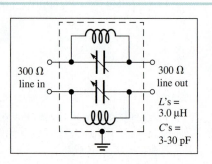

SYSTEM NOTE

EXAMPLE 13–24

Figure 13–60 shows the tank circuit from Example 13–23, which is used in a band-stop filter. Determine f_r and V_{out} at resonance. The winding resistance of the inductor is 10.5 Ω.

FIGURE 13–60

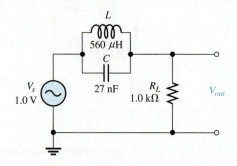

SOLUTION

The resonant frequency is

$$f_r = \frac{1}{2\pi\sqrt{LC}} = \frac{1}{2\pi\sqrt{(560\ \mu H)(27\ nF)}} = \textbf{40.9 kHz}$$

Determine V_{out} as follows:

$$X_L = 2\pi fL = 2\pi(40.9\ \text{kHz})(560\ \mu H) = 144\ \Omega$$

$$Q = \frac{X_L}{R_W} = \frac{144\ \Omega}{10.5\ \Omega} = 13.7$$

Applying Equation 13–15,

$$R_{p(eq)} = R_W(Q^2 + 1) = 10.5\ \Omega(13.7 + 1) = 1.98\ \text{k}\Omega$$

At resonance,

$$V_{out} = \left(\frac{R_L}{R_L + R_{p(eq)}}\right)V_s = \left(\frac{1.0\ \text{k}\Omega}{1.0\ \text{k}\Omega + 1.98\ \text{k}\Omega}\right)1.0\ \text{V} = \textbf{335 mV}$$

RELATED PROBLEM

What is the output at the resonant frequency if the load resistor is 470 Ω?

MULTISIM

Open the Multisim file E13-24. Compare results in Multisim with the calculations. You can read the resonant frequency and output voltage on the Bode plotter.

SECTION 13–7 CHECKUP

1. How can the bandwidth of a parallel resonant filter be increased?

2. The resonant frequency of a certain high-Q ($Q > 10$) filter is 5 kHz. If the Q is lowered to 2, does f_r change and, if so, to what value?

3. What is the impedance of a tank circuit at its resonant frequency if $R_W = 75\ \Omega$ and $Q = 25$?

13–8 RESONANT CIRCUIT APPLICATIONS

Resonant circuits are used in a wide variety of applications, particularly in communication systems. In this section, we will look at a few system applications to illustrate the importance of resonant circuits in electronic communication.

After completing this section, you should be able to

- **Discuss some applications of resonant circuits**
  - **Describe a tuned amplifier and double-tuned transformer coupling**
  - **Describe antenna coupling**
  - **Describe an AM radio receiver**

Tuned Amplifiers

A *tuned amplifier* is a circuit that amplifies signals within a specified band. Typically, a parallel resonant circuit is used in conjunction with an amplifier to achieve the selectivity. In terms of the general operation, input signals with frequencies that range over a wide band are accepted on the amplifier's input and are amplified. The resonant circuit allows only a relatively narrow band of those frequencies to be passed on. The variable capacitor allows tuning over the range of input frequencies so that a desired frequency can be selected, as indicated in Figure 13–61.

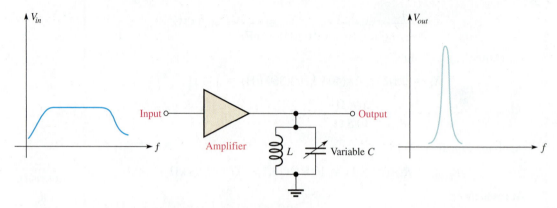

FIGURE 13–61 **A basic tuned band-pass amplifier.**

Double-Tuned Transformer Coupling in a Receiver

In some types of communication receivers, tuned amplifiers are transformer-coupled together to increase the amplification. Capacitors can be placed in parallel with the primary and secondary windings of the transformer, effectively creating two parallel resonant band-pass filters that are coupled together. This technique, illustrated in Figure 13–62, can result in a wider bandwidth and steeper slopes on the response curve, thus increasing the selectivity for a desired band of frequencies.

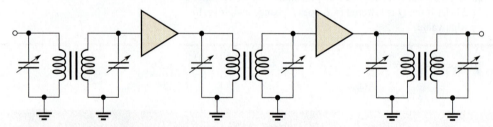

FIGURE 13–62 **Double-tuned amplifiers.**

Antenna Input to a Receiver

Radio signals are sent out from a transmitter via electromagnetic waves that propagate through the atmosphere. When the electromagnetic waves cut across the receiving antenna, small voltages are induced. Out of the wide range of electromagnetic frequencies, only one frequency or a limited band of frequencies must be extracted. Figure 13–63 shows a typical arrangement of an antenna coupled to the receiver input by a transformer (covered in

FIGURE 13–63 **Resonant coupling from an antenna.**

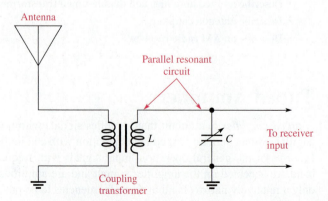

Chapter 14). A variable capacitor is connected across the transformer secondary to form a parallel resonant circuit.

Superheterodyne Receiver

Another example of resonant circuit application is in the common AM (amplitude modulation) receiver. The AM broadcast band ranges from 535 kHz to 1605 kHz. Each AM station is assigned a 10 kHz bandwidth within that range. The tuned circuits are designed to pass only the signals from the desired radio station, rejecting all others. To reject stations outside the one that is tuned, the tuned circuits must be selective, passing on only the signals in the 10 kHz band and rejecting all others. Too much selectivity is not desirable either however. If the bandwidth is too narrow, some of the higher frequency modulated signals will be rejected, resulting in a loss of fidelity. Ideally, the resonant circuit must reject signals that are not in the desired passband. A simplified block diagram of a superheterodyne AM receiver is shown in Figure 13–64.

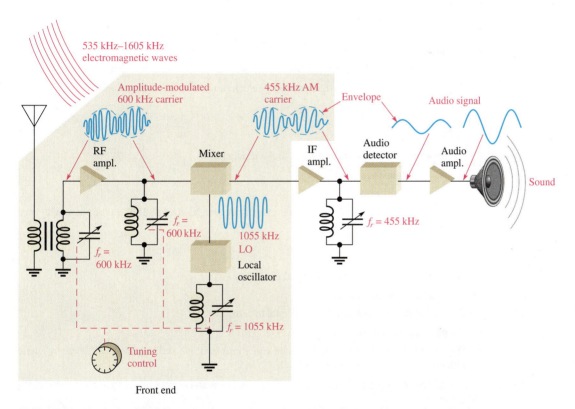

FIGURE 13–64 A simplified diagram of a superheterodyne AM radio broadcast receiver showing the application of tuned resonant circuits.

There are basically three parallel resonant circuits in the front end of the receiver. Each of these resonant circuits is gang-tuned by capacitors; that is, the capacitors are mechanically or electronically linked together so that they change together as a station is selected. The front end is tuned to receive a desired station, for example, one that transmits at 600 kHz. The input resonant circuit from the antenna and the RF (radio frequency) amplifier resonant circuit select only a frequency of 600 kHz out of all the frequencies crossing the antenna.

The actual audio (sound) signal is carried by the 600 kHz carrier frequency by modulating the amplitude of the carrier so that it follows the audio signal as indicated. The variation in the amplitude of the carrier corresponding to the audio signal is called the *envelope*. The 600 kHz is then applied to a circuit called the *mixer*.

The local oscillator (LO) is tuned to a frequency that is 455 kHz above the selected frequency (1055 kHz, in this case). By a process called *heterodyning* or *beating,* the AM signal and the local oscillator signal are mixed together, and the 600 kHz AM signal is converted by the mixer to a 455 kHz AM signal (1055 kHz − 600 kHz = 455 kHz).

The 455 kHz is the intermediate frequency (IF) for standard AM receivers. No matter which station within the broadcast band is selected, its frequency is always converted to the 455 kHz IF. The amplitude-modulated IF is amplified by the IF amplifier which is tuned to 455 kHz. The output of the IF amplifier is applied to an audio detector which removes the IF, leaving only the envelope which is the audio signal. The audio signal is then amplified and applied to the speaker.

SECTION 13–8 CHECKUP

1. Generally, why is a tuned resonant circuit necessary when a signal is coupled from an antenna to the input of a receiver?

2. What is the advantage of using an intermediate frequency in an AM radio?

3. What is meant by *ganged tuning?*

SUMMARY

- X_L and X_C have opposing effects in an *RLC* circuit.
- In a series *RLC* circuit, the larger reactance determines the net reactance of the circuit.
- In a parallel *RLC* circuit, the smaller reactance determines the net reactance of the circuit.

SERIES RESONANCE

- The reactances are equal.
- The impedance is minimum and equal to the resistance.
- The current is maximum.
- The phase angle is zero.
- The voltages across *L* and *C* are equal in magnitude and, as always, 180° out of phase with each other and thus they cancel.

PARALLEL RESONANCE

- The reactances are approximately equal for $Q \geq 10$.
- The impedance is maximum.
- The current is minimum and, ideally, equal to zero.
- The phase angle is zero.
- The currents in the *L* and *C* branches are equal in magnitude and, as always, 180° out of phase with each other and thus they cancel.
- A band-pass filter passes frequencies between the lower and upper critical frequencies and rejects all others.
- A band-stop filter rejects frequencies between its lower and upper critical frequencies and passes all others.
- The bandwidth of a resonant filter is determined by the quality factor (*Q*) of the circuit and the resonant frequency.
- Cutoff frequencies are also called −3 dB frequencies or critical frequencies.
- The output voltage is 70.7% of its maximum at the cutoff frequencies.

KEY TERMS

Key terms and other bold terms in the chapter are defined in the end-of-book glossary.

Band-pass filter   A resonant circuit that passes a range of frequencies lying between two cutoff frequencies and rejects frequencies above and below the range.

Band-stop filter   A resonant circuit that rejects a range of frequencies lying between two cutoff frequencies and passes frequencies above and below the range.

Cutoff frequency (f_c)   The frequency at which the output voltage of a filter is 70.7% of the maximum output voltage.

Decibel (dB)   A unit of logarithmic expression ten times the logarithm of the ratio of two powers or twenty times the logarithm of the ratio of two voltages measured across equal resistances.

Half-power frequency   The frequency at which the output power of a filter is 50% of the maximum (the output voltage is 70.7% of maximum); another name for *critical frequency* or *cutoff frequency*.

Parallel resonance   In a parallel *RLC* circuit, the condition where the impedance is maximum and the reactances are equal.

Resonant frequency   The frequency at which a resonant condition occurs in a series or parallel *RLC* circuit.

Selectivity   A measure of how effectively a resonant circuit passes certain frequencies and rejects others. The narrower the bandwidth, the greater the selectivity.

Series resonance   In a series *RLC* circuit, the condition where the impedance is minimum and the reactances are equal.

KEY FORMULAS

(13–1)	$X_{tot} = |X_L - X_C|$	Total series reactance (absolute value)
(13–2)	$Z_{tot} = \sqrt{R^2 + X_{tot}^2}$	Total series *RLC* impedance
(13–3)	$\theta = \tan^{-1}\left(\dfrac{X_{tot}}{R}\right)$	Series *RLC* phase angle
(13–4)	$f_r = \dfrac{1}{2\pi\sqrt{LC}}$	Ideal series resonant frequency
(13–5)	$BW = f_2 - f_1$	Bandwidth
(13–6)	$dB = 10\log\left(\dfrac{P_{out}}{P_{in}}\right)$	Decibel formula for voltage ratio
(13–7)	$dB = 20\log\left(\dfrac{V_{out}}{V_{in}}\right)$	Decibel formula for power ratio
(13–8)	$Q = \dfrac{X_L}{R}$	Series resonant quality factor
(13–9)	$BW = \dfrac{f_r}{Q}$	Bandwidth
(13–10)	$Y = \sqrt{G^2 + B_{tot}^2}$	Parallel *RLC* admittance
(13–11)	$\theta = \tan^{-1}\left(\dfrac{B_{tot}}{G}\right)$	Parallel *RLC* phase angle
(13–12)	$I_{tot} = \sqrt{I_R^2 + I_{CL}^2}$	Total parallel *RLC* current
(13–13)	$\theta = \tan^{-1}\left(\dfrac{I_{CL}}{I_R}\right)$	Parallel *RLC* phase angle
(13–14)	$L_{eq} = L\left(\dfrac{Q^2 + 1}{Q^2}\right)$	Equivalent parallel inductance
(13–15)	$R_{p(eq)} = R_W(Q^2 + 1)$	Equivalent parallel resistance
(13–16)	$f_r = \dfrac{1}{2\pi\sqrt{LC}}$	Ideal parallel resonant frequency
(13–17)	$I_{tot} = |I_L - I_C|$	Total parallel *LC* current (absolute value)
(13–18)	$I_{tot} = \dfrac{V_s}{Z_r}$	Total current at parallel resonance

$$(13\text{–}19) \quad f_r = \frac{\sqrt{1 - (R_W^2 C/L)}}{2\pi\sqrt{LC}}$$

Parallel resonant frequency (exact)

$$(13\text{–}20) \quad R_{p(tot)} = R_L \parallel R_{p(eq)}$$

Total parallel resistance

$$(13\text{–}21) \quad Q_O \cong \frac{R_{p(tot)}}{X_{L(eq)}}$$

Overall *Q* of a parallel *RLC* circuit

TRUE/FALSE QUIZ

Answers are at the end of the chapter.

1. A series *RLC* circuit can have a higher voltage than the source voltage across the resistor.

2. The impedance of a series *RLC* circuit is dependent on the source voltage.

3. Above the resonant frequency, a series resonant circuit will look inductive and current will lag the voltage.

4. A band-pass filter can be constructed from a *RLC* circuit.

5. A parallel resonant band-stop filter has minimum impedance at the resonant frequency.

6. The upper and lower cutoff frequencies of a band-stop filter determine the bandwidth.

7. The *Q* of an inductor is dependent on the frequency at which it is measured.

8. The *Q* of a band-pass filter does not affect the bandwidth.

9. In a parallel *RLC* circuit, the total impedance is always greater than the resistance.

10. At resonance, the current in a parallel resonant circuit is the same in all components.

SELF-TEST

Answers are at the end of the chapter.

1. The total reactance of a series *RLC* circuit at resonance is
   (a) zero   (b) equal to the resistance   (c) infinity   (d) capacitive

2. The phase angle of a series *RLC* circuit at resonance is
   (a) $-90°$ (b) $+90°$ (c) $0°$ (d) dependent on the reactance

3. The impedance at the resonant frequency of a series *RLC* circuit with $L = 15\ \text{mH}$, $C = 0.015\ \mu\text{F}$, and $R_W = 80\ \Omega$ is
   (a) $15\ \text{k}\Omega$ (b) $80\ \Omega$ (c) $30\ \Omega$ (d) $0\ \Omega$

4. In a series *RLC* circuit that is operating below the resonant frequency, the current
   (a) is in phase with the source voltage
   (b) lags the source voltage
   (c) leads the source voltage

5. If the value of *C* in a series *RLC* circuit is increased, the resonant frequency
   (a) is not affected   (b) increases
   (c) remains the same   (d) decreases

6. In a certain series resonant circuit, $V_C = 150\ \text{V}$, $V_L = 150\ \text{V}$, and $V_R = 50\ \text{V}$. The value of the source voltage is
   (a) $150\ \text{V}$ (b) $300\ \text{V}$ (c) $50\ \text{V}$ (d) $350\ \text{V}$

7. A certain series resonant band-pass filter has a bandwidth of 1 kHz. If the existing coil is replaced with one having a lower value of *Q*, the bandwidth will
   (a) increase   (b) decrease   (c) remain the same   (d) be more selective

8. At frequencies below resonance in a parallel *RLC* circuit, the current
   (a) leads the source voltage
   (b) lags the source voltage
   (c) is in phase with the source voltage

9. The total current into the L and C branches of a parallel circuit at resonance is ideally
   (a) maximum   (b) low   (c) high   (d) zero

10. To tune a parallel resonant circuit to a lower frequency, the capacitance should be
    (a) increased   (b) decreased   (c) left alone   (d) replaced with inductance

11. The resonant frequency of a parallel circuit is the same as a series circuit using the same components when
    (a) the Q is very low   (b) the Q is very high   (c) there is no resistance

12. If the resistance in parallel with a parallel resonant filter is reduced, the bandwidth
    (a) disappears   (b) decreases   (c) becomes sharper   (d) increases

PROBLEMS

Answers to odd-numbered problems are at the end of the book.

BASIC PROBLEMS

SECTION 13–1   Impedance and Phase Angle of Series *RLC* Circuits

1. A certain series *RLC* circuit operates at a frequency of 5 kHz and has the following values: $R = 10 \ \Omega$, $C = 0.047 \ \mu F$, and $L = 5$ mH. Determine the impedance and phase angle. What is the total reactance?

2. Find the impedance in Figure 13–65.

3. If the frequency of the source voltage in Figure 13–65 is doubled from the value that produces the indicated reactances, how does the impedance change?

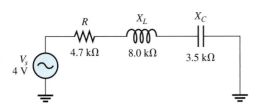

FIGURE 13–65

SECTION 13–2   Analysis of Series *RLC* Circuits

4. For the circuit in Figure 13–65, find I_{tot}, V_R, V_L, and V_C.

5. Draw the voltage phasor diagram for the circuit in Figure 13–65.

6. Analyze the circuit in Figure 13–66 for the following ($f = 25$ kHz):
   (a) I_{tot} (b) P_{true} (c) P_r (d) P_a

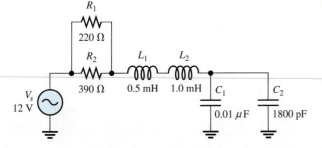

FIGURE 13–66

SECTION 13–3   Series Resonance

7. For the circuit in Figure 13–65, is the resonant frequency higher or lower than the value indicated by the reactances?

8. For the circuit in Figure 13–67, determine the voltage across R at resonance.

FIGURE 13–67

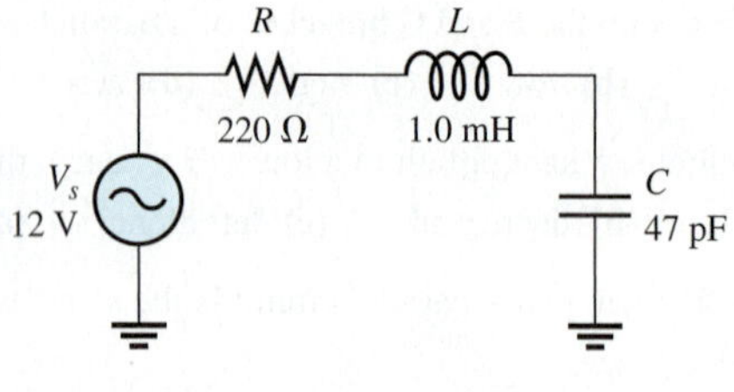

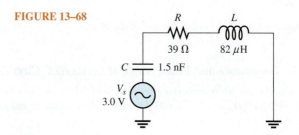

9. Find X_L, X_C, Z, and I at the resonant frequency in Figure 13–67.

10. A certain series resonant circuit has a maximum current of 50 mA and a V_L of 100 V. The source voltage is 10 V. What is Z? What are X_L and X_C?

11. For the *RLC* circuit in Figure 13–68, determine the resonant frequency and the cutoff frequencies.

12. What is the value of the current at the half-power points in Figure 13–68?

FIGURE 13–68

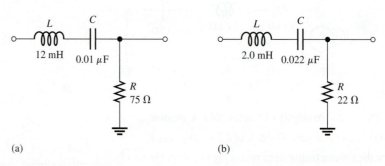

SECTION 13–4   Series Resonant Filters

13. Determine the resonant frequency for each filter in Figure 13–69. Are these filters band-pass or band-stop types?

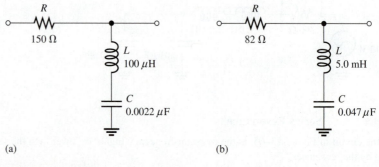

(a)   (b)

FIGURE 13–69

14. Assuming that the coils in Figure 13–69 have a winding resistance of 10 Ω, find the bandwidth for each filter.

15. Determine f_r and *BW* for each filter in Figure 13–70.

(a)   (b)

FIGURE 13–70

SECTION 13–5   Parallel *RLC* Circuits

16. Find the total impedance of the circuit in Figure 13–71.

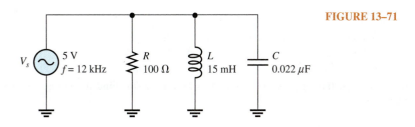

FIGURE 13–71

17. Is the circuit in Figure 13–71 capacitive or inductive? Explain.
18. For the circuit in Figure 13–71, find all the currents and voltages.
19. Find the total impedance for the circuit in Figure 13–72.

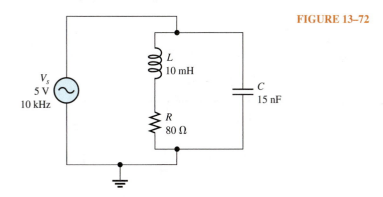

FIGURE 13–72

SECTION 13–6   Parallel Resonance

20. What is the impedance of an ideal parallel resonant circuit (no resistance in either branch)?
21. Find Z at resonance and f_r for the tank circuit in Figure 13–73.
22. How much current is drawn from the source in Figure 13–73 at resonance? What are the inductive current and the capacitive current at the resonant frequency?

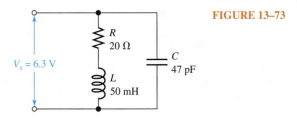

FIGURE 13–73

SECTION 13–7   Parallel Resonant Filters

23. At resonance, $X_L = 2\ k\Omega$ and $R_W = 25\ \Omega$ in a parallel resonant band-pass filter. The resonant frequency is 5 kHz. Determine the bandwidth.

24. If the lower cutoff frequency is 2400 Hz and the upper cutoff frequency is 2800 Hz, what is the bandwidth?

25. In a certain resonant circuit, the power to the load at resonance is 2.75 W. What is the power at the lower and upper cutoff frequencies?

26. What values of L and C should be used in a tank circuit to obtain a resonant frequency of 8 kHz? The bandwidth must be 800 Hz. The winding resistance of the coil is 10 Ω.

27. A parallel resonant circuit has a Q of 50 and a BW of 400 Hz. If Q is doubled, what is the bandwidth for the same f_r?

28. A parallel resonant band-stop filter is needed to reject 60 Hz power line noise. What size should the inductor be if the capacitor is 200 μF?

29. Draw the circuit described in Problem 28 if the output is taken across a 220 Ω resistor.

ADVANCED PROBLEMS

30. For each following case, express the voltage ratio in decibels:
    (a) V_{in} = 1 V, V_{out} = 1 V   (b) V_{in} = 5 V, V_{out} = 3 V
    (c) V_{in} = 10 V, V_{out} = 7.07 V   (d) V_{in} = 25 V, V_{out} = 5 V

31. Find the current through each component in Figure 13–74. Find the voltage across each component.

FIGURE 13–74

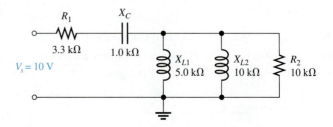

32. Determine whether there is a value of *C* that will make V_{ab} = 0 V in Figure 13–75. If not, explain.

FIGURE 13–75

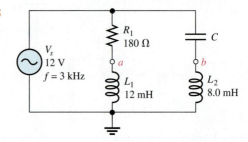

33. If the value of *C* is 0.22 μF, how much current is through each branch in Figure 13–75? What is the total current?

34. (a) Determine the frequency of oscillator-1 for the circuit in Figure 13–52. Assume L_1 is 300 μH.
    (b) What is the cutoff frequency for the low-pass filter in Figure 13–52?

35. Design a band-pass filter using a parallel resonant circuit to meet the following specifications: BW = 500 Hz, Q = 40, $I_{C(max)}$ = 20 mA, $V_{C(max)}$ = 2.5 V.

36. Design a circuit in which the following series resonant frequencies are switch-selectable: 500 kHz, 1000 kHz, 1500 kHz, 2000 kHz.

37. Design a parallel-resonant network using a single coil and switch-selectable capacitors to produce the following resonant frequencies: 8 MHz, 9 MHz, 10 MHz, and 11 MHz. Assume a 10 μH coil with a winding resistance of 5 Ω.

MULTISIM

MULTISIM TROUBLESHOOTING PROBLEMS

38. Open file P13-38; files are found at www.pearsonhighered.com/floyd. Determine if there is a fault and, if so, identify it.

39. Open file P13-39. Determine if there is a fault and, if so, identify it.

40. Open file P13-40. Determine if there is a fault and, if so, identify it.

41. Open file P13-41. Determine if there is a fault and, if so, identify it.

42. Open file P13-42. Determine if there is a fault and, if so, identify it.

43. Open file P13-43. Determine if there is a fault and, if so, identify it.

ANSWERS TO SECTION CHECKUPS

SECTION 13–1 Impedance and Phase Angle of Series *RLC* Circuits

1. The circuit is inductive if $X_L > X_C$ and capacitive if $X_C > X_L$.
2. $X_{tot} = |X_L - X_C| = 70 \ \Omega$; The circuit is capacitive.
3. $Z = \sqrt{R^2 + X_{tot}^2} = 83.2 \ \Omega$; $\theta = \tan^{-1}(X_{tot}/R) = 57.3°$; Current is leading V_s.

SECTION 13–2 Analysis of Series *RLC* Circuits

1. $V_s = \sqrt{V_R^2 + (V_C - V_L)^2} = 38.4 \ V$
2. Current leads V_s because the circuit is capacitive.
3. $X_{tot} = |X_L - X_C| = 6 \ k\Omega$

SECTION 13–3 Series Resonance

1. Series resonance occurs when $X_L = X_C$.
2. The current is maximum because the impedance is minimum.
3. $f_r = 1/(2\pi\sqrt{LC}) = 159 \ kHz$
4. It is capacitive because $X_C > X_L$.

SECTION 13–4 Series Resonant Filters

1. $V_{out} = 0.707(15 \ V) = 10.6 \ V$
2. $BW = f_r/Q = 10 \ kHz$
3. Current is maximum; output voltage is minimum.

SECTION 13–5 Parallel *RLC* Filters

1. $I_R = V_s/R = 8.0 \ mA$; $I_C = V_s/X_C = 12.0 \ mA$; $I_L = V_s/X_L = 24.0 \ mA$
2. The circuit is inductive ($X_L < X_C$).
3. $L_{eq} = L[(Q^2 + 1)/Q^2] = 20.1 \ mH$; $R_{p(eq)} = R_W(Q^2 + 1) = 1589 \ \Omega$

SECTION 13–6 Parallel Resonance

1. The impedance is maximum.
2. The current is minimum.
3. At ideal parallel resonance, $X_C = X_L = 1.5 \ k\Omega$.
4. $f_r = \sqrt{1 - (R_W^2 C/L)}/2\pi\sqrt{LC} = 33.7 \ kHz$; $Z_r = R_W(Q^2 + 1) = 440 \ \Omega$
5. $f_r = 1/(2\pi\sqrt{LC}) = 22.5 \ kHz$
6. $f_r = \sqrt{Q^2/(Q^2 + 1)}/2\pi\sqrt{LC} = 20.9 \ kHz$
7. $Z_r = R_W(Q^2 + 1) = 8.02 \ k\Omega$

SECTION 13–7 Parallel Resonant Filters

1. The bandwidth can be increased by reducing the parallel resistance.
2. f_r changes to 4.47 kHz.
3. $Z_r = R_W(Q^2 + 1) = 47.0 \ k\Omega$

SECTION 13–8 Resonant Circuit Applications

1. A tuned resonant circuit is used to select a narrow band of frequencies.
2. The same tuned circuits can be used no matter what station is selected.
3. Several variable capacitors (or inductors) whose values can be varied simultaneously with a common control is an example of ganged tuning.

ANSWERS TO RELATED PROBLEMS FOR EXAMPLES

13–1. 1.25 kΩ; 63.4°

13–2. 4.71 kΩ

13–3. Current increases, reaches a maximum at resonance, and then decreases.

13–4. More capacitive

13–5. 67.3 kHz

13–6. 2.27 mA

13–7. Z increases; Z increases

13–8. No

13–9. 200 kHz

13–10. -14 dB

13–11. 7.48

13–12. 322 Hz

13–13. V_{out} increases; V_{out} increases

13–14. Inductive

13–15. Increase. X_C will decrease and approach 0.

13–16. $R_{p(eq)} = 25$ kΩ; $L_{eq} = 5$ mH; $C = 0.022\ \mu$F, $R_{p(tot)} = 4.24$ kΩ

13–17. $I_L = 0.996$ mA, $I_C = 1.06$ mA

13–18. 80.0 kΩ

13–19. The differences are negligible.

13–20. 7.07 V

13–21. 530 kHz

13–22. 25

13–23. A larger-value load resistor has less effect on Q_O.

13–24. 185 mV

ANSWERS TO TRUE/FALSE QUIZ

1. T   **2.** F   **3.** T   **4.** T   **5.** F   **6.** T   **7.** T   **8.** F   **9.** F   **10.** F

ANSWERS TO SELF-TEST

1. (a)   **2.** (c)   **3.** (b)   **4.** (c)   **5.** (d)   **6.** (c)

7. (a)   **8.** (b)   **9.** (d)   **10.** (a)   **11.** (b)   **12.** (d)

CHAPTER 14

TRANSFORMERS

OUTLINE

OBJECTIVES

- Explain mutual inductance
- Describe how a transformer is constructed and how it works
- Describe how transformers increase and decrease voltage
- Discuss the effect of a resistive load across the secondary
- Discuss the concept of a reflected load in a transformer
- Discuss impedance matching with transformers
- Describe practical transformer ratings
- Describe several types of transformers
- Troubleshoot transformers

KEY TERMS

- Electrical isolation
- Mutual inductance (L_M)
- Transformer
- Primary winding
- Secondary winding
- Magnetic coupling
- Turns ratio (n)
- Reflected resistance
- Impedance matching
- Apparent power rating
- Center tap

INTRODUCTION

In Chapter 11, you learned about self-inductance. In this chapter, you will study mutual inductance, which is the basis for the operation of transformers. Transformers are an important component in many systems, such as power supplies, electrical power distribution, and signal coupling in communications systems.

The operation of the transformer is based on the principle of mutual inductance, which occurs when two or more coils are in close proximity. A simple transformer is actually two coils that are electromagnetically coupled by their mutual inductance. Because there is no electrical contact between two magnetically coupled coils, the transfer of energy from one coil to the other can be achieved in a situation of complete electrical isolation. In relation to transformers, the term *winding* or *coil* is commonly used to describe the primary or secondary.

VISIT THE WEBSITE
Study aids for this chapter are available at
http://pearsonhighered.com/floyd

14–1 MUTUAL INDUCTANCE

When two coils are placed close to each other, a changing electromagnetic field produced by the current in one coil will cause an induced voltage in the second coil because of the mutual inductance between the two coils.

After completing this section, you should be able to

- **Explain mutual inductance**
  - **Discuss magnetic coupling**
  - **Define** *electrical isolation*
  - **Define** *coefficient of coupling*
  - **Identify the factors that affect mutual inductance and state the formula**

Recall that the electromagnetic field surrounding a coil of wire expands, collapses, and reverses as the current increases, decreases, and reverses. When a second coil is placed very close to the first coil so that the changing magnetic lines of force cut through the second coil, the coils are magnetically coupled and a voltage is induced, as indicated in Figure 14–1.

FIGURE 14–1 **A voltage is induced in the second coil as a result of the changing current in the first coil, producing a changing magnetic field that links the second coil.**

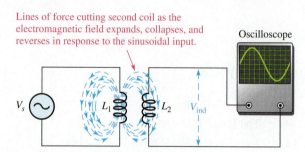

Lines of force cutting second coil as the electromagnetic field expands, collapses, and reverses in response to the sinusoidal input.

Oscilloscope

When two coils are magnetically coupled, they provide electrical isolation because there is no electrical connection between them, only a magnetic link. **Electrical isolation** is the condition in which two circuits have no common conductive path between them. If the current in the first coil is a sine wave, the voltage induced in the second coil is also a sine wave. The amount of voltage induced in the second coil as a result of the current in the first coil is dependent on the **mutual inductance, L_M**.

The mutual inductance is established by the inductance of each coil (L_1 and L_2) and by the amount of coupling (k) between the two coils. To maximize coupling, the two coils are wound on a common core. The three factors that influence mutual inductance (k, L_1, and L_2) are shown in Figure 14–2. The formula for mutual inductance is

$$L_M = k\sqrt{L_1 L_2} \qquad (14\text{–}1)$$

FIGURE 14–2 **The mutual inductance of two coils.**

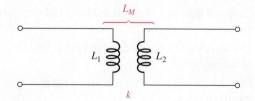

Coefficient of Coupling

The **coefficient of coupling, k**, between the two coils of a transformer is the ratio of the flux (lines of force) produced by coil 1 linking coil 2 ($\phi_{1\text{-}2}$) to the total flux produced by coil 1 (ϕ_1).

$$k = \frac{\phi_{1\text{-}2}}{\phi_1} \qquad (14\text{–}2)$$

For example, if half of the total flux produced by coil 1 links coil 2, then $k = 0.5$. A greater value of k means that more voltage is induced in coil 2 for a certain rate of change of current in coil 1. Note that k has no units. Recall that the unit of magnetic lines of force (flux) is the weber, abbreviated Wb.

The coefficient of coupling, k, depends on the physical proximity of the coils and the type of core material on which they are wound. Also, the construction and shape of the cores are factors.

EXAMPLE 14–1

Two coils are wound on a single core, and the coefficient of coupling is 0.3. The inductance of coil 1 is 10 μH, and the inductance of coil 2 is 15 μH. What is L_M?

SOLUTION

$$L_M = k\sqrt{L_1 L_2} = 0.3\sqrt{(10\,\mu\text{H})(15\,\mu\text{H})} = \mathbf{3.67\,\mu H}$$

RELATED PROBLEM*

Determine the mutual inductance when $k = 0.5$, $L_1 = 1$ mH, and $L_2 = 600\,\mu$H.

*Answers are at the end of the chapter.

EXAMPLE 14–2

One coil produces a total magnetic flux of 50 μWb, and 20 μWb link coil 2. What is k?

SOLUTION

$$k = \frac{\phi_{1\text{-}2}}{\phi_1} = \frac{20\,\mu\text{Wb}}{50\,\mu\text{Wb}} = \mathbf{0.4}$$

RELATED PROBLEM

Determine k when $\phi_1 = 500\,\mu$Wb and $\phi_{1\text{-}2} = 375\,\mu$Wb.

SECTION 14–1 CHECKUP*

1. Define *mutual inductance*.

2. Two 50 mH coils have $k = 0.9$. What is L_M?

3. If k is increased, what happens to the voltage induced in one coil as a result of a current change in the other coil?

*Answers are at the end of the chapter.

14–2 THE BASIC TRANSFORMER

A **transformer** is an electrical device constructed of two or more coils of wire (windings) electromagnetically coupled to each other so that there is a mutual inductance for the transfer of power from one winding to the other. Although many transformers have more than two windings, we will restrict our discussion in this section to a basic two-winding transformer. Later, more complicated transformers are introduced.

After completing this section, you should be able to

- Describe how a transformer is constructed and how it works
- Identify the parts of a basic transformer
- Discuss the importance of the core material
- Define *primary winding* and *secondary winding*
- Define *turns ratio*
- Discuss how the direction of windings affects voltage polarities

A schematic of a transformer is shown in Figure 14–3(a). One coil is called the **primary winding** and the other coil is called the **secondary winding** as indicated. For standard operation, the source voltage is applied to the primary winding, and a load is connected to the secondary winding, as shown in Figure 14–3(b). The primary winding is the input winding, and the secondary winding is the output winding. It is common to refer to the side of the transformer that has the source voltage as the *primary,* and the side that has the induced voltage as the *secondary.*

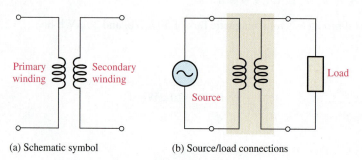

(a) Schematic symbol   (b) Source/load connections

FIGURE 14–3   The basic transformer.

The windings of a transformer are formed around the core. The core provides both a physical structure for placement of the windings and a magnetic path so that the magnetic flux is concentrated close to the coils. Three general categories of core material are air, ferrite, and iron. The schematic symbol for each type is shown in Figure 14–4.

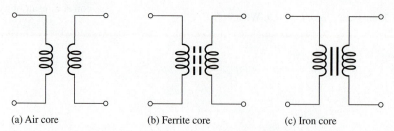

(a) Air core   (b) Ferrite core   (c) Iron core

FIGURE 14–4   Schematic symbols specify the type of core.

Iron-core transformers generally are used for audio frequency (AF) and power applications. These transformers consist of windings on a core constructed from laminated sheets of ferromagnetic material insulated from each other, as shown in Figure 14–5. This

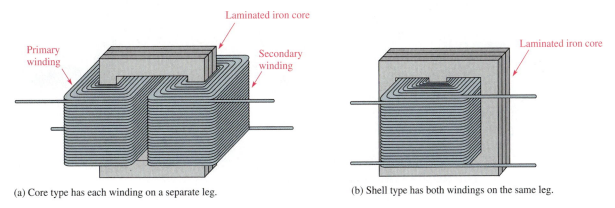

(a) Core type has each winding on a separate leg.

(b) Shell type has both windings on the same leg.

FIGURE 14–5   **Iron-core transformer construction with multilayer windings.**

construction provides an easy path for the magnetic flux and increases the amount of coupling between the windings. The figure also shows the basic construction of two major configurations of iron-core transformers. In the core-type construction, shown in Figure 14–5(a), the windings are on separate legs of the laminated core. In the shell-type construction, shown in part (b), both windings are on the same leg. Each type has certain advantages. In general, the core type has more room for insulation and can handle higher voltages. The shell type can produce higher core flux, so fewer turns are required.

Air-core and ferrite-core transformers generally are used for high-frequency applications and consist of windings on an insulating shell that is hollow (air) or constructed of ferrite, such as depicted in Figure 14–6. The wire is typically covered by a varnish-type coating to prevent the windings from shorting together. The amount of **magnetic coupling** between the primary winding and the secondary winding is set by the type of core material and by the relative positions of the windings. In Figure 14–6(a), the windings are loosely coupled because they are separated, and in part (b) they are tightly coupled because they are overlapping. The tighter the coupling, the greater the induced voltage in the secondary for a given current in the primary. Another common high-frequency transformer is the intermediate frequency (IF) transformer shown in Figure 14–6(c). The ferrite core can be adjusted with a screw that is used to tune the transformer to a specific frequency.

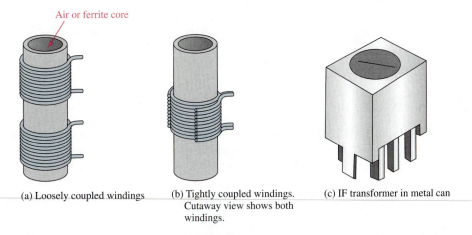

(a) Loosely coupled windings

(b) Tightly coupled windings. Cutaway view shows both windings.

(c) IF transformer in metal can

FIGURE 14–6   **Transformers with cylindrical-shaped cores.**

High-frequency transformers tend to have fewer windings and smaller inductances than power transformers. A type of high-frequency transformer that has become popular in recent years is the planar transformer. A representative planar transformer is shown in Figure 14–7(a). Planar transformers are constructed with printed circuit (PC) board assembly methods (rather than wire winding) that enable them to be produced

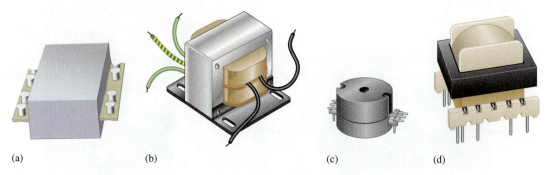

(a)　　　　　　　(b)　　　　　　　(c)　　　　　　　(d)

FIGURE 14–7 **Some common types of transformers.**

with high precision and low cost. The windings are actually traces laid out on stacked PC boards. Planar transformers are available in a variety of sizes and wattage ratings. The low profile of a planar transformer (typically less than 0.5 inches) makes it particularly suited to cases where space is critical. Figure 14–7(b) shows a low-voltage transformer commonly used in power supplies. Parts (c) and (d) show other common types of small transformers.

Intermediate Frequency Transformers

Communication systems and certain measurement systems use tuned transformers to couple a high-frequency signal from one stage to another. The signal is generally a predetermined frequency called the intermediate frequency (IF), which is a fixed frequency used in the system. The transistor circuit shown here is an intermediate frequency amplifier with parallel resonant tuned circuits shown in the yellow highlighted boxes on the input (C_1 and the secondary of T_1) and

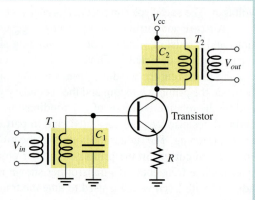

the output (C_2 and the primary of T_2). This causes the amplifier to selectively amplify the intermediate frequency.

SYSTEM NOTE

Turns Ratio

A transformer parameter that is useful in understanding how a transformer operates is the turns ratio. In this text, the **turns ratio, (n)** is defined as the ratio of the number of turns in the secondary winding (N_{sec}) to the number of turns in the primary winding (N_{pri}).

$$n = \frac{N_{sec}}{N_{pri}}$$ (14–3)

This definition of turns ratio is based on the IEEE standard for electronics power transformers as specified in the IEEE dictionary. Other categories of transformer may have a different definition, so some sources define the turns ratio as N_{pri}/N_{sec}. Either definition is correct as long as it is clearly stated and used consistently. The turns ratio of a transformer is rarely if ever given as a transformer specification. Generally, the input and output voltages and the power rating are the key specifications. However, the turns ratio is useful in studying the operating principle of a transformer.

EXAMPLE 14–3

A certain transformer used in a radar system has a primary winding with 100 turns and a secondary winding with 400 turns. What is the turns ratio?

SOLUTION

$N_{sec} = 400$ and $N_{pri} = 100$; therefore, the turns ratio is

$$n = \frac{N_{sec}}{N_{pri}} = \frac{400}{100} = 4$$

A turns ratio of 4 can be expressed as 1:4 on a schematic.

RELATED PROBLEM

A certain transformer has a turns ratio of 10. If $N_{pri} = 500$, what is N_{sec}?

Direction of Windings

Another important transformer parameter is the direction in which the windings are placed around the core. As illustrated in Figure 14–8, the direction of the windings determines the polarity of the voltage across the secondary winding (secondary voltage) with respect to the voltage across the primary winding (primary voltage). Phase dots can be used on the schematic symbols to indicate polarities, as shown in Figure 14–9.

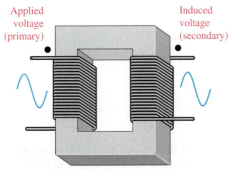

Applied voltage (primary)   Induced voltage (secondary)

(a) The primary and secondary voltages are in phase when the windings are in the same effective direction around the magnetic path.

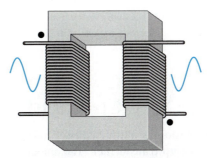

(b) The primary and secondary voltages are 180° out of phase when the windings are in the opposite direction.

FIGURE 14–8   The direction of the windings determines the relative polarities of the voltages.

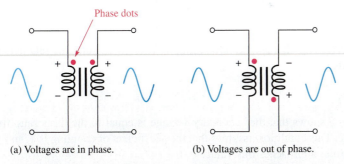

Phase dots

(a) Voltages are in phase.   (b) Voltages are out of phase.

FIGURE 14–9   Phase dots indicate corresponding polarities of primary and secondary voltages.

SECTION 14–2 CHECKUP

1. Upon what principle is the operation of a transformer based?

2. Define *turns ratio*.

3. Why are the directions of the windings of a transformer important?

4. A certain transformer has a primary winding with 500 turns and a secondary winding with 250 turns. What is the turns ratio?

5. How do the windings in a planar transformer differ from other transformers?

14–3 STEP-UP AND STEP-DOWN TRANSFORMERS

A step-up transformer has more turns in its secondary winding than in its primary winding and is used to increase ac voltage. A step-down transformer has more turns in its primary winding than in its secondary winding and is used to decrease ac voltage.

After completing this section, you should be able to

- **Describe how transformers increase and decrease voltage**
  - **Explain how a step-up transformer works**
  - **Identify a step-up transformer by its turns ratio**
  - **State the relationship between primary and secondary voltages and the turns ratio**
  - **Explain how a step-down transformer works**
  - **Identify a step-down transformer by its turns ratio**
  - **Explain the purpose of ac line conditioning**

The Step-Up Transformer

A transformer in which the secondary voltage is greater than the primary voltage is called a **step-up transformer**. The amount that the voltage is stepped up depends on the turns ratio. For any transformer,

The ratio of secondary voltage (V_{sec}) to primary voltage (V_{pri}) is equal to the ratio of the number of turns in the secondary winding (N_{sec}) to the number of turns in the primary winding (N_{pri}).

$$\frac{V_{sec}}{V_{pri}} = \frac{N_{sec}}{N_{pri}} \qquad (14\text{–}4)$$

Recall that N_{sec}/N_{pri} defines the turns ratio, n. Therefore, from this relationship, V_{sec} can be expressed as

$$V_{sec} = nV_{pri} \qquad (14\text{–}5)$$

Equation 14–5 shows that the secondary voltage is equal to the turns ratio times the primary voltage. This condition assumes that the coefficient of coupling is 1, and a good iron-core transformer approaches this value.

The turns ratio for a step-up transformer is always greater than 1 because the number of turns in the secondary winding (N_{sec}) is always greater than the number of turns in the primary winding (N_{pri}).

E X A M P L E 1 4 – 4

The transformer in Figure 14–10 has a turns ratio of 3. What is the voltage across the secondary? Voltages are in rms unless otherwise stated.

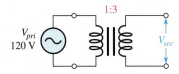

FIGURE 14–10

SOLUTION

The secondary voltage is

$$V_{sec} = nV_{pri} = 3(120 \text{ V}) = \textbf{360 V}$$

Note that the turns ratio of 3 is indicated on the schematic as 1:3, meaning that there are three secondary turns for each primary turn.

RELATED PROBLEM

The transformer in Figure 14–10 is changed to one with a turns ratio of 4. Determine V_{sec}.

MULTISIM

Open Multisim file E14-04; files are found at www. pearsonhighered.com/floyd. Measure the secondary voltage.

The Step-Down Transformer

A transformer in which the secondary voltage is less than the primary voltage is called a **step-down transformer**. The amount by which the voltage is stepped down depends on the turns ratio. Equation 14–5 also applies to a step-down transformer.

The turns ratio of a step-down transformer is always less than 1 because the number of turns in the secondary winding (N_{sec}) is always less than the number of turns in the primary winding (N_{pri}).

E X A M P L E 1 4 – 5

The transformer in Figure 14–11 is part of a laboratory power supply and has a turns ratio of 0.2. What is the secondary voltage?

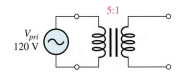

FIGURE 14–11

SOLUTION

The secondary voltage is

$$V_{sec} = nV_{pri} = 0.2(120 \text{ V}) = \textbf{24 V}$$

RELATED PROBLEM

The transformer in Figure 14–11 is changed to one with a turns ratio of 0.48. Determine the secondary voltage.

MULTISIM

Open Multisim file E14-05 and measure the secondary voltage.

SYSTEM EXAMPLE 14–1

A BASIC POWER SUPPLY

One of the most important applications for transformers is in power supplies. Power supplies convert ac to dc, which is needed by nearly every electronic system. The transformer converts the ac to a higher or lower value before the next step of changing it to pulsating dc, which is then smoothed and regulated in the final step. Many systems use low-voltage dc for control circuits.

The basic power supply shown in Figure 14–12 provides a constant 5 V dc output over a range of loads and variations in the line voltage. This is accomplished by the 7805 voltage regulator, an integrated circuit regulator that senses the output, compares it to a reference, and adjusts it to keep it constant. The 7805 is designed for positive 5.0 V dc output.

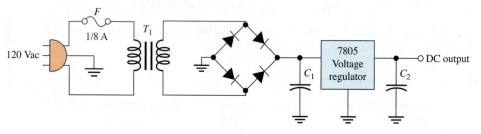

FIGURE 14–12   A basic 5.0 V dc power supply.

The operation of the power supply is as follows. The plug is connected to a wall socket that connects 120 V ac to the input of the transformer through a series fuse. The fuse is rated for a much smaller current than the maximum output current because the transformer will step *down* the voltage but step *up* the current. For the basic power supply here, the transformer output is about 6 V rms (8.5 V peak). The four diodes, which are arranged in a bridge configuration, change the voltage to a pulsating dc, which is smoothed by C_1. The 7805 regulator provides additional smoothing and keeps the output nearly constant. Final smoothing and filtering are provided by C_2.

If you work on any power supply, be aware that there are dangerous voltages present when a circuit is connected to ac; even low-voltage supplies can be lethal when they are connected to ac. In the case of high-voltage supplies, extra caution should be taken; even circuits that have been turned off remain dangerous because capacitors can retain a significant charge. Precautions for working safely around voltage sources and high energy sources (like batteries) are discussed in Chapter 1, Section 1–8, and should always be kept in mind.

DC Isolation

If there is dc through the primary circuit of a transformer, nothing happens in the secondary circuit, as indicated in Figure 14–13(a). A changing current in the primary winding is necessary in order to create a changing magnetic field. This will cause voltage to be induced in the secondary circuit, as indicated in Figure 14–13(b). Therefore, the transformer isolates

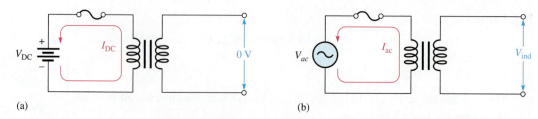

(a)   (b)

FIGURE 14–13   DC isolation and ac coupling.

the secondary circuit from any dc in the primary circuit. A transformer that is used strictly for isolation has a turns ratio of 1.

Isolation transformers are often packaged as part of a total ac line-conditioning device. In addition to the isolation transformer, the line conditioning includes surge protection, filters to eliminate interference, and sometimes automatic voltage regulation. Line conditioning is useful to isolate sensitive equipment such as microprocessor-based controllers. Specialized line conditioners for hospitals for patient-monitoring equipment provide a high degree of electrical isolation and protection from shock.

Small transformers that are used to isolate the dc bias from one stage of an amplifier to next stage are called coupling transformers because the ac signal is passed ("coupled") but the dc is blocked. Coupling transformers are widely used at high frequencies where they are designed to pass only a selected band of frequencies by making the primary and secondary coils part of a parallel resonant circuit. (Resonant circuits were discussed in Chapter 13.) A typical coupling transformer arrangement is shown in Figure 14–14 where the transformer is part of a resonant circuit on the input and output. Frequently, the core of a coupling transformer can be adjusted to fine-tune the frequency response. At audio frequencies, coupling transformers are primarily used to couple a signal from an amplifier to a speaker and to provide maximum power transfer. This type of transformer is called an impedance-matching transformer and is discussed in Section 14–6.

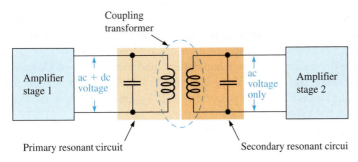

FIGURE 14–14   **A coupling transformer used to pass high frequencies in the band determined by the resonant circuit. The dc is not passed to the secondary.**

SECTION 14–3 CHECKUP

1. What does a step-up transformer do?

2. If the turns ratio is 5, how much greater is the secondary voltage than the primary voltage?

3. When 240 V ac are applied to a transformer with a turns ratio of 10, what is the secondary voltage?

4. What does a step-down transformer do?

5. A voltage of 120 V ac is applied to the primary winding of a transformer with a turns ratio of 0.5. What is the secondary voltage?

6. A primary voltage of 120 V ac is reduced to 12 V ac. What is the turns ratio?

7. What are typical features provided in an ac line conditioner?

When a resistive load is connected to the secondary winding of a transformer, the relationship of the load (secondary) current and the current in the primary circuit is determined by the turns ratio.

After completing this section, you should be able to

- Discuss the effect of a resistive load across the secondary
- Discuss power in a transformer
- Determine the current delivered by the secondary when a step-up transformer is loaded
- Determine the current delivered by the secondary when a step-down transformer is loaded

If a transformer is operated without a load, the primary acts like an inductor. Ideally, the current in an inductor will lag the voltage by 90° and the power factor is 0. When a resistive load is connected to the secondary of the transformer, the primary no longer acts as an ideal inductor. The phase angle decreases and the power factor increases. The load has caused the primary to look resistive because the primary current and voltage will be nearly in phase. For purposes of discussion, we will assume this ideal condition when the transformer is loaded with a resistive load.

The power delivered to a load can never be greater than the power delivered to the primary winding. For an ideal transformer, the power delivered by the secondary winding (P_{sec}) equals the power delivered to the primary winding (P_{pri}). When losses are considered, the power delivered by the secondary is always less.

Power is dependent on voltage and current, and there can be no increase in power in a transformer. Therefore, if the voltage is stepped up, the current is stepped down and vice versa. In an ideal transformer, the power delivered by the secondary to the load is the same as the power delivered to the primary regardless of the turns ratio.

The power delivered to the primary is

$$P_{pri} = V_{pri}I_{pri}$$

The power delivered by the secondary is

$$P_{sec} = V_{sec}I_{sec}$$

Ideally, $P_{pri} = P_{sec}$; therefore,

$$V_{pri}I_{pri} = V_{sec}I_{sec}$$

Transposing terms,

$$\frac{I_{pri}}{I_{sec}} = \frac{V_{sec}}{V_{pri}}$$

From Equation 14–4,

$$\frac{V_{sec}}{V_{pri}} = \frac{N_{sec}}{N_{pri}}$$

Therefore, since N_{sec}/N_{pri} equals the turns ratio, n, the relationship of primary current to secondary current in a transformer is

$$\frac{I_{pri}}{I_{sec}} = n \qquad\qquad (14\text{–}6)$$

Inverting both sides of Equation 14–6 and solving for I_{sec},

$$I_{sec} = \left(\frac{1}{n}\right)I_{pri} \qquad\qquad (14\text{–}7)$$

Figure 14–15 illustrates the effects of voltages and currents in a transformer. In part (a), for a step-up transformer in which n is greater than 1, the secondary current is less than the primary current because $1/n$ is less than 1. For a step-down transformer, shown in Figure 14–15(b), n is less than 1, and I_{sec} is greater than I_{pri} because $1/n$ is greater than 1.

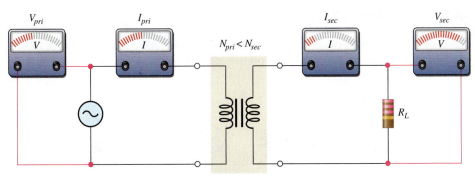

FIGURE 14–15 Illustration of voltages and currents in a transformer with a loaded secondary.

(a) Step-up transformer: $V_{sec} > V_{pri}$ and $I_{sec} < I_{pri}$

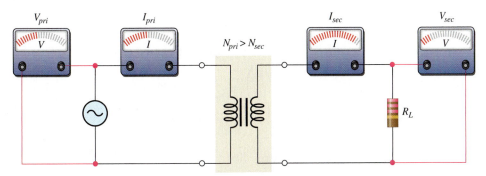

(b) Step-down transformer: $V_{sec} < V_{pri}$ and $I_{sec} > I_{pri}$

Current Transformers

A current transformer operates like a voltage transformer in that energy is transferred from the primary to the secondary through induction. Current transformers are often used to sense and monitor a large alternating current without having to open the line to make the measurement. One common application is to monitor current and supply the result to the watt-hour meter for customers with three-phase service or large-use single-phase customers.

One important consideration with current transformers is that the load must not be disconnected when current is in the primary. If this happens, the sudden opening will produce a very high voltage across the open secondary, leading to arcing and can permanently damage the transformer and equipment connected to it. In addition, it is a shock hazard to personnel working on it.

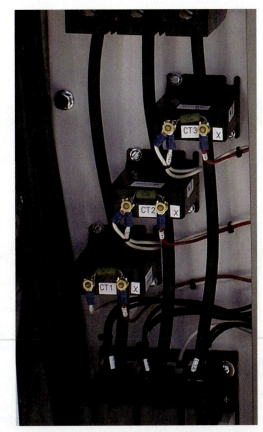

Photo courtesy of Thomas Kissell.

SYSTEM NOTE

EXAMPLE 14–6

The two ideal transformers shown in Figure 14–16 have loaded secondaries. If, as a result of the loaded secondary, the primary current is 100 mA in each case, what is the current through the load?

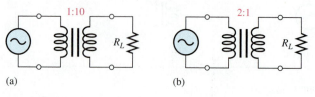

FIGURE 14–16

SOLUTION

In part (a), the turns ratio is 10. Therefore, the secondary load current is

$$I_L = I_{sec} = \left(\frac{1}{n}\right)I_{pri} = \left(\frac{1}{10}\right)I_{pri} = 0.1(100 \text{ mA}) = \textbf{10 mA}$$

In part (b), the turns ratio is 0.5. Therefore, the secondary load current is

$$I_L = I_{sec} = \left(\frac{1}{n}\right)I_{pri} = \left(\frac{1}{0.5}\right)I_{pri} = 2(100 \text{ mA}) = \textbf{200 mA}$$

RELATED PROBLEM

What is the secondary current in Figure 14–16(a) if the turns ratio is doubled? What is the secondary current in Figure 14–16(b) if the turns ratio is halved? Assume the load resistances are changed so that I_{pri} remains at 100 mA in both cases.

SECTION 14–4 CHECKUP

1. If the turns ratio of a transformer is 2, is the secondary current greater than or less than the primary current? By how much?

2. A transformer has 1000 primary turns and 250 secondary turns, and I_{pri} is 0.5 A. What is the turns ratio? What is the value of I_{sec}?

3. In Question 2, what is the primary current when there is a secondary load current of 10 A?

4. Why is it unsafe to disconnect the load on a current transformer when there is current in the primary?

14–5 REFLECTED LOAD

From the viewpoint of the primary, a load connected across the secondary winding of a transformer appears to have a resistance that is not necessarily equal to the actual resistance of the load. The actual load is "reflected" into the primary as determined by the turns ratio. This reflected load is what the source effectively sees, and it determines the amount of primary current.

After completing this section, you should be able to

• Discuss the concept of a reflected load in a transformer
  • Define *reflected resistance*
  • Explain how the turns ratio affects the reflected resistance
  • Calculate reflected resistance

The concept of a reflected load is illustrated in Figure 14–17. The load (R_L) in the secondary of a transformer is reflected into the primary by transformer action. Ideally, the load appears to the source in the primary to be a resistance (R_{pri}) with a value determined by the turns ratio and the actual value of the load resistance. The resistance R_{pri} is called the **reflected resistance**.

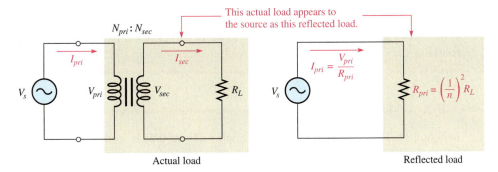

FIGURE 14–17   **Reflected load in a transformer circuit.**

The resistance in the primary of Figure 14–17 is $R_{pri} = V_{pri}/I_{pri}$. The resistance in the secondary is $R_L = V_{sec}/I_{sec}$. From Equations 14–4 and 14–6, you know that $V_{sec}/V_{pri} = n$ and $I_{pri}/I_{sec} = n$. Using these relationships, a formula for R_{pri} in terms of R_L is determined as follows:

$$\frac{R_{pri}}{R_L} = \frac{V_{pri}/I_{pri}}{V_{sec}/I_{sec}} = \left(\frac{V_{pri}}{V_{sec}}\right)\left(\frac{I_{sec}}{I_{pri}}\right) = \left(\frac{1}{n}\right)\left(\frac{1}{n}\right) = \left(\frac{1}{n}\right)^2$$

Solving for R_{pri} yields

$$R_{pri} = \left(\frac{1}{n}\right)^2 R_L \qquad (14–8)$$

Equation 14–8 shows that the resistance reflected into the primary is the square of the reciprocal of the turns ratio times the load resistance.

In a step-up transformer ($n > 1$), the reflected resistance is less than the actual load resistance; in a step-down transformer ($n < 1$), the reflected resistance is greater than the load resistance. This is illustrated in Examples 14–7 and 14–8, respectively.

EXAMPLE 14–7

Figure 14–18 shows a source that is transformer-coupled to a load resistor of 100 Ω. The transformer has a turns ratio of 4. What is the reflected resistance seen by the source?

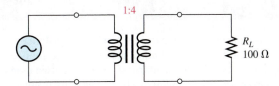

FIGURE 14–18

SOLUTION

The reflected resistance is

$$R_{pri} = \left(\frac{1}{n}\right)^2 R_L = \left(\frac{1}{4}\right)^2 100\ \Omega = \left(\frac{1}{16}\right)100\ \Omega = \mathbf{6.25\ \Omega}$$

The source sees a resistance of 6.25 Ω just as if it were connected directly, as shown in the equivalent circuit of Figure 14–19.

FIGURE 14–19

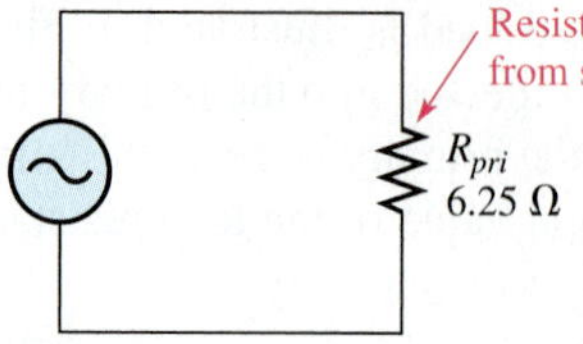

Resistance "reflected"
from secondary

R_{pri}
6.25 Ω

RELATED PROBLEM

If the turns ratio in Figure 14–18 is 10 and R_L is 600 Ω, what is the reflected resistance?

EXAMPLE 14–8

In Figure 14–18, if a transformer that has a turns ratio of 0.25 is used, what is the reflected resistance?

SOLUTION

The reflected resistance is

$$R_{pri} = \left(\frac{1}{n}\right)^2 R_L = \left(\frac{1}{0.25}\right)^2 100 \text{ Ω} = (4)^2 100 \text{ Ω} = \textbf{1600 Ω}$$

RELATED PROBLEM

To achieve a reflected resistance of 800 Ω, what turns ratio is required in Figure 14–18?

SECTION 14–5 CHECKUP

1. Define *reflected resistance.*

2. What transformer characteristic determines the reflected resistance?

3. A given transformer has a turns ratio of 10, and the load is 50 Ω. How much resistance is reflected into the primary?

4. What is the turns ratio required to reflect a 4 Ω load resistance into the primary as 400 Ω?

14–6 IMPEDANCE MATCHING

One application of transformers is in the matching of a load impedance to a source impedance to achieve maximum transfer of power or other results. This technique is called *impedance matching*. In audio systems, special wide-band transformers are often used to get the maximum amount of available power from the amplifier to the speaker by proper selection of the turns ratio. Transformers designed specifically for impedance matching usually show the input and output impedance they are designed to match.

After completing this section, you should be able to

- Discuss impedance matching with transformers
  - Discuss the maximum power transfer theorem
  - Define *impedance matching*
  - Explain the purpose of impedance matching
  - Describe a balun transformer

Recall that the maximum power transfer theorem states that maximum power is transferred from a resistive source to a resistive load when the load resistance is equal to the source resistance (see Section 6–7). In ac circuits, the total opposition to current is called *impedance* and the process of making maximum power transfer between a source and load is called **impedance matching**. The simplest and most common case for impedance matching is when the source and load are both resistive, so we will restrict our discussion to this case.

Figure 14–20(a) shows an ac source with a fixed internal resistance. Some fixed internal resistance is inherent in all sources due to their internal circuitry. Part (b) shows a load connected to the source. In this case, the objective is often to transfer as much power to the load as possible.

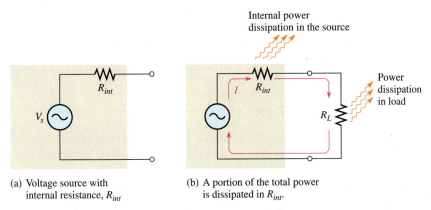

(a) Voltage source with internal resistance, R_{int}

(b) A portion of the total power is dissipated in R_{int}.

FIGURE 14–20   **Power transfer from a nonideal voltage source to a load.**

In most practical situations, the internal resistance of various types of sources is fixed. Also, in many cases, the resistance of a device that acts like a load is fixed and cannot be altered. If you need to connect a given source to a given load, remember that only by chance will their resistances match. In this situation, a special type of wide-band transformer comes in handy. You can use the reflected-resistance characteristic provided by a transformer to make the load resistance appear to have the same resistance as the source resistance. This technique is called *impedance matching*, and the transformer is called an *impedance-matching transformer*.

Figure 14–21 illustrates a specific example of an impedance-matching transformer. In this example, the source resistance is driving a 300 Ω load. The impedance-matching transformer needs to make the load resistance look like a 75 Ω resistance to the source, thus delivering maximum power to the load. To select the right transformer, you need to know how the turns ratio affects the impedance. You can use Equation 14–8 to determine the turns ratio, n, when you know the value for R_L and R_{pri}.

$$R_{pri} = \left(\frac{1}{n}\right)^2 R_L$$

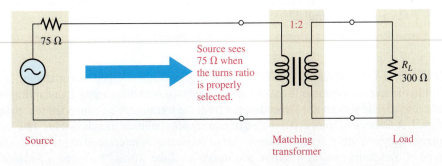

FIGURE 14–21   **Example of a load matched to a source by transformer coupling for maximum power transfer.**

Transpose terms and divide both sides by R_L.

$$\left(\frac{1}{n}\right)^2 = \frac{R_{pri}}{R_L}$$

Then take the square root of both sides.

$$\frac{1}{n} = \sqrt{\frac{R_{pri}}{R_L}}$$

Invert both sides to get the following formula for the turns ratio:

$$n = \sqrt{\frac{R_L}{R_{pri}}} \qquad\qquad (14\text{–}9)$$

Finally, solve for the particular turns ratio to match a 300 Ω load to a 75 Ω source.

$$n = \sqrt{\frac{300\ \Omega}{75\ \Omega}} = \sqrt{4} = 2$$

Therefore, a matching transformer with a turns ratio of 2 must be used in this application.

EXAMPLE 14–9

An amplifier has an 800 Ω internal resistance. In order to provide maximum power to an 8 Ω speaker, what turns ratio must be used in the coupling transformer?

SOLUTION

The reflected resistance must equal 800 Ω. Thus, from Equation 14–9, the turns ratio can be determined.

$$n = \sqrt{\frac{R_L}{R_{pri}}} = \sqrt{\frac{8\ \Omega}{800\ \Omega}} = \sqrt{0.01} = \mathbf{0.1}$$

The diagram and its equivalent reflected circuit are shown in Figure 14–22.

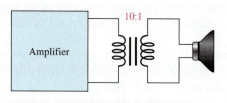
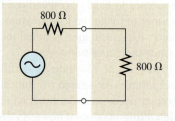

Amplifier equivalent circuit   Speaker/transformer equivalent

FIGURE 14–22

RELATED PROBLEM

What must be the turns ratio in Figure 14–22 to provide maximum power to two 8 Ω speakers in parallel?

THE BALUN TRANSFORMER   An application for impedance matching is used in high-frequency antennas. In addition to impedance matching, many transmitting antennas also require that an unbalanced signal from the transmitter be converted to a balanced signal. A *balanced signal* is composed of two equal-amplitude signals that are 180° out-of-phase with each other. An *unbalanced signal* is one that is referenced to ground. A special type of transformer called a **balun**, a contraction of the words "*bal*anced-*un*balanced" is used to convert the unbalanced signal from the transmitter to a balanced signal at the antenna as shown in Figure 14–23.

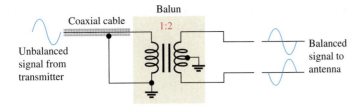

FIGURE 14–23   **Illustration of a balun transformer converting an unbalanced signal to a balanced signal.**

The transmitter is usually connected to the balun with a coaxial cable (coax). A coax is basically a conductor surrounded by an insulating region and an outer conductor called the shield. In Figure 14–23, the signal from the transmitter is referenced to ground, which is connected to the outer shield of the coax, so it is an unbalanced signal. The shield on the coax minimizes pickup of radiated noise.

The coax has a certain characteristic impedance associated with it. The turns ratio of the balun is set to match the coax impedance to the antenna impedance. For example, if the transmitting antenna represents a 300 Ω impedance and the coax has a characteristic impedance of 75 Ω, the balun can provide the impedance match with a turns ratio of two, as was shown earlier. Baluns can also be used to convert from balanced signals to unbalanced signals.

SECTION 14–6 CHECKUP

1. What does impedance matching mean?

2. What is the advantage of matching the load resistance to the resistance of a source?

3. A transformer has a turns ratio of 0.5. What is the reflected resistance with 100 Ω across the secondary?

4. What is the purpose of a balun transformer?

14–7 TRANSFORMER RATINGS AND CHARACTERISTICS

Transformer operation has been discussed from an ideal point of view. That is, the winding resistance, the winding capacitance, and nonideal core characteristics were all neglected and the transformer was treated as if it had an efficiency of 100%. For studying the basic concepts and in many applications, the ideal model is valid. However, the practical transformer has several nonideal characteristics of which you should be aware.

After completing this section, you should be able to

- **Describe practical transformer ratings**
  - **List and describe the nonideal characteristics**
  - **Explain power rating of a transformer**
  - **Define *efficiency* of a transformer**

Ratings

POWER RATING   A power transformer is typically rated in volt-amperes (VA), primary/secondary voltage, and operating frequency. For example, a given transformer rating may be specified as 2 kVA, 500/50, 60 Hz. The 2 kVA value is the **apparent power rating**. The 500 and the 50 can be either secondary or primary voltages. The 60 Hz is the operating frequency.

The transformer rating can be helpful in selecting the proper transformer for a given application. Let's assume, for example, that 50 V is the secondary voltage. In this case the load current is

$$I_L = \frac{P_{sec}}{V_{sec}} = \frac{2 \text{ kVA}}{50 \text{ V}} = 40 \text{ A}$$

On the other hand, if 500 V is the secondary voltage, then

$$I_L = \frac{P_{sec}}{V_{sec}} = \frac{2 \text{ kVA}}{500 \text{ V}} = 4 \text{ A}$$

These are the maximum currents that the secondary can handle in either case.

The reason that the power rating is in volt-amperes (apparent power) rather than in watts (true power) is as follows: If the transformer load is purely capacitive or purely inductive, the true power (watts) delivered to the load is ideally zero. However, the current for $V_{sec} = 500$ V and $X_C = 100$ Ω at 60 Hz, for example, is 5 A. This current exceeds the maximum of 4 A that the 2 kVA secondary can handle, and even though the true power is zero, the transformer may be damaged. So it is meaningless to specify power in watts for transformers.

Resonant Charging

Resonant circuits can be used for charging by transformer action. Researchers are investigating wireless transfer of power by high-frequency magnetic coupling between two resonant coils. The idea dates back to Nicholas Tesla. It may be that electric cars would have a coil on board and when parked in the garage, would automatically be charged from a coil in the floor. Researchers still need to overcome efficiency problems, due to transformer losses, before it can be a viable charging method for electric cars.

 SYSTEM NOTE

VOLTAGE AND FREQUENCY RATINGS   In addition to the apparent power rating, most power transformers will have voltage and frequency ratings placed on the transformer. The voltage ratings include the primary voltage for which it is designed and the secondary voltage that is present when the rated load is connected to the secondary and the primary is connected to the rated input voltage. Frequently, there will be a small schematic drawing showing the windings and voltage rating for each winding. The frequency for which the transformer is designed will also be specified. A transformer operated on the wrong frequency can be damaged, so it important to note the frequency specification. These are minimum specifications you need to know in order to select a power transformer for an application.

Characteristics

WINDING RESISTANCE   Both the primary and the secondary windings of a practical transformer have winding resistance. (You learned about the winding resistance of inductors in Chapter 11.) The winding resistances of a practical transformer are represented as resistors in series with the windings as shown in Figure 14–24.

Winding resistance in a practical transformer results in less voltage across a secondary load. Voltage drops due to the winding resistance effectively subtract from the primary and secondary voltages and result in a load voltage that is less than that predicted by the relationship $V_{sec} = nV_{pri}$. In most cases the effect is relatively small and can be neglected.

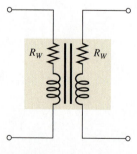

FIGURE 14–24 Winding resistance in a practical transformer.

LOSSES IN THE CORE   There is always some energy conversion in the core material of a practical transformer. This conversion is seen as a heating of ferrite and iron cores, but conversion does not occur in air cores. Part of this energy conversion is because of the continuous reversal of the magnetic field due to the changing direction of the primary current; this component of the energy conversion is called *hysteresis loss.* The rest of the energy conversion to heat is caused by eddy currents produced when voltage is induced in the core material by the changing magnetic flux, according to Faraday's law. The eddy currents occur in circular patterns in the core resistance, thus producing heat. This conversion to heat is greatly reduced by the use of laminated construction of iron cores. The thin layers of ferromagnetic material are insulated from each other to minimize the buildup of eddy currents by confining them to a small area and to keep core losses to a minimum.

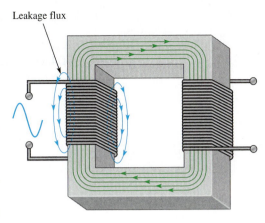

Leakage flux

MAGNETIC FLUX LEAKAGE   In an ideal transformer, all the magnetic flux produced by the primary current is assumed to pass through the core to the secondary winding and vice versa. In a practical transformer, some of the magnetic flux lines break out of the core and pass through the surrounding air back to the other end of the winding, as illustrated in Figure 14–25 for the magnetic field produced by the primary current. Magnetic flux leakage results in a reduced secondary voltage.

FIGURE 14–25   **Flux leakage in a practical transformer.**

The percentage of magnetic flux that actually reaches the secondary winding determines the coefficient of coupling of the transformer. For example, if nine out of ten flux lines remain inside the core, the coefficient of coupling is 0.90 or 90%. Most iron-core transformers have very high coefficients of coupling (greater than 0.99), whereas ferrite-core and air-core devices have lower values.

WINDING CAPACITANCE   As you learned in Chapter 11, there is always some stray capacitance between adjacent turns of a winding. These stray capacitances result in an effective capacitance in parallel with each winding of a transformer, as indicated in Figure 14–26.

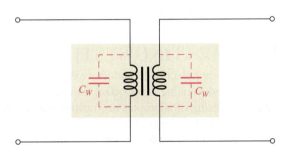

FIGURE 14–26   **Winding capacitance in a practical transformer.**

These stray capacitances have very little effect on the transformer's operation at low frequencies because the reactances (X_C) are very high. However, at higher frequencies, the reactances decrease and begin to produce a bypassing effect across the primary winding and across the secondary load. As a result, less of the total primary current is through the primary winding, and less of the total secondary current is through the load. This effect reduces the load voltage as the frequency goes up.

TRANSFORMER EFFICIENCY   Recall that the secondary power is equal to the primary power in an ideal transformer. Because the nonideal characteristics just discussed result in a power loss in the transformer, the secondary (output) power is always less than the primary (input) power. The **efficiency**, symbolized by the Greek letter eta (η), of a transformer is a measure of the percentage of the input power that is delivered to the output.

$$\eta = \left(\frac{P_{out}}{P_{in}}\right)100\%$$ (14–10)

Most power transformers have efficiencies in excess of 95%.

EXAMPLE 14–10

A certain type of transformer has a primary current of 5 A and a primary voltage of 4800 V. The secondary current is 95 A and the secondary voltage is 240 V. Determine the efficiency of this transformer.

SOLUTION

The input power is

$$P_{in} = V_{pri}I_{pri} = (4800 \text{ V})(5 \text{ A}) = 24 \text{ kVA}$$

The output power is

$$P_{out} = V_{sec}I_{sec} = (240 \text{ V})(95 \text{ A}) = 22.8 \text{ kVA}$$

The efficiency is

$$\eta = \left(\frac{P_{out}}{P_{in}}\right)100\% = \left(\frac{22.8 \text{ kVA}}{24 \text{ kVA}}\right)100\% = \mathbf{95\%}$$

RELATED PROBLEM

A transformer has a primary current of 9 A with a primary voltage of 440 V. The secondary current is 30 A and the secondary voltage is 120 V. What is the efficiency?

SYSTEM EXAMPLE 14–2

AN INDUCTION HEATING SYSTEM

A widely used method for heating electrical conductors is to use a coil that surrounds a conductive material to be heated. Heating is important for operations like brazing, soldering, heat treating, annealing, drying of adhesives and coatings, preheating metals prior to rubber molding, and many other processes. The hot coil is like the primary of a transformer; the conductive material to be heated acts as a secondary. Alternating current at frequencies from 60 Hz to over 1 MHz are used to induce a current in the material to be heated by transformer action. The heating element looks like a shorted secondary, which induces large eddy currents that are responsible for the heating.

Figure 14–27 shows a basic induction heating unit; heating units vary widely in size and heating characteristics that depend on the requirements. An important advantage of induction heating is that the material can be heated almost instantly (2000° F in less than 1 s). The heating pattern is repeatable and can provide a custom temperature profile depending on the application. For example, a soldering operation, such as done with cans, can be accomplished quickly in a continuous process, a major advantage in production lines.

Figure 14–28 shows a block diagram of a basic heating system, which is part of a larger process (not shown). The ac input is converted to dc by a power supply, which is then converted to high-frequency ac (the frequency selected depends on the material and depth of heating required.) A controller senses when a part is ready for heat; it then closes a switch that applies power to the coil through an impedance-matching transformer.

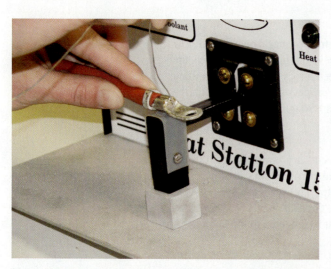

FIGURE 14–27   **A basic induction heating unit used for soldering applications**   (Photo courtesy of MagneForce, Inc.).

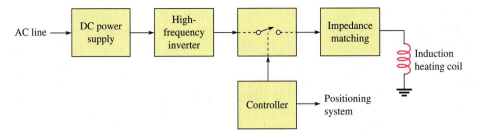

FIGURE 14–28   Block diagram of heating unit.

SECTION 14–7 CHECKUP

1. Explain how a practical transformer differs from the ideal model.

2. The coefficient of coupling of a certain transformer is 0.85. What does this mean?

3. A certain transformer has a rating of 10 kVA. If the secondary voltage is 250 V, how much load current can the transformer handle?

14–8 TAPPED AND MULTIPLE-WINDING TRANSFORMERS

The basic transformer has several important variations. They include tapped transformers, multiple-winding transformers, and autotransformers. Three-phase transformers, which are basically multiple-winding transformers, are covered in this section.

After completing this section, you should be able to

- Describe several types of transformers
  - Describe center-tapped transformers
  - Describe multiple-winding transformers
  - Describe autotransformers
  - Explain how three-phase transformers are connected

Tapped Transformers

A schematic of a transformer with a center-tapped secondary winding is shown in Figure 14–29(a). The **center tap (CT)** is equivalent to two secondary windings with half the total voltage across each.

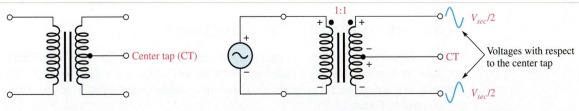

(a) Center-tapped transformer

(b) Output voltages with respect to the center tap are 180° out of phase with each other and are one-half the magnitude of the secondary voltage.

FIGURE 14–29   Operation of a center-tapped transformer.

The voltages between either end of the secondary winding and the center tap are, at any instant, equal in magnitude but opposite in polarity, as illustrated in Figure 14–29(b). Here, for example, at some instant on the sinusoidal voltage, the polarity across the entire secondary winding is as shown (top end +, bottom −). At the center tap, the voltage is less positive than the top end but more positive than the bottom end of the secondary. Therefore, measured with respect to the center tap, the top end of the secondary is positive, and the bottom end is negative. This center-tapped feature is used in many power supply rectifiers in which the ac voltage is converted to dc, as illustrated in Figure 14–30, and also in impedance-matching transformers.

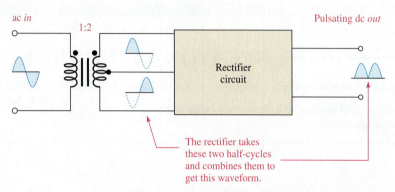

FIGURE 14–30 **Application of a center-tapped transformer in ac-to-dc conversion.**

Some tapped transformers have taps on the secondary winding at points other than the electrical center. Also, multiple primary and secondary taps are sometimes used in certain applications. Examples of these types of transformers are shown in Figure 14–31.

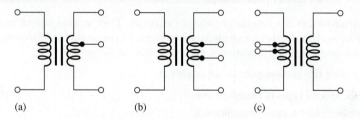

(a)   (b)   (c)

FIGURE 14–31 **Tapped transformers.**

Utility companies use many tapped transformers in distribution systems. Normally, power is generated and transmitted as three-phase power. At some point, the three-phase power is converted to single-phase for residential use. An example of a utility-pole transformer is shown in Figure 14–32 for the case where the high-voltage three-phase power has previously been converted to single-phase power (by tapping one of the three phases). It still needs to be converted to 120 V/240 V for residential customers, so a single-phase tapped transformer is used. By selecting the appropriate tap on the primary side, the utility company can make minor adjustments in the voltage that is delivered to the customer. The center tap on the secondary side is the neutral (usually uninsulated) conductor.

Multiple-Winding Transformers

Some transformers are designed to operate from either 120 V ac or 240 V ac lines. These transformers usually have two primary windings, each of which is designed for 120 V ac. When the two are connected in series, the transformer can be used for 240 V ac operation, as illustrated in Figure 14–33.

More than one secondary winding can be wound on a common core. Transformers with several secondary windings are often used to achieve several voltages by either stepping up or stepping down the primary voltage. These types are commonly used in power

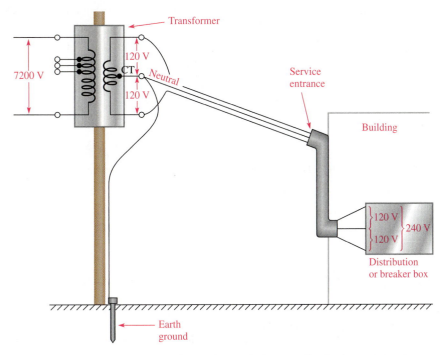

FIGURE 14–32   **Utility-pole transformer in a typical power distribution system.**

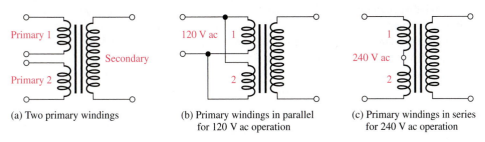

(a) Two primary windings

(b) Primary windings in parallel for 120 V ac operation

(c) Primary windings in series for 240 V ac operation

FIGURE 14–33   **Multiple-primary transformers.**

supply applications in which several voltage levels are required for the operation of an electronic system.

A typical schematic of a transformer with multiple secondary windings is shown in Figure 14–34; this transformer has three secondary windings. Sometimes you will find combinations of multiple primary windings, multiple secondary windings, and tapped transformers all in one unit.

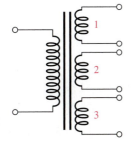

FIGURE 14–34   **A multiple-secondary transformer.**

EXAMPLE 14–11

The transformer shown in Figure 14–35 has the turns ratios for each secondary relative to the primary as indicated. One of the secondary windings is also center tapped. If 120 V ac are connected to the primary winding, determine each secondary voltage and the voltages with respect to the center tap (CT) on the middle secondary winding.

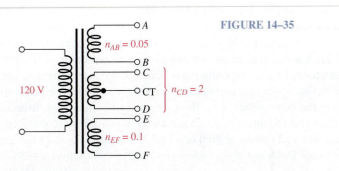

FIGURE 14–35

SOLUTION

$$V_{AB} = n_{AB}V_{pri} = (0.05)120 \text{ V} = \mathbf{6.0 \text{ V}}$$

$$V_{CD} = n_{CD}V_{pri} = (2)120 \text{ V} = \mathbf{240 \text{ V}}$$

$$V_{(CT)C} = V_{(CT)D} = \frac{240 \text{ V}}{2} = \mathbf{120 \text{ V}}$$

$$V_{EF} = n_{EF}V_{pri} = (0.1)120 \text{ V} = \mathbf{12 \text{ V}}$$

RELATED PROBLEM

Repeat the calculations if the primary winding is halved.

Autotransformers

Applications of the autotransformer include starting industrial induction motors and regulating transmission line voltages. In an **autotransformer**, one winding serves as both the primary and the secondary windings. The winding is tapped at the proper points to achieve the desired turns ratio for stepping up or stepping down the voltage.

Autotransformers differ from conventional transformers in that there is no electrical isolation between the primary and the secondary circuits because both are on one winding. Autotransformers normally are smaller and lighter than equivalent conventional transformers because they require a much lower kVA rating for a given load. Many autotransformers provide an adjustable tap using a sliding contact mechanism so that the output voltage can be varied (these are often called *variacs*.) Figure 14–36 shows schematic symbols for various types of autotransformers.

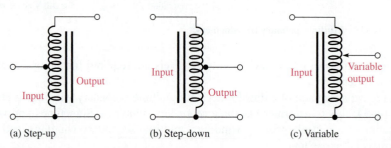

(a) Step-up   (b) Step-down   (c) Variable

FIGURE 14–36   **Variable autotransformers.**

Three-Phase Transformers

Three-phase power was introduced in Chapter 8 in relation to generators and motors. Three-phase transformers are widely used in power distribution systems. Three-phase is the most common way in which power is produced, transmitted, and used. Three-phase power is generally not used in residential applications, but it is particularly useful in industry because three-phase motors are more efficient.

A three-phase transformer consists of three sets of primary and secondary windings. Each set is wound on one leg of an iron-core assembly. Basically it is three single-phase transformers sharing a common core, as shown in Figure 14–37. It is possible (but more expensive) to connect three single-phase transformers together to achieve the same result. In a three-phase transformer, the three identical primary windings and the three identical secondary windings are connected in either of two ways, a delta (Δ) configuration or a wye (Y) configuration, to form a complete transformer unit. Delta and wye connections are shown in Figure 14–38.

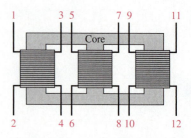

FIGURE 14–37   **Three-phase transformer.**

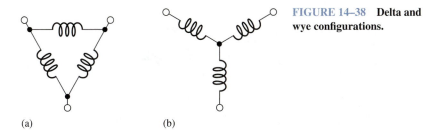

FIGURE 14–38 Delta and wye configurations.

(a)   (b)

In a three-phase transformer, the possible combinations of delta and wye configurations are

1. **Delta-to-wye (Δ-Y).** The primary winding is a delta and the secondary winding is a wye. This is the most common configuration used in commercial and industrial applications.

2. **Delta-to-delta (Δ-Δ).** The primary and secondary windings are both connected in the delta configuration. This is also commonly used in industrial applications.

3. **Wye-to-delta (Y-Δ).** The primary winding is a wye and the secondary winding is a delta. This is used in high-voltage transmission applications.

4. **Wye-to-wye (Y-Y).** The primary and secondary windings are both connected in the wye configuration. It is used in high-voltage, low kVA applications.

The delta-to-wye connections are shown in Figure 14–39. The proper winding phasing must be observed when connecting the transformers. The windings in the delta must be + to −; the polarity for each winding in the wye must be the same at the center point.

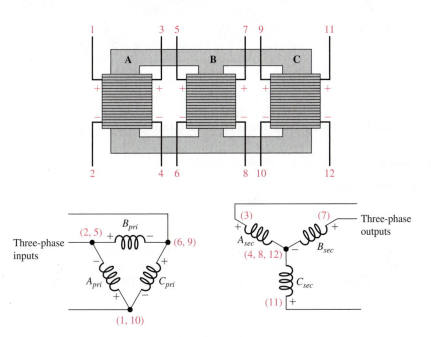

FIGURE 14–39 **Connections for a delta-to-wye transformer. The primary windings are designated A_{pri}, B_{pri} and C_{pri}; the secondary windings are designated A_{sec}, B_{sec}, and C_{sec}. The numbers in parentheses correspond to the transformer leads.**

The wye configuration has an advantage in that a neutral connection can be made at the center junction point. The delta configuration normally does not have a neutral. One exception is the special case for changing three-phase transmission-line voltage to single-phase power for residential use. In this case, a Y-Δ transformer with a center-tapped delta configuration is used, as shown in Figure 14–40. This configuration is referred to as a four-wire delta and can be used in cases where single phase is not available.

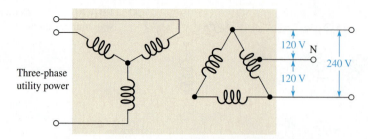

FIGURE 14–40 A tapped wye-delta transformer that can be used to convert three-phase utility voltages to single-phase residential voltages.

SECTION 14–8 CHECKUP

1. A certain transformer has two secondary windings. The turns ratio from the primary winding to the first secondary winding is 10. The turns ratio from the primary winding to the other secondary winding is 0.2. If 240 V ac are applied to the primary winding, what are the secondary voltages?

2. Name one advantage and one disadvantage of an autotransformer over a conventional transformer.

3. What is the most common configuration for a three-phase transformer?

14–9 TROUBLESHOOTING

Transformers are reliable devices when operated within their specified range. Common failures in a transformer are opens in either the primary or the secondary windings. One cause of an open is the operation of the device under conditions that exceed its ratings. Normally, when a transformer fails, it is very difficult to repair, and therefore the simplest procedure is to replace it.

After completing this section, you should be able to

- **Troubleshoot transformers**
  - **Find an open primary or secondary winding**

When there is an open primary winding, there is no primary current and, therefore, no induced voltage or current in the secondary. This condition is illustrated in Figure 14–41(a), and the method of checking with an ohmmeter is shown in part (b).

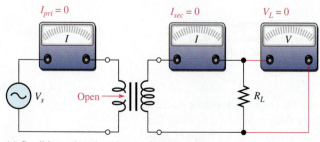

(a) Conditions when the primary winding is open

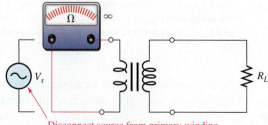

(b) Checking the primary winding with an ohmmeter

FIGURE 14–41 Open primary winding.

When there is an open secondary winding, there is no current in the secondary circuit and, as a result, no voltage across the load. Also, an open secondary winding causes the primary current to be very small (there is only a small magnetizing current). In fact, the primary current may be practically zero. This condition is illustrated in Figure 14–42(a), and the ohmmeter check is shown in part (b).

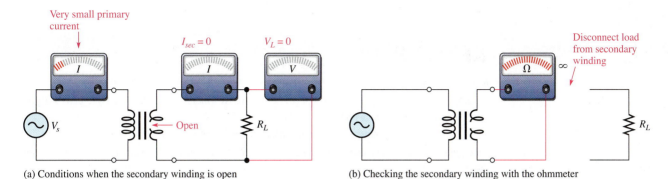

(a) Conditions when the secondary winding is open

(b) Checking the secondary winding with the ohmmeter

FIGURE 14–42   Open secondary winding.

Shorted windings are very rare and if they do occur are difficult to find unless there is a visual indication or a large number of windings are shorted. A completely shorted primary winding will draw excessive current from the source; and unless there is a breaker or a fuse in the circuit, either the source or the transformer or both will burn out. A partial short in the primary winding can cause higher than normal or even excessive primary current.

SECTION 14–9 CHECKUP

1. What is the most probable failure in a transformer?

2. What is often the cause of transformer failure?

SUMMARY

- A transformer generally consists of two or more coils that are magnetically coupled on a common core.
- There is mutual inductance between two magnetically coupled coils.
- When current in one coil changes, voltage is induced in the other coil.
- The primary is the winding connected to the source, and the secondary is the winding connected to the load.
- The number of turns in the primary winding and the number of turns in the secondary winding determine the turns ratio.
- The relative polarities of the primary and secondary voltages are determined by the direction of the windings around the core.
- A step-up transformer has a turns ratio greater than 1.
- A step-down transformer has a turns ratio less than 1.
- A transformer cannot increase power.
- In an ideal transformer, the power from the source (input power) is equal to the power delivered to the load (output power).
- If the voltage is stepped up, the current is stepped down, and vice versa.
- A load connected across the secondary winding of a transformer appears to the source as a reflected load having a value dependent on the reciprocal of the turns ratio squared.
- An impedance-matching transformer can match a load resistance to an internal source resistance to achieve maximum power transfer to the load by selection of the proper turns ratio.

- A balun is a type of transformer used to convert a balanced line (such as twisted pair wiring) to an unbalanced line (such as coaxial cable) or vice-versa.
- A transformer does not respond to constant dc.
- Conversion of electrical energy to heat in an actual transformer results from winding resistances, hysteresis loss in the core, eddy currents in the core, and flux leakage.
- Three-phase transformers are commonly used in power distribution applications.

KEY TERMS

Key terms and other bold terms in the chapter are defined in the end-of-book glossary.

Apparent power rating   The method of rating transformers in which the power capability is expressed in volt-amperes (VA).

Center tap (CT)   A connection at the midpoint of a winding in a transformer.

Electrical isolation   The condition in which two circuits have no common conductive path between them.

Impedance matching   A technique used to match a load resistance to a source resistance in order to achieve maximum transfer of power.

Magnetic coupling   The magnetic connection between two coils as a result of the changing magnetic flux lines of one coil cutting through the second coil.

Mutual inductance (L_M)   The inductance between two separate coils, such as in a transformer.

Primary winding   The input winding of a transformer; also called *primary*.

Reflected resistance   The resistance in the secondary circuit reflected into the primary circuit.

Secondary winding   The output winding of a transformer; also called *secondary*.

Transformer   An electrical device constructed of two or more coils of wire (windings) that are electromagnetically coupled to each other so that there is mutual inductance from one winding to another.

Turns ratio (n)   The ratio of turns in the secondary winding to turns in the primary winding.

KEY FORMULAS

(14–1)	$L_M = k \sqrt{L_1 L_2}$	Mutual inductance
(14–2)	$k = \dfrac{\phi_{1\text{-}2}}{\phi_1}$	Coefficient of coupling
(14–3)	$n = \dfrac{N_{sec}}{N_{pri}}$	Turns ratio
(14–4)	$\dfrac{V_{sec}}{V_{pri}} = \dfrac{N_{sec}}{N_{pri}}$	Voltage ratio
(14–5)	$V_{sec} = n V_{pri}$	Secondary voltage
(14–6)	$\dfrac{I_{pri}}{I_{sec}} = n$	Current ratio
(14–7)	$I_{sec} = \left(\dfrac{1}{n}\right) I_{pri}$	Secondary current
(14–8)	$R_{pri} = \left(\dfrac{1}{n}\right)^2 R_L$	Reflected resistance
(14–9)	$n = \sqrt{\dfrac{R_L}{R_{pri}}}$	Turns ratio for impedance matching
(14–10)	$\eta = \left(\dfrac{P_{out}}{P_{in}}\right) 100\%$	Transformer efficiency

TRUE/FALSE QUIZ

Answers are at the end of the chapter.

1. An ideal transformer delivers the same power to the load as is delivered to the primary winding.

2. Dots on the schematic symbol of a transformer show the phase relationship between the input and the output.

3. A step-down transformer has more turns in the primary winding than in the secondary winding.

4. The primary current in a transformer is always larger than the secondary current.

5. When there is no load on a transformer, the power factor is 1.

6. Impedance-matching transformers allow maximum voltage to be passed from the source to the load.

7. Reflected resistance is the same as winding resistance.

8. A balun is a form of impedance-matching transformer.

9. Power transformers are typically rated in VA rather than in watts.

10. Transformer efficiency is the ratio of output voltage divided by input voltage.

SELF-TEST

Answers are at the end of the chapter.

1. A transformer is used for
   (a) dc voltages   (b) ac voltages   (c) both dc and ac

2. Which one of the following is affected by the turns ratio of a transformer?
   (a) primary voltage   (b) dc voltage   (c) secondary voltage   (d) none of these

3. If the windings of a certain transformer with a turns ratio of 1 are in opposite directions around the core, the secondary voltage is
   (a) in phase with the primary voltage   (b) less than the primary voltage
   (c) greater than the primary voltage   (d) out of phase with the primary voltage

4. When the turns ratio of a transformer is 10 and the primary ac voltage is 6 V, the secondary voltage is
   (a) 60 V   (b) 0.6 V   (c) 6 V   (d) 36 V

5. When the turns ratio of a transformer is 0.5 and the primary ac voltage is 100 V, the secondary voltage is
   (a) 200 V   (b) 50 V   (c) 10 V   (d) 100 V

6. A certain transformer has 500 turns in the primary winding and 2500 turns in the secondary winding. The turns ratio is
   (a) 0.2   (b) 2.5   (c) 5   (d) 0.5

7. If 10 W of power are applied to the primary winding of an ideal transformer with a turns ratio of 5, the power delivered to the secondary load is
   (a) 50 W   (b) 0.5 W   (c) 0 W   (d) 10 W

8. In a certain loaded transformer, the secondary voltage is one-third the primary voltage. Ideally, the secondary current is
   (a) one-third the primary current   (b) three times the primary current
   (c) equal to the primary current   (d) less than the primary current

9. When a 1.0 kΩ load resistor is connected across the secondary winding of a transformer with a turns ratio of 2, the source "sees" a reflected load of
   (a) 250 Ω   (b) 2 kΩ   (c) 4 kΩ   (d) 1.0 kΩ

10. In Question 9, if the turns ratio is 0.5, the source "sees" a reflected load of
    (a) 1.0 kΩ   (b) 2 kΩ   (c) 4 kΩ   (d) 500 Ω

11. The turns ratio required to match a 50 Ω source to a 200 Ω load is
    (a) 0.25 (b) 0.5 (c) 4 (d) 2

12. Maximum power is transferred from a source to a load when
    (a) $R_L > R_{int}$ (b) $R_L < R_{int}$ (c) $R_L = R_{int}$ (d) $R_L = nR_{int}$

13. When a 12 V battery is connected across the primary winding of a transformer with a turns ratio of 4, the secondary voltage is
    (a) 0 V (b) 12 V (c) 48 V (d) 3 V

14. A certain transformer has a turns ratio of 1 and a 0.95 coefficient of coupling. When 1 V ac is applied to the primary, the secondary voltage is
    (a) 1 V (b) 1.95 V (c) 0.95 V

PROBLEMS

Answers to odd-numbered problems are at the end of the book.

BASIC PROBLEMS

SECTION 14–1 Mutual Inductance

1. What is the mutual inductance when $k = 0.75$, $L_1 = 1\ \mu H$, and $L_2 = 4\ \mu H$?

2. Determine the coefficient of coupling when $L_M = 1\ \mu H$, $L_1 = 8\ \mu H$, and $L_2 = 2\ \mu H$.

SECTION 14–2 The Basic Transformer

3. What is the turns ratio of a transformer having 120 turns in its primary winding and 360 turns in its secondary winding?

4. (a) What is the turns ratio of a transformer having 250 turns in its primary winding and 1000 turns in its secondary winding?
   (b) What is the turns ratio when the primary winding has 400 turns and the secondary winding has 100 turns?

5. Determine the phase of the secondary voltage with respect to the primary voltage for each transformer in Figure 14–43.

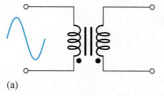 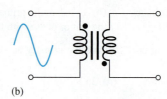 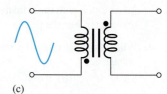

(a)   (b)   (c)

FIGURE 14–43

SECTION 14–3 Step-Up and Step-Down Transformers

6. If 120 V ac are connected to the primary of a transformer with a turns ratio of 1.5, what is the secondary voltage?

7. A certain transformer has 250 turns in its primary winding. In order to double the secondary voltage, how many turns must be in the secondary winding?

8. How many primary volts must be applied to a transformer with a turns ratio of 10 to obtain a secondary voltage of 60 V ac?

9. For each transformer in Figure 14–44, draw the secondary voltage showing its relationship to the primary voltage. Also indicate the amplitude.

FIGURE 14–44

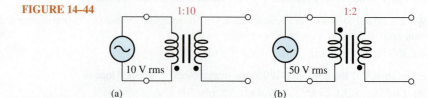

(a)   (b)

10. To step 120 V down to 30 V, what must be the turns ratio?

11. The primary winding of a transformer has 1200 V across it. What is the secondary voltage if the turns ratio is 0.2?

12. How many primary volts must be applied to a transformer with a turns ratio of 0.1 to obtain a secondary voltage of 6 V ac?

13. What is the voltage across the load in each circuit in Figure 14–45?

14. If the bottom of each secondary winding in Figure 14–45 were grounded, would the values of the load voltages be changed?

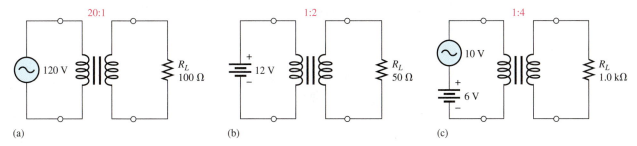

FIGURE 14–45

15. Determine the unspecified meter readings in Figure 14–46.

16. If R_L is doubled in the circuit of Figure 14–46(a), what would the secondary meter read?

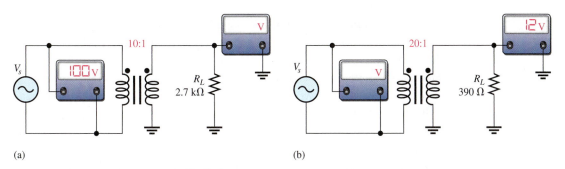

FIGURE 14–46

SECTION 14–4   Loading the Secondary

17. Determine I_{sec} in Figure 14–47.

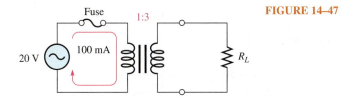

FIGURE 14–47

18. Determine the following quantities in Figure 14–48:
    (a) secondary voltage   **(b)** secondary current
    (c) primary current   **(d)** power in the load

FIGURE 14–48

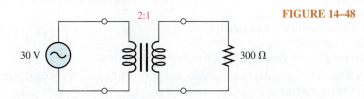

SECTION 14–5   Reflected Load

19. What is the load resistance as seen by the source in Figure 14–49?

20. What is the resistance reflected into the primary circuit in Figure 14–50?

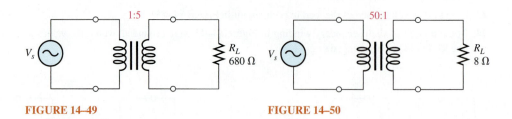

FIGURE 14–49   **FIGURE 14–50**

21. What is the primary current (rms) in Figure 14–50 if the rms source voltage is 120 V?

22. What must be the turns ratio in Figure 14–51 in order to reflect 300 Ω into the primary circuit?

FIGURE 14–51

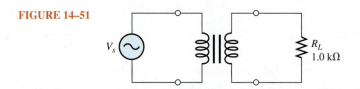

SECTION 14–6   Impedance Matching

23. For the circuit in Figure 14–52, find the turns ratio required to deliver maximum power to the 4 Ω speaker.

24. In Figure 14–52, what is the maximum power in watts delivered to the speaker?

25. Determine the value to which R_L must be adjusted in Figure 14–53 for maximum power transfer. The internal source resistance is 50 Ω.

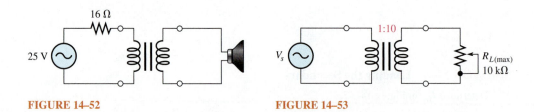

FIGURE 14–52   **FIGURE 14–53**

26. Plot the power curve for values of R_L in Figure 14–53 ranging from 1 kΩ to 10 kΩ in 1 kΩ increments. $R_L = 50$ Ω and $V_s = 10$ V.

SECTION 14–7   Transformer Ratings and Characteristics

27. In a certain transformer, the input power to the primary is 100 W. If 5.5 W are dissipated in the winding resistances, what is the output power to the load, neglecting any other losses?

28. What is the efficiency of the transformer in Problem 27?

29. Determine the coefficient of coupling for a transformer in which 2% of the total flux generated in the primary does not pass through the secondary.

30. A certain transformer is rated at 1 kVA. It operates on 60 Hz, 120 V ac. The secondary voltage is 600 V.
    (a) What is the maximum load current?
    (b) What is the smallest R_L that you can drive?
    (c) What is the largest capacitor that can be connected as a load?

31. What kVA rating is required for a transformer that must handle a maximum load current of 10 A with a secondary voltage of 2.5 kV?

SECTION 14–8   Tapped and Multiple-Winding Transformers

32. Determine each unknown voltage indicated in Figure 14–54.

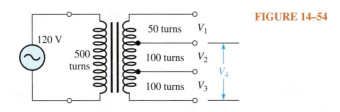

FIGURE 14–54

33. Using the indicated secondary voltages in Figure 14–55, determine the turns ratio of the primary winding to each tapped section.

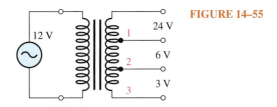

FIGURE 14–55

34. In Figure 14–56, each primary winding can accommodate 120 V ac. Show how the primaries should be connected for 240 V ac operation. Determine each secondary voltage.

35. Determine the turns ratios from each primary to each secondary in Figure 14–56.

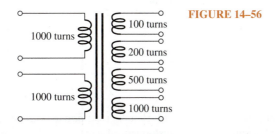

FIGURE 14–56

SECTION 14–9   Troubleshooting

36. When you apply 120 V ac across the primary winding of a transformer and check the voltage across the secondary winding, you get 0 V. Further investigation shows no primary or secondary current. List the possible faults. What is your next step in isolating the problem?

37. What is likely to happen if the primary winding of a transformer shorts?

ADVANCED PROBLEMS

38. The power supply in Figure 14–12 is designed to provide +5.0 V at 1 A, yet is fused for only ⅛ A. Explain why the fuse does not blow at the rated output.

39. For the loaded, tapped-secondary transformer in Figure 14–57, determine the following:
   (a) all load voltages and currents
   (b) the resistance looking into the primary

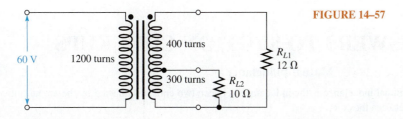

FIGURE 14–57

40. A certain transformer is rated at 5 kVA, 2400/120 V, at 60 Hz.
    (a) What is the turns ratio if the 120 V is the secondary voltage?
    (b) What is the current rating of the secondary if 2400 V is the primary voltage?
    (c) What is the current rating of the primary if 2400 V is the primary voltage?

41. Determine the voltage measured by each voltmeter in Figure 14–58. The bench-type meters have one terminal that connects to ground as indicated.

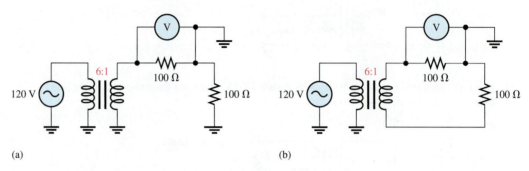

(a)   (b)

FIGURE 14–58

42. Find the appropriate turns ratio for each switch position in Figure 14–59 in order to transfer the maximum power to each load when the internal source resistance is 10 Ω. Specify the number of turns for the secondary winding if the primary winding has 100 turns.

FIGURE 14–59

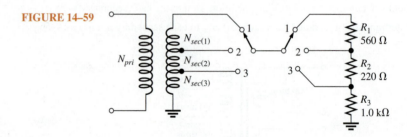

43. What must the turns ratio be in Figure 14–50 to limit the primary current to 3 mA with a source voltage of 120 V? Assume an ideal transformer and source.

44. Assume 120 V is applied to the primary of a transformer with a VA rating of 10 VA. The output voltage is 12.6 V. What is the smallest resistor that can be connected across the secondary?

45. A step-down transformer uses 120 V on the primary and 10 V on the secondary. If the secondary is rated for a maximum of 1 A, what fuse rating should be selected on the primary side?

MULTISIM TROUBLESHOOTING PROBLEMS

46. Open file P14-46; files are found at www.pearsonhighered.com/floyd. Test the circuit. If there is a fault, identify it.

47. Open file P14-47 and test the circuit. If there is a fault, identify it.

48. Determine if there is a fault in the circuit in file P14-48. If so, identify it.

49. Find and specify any faulty component in the circuit in file P14-49.

ANSWERS TO SECTION CHECKUPS

SECTION 14–1   Mutual Inductance

1. Mutual inductance is the inductance between two coils, established by the amount of coupling between the coils.

2. $L_M = k\sqrt{L_1 L_2} = 45$ mH

3. Induced voltage increases when k is increased.

SECTION 14–2   The Basic Transformer

1. Transformer operation is based on the principle of mutual inductance.

2. Turns ratio is the ratio of turns in the secondary to turns in the primary.

3. Directions of the windings determine the relative voltage polarities.

4. $n = N_{sec}/N_{pri} = 0.5$

5. The windings are formed on a printed circuit board.

SECTION 14–3   Step-Up and Step-Down Transformers

1. A step-up transformer increases voltage.

2. The secondary voltage is five times greater.

3. $V_{sec} = nV_{pri} = 2400$ V

4. A step-down transformer decreases voltage.

5. $V_{sec} = nV_{pri} = 60$ V

6. $n = 12\,\text{V}/120\,\text{V} = 0.1$

7. Electrical isolation, surge protection, and filters to eliminate interference

SECTION 14–4   Loading the Secondary

1. The secondary current is half the primary current.

2. $n = 0.25; I_{sec} = (1/n)I_{pri} = 2$ A

3. $I_{pri} = nI_{sec} = 2.5$ A

4. Disconnecting the load can lead to arcing, which can damage the transformer and equipment and present a shock hazard.

SECTION 14–5   Reflected Load

1. Reflected resistance is the resistance in the secondary circuit as seen from the primary circuit as a function of the turns ratio.

2. The reciprocal of the turns ratio determines reflected resistance.

3. $R_{pri} = (1/n)^2 R_L = 0.5\ \Omega$

4. $n = \sqrt{R_L/R_{pri}} = 0.1$

SECTION 14–6   Impedance Matching

1. Impedance matching is making maximum power transfer between a source and a load.

2. Maximum power is delivered to the load when $R_L = R_{int}$.

3. $R_{pri} = (1/n)^2 R_L = 400\ \Omega$

4. To connect unbalanced signals to balanced signals (and vice versa) and provide impedance matching.

SECTION 14–7   Transformer Ratings and Characteristics

1. In a practical transformer, conversion of electrical energy to heat reduces the efficiency. An ideal transformer has an efficiency of 100%.

2. When the coefficient of coupling is 0.85, 85% of the magnetic flux generated in the primary winding passes through the secondary winding.

3. $I_L = 10\,\text{kVA}/250\,\text{V} = 40$ A

SECTION 14–8   Tapped and Multiple-Winding Transformers

1. $V_{sec} = 10(240\,\text{V}) = 2400\,\text{V}; V_{sec} = 0.2(240\,\text{V}) = 48$ V

2. Autotransformers are smaller and lighter for the same rating. Autotransformers provide no electrical isolation.

3. The delta-to-wye configuration.

SECTION 14–9 **Troubleshooting**

1. The most probable failure is an open winding.
2. Operating above rated values causes transformer failure.

ANSWERS TO RELATED PROBLEMS FOR EXAMPLES

14–1 387 μH

14–2 0.75

14–3 5000 turns

14–4 480 V

14–5 57.6 V

14–6 5 mA; 400 mA

14–7 6 Ω

14–8 0.354

14–9 0.0707 or 14.14:1

14–10 91%

14–11 $V_{AB} = 12$ V; $V_{CD} = 480$ V; $V_{(CT)C} = V_{(CT)D} = 240$ V; $V_{EF} = 24$ V

ANSWERS TO TRUE/FALSE QUIZ

1. T **2.** T **3.** T **4.** F **5.** F **6.** F **7.** F **8.** T **9.** T **10.** F

ANSWERS TO SELF-TEST

1. (b) **2.** (c) **3.** (d) **4.** (a) **5.** (b) **6.** (c) **7.** (d)

8. (b) **9.** (a) **10.** (c) **11.** (d) **12.** (c) **13.** (a) **14.** (c)

CHAPTER 15

TIME RESPONSE OF REACTIVE CIRCUITS

OUTLINE

OBJECTIVES

- Explain the operation of an *RC* integrator
- Analyze an *RC* integrator with a single-pulse input
- Analyze an *RC* integrator with a repetitive-pulse input
- Analyze an *RC* differentiator with a single-pulse input
- Analyze an *RC* differentiator with a repetitive-pulse input
- Analyze the operation of an *RL* integrator
- Analyze the operation of an *RL* differentiator
- Discuss some applications of integrators and differentiators
- Troubleshoot *RC* integrators and *RC* differentiators

KEY TERMS

Integrator	Steady state
Transient time	Differentiator

INTRODUCTION

In Chapters 10 and 12, the frequency response of *RC* and *RL* circuits was covered. In this chapter, the time response of *RC* and *RL* circuits with pulse inputs is examined.

Before starting this chapter, you should review the material in Sections 9–5 and 11–4. Understanding exponential changes in voltages and currents in capacitors and inductors is crucial to the study of the time response of reactive circuits. Exponential formulas that were given in Chapters 9 and 11 are used throughout this chapter.

With pulse inputs, the time response of the circuits is of primary importance. In the areas of pulse and digital circuits, technicians are often concerned with how a circuit responds over an interval of time to rapid changes in voltage or current. The relationship of the circuit time constant to the input pulse characteristics, such as pulse width and period, determines the wave shapes of the voltages in the circuit.

The terms *integrator* and *differentiator* refer to circuits that can, under certain conditions, approximate the mathematical process of integration and differentiation. Mathematical integration is a summation process. Under certain conditions, integrators are capable of averaging a waveform, as you will see in this chapter. Mathematical differentiation refers to the process of finding the rate of change of a quantity. Differentiator circuits produce an output that represents the rate of change of the input. Differentiator circuits are common in systems that require timing triggers, such as radar systems.

VISIT THE WEBSITE

Study aids for this chapter are available at
http://pearsonhighered.com/floyd

In terms of time response, a series *RC* circuit in which the output voltage is taken across the capacitor is known as an **integrator**. Recall that in terms of frequency response, this particular series *RC* circuit is a low-pass filter. The term, *integrator,* is derived from a mathematical function which this type of circuit approximates under certain conditions.

After completing this section, you should be able to

- **Explain the operation of an *RC* integrator**
  - **Describe how the capacitor charges and discharges**
  - **Explain how a capacitor reacts to an instantaneous change in voltage or current**
  - **Describe the basic output voltage waveform**

Charging and Discharging of a Capacitor

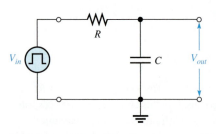

FIGURE 15–1   An *RC* integrator with a pulse generator connected.

When a pulse generator is connected to the input of an *RC* integrator, as shown in Figure 15–1, the capacitor will charge and discharge in response to the pulses. When the input goes from its low level to its high level, the capacitor charges toward the high level of the pulse through the resistor. This charging action is analogous to connecting a battery through a switch to the *RC* circuit, as illustrated in Figure 15–2(a). When the pulse goes from its high level back to its low level, the capacitor discharges back through the source. Compared to the resistance of the resistor, *R*, the resistance of the source is assumed to be negligible. This discharging action is analogous to replacing the source with a closed switch, as illustrated in Figure 15–2(b).

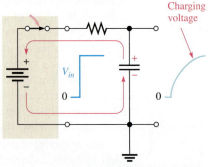 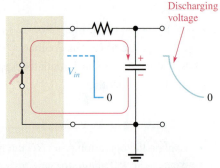

(a) When the input pulse goes HIGH, the source effectively acts as a battery in series with a closed switch, thereby charging the capacitor.

(b) When the input pulse goes back LOW, the source effectively acts as a closed switch, providing a discharge path for the capacitor.

FIGURE 15–2   The equivalent action when a pulse source charges and discharges the capacitor.

A capacitor will charge and discharge following an exponential curve. Its rate of charging and discharging depends on the *RC* **time constant**, a fixed time interval determined by *R* and *C* ($\tau = RC$).

For an ideal pulse, both edges are considered to be instantaneous. Two basic rules of capacitor behavior help in understanding the response of *RC* circuits to pulse inputs.

1. The capacitor appears as a short to an instantaneous change in current and as an open to dc.

2. The voltage across the capacitor cannot change instantaneously—it can change only exponentially.

Capacitor Voltage

In an *RC* integrator, the output is the capacitor voltage. The capacitor charges during the time that the input pulse is high. If the pulse is at its high level long enough, the capacitor will fully charge to the voltage amplitude of the pulse, as illustrated in Figure 15–3. The capacitor discharges during the time that the pulse is low. If the low time between pulses is long enough, the capacitor will fully discharge to zero, as shown in the figure. Then when the next pulse occurs, it will charge again.

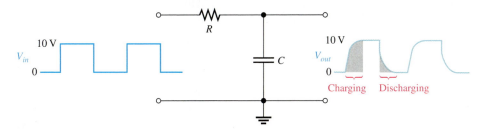

FIGURE 15–3   **Illustration of a capacitor fully charging and discharging in response to a pulse input. A pulse generator is connected to the input, but the symbol is not shown, only the waveform.**

SECTION 15–1 CHECKUP*

1. Define the term *integrator* in relation to an *RC* circuit.

2. What causes a capacitor in an *RC* circuit to charge and discharge?

*Answers are at the end of the chapter.

15–2 RESPONSE OF *RC* INTEGRATORS TO A SINGLE PULSE

From the previous section, you have a general idea of how an *RC* integrator responds over time to a pulse input. In this section, the time response to a single pulse is examined in detail.

After completing this section, you should be able to

- **Analyze an *RC* integrator with a single-pulse input**
  - **Discuss the importance of the circuit time constant**
  - **Define *transient time***
  - **Determine the response when the pulse width is equal to or greater than five time constants**
  - **Determine the response when the pulse width is less than five time constants**

Two conditions of response to a single-pulse input must be considered:

1. When the input pulse width (t_W) is equal to or greater than five time constants ($t_W \geq 5\tau$)

2. When the input pulse width is less than five time constants ($t_W < 5\tau$)

Recall that five time constants is accepted as the time a capacitor needs to fully charge or fully discharge; this time is often called the **transient time**. A capacitor will fully charge if the pulse width is equal to or greater than five time constants (5τ). This condition is expressed as $t_W \geq 5\tau$. At the end of the pulse, the capacitor fully discharges back through the source.

Figure 15–4 illustrates the output waveforms for various *RC* transient times and a fixed input pulse width. Notice that as the transient time becomes shorter, compared to the pulse width, the shape of the output pulse approaches that of the input. In each case, the output reaches the full amplitude of the input.

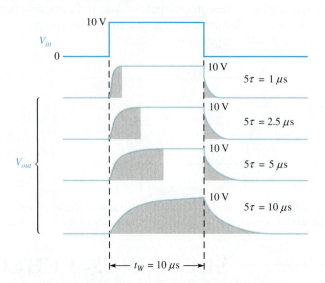

FIGURE 15–4 Variation of an integrator's output pulse shape with transient time. The shaded areas indicate when the capacitor is charging and discharging.

Figure 15–5 shows how a fixed time constant and a variable input pulse width affect the integrator output. Notice that as the pulse width is increased, the shape of the output pulse approaches that of the input. Again, this means that the transient time is short compared to the pulse width. The rise and fall times of the output remain constant.

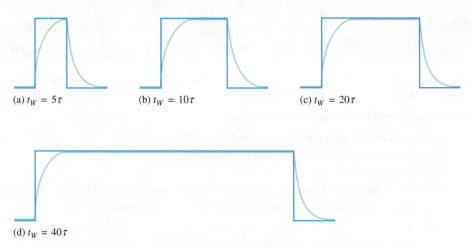

FIGURE 15–5 Variation of an integrator's output pulse shape with input pulse width (the time constant is fixed). Dark blue is input and light blue is output.

Now let's examine the case in which the width of the input pulse is less than five time constants of the *RC* integrator. This condition is expressed as $t_W < 5\tau$. As you know, a capacitor charges for the duration of the pulse and the pulse width is the time it has for charging. However, because the pulse width is less than the time the capacitor needs to fully charge (5τ), the output voltage will *not* reach the full input voltage before the end of the pulse. The capacitor only partially charges, as illustrated in Figure 15–6 for several values of *RC* time constants. Notice that for longer time constants, the output reaches a

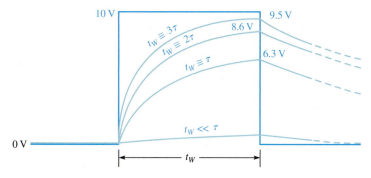

FIGURE 15–6   Capacitor voltage for various time constants that are longer than the input pulse width. Dark blue is input and light blue is output.

lower voltage because the capacitor cannot charge as much during the pulse width. Of course, in the examples with a single-pulse input, the capacitor fully discharges after the pulse ends.

When the time constant is much greater than the input pulse width, the capacitor charges very little, and, as a result, the output voltage becomes a very small almost constant value, as indicated in Figure 15–6.

Figure 15–7 illustrates the effect of reducing the input pulse width for a fixed time constant value. As the pulse width is reduced, the output voltage becomes smaller because the capacitor has less time to charge. However, it takes the capacitor the same length of time (5τ) to discharge back to zero for each condition after the pulse is removed.

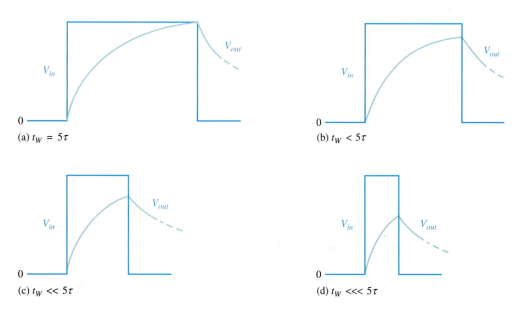

FIGURE 15–7   The capacitor charges less and less as the input pulse width is reduced. The time constant is fixed.

EXAMPLE 15–1

A single 10 V pulse with a width of 100 μs is applied to the *RC* integrator in Figure 15–8. The source resistance is assumed to be zero.

(a) To what voltage will the capacitor charge?

(b) How long will it take the capacitor to discharge?

(c) Show the output voltage waveform.

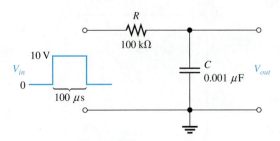

FIGURE 15–8

SOLUTION

(a) The circuit time constant is

$$\tau = RC = (100\,\text{k}\Omega)(0.001\,\mu\text{F}) = 100\,\mu\text{s}$$

Notice that the pulse width is exactly equal to one time constant. Thus, the capacitor will charge approximately 63% of the full input amplitude in one time constant, so the output will reach a maximum voltage of

$$V_{out} = (0.63)10\,\text{V} = \textbf{6.3 V}$$

(b) The capacitor discharges back through the source when the pulse ends. The total discharge time is

$$5\tau = 5(100\,\mu\text{s}) = \textbf{500}\,\mu\text{s}$$

(c) The output charging and discharging curve is shown in Figure 15–9.

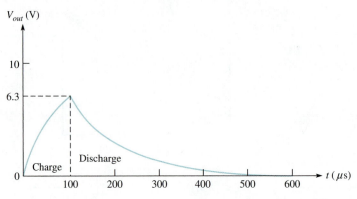

FIGURE 15–9

RELATED PROBLEM*

If the input pulse width in Figure 15–8 is increased to 200 μs, to what voltage will the capacitor charge?

Answers are at the end of the chapter.

MULTISIM

Open Multisim file E15-01; files are found at www. pearsonhighered.com/floyd. Verify the calculated results in this example and the related problem.

EXAMPLE 15–2

Determine to what voltage the capacitor in Figure 15–10 will charge when a single pulse is applied to the input. Assume C is initially uncharged and the source resistance is zero.

SOLUTION

Calculate the time constant.

$$\tau = RC = (2.2\,\text{k}\Omega)(1\,\mu\text{F}) = 2.2\,\text{ms}$$

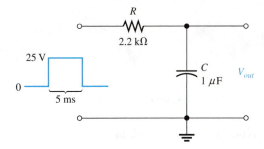

FIGURE 15–10

Because the pulse width is 5 ms, the capacitor charges for approximately 2.27 time constants ($5\text{ ms}/2.2\text{ ms} = 2.27$). Use the exponential formula, Equation 9–15, to find the voltage to which the capacitor will charge. With $V_F = 25$ V and $t = 5$ ms, the calculation is as follows:

$$v = V_F(1 - e^{-t/RC}) = (25\text{ V})(1 - e^{-5\text{ms}/2.2\text{ms}}) = \textbf{22.4 V}$$

These calculations show that the capacitor charges to 22.4 V during the 5 ms duration of the input pulse. It will discharge back to zero when the pulse goes back to zero.

RELATED PROBLEM

Determine how much C will charge if the pulse width is increased to 10 ms.

SECTION 15–2 CHECKUP

1. When an input pulse is applied to an *RC* integrator, what condition must exist in order for the output voltage to reach the full amplitude of the input?

2. For the circuit in Figure 15–11, which has a single-pulse input, find the maximum output voltage and determine how long the capacitor will discharge.

3. For Figure 15–11, draw the approximate shape of the output voltage with respect to the input pulse.

4. If an integrator time constant equals the input pulse width, will the capacitor fully charge?

5. Describe the condition in an integrator under which the output voltage has the approximate shape of a rectangular input pulse.

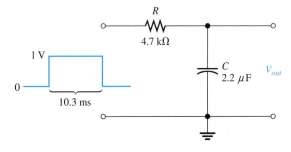

FIGURE 15–11

15–3 RESPONSE OF *RC* INTEGRATORS TO REPETITIVE PULSES

In electronic systems, you will encounter waveforms with repetitive pulses much more often than with single pulses. However, an understanding of the integrator's response to single pulses is necessary in order to understand how these circuits respond to repeated pulses.

After completing this section, you should be able to

- Analyze an *RC* integrator with a repetitive-pulse input
  - Determine the response when the capacitor does not fully charge or discharge
  - Define *steady state*
  - Describe the effect of a change in time constant on circuit response

If a periodic pulse waveform is applied to an *RC* integrator, as shown in Figure 15–12, the output waveshape depends on the relationship of the circuit time constant and the frequency of the input pulses. The capacitor, of course, charges and discharges in response to a pulse input. The amount of charge and discharge of the capacitor depends both on the circuit time constant and on the input frequency, as mentioned.

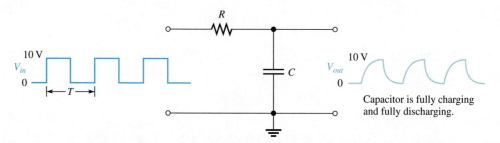

FIGURE 15–12   *RC* integrator with a repetitive-pulse waveform input.

If the pulse width and the time between pulses are each equal to or greater than five time constants, the capacitor will fully charge and fully discharge during each period of the input waveform. This case is shown in Figure 15–12.

When the pulse width and the time between pulses are shorter than 5 time constants, as illustrated in Figure 15–13 for a square wave, the capacitor will *not* completely charge or discharge. We will now examine the effects of this situation on the output voltage of the *RC* integrator.

For illustration, let's use an *RC* integrator with a charging and discharging time constant equal to the pulse width of a 10 V square wave input, as in Figure 15–14. This choice will simplify the analysis and will demonstrate the basic action of the integrator under these conditions. At this point, we really do not care what the exact time constant value is because we know that an *RC* circuit charges 63.2% during one time constant interval.

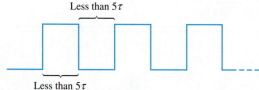

FIGURE 15–13   **Input waveform that does not allow full charge or discharge of the integrator capacitor.**

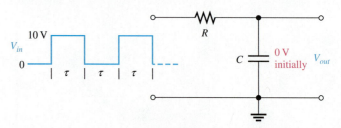

FIGURE 15–14   **Integrator with a square wave input having a period equal to two time constants ($T = 2\tau$).**

Let's assume that the capacitor in Figure 15–14 begins initially uncharged and examine the output voltage on a pulse-by-pulse basis. The results of this analysis are shown in Figure 15–15.

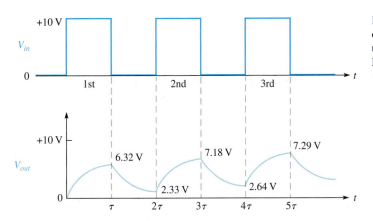

FIGURE 15–15 **Input and output for the initially uncharged integrator in Figure 15–14.**

First pulse During the first pulse, the capacitor charges. The output voltage reaches 6.32 V, which is 63.2% of 10 V, as shown in Figure 15–15.

Between first and second pulses The capacitor discharges, and the voltage decreases to 36.8% of the voltage at the beginning of this interval: 0.368(6.32 V) = 2.33 V.

Second pulse The capacitor voltage begins at 2.33 V and increases 63.2% of the way to 10 V. The total charging range is 10 V − 2.33 V = 7.67 V. The capacitor voltage will increase an additional 63.2% of 7.67 V, which is 4.85 V. Thus, at the end of the second pulse, the output voltage is 2.33 V + 4.85 V = 7.18 V, as indicated in Figure 15–15. Notice that the average is building up.

Between second and third pulses The capacitor discharges during this time, and therefore the voltage decreases to 36.8% of the initial voltage by the end of the second pulse: 0.368(7.18 V) = 2.64 V.

Third pulse At the start of the third pulse, the capacitor voltage is 2.64 V. The capacitor charges 63.2% of the way from 2.64 V to 10 V: 0.632(10 V − 2.64 V) = 4.65 V. Therefore, the voltage at the end of the third pulse is 2.64 V + 4.65 V = 7.29 V.

HANDS ON TIP
When you view pulse waveforms on an oscilloscope, set the scope to be DC coupled, not AC coupled. The reason is that AC coupling places a capacitor in series and can cause distortion in low-frequency pulse signals. Also, DC coupling allows you to observe the dc level of the pulse.

Steady-State Time Response

In the preceding discussion, the output voltage gradually built up and then began leveling off. It takes approximately 5τ for the output voltage to build up to 99% of its final value, regardless of the number of pulses that may occur during that time interval. This interval is the transient time of the circuit. Once the output voltage reaches the average value of the input voltage, a **steady-state** condition is reached, which continues as long as the periodic input continues. This condition is illustrated in Figure 15–16 based on the approximate values obtained after the first three pulses.

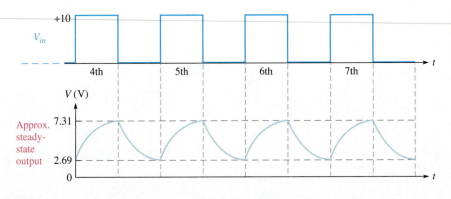

FIGURE 15–16 **Output reaches steady state after 5τ and stabilizes at the values indicated.**

The transient time for our example circuit is the time from the beginning of the first pulse to the end of the third pulse. The capacitor voltage at the end of the third pulse is 7.29 V, which is about 99% of its final value.

The Effect of an Increase in Time Constant

What happens to the output voltage if the RC time constant of the integrator is increased with a variable resistor, as indicated in Figure 15–17? As the time constant is increased, the capacitor charges less during a pulse and discharges less between pulses. The result is a smaller fluctuation in the output voltage for increasing values of time constant, as shown in Figure 15–18.

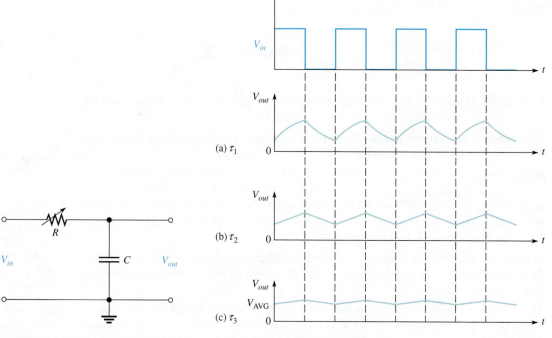

FIGURE 15–17 Integrator with a variable resistor to control the time constant.

FIGURE 15–18 Effect of increasing time constants on the output of an integrator ($\tau_3 > \tau_2 > \tau_1$).

As the time constant becomes extremely long compared to the pulse width, the output voltage approaches a constant dc voltage, as shown in Figure 15–18(c). This value is the average value of the input. For a square wave, it is one-half the amplitude.

EXAMPLE 15–3

Determine the output voltage waveform for the first two pulses applied to the integrator circuit in Figure 15–19. Assume that the capacitor is initially uncharged.

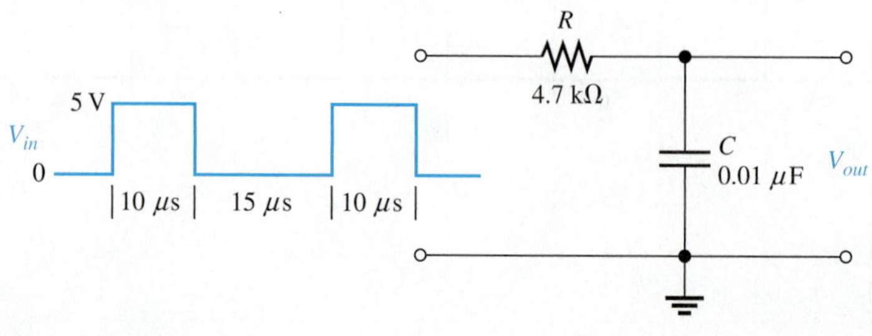

FIGURE 15–19

SOLUTION

First, calculate the circuit time constant.

$$\tau = RC = (4.7\ \text{k}\Omega)(0.01\ \mu\text{F}) = 47\ \mu\text{s}$$

Obviously, the time constant is much longer than the input pulse width or the interval between pulses (notice that the input is not a square wave). Thus, in this case, the exponential formulas must be applied, and the analysis is relatively difficult. Follow the solution carefully.

1. *Calculation for first pulse:* Use the equation for an increasing exponential (Eq. 9–15) because *C* is charging. Note that V_F is 5 V, and *t* equals the pulse width of 10 μs. Therefore,

$$v_C = V_F(1 - e^{-t/RC}) = (5\ \text{V})(1 - e^{-10\mu s/47\mu s}) = 958\ \text{mV}$$

This result is plotted in Figure 15–20(a).

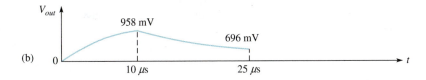

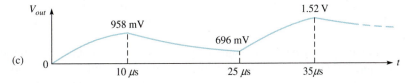

FIGURE 15–20

2. *Calculation for interval between first and second pulse:* Use the equation for a decreasing exponential (Eq. 9–16) because *C* is discharging. Note that V_i is 958 mV because *C* begins to discharge from this value at the end of the first pulse. The discharge time is 15 μs. Therefore,

$$v_C = V_i e^{-t/RC} = (958\ \text{mV})e^{-15\mu s/47\mu s} = 696\ \text{mV}$$

This result is shown in Figure 15–20(b).

3. *Calculation for second pulse:* At the beginning of the second pulse, the output voltage is 696 mV. During the second pulse, the capacitor will again charge. In this case, it does not begin at zero volts. It already has 696 mV from the previous charge and discharge. To handle this situation, you must use the general exponential formula (Eq. 9–13).

$$v = V_F + (V_i - V_F)e^{-t/\tau}$$

Using this equation, you can calculate the voltage across the capacitor at the end of the second pulse.

$$v_C = V_F + (V_i - V_F)e^{-t/RC} = 5\ \text{V} + (696\ \text{mV} - 5\ \text{V})e^{-10\mu s/47\mu s} = 1.52\ \text{V}$$

This result is shown in Figure 15–20(c).

Notice that the output waveform builds up on successive input pulses. After approximately 5τ, it will reach its steady state and will fluctuate between a constant maximum and a constant minimum, with an average equal to the average value of the input. You can demonstrate this pattern by carrying the analysis in this example further.

RELATED PROBLEM

Determine V_{out} at the beginning of the third pulse.

SYSTEM EXAMPLE 15–1

A SWEPT-FREQUENCY TEST

Communication systems have specific requirements for the frequency response of the transmitter and receiver. To test the response of the transmitter or receiver, an oscilloscope can be set up to observe the frequency response in real time. The oscilloscope's horizontal axis (the *time base*) is converted to a frequency base by putting the scope in *x-y* mode. The oscilloscope can then plot the response of the unit under test to a test signal. A block diagram of the swept-frequency test system is shown in Figure 15–21.

The swept-frequency generator supplies the test signal. It consists of three blocks that generate a frequency-modulated (FM) signal. Pulses from a square wave generator are converted to a triangle wave by an integrator (an *RC* circuit in which $\gg T$.) The triangle wave is amplified and used to change the frequency of a voltage-controlled oscillator (VCO). Simultaneously, the triangle wave moves the *x*-axis of the scope in sync with the frequency change. This causes the oscilloscope time base to become a frequency base. This signal varies continuously between two selected radio frequencies (RF) that are used for the test. The unit under test is the portion of a communication system that is being aligned.

The unit under test is part of a communication system. It responds to the frequency modulated signal as determined by its frequency response. The output is detected (the rf signal is removed), and the envelope is observed on the oscilloscope, which shows the frequency response. The unit under test can be adjusted while observing the response in real time.

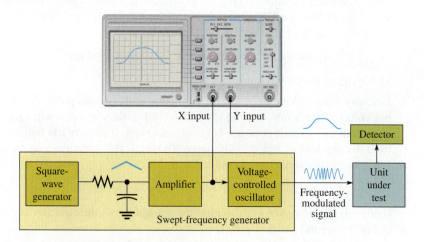

FIGURE 15–21   A swept-frequency test.

SECTION 15–3 CHECKUP

1. What conditions allow an *RC* integrator capacitor to fully charge and discharge when a periodic pulse waveform is applied to the input?

2. What will the output waveform look like if the circuit time constant is extremely small compared to the pulse width of a square wave input?

3. What is the time called that is required for the output voltage to build up to a constant average value when 5τ is greater than the pulse width of an input square wave?

4. Define *steady-state response*.

5. Describe the output of an *RC* integrator when the input is a square wave that has a period that is much less than 1τ.

6. What is the purpose of the integrator in the swept frequency measurement in System Example 15-1?

15–4 RESPONSE OF *RC* DIFFERENTIATORS TO A SINGLE PULSE

In terms of time response, a series *RC* circuit in which the output voltage is taken across the resistor is known as a **differentiator**. Recall that in terms of frequency response, this particular series *RC* circuit is a high-pass filter. The term, *differentiator*, is derived from a mathematical function which this type of circuit approximates under certain conditions.

After completing this section, you should be able to

- **Analyze an *RC* differentiator with a single-pulse input**
- **Describe the response at the rising edge of the input pulse**
- **Determine the response during and at the end of a pulse for various relationships of the pulse width and the time constant**

Figure 15–22 shows an *RC* differentiator with a pulse input. The same action occurs in the differentiator as in the integrator, except the output voltage is taken across the resistor rather than the capacitor. The capacitor charges exponentially at a rate depending on the *RC* time constant. The shape of the differentiator's resistor voltage is determined by the charging and discharging action of the capacitor.

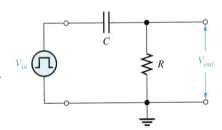

FIGURE 15–22   An *RC* differentiator with a pulse generator connected.

Pulse Response

To understand how the output voltage is shaped by a differentiator, you must consider the following:

1. The response to the rising pulse edge
2. The response between the rising and falling edges
3. The response to the falling pulse edge

Let's assume that the capacitor is initially uncharged prior to the rising pulse edge. Prior to the pulse, the input is zero volts. Thus, there are zero volts across the capacitor and also zero volts across the resistor, as indicated in Figure 15–23(a).

RESPONSE TO THE RISING EDGE OF THE INPUT PULSE   Let's also assume that a 10 V pulse is applied to the input. When the rising edge occurs, the voltage at point *A* goes to +10 V. Recall that the voltage across a capacitor cannot change instantaneously, and thus the capacitor appears instantaneously as a short. Therefore, if the

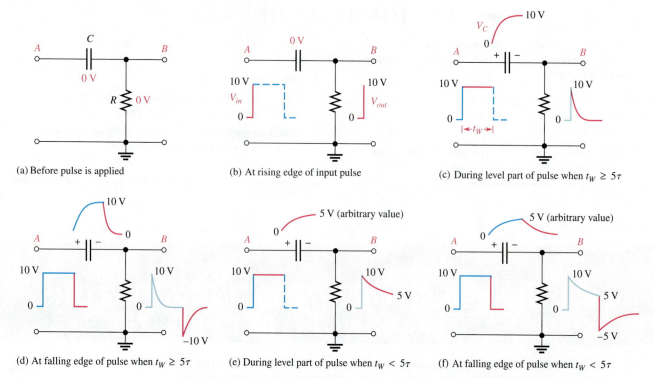

(a) Before pulse is applied

(b) At rising edge of input pulse

(c) During level part of pulse when $t_W \geq 5\tau$

(d) At falling edge of pulse when $t_W \geq 5\tau$

(e) During level part of pulse when $t_W < 5\tau$

(f) At falling edge of pulse when $t_W < 5\tau$

FIGURE 15–23   Examples of the response of a differentiator to a single-pulse input under two conditions: $t_W \geq 5\tau$ and $t_W < 5\tau$. A pulse generator is connected to the input, but the symbol is not shown, only the pulses.

voltage at point A goes instantly to $+10$ V, then the voltage at point B *must* also go instantly to $+10$ V, keeping the capacitor voltage zero for the instant of the rising edge. The capacitor voltage is the voltage from point A to point B.

The voltage at point B with respect to ground is the voltage across the resistor (and the output voltage). Thus, the output voltage suddenly goes to $+10$ V in response to the rising pulse edge, as indicated in Figure 15–23(b).

RESPONSE DURING PULSE WHEN $t_W \geq 5\tau$ While the pulse is at its high level between the rising edge and the falling edge, the capacitor is charging. When the pulse width is equal to or greater than five time constants ($t_W \geq 5\tau$), the capacitor has time to fully charge.

As the voltage across the capacitor builds up exponentially, the voltage across the resistor decreases exponentially until it reaches zero volts at the time the capacitor reaches full charge ($+10$ V in this case). This decrease in the resistor voltage occurs because the sum of the capacitor voltage and the resistor voltage at any instant must be equal to the source voltage, in compliance with Kirchhoff's voltage law ($v_C + v_R = v_{in}$). This part of the response is illustrated in Figure 15–23(c).

RESPONSE TO FALLING EDGE WHEN $t_W \geq 5\tau$ Let's examine the case in which the capacitor is fully charged at the end of the pulse ($t_W \geq 5\tau$). Refer to Figure 15–23(d). On the falling edge, the input pulse suddenly goes from $+10$ V back to zero. An instant before the falling edge, the capacitor is charged to 10 V, so the voltage at point A is $+10$ V and the voltage at point B is 0 V. The voltage across a capacitor cannot change instantaneously, so when the voltage at point A makes a transition from $+10$ V to zero on the falling edge, the voltage at point B *must* also make a 10 V transition from zero to -10 V. This keeps the voltage across the capacitor at 10 V for the instant of the falling edge.

The capacitor now begins to discharge exponentially. As a result, the resistor voltage goes from -10 V to zero in an exponential curve, as indicated in Figure 15–23(d).

RESPONSE DURING PULSE WHEN $t_W < 5\tau$ When the pulse width is less than five time constants ($t_W < 5\tau$), the capacitor does not have time to fully charge. Its partial charge depends on the relationship of the time constant and the pulse width.

Because the capacitor does not reach the full $+10$ V, the resistor voltage will not reach zero volts by the end of the pulse. For example, if the capacitor charges to $+5$ V during the pulse interval, the resistor voltage will decrease to $+5$ V, as illustrated in Figure 15–23(e).

RESPONSE TO FALLING EDGE WHEN $t_W < 5\tau$ Now, let's examine the case in which the capacitor is only partially charged at the end of the pulse ($t_W < 5\tau$). For example, if the capacitor charges to $+5$ V, the resistor voltage at the instant before the falling edge is also $+5$ V because the capacitor voltage plus the resistor voltage must add up to $+10$ V, as illustrated in Figure 15–23(e).

When the falling edge occurs, the voltage at point A goes from $+10$ V to zero. As a result, the voltage at point B goes from $+5$ V to -5 V, as illustrated in Figure 15–23(f). This decrease occurs, of course, because the capacitor voltage cannot change at the instant of the falling edge. Immediately after the falling edge, the capacitor begins to discharge to zero. As a result, the resistor voltage goes from -5 V to zero, as shown.

Summary of *RC* Differentiator Response to a Single Pulse

A good way to summarize this section is to look at the general output waveforms of a differentiator as the time constant is varied from one extreme, when 5τ is much less than the pulse width, to the other extreme, when 5τ is much greater than the pulse width. These situations are illustrated in Figure 15–24. In part (a), the output consists of narrow positive and negative "spikes." These short pulses are useful in many systems to provide a timing signal called a "trigger." In part (e), the output approaches the shape of the input. Various conditions between these extremes are illustrated in parts (b), (c), and (d).

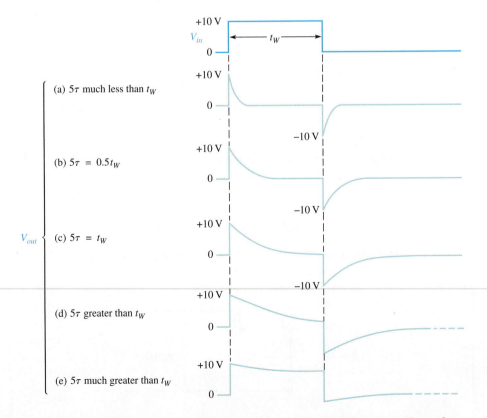

FIGURE 15–24   Effects of a change in time constant on the shape of the output voltage of an *RC* differentiator.

EXAMPLE 15–4

Show the output voltage for the circuit in Figure 15–25.

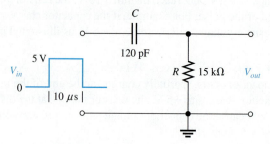

FIGURE 15–25

SOLUTION

First , calculate the time constant.

$$\tau = RC = (15\ k\Omega)(120\ pF) = 1.8\ \mu s$$

In this case, $t_W > 5\tau$, so the capacitor reaches full charge in 9 μs (before the end of the pulse).

On the rising edge, the resistor voltage jumps to +5 V and then decreases exponentially to zero before the end of the pulse. On the falling edge, the resistor voltage jumps to −5 V and then goes back to zero exponentially. The resistor voltage is, of course, the output, and its shape is shown in Figure 15–26.

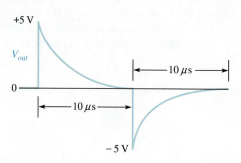

FIGURE 15–26

RELATED PROBLEM

Show the output voltage if $R = 18\ k\Omega$ and $C = 47\ pF$ in Figure 15–25.

EXAMPLE 15–5

Determine the output voltage waveform for the differentiator in Figure 15–27.

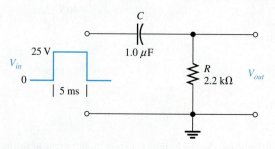

FIGURE 15–27

SOLUTION

First, calculate the time constant.

$$\tau = (2.2 \text{ k}\Omega)(1.0 \text{ }\mu\text{F}) = 2.2 \text{ ms}$$

On the rising edge, the resistor voltage immediately jumps to +25 V. Because the pulse width is 5 ms, the capacitor charges for approximately 2.27 time constants during the pulse, and therefore does not reach full charge. Thus, you must use the formula for a decreasing exponential (Eq. 9–16) in order to calculate to what voltage the output decreases by the end of the pulse.

$$v_{out} = V_i e^{-t/RC} = (25 \text{ V})e^{-5ms/2.2ms} = 2.58 \text{ V}$$

where $V_i = 25$ V and $t = 5$ ms. This calculation gives the resistor voltage at the end of the 5 ms pulse width interval.

On the falling edge, the resistor voltage immediately jumps from +2.58 V down to −22.4 V (a 25 V transition). The resulting waveform of the output voltage is shown in Figure 15–28.

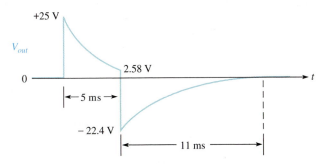

FIGURE 15–28

RELATED PROBLEM

Determine the voltage at the end of the pulse in Figure 15–27 if *R* is 1.5 kΩ.

SECTION 15–4 CHECKUP

1. Show the output of a differentiator for a 10 V input pulse when $5\tau = 0.5t_W$.

2. Under what condition does the output pulse shape most closely resemble the input pulse for a differentiator?

3. What does the differentiator output look like when 5τ is much less than the pulse width of the input?

4. If the resistor voltage in a differentiating circuit is down to +5 V at the end of a 15 V input pulse, to what negative value will the resistor voltage go in response to the falling edge of the input?

The *RC* differentiator response to a single pulse, covered in the last section, is extended in this section to repetitive pulses.

After completing this section, you should be able to

- Analyze an *RC* differentiator with a repetitive-pulse input
  - Determine the response when the pulse width is less than five time constants

If a periodic pulse waveform is applied to an *RC* differentiating circuit, two conditions again are possible: $t_W \geq 5\tau$ or $t_W < 5\tau$. Figure 15–29 shows the output when $t_W = 5\tau$. As the time constant is reduced, both the positive and the negative portions of the output become narrower. Notice that the average value of the output is zero; the waveform has equal positive and negative portions. The average value of a waveform is its **dc component**. Because a capacitor blocks dc, the dc component of the input is prevented from passing through to the output.

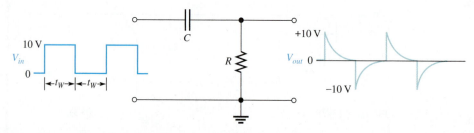

FIGURE 15–29   Example of differentiator response when $t_W = 5\tau$.

Figure 15–30 shows the steady-state output when $t_W < 5\tau$. As the time constant is increased, the positively and negatively sloping portions become flatter. For a very long time constant, the output approaches the shape of the input, but with an average value of zero.

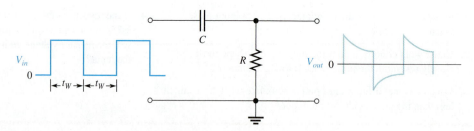

FIGURE 15–30   Example of differentiator response when $t_W < 5\tau$.

Analysis of a Repetitive Waveform

Like the integrator, the differentiator output takes time (5τ) to reach steady state. To illustrate the response, let's take an example in which the time constant equals the input pulse width. At this point, we do not care what the value of the circuit time constant is because we know that the resistor voltage will decrease to 36.8% of its maximum value during one pulse (1τ). Let's assume that the capacitor in Figure 15–31 begins initially uncharged and examine the output voltage on a pulse-by-pulse basis. The results of this analysis are shown in Figure 15–32.

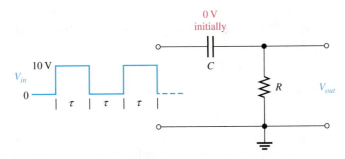

FIGURE 15–31   *RC* **differentiator with** $T = 2\tau$.

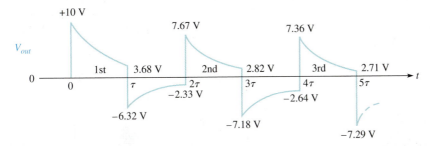

FIGURE 15–32   **Differentiator output waveform during transient time for the circuit in Figure 15–31.**

First pulse   On the rising edge, the output instantaneously jumps to +10 V. Then the capacitor partially charges to 63.2% of 10 V, which is 6.32 V. Thus, the output voltage must decrease to 3.68 V, as shown in Figure 15–32.

On the falling edge, the output instantaneously makes a negative-going 10 V transition to −6.32 V because 3.68 V − 10 V = −6.32 V.

Between first and second pulses   The capacitor discharges to 36.8% of 6.32 V, which is 2.33 V. Thus, the resistor voltage, which starts at −6.32 V, must increase to −2.33 V. Why? Because at the instant prior to the next pulse, the input voltage is zero. Therefore, the sum of v_C and v_R must be zero (+2.33 V − 2.33 V = 0). Remember that $v_C + v_R = v_{in}$ at all times, in accordance with Kirchhoff's voltage law.

Second pulse   On the rising edge, the output makes an instantaneous, positive-going, 10 V transition from −2.33 V to 7.67 V. Then by the end of the pulse the capacitor charges 0.632 (10 V − 2.33 V) = 4.85 V. Thus, the capacitor voltage increases from 2.33 V to 2.33 V + 4.85 V = 7.18 V. The output voltage drops to 0.368(7.67 V) = 2.82 V.

On the falling edge, the output instantaneously makes a negative-going transition from 2.82 V to −7.18 V, as shown in Figure 15–32.

Between second and third pulses   The capacitor discharges to 36.8% of 7.18 V, which is 2.64 V. Thus, the output voltage starts at −7.18 V and increases to −2.64 V because the capacitor voltage and the resistor voltage must add up to zero at the instant prior to the third pulse (the input is zero).

Third pulse   On the rising edge, the output makes an instantaneous 10 V transition from −2.64 V to +7.36 V. Then the capacitor charges 0.632(10 V − 2.64 V) = 4.65 V to 2.64 V + 4.65 V = +7.29 V. As a result, the output voltage drops to 0.368(7.36 V) = 2.71 V. On the falling edge, the output instantly goes from +2.71 V down to −7.29 V.

After the third pulse, five time constants have elapsed, and the output voltage is close to its steady state. Thus, the waveform in Figure 15–32 will continue to vary from a positive maximum of about +7.3 V to a negative maximum of about −7.3 V, with an average value of zero.

SYSTEM EXAMPLE 15–2

A WATCHDOG TIMER

Many systems require constant monitoring in order to take immediate action in case a power failure should occur. The circuit that monitors for a particular condition is called a *watchdog timer*, a colorful description of a monitoring circuit, usually one that can take a corrective action. For the case of a power failure monitor, there is usually enough stored power in the capacitors to take the action before the system shuts down. The action can be anything from switching to backup power, saving files, or sounding a battery-powered alarm. There is a wide variety of watchdog timer circuits, including some that are built into integrated circuits.

In this example, a watchdog timer is monitoring the input power and outputs a negative-going trigger signal when a failure occurs. The trigger can be used to start a corrective action. A key part of the circuit is two differentiating circuits. Figure 15–33 shows the basic circuit for this watchdog timer.

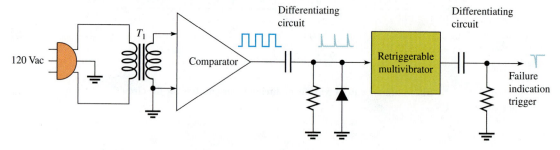

FIGURE 15–33   A watchdog timer circuit for monitoring power.

The circuit operation is to convert ac to a low voltage by T_1 and compare the low-voltage ac signal with ground. The comparator outputs a high signal when the ac is above ground and a low signal when it is below ground, thus producing a square wave. The first differentiating circuit and diode produce a series of triggers, which occur at the power line frequency. The triggers are sent to a *retriggerable multivibrator*, a circuit that has a high dc level output as long as triggers appear regularly at the input. If a power failure occurs, the dc drops to ground level and a second differentiating circuit converts this drop to a negative-going trigger, which can be used to take a corrective action.

SECTION 15–5 CHECKUP

1. What conditions allow an *RC* differentiator to fully charge and discharge when a periodic pulse waveform is applied to the input?

2. What will the output waveform look like if the circuit time constant is extremely small compared to the pulse width of a square wave input?

3. What does the average value of the differentiator output voltage equal during steady state?

4. Explain the purpose of each differentiator in the watchdog timer.

A series *RL* circuit in which the output voltage is taken across the resistor is known as an integrator in terms of time response. The response to a single pulse is discussed but can be extended to repetitive pulses as was done for the case of *RC* integrators. Although the *RL* integrator has waveforms similar to the *RC* integrator, the *RL* integrator is not as commonly used; inductors tend to be more expensive than capacitors and have winding resistance that can adversely affect the circuit.

After completing this section, you should be able to

- **Analyze the operation of an *RL* integrator**
  - Determine the response to a single-pulse input

Figure 15–34 shows an *RL* integrator. The output waveform is taken across the resistor and, under equivalent conditions, is the same shape as that for the *RC* integrator. Recall that in the *RC* case, the output was across the capacitor.

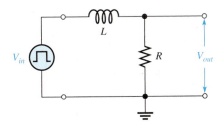

FIGURE 15–34   An *RL* integrator with a pulse generator connected.

As you know, each edge of an ideal pulse is considered to occur instantaneously. Two basic rules for inductor behavior will aid in analyzing *RL* circuit responses to pulse inputs:

1. The inductor appears as an open to an instantaneous change in current and as a short (ideally) to dc.

2. The current in an inductor cannot change instantaneously—it can change only exponentially when the input changes levels.

Response of the *RL* Integrator to a Single Pulse

When a pulse generator is connected to the input of an *RL* integrator and the voltage pulse goes from its low level to its high level, the inductor prevents a sudden change in current. As a result, the inductor acts as an open, and all the input voltage is across it at the instant of the rising pulse edge. This situation is indicated in Figure 15–35(a).

After the rising edge, the current builds up, and the output voltage follows the current as it increases exponentially, as shown in Figure 15–35(b). The current can reach a maximum of V_p/R if the transient time is shorter than the pulse width ($V_p = 10$ V in this example where V_p is pulse amplitude).

When the pulse goes from its high level to its low level, an induced voltage with reversed polarity is created across the coil in an effort to keep the current equal to V_p/R. The output voltage begins to decrease exponentially, as shown in Figure 15–35(c).

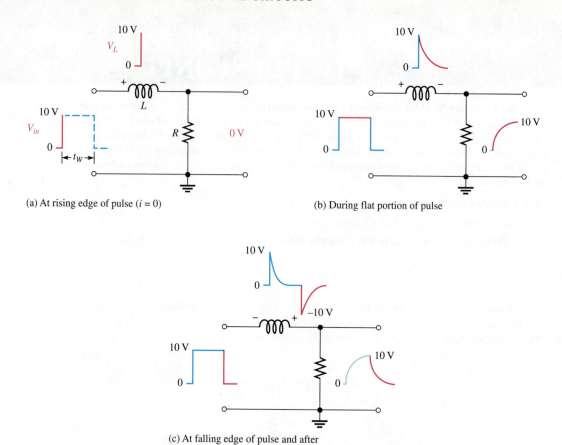

(a) At rising edge of pulse ($i = 0$)

(b) During flat portion of pulse

(c) At falling edge of pulse and after

FIGURE 15–35   Illustration of the pulse response of an *RL* integrator ($t_W > 5\tau$). A pulse generator is connected to the input but its symbol is not shown, only the pulses.

The exact shape of the output depends on the L/R time constant as summarized in Figure 15–36 for various relationships of the time constant and the pulse width. You should note that the response of this *RL* circuit in terms of the shape of the output is identical to that of the *RC* integrator. The relationship of the L/R time constant to the input pulse width has the same effect as the *RC* time constant that we discussed earlier in this chapter. For example, when $t_W < 5\tau$, the output voltage will not reach its maximum possible value.

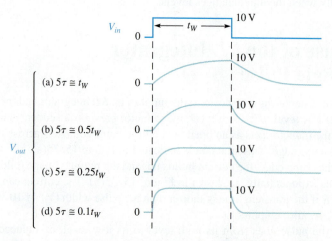

FIGURE 15–36   Illustration of the variation in integrator output pulse shape with time constant.

EXAMPLE 15–6

Determine the maximum output voltage for the integrator in Figure 15–37 when a single pulse is applied as shown.

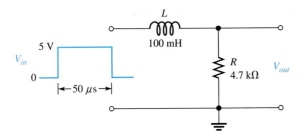

FIGURE 15–37

SOLUTION

Calculate the time constant.

$$\tau = \frac{L}{R} = \frac{100 \text{ mH}}{4.7 \text{ k}\Omega} = 21.3 \text{ } \mu s$$

Because the pulse width is 50 μs, the inductor's magnetic field increases (energizes) for approximately 2.35τ; (50 μs/21.3 μs = 2.35). Use Equation 11–6 to calculate the voltage.

$$v = V_F + (V_i - V_F)e^{-Rt/L}$$

V_i is zero. The maximum output voltage occurs at the end of the pulse. Therefore,

$$v_L = V_F(1 - e^{-t/\tau}) = 5 \text{ V}(1 - e^{-50 \text{ } \mu s/21.3 \text{ } \mu s}) = \textbf{4.52 V}$$

RELATED PROBLEM

What value must R be for the output voltage to reach 5 V by the end of the pulse?

EXAMPLE 15–7

A pulse is applied to the *RL* integrator in Figure 15–38. Determine the waveforms and the values for I, V_R, and V_L.

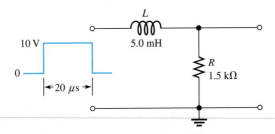

FIGURE 15–38

SOLUTION

The circuit time constant is

$$\tau = \frac{L}{R} = \frac{5.0 \text{ mH}}{1.5 \text{ k}\Omega} = 3.33 \text{ } \mu s$$

Since $5\tau = 16.7\ \mu$s is less than t_W, the current will reach its maximum value and remain there until the end of the pulse.

At the rising edge of the pulse,

$$i = 0\ \text{A}$$
$$v_R = 0\ \text{V}$$
$$v_L = 10\ \text{V}$$

The inductor appears as an open, so all of the input voltage appears across L.

During the pulse,

$$i \text{ increases exponentially to } \frac{V_p}{R} = \frac{10\ \text{V}}{1.5\ \text{k}\Omega} = 6.67\ \text{mA in } 16.7\ \mu\text{s}$$

v_R increases exponentially to 10 V in 16.7 μs

v_L increases exponentially to zero in 16.7 μs

At the falling edge of the pulse,

$$i = 6.67\ \text{mA}$$
$$v_R = 10\ \text{V}$$
$$v_L = -10\ \text{V}$$

After the pulse,

i decreases exponentially to zero in 16.7 μs

v_R decreases exponentially to zero in 16.7 μs

v_L decreases exponentially to zero in 16.7 μs

The waveforms are shown in Figure 15–39.

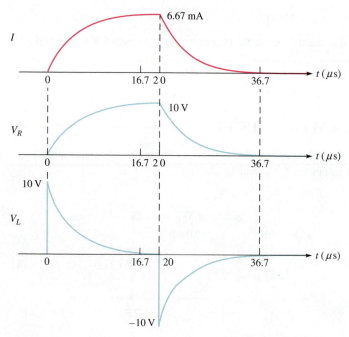

FIGURE 15–39

RELATED PROBLEM

What will be the maximum output voltage if the amplitude of the input pulse is increased to 20 V in Figure 15–38?

EXAMPLE 15–8

A 10 V pulse with a width of 10 μs is applied to the integrator in Figure 15–40. Determine the voltage level that the output will reach during the pulse. If the source has an internal resistance of 300 Ω, how long will it take the output to decrease to zero? Show the output voltage waveform.

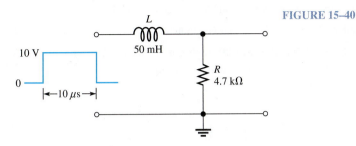

FIGURE 15–40

SOLUTION

The coil energizes through the 300 Ω source resistance plus the 4.7 kΩ external resistor. The time constant is

$$\tau = \frac{L}{R_{tot}} = \frac{50\ \text{mH}}{4700\ \Omega\ +\ 300\ \Omega} = \frac{50\ \text{mH}}{5.0\ \text{k}\Omega} = 10\ \mu\text{s}$$

Notice that in this case the pulse width is exactly equal to τ. Thus, the output V_R will reach approximately 63% of the full input amplitude in 1τ. Therefore, the output voltage gets to **6.3 V** at the end of the pulse.

After the pulse is gone, the inductor de-energizes back through the 300 Ω source resistance and the 4.7 kΩ resistor. The output voltage takes 5τ to completely decrease to zero.

$$5\tau = 5(10\ \mu\text{s}) = \textbf{50}\ \mu\text{s}$$

The output voltage is shown in Figure 15–41.

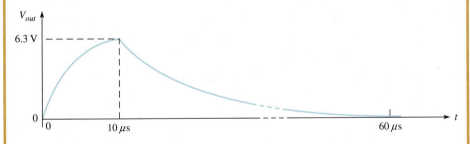

FIGURE 15–41

RELATED PROBLEM

To what approximate maximum value must R be changed in Figure 15–40 to allow the output voltage to reach the input level during the pulse?

SECTION 15–6 CHECKUP

1. In an *RL* integrator, across which component is the output voltage taken?

2. When a pulse is applied to an *RL* integrator, what condition must exist in order for the output voltage to reach the amplitude of the input?

3. Under what condition will the output voltage have the approximate shape of the input pulse?

A series *RL* circuit in which the output voltage is taken across the inductor is known as a differentiator in terms of time response. The response to a single pulse is discussed but can be extended to repetitive pulses as was done for the case of *RC* differentiators.

After completing this section, you should be able to

- Analyze the operation of an *RL* differentiator
  - Determine the response to a single-pulse input

Response of the *RL* Differentiator to a Single Pulse

Figure 15–42 shows an *RL* differentiator with a pulse generator connected to the input.

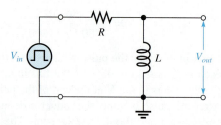

FIGURE 15–42 An *RL* differentiator with a pulse generator connected.

Initially, before the pulse, there is no current in the circuit. When the input pulse goes from its low level to its high level, the inductor prevents a sudden change in current. It does this, as you know, with an induced voltage equal and opposite to the input. As a result, *L* looks like an open, and all of the input voltage appears across it at the instant of the rising edge, as shown in Figure 15–43(a) with a 10 V pulse.

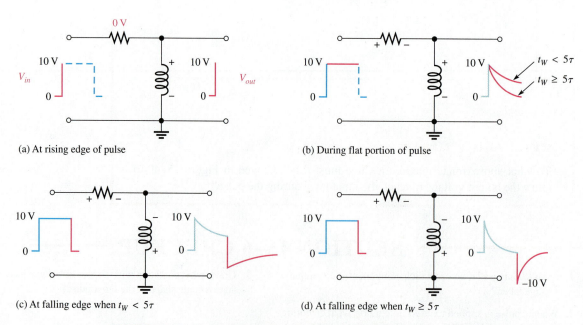

(a) At rising edge of pulse

(b) During flat portion of pulse

(c) At falling edge when $t_W < 5\tau$

(d) At falling edge when $t_W \geq 5\tau$

FIGURE 15–43 Illustration of the response of an *RL* differentiator for both time constant conditions. The pulse generator is connected to the input, but its symbol is not shown, only the pulses.

During the pulse, the current exponentially builds up. As a result, the inductor voltage decreases, as shown in Figure 15–43(b). The rate of decrease, as you know, depends on the L/R time constant. When the falling edge of the input occurs, the inductor reacts to keep the current as is, by creating an induced voltage in a direction as indicated in Figure 15–43(c). This reaction is seen as a sudden negative-going transition of the inductor voltage, as indicated in parts (c) and (d).

Two conditions are possible, as indicated in Figure 15–43(c) and (d). In part (c), 5τ is greater than the input pulse width, and the output voltage does not have time to decrease to zero. In part (d), 5τ is less than or equal to the pulse width, and so the output decreases to zero before the end of the pulse. In this case a full -10 V transition occurs at the falling edge.

Keep in mind that as far as the input and output waveforms are concerned, the *RL* integrator and differentiator perform the same as their *RC* counterparts.

A summary of the *RL* differentiator response for various relationships of time constants and pulse widths is shown in Figure 15–44. Compare this figure to Figure 15–24 for an *RC* differentiator and you will see that the responses are identical.

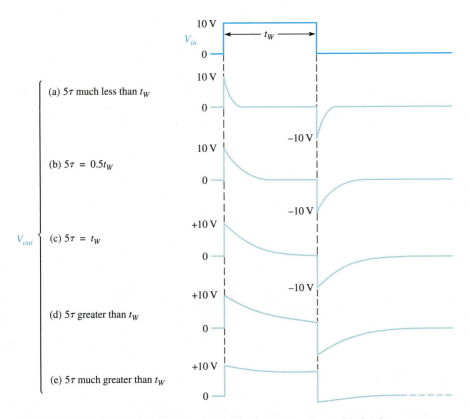

(a) 5τ much less than t_W

(b) $5\tau = 0.5t_W$

(c) $5\tau = t_W$

(d) 5τ greater than t_W

(e) 5τ much greater than t_W

V_{out}

FIGURE 15–44 **Illustration of the variation in output pulse shape with the time constant.**

EXAMPLE 15–9

Show the output voltage waveform for the *RL* differentiator in Figure 15–45.

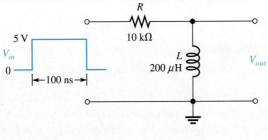

FIGURE 15–45

SOLUTION

First, calculate the time constant.

$$\tau = \frac{L}{R} = \frac{200\ \mu\text{H}}{10\ \text{k}\Omega} = 20\ \text{ns}$$

In this case, $t_W = 5\tau$, so the output will decay to zero at the end of the pulse.

On the rising edge, the inductor voltage jumps to +5 V and then decays exponentially to zero. It reaches approximately zero at the instant of the falling edge. On the falling edge of the input, the inductor voltage jumps to −5 V and then goes back to zero. The output waveform is shown in Figure 15–46.

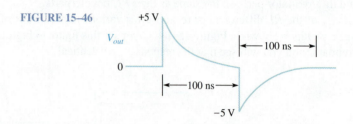

FIGURE 15–46

RELATED PROBLEM

Show the output voltage if the input pulse width is reduced to 50 ns in Figure 15–45.

EXAMPLE 15–10

Determine the output voltage waveform for the *RL* differentiator in Figure 15–47.

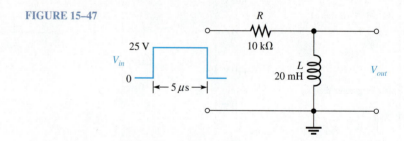

FIGURE 15–47

SOLUTION

First, calculate the time constant.

$$\tau = \frac{L}{R} = \frac{20\ \text{mH}}{10\ \text{k}\Omega} = 2\ \mu\text{s}$$

On the rising edge, the inductor voltage immediately jumps to +25 V, which is the initial value for the exponential decay. The final value is zero (assuming it could continue to decay). The exponential formula for the voltage at any instant on an inductor is Equation 11–6, which is

$$v = V_F + (V_i - V_F)e^{-Rt/L}$$

Substituting 0 for the final value gives the exponential decay formula:

$$v_L = V_i e^{-t/\tau} = (25\ \text{V})e^{-5\mu s/2/\mu s} = (25\ \text{V})e^{-2.5} = 2.05\ \text{V}$$

This result is the inductor voltage at the end of the 5 μs input pulse.

On the falling edge, the output immediately drops from +2.05 V to −22.95 V (a 25 V negative-going transition). The complete output waveform is shown in Figure 15–48.

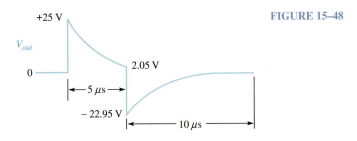

FIGURE 15–48

RELATED PROBLEM

What must be the value of R for V_{out} to reach zero by the end of the input pulse in Figure 15–47?

SECTION 15–7 CHECKUP

1. In an RL differentiator, across which component is the output taken?

2. Under what condition does the output pulse shape most closely resemble the input pulse?

3. If the inductor voltage in an RL differentiator is down to $+2$ V at the end of a $+10$ V input pulse, to what negative voltage will the output voltage go in response to the falling edge of the input?

15–8 INTEGRATOR AND DIFFERENTIATOR APPLICATIONS

Although integrators and differentiators are found in many different kinds of applications, only three basic ones are discussed in this section.

After completing this section, you should be able to

- **Discuss some applications of integrators and differentiators**
  - **Describe how RC integrators are used in timing circuits**
  - **Describe how an integrator can be used for converting pulse waveforms to dc**
  - **Describe how a differentiator can be used as a trigger pulse generator**

Timing Circuits

RC integrators can be used in timing circuits to set a specified time interval for various purposes. By changing the time constant, the interval can be adjusted. For example, the RC integrator circuit in Figure 15–49(a) is used to provide a delay between the time a power-on switch is closed and the time a certain event is activated by a threshold circuit. The threshold circuit is designed to respond to the input voltage only when the input reaches a certain specified level. Don't worry how the threshold circuit works at this point; you will learn about this type of circuit in a later course.

When the switch is thrown, the capacitor begins to charge at a rate set by the RC time constant. After a certain time, the capacitor voltage reaches the threshold value and triggers (turns on) the circuit to activate a circuit or a device such as a motor, relay, or lamp, depending on the particular application. A variable delay can be created by making R a rheostat. A circuit like this is useful in systems like radars to adjust certain features like "clutter filters." This action is illustrated by the waveforms in Figure 15–49(b).

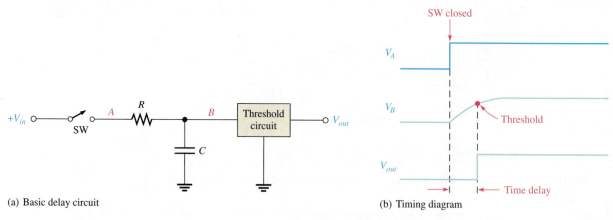

(a) Basic delay circuit

(b) Timing diagram

FIGURE 15–49 A basic time delay application of an *RC* integrator.

EXAMPLE 15–11

What is the time delay in the circuit of Figure 15–49(a) with $V_{in} = 9$ V, $R = 10$ MΩ, and $C = 0.47$ μF? Assume the threshold voltage is 5 V.

SOLUTION

Use the exponential formula, Equation 9–15, and solve for time as follows:

$$v = V_F(1 - e^{-t/RC}) = V_F - V_F e^{-t/RC}$$
$$V_F - v = V_F e^{-t/RC}$$
$$e^{-t/RC} = \frac{V_F - v}{V_F}$$

Taking the natural log (ln) of both sides of the equation yields

$$-\frac{t}{RC} = \ln\left(\frac{V_F - v}{V_F}\right)$$
$$t = -RC \ln\left(\frac{V_F - v}{V_F}\right)$$

V_F is the final voltage to which C will charge and equals V_{in}. Substituting values and solving for t yields

$$t = -(10\text{ M}\Omega)(0.47\text{ μF})\ln\left(\frac{9\text{ V} - 5\text{ V}}{9\text{ V}}\right) = -(10\text{ M}\Omega)(0.47\text{ μF})\ln\left(\frac{4\text{ V}}{9\text{ V}}\right) = \textbf{3.8 s}$$

RELATED PROBLEM

Find the time delay for the circuit in Figure 15–49 for $V_{in} = 24$ V, $R = 10$ kΩ, and $C = 1500$ pF.

A Pulse Waveform-to-DC Converter

An integrating circuit can be used to convert a pulse waveform to a constant dc value equal to the average of the waveform. This is accomplished by using a time constant that is extremely long compared to the period of the pulse waveform, as illustrated in Figure 15–50(a). Actually, there will be a slight ripple on the output as the capacitor charges and discharges a little. This ripple becomes less as the time constant becomes longer and can approach a constant level, as shown in Figure 15–50(b). In terms of sinusoidal response, this circuit is a low-pass filter.

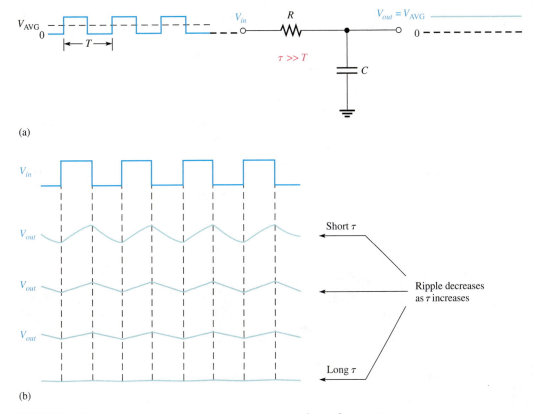

(a)

(b)

FIGURE 15–50   A long time constant integrator acts as a pulse-to-dc converter.

Trigger Pulse Generator and Waveshaping

A differentiator circuit can be used to produce very short duration pulses (spikes) with both positive and negative polarities, as you saw in System Example 15–2. When the time constant is short compared to the pulse width of the input, the differentiator produces a positive spike on each positive-going edge of the input and a negative spike on each negative-going edge, as shown in Figure 15–51. A diode can be used to remove either the positive trigger or the negative trigger.

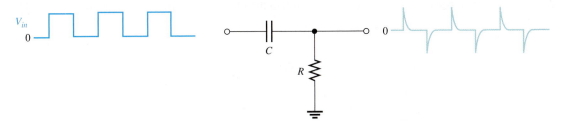

FIGURE 15–51   A very short time constant differentiator generates positive and negative spikes.

By changing the input to a triangle wave with a long period compared to the time constant, a wave-shaping circuit is produced. This time the output is a square wave that represents the rate of rise of the triangle. This idea is shown in Figure 15–52. There is a

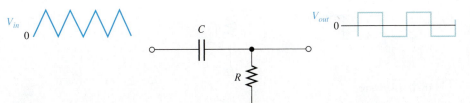

FIGURE 15–52   Triangle to square wave circuit. The time constant of the circuit is long compared to the period.

trade-off in this circuit between the rise time of the output square wave and the amplitude of the square wave. If the time constant is short, the square wave rises rapidly, but the amplitude is lower.

SECTION 15–8 CHECKUP

1. How can the delay time in the circuit of Figure 15–49 be increased?

2. Reducing the value of R in Figure 15–50 will (increase, decrease) the amount of ripple on the output.

3. What characteristic must the differentiator in Figure 15–51 have in order to produce very short duration spikes on its output?

15–9 TROUBLESHOOTING

In this section, RC circuits with pulse inputs are used to demonstrate the effects of common component failures in selected cases. The concepts can then be easily related to RL circuits.

After completing this section, you should be able to

- **Troubleshoot RC integrators and RC differentiators**
  - **Recognize the effect of an open capacitor**
  - **Recognize the effect of a shorted capacitor**
  - **Recognize the effect of an open resistor**

Open Capacitor

If the capacitor in an RC integrator opens, the output has the same waveshape as the input, as shown in Figure 15–53(a). If the capacitor in an RC differentiator opens, the output is zero because it is held at ground through the resistor, as illustrated in part (b).

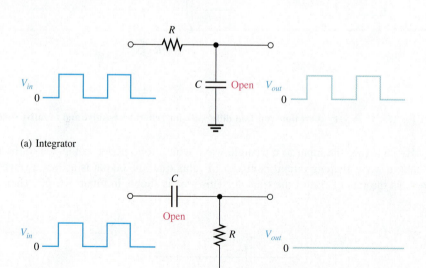

(a) Integrator

(b) Differentiator

FIGURE 15–53 Examples of the effect of an open capacitor.

Shorted Capacitor

If the capacitor in an *RC* integrator shorts, the output is at ground as shown in Figure 15–54(a). If the capacitor in an *RC* differentiator shorts, the output is the same as the input, as shown in part (b).

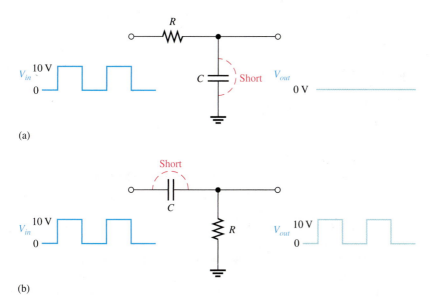

(a)

(b)

FIGURE 15–54   **Examples of the effect of a shorted capacitor.**

Open Resistor

If the resistor in an *RC* integrator opens, the capacitor has no discharge path, and, ideally, it will hold any charge it may have. In an actual situation, any charge will gradually leak off or the capacitor will discharge slowly through a measuring instrument connected to the output. This is illustrated in Figure 15–55(a).

If the resistor in an *RC* differentiator opens, the output looks like the input except for the dc level because the capacitor now must charge and discharge through the extremely high resistance of the oscilloscope, as shown in Figure 15–55(b).

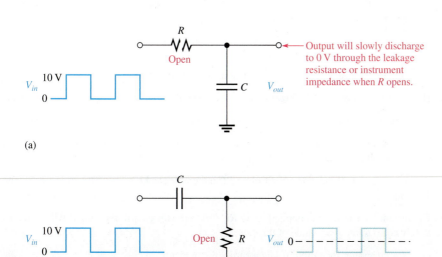

(a)

(b)

FIGURE 15–55   **Examples of the effect of an open resistor.**

SECTION 15–9 CHECKUP

1. An *RC* integrator has a zero output with a square wave input. What are the possible causes of this problem?

2. If the integrator capacitor is open, what is the output for a square wave input?

3. If the capacitor in a differentiator is shorted, what is the output for a square wave input?

SUMMARY

- In an *RC* integrator, the output voltage is taken across the capacitor.
- In an *RC* differentiator, the output voltage is taken across the resistor.
- In an *RL* integrator, the output voltage is taken across the resistor.
- In an *RL* differentiator, the output voltage is taken across the inductor.
- In an integrator, when the pulse width (t_W) of the input is much less than the transient time, the output voltage approaches a constant level equal to the average value of the input.
- In an integrator, when the pulse width of the input is much greater than the transient time, the output voltage approaches the shape of the input.
- In a differentiator, when the pulse width of the input is much less than the transient time, the output voltage approaches the shape of the input but with an average value of zero.
- In a differentiator, when the pulse width of the input is much greater than the transient time, the output voltage consists of narrow, positive-going and negative-going spikes occurring on the leading and trailing edges of the input pulses.
- *RC* integrators can be used to set up a specified time delay.

KEY TERMS

Key terms and other bold terms in the chapter are defined in the end-of-book glossary.

Differentiator   A circuit producing an output that approaches the mathematical derivative of the input.

Integrator   A circuit producing an output that approaches the mathematical integral of the input.

Steady state   The equilibrium condition of a circuit that occurs after an initial transient time.

Transient time   An interval equal to approximately five time constants.

TRUE/FALSE QUIZ

Answers are at the end of the chapter.

1. Transient time for an *RC* circuit is the same as the time constant.

2. The capacitor in a pulsed *RC* circuit will fully charge if the pulse width is equal or greater than 5τ.

3. If the source voltage is increased in a pulsed *RC* circuit, the time constant will increase.

4. Under certain conditions, the output of an *RC* integrator can have the same maximum and minimum voltage as the input.

5. Under certain conditions, the output of an *RC* integrator can be equal to the average voltage of the input.

6. The instant after a pulse is applied to an *RC* differentiator circuit, the output will be approximately equal to the peak value of the pulse.

7. In an *RL* integrator, the output is taken across the inductor.

8. If the inductor in an *RL* differentiator is open, the output will be zero.

9. The time constant for an *RL* differentiator is inversely proportional to the resistance.

10. A circuit that can develop positive and negative triggers from a square wave is a differentiator.

SELF-TEST

Answers are at the end of the chapter.

1. The output of an *RC* integrator is taken across the
   (a) resistor   (b) capacitor   (c) source   (d) coil

2. When a 10 V input pulse with a width equal to one time constant is applied to an *RC* integrator, the capacitor charges to
   (a) 10 V   (b) 5 V   (c) 6.3 V   (d) 3.7 V

3. When a 10 V input pulse with a width equal to one time constant is applied to an *RC* differentiator, the capacitor charges to
   (a) 6.3 V   (b) 10 V   (c) 5 V   (d) 3.7 V

4. In an *RC* integrator, the output pulse closely resembles the input pulse when
   (a) τ is much larger than the pulse width
   (b) τ is equal to the pulse width
   (c) τ is less than the pulse width
   (d) τ is much less than the pulse width

5. In an *RC* differentiator, the output pulse closely resembles the input when
   (a) τ is much larger than the pulse width
   (b) τ is equal to the pulse width
   (c) τ is less than the pulse width
   (d) τ is much less than the pulse width

6. The positive and negative portions of a differentiator's output voltage are maximum when
   (a) $5\tau < t_W$ (b) $5\tau > t_W$ (c) $5\tau = t_W$ (d) $5\tau > 0$

7. The output of an *RL* integrator is taken across the
   (a) resistor   (b) coil   (c) source   (d) capacitor

8. The maximum current in an *RL* differentiator is
   (a) $I = \dfrac{V_p}{X_L}$ (b) $I = \dfrac{V_p}{Z}$ (c) $I = \dfrac{V_p}{R}$

9. The current in an *RL* differentiator reaches its maximum possible value when
   (a) $5\tau = t_W$ (b) $5\tau < t_W$ (c) $5\tau > t_W$ (d) $\tau = 0.5t_W$

10. If you have an *RC* and an *RL* differentiator with equal time constants sitting side-by-side and you apply the same input pulse to both,
    (a) the *RC* has the widest output pulse
    (b) the *RL* has the most narrow spikes on the output
    (c) the output of one is an increasing exponential and the output of the other is a decreasing exponential
    (d) you can't tell the difference by observing the output waveforms

PROBLEMS

Answers to odd-numbered problems are at the end of the book.

BASIC PROBLEMS

SECTION 15–1   The *RC* Integrator

1. An integrator has $R = 2.2\ \text{k}\Omega$ in series with $C = 0.047\ \mu\text{F}$. What is the time constant?

2. Determine how long it takes the capacitor in an integrating circuit to reach full charge for each of the following series *RC* combinations:
   (a) $R = 47\ \Omega, C = 47\ \mu\text{F}$
   (b) $R = 3300\ \Omega, C = 0.015\ \mu\text{F}$
   (c) $R = 22\ \text{k}\Omega, C = 100\ \text{pF}$
   (d) $R = 4.7\ \text{M}\Omega, C = 10\ \text{pF}$

3. It is required that an integrator have a time constant of approximately 6 ms. If $C = 0.22\ \mu\text{F}$, what standard value of R must be used?

4. For the capacitor in the integrator of Problem 3 to fully charge during a pulse, what must be the minimum width of the pulse?

SECTION 15–2   Response of *RC* Integrators to a Single Pulse

5. A 20 V pulse is applied to an *RC* integrator. The pulse width equals one time constant. To what voltage does the capacitor charge during the pulse? Assume that it is initially uncharged.

6. Repeat Problem 5 for the following values of t_W:
   (a) 2τ **(b)** 3τ **(c)** 4τ **(d)** 5τ

7. Show the approximate shape of an integrator output voltage where 5τ is much less than the pulse width of a 10 V square wave input. Repeat for the case in which 5τ is much larger than the pulse width.

8. Determine the output voltage for an integrator with a single-pulse input, as shown in Figure 15–56. For repetitive pulses, how long will it take this circuit to reach steady state?

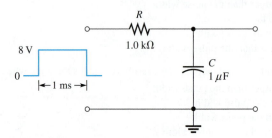

FIGURE 15–56

SECTION 15–3   Response of *RC* Integrators to Repetitive Pulses

9. Draw the integrator output in Figure 15–57 showing maximum voltages.

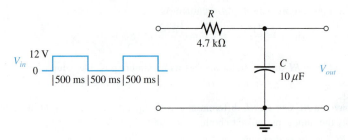

FIGURE 15–57

10. A 1 V, 10 kHz pulse waveform with a duty cycle of 25% is applied to an integrator with $\tau = 25\ \mu s$. Graph the output voltage for three initial pulses. *C* is initially uncharged.

11. What is the steady-state output voltage of the *RC* integrator with a square wave input shown in Figure 15–58?

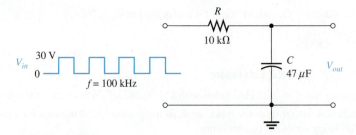

FIGURE 15–58

SECTION 15–4   Response of *RC* Differentiators to a Single Pulse

12. Repeat Problem 7 for an *RC* differentiator.

13. Redraw the circuit in Figure 15–56 to make it a differentiator, and repeat Problem 8.

SECTION 15–5   Response of *RC* Differentiators to Repetitive Pulses

14. Draw the differentiator output in Figure 15–59, showing maximum voltages.

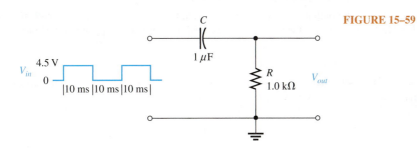

FIGURE 15–59

15. What is the steady-state output voltage of the differentiator with the square wave input shown in Figure 15–60?

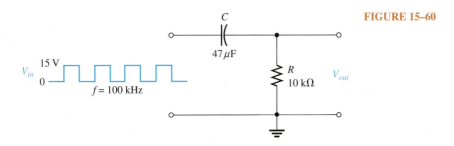

FIGURE 15–60

SECTION 15–6   Response of *RL* Integrators to Pulse Inputs

16. Determine the output voltage for the circuit in Figure 15–61. A single-pulse input is applied as shown.

17. Draw the integrator output in Figure 15–62, showing maximum voltages.

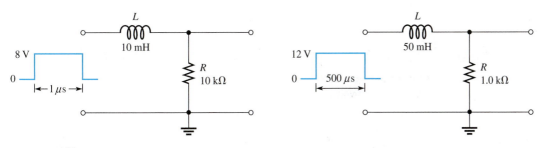

FIGURE 15–61   **FIGURE 15–62**

SECTION 15–7   Response of *RL* Differentiators to Pulse Inputs

18. (a) What is τ in Figure 15–63?
      (b) Draw the output voltage.

19. Draw the output waveform if a periodic pulse waveform with $t_W = 250$ ns and $T = 600$ ns is applied to the circuit in Figure 15–63.

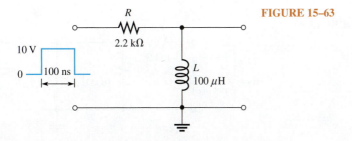

FIGURE 15–63

SECTION 15–8   Integrator and Differentiator Applications

20. What is the instantaneous voltage at point B in Figure 15–49 440 μs after the switch is closed for $R = 22$ kΩ and $C = 0.001$ μF? $V_{in} = 10$ V.

21. Ideally, what is the output of an RC integrator when the input is a square wave with an amplitude of 12 V? Assume that the time constant is much greater than the period of the input signal?

SECTION 15–9   Troubleshooting

22. Determine the most likely fault(s) in the circuit of Figure 15–64(a) for each set of waveforms in parts (b) and (c). V_{in} is a square wave with a period of 8 ms.

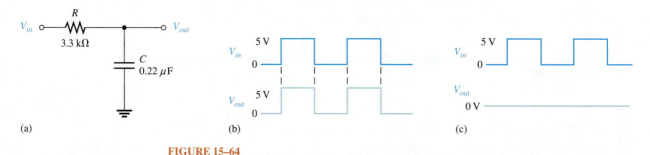

FIGURE 15–64

23. Determine the most likely fault(s), if any, in the circuit of Figure 15–65(a) for each set of waveforms in parts (b) and (c). V_{in} is a square wave with a period of 8 ms.

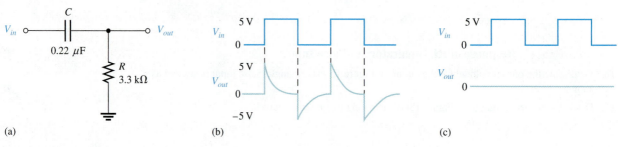

FIGURE 15–65

ADVANCED PROBLEMS

24. (a) What is τ in Figure 15–66?
(b) Draw the output voltage.

FIGURE 15–66

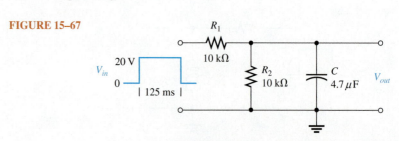

25. (a) What is τ in Figure 15–67?
(b) Draw the output voltage.

FIGURE 15–67

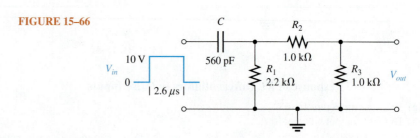

26. Determine the time constant in Figure 15–68. Is this circuit an integrator or a differentiator?

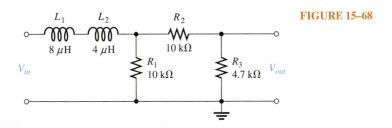

FIGURE 15–68

L_1 8 μH L_2 4 μH R_2 10 kΩ

V_{in}

R_1 10 kΩ R_3 4.7 kΩ V_{out}

27. In a time-delay circuit like that of Figure 15–49, what time constant will produce a 1 s delay if the threshold of the circuit is 2.5 V and the amplitude of the input is 5 V?

28. Draw the schematic for the circuit in Figure 15–69 and determine if the oscilloscope presentation is correct.

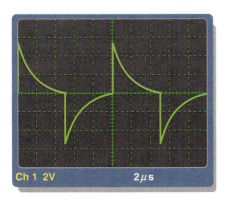

(a) Oscilloscope display

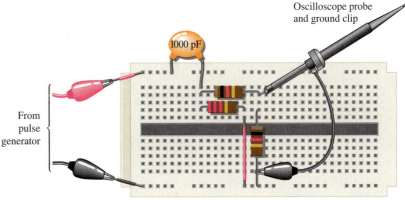

Oscilloscope probe and ground clip

1000 pF

From pulse generator

(b) Circuit board with instrument leads connected

FIGURE 15–69

29. Suggest a method that you could use to calibrate the time base for frequency in System Example 15–1.

30. Explain why the final output of System Example 15–2 is a negative-going trigger and not a positive-going trigger.

MULTISIM TROUBLESHOOTING PROBLEMS

MULTISIM

31. Open file P15-31; files are found at www.pearsonhighered.com/floyd. Test the circuit. If there is a fault, identify it.

32. Open file P15-32 and test the circuit. If there is a fault, identify it.

33. Determine if there is a fault in the circuit in file P15-33. If so, identify it.

34. Identify any faulty component in the circuit in file P15-34.

ANSWERS TO SECTION CHECKUPS

SECTION 15–1   The *RC* Integrator

1. An integrator is a series *RC* circuit with a pulse input in which the output is across the capacitor.

2. A voltage applied to the input causes the capacitor to charge. Zero volts across the input causes the capacitor to discharge.

SECTION 15–2   Response of *RC* Integrators to a Single Pulse

1. The output of an integrator reaches full amplitude when $5\tau \leq t_W$.

2. $V_{out} = (0.632)1\ \text{V} = 0.632\ \text{V}; t_{\text{disch}} = 5\tau = 51.7\ \text{ms}.$

3. See Figure 15–70.

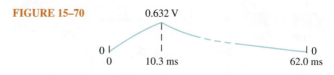

FIGURE 15–70

4. No, C will not fully charge.

5. The output is approximately the shape of the input when $5\tau \ll t_W$ (5τ much less than t_W).

SECTION 15–3   Response of *RC* Integrators to Repetitive Pulses

1. An integrator capacitor will fully charge and discharge when $5\tau \le t_W$ and $5\tau \le$ time between pulses.

2. The output looks like the input.

3. transient time

4. Steady state is the response after the transient time has passed.

5. The output is a dc voltage that is the average of the input.

6. The integrator converts the square wave to a triangle wave, which moves the beam in the x-axis.

SECTION 15–4   Response of *RC* Differentiators to a Single Pulse

1. See Figure 15–71.

FIGURE 15–71

2. The output approximates the input when $5\tau \gg t_W$.

3. The output consists of a positive and a negative spike.

4. $V_R = +5\ \text{V} - 15\ \text{V} = -10\ \text{V}$

SECTION 15–5   Response of *RC* Differentiators to Repetitive Pulses

1. C fully charges and discharges when $5\tau \le t_W$ and $5\tau \le$ time between pulses.

2. The output consists of positive and negative spikes.

3. $V_{out} = 0\ \text{V}$

4. The first differentiator converts the comparator square wave output to a series of short pulses (spikes) which continuously retrigger the one-shot as long as power is on. The second differentiator detects when the output of the one-shot goes low as a result of power loss.

SECTION 15–6   Response of *RL* Integrators to Pulse Inputs

1. The output is across the resistor.

2. The output reaches the input amplitude when $5\tau \le t_W$.

3. The output approximates the input when $5\tau \ll t_W$.

SECTION 15–7   Response of *RL* Differentiators to Pulse Inputs

1. The output is across the inductor.

2. The output approximates the input when $5\tau \gg t_W$.

3. $V_{out} = 2\ \text{V} - 10\ \text{V} = -8\ \text{V}$

SECTION 15–8   Integrator and Differentiator Applications

1. Increase the time constant to increase delay time.

2. Reducing R will increase ripple.

3. A very short time constant produces very short duration spikes.

SECTION 15–9   Troubleshooting

1. Zero output of *RC* integrator indicates shorted capacitor, open resistor, no source voltage, or open contact.
2. An open integrator capacitor results in an output exactly like the input.
3. A shorted differentiator capacitor results in an output exactly like the input.

ANSWERS TO RELATED PROBLEMS FOR EXAMPLES

15–1   8.65 V

15–2   24.7 V

15–3   1.10 V

15–4   See Figure 15–72.

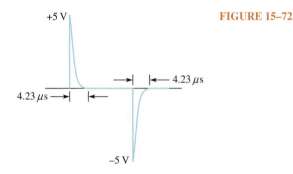

FIGURE 15–72

15–5   892 mV

15–6   10 kΩ

15–7   20 V

15–8   24.7 kΩ (assuming $R_s = 300\ \Omega$)

15–9   See Figure 15–73.

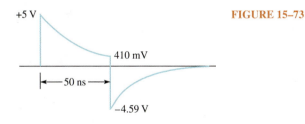

FIGURE 15–73

15–10   20 kΩ

15–11   3.5 µs

ANSWERS TO TRUE/FALSE QUIZ

1. F   **2.** T   **3.** F   **4.** T   **5.** T   **6.** T   **7.** F
8. F   **9.** T   **10.** T

ANSWERS TO SELF-TEST

1. (b)   **2.** (c)   **3.** (a)   **4.** (d)   **5.** (a)   **6.** (a)   **7.** (a)
8. (c)   **9.** (b)   **10.** (d)

Table of Standard Resistor Values

Resistance Tolerance (±%)

0.1% 0.25% 0.5%	1%	2% 5%	10%	0.1% 0.25% 0.5%	1%	2% 5%	10%	0.1% 0.25% 0.5%	1%	2% 5%	10%	0.1% 0.25% 0.5%	1%	2% 5%	10%	0.1% 0.25% 0.5%	1%	2% 5%	10%	0.1% 0.25% 0.5%	1%	2% 5%	10%
10.0	10.0	10	10	14.7	14.7	—	—	21.5	21.5	—	—	31.6	31.6	—	—	46.4	46.4	—	—	68.1	68.1	68	68
10.1	—	—	—	14.9	—	—	—	21.8	—	—	—	32.0	—	—	—	47.0	—	47	47	69.0	—	—	—
10.2	10.2	—	—	15.0	15.0	15	15	22.1	22.1	22	22	32.4	32.4	—	—	47.5	47.5	—	—	69.8	69.8	—	—
10.4	—	—	—	15.2	—	—	—	22.3	—	—	—	32.8	—	—	—	48.1	—	—	—	70.6	—	—	—
10.5	10.5	—	—	15.4	15.4	—	—	22.6	22.6	—	—	33.2	33.2	33	33	48.7	48.7	—	—	71.5	71.5	—	—
10.6	—	—	—	15.6	—	—	—	22.9	—	—	—	33.6	—	—	—	49.3	—	—	—	72.3	—	—	—
10.7	10.7	—	—	15.8	15.8	—	—	23.2	23.2	—	—	34.0	34.0	—	—	49.9	49.9	—	—	73.2	73.2	—	—
10.9	—	—	—	16.0	—	16	—	23.4	—	—	—	34.4	—	—	—	50.5	—	—	—	74.1	—	—	—
11.0	11.0	11	—	16.2	16.2	—	—	23.7	23.7	—	—	34.8	34.8	—	—	51.1	51.1	51	—	75.0	75.0	75	—
11.1	—	—	—	16.4	—	—	—	24.0	—	24	—	35.2	—	—	—	51.7	—	—	—	75.9	—	—	—
11.3	11.3	—	—	16.5	16.5	—	—	24.3	24.3	—	—	35.7	35.7	—	—	52.3	52.3	—	—	76.8	76.8	—	—
11.4	—	—	—	16.7	—	—	—	24.6	—	—	—	36.1	—	36	—	53.0	—	—	—	77.7	—	—	—
11.5	11.5	—	—	16.9	16.9	—	—	24.9	24.9	—	—	36.5	36.5	—	—	53.6	53.6	—	—	78.7	78.7	—	—
11.7	—	—	—	17.2	—	—	—	25.2	—	—	—	37.0	—	—	—	54.2	—	—	—	79.6	—	—	—
11.8	11.8	—	—	17.4	17.4	—	—	25.5	25.5	—	—	37.4	37.4	—	—	54.9	54.9	—	—	80.6	80.6	—	—
12.0	—	12	12	17.6	—	—	—	25.8	—	—	—	37.9	—	—	—	56.2	—	—	—	81.6	—	—	—
12.1	12.1	—	—	17.8	17.8	—	—	26.1	26.1	—	—	38.3	38.3	—	—	56.6	56.6	56	56	82.5	82.5	82	82
12.3	—	—	—	18.0	—	18	18	26.4	—	—	—	38.8	—	—	—	56.9	—	—	—	83.5	—	—	—
12.4	12.4	—	—	18.2	18.2	—	—	26.7	26.7	—	—	39.2	39.2	39	39	57.6	57.6	—	—	84.5	84.5	—	—
12.6	—	—	—	18.4	—	—	—	27.1	—	27	27	39.7	—	—	—	58.3	—	—	—	85.6	—	—	—
12.7	12.7	—	—	18.7	18.7	—	—	27.4	27.4	—	—	40.2	40.2	—	—	59.0	59.0	—	—	86.6	86.6	—	—
12.9	—	—	—	18.9	—	—	—	27.7	—	—	—	40.7	—	—	—	59.7	—	—	—	87.6	—	—	—
13.0	13.0	13	—	19.1	19.1	—	—	28.0	28.0	—	—	41.2	41.2	—	—	60.4	60.4	—	—	88.7	88.7	—	—
13.2	—	—	—	19.3	—	—	—	28.4	—	—	—	41.7	—	—	—	61.2	—	—	—	89.8	—	—	—
13.3	13.3	—	—	19.6	19.6	—	—	28.7	28.7	—	—	42.2	42.2	—	—	61.9	61.9	62	—	90.9	90.9	91	—
13.5	—	—	—	19.8	—	—	—	29.1	—	—	—	42.7	—	—	—	62.6	—	—	—	92.0	—	—	—
13.7	13.7	—	—	20.0	20.0	20	—	29.4	29.4	—	—	43.2	43.2	43	—	63.4	63.4	—	—	93.1	93.1	—	—
13.8	—	—	—	20.3	—	—	—	29.8	—	—	—	43.7	—	—	—	64.2	—	—	—	94.2	—	—	—
14.0	14.0	—	.	20.5	20.5	—	—	30.1	30.1	30	—	44.2	44.2	—	—	64.9	64.9	—	—	95.3	95.3	—	—
14.2	—	—	—	20.8	—	—	—	30.5	—	—	—	44.8	—	—	—	65.7	—	—	—	96.5	—	—	—
14.3	14.3	—	—	21.0	21.0	—	—	30.9	30.9	—	—	45.3	45.3	—	—	66.5	66.5	—	—	97.6	97.6	—	—
14.5	—	—	—	21.3	—	—	—	31.2	—	—	—	45.9	—	—	—	67.3	—	—	—	98.8	—	—	—

NOTE: These values are generally available in multiples of 0.1, 1, 10, 100, 1 k, and 1 M.

APPENDIX B

Capacitor Color Coding and Marking

Capacitor Color

Some capacitors have color-coded designations. The color code used for capacitors is basically the same as that used for resistors. Some variations occur in tolerance designation. The basic color codes are shown in Table B–1, and some typical color-coded capacitors are illustrated in Figure B–1.

TABLE B–1 • Typical composite color codes for capacitors (picofarads).

COLOR	DIGIT	MULTIPLIER	TOLERANCE
Black	0	1	20%
Brown	1	10	1%
Red	2	100	2%
Orange	3	1000	3%
Yellow	4	10000	
Green	5	100000	5% (EIA)
Blue	6	1000000	
Violet	7		
Gray	8		
White	9		
Gold		0.1	5% (JAN)
Silver		0.01	10%
No color			20%

NOTE: EIA stands for Electronic Industries Association, and JAN stands for Joint Army-Navy, a military standard.

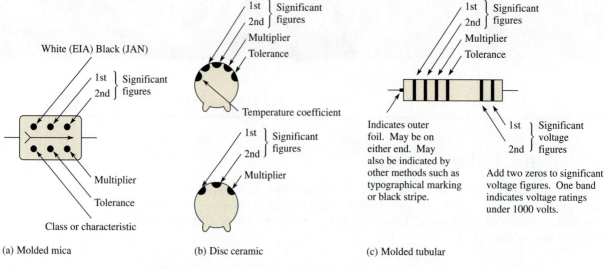

FIGURE B–1   Typical color-coded capacitors.

Marking Systems

A capacitor, as shown in Figure B–2, has certain identifying features.

- Body of one solid color (off-white, beige, gray, tan or brown).
- End electrodes completely enclose ends of part.
- Many different sizes:
  1. Type 1206: 0.125 inch long by 0.063 inch wide (3.2 mm × 1.6 mm) with variable thickness and color.
  2. Type 0805: 0.080 inch long by 0.050 inch wide (2.0 mm × 1.25 mm) with variable thickness and color.
  3. Variably sized with a single color (usually translucent tan or brown). Sizes range from 0.059 inch (1.5 mm) to 0.220 inch (5.6 mm) in length and in width from 0.032 inch (0.8 mm) to 0.197 inch (5.0 mm).
- Three different marking systems:
  1. Two place (letter and number only).
  2. Two place (letter and number or two numbers).
  3. One place (letter of varying color).

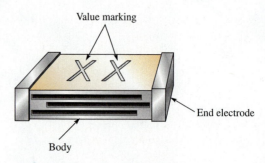

FIGURE B–2   Capacitor marking.

Standard Two-Place Code

Refer to Table B–2.

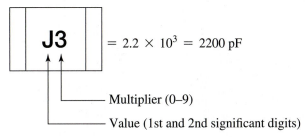

$= 2.2 \times 10^3 = 2200 \text{ pF}$

Multiplier (0–9)

Value (1st and 2nd significant digits)

Examples: $\text{S2} = 4.7 \times 100 = 470 \text{ pF}$
$\text{b0} = 3.5 \times 1.0 = 3.5 \text{ pF}$

TABLE B–2						MULTIPLIER
		VALUE*				**MULTIPLIER**
A	1.0	L	2.7	T	5.1	$0 = \times 1.0$
B	1.1	M	3.0	U	5.6	$1 = \times 10$
C	1.2	N	3.3	m	6.0	$2 = \times 100$
D	1.3	b	3.5	V	6.2	$3 = \times 1000$
E	1.5	P	3.6	W	6.8	$4 = \times 10000$
F	1.6	Q	3.9	n	7.0	$5 = \times 100000$
G	1.8	d	4.0	X	7.5	etc.
H	2.0	R	4.3	t	8.0	
J	2.2	e	4.5	Y	8.2	
K	2.4	S	4.7	y	9.0	
a	2.5	f	5.0	Z	9.1	

*Note uppercase and lowercase letters.

Alternate Two-Place Code

Refer to Table B–3.

- Values below 100 pF—Value read directly

`05` $= 5 \text{ pF}$ `82` $= 82 \text{ pF}$

- Values 100 pF and above—Letter/Number code

`A1` $= 10 \times 10 = 100 \text{ pF}$ `N3` $= 33 \times 1000$
$= 33000 \text{ pF} = .033 \ \mu\text{F}$

Multiplier (1–9)

Value (1st and 2nd significant digits)

TABLE B–3						MULTIPLIER
		VALUE*				**MULTIPLIER**
A	10	J	22	S	47	1 = ×10
B	11	K	24	T	51	2 = ×100
C	12	L	27	U	56	3 = ×1000
D	13	M	30	V	62	4 = ×10000
E	15	N	33	W	68	5 = ×100000
F	16	P	36	X	75	etc.
G	18	Q	39	Y	82	
H	20	R	43	Z	91	

*Note uppercase letters only.

Standard Single-Place Code

Refer to Table B–4.

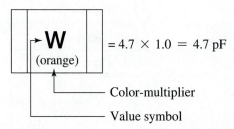

= 4.7 × 1.0 = 4.7 pF

Color-multiplier

Value symbol

Examples: R (Green) = 3.3 × 100 = 330 pF
7 (Blue) = 8.2 × 1000 = 8200 pF

TABLE B–4						MULTIPLIER (COLOR)
		VALUE				**MULTIPLIER (COLOR)**
A	1.0	K	2.2	W	4.7	Orange = ×1.0
B	1.1	L	2.4	X	5.1	Black = ×10
C	1.2	N	2.7	Y	5.6	Green = ×100
D	1.3	O	3.0	Z	6.2	Blue = ×1000
E	1.5	R	3.3	3	6.8	Violet = ×10000
H	1.6	S	3.6	4	7.5	Red = ×100000
I	1.8	T	3.9	7	8.2	
J	2.0	V	4.3	9	9.1	

APPENDIX C

Norton's Theorem and Millman's Theorem

Norton's Theorem

Like Thevenin's theorem, Norton's theorem provides a method of reducing a more complex circuit to a simpler form. The basic difference is that Norton's theorem gives an equivalent current source in parallel with an equivalent resistance. The form of Norton's equivalent circuit is shown in Figure C–1. Regardless of how complex the original circuit is, it can always be reduced to this equivalent form. The equivalent current source is designated I_N, and the equivalent resistance is designated R_N.

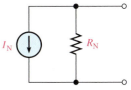

FIGURE C–1 The form of Norton's equivalent circuit.

To apply Norton's theorem, you must know how to find the two quantities I_N and R_N. Once you know them for a given circuit, simply connect them in parallel to get the complete Norton circuit.

Norton's Equivalent Current (I_N)   As stated, I_N is one part of the complete Norton equivalent circuit; R_N is the other part. I_N is defined to be the short-circuit current between two points in a circuit. Any component connected between these two points effectively "sees" a current source of value I_N in parallel with R_N.

To illustrate, suppose that a resistive circuit of some kind has a resistor connected between two points in the circuit, as shown in Figure C–2(a). We wish to find the Norton circuit that is equivalent to the one shown as "seen" by R_L. To find I_N, calculate the current between points A and B with these two points shorted, as shown in Figure C–2(b). Example C–1 demonstrates how to find I_N.

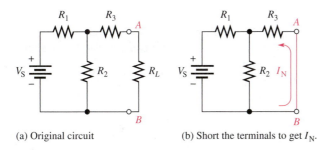

(a) Original circuit

(b) Short the terminals to get I_N.

FIGURE C–2 Determining the Norton equivalent current, I_N.

EXAMPLE C–1

Determine I_N for the circuit within the shaded area in Figure C–3(a).

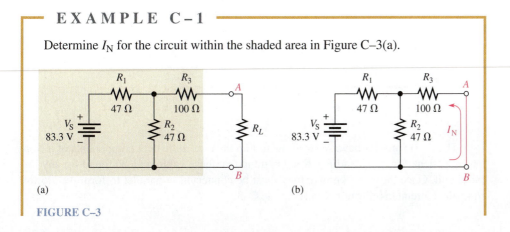

(a)

(b)

FIGURE C–3

SOLUTION

Short terminals A and B as shown in Figure C–3(b). I_N is the current through the short and is calculated as follows: First, the total resistance seen by the voltage source is

$$R_T = R_1 + \frac{R_2 R_3}{R_2 + R_3} = 47 \ \Omega + \frac{(47 \ \Omega)(100 \ \Omega)}{147 \ \Omega} = 79 \ \Omega$$

The total current from the source is

$$I_T = \frac{V_S}{R_T} = \frac{83.3 \ \text{V}}{79 \ \Omega} = 1.05 \ \text{A}$$

Now apply the current-divider formula to find I_N (the current through the short).

$$I_N = \left(\frac{R_2}{R_2 + R_3} \right) I_T = \left(\frac{47 \ \Omega}{147 \ \Omega} \right) 1.05 \ \text{A} = \mathbf{336 \ mA}$$

This is the value for the equivalent Norton current source.

Norton's Equivalent Resistance (R_N) We define R_N in the same way as R_{TH}: it is the total resistance appearing between two terminals in a given circuit with all sources replaced by their internal resistances. Example C–2 demonstrates how to find R_N.

EXAMPLE C–2

Find R_N for the circuit within the shaded area of Figure C–3(a) (see Example C–1).

SOLUTION

First reduce V_S to zero by shorting it, as shown in Figure C–4.

Looking in at terminals A and B, you can see that the parallel combination of R_1 and R_2 is in series with R_3. Thus,

$$R_N = R_3 + \frac{R_1}{2} = 100 \ \Omega + \frac{47 \ \Omega}{2} = \mathbf{124 \ \Omega}$$

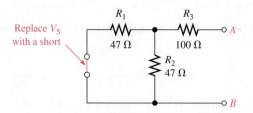

FIGURE C–4

The two examples have shown how to find the two equivalent components of a Norton equivalent circuit, I_N and R_N. Keep in mind that these values can be found for any linear circuit. Once these are known, they must be connected in parallel to form the Norton equivalent circuit, as illustrated in Example C–3.

EXAMPLE C – 3

Draw the complete Norton circuit for the original circuit in Figure C–3 (Example C–1).

SOLUTION

In Examples C–1 and C–2 you found that $I_N = 336$ mA and $R_N = 124 \,\Omega$. The Norton equivalent circuit is shown in Figure C–5.

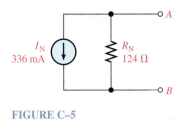

FIGURE C–5

Summary of Norton's Theorem   Any load resistor connected between the terminals of a Norton equivalent circuit will have the same current through it and the same voltage across it as if it were connected to the terminals of the original circuit. A summary of steps for theoretically applying Norton's theorem is as follows:

1. Short the two terminals between which you want to find the Norton equivalent circuit.

2. Determine the current (I_N) through the shorted terminals.

3. Determine the resistance (R_N) between the two terminals (opened) with all voltage sources shorted and all current sources opened ($R_N = R_{TH}$).

4. Connect I_N and R_N in parallel to produce the complete Norton equivalent for the original circuit.

Norton's equivalent circuit can also be derived from Thevenin's equivalent circuit by use of the source conversion method.

Millman's Theorem

Millman's theorem permits any number of parallel voltage sources to be reduced to a single equivalent voltage source. It simplifies finding the voltage across or current through a load. Millman's theorem gives the same results as Thevenin's theorem for the special case of parallel voltage sources. A conversion by Millman's theorem is illustrated in Figure C–6.

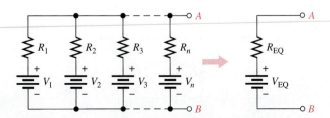

FIGURE C–6   **Reduction of parallel voltage sources to a single equivalent voltage source.**

Millman's Equivalent Voltage (V_{EQ}) and Equivalent Resistance (R_{EQ})

Millman's theorem gives us a formula for calculating the equivalent voltage, V_{EQ}. To find V_{EQ}, convert each of the parallel voltage sources into the current sources, as shown in Figure C–7.

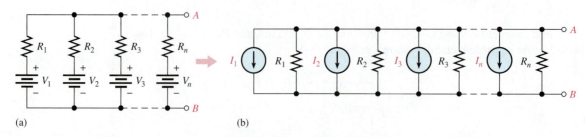

(a) (b)

FIGURE C–7 Parallel voltage sources converted to current sources.

In Figure C–7 (b), the total current from the parallel current sources is

$$I_T = I_1 + I_2 + I_3 + \cdots + I_n$$

The total conductance between terminals A and B is

$$G_T = G_1 + G_2 + G_3 + \cdots + G_n$$

where $G_T = 1/R_T$, $G_1 = 1/R_1$, and so on. Remember, the current sources are effectively open. Therefore, by Millman's theorem, the equivalent resistance, R_{EQ}, is the total resistance R_T.

$$R_{EQ} = \frac{1}{G_T} = \frac{1}{(1/R_1) + (1/R_2) + (1/R_3) + \cdots + (1/R_n)} \tag{C–1}$$

By Millman's theorem, the equivalent voltage is $I_T R_{EQ}$, where I_T is expressed as follows:

$$I_T = I_1 + I_2 + I_3 + \cdots + I_n = \frac{V_1}{R_1} + \frac{V_2}{R_2} + \frac{V_3}{R_3} + \cdots + \frac{V_n}{R_n}$$

The following is the formula for the equivalent voltage:

$$V_{EQ} = \frac{(V_1/R_1) + (V_2/R_2) + (V_3/R_3) + \cdots + (V_n/R_n)}{(1/R_1) + (1/R_2) + (1/R_3) + \cdots + (1/R_n)} \tag{C–2}$$

Equations C–1 and C–2 are the two Millman formulas. The equivalent voltage source has a polarity such that the total current through a load will be in the same direction as in the original circuit.

EXAMPLE C–4

Use Millman's theorem to find the voltage across R_L and the current through R_L in Figure C–8.

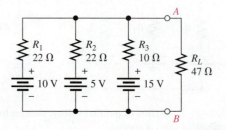

FIGURE C–8

SOLUTION

Apply Millman's theorem as follows:

$$R_{EQ} = \frac{1}{(1/R_1) + (1/R_2) + (1/R_3)}$$

$$= \frac{1}{(1/22\ \Omega) + (1/22\ \Omega) + (1/10\ \Omega)} = \frac{1}{0.19} = 5.24\ \Omega$$

$$V_{EQ} = \frac{(V_1/R_1) + (V_2/R_2) + (V_3/R_3)}{(1/R_1) + (1/R_2) + (1/R_3)}$$

$$= \frac{(10\ V/22\ \Omega) + (5\ V/22\ \Omega) + (15\ V/10\ \Omega)}{(1/22\ \Omega) + (1/22\ \Omega) + (1/10\ \Omega)} = \frac{2.18\ A}{0.19\ S} = 11.5\ V$$

The single equivalent voltage source is shown in Figure C–9.
Now calculate I_L and V_L for the load resistor.

$$I_L = \frac{V_{EQ}}{R_{EQ} + R_L} = \frac{11.5\ V}{52.2\ \Omega} = \textbf{220 mA}$$

$$V_L = I_L R_L = (220\ mA)(47\ \Omega) = \textbf{10.3 V}$$

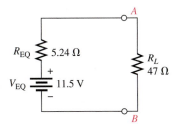

FIGURE C–9

APPENDIX D

NI Multisim for Circuit Simulation

Simulate, Prototype, and Test Circuits

Theory, Design, and Prototype

As electronic circuits and systems become more advanced, circuit designers rely on computers in the design process. It is essential for engineers and technicians to systematically design, simulate, prototype, and lay out the circuit. Students can also use the engineering and design process to reinforce concepts and theory in the classroom. The three stages of the student design process are illustrated in Figure D–1.

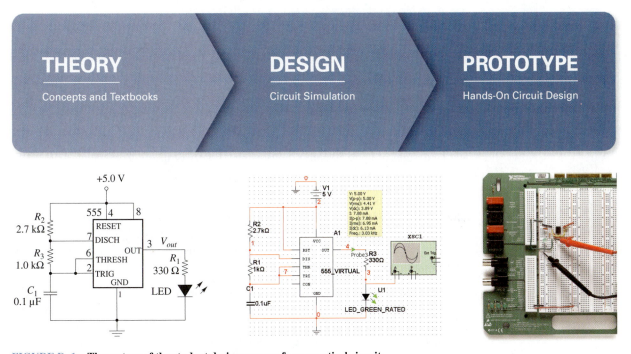

THEORY
Concepts and Textbooks

DESIGN
Circuit Simulation

PROTOTYPE
Hands-On Circuit Design

FIGURE D–1 **Three steps of the student design process for a practical circuit.**

The National Instruments Electronics Education Platform is an end-to-end tool-chain designed to meet student and educator needs. The platform consists of NI Multisim simulation software, the NI Educational Laboratory Virtual Instrumentation Suite (NI ELVIS) prototyping workstation, and the NI LabVIEW graphical programming environment. NI Multisim provides intuitive schematic capture, SPICE simulation, and integration with the NI ELVIS to help students explore circuit theory and design circuits to investigate behavior. NI ELVIS is a prototyping platform that allows students to quickly and easily create their circuits. With NI LabVIEW, students can measure real-world signals and compare simulation results.

1. Investigate Theory   Learn the fundamental theory of circuit design through the *Electric Circuits Fundamentals* textbook and course lectures. Reinforce important concepts in the easy-to-use Multisim environment by downloading the Multisim circuit files for the textbook. The Multisim circuit files help build the foundation for an in-depth understanding of circuit behavior. Using the prebuilt circuit files, simulate and analyze the circuit behavior of examples and problems in each chapter.

2. Design and Simulate   Circuit simulation provides an interactive view into a circuit. Build a circuit from scratch and learn about the design performance using built-in circuit instruments and probes in an ideal pre-laboratory environment. With the 3D breadboard in Multisim, take the leap from circuit diagram to real-world physical implementation.

3. Prototype, Measure, and Compare   Hands-on experience building physical circuitry is essential. Move from the 3D breadboard in Multisim to a real-world breadboard on NI ELVIS to seamlessly complete the entire design process by prototyping the circuit. Within the LabVIEW environment, compare real-world measurements to the simulation values to reinforce theory, fully understand circuit behavior, and build the foundations of professional engineering analysis.

NI Multisim

Multisim software integrates powerful SPICE simulation and schematic capture into a highly intuitive electronics lab on the computer. Use the Multisim Student Edition to enhance learning by:

Building circuits in a simple, easy-to-learn environment

Simulating and analyzing circuits for homework and prelab assignments

Breadboarding in 3D at home before lab sessions

Creating designs of up to 50 components using a library of nearly 4,000 devices

Integrating with the NI ELVIS prototyping workstation

While using industry-standard SPICE in the background, you can take advantage of the Multisim drag-and-drop interface to make circuit drawing, wiring, and analysis simple and easy to use. You can build circuits from scratch and learn about design performances using built-in virtual instruments and probes in an ideal environment for experimentation. With the 3D breadboard, you can easily take the leap from circuit diagram to real-world implementation.

Using Multisim with Electric Circuits Fundamentals   Create an example from the textbook to get familiar with the Multisim environment. First, launch Multisim and open a new schematic window (**File ≫ New ≫ Schematic Capture**). Design circuits in the circuit window by placing components from the component toolbar. Clicking on the component toolbar opens the component browser. Choose the family of components and select an individual component to place on the circuit window by double-clicking on it.

Once you have selected a component, it attaches itself and "ghosts" the mouse cursor. Place the component by clicking again on the desired location in the schematic. If you are new to Multisim, you should use the **BASIC_VIRTUAL** family of components, which you can assign any arbitrary value. As an example of a simple Multisim circuit, an *RLC* circuit is shown in Figure D–2. The circuit connects a 1 kΩ resistor, a 47 mH inductor, and a 10 nF capacitor. All these components are found in the Basic Group of the Component Database.

The next step is to wire the components together. Simply left-click on the source terminal and left-click on the destination terminal. Multisim automatically chooses the best path for the virtual wire between the two terminals. Always be sure you have fully captured your circuit; then you can simulate it. You can use your simulation results as a comparison with the physical circuit.

To analyze the circuit, use a measurement probe to measure the voltages and other characteristics from the circuit while the simulation is running. Use the virtual oscilloscope to analyze the output signal from our sample *RLC* circuit.

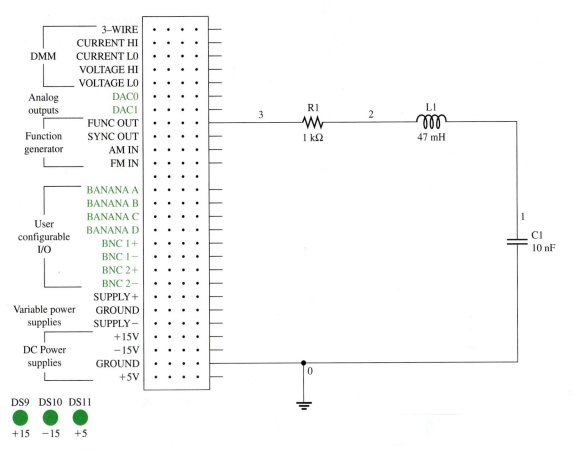

NI ELVIS

NI ELVIS is a LabVIEW-based design and prototyping environment based on LabVIEW for university science and engineering laboratories. NI ELVIS consists of virtual instruments, or functions, based on LabVIEW; a multifunction data acquisition device; and a custom-designed benchtop workstation and prototyping board. This combination provides a ready-to-use suite of instruments for educational laboratories. The NI ELVIS workstation is shown in Figure D–3.

FIGURE D–3   **The NI ELVIS workstation.**

NI ELVIS is an ideal companion to any electronics lab that uses Multisim. It features USB connectivity for easy setup, maintenance, and portability, as well as a portable breadboard for real-world circuit prototyping.

NI ELVIS features a breadboard prototyping environment with built-in instruments including a function generator, digital multimeter (DMM), oscilloscope, and variable power supplies. The breadboard is detachable, so you can work on your projects and labs independently of the NI ELVIS unit. NI ELVIS provides software based on LabVIEW for interacting with virtual instruments.

Using NI ELVIS with Electric Circuits Fundamentals   Returning to the *RLC* circuit introduced with Multisim as an example, you can continue the prototyping step. With prototyping hardware, you can quickly construct a circuit on a standard protoboard and test it with the laboratory instruments to complete the design.

Alternatively, you can repeat the simulation step but this time as a prototype using the Virtual 3D NI ELVIS from within Multisim. To construct a 3D NI ELVIS prototype, open the 3D breadboard by clicking **Tools ≫ Show Breadboard**. Place components and wires to build up your circuit. The corresponding connection points and symbols on the NI ELVIS schematic turn green, indicating the 3D connections are correct. If you create a traditional schematic, you see a standard breadboard. Figure D–4 shows the circuit on 3D Virtual NI ELVIS prototyped in Multisim and ready for construction. Once you have verified the layout of your circuit using the 3D virtual environment, you can physically build it on NI ELVIS.

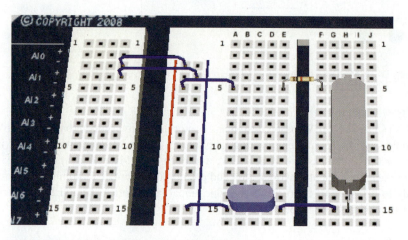

FIGURE D–4   Prototyped circuit in Multisim.

After constructing the circuit, you can use the NI ELVIS instrument launcher to measure the output signal of this circuit. Figure D–5 shows the simulated signal of the *RLC*

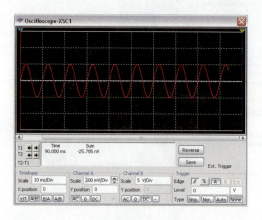

FIGURE D–5   Multisim simulated signal.

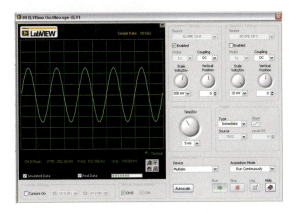

FIGURE D–6   **NI ELVIS measured signal.**

circuit on the Multisim oscilloscope. Figure D–6 shows the measured signal using the ELVIS II Oscilloscope. The most important step in the process is to compare the measurement of the prototyped circuit to the simulation. This helps you determine where potential errors exist in the design. After comparing the measurement with the theoretical values, you can revisit your design to improve it or prepare it for layout on a PCB board, using software such as NI Ultiboard.

NI Multisim Circuit Files   Download the Multisim circuit files to develop in-depth understanding of circuit behavior. To download the pre-built circuit files and Multisim resources, visit: ni.com/academic/floyd

NI Multisim Resources   The link provides references to the following resources to help you get started with NI Multisim:

Download the free Multisim 30-day evaluation version

View the Getting Started Guide to Multisim

Learn Multisim in 3-Hours tutorial

Discuss Multisim in an online discussion forum

ANSWERS TO ODD-NUMBERED PROBLEMS

Chapter 1

1. The circuit is first tested with a computer design and simulation program, which can simulate the performance and look for potential problems. When the simulation is satisfactory, a prototype circuit is constructed, tested, and modified as needed before putting it into production.

3. Electronic assemblies have become more complex but also more reliable, so there is less need for repair. It is generally cheaper for manufacturers to replace a board than troubleshoot it to the component level. Skills needed by technicians tend to be broader skills than in the past.

5. Advantages are that the digital signal can be processed and stored easily; it is also less subject to noise.

7. **(a)** An electronic oscillator generates a repetitive electronic signal.

   (b) An oscillator does not have a signal input.

9. A carrier is a high frequency radio wave that can be modulated (changed) by a lower frequency signal.

11. **(a)** 3×10^3 **(b)** 7.5×10^4 **(c)** 2×10^6

13. **(a)** 8.4×10^3 **(b)** 9.9×10^4 **(c)** 2×10^5

15. **(a)** 0.0000025 **(b)** 500 **(c)** 0.39

17. **(a)** 4.32×10^7

   (b) 5.00085×10^3

   (c) 6.06×10^{-8}

19. **(a)** 2.0×10^9 **(b)** 3.6×10^{14} **(c)** 1.54×10^{-14}

21. **(a)** 89×10^3 **(b)** 450×10^3 **(c)** 12.04×10^{12}

23. **(a)** 345×10^{-6} **(b)** 25×10^{-3} **(c)** 1.29×10^{-9}

25. **(a)** 7.1×10^{-3} **(b)** 101×10^6 **(c)** 1.50×10^6

27. **(a)** 22.7×10^{-3} **(b)** 200×10^6 **(c)** 848×10^{-3}

29. **(a)** $345\,\mu A$ **(b)** $25\,mA$ **(c)** $1.29\,nA$

31. **(a)** $3\,\mu F$ **(b)** $3.3\,M\Omega$ **(c)** $350\,nA$

33. **(a)** $5000\,\mu A$ **(b)** $3.2\,mW$

   (c) $5\,MV$ **(d)** $10,000\,kW$

35. **(a)** $50.68\,mA$ **(b)** $2.32\,M\Omega$ **(c)** $0.0233\,\mu F$

37. **(a)** 3 **(b)** 2 **(c)** 5 **(d)** 2 **(e)** 3 **(f)** 2

Chapter 2

1. $80 \times 10^{12}\,C$

3. $4.64 \times 10^{-18}\,C$

5. **(a)** $10\,V$ **(b)** $2.5\,V$ **(c)** $4\,V$

7. $20\,V$

9. $12.5\,V$

11. **(a)** 75 A **(b)** 20 A **(c)** 2.5 A

13. 2 s

15. A: $6800\,\Omega \pm 10\%$

   B: $33\,\Omega \pm 10\%$

   C: $47,000\,\Omega \pm 5\%$

17. **(a)** Red, violet, brown, gold

   (b) B: $330\,\Omega$, D: $2.2\,k\Omega$, A: $39\,k\Omega$, L: $56\,k\Omega$, F: $100\,k\Omega$

19. **(a)** $10\,\Omega \pm 5\%$

   (b) $5.1\,M\Omega \pm 10\%$

   (c) $68\,\Omega \pm 5\%$

21. **(a)** $28.7\,k\Omega \pm 1\%$

   (b) $60.4\,\Omega \pm 1\%$

   (c) $9.31\,k\Omega \pm 1\%$

23. **(a)** $22\,\Omega$ **(b)** $4.7\,k\Omega$ **(c)** $82\,k\Omega$

   (d) $3.3\,k\Omega$ **(e)** $56\,\Omega$ **(f)** $10\,M\Omega$

25. There is current through lamp 2.

27. Ammeter in series with resistors, with its negative terminal to the negative terminal of source and its positive terminal to one side of R_1. Voltmeter placed across (in parallel with) the source (negative to negative, positive to positive).

29. Position 1: $V1 = 0$, $V2 = V_S$

   Position 2: $V1 = V_S$, $V2 = 0$

31. 250 V

33. **(a)** $200\,\Omega$ **(b)** $150\,M\Omega$ **(c)** $4500\,\Omega$

35. 33.3 V

37. AWG #27

39. Circuit (b)

41. One ammeter in series with battery. One ammeter in series with each resistor (seven total).

43. See Figure P–1.

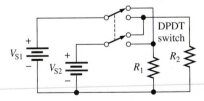

FIGURE P–1

Chapter 3

1. **(a)** 3 A **(b)** 0.2 A **(c)** 1.5 A

3. 15 mA

5. **(a)** $3.33\,mA$ **(b)** $550\,\mu A$ **(c)** $588\,\mu A$

   (d) $500\,mA$ **(e)** $6.60\,mA$

7. (a) 2.50 mA   (b) 2.27 μA   (c) 8.33 mA

9. $I = 0.642$ A, so the 0.5 A fuse will blow.

11. (a) 10 mV   (b) 1.65 V   (c) 14.1 kV

    (d) 3.52 V   (e) 250 mV   (f) 750 kV

    (g) 8.5 kV   (h) 3.53 mV

13. (a) 81 V   (b) 500 V   (c) 117.5 V

15. (a) 2 kΩ   (b) 3.5 kΩ   (c) 2 kΩ

    (d) 100 kΩ   (e) 1 MΩ

17. (a) 4 Ω   (b) 3 kΩ   (c) 200 kΩ

19. 2.6 W

21. 417 mW

23. (a) 1 MW   (b) 3 MW   (c) 150 MW   (d) 8.7 MW

25. (a) 2,000,000 μW   (b) 500 μW

    (c) 250 μW   (d) 6.67 μW

27. $P = W/t$ in watts; $V = W/Q$; $I = Q/t$. $P = VI = W/t$, so
    (1 V)(1 A) = 1 watt

29. 16.5 mW

31. 1.18 kW

33. 5.81 W

35. 25 Ω

37. 0.00186 kWh

39. 156 mW

41. 1 W

43. (a) positive at top   (b) positive at bottom

    (c) positive at right

45. 36 Ah

47. 13.5 mA

49. 4.25 W

51. Five

53. 150 Ω

55. $V = 0$ V, $I = 0$ A; $V = 10$ V, $I = 100$ mA;
    $V = 20$ V, $I = 200$ mA; $V = 30$ V, $I = 300$ mA;
    $V = 40$ V, $I = 400$ mA; $V = 50$ V, $I = 500$ mA;
    $V = 60$ V, $I = 600$ mA; $V = 70$ V, $I = 700$ mA;
    $V = 80$ V, $I = 800$ mA; $V = 90$ V, $I = 900$ mA;
    $V = 100$ V, $I = 1$ A

57. $R_1 = 0.5$ Ω; $R_2 = 1$ Ω; $R_3 = 2$ Ω

59. 10 V; 30 V

61. 216 kWh

63. 12 W

65. 2.5 A

67. Power will increase by four times.

69. See Figure P–2.

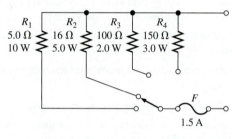

FIGURE P–2

71. No fault

73. Lamp 4 is shorted.

Chapter 4

1. See Figure P–3.

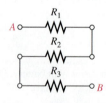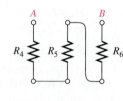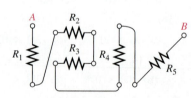

FIGURE P–3

3. 170 kΩ

5. 138 Ω

7. (a) 7.9 kΩ   (b) 33 Ω   (c) 13.24 MΩ

   The series circuit is disconnected from the source, and the
   ohmmeter is connected across the circuit terminals.

9. 1126 Ω

11. (a) 170 kΩ   (b) 50 Ω

    (c) 12.4 kΩ   (d) 1.97 kΩ

13. 0.1 A

15. (a) 625 μA

    (b) 4.26 μA. The ammeter is connected in series.

17. (a) 34.0 mA   (b) 16 V   (c) 0.543 W

19. See Figure P–4.

FIGURE P–4

21. 26 V

23. (a) $V_2 = 6.8$ V

    (b) $V_R = 8$ V, $V_{2R} = 16$ V, $V_{3R} = 24$ V, $V_{4R} = 32$ V

   The voltmeter is connected across (in parallel with) each
   resistor for which the voltage is unknown.

25. (a) 3.84 V   (b) 6.77 V

27. 3.80 V; 9.38 V

29. $V_{5.6\,k\Omega} = 10$ V; $V_{1\,k\Omega} = 1.79$ V; $V_{560\,\Omega} = 1$ V;
$V_{10\,k\Omega} = 17.9$ V

31. 55 mW

33. Measure V_A and V_B individually with respect to ground;
then $V_{R2} = V_A - V_B$.

35. 4.27 V

37. **(a)** R_4 open   **(b)** R_4 and R_5 shorted

39. 780 Ω

41. $V_A = 10$ V; $V_B = 7.72$ V; $V_C = 6.68$ V;
$V_D = 1.81$ V; $V_E = 0.57$ V; $V_F = 0$ V

43. 500 Ω

45. **(a)** 19.1 mA   **(b)** 45.8 V

   (c) $R(\frac{1}{8}\,W) = 343\,\Omega$, $R(\frac{1}{4}\,W) = 686\,\Omega$,
   $R(\frac{1}{2}\,W) = 1371\,\Omega$

47. See Figure P–5.

FIGURE P–5

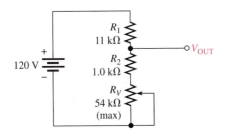

49. $R_1 + R_7 + R_8 + R_{10} = 4.23$ kΩ;
$R_2 + R_4 + R_6 + R_{11} = 23.6$ kΩ;
$R_3 + R_5 + R_9 + R_{12} = 19.9$ kΩ

51. A: 5.45 mA; B: 6.06 mA; C: 7.95 mA; D: 12 mA

53. A: $V_1 = 6.03$ V, $V_2 = 3.35$ V, $V_3 = 2.75$ V,
   $V_4 = 1.88$ V, $V_5 = 4.0$ V;

   B: $V_1 = 6.71$ V, $V_2 = 3.73$ V, $V_3 = 3.06$ V, $V_5 = 4.5$ V;

   C: $V_1 = 8.1$ V, $V_2 = 4.5$ V, $V_5 = 5.4$ V;

   D: $V_1 = 10.8$ V, $V_5 = 7.2$ V

55. Yes, R_3 and R_5 are shorted.

57. **(a)** R_{11} has burned open due to excessive power.

   (b) Replace R_{11} (10 kΩ).   **(c)** 338 V

59. $I = 375$ mA

61. $I_T = 5.0$ A

63. R_6 is shorted.

65. Lamp 4 is open.

67. The 82 Ω resistor is shorted.

Chapter 5

1. See Figure P–6.

FIGURE P–6

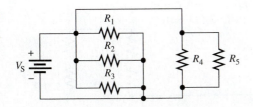

3. 3.43 kΩ

5. **(a)** 25.6 Ω **(b)** 359 Ω **(c)** 819 Ω **(d)** 996 Ω

7. 2 kΩ

9. 12 V; 5 mA

11. **(a)** 909 μA   **(b)** 76 mA

13. **(a)** $I_1 = 179\,\mu$A; $I_2 = 455\,\mu$A

   (b) $I_1 = 444\,\mu$A; $I_2 = 80\,\mu$A

15. 1350 mA

17. $I_2 = I_3 = 7.5$ mA. Connect an ammeter in series with each
resistor in each branch.

19. 6.4 A; 6.4 A

21. $I_1 = 2.19$ A; $I_2 = 811$ mA

23. 200 mW

25. 0.625 A; 3.75 A

27. The 1.0 kΩ resistor is open.

29. R_2 is open.

31. $R_2 = 25\,\Omega$; $R_3 = 100\,\Omega$; $R_4 = 12.5\,\Omega$

33. $I_R = 4.8$ A; $I_{2R} = 2.4$ A; $I_{3R} = 1.6$ A; $I_{4R} = 1.2$ A

35. **(a)** $R_1 = 100\,\Omega$, $R_2 = 200\,\Omega$, $I_2 = 50$ mA

   (b) $I_1 = 125$ mA, $I_2 = 74.9$ mA, $R_1 = 80\,\Omega$,
   $R_2 = 134\,\Omega$, $V_S = 10$ V

   (c) $I_1 = 253$ mA, $I_2 = 147$ mA, $I_3 = 100$ mA,
   $R_1 = 395\,\Omega$

37. 53.7 Ω

39. Yes; total current = 14.7 A

41. $R_1 \| R_2 \| R_5 \| R_9 \| R_{10} \| R_{12} = 100$ k$\Omega \| 220$ k$\Omega \| 560$ k$\Omega \|$
390 k$\Omega \| 1.2$ M$\Omega \| 100$ k$\Omega = 33.6$ kΩ

   $R_4 \| R_6 \| R_7 \| R_8 = 270$ k$\Omega \| 1.0$ M$\Omega \| 820$ k$\Omega \| 680$ k$\Omega =$
   135.2 kΩ

   $R_3 \| R_{11} = 330$ k$\Omega \| 1.8$ M$\Omega = 278.9$ kΩ

43. $R_2 = 750\,\Omega$; $R_4 = 423\,\Omega$

45. The 4.7 kΩ resistor is open.

47. **(a)** 75.0 Ω **(b)** 26.6 W

49. 8.9 Ω

51. R_3 is open.

53. **(a)** R from pin 1 to pin 4 agrees with calculated value.

   (b) R from pin 2 to pin 3 agrees with calculated value.

Chapter 6

1. R_2, R_3, and R_4 are in parallel, and this parallel combination
is in series with both R_1 and R_5.

3. See Figure P–7.

5. 2003 Ω

7. **(a)** 128 Ω **(b)** 791 Ω

9. **(a)** $I_1 = I_4 = 11.7$ mA, $I_2 = I_3 = 5.85$ mA;
   $V_1 = 655$ mV, $V_2 = V_3 = 585$ mV, $V_4 = 257$ mV

   (b) $I_1 = 3.8$ mA, $I_2 = 618\,\mu$A, $I_3 = 1.27$ mA,
   $I_4 = 1.91$ mA; $V_1 = 2.58$ V,
   $V_2 = V_3 = V_4 = 420$ mV

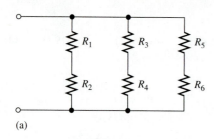

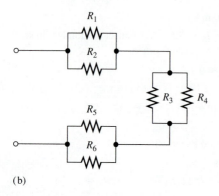

(a)

(b)

FIGURE P–7

11. 2.22 mA

13. 7.5 V unloaded; 7.29 V loaded

15. 56 kΩ load

17. 22 kΩ

19. 2 V

21. 33%

23. 360 Ω

25. 7.33 kΩ

27. $R_{TH} = 18\ k\Omega$; $V_{TH} = 2.7\ V$

29. 1.06 V; 226 μA

31. 75 Ω

33. 21 mA

35. No, the meter should read 4.39 V. The 680 Ω resistor is open.

37. The 7.62 V and 5.24 V readings are incorrect, indicating that the 3.3 kΩ resistor is open.

39. (a) $V_1 = -10\ V$, all others 0 V

    (b) $V_1 = -2.33\ V$, $V_4 = -7.67\ V$, $V_2 = -7.67\ V$, $V_3 = 0\ V$

    (c) $V_1 = -2.33\ V$, $V_4 = -7.67\ V$, $V_2 = 0\ V$, $V_3 = -7.67\ V$

    (d) $V_1 = -10\ V$, all others 0 V

41. See Figure P–8.

43. $R_T = 5.76\ k\Omega$; $V_A = 3.3\ V$; $V_B = 1.7\ V$; $V_C = 850\ mV$

45. $V_1 = 1.61\ V$; $V_2 = 6.77\ V$; $V_3 = 1.72\ V$; $V_4 = 3.33\ V$;
$V_5 = 378\ mV$; $V_6 = 2.57\ V$; $V_7 = 378\ mV$; $V_8 = 1.72\ V$;
$V_9 = 1.61\ V$

47. 110 kΩ

49. $R_1 = 180\ \Omega$; $R_2 = 60\ \Omega$. Output across R_2.

51. 845 μA

53. Pos 1: $V_1 = 88.0\ V$, $V_2 = 58.7\ V$, $V_3 = 29.3\ V$

    Pos 2: $V_1 = 89.1\ V$, $V_2 = 58.2\ V$, $V_3 = 29.1\ V$

    Pos 3: $V_1 = 89.8\ V$, $V_2 = 59.6\ V$, $V_3 = 29.3\ V$

55. $V_G = 0\ V$

57. The 2.2 kΩ resistor is open.

59. $V_B = 12.0\ V$

61. $V_{4mA} = 0.60\ V$; $V_{20mA} = 3.0\ V$

63. Three load cells can be evenly loaded because three points define a plane.

65. R_2 is shorted.

67. No fault

69. R_4 is shorted.

71. R_5 is shorted.

Chapter 7

1. Decrease

3. 37.5 μWb

5. 1000 G

7. 597

9. 1500 At

11. (a) Electromagnetic force   (b) Spring force

13. Electromagnetic force

15. Change the current.

17. Material A

19. 1 mA

21. 3.02 m/s

23. 56.3 W

25. (a) 168 W   (b) 14 W

27. 80.6%

29. The output voltage has a 10 V dc peak with a 120 Hz ripple.

31. The coil is energized because the movable contacts will be in the lower position for the polarity to be reversed.

FIGURE P–8

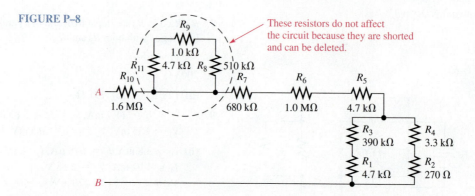

33. Design is flawed. Twelve volts is not enough voltage to operate two 12 V relays in series; 24 V is too much voltage to operate a 12 V lamp. Install a separate 12 V power supply for the lamp and change the 12 V to 24 V for the relays.

Chapter 8

1. (a) 1 Hz **(b)** 5 Hz **(c)** 20 Hz

(d) 1 kHz **(e)** 2 kHz **(f)** 100 kHz

3. $2 \, \mu s$

5. 10 ms

7. (a) 7.07 mA **(b)** 4.5 mA **(c)** 14.14 mA

9. (a) 17.7 V **(b)** 25 V **(c)** 0 V **(d)** -17.7 V

11. 15°, A leads.

13. See Figure P–9.

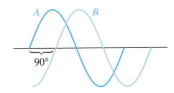

FIGURE P–9

15. (a) 22.5° **(b)** 60° **(c)** 90°

(d) 108° **(e)** 216° **(f)** 324°

17. (a) 57.4 mA **(b)** 99.6 mA **(c)** -17.4 mA

(d) -57.4 mA **(e)** -99.6 mA **(f)** 0 mA

19. 30°: 13.0 V; 45°: 14.5 V; 90°: 13.0 V; 180°: -7.5 V; 200°: -11.5 V; 300°: -7.5 V

21. (a) 7.07 mA **(b)** 0 A **(c)** 10 mA

(d) 20 mA **(e)** 10 mA

23. 7.38 V

25. 4.24 V

27. 250 Hz

29. 200 rps

31. A one phase motor requires a starting winding or other means to produce torque for starting the motor, whereas a three phase motor is self starting.

33. $t_r \cong 3.0$ ms; $t_f \cong 3.0$ ms; $t_W \cong 12.0$ ms; Ampl. $= 5$ V

35. (a) -0.375 V **(b)** 3.01 V

37. (a) 50 kHz **(b)** 10 Hz

39. 25 kHz

41. 0.424 V; 2 Hz

43. 1.4 V; 120 ms; 30%

45. A modulating signal is a lower frequency waveform that modifies a higher frequency waveform. It can be applied from an external source or it can be generated internally.

47. Burst mode outputs a specific number of cycles of a waveform; gated mode outputs the waveform for a set period of time as set by a gate signal.

49. $I_{max} = 2.38$ A; $V_{avg} = 136$ V; See Figure P–10.

FIGURE P–10 V_L (V)

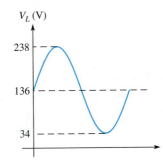

51. (a) 2.5 **(b)** 3.96 V **(c)** 12.5 kHz

53. See Figure P–11.

FIGURE P–11

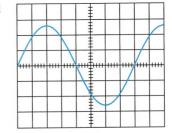

55. $V_{p(in)} = 4.44$ V; $f_{in} = 2$ Hz

57. R_3 is open.

59. 5 V; 1 ms

Chapter 9

1. (a) $5 \, \mu F$ **(b)** $1 \, \mu C$ **(c)** 10 V

3. (a) $0.001 \, \mu F$ **(b)** $0.0035 \, \mu F$ **(c)** $0.00025 \, \mu F$

5. $2 \, \mu F$

7. 88.5 pF

9. $0.0249 \, \mu F$

11. A 12.5 pF increase

13. Ceramic

15. (a) $0.022 \, \mu F$ **(b)** $0.047 \, \mu F$

(c) $0.001 \, \mu F$ **(d)** 22 pF

17. (a) Encapsulation **(b)** Dielectric (ceramic disk)

(c) Plate (metal disk) **(d)** Leads

19. (a) $0.69 \, \mu F$ **(b)** 69.7 pF **(c)** $2.6 \, \mu F$

21. $V_1 = 2.13$ V; $V_2 = 10$ V; $V_3 = 4.55$ V; $V_4 = 1$ V

23. $5.5 \, \mu F$, $27.5 \, \mu C$

25. (a) $100 \, \mu s$ **(b)** $560 \, \mu s$ **(c)** $22.1 \, \mu s$ **(d)** 15 ms

27. (a) 9.48 V **(b)** 13.0 V **(c)** 14.3 V

(d) 14.7 V **(e)** 14.9 V

29. (a) 2.72 V **(b)** 5.90 V **(c)** 11.7 V

31. (a) $339 \, k\Omega$ **(b)** $13.5 \, k\Omega$

(c) $677 \, \Omega$ **(d)** $33.9 \, \Omega$

33. $X_{C1} = 1.42 \, k\Omega$; $X_{C2} = 970 \, \Omega$; $X_{CT} = 2.39 \, k\Omega$ $V_1 = 5.94$ V; $V_2 = 4.06$ V

35. 200 Ω

37. $P_{true} = 0$ W; $P_r = 3.39$ mVAR

39. 0 Ω

41. 3.18 ms

43. 3.24 μs

45. **(a)** Charges to 3.32 V in 10 ms, then discharges to 0 V in 215 ms.

  (b) Charges to 3.32 V in 10 ms, then discharges to 2.96 V in 5 ms, then charges toward 20 V.

47. $C_1 = 1.25$ nF

49. 16 Hz

51. C_2 is open.

53. No fault

Chapter 10

1. 8 kHz; 8 kHz

3. **(a)** 288 Ω   **(b)** 1209 Ω

5. **(a)** 726 kΩ   **(b)** 155 kΩ

  (c) 91.5 kΩ   **(d)** 63.0 kΩ

7. **(a)** 34.7 mA   **(b)** 4.14 mA

9. $I_{tot} = 12.3$ mA; $V_{C1} = 1.31$ V; $V_{C2} = 0.595$ V; $V_R = 0.616$ V; $\theta = 72.0°$ (V_s lagging I_{tot})

11. 808 Ω; −36.1°

13. **(a)** 90°   **(b)** 86.4°   **(c)** 57.8°   **(d)** 9.04°

15. 326 Ω; 64.3°

17. 245 Ω; 80.5°

19. $I_{C1} = 118$ mA; $I_{C2} = 55.3$ mA; $I_{R1} = 36.4$ mA; $I_{R2} = 44.4$ mA; $I_{tot} = 191$ mA; $\theta = 65.0°$ (V_s lagging I_{tot})

21. **(a)** 3.86 kΩ   **(b)** 21.3 μA   **(c)** 14.8 μA

  (d) 25.9 μA   **(e)** 34.8° (V_s lagging I_{tot})

23. $V_{C1} = 8.74$ V; $V_{C2} = 3.26$ V; $V_{C3} = 3.26$ V; $V_{R1} = 2.11$ V; $V_{R2} = 1.15$ V;

25. $I_{tot} = 82.4$ mA; $I_{C2} = 14.4$ mA; $I_{C3} = 67.6$ mA; $I_{R1} = I_{R2} = 6.39$ mA

27. 4.03 VA

29. 0.915

31. Use the formula, $V_{out} = \left(\dfrac{X_C}{Z_{tot}}\right) 1$ V. See Figure P–12.

FREQUENCY (KHZ)	X_C (kΩ)	Z_{tot} (kΩ)	V_{out} (V)
0			1.000
1	4.08	5.64	0.723
2	2.04	4.40	0.464
3	1.36	4.13	0.329
4	1.02	4.03	0.253
5	0.816	3.98	0.205
6	0.680	3.96	0.172
7	0.583	3.94	0.148
8	0.510	3.93	0.130
9	0.453	3.93	0.115
10	0.408	3.92	0.104

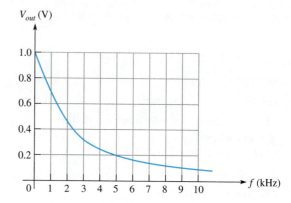

FIGURE P–12

33. See Figure P–13.

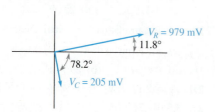

(a) For Figure 10–67

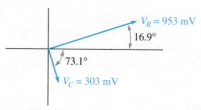

(b) For Figure 10–68

FIGURE P–13

35. Figure 10–67: 1.05 kHz; Figure 10–68: 1.59 kHz

37. No leakage: $V_{out} = 3.21$ V; $\theta = 18.7°$;

  Leakage: $V_{out} = 2.83$ V, $\theta = 33.3°$

39. **(a)** 0 V   **(b)** 0.321 V

  (c) 0.5 V   **(d)** 0 V

41. **(a)** $I_{L(A)} = 4.8$ A; $I_{L(B)} = 3.33$ A

  (b) $P_{r(A)} = 606$ VAR; $P_{r(B)} = 250$ VAR

  (c) $P_{true(A)} = 979$ W; $P_{true(B)} = 758$ W

  (d) $P_{a(A)} = 1151$ VA; $P_{a(B)} = 798$ VA

43. 11.4 kΩ

45. $P_r = 1.32$ kVAR; $P_a = 2$ kVA

47. 0.103 μF

49. C is leaky.

51. No fault

53. R_2 is open.

55. Phase shift with C_1 shorted $= 13.7°$

Chapter 11

1. **(a)** 1000 mH   **(b)** 0.25 mH

   (c) 0.01 mH   **(d)** 0.5 mH

3. 3450 turns

5. 50 mJ

7. 155 μH

9. 7.14 μH

11. **(a)** 50 mH   **(b)** 57 μH

13. **(a)** 1 μs   **(b)** 2.13 μs   **(c)** 200 μs

15. **(a)** 5.52 V   **(b)** 2.03 V   **(c)** 0.747 V

   (d) 0.275 V   **(e)** 0.101 V

17. **(a)** 157 kΩ **(b)** 179 Ω

19. $I_{tot} = 10.1$ mA; $I_{L2} = 6.7$ mA; $I_{L3} = 3.37$ mA

21. 101 mVAR

23. **(a)** -3.35 V   **(b)** -1.12 V   **(c)** -0.37 V

25. **(a)** 0.427 mA   **(b)** 0.569 mA

27. $X_{L1} = 69\ \Omega; X_{L2} = 1037\ \Omega; X_{LT} = 1106\ \Omega$

29. L_3 is open.

31. No fault

33. L_3 is shorted.

Chapter 12

1. 15 kHz

3. **(a)** 1.12 kΩ **(b)** 1.8 kΩ

5. **(a)** 17.4 Ω **(b)** 64 Ω **(c)** 127 Ω **(d)** 251 Ω

7. 100 kHz

9. **(a)** 8.94 mA   **(b)** 2.77 mA

11. 38.7°

13. See Figure P–14.

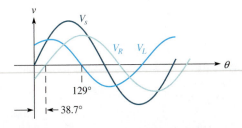

FIGURE P–14

15. 7.69 Ω

17. 537 μS

19. 2.39 kHz

21. **(a)** 274 Ω **(b)** 89.3 mA   **(c)** 159 mA

   (d) 183 mA   **(e)** 60.7° (I_{tot} lagging V_s)

23. $V_{R1} = 7.92$ V; $V_{R2} = V_L = 20.8$ V

25. $I_{tot} = 36$ mA; $I_L = 33.2$ mA; $I_{R2} = 13.9$ mA

27. 13.0 mW; 10.4 mVAR

29. $PF = 0.386; P_{true} = 347$ mW; $P_r = 692$ mVAR;
   $P_a = 900$ mVA

31. See Figure P–15.

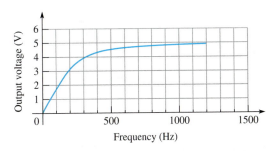

FIGURE P–15

33. $V_{R1} = V_{L1} = 18$ V; $V_{R2} = V_{R3} = V_{L2} = 0$ V

35. 5.57 V

37. 343 mA

39. **(a)** 405 mA   **(b)** 228 mA

   (c) 333 mA   **(d)** 335 mA

41. L_2 is open.

43. R_2 is open.

45. L_1 is shorted.

Chapter 13

1. 520 Ω, 88.9° (V_s lagging I); 520 Ω capacitive

3. Impedance increases.

5. See Figure P–16.

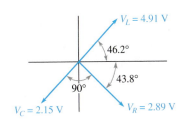

FIGURE P–16

7. f_r is lower.

9. $X_L = 4.61$ k$\Omega; X_C = 4.61$ k$\Omega; Z = 220\ \Omega; I = 54.5$ mA

11. $f_r = 454$ kHz; $f_1 = 416$ kHz; $f_2 = 492$ kHz

13. **(a)** 14.5 kHz; band-pass   **(b)** 24.0 kHz; band-pass

15. **(a)** $f_r = 339$ kHz, $BW = 239$ kHz

   (b) $f_r = 10.4$ kHz, $BW = 2.61$ kHz

17. Capacitive, $X_C < X_L$.

19. 1.47 kΩ

21. 53.1 MΩ; 104 kHz

23. 62.5 Hz

25. 1.38 W

27. 200 Hz

29. See Figure P–17.

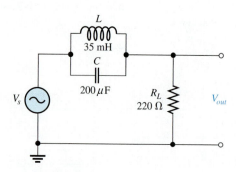

FIGURE P–17

31. $I_{R1} = I_C = 2.11$ mA; $I_{L1} = 1.33$ mA; $I_{L2} = 667$ μA;
$I_{R2} = 667$ μA; $V_{R1} = 6.96$ V; $V_C = 2.11$ V;
$V_{L1} = V_{L2} = V_{R2} = 6.67$ V

33. $I_{R1} = I_{L1} = 41.5$ mA; $I_C = I_{L2} = 133$ mA; $I_{tot} = 104$ mA

35. $L = 989$ μH; $C = 0.064$ μF

37. Neglecting R_W, 8 MHz: $C = 40$ pF; 9 MHz: $C = 31$ pF;
10 MHz: $C = 25$ pF; 11 MHz: $C = 21$ pF

39. No fault

41. L is shorted.

43. L is shorted.

Chapter 14

1. 1.5 μH

3. 3

5. **(a)** In phase   **(b)** Out of phase   **(c)** Out of phase

7. 500 turns

9. **(a)** Same polarity, 100 V rms

   (b) Opposite polarity, 100 V rms

11. 240 V

13. **(a)** 6 V   **(b)** 0 V   **(c)** 40 V

15. **(a)** 10 V   **(b)** 240 V

17. 33.3 mA

19. 27.2 Ω

21. 6.0 mA

23. 0.5

25. 5 kΩ

27. 94.5 W

29. 0.98

31. 25 kVA

33. Secondary 1: 2; Secondary 2: 0.5; Secondary 3: 0.25

35. Top secondary: $n = 100/1000 = 0.1$

   Next secondary: $n = 200/1000 = 0.2$

   Next secondary: $n = 500/1000 = 0.5$

   Bottom secondary: $n = 1000/1000 = 1$

37. Excessive primary current is drawn, potentially burning
out the source and/or the transformer, unless the primary is
protected by a fuse.

39. **(a)** $V_{L1} = 35$ V, $I_{L1} = 2.92$ A, $V_{L2} = 15$ V, $I_{L2} = 1.5$ A

   (b) 28.9 Ω

41. **(a)** 20 V   **(b)** 10 V

43. 0.0141 (70.7:1)

45. 0.1 A

47. Secondary is open.

49. Primary is open.

Chapter 15

1. 103 μs

3. 27 kΩ

5. 12.6 V

7. See Figure P–18.

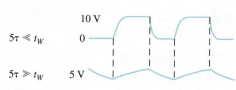

FIGURE P–18

9. See Figure P–19.

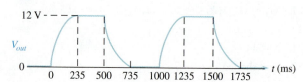

FIGURE P–19

11. 15 V dc level with a very small charge/discharge fluctuation

13. Exchange positions of R and C. See Figure P–20 for the
output voltage. $5\tau = 5$ ms (repeating Problem 8)

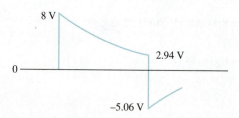

FIGURE P–20

15. Approximately the same shape as input, but with an
average value of 0 V

17. See Figure P–21.

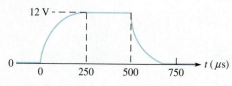

FIGURE P–21

19. See Figure P–22.

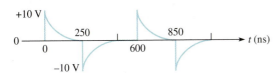

FIGURE P–22

21. A 6 V dc level

23. **(b)** No fault   **(c)** Open C or shorted R

25. **(a)** 23.5 ms   **(b)** See Figure P–23.

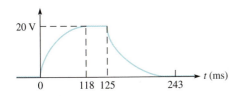

FIGURE P–23

27. 1.44 s

29. One method is as follows: Replace the input to the amplifier with a dc power supply set to the lowest voltage of the triangle wave. The VCO will have a constant output frequency, which is measured with a separate instrument and the oscilloscope horizontal axis will not move, so note the position. Increase the power supply voltage noting the frequency and the horizontal location on the oscilloscope (it will be in a new location on the screen). You can choose several points along the axis to calibrate the frequency as a function of horizontal position.

31. Capacitor is open.

33. No fault

GLOSSARY

accuracy An indication of the range of error in a measurement.

active component A component that requires electrical power to function correctly; active components can deliver more signal power than they receive from the input signal.

admittance (Y) A measure of the ability of a reactive circuit to permit current; the reciprocal of impedance. The unit is the siemens (S).

ah rating A capacity rating for batteries determined by multiplying the current in amps (A) times the length of time in hours (h) a battery can deliver that current to a load.

alternator An ac generator; alternators convert mechanical energy to electrical energy.

ammeter An electrical instrument used to measure current.

ampere (A or amp) The unit of electrical current.

ampere-turn The SI unit of magnetomotive force (mmf).

amplitude The maximum value of a voltage or current.

analog signal A continuous signal.

anode The terminal of a polarized device that electrons flow from. For a battery that is supplying current, this is the negative terminal.

apparent power (P_a) The product of the voltage times the current, expressed in volt-amperes (VA). In a purely resistive circuit it is the same as the true power.

apparent power rating The method of rating transformers in which the power capability is expressed in volt-amperes (VA).

armature The power-producing coil in an alternator, generator or motor. In dc generators, the armature is the rotor, but in alternators it can be either the rotor or the stator. In motors, the armature is the rotor.

atom The smallest particle of an element possessing the unique characteristics of that element.

atomic number The number of protons in the nucleus of an atom.

autotransformer A transformer in which the primary and secondary windings are in a single winding.

average value (1) The average value of a sine wave determined over one half-cycle as 0.637 times the peak value. (2) The average value of a pulse waveform is equal to its baseline value plus the product of its duty cycle and its amplitude.

AWG (American wire gauge) A standardization based on wire diameter.

back emf (electromotive force) A voltage developed across a spinning armature that opposes the original applied voltage.

balanced bridge A bridge circuit that is in the balanced state as indicated by zero volts across the bridge.

balun A type of transformer used to convert a balanced line (such as twisted pair wiring) to an unbalanced line (such as coaxial cable) or vice-versa.

band-pass filter A resonant circuit that passes a range of frequencies lying between two cutoff frequencies and rejects frequencies above and below the range.

band-stop filter A resonant circuit that rejects a range of frequencies lying between two cutoff frequencies and passes frequencies above and below the range.

bandwidth (BW) The characteristic of certain electronic circuits that specifies the usable range of frequencies for which signals pass from input to output without significant reduction in amplitude.

baseline The normal level of a pulse waveform; the voltage level in the absence of a pulse.

battery An energy source that uses a chemical reaction to convert chemical energy into electrical energy.

bias The application of a dc voltage to a diode or other electronic device to produce a desired mode of operation.

bleeder current The current left after the total load current is subtracted from the total current into the circuit.

block diagram A model of a system that represents its structure in a graphical format using labeled blocks to represent functions and lines to represent the signal flow.

boundary The dividing line between what is part of the system and its environment.

branch One current path in a parallel circuit.

bypass A capacitor connected from a point to ground to remove the ac signal without affecting the dc voltage. A special case of decoupling.

capacitance The ability of a capacitor to store electrical charge.

capacitive reactance The opposition of a capacitor to sinusoidal current. The unit is the ohm (Ω).

capacitive susceptance (B_C) The ability of a capacitor to permit current; the reciprocal of capacitive reactance. The unit is the siemens (S).

capacitor An electrical device that has the ability to store charge. It consists of one or more conductors separated by an insulating material called the *dielectric*.

carrier A high-frequency radio wave that can be modulated (changed) by a lower frequency signal.

cathode The terminal of a polarized device that electrons flow to. For a battery that is supplying current, this is the positive terminal.

center tap (CT) A connection at the midpoint of the secondary winding of a transformer.

charge An electrical property of matter that exists because of an excess or a deficiency of electrons. Charge can be either positive or negative.

charging For a capacitor, the process in which a current removes charge from one plate of a capacitor and deposits it on the other plate, making one plate more positive than the other.

choke An inductor. The term is used more commonly in connection with inductors used to block or choke off high frequencies.

circuit An interconnection of electrical components designed to produce a desired result. A basic circuit consists of a source, a load, and an interconnecting current path.

circuit breaker A resettable protective device used for interrupting excessive current in an electric circuit.

circular mil (CM) The unit of the cross-sectional area of a wire.

closed circuit A circuit with a complete current path.

coefficient of coupling (k) A constant associated with transformers that is the ratio of secondary magnetic flux to primary magnetic flux. The ideal value of 1 indicates that all the flux in the primary winding is coupled into the secondary winding.

coil A common term for an inductor.

color code A system of color bands or dots that identify the value of a resistor or other component.

common A reference point in a circuit; also called *reference ground*.

conductance (G) The ability of a circuit to allow current; the reciprocal of resistance. The unit is the siemens (S).

conductor A material in which electrical current is established with relative ease. An example is copper.

core The structure around which the winding of an inductor is formed. The core material influences the electromagnetic characteristics of the inductor.

coulomb (C) The unit of electrical charge; the total charge possessed by 6.25×10^{18} electrons.

Coulomb's law A law that states a force exists between two charged bodies that is directly proportional to the product of the two charges and inversely proportional to the square of the distance between them.

coupling The method of connecting a capacitor from between two points in a circuit to allow ac to pass from one point to the other while blocking dc.

covalent Related to the bonding of two or more atoms by the interaction of their valence electrons.

crystal The pattern or arrangement of atoms forming a solid material.

current The rate of flow of charge (free electrons).

current divider A parallel circuit in which the total current divides among the branches.

current source A device that produces a constant current for a varying load.

cycle One repetition of a periodic waveform.

dc component The average value of a pulse waveform.

dc power supply An electronic instrument that produces voltage, current, and power from the ac power line or batteries in a form suitable for use in powering electronic equipment.

decade A tenfold change in the value of a quantity. When a quantity becomes ten times less or ten times greater, it has changed a decade.

decibel (dB) A unit of logarithmic expression ten times the logarithm of the ratio of two powers or twenty times the logarithm of the ratio of two voltages measured across equal resistances.

decoupling The method of connecting a capacitor from one point, usually the dc power supply line, to ground to short ac to ground without affecting the dc voltage.

degree The unit of angular measure corresponding to 1/360 of a complete revolution.

dielectric The insulating material between the plates of a capacitor.

dielectric constant A measure of the ability of a dielectric material to establish an electric field.

dielectric strength A measure of the ability of a dielectric material to withstand voltage without breaking down.

digital signal A signal that has discrete levels.

DMM Digital multimeter; an electronic instrument that combines meters for the measurement of voltage, current, and resistance.

dummy load A precision high-power resistor.

duty cycle A characteristic of a pulse waveform that indicates the percentage of time that a pulse is present during a cycle; the ratio of pulse width to period.

effective value A measure of the heating effect of a sine wave; also known as the *rms (root mean square) value*.

efficiency The ratio of the output power to the input power of a circuit, usually expressed as a percentage.

electrical Related to the use of electrical voltage and current to achieve desired results.

electrical isolation The condition in which two circuits have no common conductive path between them.

electrical shock The physical sensation resulting from current through the body.

electromagnetic field A formation of a group of magnetic lines of force surrounding a conductor created by electrical current in the conductor.

electromagnetic induction The phenomenon or process by which a voltage is produced in a conductor when there is relative motion between the conductor and a magnetic or electromagnetic field.

electromagnetism The production of a magnetic field by current in a conductor.

electron A basic particle of electrical charge in matter. The electron possesses negative charge.

electronic Related to the movement and control of free electrons in semiconductors or vacuum devices.

element One of the unique substances that make up the known universe. Each element is characterized by a unique atomic structure.

energy The ability to do work. The unit is the joule (J).

engineering notation A system for representing any number as a one-, two-, or three-digit number times a power of ten with an exponent that is a multiple of 3.

error The difference between the true or best-accepted value of some quantity and the measured value.

exciter A separate dc generator used supply current to the field coils of a large generator or alternator. Normally, excitation current is controlled automatically.

exponent The number to which a base number is raised.

exponential A mathematical function described by a natural logarithm (base). Increasing and decreasing current and voltage for a capacitor or inductor are described by exponential functions.

falling edge The negative-going transition of a pulse.

fall time (t_f) The time interval required for a pulse to change from 90% to 10% of its amplitude.

farad (F) The unit of capacitance.

Faraday's law A law stating that the voltage induced across a coil of wire equals the number of turns in the coil times the rate of change of the magnetic flux.

filter A type of circuit that passes certain frequencies and rejects all others.

free electron A valence electron that has broken away from its parent atom and is free to move from atom to atom within the atomic structure of a material.

frequency A measure of the rate of change of a periodic function; the number of cycles completed in 1 s. The unit of frequency is the hertz.

frequency response In electrical circuits, the variation in the output voltage (or current) over a specified range of frequencies.

fuel cell A device that converts electrochemical energy from an external source into dc voltage. The hydrogen fuel cell is the most common type.

function generator An instrument that produces more than one type of waveform.

fundamental frequency The repetition rate of a waveform.

fuse A protective device that burns open when there is excessive current in a circuit.

gain The ratio of the output to the input for an amplifier.

gauss A CGS unit of flux density.

generator An energy source that produces electrical signals.

ground The common or reference point in a circuit.

ground plane A conducting surface used as a reference point for circuit returns or in antenna systems to image a radiating structure.

half-power frequency The frequency at which the output power of a filter is 50% of the maximum (the output voltage is 70.7% of maximum); another name for *critical* frequency or *cut-off* frequency.

half-splitting A troubleshooting procedure where one starts in the middle of a circuit or system and, depending on the first measurement, works toward the output or toward the input to find the fault.

Hall effect A change in current density across a conductor or semiconductor when current in the material is perpendicular to a magnetic field. The change in current density produces a small transverse voltage in the material, called the Hall voltage.

harmonics The frequencies contained in a composite waveform, which are integer multiples of the repetition frequency (fundamental).

henry (H) The unit of inductance.

hertz (Hz) The unit of frequency. One hertz equals one cycle per second.

high-pass circuit A certain type of circuit whereby higher frequencies are passed and lower frequencies are rejected.

horizontal organization A business structure that is decentralized, to allow managers to focus on a specialty and streamline decision making.

hypotenuse The longest side of a right triangle.

hysteresis A characteristic of a magnetic material whereby a change in magnetization lags the application of a magnetic force.

impedance (Z) The total opposition to sinusoidal current expressed in ohms.

impedance matching A technique used to match a load resistance to an internal source resistance in order to achieve a maximum transfer of power.

induced current (i_{ind}) A current induced in a conductor as a result of a changing magnetic field.

induced voltage (v_{ind}) Voltage produced as a result of a changing magnetic field.

inductance (L) The property of an inductor that produces an opposition to any change in current.

induction motor An ac motor that achieves excitation to the rotor by transformer action.

inductive reactance (X_L) The opposition of an inductor to sinusoidal current. The unit is the ohm.

inductive susceptance (B_L) The ability of an inductor to permit current; the reciprocal of inductive reactance. The unit is the siemens.

inductor A passive electrical component formed by a coil of wire and which exhibits the property of inductance.

input The voltage, current, or power applied to an electrical circuit to produce a desired result.

instantaneous power The value of power in a circuit at any given instant of time.

instantaneous value The voltage or current value of a waveform at a given instant in time.

insulator A material that does not allow current under normal conditions.

integrated circuit An active complex circuit consisting of resistors, transistors, and other components fabricated as a single unit that performs the function of multiple discrete components.

integrator A circuit that produces an output that approaches the mathematical integral of the input.

interface To make the output of one type of circuit compatible with the input of another so that they can operate properly together.

ion An atom or group of atoms with a net charge.

joule (J) The SI unit of energy.

kilowatt-hour (kWh) A large unit of energy used mainly by utility companies.

Kirchhoff's current law A law stating that the total current into a node equals the total current out of the node.

Kirchhoff's voltage law A law stating that (1) the sum of the voltage drops around a single closed loop (path) equals the source voltage in that loop or (2) the algebraic sum of all the voltages around a single closed path in a circuit is zero.

ladder diagram An electrical schematic of a parallel circuit which resembles a ladder. Two rails represent the two nodes that have the applied voltage; any number of "rungs", which are loads, complete the ladder.

ladder logic A graphical computer language used for programming PLCs. Programming is accomplished by placing simulated relays, contacts, switches, timers, and other control elements in "rungs" to allow the PLC to control an industrial process.

lag To be behind; describes a condition of the phase or time relationship of waveforms in which one waveform is behind the other in phase or time.

lead To be ahead; describes a condition of the phase or time relationship of waveforms in which one waveform is ahead of the other in phase or time; also, a wire or cable connection to a device or instrument.

leading edge The first step or transition of a pulse.

LED (light-emitting diode) A type of diode that emits light when there is forward current.

Lenz's law A law that states when the current through a coil changes, the polarity of the induced voltage created by the changing magnetic field is such that it always opposes the change that caused it. The current cannot change instantaneously.

linear Characterized by a straight-line relationship.

lines of force Magnetic flux lines that exist in a magnetic field radiating from the north pole to the south pole.

load An element (resistor or other component) connected across the output terminals of a circuit that draws current from the source and upon which work is done.

load cell A transducer that uses strain gauges to convert mechanical force into an electrical signal.

load current The output current of a circuit supplied to a load.

loading The effect on a circuit when an element that draws current from the circuit is connected across the output terminals.

low-pass circuit A certain type of circuit in which lower frequencies are passed and higher frequencies are rejected.

magnetic coupling The magnetic connection between two coils as a result of the changing magnetic flux lines of one coil cutting through the second coil.

magnetic field A force field radiating from the north pole to the south pole of a magnet.

magnetic field intensity The amount of mmf per unit length of magnetic material. The unit is the ampere-turns/meter (At/m); also called *magnetizing force*.

magnetic flux The lines of force between the north and south poles of a permanent magnet or an electromagnet.

magnetic flux density The number of lines of force per unit area perpendicular to a magnetic field.

magnetomotive force The cause of a magnetic field; measured in ampere-turns.

magnitude The value of a quantity, such as the number of volts of voltage or the number of amperes of current.

maximum power transfer The condition, when the load resistance equals the source resistance, under which maximum power is transferred from source to load.

maximum power transfer theorem A theorem that states the maximum power is transferred from a source to a load when the load resistance equals the internal source resistance.

mechatronics A synergistic combination of mechanics and electronics including instrumentation and control systems.

metric prefix A symbol that is used to replace the power of ten in numbers expressed in engineering notation.

modulate To vary a characteristic (amplitude, frequency or phase) of an electromagnetic wave.

multimeter An instrument that measures voltage, current, and resistance.

mutual inductance (L_M) The inductance between two separate coils, such as in a transformer.

negative ion An atom or group of atoms with a net negative charge.

neutron An atomic particle having no electrical charge.

node A point or junction in a circuit at which two or more components are connected.

nucleus The central part of an atom containing protons and neutrons.

ohm (Ω) The unit of resistance.

ohmmeter An instrument for measuring resistance.

Ohm's law A law stating that current is directly proportional to voltage and inversely proportional to resistance.

open A circuit condition in which the current path is interrupted.

open circuit A circuit in which there is not a complete current path.

orbit The path an electron takes as it circles around the nucleus of an atom.

oscilloscope A measurement instrument that displays signal waveforms on a screen.

oscillator An electronic circuit that produces a repetitive waveform on its output with only the dc supply voltage as an input.

output The result obtained from the system after processing the inputs.

parallel The relationship between two circuit components that exists when they are connected between the same pair of nodes.

parallel resonance In a parallel *RLC* circuit, the condition where the impedance is maximum and the reactances are equal.

passband The range of frequencies passed by a filter.

passive component A component that does not require power; passive components cannot increase signal power.

peak-to-peak value The voltage or current value of a waveform measured from its minimum to its maximum points.

peak value The voltage or current value of a waveform at its maximum positive or negative points.

period (T) The time interval of one complete cycle of a given sine wave or any periodic waveform.

periodic Characterized by a repetition at fixed time intervals.

permeability The measure of ease with which a magnetic field can be established in a material.

phase The relative displacement of a time-varying waveform in terms of its occurrence with respect to a reference.

phase angle The angle between the source voltage and the total current in a reactive circuit.

phasor A representation of a sine wave in terms of both magnitude and phase angle.

photoconductive cell A type of variable resistor that is light-sensitive.

photodiode A diode whose reverse resistance changes with incident light.

photon A discrete bundle of electromagnetic energy. In a vacuum, photons travel at the speed of light and have no rest mass.

photovoltaic effect The process whereby light energy is converted directly into electrical energy.

piezoelectric effect The property of a crystal whereby a changing mechanical stress produces a voltage across the crystal.

pole In practical terms, a single *RC* circuit in a filter or amplifier that causes the response to change at a 20 dB per decade rate above or below a certain frequency.

positive ion An atom or group of atoms with a net positive charge.

potentiometer A three-terminal variable resistor.

power The rate of energy usage. The unit is the watt (W).

power factor The relationship between volt-amperes and true power or watts. Volt-amperes multiplied by the power factor equals true power.

power of ten A numerical representation consisting of a base of 10 and an exponent; the number 10 raised to a power.

power rating The maximum amount of power that a resistor can dissipate without being damaged by excessive heat buildup.

power supply A device that converts the ac utility voltage into a constant dc.

precision A measure of the repeatability (or consistency) of a series of measurements.

primary winding The input winding of a transformer; also called *primary*.

probe An accessory used to connect a voltage to the input of an oscilloscope or other instrument.

proton A positively charged atomic particle.

pulse A type of waveform that consists of two equal and opposite steps in voltage or current separated by a time interval.

pulse repetition frequency The fundamental frequency of a repetitive pulse waveform; the rate at which the pulses repeat expressed in either hertz or pulses per second.

pulse width (t_W) The time interval between the opposite steps of an ideal pulse. For a nonideal pulse, the time between the 50% points on the leading and trailing edges.

quality factor (Q) The ratio of reactive power to true power in an inductor or a resonant circuit.

radian A unit of angular measurement. There are 2π radians in one complete revolution. One radian equals 57.3°.

Radiometric image A thermal picture in which colors represent temperature.

ramp A type of waveform characterized by a linear increase or decrease in voltage or current.

RC lag circuit A phase shift circuit in which the output voltage, taken across the capacitor, lags the input voltage by a specified angle.

RC lead circuit A phase shift circuit in which the output voltage, taken across the resistor, leads the input voltage by a specified angle.

RC time constant A fixed time interval set by the values of *R* and *C* that determines the time response of a series *RC* circuit. It equals the product of the resistance and the capacitance.

reactive power (P_r) The rate at which energy is stored and alternately returned to the source by a capacitor or inductor. The unit is the VAR.

reference ground The metal chassis that houses the assembly or a large conducive area on a printed circuit board used as the common or reference point.

reflected load The load as it appears to the source in the primary of a transformer.

reflected resistance The resistance in the secondary circuit reflected into the primary circuit.

regulated power supply A type of power supply in which the output is automatically adjusted to keep a constant value.

relay An electromagnetically controlled mechanical device in which electrical contacts are open or closed by a magnetizing current.

reluctance The opposition to the establishment of a magnetic field in a material.

resistance Opposition to current. The unit is the ohm (Ω).

resistor An electrical component designed specifically to have a certain amount of resistance.

resolution The smallest increment of a quantity that a meter can measure.

resonant frequency The frequency at which a resonant condition occurs in a series or parallel *RLC* circuit.

retentivity The ability of a material, once magnetized, to maintain a magnetized state without the presence of a magnetizing force.

rheostat A two-terminal variable resistor.

right triangle A triangle with a 90° angle.

ripple voltage The small variation in the dc voltage on the output of a filtered rectifier caused by the slight charging and discharging action of the filter capacitor.

rise time (t_r) The time interval required for a pulse to change from 10% to 90% of its amplitude.

rising edge The positive-going transition of a pulse.

RL lag circuit A phase shift circuit in which the output voltage, taken across the resistor, lags the input voltage by a specified angle.

RL lead circuit A phase shift circuit in which the output voltage, taken across the inductor, leads the input voltage by a specified angle.

RL time constant A fixed time interval set by the values of *R* and *L* that determines the time response of a circuit and is equal to L/R.

rms value The value of a sine wave that indicates its heating effect, also known as the *effective value*. It is equal to 0.707 times the peak value. RMS stands for root mean square.

roll-off The decrease in the response of a filter below or above a critical frequency.

rotor The moving part of a generator, alternator, or motor.

round off The process of dropping one or more digits to the right of the last significant digit in a number.

sawtooth waveform A type of electrical waveform composed of ramps; a special case of a triangular waveform in which one ramp is much shorter than the other.

schematic A symbolized diagram of an electrical or electronic circuit.

scientific notation A system for representing any number as a number between 1 and 10 times an appropriate power of ten.

secondary winding The output winding of a transformer; also called *secondary*.

selectivity A measure of how effectively a resonant circuit passes certain frequencies and rejects others. The narrower the bandwidth, the greater the selectivity.

self-excited generator A generator in which the field windings derive their current from the output.

semiconductor A material that has a conductance value between that of a conductor and that of an insulator. Silicon and germanium are examples.

sensitivity The ohms-per-volt rating of a voltmeter.

series In an electric circuit, a relationship of components in which the components are connected such that they provide a single current path between two points.

series-aiding An arrangement of two or more series voltage sources with polarities in the same direction.

series-opposing An arrangement of two series voltage sources with polarities in the opposite direction; not a normal configuration.

series resonance In a series *RLC* circuit, the condition where the impedance is minimum and the reactances are equal.

shell An energy band in which electrons orbit the nucleus of an atom.

short A circuit condition in which there is a zero or abnormally low resistance between two points; usually an inadvertent condition.

SI Standardized international system of units used for all engineering and scientific work; abbreviation for French *Le Système International d'Unites*.

siemens The unit of conductance.

significant digit A digit known to be correct in a number.

sine wave A type of electrical waveform that is based on the mathematical sine function.

slip The difference between the synchronous speed of the stator field and the rotor speed in an induction motor.

solenoid An electromagnetically controlled device in which the mechanical movement of a shaft or plunger is activated by a magnetizing current.

solenoid valve An electrically controlled valve for control of air, water, steam, oils, refrigerants, and other fluids.

source Any device that produces energy.

speaker An electromagnetic device that converts electrical signals to sound waves.

squirrel cage An aluminum frame within the rotor of an induction motor that forms the electrical conductors for a rotating current.

stator The fixed part of a generator, alternator, or motor.

steady state The equilibrium condition of a circuit that occurs after an initial transient time.

step-down transformer A transformer in which the secondary voltage is less than the primary voltage.

step-up transformer A transformer in which the secondary voltage is greater than the primary voltage.

stiff voltage divider A voltage divider that has a load resistor that is large compared to the resistors in the divider, so that the loaded and unloaded output voltages are nearly the same.

stopband The range of frequencies between the upper and lower cutoff points.

strain gauge A variable resistor formed by a resistive material in which a lengthening or shortening due to stress produces a proportional change in resistance.

superposition A method for analyzing circuits with two or more sources by examining the effects of each source by itself and then combining the effects.

switch An electrical or electronic device for opening and closing a current path.

synchronous motor An ac motor in which the rotor moves at the same rate as the rotating magnetic field of the stator.

system A group of interrelated parts that perform a specific function.

tank circuit A parallel resonant circuit.

tapered Nonlinear, such as a tapered potentiometer.

temperature coefficient A constant specifying the amount of change in the value of a quantity for a given change in temperature.

terminal An external contact point on an electric circuit or an electronic device.

terminal equivalency A condition that occurs when two circuits produce the same load voltage and load current where the same value of load resistance is connected to either circuit.

tesla The SI unit of flux density.

thermistor A type of variable resistor in which resistance depends on temperature.

thermocouple A type of temperature transducer formed by the junction of two dissimilar metals that produces a voltage proportional to temperature.

Thevenin's theorem A circuit theorem that provides for reducing any two-terminal resistive circuit to a single equivalent voltage source in series with an equivalent resistance.

time constant A fixed-time interval set by *R* and *C*, or *R* and *L* values, that determines the time response of a circuit.

tolerance The limits of variation in the value of a component.

trailing edge The second step or transition of a pulse.

transducer A device that transforms energy from one form to another.

transfer curve A plot showing the ratio of the output to the input.

transformer An electrical device constructed of two or more coils (windings) that are electromagnetically coupled to each other to provide a transfer of power from one coil to another.

transient A temporary passing condition in a circuit; a sudden or temporary change in circuit conditions.

transient time An interval equal to approximately five time constants.

triangular waveform A type of electrical waveform that consists of two ramps.

triangular-wave oscillator An electronic oscillator that uses a comparator with two trip points and an integrator to produce triangular waves.

trigger The activating mechanism of some electronic devices or instruments.

trimmer A small variable capacitor.

troubleshooting A systematic process of isolating, identifying, and correcting a fault in a circuit or system; the application of logical thinking combined with a thorough knowledge of circuit or system operation to find and correct a malfunction.

true power The power that is dissipated in a circuit, usually in the form of heat.

turns ratio The ratio of turns in the secondary winding to turns in the primary winding.

unbalanced bridge A bridge circuit that is in the unbalanced state as indicated by a voltage across the bridge proportional to the amount of deviation from the balanced state.

valence Related to the outer shell or orbit of an atom.

valence electron An electron that is present in the outermost shell of an atom.

VAR (volt-ampere reactive) The unit of reactive power.

varactor A diode that is used as a voltage-variable capacitor.

vertical organization A centralized business structure that has an organized top-down approach. The tendency is less toward a specialization.

volt The unit of voltage or electromotive force.

voltage The amount of energy available to move a certain number of electrons from one point to another in an electric circuit.

voltage divider A circuit consisting of series resistors across which one or more output voltages are taken.

voltage drop The decrease in the voltage across a resistor due to loss of energy.

voltage-follower A closed-loop, noninverting op-amp with a voltage gain of 1.

voltage gain The ratio of output voltage to input voltage.

voltage regulation The process of maintaining an essentially constant output voltage over variations in input voltage or load.

voltage source A device that produces a constant voltage for a varying load.

voltmeter An instrument used to measure voltage.

watt (W) The unit of power.

Watt's law A law that states the relationships of power to current, voltage, and resistance.

waveform The pattern of variations of a voltage or current showing how the quantity changes with time.

weber (Wb) The SI unit of magnetic flux, which represents 10^8 lines.

Wheatstone bridge A 4-legged type of bridge circuit with which an unknown resistance can be accurately measured using the balanced state of the bridge. Deviations in resistance can be measured using the unbalanced state.

winding The loops or turns of wire in an inductor.

winding resistance The resistance of the length of wire that makes up a coil.

wiper The sliding contact in a potentiometer.

INDEX